Elements of Signal
Detection and Estimation

Elements of Signal
Detection and Estimation

CARL W. HELSTROM

University of California
San Diego

PTR PRENTICE HALL, Englewood Cliffs, New Jersey 07632

Library of Congress Cataloging-in-Publication Data

Helstrom, Carl W.
    Elements of signal detection and estimation / Carl W. Helstrom.
        p.   cm.
    Includes bibliographical references and index.
    ISBN 0-13-808940-X
    1. Radar.   2. Sonar.   3. Signal detection.   I. Title.
TK6588.H45   1995
621.382′23—dc20   94-16150
                                                                    CIP

Acquisitions editor: *Karen Gettman*
Editorial/production supervision: *Raeia Maes*
Cover design: *Joe Di Domenico*
Manufacturing manager: *Alexis R. Heydt*

© 1995 by PTR Prentice Hall
Prentice-Hall, Inc.
A Paramount Communications Company
Englewood Cliffs, NJ 07632

The publisher offers discounts on this book when ordered
in bulk quantities. For more information, contact:

    Corporate Sales Department
    PTR Prentice Hall
    113 Sylvan Avenue
    Englewood Cliffs, NJ 07632

    Phone: 201-592-2863
    Fax: 201-592-2249

Printed in the United States of America

10 9 8 7 6 5 4 3 2 1

ISBN 0-13-808940-X

PRENTICE-HALL INTERNATIONAL (UK) LIMITED, *London*
PRENTICE-HALL OF AUSTRALIA PTY. LIMITED, *Sydney*
PRENTICE-HALL CANADA INC., *Toronto*
PRENTICE-HALL HISPANOAMERICANA, S.A., *Mexico*
PRENTICE-HALL OF INDIA PRIVATE LIMITED, *New Delhi*
PRENTICE-HALL OF JAPAN, INC., *Tokyo*
SIMON & SCHUSTER ASIA PTE. LTD., *Singapore*
EDITORA PRENTICE-HALL DO BRASIL, LTDA., *Rio de Janeiro*

To Barbro, Lars, and Stefan

To Barbra, Lars, and Stefan

Contents

PREFACE xv

1 THE STATISTICAL FOUNDATION 1

    1.1   Decision Theory   1

          1.1.1   *Bayes's Rule, 3*
          1.1.2   *Minimizing Average Cost, 6*

    1.2   Binary Decisions   8

          1.2.1   *Bayes Strategy, 8*
          1.2.2   *Neyman–Pearson Criterion, 10*
          1.2.3   *Operating Characteristic, 12*
          1.2.4   *Sufficient Statistics, 14*
          1.2.5   *Decisions Based on Discrete Random Variables, 18*

          Problems   23

2 DETECTION OF A KNOWN SIGNAL 27

    2.1   Gaussian Noise   27

          2.1.1   *The Density Functions of Gaussian Noise, 28*
          2.1.2   *Stationary Noise, 30*
          2.1.3   *Sampling by Orthonormal Functions, 31*
          2.1.4   *Gram–Schmidt Orthogonalization, 33*

2.1.5   White Noise, 34
2.1.6   The Karhunen–Loève Expansion, 37
2.1.7   Reproducing-kernel Hilbert Space, 40

2.2   Detection in Gaussian Noise   42

2.2.1   The Likelihood Ratio, 42
2.2.2   The Sufficient Statistic, 44
2.2.3   The Matched Filter, 49
2.2.4   The Irrelevance Proof, 57
2.2.5   Discrete-time Processing, 58

2.3   Solution of the Integral Equations   60

2.3.1   Inhomogeneous Equations, 60
2.3.2   Homogeneous Equations, 65

2.4   Physical Interpretation of the Signal-to-noise Ratio,   68

2.4.1   The Transmission-line Model, 69
2.4.2   The Circuit Model, 71

2.5   Decisions among a Number of Known Signals   74

2.5.1   Signal Space, 74
2.5.2   Displacement of the Signal Vectors, 79
2.5.3   Orthogonal and Simplex Signals, 79

Problems   81

3   NARROWBAND SIGNALS AND THEIR DETECTION   88

3.1   Narrowband Signals and Filters   88

3.2   Narrowband Noise   95

3.2.1   The Complex Representation, 95
3.2.2   The Complex Autocovariance Function, 97
3.2.3   Circular Gaussian Random Processes, 100
3.2.4   Narrowband White Noise, 103

3.3   Detection of a Signal of Random Phase   104

3.3.1   The Likelihood Functional for a Narrowband Input, 104
3.3.2   Signals of Random Phase, 107

3.4   The Detectability of Signals of Unknown Phase   110

3.5   Narrowband Signals in Communications   113

3.5.1   The Binary Incoherent Channel, 113
3.5.2   The Balanced Binary Incoherent Channel, 116
3.5.3   The Unilateral Binary Incoherent Channel, 117
3.5.4   The Incoherent M-ary Channel, 118

3.6   Testing Composite Hypotheses   119

3.6.1   The Bayes Criterion, 119

Contents

3.6.2 *The Extended Neyman–Pearson Criterion, 123*
3.6.3 *Detection of Signals of Unknown Amplitude, 125*
3.6.4 *Maximum-likelihood Detection, 128*

Problems   132

4   DETECTION IN MULTIPLE OBSERVATIONS 136

4.1 Optimum Detector   136

   4.1.1 *Complete Coherence, 138*
   4.1.2 *Incoherent Signals, 140*

4.2 The Threshold Detector   142

   4.2.1 *The Weak-signal Approximation, 142*
   4.2.2 *Performance of the Quadratic Threshold Detector, 144*
   4.2.3 *Detection Probability for Rayleigh Fading, 147*
   4.2.4 *Other Types of Fading Signals, 151*

4.3 Beamforming   156

   4.3.1 *Detection by an Array of Transducers, 156*
   4.3.2 *Elimination of Noise from a Point Source, 161*

4.4 Comparison of Receiver Performance   165

   4.4.1 *Asymptotic Relative Efficiency, 165*
   4.4.2 *Threshold Detection, 169*
   4.4.3 *Comparison of the Linear and the Quadratic Detector, 171*
   4.4.4 *Invariable Parameters Unknown, 172*

4.5 Distributed Detection   174

   4.5.1 *Identical Independent Sensors, 175*
   4.5.2 *Nonidentical Independent Sensors, 180*

   Problems   185

5   EVALUATING SIGNAL DETECTABILITY 188

5.1 The Edgeworth Series   188

   5.1.1 *The Moment-generating Function, 188*
   5.1.2 *The Gram–Charlier and Edgeworth Series, 190*

5.2 Numerical Laplace Inversion   195

   5.2.1 *Integration through a Saddlepoint, 195*
   5.2.2 *Integration on a Curved Path, 201*
   5.2.3 *Integer-valued Random Variables, 207*

5.3 Approximations   211

   5.3.1 *The Chernoff Bound, 211*
   5.3.2 *The Saddlepoint Approximation, 213*

5.3.3   *Calculating Approximate Decision Levels for the*
            Neyman–Pearson Criterion, 215
5.3.4   *Calculating Decision Levels for the Bayes Criterion, 217*
5.3.5   *The Uniform Asymptotic Expansion, 217*

Problems   218

6   ESTIMATION OF SIGNAL PARAMETERS   220

6.1   The Theory of Estimation   220

      6.1.1   *Maximum-a-posteriori-probability Estimators, 222*
      6.1.2   *Maximum-likelihood Estimators, 223*
      6.1.3   *Estimating the Mean of a Gaussian Distribution, 224*
      6.1.4   *Jointly Gaussian Parameters and Data, 226*
      6.1.5   *Bayes Estimates, 227*
      6.1.6   *The Quadratic Cost Function, 228*
      6.1.7   *The Principle of Orthogonality, 230*

6.2   Estimation of Signal Parameters   232

      6.2.1   *Maximum-likelihood Estimators, 232*
      6.2.2   *Estimation of Arrival Time, 233*
      6.2.3   *Asymptotic Variance of Maximum-likelihood Estimators, 236*
      6.2.4   *The Cramér–Rao Inequality, 239*
      6.2.5   *Estimation of Signal Parameters in Gaussion Noise, 242*
      6.2.6   *The Ziv–Zakai Bound, 244*

6.3   Estimation of Parameters of a Narrowband Signal   246

      6.3.1   *Arrival Time, 246*
      6.3.2   *Signal Arrival Time and Carrier Frequency, 249*
      6.3.3   *The Complex Ambiguity Function, 253*
      6.3.4   *Calculation of the Error Covariances, 255*
      6.3.5   *Coherent Pulse Trains, 259*

Problems   262

7   DETECTION OF SIGNALS WITH UNKNOWN PARAMETERS   266

7.1   Unknown Arrival Time: The Threshold Detector   266

7.2   Unknown Arrival Time: The Maximum-likelihood Detector   272

7.3   False-alarm Rate   273

      7.3.1   *The First-passage Time Problem, 273*
      7.3.2   *The Crossing-rate Approximation, 275*
      7.3.3   *The Crossing Rate of a Stochastic Process, 276*
      7.3.4   *The Crossing Rate of the Rectified Process, 278*

7.4   Unknown Arrival Time: The Probability of Detection   281

7.5   Signals of Unknown Arrival Time and Carrier Frequency   282

7.6   The Least Favorable Distribution   286

     7.6.1   *The Extended Neyman–Pearson Criterion, 286*
     7.6.2   *The Threshold Detector, 296*
     7.6.3   *Narrowband Signals in Gaussian Noise, 298*

     Problems   301

8   DETECTION OF SIGNALS UNDER CONDITIONS OF UNCERTAINTY   304

8.1   Detection in Noise of Unknown Strength   304

     8.1.1   *The CFAR Receiver, 304*
     8.1.2   *False-alarm Probability, 306*
     8.1.3   *The Probability of Detection, 308*
     8.1.4   *Colored Interference of Unknown Spectral Density, 311*

8.2   Nonparametric Detection   314

     8.2.1   *Parametric and Nonparametric Hypotheses, 314*
     8.2.2   *Nonparametric Receivers, 316*
     8.2.3   *The t-Test, 318*
     8.2.4   *The Sign Test, 320*
     8.2.5   *Rank Tests, 325*
     8.2.6   *Receivers with a Reference Input, 330*
     8.2.7   *Two-input Systems, 331*

     Problems   332

9   SEQUENTIAL DETECTION   336

9.1   The Sequential Probability Ratio Test   336

9.2   Performance of the Sequential Test   341

     9.2.1   *The Detection Probability, 342*
     9.2.2   *The Average Sample Number, 344*

9.3   Sequential Detection of Signals of Random Phase   347

9.4   Sequential Detection of Targets of Unknown Distance   350

     Problems   354

10   SIGNAL RESOLUTION   356

10.1   Specification of Receivers   356

     10.1.1   *Varieties of Resolution, 356*
     10.1.2   *The Resolution of Two Signals, 357*
     10.1.3   *The Resolution of Many Signals, 363*

10.2   The Detection of Signals in Clutter   369

Contents

        10.2.1   *The Spectrum of Clutter Interference, 369*
        10.2.2   *Detection of Single Pulses in Clutter, 370*
        10.2.3   *Coherent Pulse Trains, 377*

  10.3   The Specification of Signals   379

        10.3.1   *General Properties of the Ambiguity Function, 379*
        10.3.2   *Single Pulses, 381*
        10.3.3   *Pulse Trains, 386*

        Problems   392

11 STOCHASTIC SIGNALS 394

  11.1   Structure of the Receiver   394

        11.1.1   *Types of Stochastic Signals, 394*
        11.1.2   *Vector-space Representation, 397*
        11.1.3   *The Likelihood Ratio, 400*
        11.1.4   *Realizations of the Optimum Detector, 402*
        11.1.5   *Stationarity over a Long Observation Interval, 404*
        11.1.6   *The Threshold Detector, 405*
        11.1.7   *The Radiometer, 406*
        11.1.8   *An Example, 407*
        11.1.9   *Estimator–correlator Interpretation, 409*
        11.1.10   *The Question of Singularity, 411*

  11.2   The Performance of the Receiver   413

        11.2.1   *The Moment-generating Function of the Test Statistic, 413*
        11.2.2   *Detectability in White Noise: The Residue Series, 415*
        11.2.3   *The Threshold Detector in White Noise, 418*
        11.2.4   *Application of Saddlepoint Integration, 420*
        11.2.5   *The Toeplitz Approximation, 421*
        11.2.6   *Rational Spectral Densities, 422*
        11.2.7   *Detectability of Signals with Lorentz or Rectangular Spectral Densities, 423*

  11.3   Causal Estimator–Correlator Representation   429

        11.3.1   *Causal Estimator of the Stochastic Signal, 429*
        11.3.2   *Recalculation of the Fredholm Determinant, 431*
        11.3.3   *The Mean-square Estimation Error, 433*
        11.3.4   *The Fredholm Determinant for a Rational Spectral Density, 434*
        11.3.5   *The Likelihood Functional, 437*
        11.3.6   *The Innovation Process, 439*

  11.4   State-space Formulation   441

        11.4.1   *Generation of the Signal by a Linear System, 441*
        11.4.2   *The Kalman–Bucy Equations, 445*
        11.4.3   *The Schweppe Likelihood-ratio Receiver, 449*

  11.5   Detection of a Coherent Signal and a Stochastic Signal   451

11.5.1   *The Optimum Detector, 451*
11.5.2   *The Performance of the Optimum Detector, 452*

11.6   Discrete-time Processing   456

11.6.1   *The Optimum Statistic, 456*
11.6.2   *The Kalman Method, 458*
11.6.3   *Stationary Data, 460*
11.6.4   *Performance of the Detector, 462*

Problems   465

12   DETECTION OF OPTICAL SIGNALS   470

12.1   Photoelectron Counting   470

12.1.1   *Properties of the Light and the Detector, 470*
12.1.2   *The Probability-generating Function, 472*
12.1.3   *Evaluating the Probability-generating Function, 473*
12.1.4   *A Single Temporal Mode, 475*
12.1.5   *Many Temporal Modes, 476*
12.1.6   *The Toeplitz Approximation, 477*
12.1.7   *Rectangular Spectral Density, 478*
12.1.8   *Light with a Rational Spectral Density, 479*

12.2   Detectability of Optical Signals by Photocounting   480

12.2.1   *Negligible Background Light, 480*
12.2.2   *Detection of a Coherent Signal, 482*
12.2.3   *The Poisson Limit, 484*

12.3   Photomultiplication and Shot Noise   485

12.3.1   *Single-stage and Multistage Photomultipliers, 485*
12.3.2   *Avalanche Photodiodes, 491*
12.3.3   *Shot Noise, 493*

12.4   Optical Heterodyne and Homodyne Detection   497

12.4.1   *The Heterodyne Receiver, 497*
12.4.2   *The Homodyne Receiver, 503*

12.5   Detection Based on Counting Times   504

12.5.1   *The Poisson Limit, 505*
12.5.2   *Incoherent Gaussian Light, 506*

Problems   509

**APPENDIX A: SOLUTION OF THE DETECTION
INTEGRAL EQUATIONS   511**

A.1   Inhomogeneous Equations   511

A.2   Homogeneous Equations and Their Eigenvalues   515

Problem   518

Contents   **xiii**

APPENDIX B: CIRCULAR GAUSSIAN DENSITY FUNCTIONS 519

APPENDIX C: Q FUNCTION 523

    C.1   Properties 523

    C.2 Mth-order Q Function 526

    C.3   Computation by Recurrence 528

**APPENDIX D: ERROR PROBABILITY FOR A CHANNEL
CARRYING TWO NONORTHOGONAL SIGNALS WITH
RANDOM PHASES 531**

**APPENDIX E: RECURSIVE METHODS FOR DETECTION
PROBABILITIES: FADING SIGNALS AND RANDOM
DECISION LEVELS 537**

    E.1   Fading Signals 537

    E.2   Random Decision Level, Fixed Signal Strengths 538

    E.3   Random Signal Strengths and Random Decision Level 539

    E.4   False-alarm Probability, Random Decision Level 540

APPENDIX F: PULSE REFLECTED FROM A MOVING TARGET 542

**APPENDIX G: ASYMPTOTIC RELATIVE EFFICIENCY OF THE
RANK TEST 545**

**APPENDIX H: PROBABILITY OF DETECTION: LORENTZ
SPECTRAL DENSITY 547**

    H.1   The Residue Series 547

    H.2   The Saddlepoint Method 549

    H.3   The Toeplitz Approximation 550

**APPENDIX I: DERIVATION OF THE KALMAN–BUCY
EQUATIONS 553**

**APPENDIX J: MOMENT-GENERATING FUNCTION: COHERENT
PLUS STOCHASTIC SIGNAL 555**

BIBLIOGRAPHY 558

INDEX 575

Contents

Preface

The key component of a communication, radar, or sonar system is its receiver, whose purpose is to detect information-bearing signals, signals that have been weakened and distorted during transmission and corrupted by random noise. The systematic design of receivers and the assessment of their performance are based on signal-detection theory, and that is the subject of this book.

The detection of weak signals in noise can instructively be viewed in the framework of the statistical testing of hypotheses, whose elementary foundations are presented in Chapter 1. These are applied in Chapter 2 to the basic problem of detecting a signal of known form in Gaussian noise. There we introduce the concepts of matched filtering and signal-space representation and determine the ultimate limits on signal detectability. We then treat the detection of narrowband signals of unknown carrier phase, setting up a convenient formulation for narrowband signals and noise as a natural generalization of the phasors used in analyzing alternating currents and voltages (Chapter 3). The structure of optimum and near-optimum receivers to which multiple independent inputs are available is examined in Chapter 4. In this setting we provide a brief treatment of beamforming in transducer arrays. The performance of receivers is measured by their probabilities of error, and the efficient computation of such probabilities is treated at some length in Chapter 5.

Statistical estimation theory is introduced in Chapter 6 in order to study the estimation of signal parameters such as arrival time and Doppler shift, which provide information about the location and speed of radar targets. Calculation of the mean-square errors in such estimates is emphasized. Maximum-likelihood detection, which draws on the estimation of signal parameters, figures in the design of receivers of radar signals of unknown arrival time and Doppler shift (Chapter 7). Chapter 8 describes how estimation theory plays a role in detecting signals in the presence of noise of unknown strength. A broader uncertainty about the statistics of the noise calls for the use of nonparametric detectors, and we show how their performance can be assessed in terms of their asymptotic efficiency relative to a standard detector. For the sake of efficient scanning of the sky for radar targets, the designer should consider applying sequential processing, which is the topic of Chapter 9. The resolution of signals that are close together in arrival time and carrier frequency is approached in Chapter 10 through the estimation of signal parameters, and the efficacy of receivers that attempt to resolve close radar targets is treated in terms of the ambiguity function.

Minimum-mean-square-error estimation of time series corrupted by noise is also briefly treated in Chapter 6 as preparation for the analysis in Chapter 11 of detectors of Gaussian stochastic signals and the calculation of their error probabilities. Here the Kalman–Bucy equations play an important role. The methods developed in this context figure in the treatment of the photocounting detection of optical signals in the last chapter. Photomultiplication and heterodyne detection of optical signals are also studied. Ten appendixes collect useful formulas and present calculations too tedious for inclusion in the main text.

The reader of this book should be familiar with elementary concepts of probability, such as conditional probability, distributions of random variables, expected values, and correlation. Numerous references to the widely available textbooks by Papoulis [Pap91] and the writer [Hel91] direct the reader to fuller treatments of certain details of probability theory when they arise in the course of our study. (Notations in brackets refer to the bibliography at the end of this book.) Some acquaintance with complex variables and Fourier and Laplace transforms is also presumed. The lecture notes on which this book is based were used by graduate students in a one-year course on detection theory. An array of problems will be found at the end of each chapter.

For the sake of readers who wish to delve more deeply into the topics of this book, we have provided references to papers from which a search of the literature can begin. Our references are not necessarily to either the original work or the most recent work on a topic, and the omission of references in a particular context does not imply that the results presented are our own creation. For other approaches to signal detection and for the treatment of certain specialized aspects of the subject, the reader might consult the books by Van Trees [Van 68], Whalen [Wha71], Poor [Poo88], Kassam [Kas88], and Kazakos and Papantoni-Kazakos [Kaz90].

Detection theory was the subject of an earlier work, *Statistical Theory of Signal Detection* (New York: Pergamon Press, Inc., 1st ed., 1960; 2nd ed., 1968). I am grateful to Pergamon Press for permission to use this material.

Carl W. Helstrom

Elements of Signal

Detection and Estimation

1

The Statistical Foundation

1.1 DECISION THEORY

Signal-detection theory views a receiver as a device for making decisions about the composition of its input, which is typically the voltage across the terminals of an antenna. A radar transmitter, for instance, sends out a short burst of electromagnetic energy at regular intervals, and the radar receiver must decide whether its input contains minute portions of that energy reflected from remote targets. A communication system represents each message symbol by a signal of a particular form and periodically transmits one of its "alphabet" of signals to a distant receiver. The receiver must synchronously decide which of the signals in the alphabet has appeared at its input, and the sequence of its choices constitutes the received message.

The voltage across the terminals of the antenna continually varies in an unpredictable manner because of random fluctuations in its ambient electromagnetic field. These are caused partially by chaotic thermal motions of ions and electrons in the field of view of the antenna and partially by interfering signals from power lines, communication systems, lightning, electrical machinery, and so on. The resulting fluctuations at the input to the receiver are called *noise*. The signals about which the receiver must decide appear in the midst of this noise, which causes the decisions, however they are made, occasionally to be wrong. A radar receiver may mistakenly decide that an echo signal is present when there is none, or it may overlook one that is really there. A communications receiver may decide that the symbol *F* was transmitted when what was actually sent was an *H*. Detection theory considers how to design a receiver that suffers such errors as seldom as possible.

As a random phenomenon, the combination of signals and noise must be described statistically and analyzed in the framework of the theory of probability. For definiteness let us consider a communication system dispatching messages written in an alphabet of M symbols. To each symbol corresponds a signal of a certain form, which is transmitted to the distant receiver and appears in its input attenuated, possibly distorted, and corrupted by random noise. The proposition that the jth signal was transmitted is equivalent to a hypothesis about the composition of the input $v(t)$ to the receiver during a certain interval of time; we denote this hypothesis by H_j, $1 \leq j \leq M$. The receiver must choose one of these M hypotheses on the basis of its input $v(t)$ during the *observation interval*, say $0 \leq t \leq T$.

The input $v(t)$ is a stochastic process, described in terms of probability density functions. For simplicity we suppose that the input during $(0, T)$ has been appropriately sampled and can be represented by n samples $(v_1, v_2, \dots, v_n)$. We designate these data collectively by a vector $v = (v_1, v_2, \dots, v_n)$, and we represent v as a point in an n-dimensional Cartesian space R_n. Just how the sampling can conveniently be accomplished and how it can be expanded to include all the relevant information in $v(t)$ will be treated later.

Under hypothesis H_j, that is, when the jth of the M signals has been transmitted, the n samples v are random variables having a joint probability density function

$$p_j(v) = p_j(v_1, v_2, \dots, v_n).$$

Its form depends on the properties of the received signal and the noise. We remind the reader that a probability density function such as $p_j(v)$ is a nonnegative function whose integral over the entire space R_n equals 1:

$$\int_{-\infty}^{\infty} \cdots \int_{-\infty}^{\infty} p_j(v_1, v_2, \dots, v_n)\, dv_1\, dv_2 \dots dv_n = \int_{R_n} p_j(v)\, d^n v = 1;$$

$d^n v = dv_1\, dv_2 \dots dv_n$ is the volume element in the data space. The probability under hypothesis H_j that the point v representing a particular set of samples lies in an arbitrary region Δ of that space is

$$\Pr(v \in \Delta \mid H_j) = \int_{\Delta} p_j(v)\, d^n v, \qquad j = 1, 2, \dots, M.$$

On the basis of the observed values of $(v_1, v_2, \dots, v_n)$ the receiver is to decide among the M hypotheses $H_1, H_2, \dots, H_M$. That is, it must choose which of the M probability density functions $p_j(v)$ it believes actually to characterize the input $v(t)$ during $(0, T)$. The scheme by which the receiver makes these choices is called a *strategy*. It must assign a definite selection among $H_1, H_2, \dots, H_M$ to each possible set v of samples. The strategy can be visualized as a division of the space R_n of the data $(v_1, v_2, \dots, v_n)$ into M disjoint regions $R_1, R_2, \dots, R_M$. When the point v falls into region R_j, the receiver chooses hypothesis H_j, deciding that the jth signal was transmitted. How can this decomposition of R_n best be made?

1.1.1 Bayes's Rule

In a communication system it is desirable that the receiver make its decisions with as few errors as possible in the long run. Stated otherwise, the probability Q that the receiver makes correct decisions should be as large as possible. This probability Q depends on the structure of the regions $R_1, R_2, \ldots, R_M$, on the relative frequencies of the several signals, and on the set of probability density functions $\{p_j(v)\}$.

When signal j is transmitted, the probability that the receiver makes the correct decision equals the conditional probability under hypothesis H_j that the point v falls into region R_j:

$$\Pr(\rightarrow H_j | H_j) = \Pr(v \in R_j | H_j) = \int_{R_j} p_j(v) \, d^n v. \tag{1-1}$$

Here $\rightarrow H_j$ denotes the event "hypothesis H_j is chosen."

Let $\zeta_j = \Pr(H_j)$ be the relative frequency with which the symbol j appears in the messages and signal j is transmitted:

$$\sum_{j=1}^{M} \zeta_j = 1.$$

We call ζ_j the *prior probability* of hypothesis H_j. It must be known to the designer of the receiver. The overall probability of correct decision is then the weighted sum

$$Q = \sum_{j=1}^{M} \zeta_j \Pr(v \in R_j | H_j) = \sum_{j=1}^{M} \zeta_j \int_{R_j} p_j(v) \, d^n v. \tag{1-2}$$

The probability of error is $P_e = 1 - Q$.

Denote by

$$p(v) = \sum_{j=1}^{M} \zeta_j p_j(v) \tag{1-3}$$

the overall probability density function of the data $v = (v_1, \ldots, v_n)$. The conditional or *posterior* probability that hypothesis H_j is true when a particular set v of samples has been recorded is specified by Bayes's theorem as

$$\Pr(H_j | v) = \frac{\zeta_j p_j(v)}{p(v)}, \qquad 1 \leq j \leq M, \tag{1-4}$$

[Hel91, p. 90],[1] [Pap91, p. 83]. This follows from the basic definition of a conditional probability: for two events A and B,

$$\Pr(B | A) = \frac{\Pr(A \cap B)}{\Pr(A)} = \frac{\Pr(A | B) \Pr(B)}{\Pr(A)}$$

[Hel91, p. 25], [Pap91, p. 27]. For event B put "Hypothesis H_j is true," and for event A put "The data point lies in an infinitesimal region Δ about v;" $\Pr(A) = p(v)\Delta$; $\Pr(A | B) = p_j(v)\Delta$.

---

[1] A notation such as this refers to the Bibliography at the end of the book.

For any set v of data, one of the M posterior probabilities in (1-4), say $\Pr(H_k|v)$, will in general be largest. If the receiver then decides for hypothesis H_k, always obeying the rule, "Choose that hypothesis with the greatest posterior probability, given v," it will attain the maximum probability Q of correct decision. We can symbolize this rule as

$$\{\Pr(H_k|v) > \Pr(H_j|v), \qquad \forall\, j \neq k\} \Rightarrow \longrightarrow H_k. \tag{1-5}$$

It is known as *Bayes's rule*, for it was enounced by the Reverend Thomas Bayes in a paper, "An Essay toward Solving a Problem in the Doctrine of Chances," published posthumously in 1763 [Bay63].

It is not difficult to see that using Bayes's rule indeed maximizes the probability Q of correct decision. Let us use (1-4) to write (1-2) as

$$Q = \sum_{j=1}^{M} \int_{R_j} p(v) \Pr(H_j|v)\, d^n v. \tag{1-6}$$

In order that Q be as large as possible, those points v that are to be included in each particular region R_k are those for which $\Pr(H_k|v)$ exceeds all the other posterior probabilities $\Pr(H_j|v)$; any other assignment of points to region R_k would diminish Q. Points for which two or more of the posterior probabilities are equal lie on the boundaries of adjacent regions R_j and, for the purpose of making a definite decision, can be assigned to either of them without altering the maximum value of the probability Q of correct decision. A receiver that thus maximizes the probability Q of correct decision, or minimizes the probability P_e of error, is said to be *optimum*.

In any given decision problem the prior probabilities ζ_j of the hypotheses are fixed. What is new in each observation or trial is the set v of data, and it determines the decision through the values $p_j(v)$, $1 \leq j \leq M$, of the M probability density functions at point v, which determine the posterior probabilities $\Pr(H_j|v)$ through (1-4). The primary task of the receiver is to generate those M numbers $p_j(v)$. Equivalently, it suffices for the receiver to produce M *likelihood ratios*

$$\Lambda_j(v) = \frac{p_j(v)}{p_d(v)}, \qquad 1 \leq j \leq M, \tag{1-7}$$

where $p_d(v)$ is any probability density function of the samples $(v_1, v_2, \ldots, v_n)$ that is nonzero at all points in R_n where any probability density function $p_j(v)$ is nonzero. We can think of $p_d(v)$ as the joint probability density function of the data v under a dummy hypothesis H_d. It is often convenient to take H_d as representing the absence of any signals whatever, the input to the receiver consisting of noise alone. The dummy hypothesis H_d may or may not figure among the actually possible hypotheses H_1 through H_M.

Knowing the values of the M likelihood ratios $\Lambda_j(v)$ permits determining which of the posterior probabilities $\Pr(H_j|v)$ is maximum, for

$$\Pr(H_j|v) = \frac{\zeta_j \Lambda_j(v)}{\displaystyle\sum_{m=1}^{M} \zeta_m \Lambda_m(v)}.$$

According to (1-5), the optimum receiver calculates the M products $\zeta_j \Lambda_j(v)$, $1 \le j \le M$, and selects the hypothesis corresponding to the largest among them. The region R_k of the data space is the set of points v such that

$$\zeta_k \Lambda_k(v) \ge \zeta_j \Lambda_j(v), \qquad \forall j \ne k, \qquad v \in R_k. \tag{1-8}$$

The probability of correct decision is now

$$Q = 1 - P_e = \sum_{j=1}^{M} \zeta_j \, \Pr(v \in R_j | H_j) = \sum_{j=1}^{M} \int_{R_j} \zeta_j p_j(v) \, d^n v$$

$$= \sum_{j=1}^{M} \int_{R_j} \zeta_j \Lambda_j(v) p_d(v) \, d^n v = \int_{R_n} \max_k \, [\zeta_k \Lambda_k(v)] \, p_d(v) \, d^n v \tag{1-9}$$

$$= E \left\{ \max_k \zeta_k \Lambda_k(v) | H_d \right\},$$

where E indicates an expected value, here with respect to the distribution of the data v under the dummy hypothesis H_d that they are distributed with the probability density function $p_d(v)$. As we shall see in Chapter 2, the use of such likelihood ratios $\Lambda_j(v)$ facilitates expanding the set v of samples to include all the information in the input $v(t)$ relevant to choosing among the M hypotheses $H_1, H_2, \ldots, H_M$ in the optimum fashion.

Example 1-1   Gaussian datum with M unequal expected values

A single datum v has a Gaussian (or normal) distribution under each of M hypotheses H_j; its probability density functions are

$$p_j(v) = \frac{1}{\sqrt{2\pi\sigma^2}} \exp\left[-\frac{(v - a_j)^2}{2\sigma^2} \right], \qquad j = 1, 2, \ldots, M; \tag{1-10}$$

σ^2 is its variance, and its expected value under hypothesis H_j is

$$E(v | H_j) = a_j$$

[Hel91, p. 81], [Pap91, p. 74]. Assume for simplicity that these expected values are arranged in ascending order, $a_1 < a_2 < \cdots < a_M$. On the basis of a single observation of the random variable v, we are to decide among these M hypotheses with minimum probability of error. We shall see later that the design of a receiver in a coherent pulse-amplitude-modulated communication system reduces to a decision problem of this form. The expected values a_j correspond to the signal levels and the variance σ^2 to the strength of the noise.

Take the dummy hypothesis H_d as representing the absence of any signal at all:

$$p_d(v) = \frac{1}{\sqrt{2\pi\sigma^2}} \exp\left(-\frac{v^2}{2\sigma^2} \right).$$

Then the likelihood ratios in (1-7) are

$$\Lambda_j(v) = \exp\left[\frac{v^2}{2\sigma^2} - \frac{(v - a_j)^2}{2\sigma^2} \right] = \exp\left[\frac{2a_j v - a_j^2}{2\sigma^2} \right].$$

Assume that the M hypotheses are equally likely: $\zeta_j \equiv 1/M$. The optimum receiver selects that hypothesis for which $\zeta_j \Lambda_j(v) = M^{-1}\Lambda_j(v)$ is maximum, that is, the hypothesis for which $a_j v - \frac{1}{2}a_j^2$ is largest.

The data space is now the real line, and this decision rule corresponds to a division of the line into M segments R_j separated by points midway between the expected values a_j:

$$R_1: \quad -\infty < v \leq \tfrac{1}{2}(a_1 + a_2), \qquad R_2: \quad \tfrac{1}{2}(a_1 + a_2) < v \leq \tfrac{1}{2}(a_2 + a_3), \dots ,$$

$$R_j: \quad \tfrac{1}{2}(a_{j-1} + a_j) < v \leq \tfrac{1}{2}(a_j + a_{j+1}), \dots , \qquad R_M: \quad \tfrac{1}{2}(a_{M-1} + a_M) < v < \infty.$$

The probability when hypothesis H_j is true that hypothesis H_j is correctly chosen is, by (1-10),

$$\Pr(\rightarrow H_j \mid H_j) = \Pr(v \in R_j \mid H_j)$$

$$= \int_{\frac{1}{2}(a_{j-1}+a_j)}^{\frac{1}{2}(a_j+a_{j+1})} p_j(v) \, dv = \frac{1}{\sqrt{2\pi\sigma^2}} \int_{\frac{1}{2}(a_{j-1}-a_j)}^{\frac{1}{2}(a_{j+1}-a_j)} \exp\left(-\frac{x^2}{2\sigma^2}\right) dx$$

$$= \operatorname{erfc}\left(\frac{a_{j-1} - a_j}{2\sigma}\right) - \operatorname{erfc}\left(\frac{a_{j+1} - a_j}{2\sigma}\right).$$

This holds also for $j = 1$ and $j = M$ if we take $a_0 = -\infty$, $a_{M+1} = \infty$.

Here $\operatorname{erfc} x$ is the error-function integral:

$$\operatorname{erfc} x = \frac{1}{\sqrt{2\pi}} \int_x^\infty e^{-t^2/2} \, dt. \tag{1-11}$$

Advice about computing $\operatorname{erfc} x$ is given in Appendix A of [Hel91]; tables are to be found there, in [Abr70, pp. 966–72], and in handbooks of probability and statistics. In particular

$$\operatorname{erfc}(-\infty) = 1, \qquad \operatorname{erfc} 0 = \tfrac{1}{2}, \qquad \operatorname{erfc} \infty = 0.$$

Because $\operatorname{erfc}(-x) = 1 - \operatorname{erfc} x$, we can write the probability of correct decision under hypothesis H_j as

$$\Pr(\rightarrow H_j \mid H_j) = 1 - \operatorname{erfc}\left(\frac{a_j - a_{j-1}}{2\sigma}\right) - \operatorname{erfc}\left(\frac{a_{j+1} - a_j}{2\sigma}\right).$$

The overall probability of correct decision is therefore, as in (1-2),

$$Q = \frac{1}{M} \sum_{j=1}^{M} \Pr(\rightarrow H_j \mid H_j) = 1 - \frac{2}{M} \sum_{j=1}^{M-1} \operatorname{erfc}\left(\frac{a_{j+1} - a_j}{2\sigma}\right),$$

and the probability of error is $P_e = 1 - Q$. When the expected values are uniformly spaced by δ, this reduces to an error probability of

$$P_e = \frac{2(M - 1)}{M} \operatorname{erfc}\left(\frac{\delta}{2\sigma}\right). \tag{1-12}$$

1.1.2 Minimizing Average Cost

It was observed by Abraham Wald [Wal39] that in some situations in which choices must be made among statistical hypotheses, certain errors are more serious than others. It would then be more sensible to adopt a strategy minimizing not the overall probability of error, but the average cost of operation. The designer is presumed to know the cost C_{ij} incurred upon choosing hypothesis H_i when hypothesis H_j is true, $1 \le (i, j) \le M$. These costs C_{ij} will depend both on the action attendant on each decision and on the true "state of nature" when that action is taken. They can be assembled into a cost matrix $\mathbf{C} = \|C_{ij}\|$.

Suppose that the set v of n samples has been observed. The conditional risk associated with choosing hypothesis H_i is obtained by weighting the costs C_{ij} attending that choice by the posterior probabilities, given v, that the several hypotheses H_j are true:

$$C(\rightarrow H_i | v) = \sum_{j=1}^{M} C_{ij} \Pr(H_j | v); \tag{1-13}$$

the conditional probabilities $\Pr(H_j | v)$ are given in (1-4). Let us again represent the strategy by a certain division of the data space R_n into M decision regions R_j. The cost of applying it, averaged over a long series of trials, is

$$\overline{C} = \sum_{i=1}^{M} \int_{R_i} p(v) C(\rightarrow H_i | v) \, d^n v = \sum_{i=1}^{M} \sum_{j=1}^{M} \zeta_j C_{ij} \int_{R_i} p_j(v) \, d^n v \tag{1-14}$$

by (1-4). By the same argument as we used to establish Bayes's rule, we can convince ourselves that the strategy that minimizes the average cost $\overline{C}$ of operation is that for which each region R_k contains those points v where the conditional risk $C(\rightarrow H_k | v)$ is smaller than all the other conditional risks $C(\rightarrow H_j | v)$:

$$\{C(\rightarrow H_k | v) < C(\rightarrow H_j | v), \forall j \ne k\} \Rightarrow \rightarrow H_k. \tag{1-15}$$

The observer chooses that hypothesis whose conditional risk $C(\rightarrow H_k | v)$, given the data v, is least. The strategy based on this rule is usually called the *Bayes strategy*.

When we introduce a dummy hypothesis H_d under which the data v have a joint probability density function $p_d(v)$, these conditional risks can be expressed in terms of the likelihood ratios $\Lambda_j(v)$ defined in (1-7): From (1-13) and (1-4),

$$C(\rightarrow H_i | v) = \sum_{j=1}^{M} C_{ij} \frac{\zeta_j p_j(v)}{p(v)} = \sum_{j=1}^{M} C_{ij} \frac{\zeta_j \Lambda_j(v)}{L(v)},$$

where, by (1-3),

$$L(v) = \frac{p(v)}{p_d(v)} = \sum_{k=1}^{M} \zeta_k \Lambda_k(v).$$

Again the receiver can base its decisions on the set of likelihood ratios $\Lambda_j(v)$, $j = 1, 2, \dots, M$.

The Bayes strategy is equivalent to Bayes's rule (1-5) when the relative costs of all errors are the same:

$$C_{ij} = \begin{cases} A, & i = j, \\ A + c, & i \ne j, \end{cases} \tag{1-16}$$

with $c > 0$ and A arbitrary, for then, by (1-13),

$$C(\rightarrow H_i | v) = A + c - c \Pr(H_i | v)$$

is smallest when $\Pr(H_i | v)$ is largest.

Once a radar system has detected a target, the observer would like to know how far away it is and how fast it is approaching. The distance of the target determines the arrival time τ of the echo signal, and its speed determines the Doppler shift w of the carrier frequency of the echo. Quantities such as these are parameters of the joint probability density function of the samples v of the input $v(t)$ to the radar receiver, and the observer estimates them on the basis of those data v. How this is best accomplished will be treated in Chapter 6, where we shall see that parameter estimation can be considered as a choice among an infinite number of hypotheses and thus as a limiting case $M \to \infty$ of the M-ary decision theory we have introduced here.

Decision theory has been applied by statisticians in numerous contexts other than the design of receivers. The general theory is treated in such books as those by Blackwell and Girshick [Bla54], Luce and Raiffa [Luc57], Chernoff and Moses [Che59], Lehmann [Leh59], and Winkler [Win72]. The concepts of decision theory were applied to radar detection at the M.I.T. Radiation Laboratory during World War II [Law50]. Communication receivers based on conditional probability were recommended by Kotel'nikov in a dissertation written in Russia in 1947, published there in 1956, and translated into English in 1959 [Kot59]. In the meantime Woodward and Davies suggested applying conditional probability to signal detection [Woo50], [Woo53]. The analysis of detection problems in terms of statistical decision theory was developed by Middleton [Mid53] while the design of receivers on the basis of the likelihood ratio was being advanced by a group at the University of Michigan [Pet54]. Many problems in the detection of signals and the measurement of signal parameters have since then been studied by means of the theory of statistical decisions. William Root, who contributed much to detection theory, has written a broad survey of its development [Roo87].

1.2 BINARY DECISIONS

1.2.1 Bayes Strategy

In a radar system, viewed in the most elementary way, the task of the receiver is to decide whether the echo from some target is or is not present in its input $v(t)$ observed during a certain interval $(0, T)$. The receiver in effect chooses between two hypotheses: (H_0) "no signal is present and $v(t)$ consists only of random noise," and (H_1) "a signal or one of a specified class of signals is present in addition to the noise." The receiver is said to make a *binary* decision. Hypothesis H_0 is commonly called the *null hypothesis*, H_1 the *alternative hypothesis*.

In most communication systems messages are coded into binary symbols such as 0 and 1. For each 0 a certain signal, say $s_0(t)$, is sent. Sometimes this signal is identically zero. For each 1 some other signal $s_1(t)$ is transmitted. Here too the receiver makes a binary decision, choosing between hypothesis H_0, "Signal $s_0(t)$ has arrived in the midst of random noise," and hypothesis H_1, "Signal $s_1(t)$ has arrived in the midst of noise." Because of the overwhelming importance of binary decisions, the greater portion of our study will be devoted to them.

In binary decisions the data space R_n is divided into only two regions: R_0, containing sample points v inducing the choice of hypothesis H_0, and R_1, containing points v inducing the choice of H_1. They are separated by a hypersurface called the *decision surface* and denoted by D. The problem is to find the optimum location of this surface in the data space.

Denoting by C_{ij} the cost attending the choice of hypothesis H_i when H_j is true $(i, j = 0, 1)$, we can as in (1-15) express the Bayes strategy in a binary decision as

$$\{C(\rightarrow H_1| v) \leq C(\rightarrow H_0| v)\} \Rightarrow \rightarrow H_1$$

or, by (1-13),

$$\{C_{10} \Pr(H_0| v) + C_{11} \Pr(H_1| v) \leq C_{00} \Pr(H_0| v) + C_{01} \Pr(H_1| v)\} \Rightarrow \rightarrow H_1.$$

Substituting from (1-4) and manipulating the resulting inequality, we obtain the rule

$$\left\{\Lambda(v) = \frac{p_1(v)}{p_0(v)} \geq \frac{\zeta_0(C_{10} - C_{00})}{\zeta_1(C_{01} - C_{11})} = \Lambda_0\right\} \Rightarrow \rightarrow H_1. \qquad (1\text{-}17)$$

Here $\zeta_0 = \Pr(H_0)$ and $\zeta_1 = \Pr(H_1)$ are the prior probabilities of hypotheses H_0 and H_1, respectively; $\zeta_0 + \zeta_1 = 1$. The quantity $\Lambda(v) = p_1(v)/p_0(v)$ is the likelihood ratio appropriate for binary decision. The receiver computes $\Lambda(v)$ from the values $(v_1, v_2, \ldots, v_n)$ of the n samples of its input during $(0, T)$ and compares it with the *decision level* Λ_0, which incorporates the costs and the prior probabilities. If $\Lambda(v) < \Lambda_0$, hypothesis H_0 is selected, otherwise H_1.

Two kinds of error arise in binary decisions: choosing hypothesis H_1 when H_0 is true, an "error of the first kind," and choosing H_0 when H_1 is true, an "error of the second kind." In radar and sonar applications these are called a *false alarm* and a *false dismissal*, respectively. The relative costs of the two kinds of error are $C_{10} - C_{00}$ and $C_{01} - C_{11}$. When these are equal, $\Lambda_0 = \zeta_0/\zeta_1$; (1-17) is then equivalent to Bayes's rule, and the probability P_e of error is minimized.

The probability of an error of the first kind, or the *false-alarm probability* Q_0, is

$$Q_0 = \Pr(\rightarrow H_1| H_0) = \Pr(v \in R_1| H_0) = \int_{R_1} p_0(v)\, d^n v, \qquad (1\text{-}18)$$

where $p_0(v)$ is the joint probability density function of the samples $(v_1, v_2, \ldots, v_n)$ under hypothesis H_0.

The event complementary to an error of the second kind, that is, deciding that a signal is present when it is indeed at hand, is called a *detection*. The probability Q_d of detection is

$$Q_d = \Pr(\rightarrow H_1| H_1) = \Pr(v \in R_1| H_1) = \int_{R_1} p_1(v)\, d^n v, \qquad (1\text{-}19)$$

where $p_1(v)$ is the joint probability density function of the samples $(v_1, v_2, \ldots, v_n)$ under hypothesis H_1. The false-dismissal probability is $Q_1 = 1 - Q_d$. Statisticians call Q_0 the *size* and Q_d the *power* of the binary hypothesis test. In terms of these, the probability P_e of error is

$$P_e = \zeta_0 Q_0 + \zeta_1(1 - Q_d). \qquad (1\text{-}20)$$

1.2.2 Neyman–Pearson Criterion

In many situations both the prior probabilities and the costs are difficult to estimate or perhaps even to define. This is especially true in signal-detection problems in radar, where it is hard to judge the cost of failing to detect a target and where the prior probability of a signal may not even be a meaningful concept. When hypothesis H_1 is true extremely rarely, the principal factor in the average total cost is the fraction Q_0 of trials in which hypothesis H_1 is incorrectly chosen and an error of the first kind is made, whereupon some costly action is taken in vain. In a radar detection system, for instance, such a false alarm may lead to firing an expensive missile to attack a nonexistent target. Under such circumstances it is appropriate for the observer to determine an affordable value of the false-alarm probability Q_0 and to seek a decision strategy that attains this value and at the same time yields the minimum possible probability $Q_1 = 1 - Q_d$ of making an error of the second kind. This strategy, proposed by Neyman and Pearson [Ney33a], [Ney33b], is said to fulfill the *Neyman–Pearson criterion.* It corresponds in radar to maximizing the probability of detecting an echo signal while incurring a given false-alarm probability.

Adopting the Neyman–Pearson criterion, as we shall see, also calls for a strategy based on the likelihood ratio and expressed by

$$\left\{ \Lambda(v) = \frac{p_1(v)}{p_0(v)} \geq \lambda \right\} \Rightarrow \longrightarrow H_1. \tag{1-21}$$

The only difference from (1-17) is that the decision level λ is determined by the preassigned value of the false-alarm probability Q_0,

$$Q_0 = \Pr(\Lambda(v) \geq \lambda \mid H_0) = \int_\lambda^\infty P_0(\Lambda) \, d\Lambda, \tag{1-22}$$

where $P_0(\Lambda)$ is the probability density function of the random variable $\Lambda(v)$ under hypothesis H_0. [Remember that because the likelihood ratio $\Lambda(v)$ is a function of the n random variables $v_1, v_2, \ldots, v_n$, it is itself a random variable.] The maximum value of the probability Q_d of detection so attained is

$$Q_d = \Pr(\Lambda(v) \geq \lambda \mid H_1) = \int_\lambda^\infty P_1(\Lambda) \, d\Lambda, \tag{1-23}$$

where $P_1(\Lambda)$ is the probability density function of $\Lambda(v)$ under H_1.

In order to demonstrate that the rule in (1-21) is optimum, let us suppose that regions R_0 and R_1 are separated by the decision surface D given by $\Lambda(v) = \lambda$, with λ chosen to satisfy (1-22). Then the probability of detection is

$$Q_d = \int_{R_1} p_1(v) \, d^n v = \int_{R_1} \Lambda(v) \, p_0(v) \, d^n v \tag{1-24}$$

by virtue of (1-21). Now, as shown for $n = 2$ in Fig. 1-1, we deform the decision surface D in such a way that the value of

$$Q_0 = \int_{R_1} p_0(v) \, d^n v$$

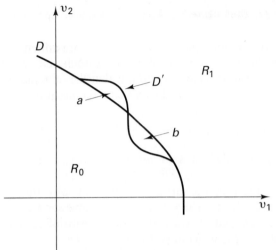

Figure 1-1. Deformation of decision surface.

remains the same. There are countless ways of doing so. Call the new surface D'. The points in region a that have been transferred from R_1 to R_0 have the same probability measure under hypothesis H_0 as those in region b that have been transferred from R_0 to R_1. If we measure the "value" of a point by $\Lambda(v)$, we can say with respect to (1-24) that in thus altering region R_1 we have exchanged points v of greater value $\Lambda(v)$—those in a—for points of lesser value—those in b. As a result, the detection probability Q_d in (1-24) has decreased, while the false-alarm probability Q_0 has remained the same. The dichotomy of the data space R_n into regions R_0 and R_1 separated by the surface $\Lambda(v) = \lambda$ must therefore be optimum under the Neyman–Pearson criterion.

Symbolically, the change δQ_d in the probability of detection when we deform the surface D into D' is

$$\delta Q_d = \int_b p_1(v)\, d^n v - \int_a p_1(v)\, d^n v$$
$$= \int_b \Lambda(v) p_0(v)\, d^n v - \int_a \Lambda(v) p_0(v)\, d^n v,$$

where $\Lambda(v)$ is the likelihood ratio defined in (1-21). In region a, $\Lambda(v) \geq \lambda$; in region b, $\Lambda(v) < \lambda$. Therefore,

$$\int_a \Lambda(v) p_0(v)\, d^n v \geq \lambda \int_a p_0(v)\, d^n v,$$
$$\int_b \Lambda(v) p_0(v)\, d^n v < \lambda \int_b p_0(v)\, d^n v,$$

and subtracting these we find

$$\delta Q_d \leq \lambda \int_b p_0(v)\, d^n v - \lambda \int_a p_0(v)\, d^n v = 0$$

because

$$\int_a p_0(v)\, d^n v = \int_b p_0(v)\, d^n v.$$

The detection probability Q_d thus decreases when the decision surface is deformed in this way.

Another way to derive the optimum strategy under the Neyman–Pearson criterion is to apply the calculus of variations. We want to minimize the probability $Q_1 = 1 - Q_d$ of a false dismissal under the constraint that the false-alarm probability Q_0 take on a preassigned value, say $\overline{Q}_0$. Introducing the Lagrange multiplier λ, we are to minimize

$$\overline{C} = Q_1 + \lambda(Q_0 - \overline{Q}_0)$$
$$= \int_{R_0} p_1(v) \, d^n v + \lambda \int_{R_1} p_0(v) \, d^n v - \lambda \overline{Q}_0 \qquad (1\text{-}25)$$

by varying the position of the decision surface D. We can solve this problem by drawing on the results of Sec. 1.1.1. The average cost of operating a strategy that chooses between hypotheses H_0 and H_1 when their prior probabilities are ζ_0 and ζ_1 ($\zeta_0 + \zeta_1 = 1$) and the costs are C_{ij} is, as in (1-14),

$$\overline{C} = \zeta_0 \left[C_{00} \int_{R_0} p_0(v) \, d^n v + C_{10} \int_{R_1} p_0(v) \, d^n v \right]$$
$$+ \zeta_1 \left[C_{01} \int_{R_0} p_1(v) \, d^n v + C_{11} \int_{R_1} p_1(v) \, d^n v \right]$$
$$= \zeta_0 C_{00} + \zeta_1 C_{11} + \zeta_0(C_{10} - C_{00}) \int_{R_1} p_0(v) \, d^n v + \zeta_1(C_{01} - C_{11}) \int_{R_0} p_1(v) \, d^n v.$$

This reduces to (1-25) if we take the costs as

$$C_{00} = -2\lambda\overline{Q}_0, \qquad C_{10} = 2\lambda(1 - \overline{Q}_0),$$
$$C_{01} = 2, \qquad C_{11} = 0,$$

and set $\zeta_0 = \zeta_1 = \frac{1}{2}$. The analysis leading to (1-17) shows that the decision surface minimizing the average cost $\overline{C}$ in (1-14) is given by the equation $\Lambda(v) = \Lambda_0 = \lambda$; we simply substitute these costs and prior probabilities into the expression for Λ_0 in (1-17). The value of the Lagrange multiplier λ is then chosen to satisfy the constraint $Q_0 = \overline{Q}_0$.

1.2.3 Operating Characteristic

As the parameter λ in (1-21) varies from 0 to ∞, the decision surface D moves through the space R_n of the data v, bounding for each value of λ the region $R_1(\lambda)$ of points v that lead to the choice of hypothesis H_1. The false-alarm probability

$$Q_0(\lambda) = \int_{R_1(\lambda)} p_0(v) \, d^n v \qquad (1\text{-}26)$$

and the detection probability

$$Q_d(\lambda) = \int_{R_1(\lambda)} p_1(v) \, d^n v \qquad (1\text{-}27)$$

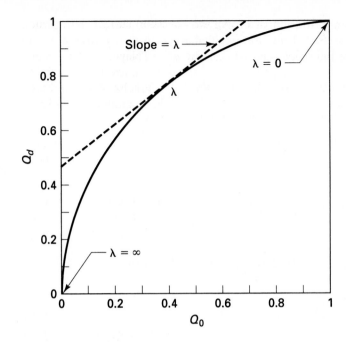

Figure 1-2. Operating characteristic. The dashed line is tangent to the operating characteristic at a point whose coordinates are $Q_0(\lambda)$, $Q_d(\lambda)$.

are in this way functions of λ, as is indeed evident also from (1-22) and (1-23). By letting λ range over $(0, \infty)$, we can plot the probability $Q_d(\lambda)$ of detection versus the probability $Q_0(\lambda)$ of a false alarm; the resulting curve is called the *operating characteristic* of the receiver. It depends only on the probability density functions of the data v under the two hypotheses and not on any costs or prior probabilities. Figure 1-2 depicts a typical operating characteristic. The parameter λ varies from $\lambda = 0$ at the upper terminus $(1, 1)$ of the curve to $\lambda = \infty$ at the lower terminus $(0, 0)$.

The slope of the operating characteristic at any point equals the value of the parameter λ at that point:

$$\frac{dQ_d}{dQ_0} = \lambda. \tag{1-28}$$

Indeed, the optimum strategy must be such as to make $\overline{C}$ in (1-25) a stationary point with respect to variations in both the position of the decision surface D and the value of λ. We find by differentiating (1-25) with respect to λ and setting the result equal to zero that

$$\frac{dQ_1}{d\lambda} + \lambda\frac{dQ_0}{d\lambda} + Q_0 - \overline{Q}_0 = 0.$$

The value of λ is chosen so that $Q_0 = \overline{Q}_0$, and because $Q_1 = 1 - Q_d$, we obtain (1-28) immediately.

When the parameter λ decreases from ∞ to 0, the slope of the tangent to the operating characteristic decreases monotonely from the origin $(0, 0)$ to the upper terminus $(1, 1)$, and the operating characteristic is therefore convex $\cap$.

In an alternative demonstration of (1-28), we consider two nearby decision surfaces in the data space R_n. They are specified respectively by $\Lambda(v) = \lambda$ and $\Lambda(v) = \lambda + d\lambda$. Designate the space between them by dR_λ. Then

$$\Pr(\lambda < \Lambda(v) < \lambda + d\lambda | \ H_1) = P_1(\lambda)d\lambda = \int_{dR_\lambda} p_1(v) \, d^n v.$$

With $d\lambda$ infinitesimal, the lamina dR_λ is so thin that within it

$$\frac{p_1(v)}{p_0(v)} = \Lambda(v) = \lambda, \qquad v \in dR_\lambda,$$

and therefore

$$P_1(\lambda)d\lambda = \int_{dR_\lambda} \Lambda(v)p_0(v) \, d^n v = \lambda \int_{dR_\lambda} p_0(v) \, d^n v.$$

Likewise,

$$\Pr(\lambda < \Lambda(v) < \lambda + d\lambda | \ H_0) = P_0(\lambda)d\lambda = \int_{dR_\lambda} p_0(v) \, d^n v.$$

Dividing, we obtain

$$\frac{P_1(\lambda)}{P_0(\lambda)} = \lambda, \qquad 0 < \lambda < \infty. \tag{1-29}$$

From (1-22) and (1-23), however,

$$\frac{dQ_d}{d\lambda} = -P_1(\lambda), \qquad \frac{dQ_0}{d\lambda} = -P_0(\lambda).$$

The slope of the operating characteristic is thus

$$\frac{dQ_d}{dQ_0} = \frac{dQ_d/d\lambda}{dQ_0/d\lambda} = \frac{P_1(\lambda)}{P_0(\lambda)} = \lambda$$

as in (1-28). Equation (1-29) is useful, for it enables us to calculate the probability density function $P_1(\Lambda)$ of the likelihood ratio under hypothesis H_1 when we know its density function $P_0(\Lambda)$ under H_0, or vice versa:

$$P_1(\Lambda) = \Lambda P_0(\Lambda). \tag{1-30}$$

1.2.4 Sufficient Statistics

The likelihood ratio $\Lambda(v) = p_1(v)/p_0(v)$ embodies all the information contained in the data $v = (v_1, v_2, \ldots, v_n)$ that is relevant to deciding between hypotheses H_0 and H_1 in the optimum fashion. If someone else measures the data v, computes the likelihood ratio $\Lambda(v)$, and tells you only the result, you can attain the same minimum Bayes cost $\overline{C}_{\min}$ by calculating the likelihood ratio

$$\Lambda = \frac{P_1(\Lambda)}{P_0(\Lambda)} \tag{1-31}$$

and comparing it with the decision level Λ_0 given by (1-17). For this reason, $\Lambda = \Lambda(v)$ is called a *sufficient statistic*. Any monotone function $G = G(\Lambda)$ of the likelihood ratio will do as well. Without loss of generality we suppose this to be an increasing function of Λ. If G exceeds a certain decision level G_0, hypothesis H_1 is chosen, otherwise H_0. Under the Neyman–Pearson criterion the value of G_0 is picked so that the false-alarm probability

$$Q_0 = \int_{G_0}^{\infty} p_0(G) \, dG \qquad (1\text{-}32)$$

takes on the preassigned value, where $p_0(G)$ is the probability density function of G under hypothesis H_0. The probability Q_d of detection is then

$$Q_d = \int_{G_0}^{\infty} p_1(G) \, dG \qquad (1\text{-}33)$$

with $p_1(G)$ the density function of G under hypothesis H_1. The quantity G is a function of the data $(v_1, v_2, \ldots, v_n)$ and, like the likelihood ratio $\Lambda(v)$, embodies all the information in them that contributes to making the decision in the best possible way under our two criteria. It too is a sufficient statistic.

A likelihood ratio can be formed with the statistic G as well,

$$\Lambda(G) = \frac{p_1(G)}{p_0(G)}, \qquad (1\text{-}34)$$

where $p_0(G)$ and $p_1(G)$ are the probability density functions of G under the two hypotheses. Under the Bayes criterion the level G_0 with which the statistic G is to be compared is given by the equation

$$\Lambda(G_0) = \frac{p_1(G_0)}{p_0(G_0)} = \Lambda_0 = \frac{\zeta_0(C_{10} - C_{00})}{\zeta_1(C_{01} - C_{11})}. \qquad (1\text{-}35)$$

The points on the operating characteristic can be indexed with the values of G_0, and the parametric equations of that curve are

$$Q_d = Q_d(G_0), \qquad Q_0 = Q_0(G_0),$$

with these functions of G_0 given by (1-32) and (1-33).

The sufficient statistic most commonly used is proportional to the logarithm of the likelihood ratio,

$$g = \ln \left[\frac{p_1(v)}{p_0(v)} \right]. \qquad (1\text{-}36)$$

When the probability density functions are jointly Gaussian, this statistic takes a particularly simple form. Furthermore, when data are taken in statistically independent batches, whatever their joint probability density functions, the logarithm of the likelihood ratio equals the sum of this statistic for each batch. The value of g for the batch comprises all the relevant information in the batch and can be measured in its stead.

Example 1-2 n Gaussian data with unequal expected values

As a simple example of these concepts, suppose that the quantities $(v_1, v_2, \ldots, v_n)$ are statistically independent and have Gaussian distributions with variance σ^2. Let the expected value of each be a_0 under hypothesis H_0 and a_1 under H_1; $a_0 < a_1$. Their joint probability density functions are then products of Gaussian density functions of the form in (1-10),

$$p_k(v) = (2\pi\sigma^2)^{-n/2} \exp\left[-\sum_{i=1}^{n} \frac{(v_i - a_k)^2}{2\sigma^2} \right], \qquad k = 0, 1, \qquad (1\text{-}37)$$

[Hel91, p. 215]. The likelihood ratio is now

$$\Lambda(v) = \frac{p_1(v)}{p_0(v)} = \exp\left[\frac{(a_1 - a_0)}{\sigma^2} \sum_{i=1}^{n} v_i - \frac{n(a_1^2 - a_0^2)}{2\sigma^2} \right]. \qquad (1\text{-}38)$$

The observer will choose hypothesis H_0 when $\Lambda(v) < \Lambda_0$, where the value of Λ_0 depends on the decision criterion used. Because the exponential function is monotone, the decision can just as well be based on the value of

$$V = \frac{1}{n} \sum_{i=1}^{n} v_i,$$

the *sample mean* of the observations, which is to be compared with the quantity V_0 given by

$$V_0 = \frac{a_0 + a_1}{2} + \frac{\sigma^2 \ln \Lambda_0}{n(a_1 - a_0)}.$$

With $a_1 > a_0$, hypothesis H_0 is chosen when $V < V_0$ and H_1 when $V \geq V_0$. If the Bayes criterion is being used, V_0 depends on the costs and prior probabilities through Λ_0, which is given by (1-17). In this example the decision surface D is the hyperplane

$$\sum_{i=1}^{n} v_i = nV_0.$$

The sample mean V is thus a sufficient statistic in the sense just explained. It is a Gaussian random variable because it is a linear combination of Gaussian random variables [Hel91, p. 245], [Pap91, p. 197]. Its expected values under hypotheses H_0 and H_1 are a_0 and a_1, respectively, and its variance is σ^2/n. Its probability density functions under the two hypotheses are therefore

$$P_k(V) = \left(\frac{2\pi\sigma^2}{n} \right)^{-1/2} \exp\left[-\frac{n(V - a_k)^2}{2\sigma^2} \right], \qquad k = 0, 1,$$

and its likelihood ratio is

$$\Lambda(V) = \frac{P_1(V)}{P_0(V)} = \exp\left[\frac{n(a_1 - a_0)}{\sigma^2} V - \frac{n\left(a_1^2 - a_0^2\right)}{2\sigma^2} \right],$$

which of course equals $\Lambda(v)$ in (1-38).

The probabilities of error of the first and second kinds are

$$Q_0 = \int_{V_0}^{\infty} P_0(V) \, dV = \operatorname{erfc}\left[\frac{V_0 - a_0}{(\sigma/\sqrt{n})} \right],$$

$$Q_1 = \int_{-\infty}^{V_0} P_1(V) \, dV = 1 - \operatorname{erfc}\left[\frac{V_0 - a_1}{(\sigma/\sqrt{n})} \right],$$

where $\operatorname{erfc}(\cdot)$ is defined in (1-11). To use the Neyman–Pearson criterion, the observer chooses the value of V_0 so that the probability Q_0 of an error of the first kind takes on the preassigned value.

Example 1-3 Gaussian data with unequal variances

Suppose the observer must decide which of two sources of Gaussian random noise is present in an input voltage, the one having a variance equal to N_0, the other a variance equal to N_1. The expected value of the noise voltage is zero in both cases. He measures the voltage at n times far enough apart that the results $(v_1, v_2, \ldots, v_n)$ are statistically independent. The two hypotheses between which he must choose are (H_0), "The variance of the voltage equals N_0," and (H_1), "The variance of the voltage equals N_1." We assume that $N_1 > N_0$, as when under hypothesis H_1 a noiselike signal is present in addition to the usual background noise, whose variance equals N_0. The joint probability density functions of the set of n measured voltages are given by

$$p_k(v) = (2\pi N_k)^{-n/2} \exp\left[-\sum_{i=1}^{n} \frac{v_i^2}{2N_k} \right], \qquad k = 0, 1. \tag{1-39}$$

The likelihood ratio for these measurements is

$$\Lambda(v) = \left[\frac{N_0}{N_1} \right]^{n/2} \exp\left[\tfrac{1}{2} \left(N_0^{-1} - N_1^{-1} \right) \sum_{i=1}^{n} v_i^2 \right]. \tag{1-40}$$

The observer computes this likelihood ratio for the outcome of the experiment and compares it with a fixed quantity Λ_0. If $\Lambda(v) < \Lambda_0$, he decides that the variance of the input voltage was N_0, otherwise that it was N_1.

The likelihood ratio in (1-40) depends on the data v only through the sum of squares

$$S = \sum_{i=1}^{n} v_i^2, \tag{1-41}$$

and the observer can base his decision just as well on the value of S, comparing it with a decision level $S_0 = r_0^2$ given by

$$S_0 = r_0^2 = \frac{2N_0 N_1}{N_1 - N_0} \ln\left[\Lambda_0 \left(\frac{N_1}{N_0} \right)^{n/2} \right];$$

if $S < r_0^2$, he chooses hypothesis H_0, otherwise H_1. The decision surface D is an n-dimensional hypersphere of radius r_0, and the regions R_0 and R_1 into which the data space R_n is divided are, respectively, the interior and exterior of this hypersphere. The sum of squares S is a sufficient statistic.

The false-alarm probability is the probability under hypothesis H_0 that the point $v = (v_1, v_2, \ldots, v_n)$ lies outside the hypersphere defined by

$$\sum_{i=1}^{n} v_i^2 = S_0 = r_0^2,$$

and this probability is calculated by integrating the joint density function $p_0(v)$ over the exterior R_1 of the hypersphere:

$$Q_0 = \Pr(S > S_0 \mid H_0) = (2\pi N_0)^{-n/2} \int_{R_1} \exp\left[-\sum_{i=1}^{n} \frac{v_i^2}{2N_0} \right] d^n v. \tag{1-42}$$

The integrand is constant on hyperspherical shells, and the volume of such a shell of radius r and thickness dr is

$$dV = \frac{2\pi^{n/2}}{\Gamma(n/2)} r^{n-1} \, dr,$$

[Edw73, p. 339], where $\Gamma(x)$ is the gamma function. [For positive integral values of x, $\Gamma(x) = (x - 1)!$ For half-integral values one can use the formulas $\Gamma(x + 1) = x\Gamma(x)$, $\Gamma(\frac{1}{2}) = \pi^{1/2}$. Thus $\Gamma(3/2) = \frac{1}{2}\pi^{1/2}$, and so forth.]

The integral in (1-42) can be written

$$Q_0(S_0) = 2(2N_0)^{-n/2} \left[\Gamma\left(\frac{n}{2}\right) \right]^{-1} \int_{r_0}^{\infty} r^{n-1} \exp\left(-\frac{r^2}{2N_0}\right) dr$$

$$= \left[\Gamma\left(\frac{n}{2}\right) \right]^{-1} \int_{x_0}^{\infty} x^{(n/2)-1} e^{-x} dx, \qquad x_0 = \frac{S_0}{2N_0}, \tag{1-43}$$

which can be evaluated by Pearson's tables of the incomplete gamma function [Pea34]. Similarly, the probability of detection is

$$Q_d(S_0) = \left[\Gamma\left(\frac{n}{2}\right) \right]^{-1} \int_{x_0'}^{\infty} x^{(n/2)-1} e^{-x} dx, \qquad x_0' = \frac{S_0}{2N_1} = \frac{x_0 N_0}{N_1}. \tag{1-44}$$

The statistic S has a scaled chi-squared distribution under each hypothesis [Hel91, p. 220], [Pap91, p. 79].

By differentiating (1-43) and (1-44) with respect to S_0 and dividing, we can verify the counterpart of (1-34):

$$\Lambda(S_0) = \frac{p_1(S_0)}{p_0(S_0)} = \left[\frac{N_0}{N_1} \right]^{n/2} \exp\left[\frac{1}{2} \left(N_0^{-1} - N_1^{-1} \right) S_0 \right] \tag{1-45}$$

for the density functions of the statistic S. This should be compared with (1-40).

The data $v_1, v_2, \ldots, v_n$ might be processed in some other way. The receiver might, for instance, use instead of S the statistic

$$S' = \sum_{k=1}^{n} v_k^4.$$

It would be compared with a decision level S_0' set to yield a preassigned false-alarm probability Q_0. Calculating the false-alarm and detection probabilities for this new statistic in order to compare its performance with that of the one defined in (1-41) would be quite difficult. Our general theory assures us, however, that when the false-alarm probabilities are set equal, the detection probability attained by the sum S' of fourth powers of the data will be less than that attained by the sum S of their squares.

1.2.5 Decisions Based on Discrete Random Variables

In the receiver of an optical communication or radar system, the entrant light may be filtered and focused onto a photoelectrically emissive surface, and decisions about the presence or absence of a signal may be based on the number k of photoelectrons emitted during some observation interval $(0, T)$. The number k, taking on only nonnegative integral values, is a discrete random variable that tends to be larger when a signal is present than when only background light enters the receiver. A more elaborate device might be designed to detect a radiant object by counting the numbers $k_1, k_2, \ldots, k_n$ of photoelectrons emitted by each of n detectors sensitive to light waves having different frequencies or arriving from different directions. In order to develop optimum strategies for such receivers, decision theory must take account of the discrete nature of data of this kind. The methods of Secs. 1.2.1 and 1.2.2 are easily modified.

A decision is to be made between two hypotheses H_0 and H_1 about a set of discrete random variables $(x_1, x_2, \ldots, x_n)$ observed in some experiment. The probability under hypothesis H_0 that the data take on the set of values $\mathbf{x} = (x_1, x_2, \ldots, x_n)$ is denoted by

$$P_0(x_1, x_2, \ldots, x_n) = P_0(\mathbf{x});$$

the probability under hypothesis H_1 that these values are observed is

$$P_1(x_1, x_2, \ldots, x_n) = P_1(\mathbf{x}).$$

One calls $P_0(\mathbf{x})$ and $P_1(\mathbf{x})$ the joint probability mass functions of the data under the two hypotheses. When these probabilities are summed over all possible values of each of the $\mathbf{x}$'s, the result is 1:

$$\sum_{\mathbf{x} \in R_n} P_k(\mathbf{x}) = 1, \qquad k = 0, 1,$$

[Hel91, p. 203]. We can again think of $(x_1, x_2, \ldots, x_n)$ as the coordinates of a point $\mathbf{x}$ in the n-dimensional Cartesian space R_n, but now all the probability is concentrated at a countable set of points. The probability that the data point falls into a region Δ of the space R_n is the sum of the probabilities attached to those points $\mathbf{x}$ in Δ; symbolically,

$$\Pr(\mathbf{x} \in \Delta \mid H_k) = \sum_{\mathbf{x} \in \Delta} P_k(\mathbf{x}), \qquad k = 0, 1. \tag{1-46}$$

A decision strategy again divides the space R_n into two regions R_0 and R_1; when the data point $\mathbf{x}$ falls into region R_k, hypothesis H_k is selected, $k = 0, 1$. The average risk associated with this strategy is given by an expression like that in (1-14), except that the integrations over R_0 and R_1 are replaced by summations over the data points $\mathbf{x}$ in those regions. Conditional risks can be defined as in (1-13) in terms of the posterior probabilities

$$\Pr(H_k \mid \mathbf{x}) = \frac{\zeta_k P_k(\mathbf{x})}{P(\mathbf{x})}, \tag{1-47}$$

where ζ_0 and ζ_1 are the prior probabilities of hypotheses H_0 and H_1, and

$$P(\mathbf{x}) = \zeta_0 P_0(\mathbf{x}) + \zeta_1 P_1(\mathbf{x}) \tag{1-48}$$

is the total probability of observing the set $\mathbf{x}$ of data. The average cost for a given strategy is then, as in (1-14),

$$\overline{C} = \sum_{\mathbf{x} \in R_0} C(\to H_0 \mid \mathbf{x}) P(\mathbf{x}) + \sum_{\mathbf{x} \in R_1} C(\to H_1 \mid \mathbf{x}) P(\mathbf{x})$$

with the conditional risks $C(\to H_0 \mid \mathbf{x})$ and $C(\to H_1 \mid \mathbf{x})$ defined as in (1-13). Again the average cost is minimum for the strategy that picks the hypothesis with the smaller conditional risk, and this implies as before that the optimum decision regions are

$$R_0 = \{\mathbf{x}: \Lambda(\mathbf{x}) < \Lambda_0\}, \qquad R_1 = \{\mathbf{x}: \Lambda(\mathbf{x}) \geq \Lambda_0\}, \tag{1-49}$$

where now the likelihood ratio

$$\Lambda(\mathbf{x}) = \frac{P_1(\mathbf{x})}{P_0(\mathbf{x})} \tag{1-50}$$

is the quotient of the *probabilities* of the data $\mathbf{x}$ under the two hypotheses. The decision level Λ_0 is again

$$\Lambda_0 = \frac{\zeta_0(C_{10} - C_{00})}{\zeta_1(C_{01} - C_{11})}. \tag{1-51}$$

The optimum strategy is, as before, to compare the likelihood ratio $\Lambda(\mathbf{x})$ with Λ_0 and to choose hypothesis H_1 if it exceeds Λ_0 and hypothesis H_0 otherwise. The false-alarm and detection probabilities are

$$Q_0 = \sum_{\mathbf{x} \in R_1} P_0(\mathbf{x}), \qquad Q_d = \sum_{\mathbf{x} \in R_1} P_1(\mathbf{x}), \tag{1-52}$$

respectively.

Example 1-4   Geometrically distributed datum

Suppose that under each hypothesis the single datum x has a geometrical distribution of nonnegative integral values,

$$P_0(x) = (1 - v_0)v_0^x, \qquad P_1(x) = (1 - v_1)v_1^x, \qquad v_1 > v_0,$$
$$v_0 = \frac{m_0}{m_0 + 1}, \qquad v_1 = \frac{m_1}{m_1 + 1}, \qquad x = 0, 1, 2, \dots, \tag{1-53}$$

where m_0 and m_1 are the expected values of the datum under hypotheses H_0 and H_1, respectively. The likelihood ratio is

$$\Lambda(x) = \frac{1 - v_1}{1 - v_0} \left(\frac{v_1}{v_0} \right)^x, \tag{1-54}$$

and hypothesis H_1 is chosen if

$$x > \left\lfloor \frac{\ln \left(\frac{1-v_0}{1-v_1} \right) + \ln \Lambda_0}{\ln (v_1/v_0)} \right\rfloor = n_0, \tag{1-55}$$

where $\lfloor u \rfloor$ denotes the greatest integer in a real number u. The false-alarm and detection probabilities are

$$Q_0 = (1 - v_0) \sum_{k=n_0+1}^{\infty} v_0^k = v_0^{n_0+1}, \qquad Q_d = v_1^{n_0+1}, \tag{1-56}$$

and the probability of error is

$$P_e = \zeta_0 Q_0 + \zeta_1(1 - Q_d).$$

This is minimum when Λ_0 is set as in (1-51), with $C_{10} - C_{00} = C_{01} - C_{11}$.

The Bayes strategy for optimum decisions among more than two hypotheses can be developed in a similar manner and is again expressed by (1-5) or (1-15). In calculating the conditional probabilities $\Pr(H_k | \mathbf{x})$, the probabilities $P_k(\mathbf{x})$ take the place of the probability density functions that figured in (1-4).

Because the datum x in Example 1-4 takes on only integral values, the decision level with which it is compared is effectively an integer, and (1-56) shows that the false-alarm probability Q_0 can take on only one of a countable set of values. If the Neyman–Pearson criterion has been adopted, it may happen that the preassigned value Q_0 of the false-alarm probability is not a member of that set. In order to achieve an arbitrary value of the false-alarm probability, it is necessary to resort to what is called *randomization*.

In Example 1-4 a randomized strategy prescribes choosing hypothesis H_1 whenever the integral datum x exceeds a certain decision level n_0, and H_0 when $x < n_0$; but when x equals n_0 exactly, hypothesis H_1 is chosen with a certain probability f and hypothesis H_0 with probability $1 - f$. The false-alarm probability is then

$$Q_0 = \sum_{x=n_0+1}^{\infty} P_0(x) + f\, P_0(n_0). \qquad (1\text{-}57)$$

The values of n_0 and f are selected to yield the desired value of this probability. Let us write (1-57) as

$$f\, P_0(n_0) = Q_0 - 1 + \sum_{x=0}^{n_0} P_0(x).$$

We sum the values of $P_0(x)$, starting with $x = 0$, until the right side first becomes positive. The value of x at which that occurs becomes the decision level n_0, and by dividing by $P_0(n_0)$, we obtain the value of the probability f. The probability of detection is then

$$Q_d = \sum_{x=n_0+1}^{\infty} P_1(x) + f\, P_1(n_0). \qquad (1\text{-}58)$$

Figure 1-3 exhibits the operating characteristic of the randomized strategy for our example of a geometrically distributed datum. It is simple to construct such an operating characteristic. The vertices of the polygon are the points at which $f = 1$ and the decision level jumps from one integer n_0 to the next; in this example they are given by (1-56) for all nonnegative integers n_0. One plots these vertices and connects them with straight lines.

In general, the likelihood ratio $\Lambda(\mathbf{x}) = P_1(\mathbf{x})/P_0(\mathbf{x})$ takes on only a countable set of values λ_k, which we can arrange in increasing order:

$$0 \le \lambda_1 < \lambda_2 < \cdots < \lambda_{k-1} < \lambda_k < \lambda_{k+1} < \cdots < \infty.$$

A number of sets $\mathbf{x}$ of data may yield the same value of $\Lambda(\mathbf{x})$. The optimum randomized strategy under the Neyman–Pearson criterion can be described as follows. The decision level is a particular one of those numbers, say λ_K. When $\Lambda(\mathbf{x}) > \lambda_K$, hypothesis H_1 is chosen, and when $\Lambda(\mathbf{x}) < \lambda_K$, hypothesis H_0 is chosen; but when $\Lambda(\mathbf{x}) = \lambda_K$, hypothesis H_1 is chosen with probability f and H_0 with probability $1 - f$. The false-alarm probability is

$$Q_0 = f\, \Pr(\Lambda(\mathbf{x}) = \lambda_K | H_0) + \sum_{k=K+1}^{\infty} \Pr(\Lambda(\mathbf{x}) = \lambda_k | H_0), \qquad (1\text{-}59)$$

and by the same technique as for (1-57), the values of f and K can be determined so that the right side of (1-59) equals the preassigned value of the false-alarm probability. The probability of detection is then

$$Q_d = f\, \Pr(\Lambda(\mathbf{x}) = \lambda_K | H_1) + \sum_{k=K+1}^{\infty} \Pr(\Lambda(\mathbf{x}) = \lambda_k | H_1). \qquad (1\text{-}60)$$

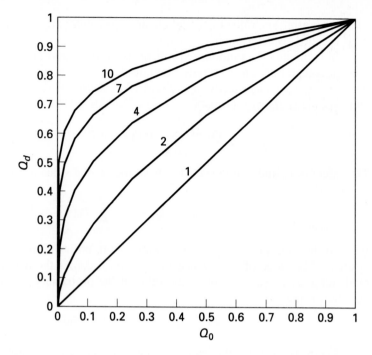

Figure 1-3. Operating characteristic: decision based on geometrically distributed datum, $m_0 = 1$. Curves are indexed with expected value m_1 under hypothesis H_1.

The chance device for making the decision whenever $\Lambda(\mathbf{x}) = \lambda_K$ can contain a random-number generator that upon that occasion produces a random number with a uniform probability distribution over the interval $(0, 1)$. If the random number lies between 0 and f, hypothesis H_1 is chosen, otherwise H_0. In practice one will ordinarily accept a pair of slightly higher or lower false-alarm and detection probabilities in order to avoid the need for such a chance device. Randomization is useful mainly for theoretical purposes: one wants to evaluate different strategies or to plot detection probabilities under different conditions of signal intensity and background illumination, and it is awkward to make comparisons unless the false-alarm probability takes a common value throughout. When the data are discrete random variables, randomization is then necessary.

In this chapter we have confined ourselves to decisions between simple hypotheses, for which the probability density functions of the data contain no unknown parameters. When unknown parameters appear in the distributions, the hypotheses are called *composite*. In Chapter 3 we shall treat the problem of deciding between a simple and a composite hypothesis. Receiver design is essentially a matter of generating likelihood ratios such as $\Lambda_j(v)$ in (1-7), $\Lambda(v)$ in (1-21), $\Lambda(\mathbf{x})$ in (1-50), their equivalents, or—in complicated situations—suitable approximations thereto. The decision levels with which they are compared depend on whether the Bayes or the Neyman–Pearson criterion has been adopted, and this aspect can in general be left up to the user of the receiver. Once a design has been chosen, its performance should

The Statistical Foundation   Chap. 1

be evaluated in terms of error probabilities or detection probabilities, and we shall give some attention to the often difficult problem of calculating them.

Problems

1-1. Find the Bayes test to choose between the hypotheses H_0 and H_1, whose prior probabilities are 0.6 and 0.4, respectively, when under H_0 the datum x has the probability density function

$$p_0(x) = \left(\frac{2}{\pi}\right)^{1/2} \exp\left(-\tfrac{1}{2}x^2\right) U(x),$$

where $U(\cdot)$ is the unit step function,

$$U(x) \equiv \begin{cases} 0, & x < 0, \\ 1, & x \geq 0. \end{cases} \tag{1-61}$$

Under H_1 x has the probability density function

$$p_1(x) = e^{-x} U(x).$$

(x is always positive.) Let the relative costs of the two kinds of errors be equal. Find the minimum attainable probability P_e of error.

1-2. Under hypothesis H_1 the random variable x has a uniform distribution over $(-1, 1)$,

$$p_1(x) \equiv \tfrac{1}{2}, \qquad -1 < x < 1; \qquad p_1(x) \equiv 0, \qquad |x| \geq 1.$$

Under hypothesis H_0 the distribution of x is triangular,

$$p_0(x) = 1 - |x|, \qquad -1 < x < 1; \qquad p_0(x) \equiv 0, \qquad |x| \geq 1.$$

We need to decide between H_1 and H_0 on the basis of the value of x in a single observation. The costs attending the decision are

$$C_{10} = 3, \qquad C_{01} = 4, \qquad C_{00} = C_{11} = 0;$$

C_{ij} is the cost of choosing hypothesis H_i when H_j is true.

Find the Bayes strategy for choosing between H_0 and H_1 for arbitrary prior probability ζ of H_0, $0 \leq \zeta \leq 1$, and calculate the minimum attainable Bayes cost $\overline{C}_{\min}(\zeta)$. Sketch the graph of that cost versus ζ. Determine the value of ζ for which it is maximum.

1-3. Consider the minimum Bayes cost $\overline{C}_{\min}$ in a binary decision as a function of the prior probability $\zeta_0 = \zeta$ of hypothesis H_0. Use (1-14) to write $\overline{C}_{\min}(\zeta)$ in terms of the false-alarm probability Q_0 and the detection probability Q_d, which are now functions of ζ through their dependence on $\lambda = \Lambda_0$ as given in (1-17). Use (1-28) to calculate the slope $d\overline{C}_{\min}/d\zeta$ of this function. Show that the function $\overline{C}_{\min}(\zeta)$ is convex $\cap$ in $0 \leq \zeta \leq 1$.

If the costs C_{ij} are known, but not the prior probabilities $\zeta_0 = \zeta$, $\zeta_1 = 1 - \zeta$, the most conservative strategy is the Bayes strategy set up for that value of ζ at which the minimum Bayes cost $\overline{C}_{\min}(\zeta)$ is greatest. It is called the *minimax* strategy; Problem 1-2 furnishes an example. Determine a linear relation between $Q_0(\zeta)$ and $Q_d(\zeta)$ from which the minimax value of ζ can be calculated. Use the operating characteristic to develop a graphical method for finding this value.

1-4. In another approach to the proof that the function $\overline{C}_{\min}(\zeta)$ introduced in Problem 1-3 is convex $\cap$, suppose that the Bayes strategy has been set up under the assumption that the prior probability of hypothesis H_0 equals $\zeta_0 = \zeta'$, but the actual relative frequency of that hypothesis equals ζ. In terms of the elements of the cost matrix **C**, write down the associated cost of operating the strategy as a function of ζ. It will depend on the false-alarm probability $Q_0(\zeta')$ and the detection probability $Q_d(\zeta')$ characterizing the

Bayes strategy adopted. This function will be linear in ζ and represented by a straight line in a graph of cost versus prior probability. Explain why that straight line must lie above the curve of the function $\overline{C}_{min}(\zeta)$ representing the average cost of the Bayes strategy, except at the point $\zeta = \zeta'$. From that fact deduce the convexity of the latter function.

1-5. Under hypothesis H_0 a datum x has the probability density function

$$p_0(x) = \begin{cases} A_0(a^2 - x^2), & |x| < a, \\ 0, & |x| \geq a, \end{cases}$$

and under hypothesis H_1 its probability density function is

$$p_1(x) = \begin{cases} A_1(b^2 - x^2), & |x| < b, \\ 0, & |x| \geq b, \end{cases}$$

with $b > a$. Calculate the constants A_0 and A_1. Determine the optimum strategy under the Neyman–Pearson criterion for deciding between the two hypotheses, and calculate the false-alarm and detection probabilities for it. With $b = 2a$, sketch the operating characteristic for this optimum strategy.

1-6. The random variables x and y are Gaussian with expected value 0 and variance 1. Their covariance $\mathrm{Cov}(x, y)$ may be either 0 or some known positive value $r > 0$. Show that the best choice between these possibilities on the basis of a measurement of x and y depends on where the point (x, y) lies with respect to a certain hyperbola in the (x, y)-plane. *Hint:* Under hypothesis H_1 the joint probability density function of x and y is

$$p_1(x, y) = \frac{1}{2\pi\sqrt{1 - r^2}} \exp\left[-\frac{x^2 + y^2 - 2rxy}{2(1 - r^2)} \right];$$

under hypothesis H_0 it has the same form, except that $r = 0$ [Hel91, p. 160], [Pap91, p. 127].

1-7. A random variable x is distributed according to a Cauchy distribution,

$$p(x) = \frac{m}{\pi(m^2 + x^2)}.$$

The parameter m can take on either of two values m_0 and m_1, $m_0 < m_1$. Design a statistical test to decide on the basis of a single measurement of x between the two hypotheses H_0 ($m = m_0$) and H_1 ($m = m_1$). Use the Neyman–Pearson criterion. For this test calculate the power $Q_d = 1 - Q_1$ as a function of the size Q_0.

1-8. A choice is to be made between hypotheses H_0 and H_1 on the basis of a single measurement of a quantity x. Under hypothesis H_0, $x = n$; under H_1, $x = s + n$. Here both s and n are independent positive random variables with the probability density functions

$$p(n) = b\, e^{-bn}\, U(n), \qquad p(s) = c\, e^{-cs}\, U(s).$$

Calculate the probability density functions of the datum x under each hypothesis. Find the decision level on x to yield a given false-alarm probability Q_0, and calculate the probability Q_d of correctly choosing hypothesis H_1.

1-9. Given are M independent data $v = (v_1, v_2, \dots, v_M)$. Under hypothesis H_0 each has a bilateral exponential, or "Laplace" distribution,

$$p_0(v) = \tfrac{1}{2}e^{-|v|}.$$

Under hypothesis H_1 the distribution of v is a shifted version of this, $p_1(v) = p_0(v - s)$, $s > 0$. Determine a sufficient statistic for deciding between these two hypotheses under the Neyman–Pearson criterion.

1-10. For a single datum v having the same distributions under hypotheses H_0 and H_1 as in Problem 1-9, determine the likelihood ratio $\Lambda(v)$ and sketch it as a function of v. For all positive values of Λ_0 in (1-17), determine the regions R_0 and R_1 on the v-axis in which hypotheses H_0 and H_1 are respectively chosen.

1-11. Under hypothesis H_0 the datum v has the Cauchy probability density function

$$p_0(v) = \frac{1}{\pi(1 + v^2)};$$

under hypothesis H_1 it has the displaced Cauchy density function

$$p_1(v) = \frac{1}{\pi\left[1 + (v - s)^2\right]}, \qquad s > 0.$$

Sketch the likelihood ratio $\Lambda(v)$ as a function of v. For all positive values of Λ_0 in (1-17), determine the regions R_0 and R_1 on the v-axis in which hypotheses H_0 and H_1 are respectively chosen.

1-12. For the logarithm g of the likelihood ratio, as defined by (1-36),

$$e^g = \Lambda(g) = \frac{p_1(g)}{p_0(g)},$$

define the moment generating functions of g under hypotheses H_0 and H_1 by

$$f_j(s) = E(e^{-gs}\,|\,H_j), \qquad j = 0, 1,$$

[Hel91, p. 276], [Pap91, p. 115]. Show that $f_1(s) = f_0(s - 1)$. Determine $f_0(s)$ and $f_1(s)$ for the logarithms of the likelihood ratios in Examples 1-2 and 1-3 in Sec. 1-2.

1-13. Assume that one has calculated the minimum error probability $P_e(\zeta)$ attainable in the choice between hypotheses H_0 and H_1, as a function of the prior probability $\zeta_0 = \zeta$ of hypothesis H_0. Show that the false-alarm and detection probabilities are then given by

$$Q_0(\zeta) = P_e + (1 - \zeta)\frac{dP_e}{d\zeta}, \qquad Q_d(\zeta) = 1 - P_e + \zeta\frac{dP_e}{d\zeta}.$$

1-14. The datum x is a nonnegative integer with Poisson probabilities under hypotheses H_0 and H_1:

$$P_k(x) = \frac{m_k^x}{x!}\exp(-m_k), \qquad x = 0, 1, 2, \ldots; \qquad k = 0, 1,$$

[Hel91, p. 43], [Pap91, p. 76]. In terms of the expected values m_0 and m_1, the decision level n_0, and the fraction f, determine the false-alarm and detection probabilities of the optimum randomized strategy for deciding between H_0 and H_1 under the Neyman–Pearson criterion, as described in Sec. 1.2.5. Show how to calculate both the decision level n_0 on x and the probability f with which H_1 is chosen when $x = n_0$. Taking $m_0 = 1$ and $m_1 = 3$, draw the operating characteristic, and assuming the relative error costs to be equal, draw the curve of the minimum attainable probability $P_e(\zeta)$ of error as a function of the prior probability $\zeta_0 = \zeta$ of hypothesis H_0.

1-15. A sequence of n independent measurements is taken of a Poisson-distributed variable x whose expected value is m_0 under hypothesis H_0 and m_1 under hypothesis H_1, as in Problem 1-14. On what combination of the measurements should a Bayes test be based, and with what decision level should its outcome be compared, for given prior probabilities ζ_0 and ζ_1 of the two hypotheses and a given cost matrix $\mathbf{C}$?

1-16. Prove from (1-59) and (1-60) that for the optimum randomized strategy with discrete data the slope of the operating characteristic equals the decision level λ on the likelihood ratio, as in (1-28).

1-17. Under hypothesis H_0 the datum x is uniformly distributed over (0, 1). Under H_1 it is uniformly distributed over $(a, a + 1)$, with $0 < a < 1$. Determine the optimum strategy under the Neyman–Pearson criterion for choosing between H_0 and H_1. Observe that as the likelihood ratio now takes on only a finite number of possible values, randomization is necessary. Calculate the operating characteristic for this hypothesis test, that is, the graph of the detection probability Q_d versus the false-alarm probability Q_0, and sketch it.

1-18. The decision about whether a certain optical signal is present or not is based on the numbers n_1 and n_2 of photoelectrons emitted during an observation interval $(0, T)$ from two separate photoelectrically emissive surfaces onto which the light from the source, along with background light, will fall. Under hypothesis H_0, "signal absent," these numbers have independent Poisson distributions,

$$\Pr(n_1 = k_1, n_2 = k_2 | H_0) = \frac{m_{01}^{k_1} m_{02}^{k_2}}{k_1! \, k_2!} \exp(-m_{01} - m_{02}),$$

and under hypothesis H_1, "signal present," their joint probability mass function is

$$\Pr(n_1 = k_1, n_2 = k_2 | H_1) = \frac{m_{11}^{k_1} m_{12}^{k_2}}{k_1! \, k_2!} \exp(-m_{11} - m_{12}), \qquad m_{11} > m_{01}, \qquad m_{12} > m_{02}.$$

Show how to process the data n_1 and n_2 to yield the maximum probability Q_d of detection for fixed false-alarm probability Q_0. Explain in detail how to calculate that maximum detection probability Q_d.

2

Detection of a Known Signal

2.1 GAUSSIAN NOISE

With the theory of statistical tests introduced in Chapter 1, we can attack the simplest signal-detection problem, that of deciding whether a signal $s(t)$ of specified form has arrived at a definite time in the midst of Gaussian noise. Gaussian noise is ubiquitous. It originates in the thermal fluctuations of all matter in the universe, which create randomly varying electromagnetic fields that excite the antenna of the receiver and generate a fluctuating voltage $n(t)$ between its terminals. The thermal fluctuations of the ions and electrons in the input resistor connected across those terminals also contribute to this *thermal noise*. As the sum of the miniscule effects of an enormous number of randomly moving charges, the noise $n(t)$ is a Gaussian stochastic process by virtue of the central-limit theorem [Hel91, pp. 260–5], [Pap91, p. 214]. Other types of noise, such as clutter in radar and reverberation in sonar, which can also be modeled as Gaussian random processes, are sometimes present as well. The detection of a known signal in Gaussian noise is a fundamental problem of signal-detection theory.

The input $v(t)$ to the receiver, which can be taken as the voltage between the terminals of its antenna, is measured during an observation interval $0 \leq t \leq T$. On the basis of this input the receiver must choose one of two hypotheses: (H_0) there is no signal present, and the input consists only of Gaussian noise with expected value zero, $v(t) = n(t)$; or (H_1) the input is the sum of the expected signal and the noise, $v(t) = s(t) + n(t)$. The receiver embodies a criterion by which its success in a large number of decisions of this kind can be evaluated; as discussed in Chapter 1, this criterion will influence how it processes the data. For example, the signal might be a rectangular pulse of duration $T' < T$:

$$s(t) \equiv A, \qquad 0 \le t_1 < t < t_1 + T' \le T,$$
$$s(t) \equiv 0, \qquad t < t_1, \qquad t > t_1 + T'.$$

It occurs at a definite time within the observation interval. A communication system might be using such pulses to convey a message that has been translated into a binary code with symbols 0 and 1. Every T seconds a pulse is or is not sent, depending on whether the current message symbol is a 1 or a 0. At the end of each interval of T seconds, the receiver decides which of the symbols was transmitted. Because of the noise it will occasionally err, and the designer's aim may be to minimize the probability of its doing so, errors in the two symbols having been judged equally expensive. The decision criterion is then the Bayes, with equal relative costs, $C_{10} - C_{00} = C_{01} - C_{11}$; the relative frequencies ζ_0 and ζ_1 with which the transmitter sends the symbols 0 and 1 are known. Alternatively, the Neyman–Pearson criterion may be adopted and the probability of detection maximized for a preassigned value of the false-alarm probability.

In order to determine the likelihood ratio on which, according to what we learned in Chapter 1, the receiver will base its decisions, we must set up an appropriate method of sampling the input $v(t)$ and then write down the joint probability density functions of the samples. To this end, we begin by reviewing the properties of Gaussian noise, forgetting the signal $s(t)$ for the present and concentrating on the probabilistic description of the noise. Out of its probability distributions we shall in Sec. 2.2 form the likelihood ratio, pass to the limit of an infinite number of samples, and show how the resulting receiver structure can be realized by a certain linear filter. Then we shall calculate the error probabilities characterizing the performance of the optimum receiver.

2.1.1 The Density Functions of Gaussian Noise

The input $v(t)$ is a stochastic process that we can assume continues throughout an infinite interval $(-\infty, \infty)$. It is defined through the array of all joint probability density functions

$$p(v_1, v_2, \dots, v_n)$$

of its samples $v_1 = v(t_1)$, $v_2 = v(t_2)$, ..., $v_n = v(t_n)$ taken at an arbitrary number n of arbitrary times $t_1, t_2, \dots, t_n$. These functions $p(v_1, v_2, \dots, v_n)$ have the basic properties of joint probability density functions mentioned at the beginning of Sec. 1.1. When as now the noise is Gaussian, the samples $v_1, v_2, \dots, v_n$ are Gaussian or normally distributed random variables; that is, their joint probability density functions have the form

$$p(v_1, v_2, \dots, v_n) = M_n \exp\left[-\tfrac{1}{2} \sum_{j=1}^{n} \sum_{k=1}^{n} \mu_{jk}^{(n)} v_j v_k \right]. \qquad (2\text{-}1)$$

The quantities $\mu_{jk}^{(n)}$ form an $n \times n$ matrix that we designate by $\mathbf{\mu}_n$. The normalization constant M_n is

$$M_n = (2\pi)^{-n/2} |\det \mathbf{\mu}_n|^{1/2}, \tag{2-2}$$

where "det" stands for the determinant [Hel91, p. 241], [Pap91, p. 197].

The expected values of all samples of the noise are zero:

$$E[v(t)| H_0] \equiv 0.$$

In particular, the first-order probability density function of a single sample $v = v(t)$ of the noise is

$$p(v) = \frac{1}{\sqrt{2\pi\sigma_t^2}} \exp\left(-\frac{v^2}{2\sigma_t^2}\right),$$

where $\sigma_t^2 = \mathrm{Var}\, v(t)$ is the variance of the noise at time t.

The matrix $\mathbf{\mu}_n$ appearing in (2-1) and (2-2) is the inverse

$$\mathbf{\mu}_n = \mathbf{\phi}_n^{-1}$$

of the symmetric $n \times n$ matrix $\mathbf{\phi}_n$ whose elements ϕ_{jk} are the covariances of the samples $v_j = v(t_j)$ and $v_k = v(t_k)$,

$$\phi_{jk} = \phi_{kj} = \mathrm{Cov}(v_j, v_k) = E[v(t_j)v(t_k)| H_0].$$

These matrix elements are determined by the autocovariance function

$$\phi(t, s) = E[v(t)v(s)| H_0] \tag{2-3}$$

of the noise [Hel91, p. 363], [Pap91, p. 289]. In particular, the variance at time t is

$$\sigma_t^2 = \phi(t, t). \tag{2-4}$$

The bivariate density function of a pair of samples $v_1 = v(t_1)$, $v_2 = v(t_2)$ of the noise, for instance, depends on the 2×2 covariance matrix

$$\mathbf{\phi}_2 = \begin{bmatrix} \phi_{11} & \phi_{12} \\ \phi_{21} & \phi_{22} \end{bmatrix} = \begin{bmatrix} \sigma_1^2 & r\sigma_1\sigma_2 \\ r\sigma_1\sigma_2 & \sigma_2^2 \end{bmatrix},$$

where $\sigma_k^2 = \mathrm{Var}\, v(t_k)$, $k = 1, 2$, and $r = r(t_1, t_2) = \phi_{12}/\sqrt{\phi_{11}\phi_{22}}$ is the correlation coefficient of the samples. The inverse of this covariance matrix is

$$\mathbf{\mu}_2 = \frac{1}{\sigma_1^2\sigma_2^2(1 - r^2)} \begin{bmatrix} \sigma_2^2 & -r\sigma_1\sigma_2 \\ -r\sigma_1\sigma_2 & \sigma_1^2 \end{bmatrix}.$$

(A factor in front of any matrix multiplies each of its elements.) Then (2-1) with $n = 2$ becomes

$$p(v_1, v_2) = \frac{1}{2\pi\sigma_1\sigma_2\sqrt{1 - r^2}} \exp\left[-\frac{\dfrac{v_1^2}{\sigma_1^2} - \dfrac{2rv_1v_2}{\sigma_1\sigma_2} + \dfrac{v_2^2}{\sigma_2^2}}{2(1 - r^2)}\right].$$

A convenient notation collects the samples $(v_1, v_2, \ldots, v_n)$ into a column vector v, whose transposed row vector is denoted by v^T, and it permits writing the joint probability density function of these samples concisely as

$$p(v) = (2\pi)^{-n/2} |\det \phi_n|^{-1/2} \exp(-\tfrac{1}{2}v^T \phi_n^{-1} v). \qquad (2\text{-}5)$$

We shall make frequent use of such matrix notation hereafter.

2.1.2 Stationary Noise

When the noise is stationary, its autocovariance function $\phi(t_1, t_2)$ depends on the times t_1 and t_2 only through the interval $\tau = t_2 - t_1$ between them, and it is written $\phi(t_2 - t_1)$. This autocovariance function is an even function

$$\phi(\tau) = \phi(-\tau),$$

and $\phi(0) = \sigma^2$ is the variance, now constant, of the noise. The autocovariance function is the Fourier transform

$$\phi(\tau) = \int_{-\infty}^{\infty} \Phi(\omega) e^{i\omega\tau} \frac{d\omega}{2\pi} \qquad (2\text{-}6)$$

of a nonnegative function $\Phi(\omega)$ known as the *spectral density* of the noise:

$$\Phi(\omega) = \int_{-\infty}^{\infty} \phi(\tau) e^{-i\omega\tau} d\tau, \qquad \Phi(\omega) = \Phi(-\omega) \geq 0, \qquad (2\text{-}7)$$

[Hel91, pp. 383][1]. The variance of the noise, in particular, is the integral of the spectral density,

$$\sigma^2 = \phi(0) = \int_{-\infty}^{\infty} \Phi(\omega) \frac{d\omega}{2\pi}. \qquad (2\text{-}8)$$

The average power in the spectral components of the process lying between positive frequencies $\omega/2\pi$ and $(\omega + d\omega)/2\pi$ (Hz) equals

$$[\Phi(\omega) + \Phi(-\omega)]\frac{d\omega}{2\pi} = 2\Phi(\omega)\frac{d\omega}{2\pi}.$$

It is measured by a spectrum analyzer [Hel91, pp. 433–9], [Pap91, pp. 438–9].

When stationary Gaussian noise $v(t)$ passes through a stationary linear filter whose impulse response is $k(\tau)$, the output

$$v_0(t) = \int_0^{\infty} k(\tau)v(t - \tau) d\tau$$

is a stationary Gaussian random process whose spectral density is

$$\Phi_0(\omega) = |y(\omega)|^2 \Phi(\omega), \qquad (2\text{-}9)$$

where

$$y(\omega) = \int_0^{\infty} k(\tau) e^{-i\omega\tau} d\tau \qquad (2\text{-}10)$$

is the transfer function of the filter [Hel91, p. 384], [Pap91, p. 324].

---

[1] Papoulis [Pap91, p. 319] defines the spectral density as the Fourier transform of the autocorrelation function. When as here the process $v(t)$ has expected value zero, this distinction is of no consequence.

2.1.3 Sampling by Orthonormal Functions

In determining optimum strategies for decisions about Gaussian random processes, as for the detection of known signals in Gaussian noise, it is convenient to sample random processes such as $v(t)$ not at particular instants of time, as in Sec. 2.1.1, but by means of their expansions in a particular kind of Fourier series. Let a random process $v(t)$ be observed during an interval $(0, T)$. Define an infinite set of functions $f_k(t)$ orthonormal over that interval in the sense that

$$\int_0^T [f_k(t)]^2 \, dt \equiv 1, \qquad k = 1, 2, 3, \ldots,$$

$$\int_0^T f_j(t) f_k(t) \, dt \equiv 0, \qquad j \neq k. \tag{2-11}$$

The most familiar set of orthonormal functions is made up of sines and cosines:

$$f_1(t) \equiv T^{-1/2},$$

$$f_{2k+1}(t) = \left(\frac{2}{T}\right)^{1/2} \cos\left[\frac{2\pi k}{T}\left(t - \frac{T}{2}\right)\right], \qquad k \geq 1,$$

$$f_{2k}(t) = \left(\frac{2}{T}\right)^{1/2} \sin\left[\frac{2\pi k}{T}\left(t - \frac{T}{2}\right)\right], \qquad k \geq 1.$$

Another set is obtained by shifting and scaling the Legendre polynomials:

$$f_k(t) = \left(\frac{2k-1}{T}\right)^{1/2} P_{k-1}\left(\frac{2t}{T} - 1\right), \qquad k = 1, 2, 3, \ldots.$$

There is no limit to the number of sets of orthonormal functions that can be constructed, and in a particular problem, one set may be more convenient than another. Indeed, given any infinitely numerous set of linearly independent functions $g_1(t), g_2(t), \ldots$, one can construct from them a set of functions $\{f_k(t)\}$ orthonormal over the interval $(0, T)$ by using what is known as the *Gram–Schmidt procedure*. It will be explained in Sec. 2.1.4.

We write the random process $v(t)$ as a Fourier series,

$$v(t) = \sum_{k=1}^{\infty} v_k f_k(t), \tag{2-12}$$

and as its "samples" we take the coefficients

$$v_k = \int_0^T f_k(s) v(s) \, ds. \tag{2-13}$$

We assume that the set of functions $f_k(t)$ is sufficiently numerous so that any realization $v(t)$ of the random process can be represented as in (2-12); it is said to be *complete*. Substituting (2-13) into (2-12) shows that

$$\sum_{k=1}^{\infty} f_k(t) f_k(s) = \delta(t - s), \tag{2-14}$$

where $\delta(t - s)$ is the Dirac delta function, which is so defined that for any function $h(t)$

$$\int_{-\infty}^{\infty} h(s)\delta(t - s) \, ds = h(t),$$

provided $h(\cdot)$ is continuous at t. Equation (2-14) is known as the *completeness relation* for the set of orthonormal functions.

When $v(t)$ is purely random noise, its expected value is zero at all times t, and hence all coefficients have zero expected values:

$$E(v_k | H_0) \equiv 0, \qquad \forall k. \tag{2-15}$$

The covariances of the samples are, by (2-3),

$$
\begin{aligned}
\phi_{jk} = E(v_j v_k | H_0) &= \int_0^T \int_0^T f_j(t) f_k(s) E[v(t)v(s)| H_0] \, dt \, ds \\
&= \int_0^T \int_0^T f_j(t)\phi(t, s) f_k(s) \, dt \, ds
\end{aligned} \tag{2-16}
$$

in terms of the autocovariance function $\phi(t, s)$ of the noise.

When $v(t)$ is a Gaussian random process, the samples v_k defined as in (2-13) are Gaussian random variables, the joint probability density function of any n of which is given by an expression like (2-1) or, in vector notation, (2-5). The reason is that any linear combination of Gaussian random variables is a Gaussian random variable [Hel91, p. 245], [Pap91, p. 197], and the quantities v_k defined by (2-13) are linear combinations of the values of $v(t)$ at all times t in $(0, T)$.

The sample v_k can be generated by passing the random process $v(t)$ through a linear filter whose impulse response is

$$h_k(\tau) = \begin{cases} f_k(T - \tau), & 0 \le \tau \le T, \\ 0, & \tau < 0, \end{cases} \tag{2-17}$$

and measuring the output of the filter at time $t = T$. The input $v(t)$ having been turned on at time $t = 0$, the output is

$$w_k(t) = \int_0^t h_k(\tau)v(t - \tau) \, d\tau,$$

and at time $t = T$, by (2-17),

$$w_k(T) = \int_0^T f_k(T - \tau)v(T - \tau) \, d\tau = \int_0^T f_k(u)v(u) \, du = v_k.$$

The filter whose impulse response is given by (2-17) is said to be *matched* to the signal $f_k(t)$ over the interval $(0, T)$. Matched filters will turn out to be most useful in constructing optimum detectors. By passing the input $v(t)$ through a bank of filters matched to the "signals" $f_1(t)$, $f_2(t)$, $f_3(t)$, ... in our orthonormal set, we could generate as many of the samples $v_1, v_2, v_3, \ldots$ as we liked. As we shall see, the optimum processing of the input $v(t)$ will not require us to do so.

The expected value of the sum of the squares of all our samples equals the average total energy of the noise $v(t)$ received during the interval $(0, T)$, for by (2-12) and the orthonormality relation (2-11)

$$E \int_0^T [v(t)]^2 \, dt = \int_0^T \phi(t, t) \, dt = \sum_{k=1}^{\infty} \sum_{m=1}^{\infty} E(v_k v_m) \int_0^T f_k(t) f_m(t) \, dt$$

$$= \sum_{k=1}^{\infty} \sum_{m=1}^{\infty} E(v_k v_m) \delta_{km} = E \sum_{k=1}^{\infty} v_k^2. \tag{2-18}$$

Here we have used the Kronecker delta,

$$\delta_{km} = \begin{cases} 1, & k = m, \\ 0, & k \neq m. \end{cases} \tag{2-19}$$

If the noise is stationary, the quantity in (2-18) equals T times the variance σ^2 of the noise as given in (2-8).

2.1.4 Gram–Schmidt Orthogonalization

The Gram–Schmidt orthogonalization procedure enables us to start with a set of linearly independent functions $g_1(t)$, $g_2(t)$, ... , and from them to construct a set of functions $f_1(t)$, $f_2(t)$, ... that are orthonormal in the sense of (2-11). It rests on the idea that functions of time t in $(0, T)$ can be thought of as vectors in a Cartesian space of infinite dimensionality.

What corresponds to the scalar product of any two functions $h(t)$ and $m(t)$ will be denoted by

$$(\mathbf{h}, \mathbf{m}) = (\mathbf{m}, \mathbf{h}) = \int_0^T h(t) m(t) \, dt. \tag{2-20}$$

The "length" of a function $h(t)$, or of the vector representing it in the Cartesian space, is $(\mathbf{h}, \mathbf{h})^{1/2}$. In terms of this notation, the orthonormality conditions (2-11) are

$$(\mathbf{f}_k, \mathbf{f}_m) = \delta_{km} = \begin{cases} 1, & k = m, \\ 0, & k \neq m. \end{cases} \tag{2-21}$$

The first element $f_1(t)$ of the new set of functions is taken proportional to $g_1(t)$:

$$f_1(t) = \beta_1 g_1(t), \qquad \beta_1 = (\mathbf{g}_1, \mathbf{g}_1)^{-1/2} = \left[\int_0^T [g_1(t)]^2 \, dt \right]^{-1/2}.$$

The constant β_1 serves to normalize the function $f_1(t)$ to unit length.

The second function will be a linear combination of $f_1(t)$ and $g_2(t)$,

$$f_2(t) = \alpha f_1(t) + \beta_2 g_2(t),$$

in which the constants α and β_2 remain to be chosen. Thinking of the functions $g_1(t)$ and $g_2(t)$ as vectors, we see that they define a plane. The new function $f_1(t)$ lies along $g_1(t)$, and the new function $f_2(t)$ lies in the same plane and is perpendicular to $f_1(t)$. Taking the scalar product with $f_1(t)$, we obtain

$$(\mathbf{f}_1, \mathbf{f}_2) = \alpha(\mathbf{f}_1, \mathbf{f}_1) + \beta_2(\mathbf{f}_1, \mathbf{g}_2) = 0,$$

whence $\alpha = -\beta_2(\mathbf{f}_1, \mathbf{g}_2)$. Furthermore,

$$1 = (\mathbf{f}_2, \mathbf{f}_2) = \alpha(\mathbf{f}_2, \mathbf{f}_1) + \beta_2(\mathbf{f}_2, \mathbf{g}_2) = \beta_2(\mathbf{f}_2, \mathbf{g}_2)$$
$$= \beta_2[\alpha(\mathbf{f}_1, \mathbf{g}_2) + \beta_2(\mathbf{g}_2, \mathbf{g}_2)]$$
$$= \beta_2^2[(\mathbf{g}_2, \mathbf{g}_2) - (\mathbf{f}_1, \mathbf{g}_2)^2].$$

Hence $\beta_2 = [(\mathbf{g}_2, \mathbf{g}_2) - (\mathbf{f}_1, \mathbf{g}_2)^2]^{-1/2}$, and

$$f_2(t) = \beta_2[g_2(t) - (\mathbf{f}_1, \mathbf{g}_2)f_1(t)].$$

As we continue through this Gram–Schmidt procedure, each new function $f_k(t)$ is a linear combination of $g_k(t)$, which has not yet been used, and of the orthonormal functions $f_i(t)$ previously formed, $1 \leq i \leq k - 1$. If for some β_k we set

$$f_k(t) = \beta_k \left[g_k(t) - \sum_{i=1}^{k-1} (\mathbf{f}_i, \mathbf{g}_k)f_i(t) \right], \tag{2-22}$$

the function $f_k(t)$ will be orthogonal to the previous members $f_j(t)$ of the set, $j = 1, 2, \ldots, k - 1$, for by (2-21)

$$(\mathbf{f}_j, \mathbf{f}_k) = \beta_k \left[(\mathbf{f}_j, \mathbf{g}_k) - \sum_{i=1}^{k-1} (\mathbf{f}_i, \mathbf{g}_k)\delta_{ij} \right] = 0.$$

The constant β_k in (2-22) is determined so that $(\mathbf{f}_k, \mathbf{f}_k) = 1$; that is, by (2-21) and (2-22),

$$(\mathbf{f}_k, \mathbf{f}_k) = \beta_k(\mathbf{g}_k, \mathbf{f}_k) = \beta_k^2 \left[(\mathbf{g}_k, \mathbf{g}_k) - \sum_{i=1}^{k-1} (\mathbf{f}_i, \mathbf{g}_k)(\mathbf{g}_k, \mathbf{f}_i) \right] = 1,$$

and the normalization constant in (2-22) is

$$\beta_k = \left[(\mathbf{g}_k, \mathbf{g}_k) - \sum_{i=1}^{k-1} (\mathbf{f}_i, \mathbf{g}_k)^2 \right]^{-1/2}.$$

In this way we can create from an infinite set of linearly independent functions $g_j(t)$ as many members $f_k(t)$ of a set of orthonormal functions as we need.

2.1.5 White Noise

The thermal noise that, as we said at the beginning of this section, is ever present at the input to a receiver possesses a spectral density much broader than the spectra of any signals one has occasion to detect, and even much broader than the passband of the input circuitry—antenna, leads or waveguides, and so on—that conducts the signals into the receiver. It is customary to model this thermal noise as a stationary random process $v(t)$ whose spectral density is uniform,

$$\Phi(\omega) \equiv \frac{N}{2}, \tag{2-23}$$

over a range of angular frequencies $-\omega_0 < \omega < \omega_0$ encompassing the spectra of the signals to be detected and extending far beyond them. The quantity N is the unilateral spectral density of the noise: the average noise power passing through a filter of unit gain and bandwidth $\Delta\nu = \Delta\omega/2\pi$ Hz, including components of both positive and negative frequencies, equals $N\Delta\nu$. [The quantity $N/2$ in (2-23) is called the *bilateral* spectral density; the noise is thought of as equally divided between positive and negative frequencies.] Because of the uniformity of its spectral density, this kind of noise is called *white noise* [Hel91, pp. 403–4], [Pap91, p. 295].

According to (2-6) the autocovariance function of white noise is

$$\phi(\tau) = \frac{N}{2}\delta(\tau). \tag{2-24}$$

This Dirac delta function $\delta(\tau)$ can be thought of as a peaked function of unit area,

$$\int_{-\infty}^{\infty} \delta(\tau)\, d\tau = 1,$$

whose duration is much shorter than any time interval our instruments can resolve.

Because (2-4) with (2-8) and (2-23) indicates that time samples of white noise would have infinite variance, temporal sampling such as we started with in Sec. 2.2.1 is unsuitable. Instead we sample white noise by means of an arbitrary complete set of orthonormal functions $\{f_k(t)\}$ as described in Sec. 2.1.3, and we find by (2-16) that the samples v_k are uncorrelated and hence statistically independent Gaussian random variables with variances equal to $N/2$; the elements of their covariance matrix are

$$\begin{aligned}
\phi_{jk} &= \frac{N}{2}\int_0^T \int_0^T f_j(t)\delta(t-s)f_k(s)\, dt\, ds \\
&= \frac{N}{2}\int_0^T f_j(t)f_k(t)\, dt = \frac{N}{2}\delta_{jk}.
\end{aligned} \tag{2-25}$$

The joint probability density function of any set $v_1, v_2, \ldots, v_n$ of n of these samples is therefore

$$\begin{aligned}
p(v_1, v_2, \ldots, v_n) &= \prod_{k=1}^{n} (\pi N)^{-1/2} \exp\left(-\frac{v_k^2}{N}\right) \\
&= (\pi N)^{-n/2} \exp\left[-\frac{1}{N}\sum_{k=1}^{n} v_k^2\right].
\end{aligned} \tag{2-26}$$

White noise can instructively be pictured as a dense succession of sharp pulses occurring at random times τ_m and having independent and identically distributed random amplitudes a_m:

$$v(t) = \sum_{m=-\infty}^{\infty} a_m \delta(t - \tau_m). \tag{2-27}$$

For the sake of definiteness, the instants τ_m are taken to constitute a Poisson point process: the number n of such instants in an interval of duration Δ has a Poisson distribution with expected value $\lambda\Delta$, where λ is the average number of pulses per unit time:

$$\Pr(n = k) = \frac{(\lambda\Delta)^k}{k!}e^{-\lambda\Delta}, \qquad k = 0, 1, 2, \ldots. \tag{2-28}$$

The numbers of pulses in any disjoint intervals are furthermore statistically independent [Hel91, p. 390], [Pap91, pp. 357–8]. The amplitudes a_m are taken to have expected value zero. The white noise is thus a kind of shot noise. By Campbell's theorem [Hel91, p. 397], [Pap91, p. 360] the autocovariance function of this random process is

$$E[v(t)v(s)] = \lambda E(a_m^2) \int_{-\infty}^{\infty} \delta(u - t)\delta(u - s) \, du$$

$$= \lambda E(a_m^2)\delta(t - s).$$

(2-29)

White noise is modeled as in (2-27) in the limit in which the rate λ grows beyond all bounds and the mean-square amplitude $E(a_m^2)$ vanishes in such a way that their product $\lambda E(a_m^2) = N/2$ remains fixed. We call this the *high-rate limit*.

When the random process of (2-27) passes into a linear filter whose impulse response is $k(\tau)$, each pulse $a_m\delta(t - \tau_m)$ causes the filter to put out a copy $a_m k(t - \tau_m)$ of that response, and the net output is the shot-noise process

$$v_0(t) = \int_{-\infty}^{\infty} k(\tau)v(t - \tau) \, d\tau = \sum_{m=-\infty}^{\infty} a_m k(t - \tau_m).$$

(2-30)

The central-limit theorem assures us that for a broad class of probability distributions of the amplitudes a_m, the output $v_0(t)$ of a linear filter, as defined by (2-30), will be a Gaussian random process in the high-rate limit. From this standpoint it is unnecessary to define the white noise itself as a Gaussian process.

The integrals for the samples v_k as defined by (2-13) will have the form of a summation like that in (2-30):

$$v_k = \sum_{\tau_m \in (0,T)} a_m f_k(\tau_m).$$

(2-31)

By virtue of the central-limit theorem, they too will be Gaussian random variables in the high-rate limit, and the joint probability density function of any number of such samples will be given by (2-26).

Let us consider the sum

$$g = \sum_{k=1}^{\infty} s_k v_k,$$

(2-32)

where the v_k's are samples of white noise in this sense, and the s_k's are the Fourier coefficients of the signal $s(t)$ with respect to the same set of orthonormal functions:

$$s(t) = \sum_{k=1}^{\infty} s_k f_k(t).$$

(2-33)

Substituting (2-31) into (2-32), we obtain

$$g = \sum_{k=1}^{\infty} \sum_{\tau_m \in (0,T)} a_m s_k f_k(\tau_m) = \sum_{\tau_m \in (0,T)} a_m s(\tau_m) = \int_0^T s(t)v(t) \, dt,$$

(2-34)

when $v(t)$ is represented as in (2-27). If we like, we can regard (2-34) as a definition of what is meant by an integral of the form

$$\int_0^T s(t)v(t)\,dt$$

when $v(t)$ represents white noise.

2.1.6 Karhunen–Loève Expansion

Setting up likelihood ratios appropriate for the detection of signals in Gaussian noise is much simpler when the samples v_k are uncorrelated and hence independent. We have seen that when the noise is white, samples defined as in (2-13) in terms of an arbitrary set of orthonormal functions $f_k(t)$ are uncorrelated. It would be convenient for the samples to be uncorrelated even when the noise is not white, but is described by an arbitrary autocovariance function $\phi(t, s)$ as in (2-3). (Such noise is said to be *colored*.) This will be so if the orthonormal functions utilized are the solutions of the homogeneous integral equation

$$\lambda_k f_k(t) = \int_0^T \phi(t, s)f_k(s)\,ds, \qquad 0 \le t \le T, \tag{2-35}$$

whose *kernel* $\phi(t, s)$ is the autocovariance function of the noise. Our observation interval remains $(0, T)$. Nonzero solutions of this equation exist only for special values of the constants λ_k; those values are called the *eigenvalues* (or "proper" or "characteristic" values) of (2-35). The associated solutions $f_k(t)$ are called the *eigenfunctions* of (2-35). (How to solve integral equations of this kind will be discussed in Sec. 2.3.) When these eigenfunctions $f_k(t)$ are used in the series (2-12),

$$v(t) = \sum_{k=1}^{\infty} v_k f_k(t),$$

the series is called the *Karhunen–Loève expansion* of the input $v(t)$ [Loè45], [Loè46], [Kar47].

The equation (2-35) is called a *homogeneous Fredholm integral equation*. For the sake of future applications, we allow its kernel $\phi(t, s)$ to be complex, but we impose on it the symmetry property

$$\phi(t, s) = \phi^*(s, t), \tag{2-36}$$

where the asterisk denotes the complex conjugate. When such a homogeneous integral equation arises in detection theory, its kernel satisfies this condition. The kernel is then said to be *Hermitian*; if it is also real, it is described as *symmetrical*.

The theory of integral equations such as (2-35) is described in a number of books: [Lov24], [Cou31], [Hil53], and [Mor53], to name a few. The theory is akin to that of linear operators, which act on vectors in a *Hilbert space* of an infinite number of dimensions; each vector corresponds to a function defined over the interval $0 \le t \le T$. The components of such a vector are the coefficients of the Fourier series for the function with respect to the particular set of orthonormal functions adopted as a basis. Thus to the function $s(t)$ corresponds the vector $(s_1, s_2, \ldots, s_k, \ldots)$, where the s_k's are the coefficients in the Fourier series (2-33) for $s(t)$.

Multiplication of a function of s by $\phi(t, s)ds$ and integration over $0 \leq s \leq T$ to yield a new function—now of t—constitute a particular type of linear operation. A linear operator rotates and stretches the vectors on which it acts in such a way that the transformed sum of two vectors is the sum of the transformed vectors, and so on. An integral like

$$\int_0^T f^*(t)g(t)\, dt$$

corresponds to the scalar product $(\mathbf{f}, \mathbf{g})$ of the vectors representing the functions $f(t)$ and $g(t)$.

First we shall prove that the eigenfunctions of (2-35) possess an orthonormality property like that in (2-11). To do so, we first multiply both sides of (2-35) by $f_m^*(t)\, dt$ and integrate over $(0, T)$:

$$\lambda_k \int_0^T f_m^*(t)f_k(t)\, dt = \int_0^T \int_0^T f_m^*(t)\phi(t, s)f_k(s)\, dt\, ds. \qquad (2\text{-}37)$$

When we take the complex conjugate of (2-35) and write it for the mth eigenfunction, it becomes

$$\lambda_m^* f_m^*(s) = \int_0^T \phi^*(s, t)f_m^*(t)\, dt = \int_0^T f_m^*(t)\phi(t, s)\, dt$$

on account of the Hermitian character (2-36) of the kernel. If we multiply both sides of this equation by $f_k(s)ds$ and integrate, it becomes

$$\lambda_m^* \int_0^T f_m^*(s)f_k(s)\, ds = \int_0^T \int_0^T f_m^*(t)\phi(t, s)f_k(s)\, dt\, ds.$$

Subtracting from (2-37), we find

$$(\lambda_k - \lambda_m^*)\int_0^T f_m^*(t)f_k(t)\, dt = 0. \qquad (2\text{-}38)$$

In most problems the eigenvalues λ_k are all distinct. Then for $k \neq m$ the associated eigenfunctions must be orthogonal in the sense appropriate for complex functions on the interval $(0, T)$:

$$\int_0^T f_m^*(t)f_k(t)\, dt = 0, \qquad k \neq m.$$

If two or more eigenvalues are identical, new eigenfunctions can be formed as linear combinations of their associated eigenfunctions in such a way that they are orthogonal; one applies the Gram–Schmidt procedure described in Sec. 2.1.4, suitably modified if necessary to accommodate complex functions.

For $k = m$, on the other hand, the integral in (2-38) is positive, and $\lambda_k - \lambda_k^* = 0$; all the eigenvalues λ_k are therefore real. From the form of (2-35) we see that it specifies an eigenfunction only up to an arbitrary multiplying constant, which can be chosen so that the integral of the absolute square of the function equals 1. Then for all indices

$$\int_0^T f_m^*(t)f_k(t)\, dt = \delta_{km} = \begin{cases} 1, & k = m, \\ 0, & k \neq m. \end{cases} \qquad (2\text{-}39)$$

The eigenfunctions form an orthonormal set in this sense. In the vector-space analogy, they correspond to a mutually orthogonal set of vectors of unit length.

In detection theory it suffices to assume that the kernel $\phi(t, s)$ is positive definite, which means that for any function $g(t)$ that is not identically zero

$$E\left[\left|\int_0^T g(t)v(t)\, dt\right|^2 \Big| H_0\right] = \int_0^T \int_0^T g^*(t)\phi(t, s)g(s)\, dt\, ds > 0. \qquad (2\text{-}40)$$

The eigenvalues λ_k are then all strictly positive, for from (2-35) and (2-39) we obtain

$$\lambda_k \int_0^T |f_k(t)|^2\, dt = \lambda_k = \int_0^T \int_0^T f_k^*(t)\phi(t, s)f_k(s)\, dt\, ds > 0.$$

The linear operator represented by the kernel $\phi(t, s)$ changes the lengths of the orthogonal vectors corresponding to the eigenfunctions $f_k(t)$, but it does not rotate them. A positive definite linear operator $\phi(t, s)$ does not nullify or reverse the direction of any of these basic orthogonal vectors.

If the kernel $\phi(t, s)$ is real and symmetrical, the eigenfunctions can also be taken to be real. Indeed, upon taking the complex conjugate of (2-35), we find that both $f_k(t)$ and $f_k^*(t)$ are eigenfunctions of the kernel $\phi(t, s)$ corresponding to the same eigenvalue. The real and the imaginary parts of $f_k(t)$, as linear combinations of these, must therefore also be eigenfunctions, and we can use them instead. Ordinarily $f_k(t) = f_k^*(t)$.

Returning now to noise with a real autocovariance function $\phi(t, s)$, we can assume that its eigenfunctions in (2-35) are real. We can then use (2-11) and (2-16) to show that the covariances of the samples v_k are

$$\begin{aligned}
\phi_{jk} = E(v_j v_k | H_0) &= \int_0^T \int_0^T f_j(t)\phi(t, s)f_k(s)\, dt\, ds \\
&= \lambda_k \int_0^T f_j(t)f_k(t)\, dt = \lambda_k \delta_{jk} = \begin{cases} \lambda_k, & j = k, \\ 0, & j \neq k. \end{cases}
\end{aligned} \qquad (2\text{-}41)$$

The variance of the sample v_k equals the associated eigenvalue λ_k, which, as we have seen, must be positive: $\lambda_k > 0$.

If we write a Fourier expansion of the type in (2-12) for $\phi(t, s)$ considered as a function of t, its kth coefficient ϕ_k must, by (2-13), equal

$$\phi_k = \int_0^T f_k(u)\phi(u, s)\, du = \lambda_k f_k(s)$$

by (2-36) and (2-35), and the Fourier expansion of $\phi(t, s)$ becomes

$$\phi(t, s) = \sum_{k=1}^{\infty} \lambda_k f_k(t)f_k(s), \qquad (2\text{-}42)$$

which is known as *Mercer's formula*.

2.1.7 Reproducing-kernel Hilbert Space[2]

In Sec. 2.1.3 we found that with white noise there is an unlimited number of sets of orthonormal functions $f_k(t)$ with which one can sample the input $v(t)$ and obtain samples that are uncorrelated and, being Gaussian random variables, independent. In Sec. 2.1.6, on the other hand, it appears as though when the noise is colored, only a single set would serve that purpose, specifically, the set of eigenfunctions of the integral equation (2-35). In order to find an analogous plenitude of sampling functions that will produce uncorrelated samples, the definition of orthogonality must be modified. It is now based on a new definition of the scalar product of two functions, and to distinguish the new scalar product from that in (2-20), we mark it with angular brackets.

For functions $h(t)$ and $m(t)$ defined in the interval $(0, T)$, this scalar product is defined by

$$\langle \mathbf{h}, \mathbf{m} \rangle = \langle \mathbf{m}, \mathbf{h} \rangle = \int_0^T H(t)m(t) \, dt = \int_0^T h(t)M(t) \, dt \qquad (2\text{-}43)$$

where $H(t)$ and $M(t)$ are solutions of the integral equations

$$h(t) = \int_0^T \phi(t, s)H(s) \, ds,$$
$$m(t) = \int_0^T \phi(t, s)M(s) \, ds, \qquad 0 \leq t \leq T. \qquad (2\text{-}44)$$

As before, $\phi(t, s)$ is the autocovariance function of the noise, and we are here assuming it to be real and as in (2-36) symmetrical: $\phi(t, s) = \phi(s, t)$. We call $H(t)$ and $M(t)$ the *cofunctions* of $h(t)$ and $m(t)$, respectively.

The *norm* of a function is its scalar product with itself:

$$\|h(t)\| = \langle \mathbf{h}, \mathbf{h} \rangle = \int_0^T h(t)H(t) \, dt = \int_0^T \int_0^T H(t)\phi(t, s)H(s) \, ds \, dt. \qquad (2\text{-}45)$$

Because $\phi(t, s)$ is positive definite—see (2-40)—, the norm is always positive unless $h(t) \equiv 0$. The theory deals only with functions of finite norm in this sense. The square root of the norm is analogous to the length of a vector, and the functionals $\langle \mathbf{h}, \mathbf{m} \rangle$ have properties analogous to scalar products in an ordinary Cartesian vector space.

The functions $h(t)$ and $m(t)$ are now termed *orthogonal* if $\langle \mathbf{h}, \mathbf{m} \rangle = 0$. A set of functions $\{f_k(t)\}$ is said to be *orthonormal* if

$$\langle \mathbf{f}_j, \mathbf{f}_k \rangle = \delta_{jk} = \begin{cases} 1, & j = k, \\ 0, & j \neq k. \end{cases} \qquad (2\text{-}46)$$

The eigenfunctions of the integral equation (2-35) are orthogonal in this new sense, and if after normalization as in (2-39) they are multiplied by $\lambda_k^{1/2}$, they acquire norms $\|f_k(t)\|$ equal to 1.

---

[2]Reading this part can be deferred until Sec. 2.2.4.

From a set of linearly independent functions $g_k(t)$ of finite norm, a set of functions $f_k(t)$ orthonormal as in (2-46) can be constructed by the Gram–Schmidt procedure. It is merely necessary to replace the scalar product $(\cdot, \cdot)$ used in Sec. 2.1.4 by the scalar product $\langle \cdot, \cdot \rangle$ introduced here. We define samples of the input $v(t)$ by

$$v'_k = \langle v, \mathbf{f}_k \rangle = \int_0^T F_k(t)v(t) \, dt, \qquad \forall k, \qquad (2\text{-}47)$$

where $F_k(t)$ is the cofunction to $f_k(t)$. These samples are uncorrelated, for their covariances are

$$\text{Cov}(v'_k, v'_m) = E(v'_k v'_m \mid H_0) = \int_0^T \int_0^T F_k(t)F_m(s)E[v(t)v(s) \mid H_0] \, dt \, ds$$

$$= \int_0^T \int_0^T F_k(t)\phi(t, s)F_m(s) \, dt \, ds = \int_0^T F_k(t) f_m(t) \, dt$$

$$= \langle \mathbf{f}_k, \mathbf{f}_m \rangle = \delta_{km},$$

and the samples v'_k, being Gaussian random variables, are statistically independent.

The autocovariance function $\phi(t, u)$ of the noise is called the *kernel* of the space of functions having finite norm in the sense of (2-45). Considered as a function only of t, with u a parameter, it has the so-called *reproducing property*

$$\langle h(t), \phi(t, u) \rangle = h(u), \qquad 0 \le u \le T, \qquad (2\text{-}48)$$

for any function in the space. To demonstrate this, we again denote the cofunction of $h(t)$ by $H(t)$, whereupon the scalar product in (2-48) is

$$\langle h(t), \phi(t, u) \rangle = \int_0^T H(t)\phi(t, u) \, dt = \int_0^T \phi(u, t)H(t) \, dt = h(u)$$

by the symmetry of $\phi(t, u)$ and by (2-44). Because of (2-48), a function space with a scalar product defined as in (2-43) is called a *reproducing kernel Hilbert space*.

As we shall see in Sec. 2.3, although cofunctions such as $H(t)$ and $M(t)$ in (2-44) often contain delta functions and derivatives of delta functions, the scalar product $\langle \mathbf{h}, \mathbf{m} \rangle$ of $h(t)$ and $m(t)$ can—for a large class of autocovariance functions—be written in a form that is free of such "pathological" entities. If one has by some means worked out the form of the scalar product $\langle \mathbf{h}, \mathbf{m} \rangle$, one needs only to verify that it possesses the reproducing property (2-48). The detection of signals in colored Gaussian noise has been analyzed in the framework of the reproducing kernel Hilbert space by Kailath and others [Kai67], [Kai71]. In particular, the latter exhibits the norms arising from a variety of autocovariance functions, and from these the form of the scalar product $\langle \mathbf{h}, \mathbf{m} \rangle$ can be directly derived by the rule

$$\|h(t) + m(t)\| = \|h(t)\| + \|m(t)\| + 2\langle \mathbf{h}, \mathbf{m} \rangle.$$

2.2 DETECTION IN GAUSSIAN NOISE

2.2.1 The Likelihood Ratio

The input $v(t)$ is observed during an interval $(0, T)$. We seek the optimum strategy for deciding between the hypotheses

$$H_0: \quad v(t) = n(t), \qquad\qquad 0 \leq t \leq T,$$

and

$$H_1: \quad v(t) = s(t) + n(t), \qquad 0 \leq t \leq T,$$

where $n(t)$ is Gaussian noise with autocovariance function

$$E[n(t)n(s)] = E[v(t)v(s)|\, H_0] = \phi(t, s),$$

and $s(t)$ is a signal of known form and amplitude. For simplicity we assume that $s(t)$ vanishes outside the interval $(0, T)$. If the noise is stationary, its autocovariance function has the form $\phi(t, s) = \phi(t - s)$, and it possesses a spectral density $\Phi(\omega)$ as in (2-7). Such noise is called colored to distinguish it from white noise, whose spectral density is uniform and whose autocovariance function is proportional to a Dirac delta function as in (2-24). Even if the noise is nonstationary and possesses no spectral density, we refer to it as colored noise.

The input $v(t)$ is sampled in terms of a set of functions $f_k(t)$ orthonormal over $(0, T)$, as described in Sec. 2.1.6, with the samples defined by

$$v_k = \int_0^T f_k(t)v(t)\, dt, \tag{2-49}$$

in which the functions $f_k(t)$ are the eigenfunctions of the integral equation (2-35), whose kernel is the autocovariance function $\phi(t, s)$ of the noise. We are expressing the input $v(t)$ in a Karhunen–Loève expansion

$$v(t) = \sum_{k=1}^{\infty} v_k f_k(t). \tag{2-50}$$

For the moment we base our decision on only the first n of these samples, $v_1, v_2, \ldots, v_n$. Later we shall let n grow beyond all bounds.

We have seen that these samples are statistically independent Gaussian random variables. As in (2-41) their variances are

$$\text{Var } v_k = \lambda_k, \tag{2-51}$$

where λ_k is the kth eigenvalue of the integral equation (2-35). Under hypothesis H_0, as in (2-15), their expected values are zero. Under hypothesis H_1, on the other hand, when the signal $s(t)$ is present, $E[v(t)|\, H_1] = s(t)$, and

$$E(v_k|\, H_1) = \int_0^T f_k(t)s(t)\, dt = s_k, \tag{2-52}$$

Detection of a Known Signal   Chap. 2

where s_k is the kth coefficient of a Fourier expansion of the signal $s(t)$ in terms of the orthonormal eigenfunctions $f_k(t)$ of (2-35):

$$s(t) = \sum_{k=1}^{\infty} s_k f_k(t). \tag{2-53}$$

The joint probability density function of the samples v_k under hypothesis H_0 is thus

$$p_0(v_1, v_2, \ldots, v_n) = p_0(v) = \prod_{k=1}^{n} \frac{1}{\sqrt{2\pi\lambda_k}} \exp\left(-\frac{v_k^2}{2\lambda_k}\right).$$

Under hypothesis H_1 the noise component of the kth sample is $v_k - s_k$, and the joint density function of the samples $v_1, v_2, \ldots, v_n$ under hypothesis H_1 is

$$p_1(v_1, v_2, \ldots, v_n) = p_1(v) = \prod_{k=1}^{n} \frac{1}{\sqrt{2\pi\lambda_k}} \exp\left[-\frac{(v_k - s_k)^2}{2\lambda_k}\right].$$

The optimum strategy, as we learned in Chapter 1, bases the decision on the likelihood ratio

$$\Lambda_n(v) = \frac{p_1(v)}{p_0(v)} = \prod_{k=1}^{n} \exp\left[\frac{v_k^2 - (v_k - s_k)^2}{2\lambda_k}\right] = \exp \sum_{k=1}^{n} \left[\frac{s_k v_k}{\lambda_k} - \frac{s_k^2}{2\lambda_k}\right].$$

The data v appear in this expression only combined as in

$$g_n = \sum_{k=1}^{n} \frac{s_k v_k}{\lambda_k}, \tag{2-54}$$

and the likelihood ratio $\Lambda_n(v)$ is a monotone increasing function of g_n. Hence, according to Sec. 1.2.4, the decision can just as well be based on the sufficient statistic g_n, which will be compared with some decision level g_{0n}; if $g_n \geq g_{0n}$, H_1 is chosen, otherwise H_0. Let us evaluate the false-alarm and detection probabilities characterizing this strategy.

Because g_n, defined by (2-54), is a linear combination of the Gaussian random variables $v_1, v_2, \ldots, v_n$, it too has a Gaussian distribution under both hypotheses. Under hypothesis H_0 the expected value of g_n is zero because $E(v_k | H_0) = 0$. Its variance is

$$\sigma_n^2 = \text{Var}(g_n | H_0) = \sum_{k=1}^{n} \left(\frac{s_k}{\lambda_k}\right)^2 \text{Var}(v_k | H_0) = \sum_{k=1}^{n} \frac{s_k^2}{\lambda_k} \tag{2-55}$$

by (2-51). Its probability density function is therefore

$$p_0(g_n) = \frac{1}{\sqrt{2\pi\sigma_n^2}} \exp\left(-\frac{g_n^2}{2\sigma_n^2}\right), \tag{2-56}$$

and the probability Q_0 of a false alarm is

$$Q_0 = \frac{1}{\sqrt{2\pi\sigma_n^2}} \int_{g_{0n}}^{\infty} \exp\left(-\frac{g_n^2}{2\sigma_n^2}\right) dg_n = \text{erfc}\left(\frac{g_{0n}}{\sigma_n}\right) \tag{2-57}$$

by (1-32), where erfc($\cdot$) is the error-function integral defined in (1-11).

Sec. 2.2   Detection in Gaussian Noise

Under hypothesis H_1 the expected value of the statistic g_n is

$$E(g_n | H_1) = \sum_{k=1}^{n} \frac{s_k}{\lambda_k} E(v_k | H_1) = \sum_{k=1}^{n} \frac{s_k^2}{\lambda_k} = \sigma_n^2$$

by (2-52). Its variance is the same as under hypothesis H_0 because g_n in (2-54) is a linear combination of the variables v_k, and the terms due to the signal will cancel out when $\mathrm{Var}(g_n | H_1)$ is calculated. Hence the probability density function of the statistic g_n under hypothesis H_1 is

$$p_1(g_n) = \frac{1}{\sqrt{2\pi\sigma_n^2}} \exp\left[-\frac{(g_n - \sigma_n^2)^2}{2\sigma_n^2} \right], \tag{2-58}$$

and the probability of detection is

$$Q_d = \mathrm{erfc}\left(\frac{g_{0n} - \sigma_n^2}{\sigma_n} \right) = \mathrm{erfc}\left(\frac{g_{0n}}{\sigma_n} - \sigma_n \right) \tag{2-59}$$

by (1-33). Given the false-alarm probability Q_0, we solve

$$Q_0 = \mathrm{erfc}\, x \tag{2-60}$$

for x, whereupon the probability of detection is

$$Q_d = \mathrm{erfc}(x - \sigma_n), \tag{2-61}$$

where σ_n is given by (2-55).

The more samples $v_1, v_2, \dots, v_n$ of the input $v(t)$ we utilize, the larger the probability Q_d of detection, for σ_n^2 as defined by (2-55) increases with n. Hence, as we might expect, the maximum probability of detection is attained by utilizing all the samples, letting n go to infinity, and basing the decision on the statistic

$$g = \sum_{k=1}^{\infty} \frac{s_k v_k}{\lambda_k}. \tag{2-62}$$

Similarly passing to the limit $n \to \infty$ in (2-55), we find that

$$\sigma_n^2 \to \mathrm{Var}\, g = d^2 = \sum_{k=1}^{\infty} \frac{s_k^2}{\lambda_k}. \tag{2-63}$$

The random variable g defined by (2-62) makes sense only if its variance d^2 is finite. The quantity d^2 is the basic signal-to-noise ratio in this detection problem.

2.2.2 The Sufficient Statistic

Just as the function $s(t)$ corresponds to the set of signal samples $\{s_k\}$ through (2-53), in which the $f_k(t)$ are the eigenfunctions in (2-35), so we can define a function $q(t)$ by the Fourier series

$$q(t) = \sum_{k=1}^{\infty} q_k f_k(t), \qquad q_k = \frac{s_k}{\lambda_k}. \tag{2-64}$$

Then by virtue of the orthonormality of the eigenfunctions, we can write our sufficient statistic as

$$g = \sum_{k=1}^{\infty} q_k v_k = \sum_{k=1}^{\infty} q_k \int_0^T f_k(t)v(t)\,dt = \int_0^T q(t)v(t)\,dt, \qquad (2\text{-}65)$$

by (2-13). Furthermore, by (2-35),

$$\int_0^T \phi(t, u)q(u)\,du = \sum_{k=1}^{\infty} q_k \int_0^T \phi(t, u)f_k(u)\,du$$

$$= \sum_{k=1}^{\infty} \lambda_k q_k f_k(t) = \sum_{k=1}^{\infty} s_k f_k(t) = s(t),$$

whence $q(t)$ is the solution of the inhomogeneous integral equation

$$s(t) = \int_0^T \phi(t, u)q(u)\,du, \qquad 0 \le t \le T. \qquad (2\text{-}66)$$

In the next section we shall consider methods of solving this fundamental integral
equation.

The statistic g can be generated by multiplying the input $v(t)$ by $q(t)$ and
integrating over the observation interval $(0, T)$. One speaks of correlating the input
with $q(t)$, and a receiver that does so has been called a *correlation receiver*. We
shall see in Sec. 2.2.3, however, that it is simpler to produce the statistic g by linear
filtering of $v(t)$.

By replacing v_k by s_k in (2-65) and comparing with (2-63), we see that we can
write the signal-to-noise ratio d^2 as

$$d^2 = \int_0^T s(t)q(t)\,dt. \qquad (2\text{-}67)$$

Under hypothesis H_0 the expected value of our statistic g equals 0; under hypothesis
H_1 it is

$$E(g| H_1) = \int_0^T q(t)E[v(t)| H_1]\,dt = \int_0^T q(t)s(t)\,dt = d^2.$$

As g is a Gaussian random variable and its variance equals d^2 under both hypotheses
by (2-63), its probability density functions under the two hypotheses are

$$p_0(g) = \frac{1}{\sqrt{2\pi d^2}} \exp\left(-\frac{g^2}{2d^2}\right), \qquad (2\text{-}68)$$

$$p_1(g) = \frac{1}{\sqrt{2\pi d^2}} \exp\left[-\frac{(g - d^2)^2}{2d^2}\right]. \qquad (2\text{-}69)$$

The likelihood ratio of our statistic g is therefore

$$\Lambda(g) = \frac{p_1(g)}{p_0(g)} = \exp(g - \tfrac{1}{2}d^2), \qquad (2\text{-}70)$$

and by (2-65) and (2-67) this can be written in terms of the input $v(t)$ as

$$\Lambda[v(t)] = \exp\left[\int_0^T q(t)v(t)\, dt - \tfrac{1}{2}\int_0^T s(t)q(t)\, dt\right]. \tag{2-71}$$

We call this the *likelihood functional* for detection of the signal $s(t)$ in Gaussian noise having autocovariance function $\phi(t, s)$.

Under the Bayes criterion, the likelihood functional is compared with the quantity Λ_0 given by (1-17) in terms of the prior probabilities ζ_0 and ζ_1 of hypotheses H_0 and H_1 and the costs C_{ij} attending the various combinations of decision and true hypothesis. Equivalently, hypothesis H_1 is chosen whenever

$$g \ge g_0 = \tfrac{1}{2}d^2 + \ln \Lambda_0.$$

By passing to the limit $n \to \infty$ in (2-57) and (2-59) or by putting (2-68) and (2-69) into (1-32) and (1-33), we find that the false-alarm and detection probabilities have the forms

$$Q_0 = \operatorname{erfc} x, \qquad x = \frac{g_0}{d}, \qquad Q_d = \operatorname{erfc}(x - d), \tag{2-72}$$

where d^2 is the signal-to-noise ratio defined in (2-67).

When the Neyman–Pearson criterion is used, the value of the false-alarm probability Q_0 is fixed in advance, usually on the basis of the relative frequency of errors of the first kind that the observer can tolerate. The decision level g_0 is then determined from (2-72) as $g_0 = dx$.

In Fig. 2-1 the probability of detection Q_d has been plotted against d for a number of values of the false-alarm probability Q_0. The pair of equations (2-72) represents in parametric form the operating characteristic of the statistical test or the detection strategy; a number of these are shown in Fig. 2-2 for various values of the signal-to-noise parameter d.

It is often convenient to describe the effectiveness of a receiver by quoting the signal-to-noise ratio of the *minimum detectable signal*, that is, the value of d^2 required to attain a certain probability Q_d of detection for a given false-alarm probability Q_0. If, for instance, the values adopted are $Q_0 = 10^{-6}$ and $Q_d = 0.99$, we obtain from (2-72) the ratio $d^2 = 50.12 \equiv 17.01$ dB.

When the noise is white with unilateral spectral density N, its autocovariance function is

$$\phi(t, s) = \frac{N}{2}\delta(t - s)$$

as in (2-24), and substitution into (2-66) shows that

$$q(t) = \frac{2}{N}s(t). \tag{2-73}$$

The likelihood functional in (2-71) becomes

$$\Lambda[v(t)] = \exp\left[\frac{2}{N}\int_0^T s(t)v(t)\, dt - \frac{1}{N}\int_0^T [s(t)]^2\, dt\right]. \tag{2-74}$$

The detection statistic can now be taken as

$$g = \int_0^T s(t)v(t)\, dt,$$

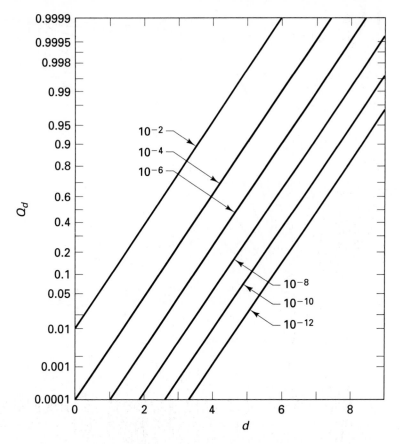

Figure 2-1. Probability of detection: completely known signal in Gaussian noise. Curves are indexed by the false-alarm probability Q_0.

the factor of $(2/N)$ being absorbed into the decision level g_0. The signal-to-noise ratio becomes

$$d^2 = \frac{2}{N} \int_0^T [s(t)]^2 \, dt = \frac{2E}{N} \tag{2-75}$$

in terms of the "energy"

$$E = \int_0^T [s(t)]^2 \, dt \tag{2-76}$$

of the signal. In Sec. 2.4 we shall present a physical interpretation of this basic signal-to-noise ratio.

That the statistic g defined in (2-65) is optimum for deciding between the hypotheses H_0 and H_1 was demonstrated by Grenander in 1950 [Gre50], but under the assumption that the solution $q(t)$ of (2-66) is square integrable over $(0, T)$. We shall see in Sec. 2.3 that that solution often contains delta functions and their derivatives, which are not square integrable. Kadota [Kad67] has shown that g is optimum even under those conditions, provided that the signal-to-noise ratio d^2 in (2-67) is finite.

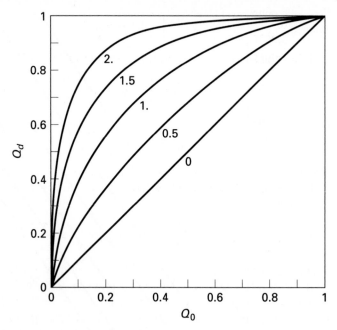

Figure 2-2. Receiver operating characteristics: completely known signal in Gaussian noise. Curves are indexed by signal-to-noise ratio d.

When, as ordinarily, a portion of the noise $n(t)$ is white and the energy of the signal is finite, the signal-to-noise ratio d^2 in (2-67) will be finite, and errors have nonzero probabilities under both hypotheses. If the white component has been neglected in modeling the noise in a detection problem, it is possible for the signal-to-noise ratio d^2 to turn out to be infinite. Examination of (2-56) and (2-58) shows that when we make the number n of samples larger and larger, the probability density functions of the statistic g_n move farther and farther apart; and if $d^2 = \infty$, we can, by setting a decision level midway between them, cause the probabilities Q_0 and $Q_1 = 1 - Q_d$ of errors of the first and second kinds to go to zero. A situation like this, which is of hardly more than mathematical interest, results in what is known as *singular detection*. Conditions under which it may occur have been extensively studied [Roo63], [Yag63], [Kai66a]. A rigorous treatment of this matter can be found in the book by Poor [Poo88]. For those who wish to delve into the literature on this topic, we mention that in its terminology the "equivalence" of the probability measures associated with $v(t)$ under the two hypotheses means that detection is imperfect; "perpendicularity" of the measures means that error probabilities are zero and the singular case is at hand. What we have called the likelihood functional is often termed the Radon–Nikodym derivative of one measure with respect to the other.

An indication of the kind of noise model that may imply singular detection can be obtained by studying the detection of a signal $s(t)$ in stationary Gaussian noise on the basis of observation of the input $v(t)$ during an infinite interval. The integral equation (2-66) then becomes

$$s(t) = \int_{-\infty}^{\infty} \phi(t - u)q_\infty(u)\, du. \tag{2-77}$$

Because it has the form of a convolution, it can be solved by Fourier transformation,

$$q_\infty(t) = \int_{-\infty}^{\infty} Q_\infty(\omega)\, e^{i\omega t} \frac{d\omega}{2\pi}, \tag{2-78}$$

where

$$Q_\infty(\omega) = \frac{S(\omega)}{\Phi(\omega)} \tag{2-79}$$

in terms of the spectral density $\Phi(\omega)$ of the noise and the spectrum $S(\omega)$ of the signal. In this limit of an infinitely long observation interval, the signal-to-noise ratio (2-67) becomes

$$d_\infty^2 = \int_{-\infty}^{\infty} s(t)q_\infty(t)\, dt = \int_{-\infty}^{\infty} S^*(\omega)Q_\infty(\omega) \frac{d\omega}{2\pi} = \int_{-\infty}^{\infty} \frac{|S(\omega)|^2}{\Phi(\omega)} \frac{d\omega}{2\pi}. \tag{2-80}$$

The longer the observation interval is, the more information in the input $v(t)$ is utilized for deciding between the two hypotheses, and the higher the probability Q_d of detection for a given false-alarm probability Q_0 must be. The effective signal-to-noise ratio d^2 must therefore increase with the length of the interval $(0, T)$, and the quantity d_∞^2 of (2-80) must represent an upper bound to the signal-to-noise ratio for any finite interval,

$$d^2 \le d_\infty^2.$$

If d_∞^2 is finite, d^2 must be finite, and detection will not be singular. This will be the case when the spectral density $\Phi(\omega)$ of the noise is nonzero for all ω and drops off to zero as $|\omega| \to \infty$ more slowly than $|S(\omega)|^2$ and, in particular, when the noise contains a white component whose spectral density is uniform at all frequencies. If, on the other hand, $d_\infty^2 = \infty$, as would happen, for instance, if the spectral density $\Phi(\omega)$ of the noise vanished in any frequency band where the spectrum $S(\omega)$ of the signal did not, detection might be singular even for a finite observation interval.

2.2.3 The Matched Filter

The sufficient statistic g in (2-65) can be generated by passing the input $v(t)$ through a filter matched to the "signal" $q(t)$. The impulse response of this filter is

$$k(\tau) = \begin{cases} q(T - \tau), & 0 \le \tau \le T, \\ 0, & \tau < 0. \end{cases} \tag{2-81}$$

It is customary, although unnecessary, to take $k(\tau) \equiv 0$ for $\tau > T$ as well. When the input $v(t)$ is turned on at time $t = 0$, the output of this filter at a later time t in $0 \le t \le T$ is

$$v_0(t) = \int_0^t k(\tau)v(t - \tau)\, d\tau = \int_0^t q(T - \tau)v(t - \tau)\, d\tau$$

$$= \int_{T-t}^T q(u)v(t - T + u)\, du,$$

and at time T the output equals

$$v_0(T) = \int_0^T q(u)v(u)\,du = g.$$

The optimum receiver therefore consists of the matched filter $k(\cdot)$ and a sampling device that samples the output $v_0(t)$ of the filter at the end of the observation interval $(0, T)$ and compares it with a decision level g_0; if g exceeds g_0, the receiver decides that a signal is present.

Among all linear filters through which the input $v(t)$ might be passed, the matched filter yields the largest output signal-to-noise ratio

$$d_{\text{out}}^2 = \frac{[s_0(T)]^2}{\text{Var } v_n(T)}. \tag{2-82}$$

Here $s_0(t)$ and $v_n(t)$ are the outputs of an arbitrary linear filter when the signal and the noise are, respectively, alone present in the input, which we now suppose turned on at time $t = 0$. (When the noise is nonwhite, that is, correlated, its values for $t < 0$ would provide information enhancing signal detectability even though the signal started after $t = 0$.) With $k(\tau)$ the impulse response of the filter, these components of the output are at time T

$$s_0(T) = \int_0^T k(\tau)s(T - \tau)\,d\tau = \int_0^T k(T - u)s(u)\,du,$$

$$v_n(T) = \int_0^T k(T - u)n(u)\,du.$$

Because the noise output $v_n(T)$ has zero expected value, its variance is

$$\begin{aligned}
\text{Var } v_n(T) &= E\{[v_n(T)]^2 \mid H_0\} \\
&= \int_0^T \int_0^T k(T - u_1)k(T - u_2)E[n(u_1)n(u_2)]\,du_1\,du_2 \\
&= \int_0^T \int_0^T k(T - u_1)\phi(u_1, u_2)k(T - u_2)\,du_1\,du_2,
\end{aligned}$$

where $\phi(\cdot, \cdot)$ is still the autocovariance function of the noise.

Let us write both $s(t)$ and $k(T - t)$ as Fourier expansions in terms of the eigenfunctions $f_k(t)$ of (2-35):

$$s(t) = \sum_{n=1}^\infty s_n f_n(t), \qquad k(T - t) = \sum_{m=1}^\infty k_m f_m(t).$$

Then when we use (2-35) and (2-39), we can write the output signal-to-noise ratio as

$$d_{\text{out}}^2 = \frac{\left[\sum_{m=1}^\infty k_m s_m\right]^2}{\sum_{m=1}^\infty \lambda_m k_m^2} = \frac{\left[\sum_{m=1}^\infty \left(\lambda_m^{1/2} k_m\right)\left(\lambda_m^{-1/2} s_m\right)\right]^2}{\sum_{m=1}^\infty \left(\lambda_m^{1/2} k_m\right)^2}.$$

The Cauchy–Schwarz inequality for sequences states that for any two sequences $(a_1, a_2, \ldots)$ and $(b_1, b_2, \ldots)$

$$\left[\sum_{m=1}^{\infty} a_m b_m\right]^2 \leq \sum_{m=1}^{\infty} a_m^2 \sum_{n=1}^{\infty} b_n^2 \qquad (2\text{-}83)$$

with equality if and only if $b_m \equiv c a_m$, $\forall m$, for some constant c. Taking $a_m = \lambda_m^{1/2} k_m$, $b_m = \lambda_m^{-1/2} s_m$, we find

$$d_{\text{out}}^2 \leq \sum_{m=1}^{\infty} \frac{s_m^2}{\lambda_m} = \sum_{m=1}^{\infty} s_m q_m = \int_0^T s(t) q(t) \, dt = d^2$$

with equality if and only if $k_m \equiv c s_m/\lambda_m = c q_m$; that is,

$$k(T - t) = c q(t).$$

The maximum possible output signal-to-noise ratio is therefore given by d^2 of (2-67) and is attained by the matched filter specified by (2-81). The concept of the matched filter was proposed by North [Nor43] and applied to this decision problem by Peterson, Birdsall, and Fox [Pet54].

2.2.3.1. The matched filter for white noise   When the noise is white, the function $q(t)$ is proportional to the signal $s(t)$, as in (2-73). The constant of proportionality can be absorbed in the decision level g_0, and the impulse response of the matched filter for detection of this signal in white noise can be taken as

$$k(\tau) = \begin{cases} s(T - \tau), & 0 \leq \tau \leq T, \\ 0, & \tau \leq 0, \quad \tau > T. \end{cases} \qquad (2\text{-}84)$$

Example 2-1   Triangular pulse in white noise

Let the signal be a triangular pulse of the form

$$s(t) = \begin{cases} t, & 0 \leq t \leq T', \\ 0, & t < 0, \quad t > T'. \end{cases}$$

It is illustrated in Fig. 2-3. The observation interval is still $(0, T)$, with $T > T'$, and the impulse response of the matched filter is

$$k(\tau) = \begin{cases} T - \tau, & T - T' \leq \tau \leq T, \\ 0, & \tau < T - T', \quad \tau > T, \end{cases}$$

as shown in the same figure. Then the output of the filter when only the signal $s(t)$ is applied to it is

$$s_0(t) = \begin{cases} 0, & t < T - T', \\ \frac{1}{6}(|T - t| + 2T')(T' - |T - t|)^2, & T - T' \leq t \leq T + T', \\ 0, & t > T + T'. \end{cases}$$

It is shown in the lowest part of Fig. 2-3. This output reaches a maximum at $t = T$, the time at which the decision is made. For a pulse of duration limited to $(0, T')$, there is no point to making the observation interval longer than T'.

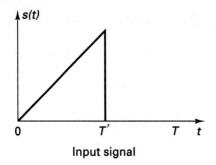

Input signal

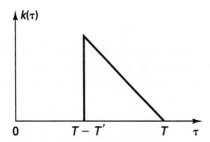

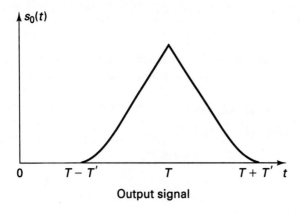

Output signal

Figure 2-3. Matched filter and its output.

If the signal is confined to the observation interval $(0, T)$, as we have supposed, the transfer function of the matched filter is proportional to the complex conjugate of the spectrum

$$S(\omega) = \int_0^T s(t) \, e^{-i\omega t} \, dt$$

of the signal. It is defined by

$$
\begin{aligned}
y(\omega) &= \int_{-\infty}^{\infty} k(\tau) \, e^{-i\omega\tau} \, d\tau = \int_0^T s(T - \tau) \, e^{-i\omega\tau} \, d\tau \\
&= \int_0^T s(u) \, e^{-i\omega(T-u)} \, du = e^{-i\omega T} S^*(\omega).
\end{aligned}
\tag{2-85}
$$

Detection of a Known Signal   Chap. 2

The factor $\exp(-i\omega T)$ corresponds to a delay of T seconds in the response of the filter.

The spectrum of the signal component $s_0(t)$ of the output of the matched filter is

$$\int_{-\infty}^{\infty} s_0(t)\, e^{-i\omega t}\, dt = y(\omega)S(\omega) = e^{-i\omega T}|S(\omega)|^2,$$

and hence that component is

$$s_0(t) = \int_{-\infty}^{\infty} |S(\omega)|^2\, e^{i\omega(t-T)}\frac{d\omega}{2\pi}.$$

Because $|S(\omega)| = |S(-\omega)|$, this is

$$s_0(t) = \int_{-\infty}^{\infty} |S(\omega)|^2 \cos\omega(t-T)\frac{d\omega}{2\pi},$$

and we see that the signal component of the output of the matched filter is an even function of $(t-T)$:

$$s_0(T-x) = s_0(T+x).$$

The matched filter can be constructed by methods that have been developed for synthesizing filters with prescribed transient response [Gui57], [Yen64]. Matched filters find extensive application in signal processing; Turin has written a thorough review of this subject [Tur60a].

2.2.3.2. The matched filter for stationary colored noise

The impulse response of the matched filter for detecting a signal in colored Gaussian noise has been given in (2-81). When the noise is stationary and the observation interval, which we now take as $(-T, T)$, is much longer than either the duration of the signal $s(t)$ or the width of the autocovariance function $\phi(\tau)$, the solution $q(t)$ of (2-66) will be close to the function $q_\infty(t)$ defined by (2-78) and (2-79), and the transfer function of the matched filter in (2-81) will be approximately

$$y_\infty(\omega) = Q_\infty^*(\omega)\, e^{-i\omega T} = e^{-i\omega T}\frac{S^*(\omega)}{\Phi(\omega)} \qquad (2\text{-}86)$$

[Dwo50]. If the spectral density of the noise can be factored as

$$\Phi(\omega) = \Gamma(\omega)\Gamma^*(\omega), \qquad (2\text{-}87)$$

where the function $\Gamma(\omega)$ contains all the poles and zeros of the spectral density $\Phi(\omega)$ lying above the real axis in the ω-plane, the transfer function $y_\infty(\omega)$ in (2-86) can be expressed as

$$y_\infty(\omega) = y_1(\omega)y_2(\omega) \qquad (2\text{-}88)$$

with

$$y_1(\omega) = \frac{1}{\Gamma(\omega)}, \qquad y_2(\omega) = e^{-i\omega T}\frac{S^*(\omega)}{\Gamma^*(\omega)}, \qquad (2\text{-}89)$$

and the matched filtering can be carried out in two stages, as shown in Fig. 2-4.

At the output of the first filter, the spectral density of the noise is uniform, $\Phi_1(\omega) = \Phi(\omega)|y_1(\omega)|^2 \equiv 1$; the noise $n_1(t)$ at that point is white. For this reason the

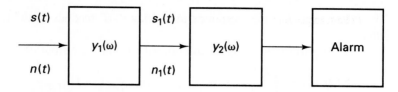

Figure 2-4. Matched filter for a long observation interval: $y_1(\omega)\,y_2(\omega) = \exp(-i\omega T)S^*(\omega)/\Phi(\omega)$.

filter whose transfer function is $y_1(\omega) = 1/\Gamma(\omega)$ is called a *whitening filter* [Bod50]. The spectrum of the signal $s_1(t)$ at that point is

$$S_1(\omega) = S(\omega)y_1(\omega) = \frac{S(\omega)}{\Gamma(\omega)}.$$

This filter is causal because all the poles and zeros of $y_1(\omega)$ lie above the Re ω-axis. Its impulse response $k_1(\tau)$ is nonzero only for $\tau \geq 0$.

The task of the second filter is to facilitate the detection of a known signal, $s_1(t)$, in white noise, and (2-85) and (2-89) show that it has indeed the proper transfer function. A system such as this can be realized only approximately and even so only by accepting a long delay T, but it serves to elucidate the results of our mathematical analysis. The delay must be long enough so that the tail of the output $s_1(t)$ of the whitening filter is negligible.

This approximate realization of the matched filter provides an alternative way of calculating the maximum attainable signal-to-noise ratio (2-80),

$$d_\infty^2 = \int_0^\infty [s_1(t)]^2\, dt, \tag{2-90}$$

which follows from (2-75) because at the output of the whitening filter the signal is $s_1(t)$, and it is immersed in white noise of unit bilateral spectral density, $N/2 \equiv 1$.

Example 2-2   Triangular pulse in colored noise

Let the signal $s(t)$ have the same triangular form as in Example 2-1: $s(t) = t, 0 \leq t \leq T'$; $s(t) \equiv 0, t < 0, t > T'$. The noise is a combination of white noise having unilateral spectral density N and noise having a Lorentz spectral density with variance ϕ_0:

$$\Phi(\omega) = \frac{N}{2} + \frac{2\mu\phi_0}{\mu^2 + \omega^2};$$

its autocovariance function is

$$\phi(\tau) = \frac{N}{2}\delta(\tau) + \phi_0\, e^{-\mu|\tau|}.$$

We write the spectral density as

$$\Phi(\omega) = \frac{N}{2}\frac{\beta^2 + \omega^2}{\mu^2 + \omega^2}, \qquad \beta^2 = \mu^2 + \frac{4\mu\phi_0}{N},$$

which can be factored as in (2-87) with

$$\Gamma(\omega) = \left(\frac{N}{2}\right)^{1/2}\frac{\beta + i\omega}{\mu + i\omega}.$$

Detection of a Known Signal   Chap. 2

The transfer function of the first filter in Fig. 2-4 is

$$y_1(\omega) = \left(\frac{2}{N}\right)^{1/2} \frac{\mu + i\omega}{\beta + i\omega} = \left(\frac{2}{N}\right)^{1/2} \left[1 - \frac{\beta - \mu}{\beta + i\omega}\right],$$

and its impulse response is

$$k_1(\tau) = \left(\frac{2}{N}\right)^{1/2} [\delta(\tau) - (\beta - \mu) e^{-\beta\tau} U(\tau)].$$

The signal component of the output of this whitening filter is

$$s_1(t) = \int_0^\infty k_1(\tau) s(t - \tau) \, d\tau = \left(\frac{2}{N}\right)^{1/2} \left[s(t) - (\beta - \mu) \int_0^\infty e^{-\beta\tau} s(t - \tau) \, d\tau\right]$$

$$\equiv 0, \qquad t < 0,$$

$$= \left(\frac{2}{N}\right)^{1/2} \beta^{-2} [\mu\beta t + (\beta - \mu)(1 - e^{-\beta t})], \qquad 0 \le t \le T',$$

$$= -\left(\frac{2}{N}\right)^{1/2} \beta^{-2} (\beta - \mu)(\beta T' - 1 + e^{-\beta T'}) e^{-\beta(t-T')}, \qquad t > T'.$$

It is sketched in Fig. 2-5(a). The second filter in Fig. 2-4 is matched to this signal $s_1(t)$ with a delay T that needs only to be long enough so that the tail of $s_1(t)$ is insignificant: $\beta(T - T') \gg 1$.

The solution $q_\infty(t)$ of (2-77) is the inverse Fourier transform of

$$Q_\infty(\omega) = \frac{2}{N} \frac{\mu^2 + \omega^2}{\beta^2 + \omega^2} S(\omega) = \frac{2}{N} \left[1 - \frac{\beta^2 - \mu^2}{\beta^2 + \omega^2}\right] S(\omega)$$

and can be considered as the output of a noncausal filter whose input is the signal $s(t)$ and whose impulse response is

$$k_\infty(\tau) = \frac{2}{N} \left[\delta(\tau) - \frac{\beta^2 - \mu^2}{2\beta} e^{-\beta|\tau|}\right].$$

For our triangular signal this is

$$q_\infty(t) = \begin{cases} -\dfrac{\beta^2 - \mu^2}{N\beta^3} \left[1 - e^{-\beta T'}(\beta T' + 1)\right] e^{\beta t}, & t < 0, \\[2ex] \dfrac{2}{N}\left\{t - \dfrac{\beta^2 - \mu^2}{2\beta^3} \left[2\beta t + e^{-\beta t} - (\beta T' + 1)e^{-\beta(T'-t)}\right]\right\}, & 0 \le t \le T', \\[2ex] -\dfrac{\beta^2 - \mu^2}{N\beta^3}(\beta T' - 1 + e^{-\beta T'})e^{-\beta(t-T')}, & t > T'. \end{cases}$$

It is sketched in Fig. 2-5(b).

The signal component of the output of the entire matched filter (2-81) when the delay T is so long that $T \gg T'$, $\beta(T - T') \gg 1$, is

$$s_0(t) = \int_{-\infty}^\infty \frac{|S(\omega)|^2}{\Phi(\omega)} e^{-i\omega(t-T)} \frac{d\omega}{2\pi},$$

which by the convolution theorem can be expressed as

$$s_0(t) = \int_{-\infty}^\infty s(v) q_\infty(T - t + v) \, dv.$$

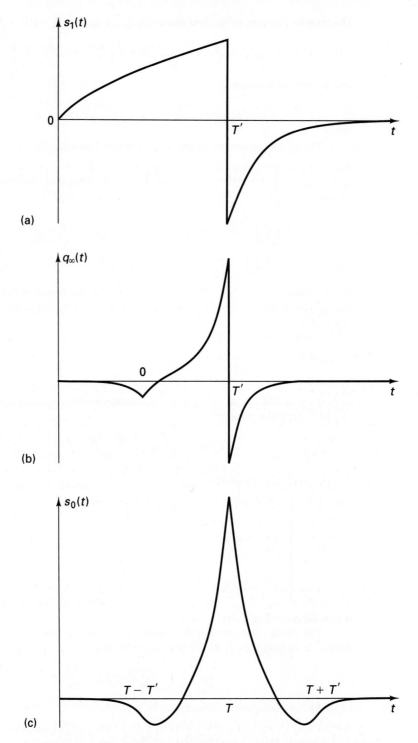

Figure 2-5. Example 2-2: (a) Output $s_1(t)$ of the whitening filter; (b) solution $q_\infty(t)$ of (2-77); (c) output $s_0(t)$ of the matched filter (2-81). Here we used the parameters $\phi_0 T'/N = 4$, $\mu T' = 2$, $\beta T' = 6$.

A somewhat tedious integration yields for this output $s_0(t) = \bar{s}(|t - T|)$, where

$$\bar{s}(t) = (N\beta^5)^{-1}\Big\{2\mu^2\beta^3\bar{s}_0(t) - (\beta^2 - \mu^2)\Big[2\beta t + e^{-\beta t}[2 - \beta^2 T'^2 - e^{-\beta T'}(1 + \beta T')]$$

$$- (\beta T' + 1)e^{\beta(t-T')}\Big]\Big\}, \qquad 0 \le t \le T',$$

$$= -(N\beta^5)^{-1}(\beta^2 - \mu^2)[1 - e^{-\beta T'}(\beta T' + 1)](\beta T' - 1 + e^{-\beta T'})e^{-\beta(t-T')}, \qquad t > T',$$

with

$$\bar{s}_0(t) = \tfrac{1}{6}(2T' + t)(T' - t)^2$$

the form of the output of the matched filter in Example 2-1 for $0 \le t \le T'$. The output signal $s_0(t)$ is sketched in Fig. 2-5(c). In particular, the upper bound on the signal-to-noise ratio in (2-80) is obtained by setting $t = T$:

$$d_\infty^2 = s_0(T) = \bar{s}(0)$$
$$= d_0^2 \beta^{-5}\Big\{\mu^2\beta^3 - \tfrac{3}{2}(\beta^2 - \mu^2)\Big[2 - \beta^2 T'^2 - 2(1 + \beta T')\,e^{-\beta T'}\Big]T'^{-3}\Big\}$$

with

$$d_0^2 = \frac{2T'^3}{3N}$$

the signal-to-noise ratio for detection in the absence of the colored component, whereupon $\beta = \mu$.

2.2.4 The Irrelevance Proof

An alternative derivation of the sufficient statistic for detecting the signal $s(t)$ in Gaussian noise having autocovariance function $\phi(t, s)$ is carried out in the framework of the reproducing-kernel Hilbert space (RKHS) introduced in Sec. 2.1.7. We saw there that from an arbitrary set of linearly independent functions $g_k(t)$ we can form a set of functions $f_k(t)$ that are orthonormal in the sense of (2-46), and that samples v_k' of the input $v(t)$ generated from these as in (2-47) are uncorrelated and hence, being Gaussian random variables, are statistically independent. These new functions $f_k(t)$ are not necessarily the eigenfunctions of (2-35).

If we now take the first member of our set as $g_1(t) = s(t)$, the first member of the set of orthonormal functions will be

$$f_1(t) = d^{-1}s(t),$$

which is normalized in accordance with (2-46) by virtue of the definition of d in (2-67). Furthermore, $E(v_k'|H_0) \equiv 0$, and by (2-47), with $F_k(t)$ the cofunction of $f_k(t)$ as defined in (2-43) and (2-44),

$$E(v_k'|H_1) = \int_0^T F_k(t)E[v(t)|H_1]\,dt = \int_0^T F_k(t)s(t)\,dt$$

$$= d\langle \mathbf{f}_k, \mathbf{f}_1\rangle = \begin{cases} d, & k = 1, \\ 0, & k \ne 1. \end{cases}$$

The samples v_k', $k \ge 2$, therefore have the same probability density functions under both hypotheses. When we form the likelihood ratio

$$\Lambda(v_1, v_2, \ldots, v_n) = \frac{p_1(v_1', v_2', \ldots, v_n')}{p_0(v_1', v_2', \ldots, v_n')} = \prod_{k=1}^{n} \frac{p_1(v_k')}{p_0(v_k')},$$

all the factors will cancel except that for v_1'. This will be the case however large the number n of samples may be. The sample v_1' is therefore a sufficient statistic. Because the cofunction $F_1(t)$ equals $d^{-1}q(t)$ by (2-44) and (2-66), $v_1' = d^{-1}g$ is proportional to the statistic g defined in (2-62), which is therefore also a sufficient statistic.

What we have done here has been to divide the input $v(t)$ into two statistically independent parts,

$$v(t) = v'(t) + v''(t),$$
$$v'(t) = d^{-2}s(t)g,$$
$$v''(t) = \sum_{k=2}^{\infty} v_k' f_k(t),$$

in such a way that the distributions of $v''(t)$ are identical under both hypotheses H_0 and H_1, and the random process $v''(\cdot)$ is uncorrelated with $v'(\cdot)$ and hence statistically independent of it. Thus $v'(t)$ contains all the information in the input relevant to making the decision between the hypotheses in the optimum fashion. The component $v''(t)$ is irrelevant, and the foregoing argument is often called the *irrelevance proof* that g is the optimum detection statistic. It is due to Kailath [Kai67], who developed the connection with reproducing-kernel Hilbert spaces in a later paper [Kai71]. The latter presents some examples of RKHS norms, and further methods of calculating such norms are outlined in [Kai72]. The signal-to-noise ratio d^2 specified by (2-67) is the RKHS norm of our signal $s(t)$ as defined in (2-45), and by working only with functions having a finite norm, this approach circumvents the question of singular detection.

2.2.5 Discrete-time Processing

In some situations it is convenient to sample the input $v(t)$ at n times $t_1, t_2, \ldots, t_n$ uniformly separated throughout the observation interval $(0, T)$, creating a set of samples $v_k = v(t_k)$ that can be collected into a row vector

$$v^T = (v_1, v_2, \ldots, v_n).$$

Preliminary filtering will have removed noise of frequencies outside the spectral band of the signal $s(t)$ to be detected. The samples are characterized by a covariance matrix

$$\phi = E(vv^T | H_0)$$

whose elements are

$$\phi_{jk} = \text{Cov}(v_j, v_k) = E[v(t_j)v(t_k) | H_0]$$

as in Sec. 2.1.1. Their expected values are zero under hypothesis H_0; under hypothesis H_1 they are temporal samples of the signal,

$$E(v_k | H_1) = s(t_k) = s_k.$$

The joint probability density function of the samples under hypothesis H_0 has the form in (2-5)

$$p_0(v) = (2\pi)^{-n/2}| \det \boldsymbol{\phi}|^{-1/2} \exp(-\tfrac{1}{2}v^T \boldsymbol{\phi}^{-1}v),$$

and under hypothesis H_1 it is

$$p_1(v) = (2\pi)^{-n/2}| \det \boldsymbol{\phi}|^{-1/2} \exp\left[-\tfrac{1}{2}(v^T - s^T)\boldsymbol{\phi}^{-1}(v - s)\right],$$

where $s^T = (s_1, s_2, \dots, s_n)$ is the row vector of the samples of the signal.

The optimum processing of these samples requires forming their likelihood ratio

$$\Lambda(v) = \frac{p_1(v)}{p_0(v)} = \exp\left[\tfrac{1}{2}v^T \boldsymbol{\phi}^{-1}v - \tfrac{1}{2}(v^T - s^T)\boldsymbol{\phi}^{-1}(v - s)\right]$$

$$= \exp\left[v^T \boldsymbol{\phi}^{-1}s - \tfrac{1}{2}s^T \boldsymbol{\phi}^{-1}s\right],$$

and we see that a sufficient statistic is the linear combination

$$g = v^T \mathbf{q} = \sum_{k=1}^{n} q_k v_k,$$

where $\mathbf{q} = \boldsymbol{\phi}^{-1}s$. The coefficients q_k are the solutions of the linear equations

$$\mathbf{s} = \boldsymbol{\phi}\mathbf{q} \quad (\text{or}) \quad s_k = \sum_{j=1}^{n} \phi_{kj}q_j,$$

and they can be determined in advance. The statistic g can be evaluated by a digital computer or analogous device when the sampling intervals T/n are sufficiently long. It is compared with a decision level g_0, and as before, the decision is for hypothesis H_1 when $g \geq g_0$ and for H_0 when $g < g_0$.

The expected values of the statistic g are

$$E(g|H_0) = 0, \qquad E(g|H_1) = s^T \mathbf{q} = d^2,$$

under the two hypotheses, and its variance is

$$\text{Var } g = E(g^2|H_0) = E(\mathbf{q}^T vv^T \mathbf{q}|H_0)$$

$$= \mathbf{q}^T \boldsymbol{\phi}\mathbf{q} = s^T \mathbf{q} = d^2$$

under both. The false-alarm and detection probabilities are again given by (2-72), and the governing signal-to-noise ratio is

$$d^2 = s^T \mathbf{q} = \sum_{k=1}^{n} s_k q_k,$$

which is the discrete-time counterpart of (2-67).

2.3 SOLUTION OF THE INTEGRAL EQUATIONS

2.3.1 Inhomogeneous Equations

To determine the impulse response of the optimum filter for detecting a known signal $s(t)$ in stationary Gaussian noise of autocovariance function $\phi(\tau)$, one must solve an integral equation of the form

$$s(t) = \int_0^T \phi(t - u)q(u) \, du, \qquad 0 \le t \le T, \qquad (2\text{-}91)$$

for the unknown function $q(t)$. This type of equation, which is called a Fredholm integral equation of the first kind, occurs frequently in detection theory and in the theory of linear prediction and filtering. A continuous solution $q(t)$ does not in general exist for continuous $s(t)$ unless the kernel $\phi(t - u)$ has some singularity or the range of integration is unbounded [Cou31, vol. 1, p. 135]. For certain types of kernel a solution in closed form can be obtained, but it involves delta functions and their derivatives. One's first thought is to treat (2-91) numerically, using a quadrature formula to replace the integral by a summation and solving the resulting set of linear simultaneous equations for the values of $q(t)$ at a finite set of points in the interval $0 \le t \le T$, but a solution involving delta-function singularities can hardly be well approximated in this way.

The situation is more favorable when, as is indeed usually the case, the noise contains a white component that has a flat spectral density, that is, when the autocovariance of the noise is of the form

$$\phi(\tau) = \frac{N}{2}\delta(\tau) + \pi(\tau),$$

where $\pi(\tau)$ is continuous, positive definite, and integrable. Then (2-91) becomes

$$s(t) = \frac{N}{2}q(t) + \int_0^T \pi(t - u)q(u) \, du, \qquad 0 \le t \le T; \qquad (2\text{-}92)$$

this is a Fredholm integral equation of the second kind, and a solution will generally exist unless $(-N/2)$ is an eigenvalue of the integral equation

$$\lambda f(t) = \int_0^T \pi(t - u)f(u) \, du$$

[Cou31, vol. 1, Ch. 3]. Because $\pi(t - u)$ is ordinarily a positive-definite kernel representing an additive colored noise component, this integral equation cannot have a negative eigenvalue, and there is no trouble about the existence of a solution that is free of singularities. If the amount of colored noise is small, a solution of (2-92) can be obtained by iteration,

$$q(t) = \frac{2}{N}\left[s(t) - \frac{2}{N}\int_0^T \pi(t - u)s(u) \, du + \cdots \right].$$

Otherwise one can calculate a solution numerically by replacing the integral by a summation and solving the resulting simultaneous equations for the values of $q(t)$

at a finite set of points in $(0, T)$. See [Bak77, Ch. 4] for an exposition of these numerical methods.

A type of autocovariance function $\phi(t - s)$ for which the integral equation (2-91) can be solved explicitly is that characterizing stationary noise whose spectral density $\Phi(\omega)$ is a rational function of the angular frequency ω. Noise of this kind can be generated by passing white noise through a linear filter composed of linear circuit elements—resistors, inductors, and capacitors. The transfer function $y(\omega)$ of such a filter is a rational function of ω, and the spectral density $\Phi(\omega)$ of the noise output, proportional to $|y(\omega)|^2$, is a rational function of ω^2, which can be written in the form

$$\Phi(\omega) = C \frac{\prod_{j=1}^{m} (\omega^2 + h_j^2)}{\prod_{k=1}^{n} (\omega^2 + m_k^2)}, \qquad C > 0, \tag{2-93}$$

with $m \leq n$. The constants h_j and m_k are either real or, if complex, occur in complex-conjugate pairs. We call noise of this kind *leucogenic*[3] because it can be thought of as arising from white noise through linear filtering. The autocovariance function of the noise is the sum of exponential functions of $|\tau|$, possibly multiplied by polynomials. Terms corresponding to complex m_k's are bilaterally damped sinusoids.

The spectral density $\Phi(\omega)$ can be written in the form

$$\Phi(\omega) = \frac{N(\omega^2)}{P(\omega^2)}, \tag{2-94}$$

where

$$N(x) = \beta_m x^m + \beta_{m-1} x^{m-1} + \cdots + \beta_1 x + \beta_0$$

is a polynomial of degree m and

$$P(x) = \alpha_n x^n + \alpha_{n-1} x^{n-1} + \cdots + \alpha_1 x + \alpha_0$$

is a polynomial of degree n; $m \leq n$. Then the autocovariance function $\phi(t - u)$ is a solution of the linear differential equation

$$P(-D^2)\phi(t - u) = N(-D^2)\delta(t - u), \qquad D = \frac{d}{dt},$$
$$D^2 = \frac{d^2}{dt^2}, \qquad -\infty < t < \infty, \tag{2-95}$$

in which the differential operators on each side are obtained by replacing x by $-D^2$ everywhere:

$$P(-D^2) = (-1)^n \alpha_n \frac{d^{2n}}{dt^{2n}} + (-1)^{n-1} \alpha_{n-1} \frac{d^{2(n-1)}}{dt^{2(n-1)}} + \cdots - \alpha_1 \frac{d^2}{dt^2} + \alpha_0,$$
$$N(-D^2) = (-1)^m \beta_m \frac{d^{2m}}{dt^{2m}} + (-1)^{m-1} \beta_{m-1} \frac{d^{2(m-1)}}{dt^{2(m-1)}} + \cdots - \beta_1 \frac{d^2}{dt^2} + \beta_0.$$

The boundary conditions on (2-95) are that $\phi(t - u)$ go to zero as t goes to ∞ and $-\infty$. Indeed, taking the Fourier transform of (2-95), with (2-7), yields (2-94) immediately.

[3] Greek λευκός = white, γένος = lineage, family, descent.

Let us now operate on both sides of (2-91) with the operator $P(-D^2)$. By virtue of (2-95) we obtain

$$P(-D^2)s(t) = P(-D^2) \int_0^T \phi(t-u)q(u) \, du = N(-D^2) \int_0^T \delta(t-u)q(u) \, du,$$

or

$$P(-D^2)s(t) = N(-D^2)q(t), \qquad 0 \leq t \leq T. \tag{2-96}$$

This is an inhomogeneous linear differential equation of degree $2m$ for $q(t)$. Its solution involves the sum of a solution of the homogeneous differential equation

$$N(-D^2)q_1(t) = 0 \tag{2-97}$$

and a particular solution $q_0(t)$ of (2-96). One particular solution can be found by taking the Fourier transform of (2-96) and is the same as $q_\infty(t)$ defined by (2-78) and (2-79). This may not be the simplest particular solution, especially if the signal $s(t)$ is undefined outside the interval $(0, T)$. It may be more convenient to solve (2-96) by Laplace transformation, and sometimes a particular solution can be written down by inspection. The solution of (2-97) will have the form

$$q_1(t) = \sum_{k=1}^{m} \{c_k \exp(-h_k t) + d_k \exp[-h_k(T-t)]\}, \tag{2-98}$$

where h_k and $-h_k$, $1 \leq k \leq m$, are the $2m$ roots of the algebraic equation

$$N(-h_k^2) = 0; \tag{2-99}$$

as in (2-93) the zeros of the spectral density $\Phi(\omega)$ are $\pm ih_k$, $1 \leq k \leq m$.

Additional terms must be included in the solution of (2-91) to account for the finite end points 0 and T of the range of integration. These have been shown to involve delta functions and their derivatives situated at those points, but standing just inside the interval $(0, T)$. The complete solution of (2-91) then has the form

$$q(t) = q_0(t) + \sum_{j=0}^{n-m-1} [a_j \delta^{(j)}(t) + b_j \delta^{(j)}(t-T)]$$

$$+ \sum_{k=1}^{m} \{c_k \exp(-h_k t) + d_k \exp[-h_k(T-t)]\}, \tag{2-100}$$

in which the a's, b's, c's, and d's are constants to be determined [Zad50], [Zad52]; $q_0(t)$ is $q_\infty(t)$ of (2-78) and (2-79) or any other convenient particular solution of (2-96). Here $\delta^{(j)}(t-u)$ is the jth derivative of the delta function, defined by

$$\int_a^b f(t)\delta^{(j)}(t-u) \, dt = (-1)^j \frac{d^j}{dt^j} f(t) \bigg|_{t=u} = (-1)^j f^{(j)}(u), \qquad a < u < b,$$

the superscript (j) indicating j-fold differentiation. When $m = n$, the terms with the delta functions do not appear in (2-100); when $m = 0$, the exponential functions are absent.

The detection statistic g is obtained by substituting $q(t)$ into (2-65):

$$g = \int_0^T q_0(t)v(t)\,dt + \sum_{j=0}^{n-m-1} (-1)^j [a_j v^{(j)}(0) + b_j v^{(j)}(T)]$$

$$+ \sum_{k=1}^{m} \int_0^T \{c_k \exp(-h_k t) + d_k \exp[-h_k(T-t)]\}v(t)\,dt.$$

(2-101)

The terms in the first summation require the input $v(t)$ to be differentiated at most $n - m - 1$ times and sampled at $t = 0$ and $t = T$. The noise in the input can indeed be differentiated as many as $n - m - 1$ times because its spectral density $\Phi(\omega)$ decreases at infinity like $|\omega|^{-2(n-m)}$, and the variance

$$\mathrm{Var}\, n^{(n-m-1)}(t) = \int_{-\infty}^{\infty} \omega^{2(n-m-1)}\Phi(\omega)\frac{d\omega}{2\pi}$$

of the $(n - m - 1)$th derivative of the noise $n(t)$ is finite. The remaining terms in (2-101) can be generated by passing the input through a filter matched to the signal

$$q_0(t) + \sum_{k=1}^{m} \{c_k \exp(-h_k t) + d_k \exp[-h_k(T-t)]\}$$

and sampling the output at the end of the observation interval. Although the solution of the integral equation contains delta functions and their derivatives, no singularities appear in the expression for the sufficient statistic g.

The solution in (2-100) involves $2n$ constants, $n - m$ each of the a's and b's, and m each of the c's and d's. To find them, one can substitute $q(t)$ from that equation into the integral equation (2-96). After the integration is carried out, one will be able to cancel $s(t)$, and there will remain $2n$ distinct functions of t, each multiplied by some linear combination of the unknown constants. If the $2n$ poles $\pm i m_k$ of $\Phi(\omega)$ are distinct, these functions will be $\exp m_k t$ and $\exp(-m_k t)$, $k = 1, 2, \ldots, n$. The coefficients of each of these functions must vanish in order for the integral equation to be satisfied, and one obtains in this way $2n$ linear equations that can be solved for the unknown constants in (2-100).

Example 2-3   Lorentz spectral density: the inhomogeneous equation

As a simple example, let the autocovariance of the noise be

$$\phi(\tau) = \phi_0 e^{-\mu|\tau|}.$$

(2-102)

The spectral density then has the "Lorentz" form,

$$\Phi(\omega) = \frac{2\mu\phi_0}{\omega^2 + \mu^2},$$

(2-103)

and in (2-93) $m = 0$, $n = 1$, $C = 2\mu\phi_0$, and $m_1 = \mu$. The particular solution $q_0(t)$ is now, by (2-78) and (2-79) and with $S(\omega)$ the spectrum of the signal $s(t)$,

$$q_0(t) = q_\infty(t) = (2\mu\phi_0)^{-1} \int_{-\infty}^{\infty} (\omega^2 + \mu^2)S(\omega)\, e^{i\omega t}\frac{d\omega}{2\pi}$$

$$= (2\mu\phi_0)^{-1}[\mu^2 s(t) - s''(t)],$$

where the primes indicate differentiation with respect to t. The solution of the integral equation will be

$$q(t) = (2\mu\phi_0)^{-1}[\mu^2 s(t) - s''(t)] + a_0\delta(t) + b_0\delta(t - T) \qquad (2\text{-}104)$$

as in (2-100). When this is substituted into the integral equation (2-91), we obtain

$$s(t) = \phi_0 \int_0^T e^{-\mu|t-u|}q(u)\,du = \phi_0[a_0 e^{-\mu t} + b_0 e^{-\mu(T-t)}]$$
$$+ (2\mu)^{-1} e^{-\mu t} \int_0^t e^{\mu u}[\mu^2 s(u) - s''(u)]\,du + (2\mu)^{-1} e^{\mu t} \int_t^T e^{-\mu u}[\mu^2 s(u) - s''(u)]\,du.$$

Integrating the terms containing $s''(u)$ twice by parts, we finally get

$$s(t) = \phi_0[a_0 e^{-\mu t} + b_0 e^{-\mu(T-t)}]$$
$$- (2\mu)^{-1}\{e^{-\mu t}[\mu s(0) - s'(0)] + e^{-\mu(T-t)}[\mu s(T) + s'(T)] - 2\mu s(t)\}.$$

In order for this equation to hold for all values of t in the interval $(0, T)$, the coefficients of $e^{\mu t}$ and $e^{-\mu t}$ must each vanish, and the coefficients a_0 and b_0 must be given by

$$a_0 = \frac{\mu s(0) - s'(0)}{2\mu\phi_0}, \qquad b_0 = \frac{\mu s(T) + s'(T)}{2\mu\phi_0}. \qquad (2\text{-}105)$$

The detection statistic g now becomes

$$g = (2\mu\phi_0)^{-1}\{[\mu s(0) - s'(0)]\,v(0) + [\mu s(T) + s'(T)]\,v(T)\}$$
$$+ (2\mu\phi_0)^{-1}\int_0^T [\mu^2 s(t) - s''(t)]v(t)\,dt. \qquad (2\text{-}106)$$

The signal-to-noise ratio d^2 that determines the probability of detection is obtained by putting $s(t)$ for $v(t)$ in g, and after an integration by parts we find

$$d^2 = \frac{1}{2\phi_0}\left\{[s(0)]^2 + [s(T)]^2 + \frac{1}{\mu}\int_0^T \{\mu^2[s(t)]^2 + [s'(t)]^2\}\,dt\right\}. \qquad (2\text{-}107)$$

This will be finite provided the signal is differentiable at least once within the interval $(0, T)$. As defined in (2-43) and (2-44), the RKHS scalar product between two functions $h(t)$ and $m(t)$, each differentiable at least once in $(0, T)$, can be written down by replacing $s(t)$ by $h(t)$ and $v(t)$ by $m(t)$ in (2-106) and then integrating by parts:

$$\langle \mathbf{h}, \mathbf{m} \rangle = (2\phi_0)^{-1}[h(0)m(0) + h(T)m(T)] + \frac{1}{2\mu\phi_0}\int_0^T [\mu^2 h(t)m(t) + h'(t)m'(t)]\,dt.$$

If the rational spectral density $\Phi(\omega)$ is any more complicated than the one in this example, the method of substituting $q(t)$ of (2-100) into the integral equation (2-96) will be extremely tedious. Formal schemes requiring less labor have been developed by Slepian and Kadota [Sle69] and by Baggeroer [Bag69], [Bag71]; see also Appendix A.

If $\phi(t, s)$ is the autocovariance function of nonstationary leucogenic noise, the integral equation (2-66) can be solved by a natural extension of the method of this section; it is described by Laning and Battin [Lan56, Sec. 8.5] and by Miller and Zadeh [Mil56]. If this kind of noise contains a component that is white, so that the kernel of (2-66) has the form

$$\phi(t, s) = R(t)\delta(t - s) + \pi(t, s),$$

the integral equation becomes

$$s(t) = R(t)q(t) + \int_0^T \pi(t, s)q(s)\,ds.$$

If the function $\pi(t, s)$ is the autocovariance of leucogenic noise $y(t)$ that can be modeled as the output of a linear system characterized by a finite number of state variables and driven by white noise, this equation can be solved by a method due to Baggeroer [Bag69], [Bag71]. One must be able to construct a linear system that depends on a finite number of state variables and is driven by white noise and whose output is the stochastic process $y(t)$. We shall examine a method of this kind in Sec. 11.4.

Kailath [Kai66] has shown how to solve the integral equation (2-66) for a kernel of the form

$$\phi(t, s) = \begin{cases} f(t)g(s), & 0 \le t \le s \le T, \\ f(s)g(t), & 0 \le s \le t \le T, \end{cases} \tag{2-108}$$

where $g(t)$ and $f(t)$ are continuous functions of t and their quotient $f(t)/g(t)$ is continuous and strictly increasing in the interval $(0, T)$. Equations with similar kernels were treated by Shinbrot [Shi57]. Kailath [Kai66b] has also solved the integral equation for a triangular kernel, $\phi(t, s) = 1 - |t - s|$, $0 \le |t - s| \le 1$, $\phi(t, s) \equiv 0$, $|t - s| > 1$, and for linear combinations of the triangular kernel and a kernel of the type of (2-108).

2.3.2 Homogeneous Equations

When the kernel $\phi(t - s)$ of the homogeneous integral equation

$$\lambda f(t) = \int_0^T \phi(t - s)f(s)\,ds \tag{2-109}$$

is the autocovariance function of noise with a rational spectral density as in (2-93), the process of solving it is much like the method of Sec. 2.3.1. The solution will in general be a combination of exponential functions

$$f(t) = \sum_{k=1}^n [g_k \exp p_k t + e_k \exp(-p_k t)]. \tag{2-110}$$

When this is substituted into (2-109), it is found that the n numbers p_k must be solutions of the algebraic equation

$$N(-p^2) - \lambda P(-p^2) = 0, \qquad p = \pm p_k, \qquad k = 1, 2, \dots, n, \tag{2-111}$$

where $N(\omega^2)$ and $P(\omega^2)$ are the polynomials in the numerator and the denominator of the spectral density $\Phi(\omega)$ as in (2-94).

At the same time, certain linear combinations of the functions $\exp m_k t$ and $\exp(-m_k t)$ appear, where im_k, $-im_k$ are the $2n$ roots of the equation $P(\omega^2) = 0$, $k = 1, 2, \dots, n$. These linear combinations must vanish in order for the integral equation to be satisfied, and in this way one obtains $2n$ homogeneous linear equations for the coefficients g_k and e_k, $k = 1, 2, \dots, n$. These linear equations have a nonzero solution only when the parameter λ is one of the eigenvalues of the integral equation

(2-109), and the vanishing of the determinant of the coefficients of the $2n$ linear equations provides a transcendental equation for the eigenvalues. General formulas that abbreviate the labor have been given by Youla [You57] and by Slepian and Kadota [Sle69]; see also Appendix A.

Example 2-4 Lorentz spectral density: the homogeneous equation

For the exponential kernel in (2-102) we can use the solution in (2-104) and (2-105) by putting $s(t) = \lambda f(t)$:

$$f(t) = \frac{\lambda}{2\mu\phi_0} \left\{ [\mu f(0) - f'(0)]\delta(t) + [\mu f(T) + f'(T)]\delta(t-T) + \mu^2 f(t) - f''(t) \right\}.$$

It is apparent from (2-109) that the solution $f(t)$ cannot contain any delta functions. The coefficients of $\delta(t)$ and $\delta(t-T)$ must therefore vanish, and we obtain the boundary conditions

$$\mu f(0) - f'(0) = 0,$$
$$\mu f(T) + f'(T) = 0,$$

on the differential equation

$$f''(t) + \Gamma^2 f(t) = 0, \qquad \Gamma^2 = 2\mu\phi_0\lambda^{-1} - \mu^2. \tag{2-112}$$

The solution of this differential equation is

$$f(t) = A \cos \Gamma t + B \sin \Gamma t,$$

and the boundary conditions give two equations for the coefficients A and B:

$$\mu A - \Gamma B = 0,$$
$$\mu(A \cos \Gamma T + B \sin \Gamma T) + \Gamma(B \cos \Gamma T - A \sin \Gamma T) = 0.$$

In order for a nonzero solution of these equations to exist, the determinant of the coefficients of A and B must vanish:

$$\begin{vmatrix} \mu & -\Gamma \\ \mu \cos \Gamma T - \Gamma \sin \Gamma T & \mu \sin \Gamma T + \Gamma \cos \Gamma T \end{vmatrix} = 0.$$

This equation determines the values of Γ and hence, from (2-112), of the eigenvalues λ. If we number the eigenfunctions starting from $k = 0$, we obtain from the determinant

$$(m^2 - g_k^2) \tan g_k + 2mg_k = 0, \qquad g_k = \Gamma_k T, \qquad m = \mu T.$$

By substituting $g_k = m \cot \theta_k$, we find $\tan g_k = \tan 2\theta_k$, whereupon

$$m \cot \theta_k = 2\theta_k + k\pi, \qquad k = 0, 1, 2, \dots . \tag{2-113}$$

This can easily be solved by Newton's method. The angles θ_k lie between 0 and $\pi/2$, decreasing toward 0 with increasing k. From (2-112) we then obtain

$$\lambda_k = \frac{2m\phi_0 T}{g_k^2 + m^2} = \frac{2\phi_0 T}{m} \sin^2 \theta_k.$$

It can be shown that for $m = \mu T \ll 1$,

$$g_0 \approx \sqrt{2m} \left(1 - \frac{m}{12}\right), \qquad \lambda_0 \approx \phi_0 T \left(1 - \frac{m}{3}\right),$$

$$g_k \approx k\pi + \frac{2m}{k\pi}, \qquad \lambda_k \approx \frac{2m}{k^2\pi^2} \phi_0 T,$$

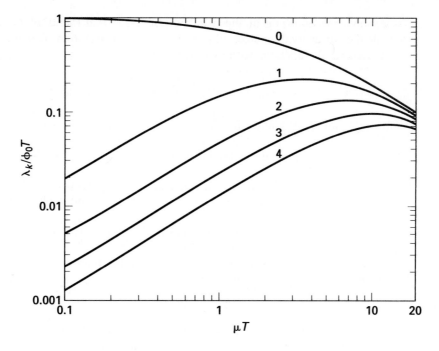

Figure 2-6. Eigenvalues of the kernel $\phi(t - u) = \phi_0\, e^{-\mu|t-u|}$.

and for $m = \mu T \gg \max(1, k\pi)$,

$$g_k \approx (k + 1)\left(1 - \frac{2}{m}\right)\pi, \qquad \lambda_k \approx \frac{2\phi_0 T}{m}.$$

The first five eigenvalues of the integral equation are plotted versus the parameter μT in Fig. 2-6. The eigenfunctions $f_k(t)$ are proportional to $\cos \Gamma_k(t - \frac{1}{2}T)$ for k even and to $\sin \Gamma_k(t - \frac{1}{2}T)$ for k odd. The number of oscillations of the eigenfunction in the interval $(0, T)$ increases with the index k as the eigenvalue λ_k decreases. The eigenfunction $f_k(t)$ in this example has exactly k zeros in the interval $(0, T)$.

If the kernel is not of the exponential type studied here, the homogeneous equation (2-35) may be very difficult to solve exactly. Many approximation techniques have been developed for calculating the eigenfunctions and eigenvalues of positive-definite linear operators, especially in connection with problems in quantum mechanics. Methods and references to the literature are given by Morse and Feshbach [Mor53, Ch. 9] and Baker [Bak77, Ch. 3]. A convenient one, known as the *Rayleigh–Ritz method*, is based on the fact that among all functions $g(t)$ that are normalized in the sense that

$$\int_0^T |g(t)|^2 \, dt = 1 \qquad (2\text{-}114)$$

the quantity

$$R = \int_0^T \int_0^T g^*(t)\phi(t, s)g(s) \, dt \, ds \qquad (2\text{-}115)$$

is stationary when $g(t)$ is an eigenfunction of (2-35), and the value it then takes on equals the corresponding eigenvalue λ. If we use for $g(t)$ a linear combination of a finite number of functions with arbitrary coefficients,

$$g(t) = \sum_{k=1}^{n} c_k h_k(t), \qquad (2\text{-}116)$$

substitute into (2-115), and vary the coefficients c_k, taking account of the constraint (2-114), until a stationary value of R is found, we determine a set of homogeneous linear simultaneous equations for the coefficients c_k:

$$\sum_{k=1}^{n} (m_{jk} - \lambda h_{jk}) c_k = 0, \qquad 1 \leq j \leq n,$$

$$m_{jk} = \int_0^T \int_0^T h_j^*(t) \phi(t, s) h_k(s) \, dt \, ds, \qquad h_{jk} = \int_0^T h_j^*(t) h_k(t) \, dt.$$

In order for these equations to have a solution other than $c_k \equiv 0$, the determinant of the coefficients must vanish:

$$\det \left(m_{jk} - \lambda h_{jk} \right) = 0.$$

Computer programs exist for finding the roots λ of this determinantal equation and for calculating the associated coefficients c_k. Each such root is an approximation to an eigenvalue of (2-35), and when the coefficients c_k thus calculated are substituted into (2-116), the resulting function $g(t)$ is an approximation to the associated eigenfunction.

The homogeneous integral equation (2-109) has been solved by Slepian and Pollak [Sle61] for a kernel that is the autocovariance function of bandlimited white noise,

$$\phi(\tau) = \frac{\sin W\tau}{\pi W \tau}.$$

The eigenfunctions are angular prolate spheroidal functions, and they possess the unusual property of being orthogonal over both the finite interval $(0, T)$ and the infinite interval $(-\infty, \infty)$. These solutions have provided the basis of an extensive treatment of the uncertainty relation for signals [Lan61], [Lan62].

If the noise whose autocovariance function $\phi(t, s)$ is the kernel of the homogeneous equation (2-35) can be modeled as the output of a linear system driven by white noise, the eigenvalues λ can be calculated numerically by a method given by Baggeroer [Bag69].

2.4 PHYSICAL INTERPRETATION OF THE SIGNAL-TO-NOISE RATIO

In Sec. 2.2 we studied the detection of a signal $s(t)$ in white Gaussian noise $n(t)$, and there we learned that the signal-to-noise ratio d^2 determining the probability of detection is equal to $2E/N$, where N is the unilateral spectral density of the noise $n(t)$ and

$$E = \int_0^T [s(t)]^2 \, dt.$$

We called E the *energy* of the signal $s(t)$.

The signal $s(t)$ is the voltage induced across the terminals of the receiving antenna by the electromagnetic pulse launched by a distant transmitter. The noise $n(t)$ is a random voltage induced across those terminals by the fluctuations of the electromagnetic field in the neighborhood of the antenna, and these are caused by thermal agitation of ions and electrons everywhere in the universe. We now want to show that this signal-to-noise ratio d^2 is equal to $2\varepsilon/kT_0$, where ε is the maximum physical energy that the receiver can extract from the electromagnetic field carrying the signal, $k = 1.38 \cdot 10^{-23}$ joule/deg is Boltzmann's constant, and T_0 is the effective absolute temperature of the fluctuating field. By effective temperature we mean that the spectral density of the noise $n(t)$ is the same as though antenna, receiver, and all were enclosed in an enormous box that is in thermal equilibrium at absolute temperature T_0. Physicists call such an enclosure a *Hohlraum*.

2.4.1 The Transmission-line Model

The simplest way to show this is based on the same model that Nyquist used to derive the spectral density of the fluctuating voltage across a resistor [Nyq28]. We imagine that a very long transmission line is connected to the terminals of the antenna and that everything is enclosed in a *Hohlraum* in thermal equilibrium at absolute temperature T_0. The transmission line is matched to the antenna so that at all frequencies it extracts the maximum power from the ambient field. The electromagnetic field of the line is decomposed into normal modes, and the amplitude of the jth mode is denoted by v_j. This modal coefficient v_j is so scaled that v_j^2 is the energy in the jth mode.

Under hypothesis H_0 v_j, as a linear functional of the Gaussian noise process $n(t)$, is a Gaussian random variable with zero expected value and a variance Var v_j equal to the average thermal energy in the jth mode. Each v_j constitutes a degree of freedom of the system. According to the theorem of equipartition of energy, each degree of freedom must hold an average energy

$$\sigma^2 = \text{Var } v_j = \tfrac{1}{2}kT_0$$

when the entire system is in thermal equilibrium at absolute temperature T_0. The modal coefficients v_j are statistically independent.

Let us suppose that the signal $s(t)$ has arrived, and let us measure the modal amplitudes v_j at such a time that the entire signal field is on its way down the transmission line, but before it has reached the end of the line. The line is assumed long enough so that this will be possible. Then under hypothesis H_1 the expected value of v_j will equal the amplitude s_j of the jth mode of the field in the line created by the signal. We base our decision about the presence or absence of the signal on the observed values of all the modal amplitudes v_j. Under the Neyman–Pearson criterion the optimum way to process these amplitudes is to form their likelihood ratio and compare it with a decision level λ set to yield a preassigned false-alarm probability Q_0, just as we did in Sec. 2.2. The entire analysis continues as in that section, with the variances λ_k in (2-51) through (2-63) replaced by σ^2. The optimum detection statistic is a linear combination of the modal amplitudes v_j, weighted by the signal amplitudes s_j. The false-alarm and detection probabilities are again given by (2-72), and by (2-63) the governing signal-to-noise ratio equals

$$d^2 = \sum_{j=1}^{\infty} \frac{s_j^2}{\sigma^2} = \frac{2\varepsilon}{kT_0} \tag{2-117}$$

because the available signal energy ε equals the sum of the squares of the signal coefficients s_j, and $\sigma^2 = \frac{1}{2}kT_0$. This is what we set out to prove.

One might object that if an average energy $\frac{1}{2}kT_0$ is associated with each variable v_j under hypothesis H_0, and if there are an infinite number of these, the total average energy in the field of the transmission line must be infinite. The resolution of this paradox rests on Max Planck's quantum hypothesis. First of all, we must assert that the modal variables actually occur in pairs, which we denote by v_j' and v_j'', and each pair is associated with a frequency v_j that is an integral multiple of c/L, where c is the velocity of propagation of waves along the line and L is its length. By virtue of Maxwell's equations, the jth such mode behaves like a harmonic oscillator of frequency v_j. We can think of v_j' and v_j'' as the real and imaginary parts of the complex amplitude

$$(v_j' + iv_j'') \exp 2\pi i v_j t$$

of that oscillator or alternatively as related to the electric and magnetic components, respectively, of the field in the line.

The total energy in the jth modal oscillator is now, by our normalization,

$$W_j = v_j'^2 + v_j''^2,$$

and because v_j' and v_j'' are Gaussian and independent, the probability density function of this energy under hypothesis H_0 is

$$P_0(W_j) = (kT_0)^{-1} \exp\left(-\frac{W_j}{kT_0}\right),$$

which is known as the *Boltzmann distribution*.

Planck postulated that each field oscillator cannot take up an arbitrary energy W_j, but only a discrete amount that is an integral multiple of hv_j; $h = 6.626 \cdot 10^{-34}$ joule-sec became known as *Planck's constant*. That energy is still, however, governed by a Boltzmann distribution, whereby for any nonnegative integer n the probability that the oscillator holds n such "quanta" is

$$\Pr(W_j = nhv_j \mid H_0) = C \exp\left(-\frac{nhv_j}{kT_0}\right), \qquad n = 0, 1, 2, \dots .$$

A simple calculation shows that because these probabilities must sum to 1, the constant C must be

$$C = 1 - w_j, \qquad w_j = \exp\left(-\frac{hv_j}{kT_0}\right).$$

The average number of quanta is then

$$E(n \mid H_0) = (1 - w_j) \sum_{n=0}^{\infty} n w_j^n = \frac{w_j}{1 - w_j} = \left[\exp\left(\frac{h\nu_j}{kT_0}\right) - 1 \right]^{-1}.$$

The average total energy in the jth modal oscillator is therefore not kT_0, but

$$E(W_j \mid H_0) = h\nu_j \left[\exp\left(\frac{h\nu_j}{kT_0}\right) - 1 \right]^{-1}.$$

This approximately equals kT_0 when $\nu_j \ll kT_0/h$, but decreases exponentially with ν_j when $\nu_j \gg kT_0/h$. (For $T_0 = 17°C = 290$ K, $kT_0/h = 6 \cdot 10^{12}$ Hz, a frequency far higher than those encountered in radio communications or in radar.) The total average thermal energy in the field of the transmission line, which is the sum over j of $E(W_j \mid H_0)$, is therefore finite.

It would be incorrect, however, to replace kT_0 in (2-117) by the value of $E(W_j \mid H_0)$ just quoted. If the frequencies in the signal $s(t)$ are so high that the average thermal energy in each mode it occupies must be so expressed, ordinary statistical decision theory is inapplicable, and the decision about the presence of the signal must be treated in a manner consistent with the laws of quantum mechanics. How to do this has been described elsewhere [Hel76], [Hol82] and lies beyond the scope of this book. The effects of the quantum nature of signals at frequencies larger than kT_0/h will show up, however, when we treat the detection of optical signals in Chapter 12.

2.4.2 The Circuit Model

An alternative derivation of (2-117), based on an analysis of a circuit model of the receiver, can be instructive. The signal is created in the receiver when the pulsed electromagnetic wave from the transmitter excites the antenna and causes a current to flow through an input load impedance. The voltage across this impedance Z_L, as shown in Fig. 2-7, is the $v(t)$ that we have taken as the input to our receiver in the previous sections. The signal component of this voltage, when present, is $s(t)$. In this model the noise component $n(t)$ arises from the random external electromagnetic fields exciting the antenna and from thermal agitation of the constituents of the load impedance Z_L. We should like to interpret the maximum attainable signal-to-noise ratio d_∞^2 in (2-80) in terms of the total energy ε delivered by the transmitter to the receiver and the effective absolute temperature T_0 of the perturbing thermal noise.

The antenna is replaced by its Thévenin equivalent circuit, consisting as shown in Fig. 2-7 of an ideal voltage generator in series with the impedance Z_A of the antenna. Its open-circuit voltage, induced by the incident signal field, equals $s_0(t)$. The voltage generator produces a pulsed voltage that induces a current in the circuit composed of Z_L and Z_A in series, and the voltage thus created across the load impedance Z_L is the signal $s(t)$. Denote the Fourier transforms of $s_0(t)$ and $s(t)$ by $S_0(\omega)$ and $S(\omega)$, respectively. Then

$$S(\omega) = \frac{Z_L}{Z_L + Z_A} S_0(\omega). \tag{2-118}$$

When the region of space toward which the beam of the antenna is pointed has an effective absolute temperature T_0, the antenna appears, according to Nyquist's

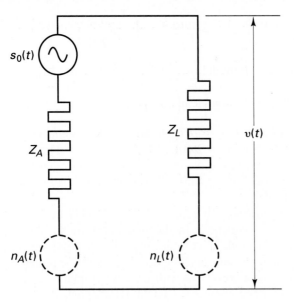

Figure 2-7. Equivalent circuit for the input to a receiver.

law, to have in series with its equivalent impedance Z_A an ideal generator of a random voltage $n_A(t)$ with spectral density

$$\Phi_0(\omega) = 2kT_0 \operatorname{Re} Z_A = 2kT_0 R_A.$$

Here k is again Boltzmann's constant and R_A is the resistive component of the impedance Z_A, which is largely the radiation resistance of the antenna [Hel91, pp. 418-31], [Pap91, pp. 351–4].

The load, at absolute temperature T_L, likewise seems to have in series with it an ideal generator of a random voltage $n_L(t)$ with spectral density

$$\Phi_L(\omega) = 2kT_L \operatorname{Re} Z_L = 2kT_L R_L.$$

The net spectral density of the noise voltage $n(t)$ observed between the terminals of the load is, by (2-9),

$$\Phi(\omega) = \Phi_0(\omega) \, |y_{A\to o}(\omega)|^2 + \Phi_L(\omega) \, |y_{L\to o}(\omega)|^2 \, ,$$

where as in (2-118)

$$y_{A\to o}(\omega) = \frac{Z_L}{Z_L + Z_A}$$

is the transfer function to the output from an ideal generator in series with Z_A, and where

$$y_{L\to o}(\omega) = \frac{Z_A}{Z_L + Z_A}$$

is the transfer function to the output from an ideal generator in series with Z_L. Hence the spectral density of the noise in the output is

$$\Phi(\omega) = \frac{2kT_0 R_A \, |Z_L|^2 + 2kT_L R_L \, |Z_A|^2}{|Z_L + Z_A|^2} .$$

Detection of a Known Signal   Chap. 2

The maximum signal-to-noise ratio at the output with respect to detection of the signal $s(t)$ is then, by (2-80),

$$d_\infty^2 = \int_{-\infty}^{\infty} \frac{|S(\omega)|^2}{\Phi(\omega)} \frac{d\omega}{2\pi} = \int_{-\infty}^{\infty} \frac{|S_0(\omega)|^2 |Z_L|^2}{2kT_0R_A |Z_L|^2 + 2kT_LR_L |Z_A|^2} \frac{d\omega}{2\pi}.$$

Diminishing the temperature T_L of the load causes d_∞^2 to increase, and we see that it is bounded above by

$$d_\infty^2 \le d_1^2 = \int_{-\infty}^{\infty} \frac{|S_0(\omega)|^2}{2kT_0R_A} \frac{d\omega}{2\pi}. \tag{2-119}$$

The smaller the term $kT_LR_L|Z_A|^2$ is in comparison with the term $kT_0R_A|Z_L|^2$, the more closely equality is approached. We now endeavor to express this maximum possible output signal-to-noise ratio in terms of the signal field striking the antenna.

The integral in (2-119) must be proportional to the energy density in the electromagnetic field of the signal as it passes the antenna, and that energy density is conveniently measured by the energy ε that would be absorbed from the signal field if the load were matched to the antenna [Hel91, pp. 429–31]. To match the antenna, the load is given an impedance $Z_L = Z_A* = R_A - iX_A$, and the absorbed energy is what is then dissipated in the load:

$$\varepsilon = \int_{-\infty}^{\infty} \frac{|S(\omega)|^2}{Z_L} \frac{d\omega}{2\pi} = \int_{-\infty}^{\infty} \frac{Z_L^* |S_0(\omega)|^2}{|Z_L + Z_A|^2} \frac{d\omega}{2\pi} = \int_{-\infty}^{\infty} \frac{|S_0(\omega)|^2}{4R_A} \frac{d\omega}{2\pi}.$$

Under the reasonable assumption that the effective temperature T_0 is uniform over the band of frequencies occupied by the signal, we can therefore write (2-119) as

$$d_\infty^2 \le d_1^2 = \frac{2\varepsilon}{kT_0}, \tag{2-120}$$

where ε is the maximum energy that the receiver can draw from the incident signal field. This energy must have been supplied by the transmitter.

The energy ε absorbed by a matched load is equal to E_1A_r, where E_1 is the total energy in the electromagnetic field of the signal passing through a unit area normal to the direction of propagation, and A_r is the effective area of the antenna for waves moving in that direction. In terms of the gain Γ of the antenna in that same direction, $A_r = \Gamma\lambda^2/4\pi$, where λ is the wavelength of the signal radiation. If the dimensions of a plane antenna are much greater than a wavelength, as shown by antenna theory [Sil49], the maximum value of the effective area A_r is equal to the geometrical area of the antenna. The energy ε is then the total energy in the signal field that passes through an area equal to that of the antenna and perpendicular to the direction of propagation. In this sense we can call ε the signal energy intercepted by the antenna, and (2-120) informs us that at best the output signal-to-noise ratio d^2 equals the intercepted signal energy divided by $\frac{1}{2}kT_0$. The quantity kT_0 is the maximum noise power per unit of frequency available from the surroundings when they are at absolute temperature T_0 [Hel91, p. 430]. Thus in the example following (2-72), the minimum detectable signal must furnish a total energy ε of at least $25.1kT_0$ during the observation interval. (For $T_0 = 17°C = 290$ K, $kT_0 = 4 \cdot 10^{-21}$ joules.)

In a practical receiver the input from the antenna must be amplified before it can be processed to form the detection statistic g for comparison with the decision level g_0. The signal and the noise are amplified together, and the amplifiers themselves add noise generated through the shot effect and other random electronic mechanisms.

How an amplifier affects the signal-to-noise ratio is often measured by its *noise figure F, F $\geq$ 1*. If G is the power gain of the amplifier, the noise power per unit bandwidth at the output of an ideal amplifier would be G times that at the input. Because the amplifier itself adds noise, the actual noise power density at its output is not G, but FG times that at the input. The effective signal-to-noise ratio at the output is therefore

$$d_{out}^2 = \frac{2G\varepsilon}{FGkT_0} = \frac{d_{in}^2}{F}$$

in terms of the effective input signal-to-noise ratio $d_{in}^2 = 2\varepsilon/kT_0$.

It is shown in texts on radio and radar that for two networks in cascade having noise figures F_1 and F_2 and power gains G_1 and G_2, in that order, the noise figure of the combination is given by

$$F = F_1 + \frac{F_2 - 1}{G_1}$$

[Hel91, p. 432]. Hence if the first amplifier has such a large gain that $G_1 \gg F_2 - 1$, its noise figure effectively determines that of the combination. Extension to a chain of many amplifiers is immediate. For further discussion of these matters, see [Dav58, Ch. 10] and [Hau58].

The signal $s(t)$ whose detection has been the subject of this chapter has been assumed to be completely known in amplitude, form, and time of arrival. In radar practice, as often in communications, signal parameters may be known only within wide ranges, with a consequent decrease in the probability of detection or increase in the energy of the minimum detectable signal. Strategies for detecting more vaguely known signals will be discussed in later chapters.

2.5 DECISIONS AMONG A NUMBER OF KNOWN SIGNALS

2.5.1 Signal Space

A communication system might be transmitting information coded in M symbols, as described in Sec. 1.1. The jth symbol causes the transmitter to emit a signal that appears at the input to the receiver as $s_j(t)$. The receiver must decide which of the M signals $s_i(t)$ is present in its input,

$$v(t) = s_i(t) + n(t), \qquad 0 \leq t \leq T, \qquad (H_i), \quad i = 1, 2, \dots, M, \qquad (2\text{-}121)$$

during a particular observation interval; that is, it must choose one of the M hypotheses H_i. The form of each of the signals is completely known. We assume that each signal is confined to $(0, T)$ and that intersymbol interference is absent. The ith signal is transmitted with relative frequency ζ_i, which is therefore the prior probability of hypothesis H_i, and

Detection of a Known Signal   Chap. 2

$$\sum_{i=1}^{M} \zeta_i = 1.$$

The noise $n(t)$ will be taken as white and Gaussian with unilateral spectral density N. The receiver is to incur minimum probability of error in its decisions.

As in Sec. 2.1.3, we sample the input $v(t)$ by means of a set of functions $f_i(t)$, $i = 1, 2, \ldots$, that are orthonormal over the observation interval $(0, T)$ in the sense of (2-11). Remembering the irrelevance argument in Sec. 2.2.4, we create that set by applying the Gram–Schmidt procedure of Sec. 2.1.4 to a set of functions $g_1(t), g_2(t), \ldots$, that begins with the signals $s_i(t)$ to be detected. When we have utilized all the signals—there are only M—, we must bring in other functions $g_j(t)$ to complete the data space. If the M signals $s_i(t)$ are linearly independent, we can take $g_i(t) = s_i(t)$, $i = 1, 2, \ldots, M$, and by the orthogonalization procedure we can form a set of M orthonormal functions $f_i(t)$ that are linear combinations of them. The signals might, for instance, be orthogonal in the first place,

$$\int_0^T s_i(t)s_j(t)\, dt = E_i \delta_{ij}, \tag{2-122}$$

where E_i is the energy of the ith signal. Then

$$f_i(t) = E_i^{-1/2} s_i(t), \qquad 1 \le i \le M.$$

If the signals $s_i(t)$ are not linearly independent, however, we shall be able to use only some subset of $D < M$ of them before we run out of usable functions $g_i(t)$ and must seek elsewhere. The signals might, for instance, differ only in amplitude, as in a pulse-amplitude-modulated (PAM) communication system,

$$s_k(t) = A_k s(t), \qquad k = 1, 2, \ldots, M, \tag{2-123}$$

whereupon $D = 1$. As another example, consider a quaternary communication system in which the four received signals are

$$\begin{aligned} s_1(t) &= A f_1(t), & s_2(t) &= -A f_1(t), \\ s_3(t) &= A f_2(t), & s_4(t) &= -A f_2(t), \end{aligned} \tag{2-124}$$

where the functions $f_1(t)$ and $f_2(t)$ are orthonormal as in (2-11). Only two of these four signals are linearly independent, and $D = 2$.

Each signal can be written as a linear combination of the D orthonormal functions $f_i(t)$ formed by the Gram–Schmidt procedure,

$$s_k(t) = \sum_{i=1}^{D} s_{ki} f_i(t), \qquad s_{ki} = \int_0^T s_k(t) f_i(t)\, dt. \tag{2-125}$$

The D components s_{ki} form a vector

$$\mathbf{s}_k = (s_{k1}, s_{k2}, \ldots, s_{kD})$$

that represents the signal. For amplitude-modulated signals as in (2-123), these vectors have a single component proportional to A_k. For M orthogonal signals of equal energy E, each vector has M components, of which one equals $E^{1/2}$ and the

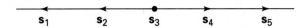

Figure 2-8. Signal vectors, PAM communication system, $M = 5$.

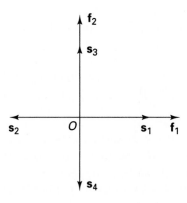

Figure 2-9. Signal vectors, quaternary communication system.

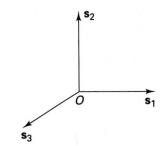

Figure 2-10. Orthogonal signal vectors, $M = 3$, $E_i \equiv E$.

rest equal 0. The M signals $s_i(t)$ are said to span a D-dimensional space, as do the D orthonormal functions $f_i(t)$ formed by combining them. This space is called the *signal space*.

Figure 2-8 shows the signal vectors $\mathbf{s}_i$ associated with the PAM signals in (2-123) when $M = 5$ and the signals have amplitudes $A_k = (k - 3)A$, $1 \le k \le 5$; $\mathbf{s}_3 = \mathbf{0}$ is a null vector. Figure 2-9 shows the two-dimensional signal space that embeds the four quaternary signals in (2-124). Figure 2-10 shows the three-dimensional signal space spanned by three orthogonal signals of equal energy.

The remaining members $f_k(t)$, $k > D$, of the set of sampling functions will be orthogonal to the first D functions $f_i(t)$, $1 \le i \le D$, and hence to all M signals, which are linear combinations of the functions $f_j(t)$ for $1 \le j \le D$.

As in Sec. 2.2.1, the receiver is to base its choice among the M hypotheses H_i on the samples

$$x_i = \int_0^T f_i(t)v(t)\, dt. \tag{2-126}$$

These data are uncorrelated because, as in Sec. 2.1.5, the noise is white and the sampling functions $f_i(t)$ are orthonormal. As the data are Gaussian random variables, they are therefore statistically independent. They have expected values

Detection of a Known Signal   Chap. 2

$$E(x_i \mid H_j) = \int_0^T f_i(t)s_j(t)\,dt = s_{ji}, \qquad 1 \le i \le D,$$

$$\equiv 0, \qquad i > D,$$

when the jth signal is present, and their variances are equal,

$$\mathrm{Var}\,(x_i \mid H_j) \equiv \frac{N}{2},$$

as can be verified by the techniques of Sec. 2.2. The joint probability density function of any number $n > D$ of the samples has the multivariate Gaussian form

$$p_j^{(n)}(\mathbf{x}) = (\pi N)^{-n/2} \exp\left[-\frac{1}{N}\sum_{i=1}^{D}(x_i - s_{ji})^2\right]\exp\left[-\frac{1}{N}\sum_{i=D+1}^{n}x_i^2\right] \qquad (2\text{-}127)$$

under hypothesis H_j that the jth signal is present.

According to what we learned in Sec. 1.1, the receiver will attain minimum probability of error if it chooses the hypothesis with the greatest posterior probability $\Pr(H_j \mid \mathbf{x})$, given the set $\mathbf{x} = (x_1, x_2, \ldots, x_D, \ldots, x_n)$. From (1-4) we see that all it needs to do is to compare the M quantities $\zeta_j p_j^{(n)}(\mathbf{x})$ and choose the largest. The rightmost exponential factor in (2-127) is common to all these quantities and cancels out from the comparisons: the data x_i for $i > D$ are irrelevant to the decision because they contain no information about which hypothesis is true. The D random variables x_i, $1 \le i \le D$, which we gather into a data vector

$$\mathbf{x} = (x_1, x_2, \ldots, x_D), \qquad D \le M,$$

contain all the information in the input $v(t)$ relevant to deciding among the M hypotheses in the optimum fashion. The receiver bases its decision on the M quantities $\zeta_j p_j(\mathbf{x})$, where

$$p_j(\mathbf{x}) = (\pi N)^{-D/2} \exp\left[-\frac{1}{N}|\mathbf{x} - \mathbf{s}_j|^2\right], \qquad j = 1, 2, \ldots, M. \qquad (2\text{-}128)$$

Here

$$|\mathbf{x} - \mathbf{s}_j| = (\mathbf{x} - \mathbf{s}_j, \mathbf{x} - \mathbf{s}_j)^{1/2} = \left[\sum_{i=1}^{D}(x_i - s_{ji})^2\right]^{1/2}$$

is the distance in the signal space from the "data point" $\mathbf{x} = (x_1, x_2, \ldots, x_D)$ to the vertex of the jth signal vector $\mathbf{s}_j$. The receiver picks that hypothesis for which $\zeta_j p_j(\mathbf{x})$ is largest. Equivalently, the receiver chooses that hypothesis H_j for which

$$\zeta_j \exp\left[\frac{2}{N}(\mathbf{s}_j, \mathbf{x}) - \frac{1}{N}(\mathbf{s}_j, \mathbf{s}_j)\right] > \zeta_i \exp\left[\frac{2}{N}(\mathbf{s}_i, \mathbf{x}) - \frac{1}{N}(\mathbf{s}_i, \mathbf{s}_i)\right], \qquad \forall i \ne j. \qquad (2\text{-}129)$$

We use the same notation as in Sec. 2.1.4;

$$(\mathbf{s}_k, \mathbf{x}) = \sum_{i=1}^{D}s_{ki}x_i = \int_0^T s_k(t)v(t)\,dt$$

is the scalar product of the kth signal vector $\mathbf{s}_k$ and the data vector $\mathbf{x}$.

The D data $x_1, x_2, \ldots, x_D$ on which the decision is based can be generated by passing the input $v(t)$ through a set of D parallel filters matched as in (2-17) to the D orthogonal "signals" $f_k(t)$, $1 \le k \le D$, and measuring the output of each filter at the end $t = T$ of the observation interval. For amplitude-modulated signals as in (2-123), we have a single datum x, the output of a filter matched to the signal $s(t)$ to which all M of the signals are proportional. When the signals are orthogonal, there are M independent data $x_1, x_2, \ldots, x_M$.

Let us designate by R_j the region in the D-dimensional signal space containing those data vectors $\mathbf{x}$ that lead to the choice of hypothesis H_j, "Signal $s_j(t)$ is present." From (2-129) we see that these decision regions are bounded by hyperplanes on which the scalar product $((\mathbf{s}_j - \mathbf{s}_i), \mathbf{x})$ is constant, and each hyperplane is perpendicular to the vector $\mathbf{s}_j - \mathbf{s}_i$ connecting the vertices of some pair $\mathbf{s}_j, \mathbf{s}_i$ of signal vectors.

The probability q_{jj} of correctly choosing hypothesis H_j is the integral over R_j of the probability density function $p_j(\mathbf{x})$ of the data $x_1, x_2, \ldots, x_D$. This density function has the multivariate Gaussian form in (2-128), and that probability is therefore

$$q_{jj} = (\pi N)^{-D/2} \int_{R_j} \int \exp\left[-\frac{1}{N} |\mathbf{x} - \mathbf{s}_j|^2\right] d^D \mathbf{x}, \qquad (2\text{-}130)$$

where $d^D \mathbf{x} = dx_1 \ldots dx_D$ is the volume element in the D-dimensional space. The probability of error is then

$$P_e = 1 - \sum_{j=1}^{M} \zeta_j q_{jj}. \qquad (2\text{-}131)$$

For an arbitrary configuration of signal vectors $\mathbf{s}_i$, evaluating the integral in (2-130) may be quite difficult.

In many communication systems information is transmitted at maximum rate by coding it into M symbols in such a way that they occur equally often, $\zeta_i \equiv M^{-1}$. Then the receiver chooses the signal $s_k(t)$ whose vector $\mathbf{s}_k$ lies closest to the data vector $\mathbf{x}$ in the sense that the distance $|\mathbf{x} - \mathbf{s}_k|$ is smallest among all the distances $|\mathbf{x} - \mathbf{s}_i|$, $1 \le i \le M$.

The probability of error is invariant to rigid rotation of the set of M signal vectors $\mathbf{s}_k$, for the surfaces bounding the decision regions R_j rotate with them, and the probability density functions move with them unchanged. This invariance is sometimes useful in calculating error probabilities. The probability of error depends only on the configuration of the M signal vectors $\mathbf{s}_k$ in the signal space, and not on the particular orthonormal functions $f_i(t)$, $i = 1, 2, \ldots, D$, used to represent the signals through (2-125). Thus a great variety of sets of received signals may incur the same probability of error.

If the noise is not white, but colored, possessing an autocovariance function $\phi(t, s)$, a signal space of the same kind can be constructed as we have shown here if one uses instead the type of scalar product introduced in Sec. 2.1.7 and designated by angular brackets $\langle \cdot, \cdot \rangle$.

2.5.2 Displacement of the Signal Vectors

Suppose that a new set of signals $s_i'(t)$ is generated by adding to each signal $s_i(t)$ a common signal $a(t)$ that lies in the same signal space and is represented there by a vector $\mathbf{a}$ as in (2-125). The new signals are represented by the vectors

$$\mathbf{s}_i' = \mathbf{s}_i + \mathbf{a}, \qquad 1 \le i \le M. \tag{2-132}$$

The decision rule for the receiver of these new signals $s_i'(t)$ is the same as in (2-129), except that $\mathbf{s}_j$ is replaced by $\mathbf{s}_j'$ and $\mathbf{s}_i$ by $\mathbf{s}_i'$. Substituting into it from (2-132), we find that the hyperplanes separating the decision regions R_i' for the new receiver are obtained by displacing all the points in R_i by the same vector $\mathbf{a}$. The decision regions R_i' for the new set of signals are simply the old regions R_i moved rigidly along with the signal points $\mathbf{s}_i$ in the displacement (2-132).

Although the energies of the new signals $s_i'(t)$ differ from those of the original ones, the minimum probability P_e' of error in deciding among them is equal to the minimum probability P_e of error in deciding among the original signals $s_i(t)$. Indeed, the probability q_{jj}' that the new receiver correctly chooses signal $s_j'(t)$ (hypothesis H_j') when H_j' is true is, as in (2-130),

$$q_{jj}' = (\pi N)^{-D/2} \int_{R_j'} \int \exp\left[-\frac{1}{N}|\mathbf{x}' - \mathbf{s}_j'|^2\right] d^D \mathbf{x}'.$$

Changing variables to $\mathbf{x} = \mathbf{x}' - \mathbf{a}$, we find that this becomes identical with the expression in (2-130) for the probability q_{jj} that the original receiver correctly chooses hypothesis H_j ("Signal $s_j(t)$ is present") when H_j is true. The probability of error in the new receiver must therefore be the same as the probability P_e of error (2-131) in the receiver of the original signals $s_i(t)$. Because of this invariance of the probability of error, when a new set of signals is formed by displacement from a set for which the probability of error is already known, the only problem remaining is to calculate the energies of the new signals.

2.5.3 Orthogonal and Simplex Signals

The simplest configuration of signals is one in which they are orthogonal, convey equal energies E to the receiver,

$$(\mathbf{s}_i, \mathbf{s}_j) = E\delta_{ij},$$

and are transmitted equally often. The signal vector $\mathbf{s}_j$ whose vertex is closest to the data point $\mathbf{x}$ will be that for which the jth component x_j is largest:

$$\{x_j > x_i, \forall i \ne j\} \Rightarrow \to H_j.$$

All probabilities q_{ii} of choosing hypothesis H_i when H_i is true are now equal, and in calculating the average probability Q_c of correct decision, we can assume that the first signal $s_1(t)$ was actually the one received and that hypothesis H_1 is true. Hypothesis H_1 is correctly selected if $x_1 > x_j, \forall j \ne 1$; and because $s_{ij} = E^{1/2}\delta_{ij}$, the probability of correct decision is

$$Q_c = q_{11} = \Pr(x_j < x_1, \, j = 2, 3, \ldots, M | H_1)$$

$$= \int_{-\infty}^{\infty} p_1(x_1) \, dx_1 \left[\int_{-\infty}^{x_1} p_0(x_j) \, dx_j \right]^{M-1},$$

where

$$p_0(x) = \frac{1}{\sqrt{\pi N}} \exp\left(-\frac{x^2}{N}\right), \qquad p_1(x) = \frac{1}{\sqrt{\pi N}} \exp\left[-\frac{(x - E)^2}{N}\right].$$

Hence

$$Q_c = \frac{1}{\sqrt{2\pi}} \int_{-\infty}^{\infty} \exp\left[-\tfrac{1}{2}(y - d)^2\right] (1 - \text{erfc } y)^{M-1} \, dy, \qquad (2\text{-}133)$$

in terms of the error-function integral (1-11). Here $d^2 = 2E/N$ is the signal-to-noise ratio for each signal. The error probabilities $P_e = 1 - Q_c$ have been tabulated and plotted by Viterbi [Vit64] for values of M that are integral powers of 2.

When M vectors $\mathbf{s}_k'$ all lie in a hyperspace of dimension $M - 1$, all of them having equal lengths $|\mathbf{s}_k'| = E'^{1/2}$ and making equal angles with each other, they are said to form a *regular simplex*. For $M = 2$ the corresponding signals $s_k'(t)$ are antipodal. The tips of the vectors $\mathbf{s}_k'$ form for $M = 3$ an equilateral triangle and for $M = 4$ a regular tetrahedron. For all M the signals add to zero:

$$\sum_{j=1}^{M} s_j'(t) \equiv 0, \qquad \sum_{j=1}^{M} \mathbf{s}_j' = \mathbf{0}.$$

We shall show that the probability of correct decision is again given by (2-133), but with the signal-to-noise ratio d^2 replaced by the slightly larger value

$$d'^2 = \left[\frac{M}{M - 1}\right] d^2. \qquad (2\text{-}134)$$

It has been conjectured that the regular simplex configuration attains the lowest error probability P_e among all sets of signals having a given total energy. For $M \gg 1$ this probability is only slightly less than that incurred by M orthogonal signals.

In order to prove that (2-133) and (2-134) determine the probability of correct decision for the simplex signals, we observe first that they are obtained from a set of M orthogonal signals $s_i(t)$ of equal energies E by adding to each the signal

$$a(t) = -\frac{1}{M} \sum_{j=1}^{M} s_j(t),$$

which is represented by the vector

$$\mathbf{a} = -\frac{1}{M} \sum_{j=1}^{M} \mathbf{s}_j.$$

According to what we said in Sec. 2.5.2, when the receiver optimum for deciding among them is used, these new signals $s_i'(t) = s_i(t) + a(t)$ result in the same probability P_e of error as the original set of M orthogonal signals, for which P_e is given by (2-133). The sum of the new signals is zero, and the reader can easily show that they all have equal energies and that their pairwise scalar products (s_i', s_j'), $i \neq j$, are all equal. The new signals have smaller energies

$$E' \equiv (s_i', s_i') = (s_i, s_i) + 2(a, s_i) + (a, a)$$
$$= E - 2M^{-1}E + EM^{-1} = E(1 - M^{-1})$$

than the original orthogonal signals from which they were derived. The signal-to-noise ratio for the original signals, which figures in (2-133), is therefore larger than that for the simplex signals by a factor $(1 - M^{-1})^{-1} = M/(M - 1)$, whence (2-134).

The problem of finding a set of signal vectors that yields low probability of error in a space of dimension D less than the number M of signals and the question of the optimality of the simplex signals are treated at length in the book by Weber [Web68]. The performance of systems involving other signal sets is analyzed in textbooks on communication theory, such as [Vit66] and [Pro89].

Problems

In these problems the noise is Gaussian and the observation interval is $(0, T)$ unless otherwise stated.

2-1. Let λ_k be the kth eigenvalue of the integral equation

$$\lambda f(t) = \int_0^T \phi(t, s) f(s) \, ds, \qquad 0 \leq t \leq T.$$

Prove that

$$\sum_{k=1}^{\infty} \lambda_k = \int_0^T \phi(t, t) \, dt, \qquad \sum_{k=1}^{\infty} \lambda_k^2 = \int_0^T \int_0^T \phi(t, u) \phi(u, t) \, dt \, du.$$

Hint: Use (2-42).

2-2. A stationary Gaussian random process $x(t)$ has expected value 0 and autocovariance function $\phi(\tau)$. The variance of the process is estimated by forming the quantity

$$Z = \frac{1}{T} \int_0^T [x(t)]^2 \, dt.$$

Find the expected value of Z and show that

$$\text{Var } Z = \frac{4}{T^2} \int_0^T (T - s)[\phi(s)]^2 \, ds.$$

Hint: Here you will need the rule

$$E(x_1 x_2 x_3 x_4) = E(x_1 x_2)E(x_3 x_4) + E(x_1 x_3)E(x_2 x_4) + E(x_1 x_4)E(x_3 x_2)$$

for Gaussian random variables with expected values zero [Hel91, p. 244], [Pap91, p. 197].

2-3. Taking as your interval $(-1, 1)$ instead of $(0, T)$, use the Gram–Schmidt procedure in Sec. 2.1.4 to generate the first six Legendre polynomials $P_0(t), \dots, P_5(t)$ from the powers $t^0, \dots, t^5$. [The conventional normalization of these polynomials is $P_k(1) = 1$.]

2-4. In [Kai71] Kailath defines the scalar product in the reproducing-kernel Hilbert space in a manner different from that in Sec. 2.1.7. Taking a random process $n(t)$ with zero expected value and autocovariance function $\phi(t, s)$, one assigns a random variable u_h to a function $h(t)$ in the interval $(0, T)$ by the relation $E[n(t)u_h] = h(t)$. To a function

$m(t)$, the random variable u_m is likewise assigned. The scalar product is then defined as

$$\langle \mathbf{h}, \mathbf{m} \rangle = E(u_h u_m).$$

Show that

$$u_h = \int_0^T H(s)n(s)\,ds,$$

where $H(t)$ is the cofunction of $h(t)$ as in (2-44); u_m is similarly specified. Then show that Kailath's definition of the scalar product is equivalent to that in (2-43).

2-5. The optimum detector for a signal $s(t)$ in white Gaussian noise has been constructed, but the signal that appears at its input is not $s(t)$, but $s_1(t)$. Calculate the probability of detecting $s_1(t)$ and show how it depends on the integral

$$\int_0^T s(t)s_1(t)\,dt.$$

2-6. The signal $s(t) = A[1 - \exp(-at)]$ is to be detected in the presence of white Gaussian noise of unilateral spectral density N. Let the observation interval be $0 \le t \le T$. Find the impulse response of the proper matched filter, and work out the output of the matched filter as a function of time when the input is the signal $s(t)$. Assume $s(t) \equiv 0$, $t > T$.

2-7. A system is to be designed to decide which of two signals, $s_0(t)$ or $s_1(t)$, has been received in the presence of white Gaussian noise of unilateral spectral density N. Show that the system can base its decision on the correlation of the input with the difference of the signals. Relate the decision level on the optimum statistic to the critical value Λ_0 of the likelihood ratio.

2-8. In Problem 2-7, find the probabilities Q_0 and Q_1 of each of the two kinds of errors. Show that for Q_0 fixed, the probability Q_1 depends on an *effective signal-to-noise ratio*

$$d^2 = \frac{2(E_1 + E_0 - 2R)}{N},$$

where E_0 and E_1 are the energies of the two signals, and R is

$$R = \int_0^T s_0(t)s_1(t)\,dt.$$

2-9. A signal

$$s(t) = A \cos\left(\frac{2\pi t}{T}\right) + B \cos\left(\frac{4\pi t}{T}\right)$$

is to be detected in Gaussian noise $n(t)$ with autocovariance function

$$\phi(t, s) = \alpha \cos(2\pi t/T)\cos(2\pi s/T) + \beta \cos(4\pi t/T)\cos(4\pi s/T), \quad \alpha > 0, \beta > 0,$$

by observing the input $v(t)$ to the receiver during the interval $(0, T)$. That is, under hypothesis H_0 $v(t) = n(t)$; under H_1 $v(t) = n(t) + s(t)$. Describe in detail the optimum detector under the Neyman–Pearson criterion and calculate the probability Q_d of detection, showing how it depends on the false-alarm probability Q_0.

2-10. A signal of the form

$$s(t) = A \cos wt, \qquad 0 \le t \le T, \qquad w = \frac{2\pi}{T},$$

is to be detected in Gaussian noise having the autocovariance function

$$\phi(t, u) = \sum_{k=1}^{n} \mu_k \cos kwt \cos kwu, \qquad w = \frac{2\pi}{T},$$

where all μ_k are positive. The input $v(t)$ to the receiver is observed during the interval $(0, T)$. Determine the optimum detector of the signal $s(t)$ under the Neyman–Pearson criterion, and calculate the false-alarm and detection probabilities Q_0 and Q_d. Express Q_d in terms of the energy of the signal $s(t)$ and other parameters in the formulation of the problem.

2-11. The signal $s(t) = At \exp(-bt)U(t)$ is received in the presence of noise of autocovariance function $\phi(\tau) = \phi_0 \exp(-\mu|\tau|)$ as in (2-102). Show how the input $v(t)$ to the receiver should be processed by a matched filter and a delay line to decide whether the signal is present. Calculate the effective signal-to-noise ratio d^2, and state how the probability of detection depends on it. *Hint:* Use the results of Sec. 2.3.

2-12. A signal $s(t)$ is to be detected in a mixture of white noise of unilateral spectral density N and correlated noise whose autocovariance function is that given in (2-102). The spectral density of the noise is thus the same as that in Example 2-2. Show how to calculate the impulse response of the detection filter by the technique described in Sec. 2.3. As an example take the signal as a constant, $s(t) \equiv A$, and work out the impulse response of the filter and the effective signal-to-noise ratio d^2.

Show how the solution $q(t)$ of (2-91) passes in the limit of vanishing white noise $(N \rightarrow 0)$ to a solution involving delta functions as in (2-104). In Fig. 2-11 we have plotted $q(t)$ for $\mu T = 2$ and for various values of the parameter $4\phi_0/\mu N$. The peaks at $t = 0$ and $t = T$ become sharper and sharper as the spectral density of the white noise decreases.

2-13. The noise at the input to a receiver is stationary and Gaussian and consists of the sum of white noise with unilateral spectral density N and noise whose autocovariance function is that given in (2-102). The receiver is to detect a signal of the form

$$s(t) = A e^{-a|t|}, \qquad -\infty < t < \infty.$$

An infinitely long observation interval is allowed. Find the optimum detection statistic under the Neyman–Pearson criterion and calculate its probability of detection for fixed false-alarm probability.

2-14. (a) A rectangular signal of duration T_1,

$$s(t) = \begin{cases} A, & 0 < t < T_1, \\ 0, & t \le 0, \ t \ge T_1, \end{cases}$$

is to be detected in the presence of Gaussian noise that is a combination of white noise and stationary noise having a Lorentz spectral density as in Example 2-2. The input to the receiver is observed during an interval $(-T, T)$ that is much longer than both the signal and the width μ^{-1} of the autocovariance function of the nonwhite component of the noise.

Determine the impulse response of the optimum linear filter for detecting this signal with maximum probability Q_d of detection for preassigned false-alarm probability Q_0. Assume that $T \gg T_1$, whereupon (2-77) is appropriate. Sketch the impulse response and describe how the receiver utilizes this filter. Sketch the signal component of the output of this filter under hypothesis H_1.

(b) Show how the optimum receiver of part (a) can be decomposed as in Fig. 2-4 into a whitening filter followed by a second linear filter. Give the transfer function and

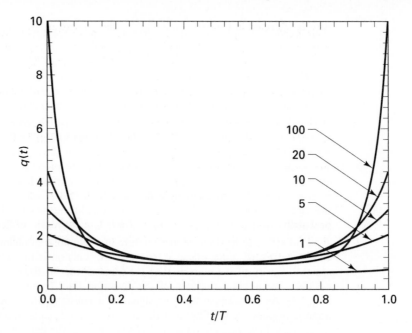

Figure 2-11. Solution $q(t)$ of the detection integral equation (2-91) with the spectral density of Problem 2-12 and a constant signal $s(t) \equiv 1$; $\mu T = 2$. Curves are indexed with values of the parameter $4\phi_0/\mu N$.

the impulse response of the whitening filter for the signal and noise specified in part (a), and calculate the signal $s_1(t)$ that appears at the output of the whitening filter when the rectangular signal $s(t)$ is present at the input. Sketch this output signal and give the impulse response of the second member of the cascade.

(c) Calculate the maximum probability Q_d of detection attained by the optimum receiver of part (a) in terms of the preassigned false-alarm probability Q_0 and an appropriate signal-to-noise ratio d_∞^2, and calculate that signal-to-noise ratio.

 Hint: This problem is most simply solved in the time domain by making use of the convolution theorem. Compare with Example 2-2.

2-15. A signal whose spectrum $S(\omega)$ is bandlimited,

$$S(\omega) \equiv A, \qquad -\pi W < \omega < \pi W,$$
$$\equiv 0, \qquad |\omega| \geq \pi W,$$

is to be detected in the presence of stationary Gaussian noise of autocovariance function

$$\phi(\tau) = \frac{\phi_0 a^2}{a^2 + \tau^2}.$$

Observation during an infinite interval is permitted. Calculate in terms of the error-function integral the maximum possible probability Q_d of detection for a fixed false-alarm probability Q_0.

2-16. A binary communication system is to transmit messages coded into 0's and 1's, which occur independently every T seconds and with equal relative frequencies. Two systems are contemplated. In system (a) the 1's are transmitted by sending a signal that is received as $af(t)$, which falls entirely within the interval $(0, T)$; for 0's nothing is sent.

This is called an *on–off* system. In system (b) the 1's are transmitted by sending a signal received as $bf(t)$, the 0's by sending one received as $-bf(t)$; these are called *antipodal* signals. The signals are received in white Gaussian noise of unilateral spectral density N. For each system determine the receiver that minimizes the average probability P_e of error. Find the ratio of the average power transmitted by system (a) to that required by system (b) in order that the error probabilities P_e be equal.

2-17. A signal $Af(t)$ of known form and amplitude is to be detected in white Gaussian noise of unilateral spectral density N. Fix the cost matrix $\mathbf{C}$ and the prior probabilities ζ_0 and ζ_1 of hypotheses H_0 and H_1. Show that the false-alarm and detection probabilities are given as in (2-72) with

$$x = \frac{1}{d}\ln \Lambda_0 + \frac{d}{2}, \qquad d = \sqrt{2E/N},$$

and with Λ_0 as given in (1-17). Find the minimum Bayes cost $\overline{C}_{\min}(d)$ as a function of the signal-to-noise ratio d, and show that as d goes to zero, $\overline{C}_{\min}(0) - \overline{C}_{\min}(d)$ is proportional, for $d \ll 1$, to

$$d^3 \exp\left[-\frac{\ln^2 \Lambda_0}{2d^2}\right].$$

This difference $\overline{C}_{\min}(0) - \overline{C}_{\min}(d)$ of Bayes costs therefore approaches zero faster than any power of d as d goes to zero. *Hint:* You will need the asymptotic form of the error-function integral in (1-11):

$$\operatorname{erfc} x \approx \frac{1}{\sqrt{2\pi}\,x} e^{-x^2/2}\left(1 - \frac{1}{x^2} + \cdots\right), \qquad x \gg 1, \qquad (2\text{-}135)$$

[Abr70, eq. (26.2.12), p. 932]. This can be derived by successively integrating (1-11) by parts.

2-18. For $a > 2T$, solve the integral equation

$$\cos wt = \phi_0 \int_{-T}^{T}\left(1 - \frac{|t-s|}{a}\right)q(s)\,ds, \qquad -T \le t \le T.$$

Hint: By differentiating the triangular kernel $\phi(t-s)$ twice, show that it satisfies the differential equation

$$\frac{d^2\phi}{dt^2} = F(t),$$

where $F(t)$ involves delta functions. Determine $F(t)$ and then use the technique of Sec. 2.3.1.

2-19. Consider a pulse-amplitude-modulated (PAM) system in which messages are coded into an odd number $M = 2n + 1$ of symbols corresponding to signals received as

$$s_j(t) = A_j f(t), \qquad \int_{0}^{T}[f(t)]^2\,dt = 1, \qquad 1 \le j \le M = 2n + 1,$$

during an observation interval $(0, T)$. The amplitudes $A_1, A_2, \ldots, A_M$ are uniformly spaced about zero: $A_{n+1} = 0$. The noise is white and Gaussian with unilateral spectral density N. The signals occur with equal relative frequencies M^{-1}. Determine the optimum receiver for deciding among these PAM signals and calculate the probability of error as a function of the average received energy E and the noise spectral density N. Observe the relevance of Example 1-1.

2-20. A ternary communication system transmits one of three signals $f(t)$, 0, or $-f(t)$ every T seconds. They are received with energy E or energy 0 in white Gaussian noise of unilateral spectral density N. At the end of each interval $(0, T)$ the output x of a filter matched to $f(t)$ is measured and compared with decision levels $+a$ and $-a$. If $x > a$, the decision is made that $+f(t)$ was sent; if $x < a$, that $-f(t)$ was sent; and if $-a < x < a$, that 0 was sent. What is the probability $Q_c = 1 - P_e$ of a correct decision as a function of a, E, and N when all three signals are sent equally often? What is the maximum possible value of Q_c and for what value of a is it attained?

2-21. A quaternary communication system transmits every T seconds one of four equally likely signals as in (2-124). The functions $f_1(t)$ and $f_2(t)$ are orthogonal, and the signals are received with equal energies E in white Gaussian noise. The receiver has filters matched to $f_1(t)$ and $f_2(t)$ and observes their outputs y_1 and y_2 at the end of each interval $(0, T)$. In terms of these, specify the strategy that minimizes the probability of error in deciding which signal has been received. Calculate that minimum probability of error as a function of the signal-to-noise ratio. *Hint*: A judiciously selected rotation of axes in the signal space much simplifies this calculation.

2-22. Every T seconds a quaternary PAM communication system transmits one of four signals,

$$A = a_1 f(t), \quad B = a_2 f(t), \quad C = -a_2 f(t), \quad D = -a_1 f(t), \quad 0 < a_2 < a_1.$$

The signals are sent with equal relative frequencies. They are received with a common attenuation μ in the presence of white Gaussian noise of spectral density N; that is, if A is sent, $\mu a_1 f(t)$ is received, and so on. At the end of each interval $(0, T)$ the output y of a filter matched to $f(t)$ is compared with three decision levels b, 0, and $-b$, and the decisions about the transmitted signals are made on the basis of the scheme

$$y > b \rightarrow A; \quad 0 < y \leq b \rightarrow B; \quad -b < y \leq 0 \rightarrow C; \quad y \leq -b \rightarrow D.$$

Calculate the average probability of error in deciding among these four signals, and choose the value of b that minimizes it. Show how you would determine the values of the amplitudes a_1 and a_2 to make this minimum probability of error as small as possible under the constraint of fixed average transmitted power.

2-23. In a communication system sending messages expressed in an alphabet of four symbols A, B, C, and D, the transmitter sends nothing for each A. For each B it sends a signal received as $s_1(t)$. For each C it sends a signal that is received as $s_2(t)$; $s_2(t)$ is orthogonal to $s_1(t)$, but does not necessarily have the same energy. For each D the transmitter sends a signal received as $s_1(t) + s_2(t)$. The received signals are confined to an observation interval $(0, T)$. Successive message symbols appear every T seconds, are statistically independent, and occur with equal relative frequencies $\frac{1}{4}$. The signals are received in white Gaussian noise with unilateral spectral density N. Describe the receiver that decides among the four possible signals with minimum probability P_e of error, and calculate that probability in terms of N and of the energies E_1 and E_2 of signals $s_1(t)$ and $s_2(t)$.

2-24. A nonary communication system sends messages coded into an alphabet of nine symbols, which occur with equal relative frequencies $1/9$. In terms of two functions $f_1(t)$ and $f_2(t)$ that are orthonormal over the observation interval $(0, T)$, the nine signals, as received, are as follows:

$$s_1(t) \equiv 0; \quad s_2(t) = Af_1(t); \quad s_3(t) = A[f_1(t) + f_2(t)];$$
$$s_4(t) = Af_2(t); \quad s_5(t) = A[-f_1(t) + f_2(t)]; \quad s_6(t) = -Af_1(t);$$
$$s_7(t) = -A[f_1(t) + f_2(t)]; \quad s_8(t) = -Af_2(t); \quad s_9(t) = A[f_1(t) - f_2(t)].$$

The signals are received in white Gaussian noise $n(t)$ during observation intervals that we denote as usual by $(0, T)$.

(a) Draw a diagram representing these nine signals in an appropriately defined signal space based on the functions $f_1(t)$ and $f_2(t)$.

(b) Determine the optimum receiver for deciding with minimum probability P_e of error which of these nine signals is present in the input

$$v(t) = n(t) + s_k(t), \qquad 1 \le k \le 9.$$

Use the signal-space diagram to indicate the decision regions.

(c) Calculate the probability P_e of error for the optimum system. Express it in terms of the average signal power received and the noise spectral density N.

(d) Taking

$$f_1(t) = \left(\frac{2}{T}\right)^{1/2} \cos wt, \qquad f_2(t) = \left(\frac{2}{T}\right)^{1/2} \sin wt, \qquad w = \frac{2\pi k}{T},$$

for some integer k, determine the nine signals in the simplest possible form.

2-25. In the quaternary communication system characterized by (2-124), the functions $f_k(t)$ are normalized as before, but they are not orthogonal:

$$\int_0^T [f_k(t)]^2 \, dt = 1, \qquad k = 1, 2;$$

$$\int_0^T f_1(t) f_2(t) \, dt = r, \qquad 0 < |r| < 1.$$

Thus the signals have equal energies, but the outputs at $t = T$ of filters matched to them are correlated. Their relative frequencies are still equal to $\frac{1}{4}$, and they are received in white Gaussian noise of unilateral spectral density N. Calculate the minimum probability P_e of error in deciding among these signals, and determine the receiver that attains it.

3

Narrowband Signals and Their Detection

3.1 NARROWBAND SIGNALS AND FILTERS

Because an antenna of convenient size radiates most efficiently at high frequencies, the signals transmitted in radio communications and radar consist of a high-frequency carrier modulated in amplitude or frequency or both. The strategy developed in Chapter 2 for detecting a signal in Gaussian noise required that its form $s(t)$ be completely known to the receiver. More often than not, however, the phase of the carrier of a received radio or radar signal is unknown, and other parameters such as its amplitude, its arrival time, and even the frequency of its carrier may also be unknown or uncertain within a wide range of possible values. We must broaden our theory to accommodate such uncertainties in the signals to be detected. Before entering on the necessary modifications to the general theory, we shall discuss its simplest extension, which treats the detection of a high-frequency signal of unknown carrier phase. First we must review the concept of modulation and introduce a concise notation for expressing modulated signals and the properties of the noise that accompanies them.

A simple radar pulse may look somewhat like the signal depicted in Fig. 3-1, for which $s(t)$ can be written as

$$s(t) = \begin{cases} A \sin \Omega t, & 0 \le t \le T, \\ 0, & t < 0, \quad t > T. \end{cases}$$

A pulse of this kind usually contains a large number of cycles of the radio-frequency carrier; in a typical such signal the carrier frequency Ω may be $2\pi \cdot 10^9$ rad/sec (a 1000-MHz carrier), and the duration T may be 1 μsec $= 10^{-6}$ sec.

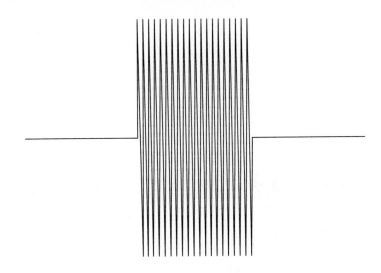

Figure 3-1. Narrowband pulse.

This kind of signal is said to be *amplitude modulated*, for only the extent of the oscillations of the carrier is affected. The outputs of certain broadcasting transmitters are also amplitude modulated and can be described by the expression

$$s(t) = A[1 + m(t)] \cos \Omega t. \qquad (3\text{-}1)$$

The function $m(t)$, which changes only slightly over a period $2\pi/\Omega$ of the carrier $\cos \Omega t$, represents the voices and music being transmitted. Care is taken to prevent the factor $[1 + m(t)]$ from becoming negative. The signal in (3-1) looks like a sinusoid with an irregularly fluctuating amplitude.

The output of a transmitter modulated in frequency can be written in the form

$$s(t) = A \cos\left[\Omega t + \int_0^t w(s)\, ds \right].$$

The instantaneous frequency of this signal is $\Omega + w(t)$, and the slowly varying function $w(t)$ carries the information being broadcasted. Such a wave has a constant amplitude, but the times at which it crosses the zero level shift about with the modulation.

A high-frequency carrier with the most general kind of modulation can be represented by the equation

$$s(t) = M(t) \cos[\Omega t + \phi(t)]. \qquad (3\text{-}2)$$

We shall call the function $M(t)$ the *amplitude modulation* and the function $\phi(t)$ the *phase modulation*; the derivative $d\phi/dt$ is the *frequency modulation* of the signal. All these modulations vary much more slowly than the radio-frequency carrier. Some radar signals are modulated in both amplitude and frequency (or phase) in this way.

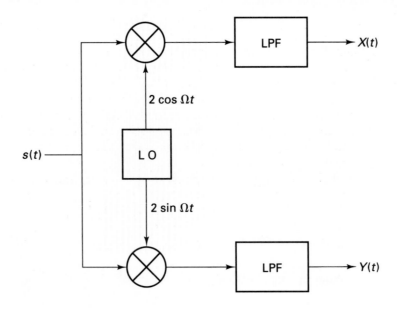

Figure 3-2. Demodulating receiver.

By expanding the cosine in (3-2) we obtain

$$s(t) = M(t)[\cos \phi(t) \cos \Omega t - \sin \phi(t) \sin \Omega t]$$
$$= X(t) \cos \Omega t - Y(t) \sin \Omega t \qquad (3\text{-}3)$$

with

$$X(t) = M(t) \cos \phi(t), \qquad Y(t) = M(t) \sin \phi(t); \qquad (3\text{-}4)$$

$X(t)$ is often called the in-phase component and $Y(t)$ the quadrature component, but we shall designate the pair of them as the *quadrature components* of the signal. They too change only slightly during one cycle of the carrier $\cos \Omega t$. The function $X(t)$ can be extracted by multiplying the signal $s(t)$ with the output $2 \cos \Omega t$ of a local oscillator (L O) and filtering off the components of the product with carrier frequency 2Ω by means of a low-pass filter (LPF):

$$2s(t) \cos \Omega t = X(t)(1 + \cos 2\Omega t) - Y(t) \sin 2\Omega t \rightarrow X(t).$$

A device accomplishing this is called a *mixer* or a *homodyne detector*. It is said to mix or beat together the signals $s(t)$ and $2 \cos \Omega t$. The other quadrature component $Y(t)$ can be extracted by mixing $s(t)$ with the signal $2 \sin \Omega t$. Figure 3-2 shows the block diagram of a receiver whose input is the modulated signal $s(t)$ and whose outputs are the quadrature components $X(t)$ and $Y(t)$.

By combining these quadrature components into a complex function of the time, we obtain a convenient representation of a modulated signal in terms of its *complex envelope*

$$F(t) = M(t) e^{i\phi(t)}, \qquad (3\text{-}5)$$

and we write the signal compactly as

$$s(t) = \text{Re } F(t) e^{i\Omega t}; \qquad (3\text{-}6)$$

Re stands for the real part of the complex number following it. The complex function $F(t)$ can be pictured as a vector at the origin of the XY-plane. The end of the vector moves about in the plane, and all the while the plane itself rotates with an angular velocity Ω. The signal $s(t)$ is the projection of this rotating vector on a fixed line. The complex envelope of a modulated signal is a natural generalization of the phasor representation of alternating currents and voltages. When the motion of the vector $F(t)$ within the rotating plane is much slower than the rate of rotation, the signal $s(t)$ is said to be *quasiharmonic*.

If the signal is a pulse of finite energy, its complex envelope possesses a Fourier transform

$$f(\omega) = \int_{-\infty}^{\infty} F(t) e^{-i\omega t} \, dt, \qquad (3\text{-}7)$$

in terms of which the spectrum $S(\omega)$ of the signal is

$$S(\omega) = \int_{-\infty}^{\infty} s(t) e^{-i\omega t} \, dt = \tfrac{1}{2} \int_{-\infty}^{\infty} [F(t) e^{i\Omega t} + F^*(t) e^{-i\Omega t}] e^{-i\omega t} \, dt$$

$$= \tfrac{1}{2}[f(\omega - \Omega) + f^*(-\omega - \Omega)]. \qquad (3\text{-}8)$$

Because the quadrature components of $F(t)$ vary much more slowly than the carrier $\cos \Omega t$, the width in frequency of the modulus $|f(\omega)|$ of its Fourier transform is much smaller than Ω. The modulus $|S(\omega)|$ of the spectrum of the signal then exhibits two narrow peaks, one near the frequency Ω and the other near $-\Omega$. Because of this structure, $s(t)$ is called a *narrowband signal*.

The spectrum in (3-8) satisfies the condition $S(-\omega) = S^*(\omega)$ imposed by the reality of the signal $s(t)$. The Fourier transform $f(\omega)$ of the complex envelope satisfies a similar condition if $F(t)$ is real and the signal is purely amplitude modulated. Only then will the modulus $|f(\omega)|$ be an even function and will the peaks of $|S(\omega)|$ be symmetrical about the carrier frequency Ω. Indeed, the carrier frequency is quite arbitrary. Shifting it by an amount k simply introduces a factor $\exp(-ikt)$ into the complex envelope,

$$F(t) e^{i\Omega t} = [F(t) e^{-ikt}] e^{i(\Omega + k)t},$$

without changing the signal $s(t)$. An appropriate choice of the carrier frequency Ω sometimes simplifies a calculation.

If a narrowband signal is not presented in the explicitly modulated form of (3-2), its complex envelope can be derived from what is called the *analytic signal*. We write $s(t)$ as the sum of a positive-frequency part $s_+(t)$ and a negative-frequency part $s_-(t)$, which are defined in terms of its spectrum $S(\omega)$ by

$$s_+(t) = \int_0^{\infty} S(\omega) e^{i\omega t} \frac{d\omega}{2\pi}, \qquad s_-(t) = s_+^*(t) = \int_{-\infty}^0 S(\omega) e^{i\omega t} \frac{d\omega}{2\pi}. \qquad (3\text{-}9)$$

For a narrowband signal the positive-frequency part of the spectrum is concentrated in the neighborhood of an angular frequency Ω, and the complex envelope can be defined as

$$F(t) = 2s_+(t) e^{-i\Omega t}.$$

The analytic signal $s_+(t)$ was introduced by Gabor [Gab46]. It is more general than the complex envelope, for it can be defined for any signal that possesses a spectrum. In terms of the signal $s(t)$ itself, it is given by the integral

$$s_+(t) = \frac{1}{2\pi i} \int_{-\infty}^{\infty} \frac{s(z)\, dz}{t - z}, \tag{3-10}$$

in which z has an infinitesimal negative imaginary part. The real and imaginary parts of the analytic signal are related by the Hilbert transform, for which we refer the reader to [McD56], [Gui63, Ch. 18], and [Sch66, Sec. 1.6]. The definitions in (3-9) and (3-10) are of little practical value, and the analytic signal is useful mainly when the signal $s(t)$ is quasiharmonic, whereupon the complex envelope serves as well. This and other definitions of complex envelopes have been reviewed by Rice [Ric82].

A filter whose output is a function of the amplitude modulation of a narrowband signal applied to its input is known as a *rectifier*. It must be nonlinear, for a linear filter could not remove the oscillations of the carrier without destroying the envelope as well. A typical rectifier whose output is related directly to the amplitude modulation of its input is the quadratic rectifier. It first squares its input, yielding by (3-3)

$$[s(t)]^2 = \tfrac{1}{2}\{[X(t)]^2 + [Y(t)]^2\}$$
$$+ \tfrac{1}{2}\{[X(t)]^2 - [Y(t)]^2\} \cos 2\Omega t - X(t)Y(t) \sin 2\Omega t, \tag{3-11}$$

after which a low-pass filter removes the terms with frequencies in the vicinity of 2Ω, so that the final output is proportional to $|F(t)|^2 = [M(t)]^2 = [X(t)]^2 + [Y(t)]^2$. This we call a *quadratic* rectifier. Rectifiers having other than a quadratic characteristic produce some other monotone function of the absolute value $|F(t)| = M(t)$; a *linear* rectifier yields $M(t)$ itself.

A device whose output is proportional to the instantaneous frequency deviation $\phi'(t)$ is known as a *discriminator*; its output can be taken as proportional to

$$\frac{d}{dt} \operatorname{Im}[\ln F(t)]$$

in terms of the complex envelope $F(t)$ of the input. Any given discriminator or rectifier circuit must of course be analyzed to determine the accuracy of these descriptions of its action on the input signal.

If we integrate (3-11) over $-\infty < t < \infty$ and recognize that for quasiharmonic signals the integrals of the terms proportional to $\cos 2\Omega t$ and $\sin 2\Omega t$ will be much smaller than the others, we obtain

$$E = \int_{-\infty}^{\infty} [s(t)]^2 \, dt \approx \tfrac{1}{2} \int_{-\infty}^{\infty} |F(t)|^2 \, dt = \tfrac{1}{2} \int_{-\infty}^{\infty} |f(\omega)|^2 \frac{d\omega}{2\pi}$$

for the energy of the signal.

Quasiharmonic signals are often transformed by means of linear filters that attenuate components of all frequencies except those in the neighborhood of the input carrier frequency. One purpose of such filtering is to eliminate noise lying outside the frequency band of the signals. The analysis of these pass filters illustrates the simplification brought about by the complex notation introduced here.

We shall deal with linear narrowband pass filters that least attenuate those components of their inputs whose frequencies lie in a range of width W about some high frequency Ω, with W much smaller than Ω, $W \ll \Omega$. It is convenient to write the transfer function $y(\omega)$ of such a filter in the form

$$y(\omega) = Y(\omega - \Omega) + Y^*(-\omega - \Omega), \qquad (3\text{-}12)$$

in which the complex function $Y(\omega)$ differs significantly from zero only over a narrow range of frequencies about $\omega = 0$. Equation (3-12) satisfies the condition of symmetry $y(-\omega) = y^*(\omega)$, which is a consequence of the reality of the impulse response

$$k(\tau) = \int_{-\infty}^{\infty} y(\omega) \, e^{i\omega\tau} \frac{d\omega}{2\pi} \qquad (3\text{-}13)$$

of the filter. If a linear narrowband pass filter consists of lumped circuit elements, the poles of its transfer function $y(\omega)$ lie in the neighborhood of $\omega = +\Omega$ and $\omega = -\Omega$. We can then decompose $y(\omega)$ as in (3-12) from its expansion in partial fractions, taking the terms with poles near $\omega = +\Omega$ into $y(\omega - \Omega)$ and leaving the rest for the term $y^*(-\omega - \Omega)$.

By means of (3-12) and (3-13) we can write the impulse response $k(\tau)$ of the narrowband filter as

$$k(\tau) = \int_{-\infty}^{\infty} [Y(\omega - \Omega) + Y^*(-\omega - \Omega)] \, e^{i\omega\tau} \frac{d\omega}{2\pi}$$

$$= \int_{-\infty}^{\infty} Y(\omega) \, e^{i(\omega+\Omega)\tau} \frac{d\omega}{2\pi} + \int_{-\infty}^{\infty} Y^*(\omega) \, e^{-i(\omega+\Omega)\tau} \frac{d\omega}{2\pi}$$

$$= 2 \, \mathrm{Re} \, K(\tau) \, e^{i\Omega\tau},$$

where

$$K(\tau) = \int_{-\infty}^{\infty} Y(\omega) \, e^{i\omega\tau} \frac{d\omega}{2\pi}. \qquad (3\text{-}14)$$

If $Y(\omega)$ is significant only over a range of frequencies of width W, the function $K(\tau)$ changes appreciably in a range of values of τ whose width is of the order of $1/W$. By analogy with the concept of the complex envelope of a narrowband signal, we can consider $K(\tau)$ as one-half the complex envelope of the impulse response of the narrowband filter. When such a filter is excited by a sharp impulse, it "rings," and its output oscillates with frequency Ω; this output decays with time in a manner described by the envelope $2K(\tau)$.

We shall now show that the envelope function $K(\tau)$ transforms the complex envelope of the input signal to produce that of the output signal in much the same way as the impulse response $k(\tau)$ acts on the signal itself,

$$s_o(t) = \int_0^{\infty} k(\tau) s_i(t - \tau) \, d\tau, \qquad (3\text{-}15)$$

the subscripts i and o denoting input and output, respectively. The spectra of the input and output signals are related by

$$S_o(\omega) = y(\omega) S_i(\omega),$$

and if we write them in the form (3-8) and use (3-12) for $y(\omega)$, we find

$$\tfrac{1}{2}[f_o(\omega - \Omega) + f_o^*(-\omega - \Omega)]$$

$$= [Y(\omega - \Omega) + Y^*(-\omega - \Omega)][\tfrac{1}{2}f_i(\omega - \Omega) + \tfrac{1}{2}f_i^*(-\omega - \Omega)]$$

$$= \tfrac{1}{2}Y(\omega - \Omega)f_i(\omega - \Omega) + \tfrac{1}{2}Y^*(-\omega - \Omega)f_i^*(-\omega - \Omega)$$

$$+ \tfrac{1}{2}Y^*(-\omega - \Omega)f_i(\omega - \Omega) + \tfrac{1}{2}Y(\omega - \Omega)f_i^*(-\omega - \Omega).$$

For quasiharmonic signals and narrowband filters the last two terms are much smaller than the first two, and we can write approximately

$$f_o(\omega - \Omega) \approx Y(\omega - \Omega)f_i(\omega - \Omega) \quad \text{or} \quad f_o(\omega) \approx Y(\omega)f_i(\omega),$$

from which, by the convolution theorem for Fourier transforms, we find for the complex envelope of the output signal

$$F_o(t) \approx \int_0^\infty K(\tau)F_i(t - \tau)\, d\tau. \tag{3-16}$$

By comparing the terms we dropped with those we retained, we see that the relative error we committed is on the order of the magnitude of $|Y(2\Omega)/Y(0)|$ or $|f_i(2\Omega)/f_i(0)|$, whichever is the greater. Within this approximation, we can use $K(\tau)$, which we call the *complex impulse response* of the narrowband filter, in the same way when dealing with complex envelopes as we ordinarily use the original impulse response $k(\tau)$ when dealing with the signals themselves.

Let us illustrate these results for a filter consisting of a simply resonant circuit, as shown in Fig. 3-3. When the input and output are measured across the terminals shown, the transfer function is

$$y(\omega) = \frac{R}{R + i(\omega L - \frac{1}{\omega C})} = \frac{2i\mu\omega}{\omega_0^2 + 2i\mu\omega - \omega^2}, \qquad \mu = \frac{R}{2L}, \qquad \omega_0^2 = \frac{1}{LC}.$$

The poles of $y(\omega)$ lie at $\omega = i\mu + \nu$ and $\omega = i\mu - \nu$, $\nu^2 = \omega_0^2 - \mu^2$. For a narrowband or "high-Q" filter, $\mu \ll \omega_0$; and we can take the pass frequency as $\Omega = \nu$, which is close to the resonant frequency ω_0. Then it is simple to show that $Y(\omega)$ satisfies (3-12) when it is defined by

$$Y(\omega) = \frac{\mu(\mu - i\nu)}{\nu(\omega - i\mu)} \approx \frac{1}{1 + i\omega/\mu}.$$

By (3-14) the complex impulse response has the simple form

$$K(\tau) = \frac{\mu}{\nu}(\nu + i\mu)\, e^{-\mu\tau}U(\tau) \approx \mu\, e^{-\mu\tau}U(\tau), \tag{3-17}$$

where $U(\cdot)$ is the unit step function. For an input consisting of a carrier of frequency near Ω whose modulation varies slowly, it is simpler to calculate the modulation of the output by means of (3-17) and (3-16) than to use the usual methods embodied in formulas like (3-15), in which $k(\tau)$ is a somewhat more complicated function.

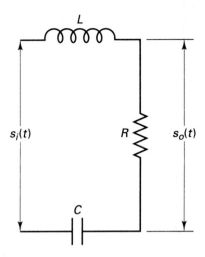

Figure 3-3. Simply resonant circuit.

3.2 NARROWBAND NOISE

3.2.1 The Complex Representation

The antenna that picks up a high-frequency radio or radar signal and the waveguide or coaxial cable that conveys it to the receiver respond significantly only over a relatively narrow range of frequencies about the carrier frequency of the signal. The input $v(t)$ to the receiver has in effect passed through a narrowband filter of the type specified by (3-12). Its passband is usually broad enough so that the filter does not appreciably distort the signal, but it converts the thermal noise into noise with a spectral density concentrated in the neighborhood of the pass frequency Ω of the input. Even if the input circuitry did not have such an effect on the noise, it would generally be convenient, before processing the input further, to remove from it the noise in frequency ranges far from that of the signal by passing the input through just such a narrowband filter as that in (3-12). The output noise $n(t)$ will then have a spectral density of the form

$$\Phi(\omega) = \tilde{\Phi}(\omega - \Omega) + \tilde{\Phi}(-\omega - \Omega), \tag{3-18}$$

in which the function $\tilde{\Phi}(\omega)$ is real and nonnegative and differs significantly from zero only over a range of angular frequencies of width $W \ll \Omega$. The autocovariance function of this noise is

$$\phi(\tau) = \int_{-\infty}^{\infty} \Phi(\omega)\, e^{i\omega\tau} \frac{d\omega}{2\pi} = \int_{-\infty}^{\infty} [\tilde{\Phi}(\omega - \Omega) + \tilde{\Phi}(-\omega - \Omega)]\, e^{i\omega\tau} \frac{d\omega}{2\pi}$$

$$= \int_{-\infty}^{\infty} [\tilde{\Phi}(u)\, e^{i(u+\Omega)\tau} + \tilde{\Phi}(u)\, e^{-i(u+\Omega)\tau}] \frac{du}{2\pi} = \mathrm{Re}[\tilde{\phi}(\tau)\, e^{i\Omega\tau}],$$

where the function $\tilde{\phi}(\tau)$, generally complex, is defined by the Fourier transform

$$\tilde{\phi}(\tau) = 2\int_{-\infty}^{\infty} \tilde{\Phi}(\omega)\, e^{i\omega\tau} \frac{d\omega}{2\pi}, \tag{3-19}$$

and

$$\tilde{\Phi}(\omega) = \tfrac{1}{2} \int_{-\infty}^{\infty} \tilde{\phi}(\tau) \, e^{-i\omega\tau} \, d\tau. \tag{3-20}$$

We call $\tilde{\Phi}(\omega)$ the *narrowband spectral density* and $\tilde{\phi}(\tau)$ the *complex autocovariance function* of the stationary narrowband noise $n(t)$.

The variance of this random process is

$$\text{Var } n(t) = \phi(0) = \tilde{\phi}(0) = 2\int_{-\infty}^{\infty} \tilde{\Phi}(\omega)\frac{d\omega}{2\pi}. \tag{3-21}$$

The factor 2 in (3-19) and (3-21) can be remembered by keeping in mind that the total power in the process is obtained by integrating the spectral density over all frequencies, both positive and negative. That spectral density (3-18) has two peaks, each of which contributes half the power.

When noise of spectral density $\Phi_i(\omega)$ passes into a narrowband filter whose transfer function is given by (3-12), the spectral density of the output is, according to (2-9),

$$\begin{aligned} \Phi_o(\omega) &= \Phi_i(\omega)|y(\omega)|^2 = \Phi_i(\omega)|Y(\omega - \Omega) + Y^*(-\omega - \Omega)|^2 \\ &\approx \Phi_i(\omega)[|Y(\omega - \Omega)|^2 + |Y(-\omega - \Omega)|^2]. \end{aligned} \tag{3-22}$$

When the absolute square in the first line of (3-22) is expanded, the cross-product $\text{Re}[Y(\omega - \Omega)Y^*(-\omega - \Omega)]$ can be neglected because for narrowband filters it is much smaller than the other terms. If the spectral density of the input noise is much broader than the transfer function of the filter, we can replace $\Phi_i(\omega)$ by its value at Ω, and by comparison with (3-18) we find

$$\tilde{\Phi}_o(\omega) = |Y(\omega)|^2 \Phi_i(\Omega) \tag{3-23}$$

for the narrowband spectral density of the output at all frequencies ω where this function is of significant magnitude. If, on the other hand, the input noise is narrowband, with a spectral density like that in (3-18), we find

$$\tilde{\Phi}_o(\omega) = |Y(\omega)|^2 \tilde{\Phi}_i(\omega) \tag{3-24}$$

by discarding all the terms in (3-22) that are small in this approximation. This equation has the same form as (2-9) and shows that we can use the narrowband spectral density and transfer function in much the same way as we use the original functions $\Phi(\omega)$ and $y(\omega)$.

In noise $n(t)$ of this kind, frequencies in the neighborhood of Ω predominate, and we should expect to write it in the quasiharmonic form

$$n(t) = \text{Re } N(t) \, e^{i\Omega t}, \tag{3-25}$$

where

$$N(t) = X(t) + iY(t) \tag{3-26}$$

is a complex envelope whose quadrature components $X(t)$ and $Y(t)$ are random processes varying much more slowly than the carrier.

Indeed, a stationary random process with spectral density $\Phi(\omega)$ can be considered as a succession of randomly occurring pulses $s(t - t_m)$ whose spectra are proportional to

$$S(\omega) = [\Phi(\omega)]^{1/2} \, e^{ig(\omega)}, \tag{3-27}$$

with $g(\omega)$ an arbitrary phase factor,

$$n(t) = \sum_{m=-\infty}^{\infty} a_m s(t - t_m); \tag{3-28}$$

the amplitudes a_m are independently random with zero expected values, and as we said in Sec. 2.1.5, the epochs t_m of the pulses form a Poisson point process. Observe that (3-28) is the same as (2-30) when we put $s(t) = k(t)$, whereupon $S(\omega) = y(\omega)$. By (2-9) with $\Phi(\omega) \equiv N/2$, the spectral density of $n(t)$ must be proportional to $|y(\omega)|^2 = |S(\omega)|^2$.

When the spectral density $\Phi(\omega)$ has the form of (3-18), the spectrum $S(\omega)$ in (3-27) will have the similar form (3-8), and the component pulses in (3-28) can be represented as

$$s(t - t_m) = \text{Re}[F(t - t_m) \exp i\Omega(t - t_m)]$$

in terms of a complex envelope $F(t)$. As a result, the noise $n(t)$ in (3-28) has the form of (3-25) with a complex envelope

$$N(t) = \sum_{m=-\infty}^{\infty} a_m \exp(i\phi_m)F(t - t_m), \qquad \phi_m = -\Omega t_m, \tag{3-29}$$

which represents a sequence of complex pulses $F(t - t_m)$ with random amplitudes a_m and random phases ϕ_m. We shall now study the properties of narrowband or quasiharmonic noise of this kind, and in particular we shall relate its autocovariance function to those of its quadrature components $X(t)$ and $Y(t)$.

3.2.2 The Complex Autocovariance Function

Quasiharmonic noise is not necessarily stationary. In a scatter-multipath communication system, for instance, narrowband transmitted pulses impinge on a myriad of scatterers randomly located in the ionosphere, each of which reradiates a pulse of the same form. These scattered pulses combine with random amplitudes and phases to produce a received signal much like that in (3-28) and (3-29), except that the epochs t_m do not stretch from $-\infty$ to ∞, but occur only during a limited interval determined by the thickness of the scattering layer and the angles of the incident and scattered beams. A similar phenomenon occurs in radar astronomy when a narrowband radar pulse is reflected by numerous randomly located points on the surface of a rough planet. In order later to treat the detection of random quasiharmonic signals of this kind, we allow the process $n(t) = \text{Re } N(t) \exp i\Omega t$ to be nonstationary.

Writing the process as

$$n(t) = \tfrac{1}{2}[N(t) e^{i\Omega t} + N^*(t) e^{-i\Omega t}],$$

we form its autocovariance function

$$\begin{aligned}
\phi(t_1, t_2) &= E[n(t_1)n(t_2)] \\
&= \tfrac{1}{4}\{E[N(t_1)N(t_2)] \exp[i\Omega(t_1 + t_2)] + \text{c.c.} \\
&\qquad + E[N(t_1)N^*(t_2)] \exp[i\Omega(t_1 - t_2)] + \text{c.c.}\},
\end{aligned} \tag{3-30}$$

where c.c. denotes the complex conjugate of the preceding term.

In naturally occurring quasiharmonic random processes, whether stationary or not, the term $E[N(t_1)N(t_2)]$ must vanish, for otherwise the process would provide information about the phase of the carrier, and it would appear as though the process were governed by an inherent "clock" with which we could synchronize our own clocks. The variance of the process, for instance, would have the form

$$\text{Var } n(t) = E\{[n(t)]\}^2 = \tfrac{1}{2}E\{|N(t)|^2\} + \Sigma \cos(2\Omega t + \theta), \qquad \Sigma = \tfrac{1}{2}|E\{[N(t)]^2\}|,$$

and the variance would pulsate at frequency 2Ω with a determinate phase θ. When we assert that

$$E[N(t_1)N(t_2)] \equiv 0, \qquad \forall(t_1, t_2), \tag{3-31}$$

we are denying the existence of a clock inherent in the ensemble of temporal functions $n(t)$ that, equipped with a probability measure, defines the random process. We assume that for each realization $\text{Re}[N'(t) \exp i\Omega t]$ in the ensemble, the ensemble also contains the process $\text{Re}[N'(t) \exp i\psi \exp i\Omega t]$ for all possible phases ψ in $(0, 2\pi)$, no phase ψ being preferred over any other. This means that when we form the expected value in (3-31), we are forming

$$E[N'(t_1)N'(t_2)\, e^{2i\psi}]$$

and taking an average over a phase ψ uniformly distributed over $(0, 2\pi)$, in addition to the average over the totality of complex envelopes $N'(t)$. Because

$$E(e^{2i\psi}) = E(\cos 2\psi) + iE(\sin 2\psi) = 0,$$

that expected value must vanish as in (3-31).

The second expected value in (3-30) is

$$\tfrac{1}{2}E[N(t_1)N^*(t_2)] = \tfrac{1}{2}E[N'(t_1)N'^*(t_2)]$$

by the same reasoning, and the phase ψ drops out. We designate this expected value, which does not vanish in general, by

$$\tilde{\phi}(t_1, t_2) = \tfrac{1}{2}E[N(t_1)N^*(t_2)], \tag{3-32}$$

and we call it the complex autocovariance function of the nonstationary complex random process $N(t)$. In terms of it the autocovariance function of the quasiharmonic random process $n(t)$ is

$$\phi(t_1, t_2) = \text{Re}[\tilde{\phi}(t_1, t_2) \exp i\Omega(t_1 - t_2)]$$

from (3-30). The complex autocovariance function $\tilde{\phi}(t_1, t_2)$ possesses the Hermitian property

$$\tilde{\phi}(t_1, t_2) = \tilde{\phi}^*(t_2, t_1), \tag{3-33}$$

which follows immediately from its definition (3-32).

The autocovariance and cross-covariance functions of the quadrature components $X(t)$ and $Y(t)$ defined by (3-26) are

$$E[X(t_1)X(t_2)] = E[Y(t_1)Y(t_2)] = \text{Re } \tilde{\phi}(t_1, t_2),$$
$$E[Y(t_1)X(t_2)] = -E[X(t_1)Y(t_2)] = \text{Im } \tilde{\phi}(t_1, t_2). \tag{3-34}$$

Unless the complex autocovariance function $\tilde{\phi}(t_1, t_2)$ happens to be real, the real and imaginary parts of the complex envelope $N(t)$ are correlated random processes, except for their values observed at the same time $t_1 = t_2$.

The complex autocovariance function $\tilde{\phi}(t_1, t_2)$ of a narrowband random process $n(t) = \mathrm{Re}\, N(t) \exp i\Omega t$ is nonnegative definite because for any complex function $g(t)$,

$$\tfrac{1}{2}E \left| \int_0^T g^*(t)N(t)\, dt \right|^2 \geq 0,$$

so that

$$\tfrac{1}{2}E \int_0^T \int_0^T g^*(t_1)N(t_1)N^*(t_2)g(t_2)\, dt_1\, dt_2 = \int_0^T \int_0^T g^*(t_1)\tilde{\phi}(t_1, t_2)g(t_2)\, dt_1\, dt_2 \geq 0,$$

which is the criterion that the kernel $\tilde{\phi}(t_1, t_2)$ of such a quadratic form be nonnegative definite. We shall assume, as in (2-40), that $\tilde{\phi}(t_1, t_2)$ is positive definite, so that for no function $g(t)$ except $g(t) \equiv 0$ does

$$E \left| \int_0^T g^*(t)N(t)\, dt \right|^2$$

vanish.

When the quasiharmonic process $n(t)$ is stationary, its complex autocovariance function $\tilde{\phi}(t_1, t_2)$ depends on the times t_1 and t_2 only through their difference $\tau = t_1 - t_2$,

$$\tilde{\phi}(t_1, t_2) \rightarrow \tilde{\phi}(t_1 - t_2) = \tilde{\phi}(\tau),$$

and this complex autocovariance function $\tilde{\phi}(\tau)$ is the same as that defined by (3-19) in terms of the narrowband spectral density $\tilde{\Phi}(\omega)$ of the process. The average power in the process

$$n(t) = X(t) \cos \Omega t - Y(t) \sin \Omega t$$
$$= X'(t) \cos(\Omega t + \psi) - Y'(t) \sin(\Omega t + \psi), \qquad N'(t) = X'(t) + iY'(t),$$

is

$$E[n(t)]^2 = \tilde{\phi}(0) = \tfrac{1}{2}E[|N(t)|^2] = \tfrac{1}{2}\{E[X(t)]^2 + E[Y(t)]^2\}.$$

Remembering that all phases ψ are equally likely, we can think of the $\frac{1}{2}$ appearing here as representing the averages of the factors $\cos^2(\Omega t + \psi)$ and $\sin^2(\Omega t + \psi)$ that figure in $[n(t)]^2$.

The Hermitian property (3-33) of complex autocovariances implies that

$$\tilde{\phi}(-\tau) = \tilde{\phi}^*(\tau).$$

The real part of $\tilde{\phi}(\tau) = \tilde{\phi}_x(\tau) + i\tilde{\phi}_y(\tau)$ is therefore an even function and the imaginary part an odd function:

$$\tilde{\phi}_x(-\tau) = \tilde{\phi}_x(\tau), \qquad \tilde{\phi}_y(-\tau) = -\tilde{\phi}_y(\tau).$$

As a quasiharmonic function, the random process $n(t)$ must have an amplitude modulation $|N(t)|$ and a phase modulation $\arg N(t)$ of the same kind as described in Sec. 3.1. From the statistics of $n(t)$ one can determine the probability distributions of the outputs of rectifiers, discriminators, and other devices whose inputs are narrowband signals and noise. The statistical properties of the envelope of narrowband noise have been treated in [Ric44], [Mid48], [Bun49], [Are57], [Dug58], and [Ree62]. When the noise is both narrowband and Gaussian, its joint probability density functions can be put into an especially simple form that facilitates calculations. We turn now to developing it.

3.2.3 Circular Gaussian Random Processes

Just as in Chapter 2 an ordinary random process was sampled by means of the coefficients in its Fourier expansion in terms of a set of orthonormal functions, so can we sample a narrowband random process $n(t)$ by a similar expansion of its complex envelope $N(t)$,

$$N(t) = \sum_{k=1}^{\infty} z_k f_k(t),$$

in which the functions $f_k(t)$ may be complex and are orthonormal in the sense of (2-39):

$$\int_0^T f_m^*(t) f_k(t)\, dt = \delta_{km}.$$

The complex samples of the envelope are then

$$z_k = x_k + i y_k = \int_0^T f_k^*(t) N(t)\, dt. \tag{3-35}$$

From (3-32) we obtain the complex autocovariance matrix $\tilde{\boldsymbol{\phi}}$ of the samples z_k; its elements are

$$\begin{aligned}
\tilde{\phi}_{km} &= \tfrac{1}{2} E(z_k z_m^*) = \tfrac{1}{2} \int_0^T \int_0^T f_k^*(t) E[N(t) N^*(s)] f_m(s)\, dt\, ds \\
&= \int_0^T \int_0^T f_k^*(t) \tilde{\phi}(t,s) f_m(s)\, dt\, ds.
\end{aligned} \tag{3-36}$$

The matrix $\tilde{\boldsymbol{\phi}}$ is Hermitian,

$$\tilde{\boldsymbol{\phi}}^+ = \tilde{\boldsymbol{\phi}},$$

where $\tilde{\boldsymbol{\phi}}^+$ indicates a matrix derived from $\tilde{\boldsymbol{\phi}}$ by transposing rows and columns and taking the complex conjugate of each element; the km element of $\tilde{\boldsymbol{\phi}}^+$ is $\tilde{\phi}_{mk}^*$, and by (3-33) and (3-36)

$$\tilde{\phi}_{mk}^* = \tilde{\phi}_{km}.$$

Furthermore, because of (3-31),

$$E(z_k z_m) = E(z_k^* z_m^*) \equiv 0, \qquad \forall (k,m).$$

From these relations we find the counterparts of (3-34):

$$\begin{aligned}
E(x_k x_m) &= E(y_k y_m) = \operatorname{Re} \tilde{\phi}_{km} = \tilde{\phi}_{x,km}, \\
E(y_k x_m) &= -E(x_k y_m) = \operatorname{Im} \tilde{\phi}_{km} = \tilde{\phi}_{y,km},
\end{aligned} \tag{3-37}$$

where $\tilde{\phi} = \tilde{\phi}_x + i\tilde{\phi}_y$. Thus the $2n \times 2n$ covariance matrix $\tilde{\Phi}$ of the $2n$ random variables $x_1, x_2, \ldots, x_n, y_1, y_2, \ldots, y_n$ can be divided into $n \times n$ blocks

$$\tilde{\Phi} = \begin{bmatrix} \tilde{\phi}_x & -\tilde{\phi}_y \\ \tilde{\phi}_y & \tilde{\phi}_x \end{bmatrix}. \tag{3-38}$$

In particular $\tilde{\phi}_{y,kk} = E(x_k y_k) \equiv 0$, and

$$E(x_k^2) = E(y_k^2) = \tfrac{1}{2}E(x_k^2 + y_k^2) = \tfrac{1}{2}E[|z_k|^2] = \tilde{\phi}_{kk} = \tilde{\phi}_{x,kk}.$$

For instance, to the 4×4 matrix

$$\tilde{\Phi} = \begin{bmatrix} 4 & 2 & 0 & -1 \\ 2 & 3 & 1 & 0 \\ 0 & 1 & 4 & 2 \\ -1 & 0 & 2 & 3 \end{bmatrix}$$

with

$$\tilde{\phi}_x = \begin{bmatrix} 4 & 2 \\ 2 & 3 \end{bmatrix}, \qquad \tilde{\phi}_y = \begin{bmatrix} 0 & 1 \\ -1 & 0 \end{bmatrix},$$

corresponds the 2×2 Hermitian matrix

$$\tilde{\phi} = \tilde{\phi}_x + i\tilde{\phi}_y = \begin{bmatrix} 4 & 2+i \\ 2-i & 3 \end{bmatrix}.$$

Its inverse is

$$\tilde{\phi}^{-1} = \tfrac{1}{7}\begin{bmatrix} 3 & -2-i \\ -2+i & 4 \end{bmatrix},$$

from which we can write down the inverse of the 4×4 matrix $\tilde{\Phi}$:

$$\tilde{\Phi}^{-1} = \tfrac{1}{7}\begin{bmatrix} 3 & -2 & 0 & 1 \\ -2 & 4 & -1 & 0 \\ 0 & -1 & 3 & -2 \\ 1 & 0 & -2 & 4 \end{bmatrix}.$$

When the narrowband noise $n(t)$ is Gaussian, the random variables x_k and y_k, $k = 1, 2, \ldots, n$, have a jointly Gaussian probability density function that is determined entirely by the covariance matrix $\tilde{\Phi}$ defined through (3-34), (3-37), and (3-38). In particular, the bivariate probability density function of the real and imaginary parts of z_k is

$$\begin{aligned} p(x_k, y_k) &= \frac{1}{2\pi\tilde{\phi}_{kk}} \exp\left[-\frac{x_k^2 + y_k^2}{2\tilde{\phi}_{kk}}\right] \\ &= \frac{1}{2\pi\tilde{\phi}_{kk}} \exp\left[-\frac{|z_k|^2}{2\tilde{\phi}_{kk}}\right] = \tilde{p}(z_k). \end{aligned} \tag{3-39}$$

Because of the circular symmetry of this probability density function $p(x_k, y_k)$, the complex random variables $z_k = x_k + iy_k$ are called *circular* complex Gaussian random variables.

The joint probability density function of the real and imaginary parts of n of these complex random variables has the *circular Gaussian* form

$$p(x_1, \ldots, x_n, y_1, \ldots, y_n) = \tilde{p}(z_1, z_2, \ldots, z_n)$$
$$= (2\pi)^{-n}(\det \tilde{\boldsymbol{\phi}})^{-1} \exp(-\tfrac{1}{2}\mathbf{z}^+ \tilde{\boldsymbol{\phi}}^{-1} \mathbf{z})$$
$$= (2\pi)^{-n}(\det \tilde{\boldsymbol{\phi}})^{-1} \exp\left[-\tfrac{1}{2}\sum_{i=1}^{n}\sum_{j=1}^{n} z_i^* \tilde{\mu}_{ij} z_j\right], \tag{3-40}$$

which is shown in Appendix B to be equivalent to the usual Gaussian density function involving the $2n \times 2n$ matrix $\tilde{\boldsymbol{\Phi}}$ of (3-38). Here $\tilde{\mu}_{ij}$ are the elements of the inverse $\tilde{\boldsymbol{\phi}}^{-1} = \tilde{\boldsymbol{\mu}}$ of the Hermitian matrix $\tilde{\boldsymbol{\phi}} = \tilde{\boldsymbol{\phi}}_x + i\tilde{\boldsymbol{\phi}}_y$; $\mathbf{z}$ is an n-element column vector of the complex numbers $z_1, z_2, \ldots, z_n$, and

$$\mathbf{z}^+ = (z_1^*, z_2^*, \ldots, z_n^*)$$

is its transposed conjugate row vector. Furthermore

$$(\det \tilde{\boldsymbol{\phi}})^2 = \det \tilde{\boldsymbol{\Phi}} = \det\begin{bmatrix} \tilde{\boldsymbol{\phi}}_x & -\tilde{\boldsymbol{\phi}}_y \\ \tilde{\boldsymbol{\phi}}_y & \tilde{\boldsymbol{\phi}}_x \end{bmatrix}$$

is the square of the determinant of the $n \times n$ Hermitian matrix $\tilde{\boldsymbol{\phi}}$. In our example, for instance, $\det \tilde{\boldsymbol{\phi}} = 7$ and $\det \tilde{\boldsymbol{\Phi}} = 49$.

Equation (3-40) is only a concise way of writing the joint probability density function of the $2n$ random variables $x_1, x_2, \ldots, x_n, y_1, y_2, \ldots, y_n$; it should not be considered as the probability density function of $z_1, z_2, \ldots, z_n$. When the real and imaginary parts of the samples z_k of the complex random process $N(t)$, defined as in (3-35), possess such a circular Gaussian probability density function, $N(t)$ is said to be a circular Gaussian random process, and the complex envelope of narrowband Gaussian noise is a process of this kind.

When a narrowband signal $s(t) = \mathrm{Re}\, S(t) \exp i\Omega t$ is present in addition to this narrowband Gaussian noise, the complex samples z_k have expected values that are not zero, but are given by

$$E(z_k) = S_k = \int_0^T f_k^*(t) S(t)\, dt,$$

and the joint probability density function of the samples x_k, y_k becomes, in place of (3-40)

$$\tilde{p}(z_1, z_2, \ldots, z_n) = (2\pi)^{-n}(\det \tilde{\boldsymbol{\phi}})^{-1} \exp\left[-\tfrac{1}{2}\sum_{i=1}^{n}\sum_{j=1}^{n}(z_i^* - S_i^*)\tilde{\mu}_{ij}(z_j - S_j)\right]. \tag{3-41}$$

The joint characteristic function of the $2n$ random variables $x_1, x_2, \ldots, x_n$, $y_1, y_2, \ldots, y_n$ can be written in a similarly concise form. It is the $2n$-dimensional Fourier transform of the joint density function in (3-41), and Appendix B shows that it can be expressed as

$$h(u_{x1}, \ldots, u_{xn}, u_{y1}, \ldots, u_{yn})$$
$$= \tilde{h}(w_1, \ldots, w_n; w_1^*, \ldots, w_n^*)$$
$$= E \exp\left[i \sum_{j=1}^{n}(u_{xj} x_j + u_{yj} y_j)\right] \tag{3-42}$$

$$= \exp\left[\frac{1}{2} i \sum_{j=1}^{n} (w_j^* S_j + w_j S_j^*) - \frac{1}{2} \sum_{k=1}^{n} \sum_{m=1}^{n} w_k^* \tilde{\phi}_{km} w_m \right]$$

with $w_k = u_{xk} + i u_{yk}$.

By writing

$$\tilde{h}(w_1, \dots, w_n; w_1^*, \dots, w_n^*) = E \exp\left[\frac{1}{2} i \sum_{j=1}^{n} (w_j^* z_j + w_j z_j^*) \right] \qquad (3\text{-}43)$$

and regarding w_j and w_j^* as mathematically distinct variables, we can derive the covariances of the samples of a circular Gaussian process by differentiating (3-42). Thus if we take $S_j \equiv 0$ for simplicity, we find, with all partial derivatives evaluated at $w_j \equiv 0, j = 1, 2, \dots, n$,

$$\frac{1}{2} E(z_k z_m^*) = -2 \frac{\partial^2 h}{\partial w_k^* \, \partial w_m} \bigg|_{w \equiv 0} = \tilde{\phi}_{km}$$

as in (3-36). Furthermore,

$$\frac{1}{4} E(z_1^* z_2^* z_3 z_4) = 4 \frac{\partial^4 h}{\partial w_1 \partial w_2 \partial w_3^* \partial w_4^*} \bigg|_{w \equiv 0} \qquad (3\text{-}44)$$
$$= \tilde{\phi}_{31} \tilde{\phi}_{42} + \tilde{\phi}_{32} \tilde{\phi}_{41} = \frac{1}{4} E(z_3 z_1^*) E(z_4 z_2^*) + \frac{1}{4} E(z_3 z_2^*) E(z_4 z_1^*).$$

Formulas for moments of higher order have been given by Reed [Ree62]; they follow the same pattern as (3-44). With $S_k \equiv 0$, the expected value of a product vanishes unless the numbers of starred and unstarred factors are equal. When those numbers are equal, one forms all possible pairs of starred and unstarred z's, as in (3-44), multiplies the expected values of the paired products, and adds.

3.2.4 Narrowband White Noise

Ordinary white noise cannot be expressed directly in terms of quadrature components, for its spectral density occupies much too wide a range of frequencies. In analyzing the detection of a narrowband signal in white noise, however, it would be convenient to represent the input noise in this way. As we said at the beginning of this section, we can always assume that the signal and the noise have passed through a filter whose passband includes the spectrum of the signal, but is much wider. Usually it will be possible to treat this new filter as narrowband and represent its transfer function as in (3-12). It can have little effect on signal detectability, however, for it cuts out only noise components with frequencies far from those of the signal, affecting the signal hardly at all. Then one can conveniently write the white noise $n(t)$ as

$$n(t) = \text{Re } N(t) \, e^{i\Omega t}, \qquad N(t) = X(t) + iY(t),$$

with

$$E[X(t_1)X(t_2)] = E[Y(t_1)Y(t_2)] = N\delta(t_1 - t_2),$$

$$E[X(t_1)Y(t_2)] = 0,$$

$$\tfrac{1}{2}E[N(t_1)N^*(t_2)] = N\delta(t_1 - t_2), \qquad E[N(t_1)N(t_2)] = 0, \qquad (3\text{-}45)$$

$$\tilde{\Phi}(\omega) \equiv \frac{N}{2}, \qquad \tilde{\phi}(\tau) = N\delta(\tau),$$

where N is the unilateral spectral density of the white noise. These relations are consistent with the definition in (3-19) and with (3-23) and (3-24), in which both $\Phi_i(\omega)$ and $\tilde{\Phi}_i(\omega)$ can be set equal to $N/2$.

There is no formal difficulty with regarding white noise as narrowband noise whose spectral density is uniform over a range of frequencies—usually those of a quasiharmonic signal—of interest in many detection problems, and a considerable simplification follows from this viewpoint. Samples x_k and y_k, $1 \le k \le n$, of white noise, defined as in (3-35), are statistically independent, and by (3-36) and (3-45) their joint probability density function has the circular-complex Gaussian form

$$\tilde{p}(z_1, z_2, \ldots, z_n) = (2\pi N)^{-n} \exp\left[-\frac{1}{2N} \sum_{j=1}^{n} |z_j|^2\right]. \qquad (3\text{-}46)$$

The Hermitian quadratic form in the exponent of (3-40) is now simply the sum of the absolute squares of the z_j's, divided by N.

3.3 DETECTION OF A SIGNAL OF RANDOM PHASE

3.3.1 The Likelihood Functional for a Narrowband Input

A narrowband signal
$$s(t) = \operatorname{Re} S(t)\, e^{i\Omega t}$$

with complex envelope $S(t)$ is to be detected in the presence of Gaussian noise $n(t)$ that for the reasons mentioned at the beginning of Sec. 3.2 we can also presume to be narrowband:
$$n(t) = \operatorname{Re} N(t)\, e^{i\Omega t}.$$

Its complex envelope $N(t)$ is a circular Gaussian random process with complex auto-covariance function $\tilde{\phi}(t_1, t_2)$ as in (3-32). When as here the input $v(t) = \operatorname{Re} V(t) \exp i\Omega t$ is narrowband, both the real and imaginary parts of its complex envelope $V(t)$ can be measured separately by mixing the input with $2\cos\Omega t$ and $2\sin\Omega t$, respectively, as described in Sec. 3.1 and illustrated in Fig. 3-2. We can therefore assume that the complex envelope $V(t)$ itself is available to the receiver and that the receiver will base its decision between the two hypotheses

$$H_0:\ V(t) = N(t),$$
$$H_1:\ V(t) = S(t) + N(t), \qquad 0 \le t \le T,$$

on the complex envelope $V(t)$.

The complex envelope $V(t)$ is sampled by determining the complex coefficients V_k of its Fourier expansion in terms of a set of functions $f_k(t)$ orthonormal over the observation interval $(0, T)$ as in (2-39):

$$V_k = \int_0^T f_k^*(t) V(t)\, dt;$$

compare (3-35). Under hypothesis H_0 the real and imaginary parts of V_k are Gaussian random variables with expected values 0 and a joint probability density function of the circular Gaussian form in (3-40). For simplicity we take the functions $f_k(t)$ as the eigenfunctions of the complex autocovariance function $\tilde{\phi}(t_1, t_2)$ of the noise:

$$\lambda_k f_k(t) = \int_0^T \tilde{\phi}(t, u) f_k(u)\, du.$$

Because as in (3-33) the complex autocovariance function is Hermitian, the eigenvalues λ_k are real; and because $\tilde{\phi}(t_1, t_2)$ is positive definite, they are positive. Then the covariance matrix $\tilde{\boldsymbol{\phi}}$ of the samples $V_1, V_2, \ldots, V_n$ is diagonal,

$$\tilde{\phi}_{km} = \lambda_k \delta_{km},$$

as is its inverse $\tilde{\boldsymbol{\phi}}^{-1} = \tilde{\boldsymbol{\mu}}$ figuring in (3-40); and the joint probability density function of the real and imaginary parts of the first n complex samples $V_1, V_2, \ldots, V_n$ takes the form

$$\tilde{p}_0(\mathbf{V}) = \prod_{k=1}^{n} \frac{1}{2\pi\lambda_k} \exp\left[-\frac{|V_k|^2}{2\lambda_k} \right]; \tag{3-47}$$

$\mathbf{V}$ stands for the set of n circular complex Gaussian random variables $V_1, V_2, \ldots, V_n$. Under hypothesis H_1 the expected value of the kth complex sample V_k is

$$E(V_k \mid H_1) = S_k = \int_0^T f_k^*(t) S(t)\, dt, \tag{3-48}$$

where $S(t)$ is the complex envelope of the signal. The covariance matrix of the samples and its inverse are unchanged, and the joint probability density function of the real and imaginary parts of the samples V_k is, as in (3-41),

$$\tilde{p}_1(\mathbf{V}) = \prod_{k=1}^{n} \frac{1}{2\pi\lambda_k} \exp\left[-\frac{|V_k - S_k|^2}{2\lambda_k} \right]. \tag{3-49}$$

The decision between the two hypotheses is optimally based on the likelihood ratio

$$\Lambda_n(\mathbf{V}) = \frac{\tilde{p}_1(\mathbf{V})}{\tilde{p}_0(\mathbf{V})} = \prod_{k=1}^{n} \exp\left[\frac{|V_k|^2 - |V_k - S_k|^2}{2\lambda_k} \right]$$

$$= \exp\left[\operatorname{Re} \sum_{k=1}^{n} \frac{S_k^* V_k}{\lambda_k} - \frac{1}{2}\sum_{k=1}^{n} \frac{|S_k|^2}{\lambda_k} \right]. \tag{3-50}$$

The data $V_1, V_2, \ldots, V_n$ appear only in the first summation, whose real part is a sufficient statistic when only n samples are included. Again we utilize all the data in the input $v(t)$ by passing to the limit $n \to \infty$, and the decision is based on the sufficient statistic

$$G = \text{Re} \sum_{k=1}^{\infty} \frac{S_k^* V_k}{\lambda_k} = \text{Re} \sum_{k=1}^{\infty} Q_k^* V_k,$$

with $Q_k = S_k/\lambda_k$. By the same procedure as we used in Sec. 2.2, we can write our sufficient statistic as

$$G = \text{Re} \int_0^T Q^*(t) V(t) \, dt, \tag{3-51}$$

where $Q(t)$ is the solution of the integral equation

$$S(t) = \int_0^T \tilde{\phi}(t, u) Q(u) \, du, \qquad 0 \le t \le T, \tag{3-52}$$

whose kernel is the complex autocovariance function of the narrowband noise. The signal-to-noise ratio is now

$$d^2 = \sum_{k=1}^{\infty} \frac{|S_k|^2}{\lambda_k} = \sum_{k=1}^{\infty} S_k^* Q_k = \int_0^T S^*(t) Q(t) \, dt, \tag{3-53}$$

and the likelihood functional in terms of the complex envelope $V(t)$ of the input is determined by taking (3-50) to the limit $n \to \infty$:

$$\Lambda[V(t)] = \exp(G - \tfrac{1}{2} d^2)$$
$$= \exp \left[\text{Re} \int_0^T Q^*(t) V(t) \, dt - \tfrac{1}{2} \int_0^T S^*(t) Q(t) \, dt \right]. \tag{3-54}$$

When the noise is white, $Q(t) = S(t)/N$ by (3-45).

The matched filter is now a narrowband filter with complex impulse response

$$K(\tau) = \begin{cases} Q^*(T - \tau), & 0 \le \tau \le T, \\ 0, & \tau < 0, \qquad \tau > T. \end{cases} \tag{3-55}$$

The output of this filter at time t is a narrowband random process $v_0(t) = \text{Re } V_0(t) \exp i\Omega t$ whose complex envelope, as in (3-16), is

$$V_0(t) = X_0(t) + i Y_0(t) = \int_0^t K(\tau) V(t - \tau) \, d\tau$$
$$= \int_0^t Q^*(T - \tau) V(t - \tau) \, d\tau = \int_0^t Q^*(T - t + u) V(u) \, du,$$

and the statistic G on which the decision is based is obtained by sampling the real part $X_0(t) = \text{Re } V_0(t)$ of this output at the end of the observation interval,

$$G = X_0(T) = \text{Re } V_0(T).$$

The real part $X_0(t)$ can be obtained by mixing the output of the filter with a locally generated signal $2 \cos \Omega t$.

3.3.2 Signals of Random Phase

The detection strategy developed in Sec. 3.3.1 requires that the phase of the carrier of the signal be known precisely. If that phase is in error by ψ, the signal, instead of being Re $S(t) \exp i\Omega t$, is

$$s(t; \psi) = \text{Re } S(t) \, e^{i\Omega t + i\psi}. \qquad (3\text{-}56)$$

The signal component of the statistic G under hypothesis H_1 will then be not d^2, but

$$E[G \mid H_1, \psi] = \text{Re } e^{i\psi} \int_0^T Q^*(t) S(t) \, dt = d^2 \cos \psi.$$

If the phase error $|\psi|$ is more than $90°$, this component will be negative, and the probability of detecting the signal will be less than the false-alarm probability Q_0.

Consider a radar system set up to determine whether a target is present at a certain distance from its transmitter. A typical signal has the form

$$s(t) = \text{Re } F(t - t_0) \exp i\Omega(t - t_0),$$

where $F(t)$ is the complex envelope of the transmitted pulse, Ω is the carrier frequency, and t_0 is the time when some distinguishing point of the echo from the target reaches the receiver. For narrowband signals the carrier frequency is so large that many cycles of the carrier occur within the duration of the pulse. The time t_0 is proportional to the distance from the target to the receiving antenna. A small change in t_0 on the order of $(1/\Omega)$, corresponding to an alteration in the distance of the target on the order of a wavelength of the radiation, makes a very small change in the envelope $F(t - t_0)$, but a large change in the carrier phase $(-\Omega t_0)$. [A typical value of the pulse duration is 10^{-6} sec; $(1/\Omega)$ may be on the order of 10^{-9} sec.] The distance to the target will seldom be specified within a fraction of a wavelength of the radiation, and hence not precisely enough to determine the phase $\psi = -\Omega t_0$ of the received echo within a fraction of 2π, although one may be able to time its arrival within a small fraction of the width of the pulse envelope $|F(t)|$. In a communication system the phase ψ of the carrier of the signal may also be uncertain by a large multiple of 2π unless the distance from transmitter to receiver is known within a fraction of a wavelength or the phase of the carrier has been tracked since the inception of transmission by some device such as a phase-lock loop.

When the phase ψ of the carrier is uncertain by a considerable fraction of 2π or more, the observer generally has no reason to assign to it one value rather than another. In effect the receiver is called on to detect any one of a class of narrowband signals $s(t; \psi)$ as in (3-56), with the phase ψ a random variable uniformly distributed over $(0, 2\pi)$. This is our first example of the detection of a signal one or more of whose parameters are unknown. A hypothesis of the form "A signal from a class of signals having parameter values lying in a certain range is present" is known as a *composite* hypothesis.

When the complex envelope of the signal is $S(t) \exp i\psi$ instead of $S(t)$, the signal samples S_k in (3-48) must be replaced by $S_k \exp i\psi$, and the joint probability density function of the real and imaginary parts of the complex samples $V_1, V_2, \dots, V_n$ under the now composite hypothesis H_1 becomes, in place of (3-49),

$$\tilde{p}_1(\mathbf{V}; \psi) = \prod_{k=1}^{n} (2\pi\lambda_k)^{-1} \exp\left[-\frac{|V_k - S_k e^{i\psi}|^2}{2\lambda_k}\right]. \qquad (3\text{-}57)$$

The actual joint probability density function of these data, taking into account the randomness of the phase ψ, is obtained by averaging (3-57) over $0 \le \psi < 2\pi$:

$$\tilde{p}_1(\mathbf{V}) = \int_0^{2\pi} \tilde{p}_1(\mathbf{V}; \psi)\frac{d\psi}{2\pi}. \qquad (3\text{-}58)$$

Under hypothesis H_0 the joint probability density function of the data remains $\tilde{p}_0(\mathbf{V})$ as in (3-47). The likelihood ratio on which the decision is based is then not (3-50), but in the limit $n \to \infty$

$$\Lambda(\mathbf{V}) = \lim_{n\to\infty} \int_0^{2\pi} \frac{\tilde{p}_1(\mathbf{V}; \psi)}{\tilde{p}_0(\mathbf{V})} \frac{d\psi}{2\pi} = \int_0^{2\pi} \Lambda[V(t)| \psi]\frac{d\psi}{2\pi},$$

where $\Lambda[V(t)| \psi]$ is a likelihood functional obtained from that in (3-54) by replacing $S(t)$ by $S(t) \exp i\psi$. By (3-52) $Q(t)$ must be replaced by $Q(t) \exp i\psi$, and the likelihood functional becomes, in place of (3-54),

$$\Lambda[V(t)] = \int_0^{2\pi} \exp\left[\operatorname{Re} e^{-i\psi} \int_0^T Q^*(t)V(t)\, dt - \tfrac{1}{2}\int_0^T S^*(t)Q(t)\, dt\right]\frac{d\psi}{2\pi}.$$

In order to evaluate this average over the phase ψ, we put

$$\int_0^T Q^*(t)V(t)\, dt = R e^{i\theta}, \qquad R = \left|\int_0^T Q^*(t)V(t)\, dt\right|. \qquad (3\text{-}59)$$

Then with (3-53),

$$\Lambda[V(t)] = \int_0^{2\pi} \exp[R \cos(\theta - \psi) - \tfrac{1}{2}d^2]\frac{d\psi}{2\pi} \qquad (3\text{-}60)$$
$$= e^{-d^2/2}I_0(R),$$

where

$$I_0(x) = \int_0^{2\pi} e^{x \cos\phi}\, \frac{d\phi}{2\pi} = \sum_{k=0}^{\infty} \frac{(\tfrac{1}{2}x)^{2k}}{(k!)^2} \qquad (3\text{-}61)$$

is the modified Bessel function of order zero.

The data $V(t)$ now appear in the likelihood functional only through the quantity $I_0(R)$, which is a monotone function of R; and R, defined in (3-59), becomes a sufficient statistic for deciding between hypothesis H_0, "No signal is present," and the composite hypothesis H_1, "A signal $s(t; \psi)$ is present with a phase ψ just as likely to have one value as another in $(0, 2\pi)$." The receiver chooses hypothesis H_1 if $R \ge R_0$ and H_0 if $R < R_0$, for some decision level R_0. Under the Bayes criterion,

$$e^{-d^2/2} I_0(R_0) = \Lambda_0,$$

with Λ_0 given by (1-17). Under the Neyman–Pearson criterion the value of R_0 is chosen so that the false-alarm probability

$$Q_0 = \Pr(R > R_0| H_0) = \int_{R_0}^{\infty} P_0(R)\, dR \qquad (3\text{-}62)$$

equals a preassigned value, $P_0(R)$ being the probability density function of the statistic R under hypothesis H_0. The probability $Q_d(\psi)$ of detecting a signal $s(t; \psi)$ with a particular phase ψ is

$$Q_d(\psi) = \Pr(R > R_0 \mid H_1, \psi) = \int_{R_0}^{\infty} P_1(R; \psi)\, dR, \qquad (3\text{-}63)$$

where $P_1(R; \psi)$ is the probability density function of R when the signal $s(t; \psi)$ of (3-56) is present in the input $v(t)$. In the next section we shall calculate these probabilities.

The statistic R can be determined by passing the input $v(t)$ through the narrowband matched filter of (3-55) and rectifying its output $v_0(t) = \mathrm{Re}\, V_0(t) \exp i\Omega t$. A linear rectifier produces $|V_0(t)|$, and the value of this at time $t = T$ is the statistic $R = |V_0(T)|$ required.

If the noise $n(t)$ is the result of passing white noise through an input filter whose passband is broad enough not to distort the signal, we can assume that $n(t)$ is "narrowband white noise" of the type discussed in Sec. 3.2.4. Its complex autocovariance function is

$$\tilde{\phi}(t, u) = N\delta(t - u)$$

as in (3-45), and when this is taken as the kernel of the integral equation (3-52), its solution is $Q(t) = S(t)/N$, whereupon the test statistic R becomes

$$R = \frac{1}{N}\left| \int_0^T S^*(t)V(t)\, dt \right|. \qquad (3\text{-}64)$$

It is proportional to the rectified output, at the end of the observation interval, of a filter matched to any of the expected signals $s(t; \psi)$. The average likelihood functional for detecting a signal of unknown phase in white noise is that in (3-60) with R now given by (3-64) and with the signal-to-noise ratio now

$$d^2 = \frac{1}{N}\int_0^T |S(t)|^2\, dt = \frac{2E}{N}. \qquad (3\text{-}65)$$

For narrowband signals the quantity d^2 in (3-53) or (3-65) is the same as that defined in (2-67) or (2-75). To show this, put

$$s(t) = \mathrm{Re}\, S(t)\, e^{i\Omega t} \quad \text{and} \quad q(t) = 2\, \mathrm{Re}\, Q(t)\, e^{i\Omega t}$$

into (2-67), expressing them in terms of their real and imaginary parts. The integrand will be found to have a group of terms with the factors $\cos 2\Omega t$ or $\sin 2\Omega t$; these oscillate much more rapidly than the other terms. With $\Omega T \gg 1$ in the quasiharmonic approximation, that group of terms contributes negligibly after integration, and we obtain the formula in (3-53) for d^2. When the noise is white, that reduces to (3-65). We are justified, therefore, in calling the quantity d^2 in (3-53) the signal-to-noise ratio for the detection of narrowband signals. When the noise is white, it becomes the familiar $2E/N$, where E is the energy of the signal and N the unilateral spectral density of the noise.

3.4 THE DETECTABILITY OF SIGNALS OF UNKNOWN PHASE

The performance of the receiver derived in the Sec. 3.3 can be evaluated on the basis of its probabilities Q_0 of a false alarm and Q_d of detection. In that receiver hypothesis H_0 is chosen when the statistic R of (3-59) is less than the decision level R_0; H_1 is chosen when it is greater. The probabilities in question are given by (3-62) and (3-63), where $P_0(R)$ is the probability density function of the statistic R when the input consists of noise alone; $P_1(R; \psi)$ is its probability density function when the input consists of noise plus the signal $s(t; \psi)$ of phase ψ. In (3-63) we allow for the possibility that the probability of detecting the signal may depend on its phase.

To determine these probabilities we write the statistic R as

$$R = |z| = |x + iy| = (x^2 + y^2)^{1/2}, \tag{3-66}$$

where x and y are the real and imaginary parts of the complex random variable

$$z = x + iy = R\,e^{i\theta} = \int_0^T Q^*(t)V(t)\,dt. \tag{3-67}$$

These components x and y are Gaussian random variables, for they are the results of linear operations on the Gaussian random processes $X(t)$ and $Y(t)$, which are the quadrature components of the input, $v(t) = \mathrm{Re}\,V(t)\exp i\Omega t$, $V(t) = X(t) + iY(t)$. Under hypothesis H_0 their expected values are zero, for $E[V(t)|\,H_0] = 0$. Their covariances can be calculated in the following way. First averaging the absolute square of $(x + iy)$ and dividing by 2, we obtain

$$\tfrac{1}{2}E(|x + iy|^2|\,H_0) = \tfrac{1}{2}[E(x^2|\,H_0)] + \tfrac{1}{2}[E(y^2|\,H_0)] = \tfrac{1}{2}E(zz^*|\,H_0)$$

$$= \tfrac{1}{2}\int_0^T\!\!\int_0^T Q^*(t)Q(u)E[V(t)V^*(u)|\,H_0]\,dt\,du \tag{3-68}$$

$$= \int_0^T\!\!\int_0^T Q^*(t)Q(u)\tilde{\phi}(t, u)\,dt\,du = \int_0^T Q^*(t)S(t)\,dt = d^2;$$

here we have used (3-32), the integral equation (3-52), and the definition (3-53). Similarly, by (3-31),

$$E[(x + iy)^2|\,H_0] = E[x^2 - y^2 + 2ixy|\,H_0] = E(z^2|\,H_0)$$

$$= \int_0^T\!\!\int_0^T Q^*(t)Q^*(u)E[V(t)V(u)|\,H_0]\,dt\,du = 0,$$

which implies that the expected values of x^2 and y^2 are equal and that $E(xy|\,H_0) = 0$. From (3-68) it then follows that

$$\mathrm{Var}\,x = \mathrm{Var}\,y = d^2, \tag{3-69}$$

and $\mathrm{Cov}(x, y) = 0$. The components x and y, as Gaussian random variables, are therefore statistically independent, and their variances are equal to d^2. Under hypothesis H_0 their joint probability density function is thus

$$p_0(x, y) = \frac{1}{2\pi d^2}\exp\left[-\frac{x^2 + y^2}{2d^2}\right].$$

The probability Q_0 of a false alarm equals the probability that the point with coordinates (x, y) lies outside a circle of radius R_0, and it can be calculated by integrating the joint density function $p_0(x, y)$ over the exterior of that circle:

$$Q_0 = \Pr[x^2 + y^2 > R_0^2 | H_0] = \frac{1}{2\pi d^2} \int_{R > R_0} \int \exp\left[-\frac{x^2 + y^2}{2d^2}\right] dx\, dy$$

$$= \frac{1}{d^2} \int_{R_0}^{\infty} R\, e^{-R^2/2d^2}\, dR = \exp\left(-\frac{R_0^2}{2d^2}\right).$$

(3-70)

The integral was evaluated by changing to polar coordinates. Incidentally we have shown that the probability density function of the statistic R under hypothesis H_0 is

$$P_0(R) = \frac{R}{d^2} \exp\left(-\frac{R^2}{2d^2}\right) U(R),$$

(3-71)

which is known as the *Rayleigh distribution*.

When under hypothesis H_1 there is a signal $s(t; \psi)$ having a particular phase ψ present, the components x and y in (3-67) are again Gaussian random variables with covariances given by (3-69). The expected value of the complex envelope $V(t)$ of the input is now $S(t) \exp i\psi$, and the expected values of the components x and y are given by

$$E(x + iy | H_1, \psi) = e^{i\psi} \int_0^T Q^*(t) S(t)\, dt = d^2 e^{i\psi},$$

$$E(x | H_1, \psi) = d^2 \cos \psi, \qquad E(y | H_1, \psi) = d^2 \sin \psi.$$

The joint probability density function of x and y under hypothesis H_1 is therefore

$$p_1(x, y; \psi) = \frac{1}{2\pi d^2} \exp\left[-\frac{(x - d^2 \cos \psi)^2 + (y - d^2 \sin \psi)^2}{2d^2}\right]$$

$$= \frac{1}{2\pi d^2} \exp\left[-\frac{x^2 + y^2 - 2d^2(x \cos \psi + y \sin \psi) + d^4}{2d^2}\right].$$

The probability that the signal $s(t; \psi)$ is detected is the probability that the point with coordinates (x, y) lies outside the circle of radius R_0 under hypothesis H_1:

$$Q_d(\psi) = \Pr[x^2 + y^2 > R_0^2 | H_1] = \int_{R > R_0} \int p_1(x, y; \psi)\, dx\, dy.$$

To evaluate this integral, we introduce polar coordinates $x = R \cos \theta$, $y = R \sin \theta$. The element of area is $dx\, dy = R\, dR\, d\theta$. Using the definition (3-61) of the modified Bessel function $I_0(x)$, we find

$$Q_d(\psi) = \frac{1}{2\pi d^2} \int_{R_0}^{\infty} \int_0^{2\pi} R \exp\left[-\frac{R^2 - 2d^2 R \cos(\theta - \psi) + d^4}{2d^2}\right] dR\, d\theta$$

$$= \frac{1}{d^2} \int_{R_0}^{\infty} R \exp\left[-\frac{R^2 + d^4}{2d^2}\right] I_0(R)\, dR.$$

(3-72)

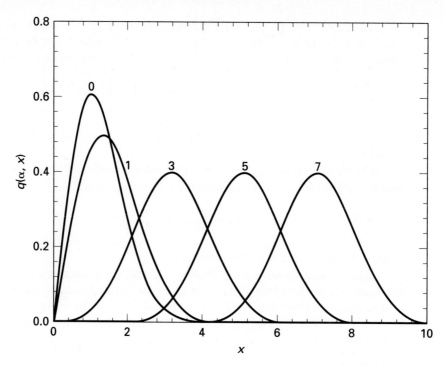

Figure 3-4. Probability density function $q(\alpha, x)$ of the output of the detector. Curves are indexed with the value of α.

The probability density function of the statistic R under hypothesis H_1 that one of the signals $s(t; \psi)$ is present is therefore

$$P_1(R; \psi) = \frac{1}{d} q\left(d, \frac{R}{d}\right),$$

$$q(\alpha, x) = x \, e^{-\frac{1}{2}(x^2 + \alpha^2)} I_0(\alpha x) U(x). \tag{3-73}$$

This density function $q(\alpha, x)$ has been plotted in Fig. 3-4 for a few values of the parameter α. The curve for $\alpha = 0$ represents the Rayleigh distribution; for $\alpha > 0$ (3-73) is called the *Rayleigh–Rice* or the *Rice–Nakagami* distribution. For large values of α the density function looks much like that for a Gaussian distribution. Indeed, with the asymptotic formula

$$I_0(x) \approx \frac{e^x}{\sqrt{2\pi x}}, \qquad x \gg 1, \tag{3-74}$$

for the modified Bessel function [Abr70, p. 377, eq. 9.7.1], the density function $q(\alpha, x)$ becomes

$$q(\alpha, x) \approx \left[\frac{x}{2\pi\alpha}\right]^{1/2} \exp[-\tfrac{1}{2}(x - \alpha)^2], \qquad \alpha x \gg 1.$$

Averaging the conditional probability density function $P_1(R; \psi)$ with respect to any distribution of the phase ψ yields the density function of the statistic R,

$$P_1(R) = \frac{R}{d^2} \exp\left[-\frac{R^2 + d^4}{2d^2}\right] I_0(R) U(R).$$

Dividing by the probability density function of R under hypothesis H_0, the Rayleigh distribution in (3-71), we obtain the likelihood ratio for the statistic R,

$$\Lambda(R) = e^{-d^2/2} I_0(R),$$

which agrees with the average likelihood functional $\Lambda[v(t)]$ in (3-60), as indeed it must, for R is a sufficient statistic.

The integral in (3-72) cannot be evaluated in closed form. We put

$$Q_d(\psi) = Q(d, R_0/d), \tag{3-75}$$

expressing it in terms of *Marcum's Q function*,

$$Q(\alpha, \beta) = \int_\beta^\infty x \, e^{-\frac{1}{2}(x^2 + \alpha^2)} I_0(\alpha x) \, dx. \tag{3-76}$$

This function has been extensively tabulated by Marcum [Mar50], and its properties have been studied by Rice [Ric44] and Marcum [Mar48]. In particular we note the initial values

$$Q(\alpha, 0) = 1, \qquad Q(0, \beta) = e^{-\beta^2/2}. \tag{3-77}$$

Various properties of the Q function and algorithms for computing it are to be found in Appendix C.

When the Neyman–Pearson criterion is being used, the decision level R_0 on the statistic R is picked so that the false-alarm probability Q_0 in (3-70) equals the preassigned value. The probability $Q_d(\psi)$ of detecting the signal $s(t; \psi)$ is then given by (3-75); it is independent of the phase ψ of the signal that happens to be present. This detection probability Q_d is a function of the parameter d in (3-53) or, when the noise is white, in (3-65); as before, we call d^2 the signal-to-noise ratio.

The signal-to-noise ratio d^2 required to attain a given probability Q_d of detection for a fixed false-alarm probability Q_0 is slightly larger when the signal phase ψ is unknown than when it is known. If we denote the former by d_1^2, the latter by d_0^2, the ratio d_1^2/d_0^2 measures how much larger the signal-to-noise ratio must be when the phase ψ is unknown than when it is known in order for the receiver to achieve the pair (Q_0, Q_d); we call this ratio the *loss* entailed by not knowing the phase ψ of the signal a priori. In Fig. 3-5 we plot this loss in decibels, that is, $10 \log_{10}(d_1^2/d_0^2)$, versus $\log_{10} Q_0$ for various values of the probability Q_d of detection. Over most of the range the loss does not exceed 1 dB.

3.5 NARROWBAND SIGNALS IN COMMUNICATIONS

3.5.1 The Binary Incoherent Channel

Digital communications most often utilize binary signals transmitted at a constant rate, and both the theory of coding binary data and that of detecting binary signals are among the most extensively developed parts of communication theory. Let us

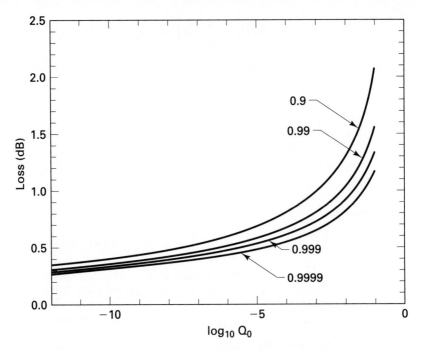

Figure 3-5. Signal-to-noise ratio loss in decibels entailed by ignorance of the phase of the signal to be detected, versus $\log_{10} Q_0$. The curves are indexed with the probability Q_d of detection.

suppose that messages to be dispatched have been coded into a stream of 0's and 1's. For each 0 the transmitter sends a signal received as $s_0(t)$ and for each 1 a signal received as $s_1(t)$. The relative frequencies of 0's and 1's are ζ_0 and ζ_1, respectively, and $\zeta_0 + \zeta_1 = 1$. One speaks then of communicating over a *binary channel*.

In treating the detection of these signals we suppose that the elements of the sequence of 0's and 1's that they represent are statistically independent. Each received signal is confined to an interval of T seconds' duration, and there is no intersymbol interference. With both the noise and the sequence of symbols taken as stationary random processes, all intervals are statistically alike, and we need to consider only a single interval $(0, T)$. On the basis of its input $v(t)$ during that interval, the receiver is to choose between two hypotheses, H_0, "Signal $s_0(t)$ was received," and H_1, "Signal $s_1(t)$ was received." When it selects H_0, the receiver issues a 0; when it selects H_1, a 1. The receiver is to be designed so that the relative frequency of errors in the stream of digits it puts out—that is, the probability of error in each decision—is as small as possible.

When the transmitted signals are narrowband pulse modulations of a carrier of frequency Ω, the received signals $s_0(t)$ and $s_1(t)$ under the two hypotheses can be taken to have the forms

$$s_j(t) = \text{Re } S_j(t) \exp(i\Omega t + i\psi_j), \qquad j = 0, 1. \tag{3-78}$$

Narrowband Signals and Their Detection   Chap. 3

Their phases ψ_0 and ψ_1 may be unknown at the receiver for several reasons. The transmitter may be pulsing a high-frequency oscillator with no attempt to keep the phases of the output coherent from one signal to the next. Transmission over a number of paths of different and variable lengths or rapidly varying delays in the propagation of the signals from transmitter to receiver may change the phases of the received signals in ways the receiver cannot follow. Synchronization with the phase of the transmitted carrier may simply be too costly, and the designer may choose to disregard phase relations between successively received signals. We speak then of *incoherent* detection.

To the phases ψ_0 and ψ_1 of the received signal we assign the uniform prior probability density function

$$z(\psi_j) \equiv \frac{1}{2\pi}, \qquad 0 \le \psi_j < 2\pi, \qquad j = 0, 1.$$

The noise is taken as stationary, narrowband, and Gaussian with autocovariance function

$$\phi(\tau) = \mathrm{Re}\ \tilde{\phi}(\tau)\ e^{i\Omega\tau}.$$

In order to minimize the probability of error, the receiver selects the hypothesis with the greater posterior probability, given its observation of the complex envelope $V(t)$ of its input $v(t) = \mathrm{Re}\ V(t) \exp i\Omega t$. If as in Sec. 3.3 the receiver has taken a set $\mathbf{V} = (V_1, V_2, \ldots, V_n)$ of complex samples of $V(t)$ by means, for instance, of a set of functions $f_j(t)$ orthonormal over $(0, T)$ as in (2-39), then according to (1-4) it chooses hypothesis H_1 if $\zeta_1\tilde{p}_1(\mathbf{V}) \ge \zeta_0\tilde{p}_0(\mathbf{V})$; otherwise it chooses H_0. Here $\tilde{p}_j(\mathbf{V})$ is the joint probability density function of the real and imaginary parts of the samples V_i under hypothesis H_j. As in (3-58),

$$\tilde{p}_j(\mathbf{V}) = \int_0^{2\pi} \tilde{p}_j(\mathbf{V}; \psi)\frac{d\psi}{2\pi}, \qquad j = 0, 1, \tag{3-79}$$

where $\tilde{p}_j(\mathbf{V}; \psi)$ is the joint probability density function of those real and imaginary parts when the signal $s_j(t)$ is on hand and the phase of its carrier is ψ. If we introduce the dummy hypothesis H_n that no signal at all is present, the receiver can just as well compare the quantities

$$\zeta_0\frac{\tilde{p}_0(\mathbf{V})}{\tilde{p}_n(\mathbf{V})} \quad \text{and} \quad \zeta_1\frac{\tilde{p}_1(\mathbf{V})}{\tilde{p}_n(\mathbf{V})},$$

where $\tilde{p}_n(\mathbf{V})$ is the joint probability density function of the data $\mathbf{V}$ when noise alone is present; it is given by the right side of (3-47). Passing to the limit of an infinite number of samples as we did in Sec. 3.3, we see that the receiver can compare the quantities

$$\zeta_0\Lambda_0[V(t)] \quad \text{and} \quad \zeta_1\Lambda_1[V(t)],$$

where by (3-59) and (3-60)

$$\Lambda_j[V(t)] = \exp(-\tfrac{1}{2}d_j^2)I_0(R_j), \qquad j = 0, 1,$$

$$R_j = \left| \int_0^T Q_j^*(t)V(t)\, dt \right|, \qquad d_j^2 = \int_0^T S_j^*(t)Q_j(t)\, dt, \tag{3-80}$$

with $Q_j(t)$ the solution of the integral equation

$$S_j(t) = \int_0^T \tilde{\phi}(t - u)Q_j(u)\, du, \qquad 0 \le t \le T, \qquad j = 0, 1, \qquad (3\text{-}81)$$

as in (3-52). Thus the receiver compares the two quantities

$$y_0 = \zeta_0 \exp(-\tfrac{1}{2}d_0^2)I_0(R_0) \quad \text{and} \quad y_1 = \zeta_1 \exp(-\tfrac{1}{2}d_1^2)I_0(R_1).$$

The receiver determines the statistic R_j, $j = 0, 1$, at the end of each observation interval $(0, T)$ by sampling the rectified output of a filter matched to the signal $\operatorname{Re} Q_j(t) \exp i\Omega t$.

If $y_0 \ge y_1$, the receiver chooses hypothesis H_0 and issues a 0; if $y_0 < y_1$, it issues a 1. Setting $y_0 = y_1$, one can compute a monotone function $f(R_0)$ such that the receiver selects hypothesis H_1 when $R_1 \ge f(R_0)$; otherwise it selects H_0. The false-alarm probability is then

$$Q_0 = \int_0^\infty dR_0 \int_{f(R_0)}^\infty p_0(R_0, R_1)\, dR_1,$$

where $p_0(R_0, R_1)$ is the joint probability density function of R_0 and R_1 under hypothesis H_1. If the signals are orthogonal in the sense that

$$\int_0^\infty S_0^*(t)Q_1(t)\, dt = \int_0^\infty S_1^*(t)Q_0(t)\, dt = 0,$$

R_0 and R_1 are independent random variables, and

$$p_0(R_0, R_1) = \frac{R_0}{d_0^2} \exp\left[-\tfrac{1}{2}d_0^2 - \frac{R_0^2}{2d_0^2}\right]I_0(R_0) \cdot \frac{R_1}{d_1^2} \exp\left[-\frac{R_1^2}{2d_1^2}\right]$$

by (3-71) and (3-73). Then

$$\Pr(R_1 \ge f(R_0)| H_0) = \exp\left[-\frac{[f(R_0)]^2}{2d_1^2}\right],$$

and the false-alarm probability is

$$Q_0 = e^{-\frac{1}{2}d_0^2} \int_0^\infty \frac{R_0}{d_0^2} \exp\left[-\frac{R_0^2}{2d_0^2} - \frac{[f(R_0)]^2}{2d_1^2}\right]I_0(R_0)\, dR_0.$$

The false-dismissal probability Q_1 can be expressed by a similar integral; both must be evaluated numerically. The overall probability of error is then

$$P_e = \zeta_0 Q_0 + \zeta_1 Q_1.$$

3.5.2 The Balanced Binary Incoherent Channel

If the 0's and 1's occur equally often, $\zeta_0 = \zeta_1 = \tfrac{1}{2}$; and if the received signal-to-noise ratios are equal, $d_0 = d_1 = d$, the binary channel is appropriately termed *balanced*. When the noise is white, both signals are being received with equal energies, $E_0 = E_1 = E$. There is complete symmetry between them. The receiver can then simply compare R_0 with R_1, deciding that symbol 0 was sent if $R_0 \ge R_1$ and that 1 was sent if $R_0 < R_1$.

The probabilities of errors of the first and second kinds are equal in the balanced channel. A rather long calculation, presented in Appendix D, shows that the probability of error is

$$P_e = \Pr(R_0 \geq R_1 | H_1) = Q(\mu_1 d, \mu_2 d) - \tfrac{1}{2} e^{-d^2/4} I_0(\tfrac{1}{4}|\lambda| d^2)$$
$$= \tfrac{1}{2}[1 - Q(\mu_2 d, \mu_1 d) + Q(\mu_1 d, \mu_2 d)], \tag{3-82}$$
$$\mu_1 = \tfrac{1}{2}\left[1 - \sqrt{1 - |\lambda|^2}\right]^{1/2}, \qquad \mu_2 = \tfrac{1}{2}\left[1 + \sqrt{1 - |\lambda|^2}\right]^{1/2},$$

where λ is the *correlation coefficient* of the two signals,

$$\lambda = \frac{1}{d^2} \int_0^T \int_0^T Q_0^*(t)\tilde{\phi}(t - u)Q_1(u)\, dt\, du = \frac{1}{d^2} \int_0^T Q_0^*(t)S_1(t)\, dt$$

and $Q(\cdot, \cdot)$ is the Q function defined in (3-76). If $\lambda = 0$, the signals are orthogonal with respect to the interval $(0, T)$ and the kernel $\tilde{\phi}(t - u)$, whereupon the probability of error becomes simply

$$P_e = Q(0, 2^{-1/2}d) - \tfrac{1}{2} e^{-d^2/4} = \tfrac{1}{2} e^{-d^2/4}.$$

3.5.3 The Unilateral Binary Incoherent Channel

In an on–off binary communication system, $s_0(t) \equiv 0$, and the receiver uses the same decision scheme as developed in Sec. 3.3.2, comparing the rectified output R of a filter matched to the signal $\text{Re}[Q_1(t) \exp i\Omega t]$ with a decision level that we here call r_0, and it issues a 1 when $R \geq r_0$ and a 0 when $R < r_0$. From (3-60) we see that the decision level r_0 is given by the equation

$$\exp(-\tfrac{1}{2}d_1^2)\, I_0(r_0) = \frac{\zeta_0}{\zeta_1} \tag{3-83}$$

in terms of the relative frequencies ζ_0 and ζ_1 of 0's and 1's; d_1^2 is as in (3-80) the signal-to-noise ratio of the received signal when a 1 is transmitted.

The average probability of error in this *unilateral* binary incoherent channel is

$$P_e = \zeta_0 \Pr(R \geq r_0 | H_0) + \zeta_1 \Pr(R < r_0 | H_1),$$

and by (3-70) and (3-76) we can write this as

$$P_e = \zeta_0 e^{-b^2/2} + \zeta_1[1 - Q(d_1, b)], \qquad b = \frac{r_0}{d_1},$$

with $Q(\cdot, \cdot)$ again Marcum's Q function. By using (C-11) and (3-83) this expression can be simplified to

$$P_e = \zeta_1 Q(b, d_1). \tag{3-84}$$

When $\zeta_0 = \zeta_1 = \tfrac{1}{2}$, pulses and blanks are being sent equally often, and the average signal-to-noise ratio, which is proportional to the average transmitted power, is $d_{av}^2 = \tfrac{1}{2}d_1^2$. In comparing the unilateral and the balanced channels, we should equate their average signal-to-noise ratios and set $d_{av}^2 = d^2$, where d^2 is the signal-to-noise ratio in Sec. 3.5.2, for the average power expended by their transmitters will then be the same. When the noise is white, $\tfrac{1}{2}d^2 = E/N$, where E is the energy of

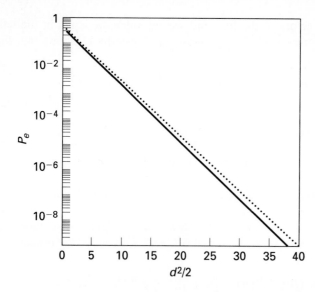

Figure 3-6. Error probabilities for the unilateral and balanced incoherent binary channels versus average signal-to-noise ratio $d^2/2$. Solid line: unilateral channel; dotted line: balanced channel.

each signal received in the balanced channel and N is the unilateral spectral density of the noise. In Fig. 3-6 we have plotted the probability P_e of error for each of these channels versus $\frac{1}{2}d^2$. The unilateral channel has the smaller probability of error.

3.5.4 The Incoherent M-ary Channel

Suppose that the transmitter is sending messages coded into an alphabet of M symbols, each invoking a different narrowband signal. The received signals are then narrowband pulses modulating a radio-frequency carrier,

$$s_j(t) = \text{Re } S_j(t) \exp(i\Omega t + i\psi_j), \qquad j = 1, 2, \dots, M,$$

and when no attempt is made to track the phase of the received carrier, the phases ψ_j must be treated as random variables. It is most reasonable to assign them the uniform distribution over $(0, 2\pi)$. The receiver is to decide which of these M signals is present during an observation interval $(0, T)$.

This problem is an extension to M hypotheses of the binary decision problem treated in Sec. 3.5.1. Again a useful dummy hypothesis H_n states that noise alone is present, and by carrying through the same analysis as in that part, we find that the optimum receiver decides that that hypothesis H_j is true for which

$$\zeta_j \Lambda_j[v(t)] = \zeta_j \exp(-\tfrac{1}{2}d_j^2)I_0(R_j), \qquad j = 1, 2, \dots, M, \qquad (3\text{-}85)$$

is largest; the statistic R_j and the signal-to-noise ratio d_j^2 are defined as in (3-80). The receiver embodies a bank of M filters in parallel, each matched to one of the signals $\text{Re } Q_j(t) \exp i\Omega t$, where $Q_j(t)$ is the solution of the integral equation (3-81). Each filter is followed by a linear rectifier, whose output, sampled at the end of the observation interval $(0, T)$, is the statistic R_j. The receiver computes by analog or digital means the M likelihood functionals $\Lambda_j[v(t)]$ weighted with the prior probabilities ζ_j and determines the largest. In general it will be difficult to calculate the error probability of such a communication system.

If the signals occur equally often, $\zeta_j \equiv M^{-1}$, arrive at the receiver with equal signal-to-noise ratios $d_j^2 \equiv d^2$, and are orthogonal in the sense of (2-46), that is,

$$\int_0^T S_i^*(t)Q_j(t)\,dt = d^2\delta_{ij},$$

the receiver simply decides that the jth signal is present when $R_j > R_i$, $\forall i \neq j$. When the noise is white, the signals must be orthogonal in the usual sense of (2-39), and $d_j^2 = 2E_j/N$, where E_j is the energy of the jth signal and N is the unilateral spectral density of the noise.

To calculate the probability Q_c of a correct decision, we assume that the first signal is the one received, and as in (2-133) we find by (3-73)

$$Q_c = \Pr(R_1 \geq R_i, i = 2, 3, \ldots, M|\,H_1)$$
$$= \int_0^\infty x(1 - e^{-x^2/2})^{M-1}\,e^{-(x^2+d^2)/2}I_0(xd)\,dx.$$

We expand the factor $(\cdot)^{M-1}$ by the binomial theorem and integrate term by term, utilizing the normalization integral for the Rayleigh–Rice distribution in (3-73)—see (C-4)—, and after a little algebra we find for the probability of error

$$P_e = 1 - Q_c = \frac{1}{M}\sum_{r=2}^M (-1)^r \binom{M}{r}\,e^{-(r-1)d^2/2r}.$$

Many other types of signal sets are used in communication systems, and their virtues and disadvantages are discussed in texts on communication theory, where one can also find calculations of the error probabilities they suffer. This large subject is beyond the scope of this book.

3.6 TESTING COMPOSITE HYPOTHESES

3.6.1 The Bayes Criterion

Chapter 1 treated strategies for choosing between two hypotheses on the basis of a number of measurements, strategies that in essence select which of two probability density functions $p_0(v)$ or $p_1(v)$ is the more consistent with the observed values $v_1, v_2, \ldots, v_n$ of n random variables. The choice is to be optimum under some criterion—Bayes or Neyman–Pearson—corresponding to a definition of long-run success. It was assumed that the two probability density functions are known in all respects. In Chapter 2 this theory was applied to the detection of a unique signal in Gaussian noise.

Signals to be detected, however, are rarely unique; seldom is the form with which they appear at the receiver completely known. Usually it is necessary to detect one of a class of signals specified by parameters taking values anywhere in more or less well-defined ranges. A narrowband radar echo, for example, has the form

$$s(t; A, t_0, \Omega) = A \operatorname{Re} S(t - t_0) \exp i\Omega(t - t_0).$$

Its amplitude A depends on the size and reflectivity of the target; the same target in different orientations may reflect the radar pulse with widely different amplitudes. The arrival time t_0 depends on the distance to the target, and the carrier frequency Ω, through the Doppler effect, depends on the component of the velocity of the target in the direction of the receiver. None of these three quantities may be known in advance with much precision. A radar system must be designed to detect echoes with a spectrum of values of A, t_0, and Ω. In a communication system utilizing quasiharmonic signals $A \operatorname{Re} F(t) \exp(i\Omega t + i\psi)$, it is only under the most carefully controlled circumstances that the amplitude A and the phase ψ of the carrier are known by the receiver. The carrier frequency Ω may also vary, as when an enemy, in order to hinder our intercepting his messages, changes his carrier frequency Ω from pulse to pulse in a way known to his receiver, but not to ours. In this chapter we have analyzed the detection of narrowband signals of unknown carrier phase ψ. Now we must consider how to adapt the general theory to accommodate uncertainties in any parameters of the signals to be detected.

Denote the unknown parameters of the signal by $\theta_1, \theta_2, \ldots, \theta_m$, supposing there to be m all told. We can represent them by a vector $\boldsymbol{\theta} = (\theta_1, \theta_2, \ldots, \theta_m)$ in an m-dimensional parameter space, which we designate by $\boldsymbol{\Theta}$. Under hypothesis H_1 these parameters appear in the probability density function $p_1(v|\boldsymbol{\theta})$ of the data v through its dependence on the signal $s(t; \boldsymbol{\theta})$. Hypothesis H_1 now asserts that one of a class of signals $s(t; \boldsymbol{\theta})$ with parameters $\boldsymbol{\theta} \in \boldsymbol{\Theta}$ is present; it is said to be a *composite* hypothesis. Hypothesis H_0, we shall presume for the most part, states that only noise is present; and if the statistical characteristics of the noise are completely known, hypothesis H_0 remains what is termed a *simple* hypothesis. Under H_0 the probability density function of the data is $p_0(v)$.

The task of the receiver is to decide whether the values of the data v actually measured were drawn from a population described by the probability density function $p_0(v)$ (hypothesis H_0) or from one described by a probability density function $p_1(v|\boldsymbol{\theta})$ for a set of parameters $\boldsymbol{\theta} \in \boldsymbol{\Theta}$ (hypothesis H_1). Again the decision strategy can be described as a division of the n-dimensional Cartesian space R_n of the observations v into two regions R_0 and R_1. Hypothesis H_0 is chosen when the point whose coordinates are the observed values $v = (v_1, v_2, \ldots, v_n)$ lies in region R_0, and H_1 when it lies in R_1. The decision surface D dividing these regions is to be selected so that the statistical test is optimum in some sense. Ideally, but rarely, all prior probabilities and costs are well defined and the Bayes criterion is applicable, a situation treated by Wald [Wal50], whose statistical decision theory was applied to signal detection by Middleton and Van Meter [Mid55]. The observer knows not only the prior probabilities ζ_0 and ζ_1 with which hypotheses H_0 and H_1, respectively, hold, but also a joint prior probability density function $z(\boldsymbol{\theta}) = z(\theta_1, \theta_2, \ldots, \theta_m)$ of the m parameters $\boldsymbol{\theta}$, which describes their relative frequencies of occurrence when hypothesis H_1 is true. As with all joint density functions, its integral over the parameter space equals 1:

$$\int_{\boldsymbol{\Theta}} z(\boldsymbol{\theta}) \, d^m\boldsymbol{\theta} = 1, \qquad d^m\boldsymbol{\theta} = d\theta_1 \, d\theta_2 \ldots d\theta_m. \tag{3-86}$$

The observer must in addition know the costs C_{00} and C_{10} of choosing hypotheses H_0 and H_1, respectively, when H_0 is true, and the costs $C_{01}(\boldsymbol{\theta})$ and $C_{11}(\boldsymbol{\theta})$ of choosing H_0 and H_1, respectively, when H_1 is true and a signal with parameters $\boldsymbol{\theta}$ is present. These costs may depend on the parameters $\boldsymbol{\theta}$ of that signal.

The average cost per decision is now, as an evident modification of (1-14),

$$\overline{C} = \zeta_0 \left[C_{00} \int_{R_0} p_0(v)\, d^n v + C_{10} \int_{R_1} p_0(v)\, d^n v \right]$$

$$+ \zeta_1 \left[\int_{R_0} d^n v \int_{\Theta} d^m \boldsymbol{\theta}\, z(\boldsymbol{\theta}) C_{01}(\boldsymbol{\theta}) p_1(v|\boldsymbol{\theta}) \right. \tag{3-87}$$

$$\left. + \int_{R_1} d^n v \int_{\Theta} d^m \boldsymbol{\theta}\, z(\boldsymbol{\theta}) C_{11}(\boldsymbol{\theta}) p_1(v|\boldsymbol{\theta}) \right].$$

The first bracket of (3-87) is the risk associated with hypothesis H_0; the second is that associated with H_1. The decision surface D separating the regions R_0 and R_1 must be situated so that $\overline{C}$ is minimum. The analysis proceeds as in Sec. 1.2, where the Bayes strategy for a choice between simple hypotheses was derived, and it shows that the decision surface D consists of those points v satisfying the equation

$$\zeta_0(C_{10} - C_{00})p_0(v) = \zeta_1 \int_{\Theta} d^m \boldsymbol{\theta}\, z(\boldsymbol{\theta})[C_{01}(\boldsymbol{\theta}) - C_{11}(\boldsymbol{\theta})]p_1(v|\boldsymbol{\theta}),$$

$$v \in D. \tag{3-88}$$

Those points v for which the left side of (3-88) is the larger make up R_0; those for which it is the smaller make up R_1. To choose between hypotheses H_0 and H_1 the observer calculates the cost-likelihood ratio

$$\Lambda_c = \frac{\zeta_1 \int_{\Theta} d^m \boldsymbol{\theta}\, z(\boldsymbol{\theta})[C_{01}(\boldsymbol{\theta}) - C_{11}(\boldsymbol{\theta})]p_1(v|\boldsymbol{\theta})}{\zeta_0(C_{10} - C_{00})p_0(v)}$$

on the basis of the observations v and decides for hypothesis H_0 if $\Lambda_c < 1$ and for H_1 if $\Lambda_c \geq 1$.

When, as we assume henceforth, the costs C_{01} and C_{11} do not depend on the values of the parameters $\boldsymbol{\theta}$, we can carry out the integration over Θ in these expressions, and by introducing the overall probability density function

$$p_1(v) = \int_{\Theta} z(\boldsymbol{\theta}) p_1(v|\boldsymbol{\theta})\, d^m \boldsymbol{\theta}$$

of the data under hypothesis H_1 we can reduce our expressions to the same form as in Sec. 1.2. The observer now forms the average likelihood ratio

$$\overline{\Lambda}(v) = \frac{p_1(v)}{p_0(v)} = \int_{\Theta} d^m \boldsymbol{\theta}\, z(\boldsymbol{\theta})\Lambda(v|\boldsymbol{\theta}), \qquad \Lambda(v|\boldsymbol{\theta}) = \frac{p_1(v|\boldsymbol{\theta})}{p_0(v)}, \tag{3-89}$$

and compares it with the quantity

$$\Lambda_0 = \frac{\zeta_0(C_{10} - C_{00})}{\zeta_1(C_{01} - C_{11})} \tag{3-90}$$

as before; if $\overline{\Lambda}(v) < \Lambda_0$, hypothesis H_0 is chosen, otherwise H_1.

The Bayes cost, which is the minimum value $\overline{C}_{\min}$ of (3-87) obtained when the decision surface is given by (3-88), depends on the prior probabilities ζ_0 and ζ_1 and on the form of the prior probability density function $z(\theta)$. If these are unknown, but the costs are defined, one might seek those prior probabilities ζ_0 and ζ_1 and that prior density function $z(\theta)$ for which $\overline{C}_{\min}$ is maximum, assuming perhaps that some adversary is picking them so as to make the observer's minimum loss as large as possible. The form of the prior density function $z(\theta)$ that with the proper values of ζ_0 and ζ_1 maximizes the Bayes cost $\overline{C}_{\min}$ is called the *least favorable distribution* of the parameters θ. In a few problems it can be found by inspection; in others it may be most difficult to calculate. The concept of a least favorable distribution will appear again when we discuss the Neyman–Pearson criterion. For an extensive treatment of the Bayes criterion in problems of conventional statistics, we refer to [Bla54].

If both hypotheses H_0 and H_1 are composite, prior probability density functions $z_0(\theta)$ and $z_1(\theta)$ of the parameters under both hypotheses must be specified. If the costs are independent of the true values of the parameters, the optimum decision under the Bayes criterion, as it is not hard to see, is made by comparing the likelihood ratio

$$\overline{\Lambda}(v) = \frac{p_1(v)}{p_0(v)}, \qquad p_j(v) = \int_\Theta z_j(\theta) p_j(v|\,\theta)\, d^m\theta, \qquad j = 0, 1, \qquad (3\text{-}91)$$

with the same decision level Λ_0 as in (3-90). Again unknown prior density functions $z_j(\theta)$ might be replaced by least favorable ones if these can be discovered.

Suppose that the receiver is to decide, on the basis of its input $v(t)$, which of two signals $s_0(t; \theta)$ and $s_1(t; \theta)$ has been added to the random noise $n(t)$ to form the input $v(t)$. If we divide $p_j(v)$ in (3-91) by $p_n(v)$, which is the probability density function of the data v when no signal at all is present, the likelihood ratio can be written

$$\overline{\Lambda}(v) = \frac{\Lambda_1(v)}{\Lambda_0(v)}, \qquad \Lambda_j(v) = \int_\Theta z_j(\theta)\Lambda_j(v|\,\theta)\, d^m\theta,$$

$$\Lambda_j(v|\,\theta) = \frac{p_j(v|\,\theta)}{p_n(v)}, \qquad j = 0, 1. \qquad (3\text{-}92)$$

Here $\Lambda_j(v|\,\theta)$ is the likelihood ratio appropriate to the decision between two simple hypotheses, H_j: "Signal $s_j(t; \theta)$ is present in the midst of random noise $n(t)$," and H_n: "Noise $n(t)$ alone is present." When the noise is Gaussian, we can as in Chapter 2 pass to the limit $n \to \infty$ of an infinite number of samples of the input, whereupon the likelihood ratio $\Lambda_j(v|\,\theta)$ goes into a likelihood functional $\Lambda_j[v(t)|\,\theta]$ similar to that in (2-71). The decision can now be based on the quantity

$$\Lambda[v(t)] = \frac{\overline{\Lambda}_1[v(t)]}{\overline{\Lambda}_0[v(t)]}, \qquad \overline{\Lambda}_j[v(t)] = \int_\Theta z_j(\theta)\Lambda_j[v(t)|\,\theta]\, d^m\theta, \qquad j = 0, 1. \qquad (3\text{-}93)$$

Similar passages to the limit $n \to \infty$ of an infinite number of data can be carried out in the other formulations of this section. Observe that we average over the parameters θ before dividing to form the ratio $\Lambda[v(t)]$ in (3-93). An example of this method has been seen in Sec. 3.5.1, where we averaged over the phase ψ of each signal $s_j(t; \psi)$, $j = 0, 1$.

3.6.2 The Extended Neyman–Pearson Criterion

We return to the choice between the simple hypothesis H_0, "No signal is present," and the composite hypothesis H_1, "One of a class of signals $s(t; \theta)$ with parameters $\theta \in \Theta$ is present." Once the decision strategy has been adopted as a dichotomy of the data space R_n into regions R_0 and R_1, the probability Q_0 of an error of the first kind, or false alarm, is

$$Q_0 = \int_{R_1} p_0(v) \, d^n v, \qquad (3\text{-}94)$$

R_1 being the portion of R_n in which hypothesis H_1 is chosen. The probability of detecting a signal $s(t; \theta)$ with a particular set θ of parameters is

$$Q_d(\theta) = \int_{R_1} p_1(v| \theta) \, d^n v. \qquad (3\text{-}95)$$

Having adopted the Neyman–Pearson criterion, one would like to find a decision surface D separating regions R_0 and R_1, yielding the maximum probability of detecting a signal with any set of parameters, and incurring the preassigned false-alarm probability Q_0. If the same surface D is thus optimum for all values of the parameters θ, the strategy so defined provides what is called a *uniformly most powerful test* of hypothesis H_1 against hypothesis H_0.

Here is an example of a uniformly most powerful test. Let the n observations v be normally distributed, independent random variables with variance σ^2, and let their expected values be zero under hypothesis H_0 and $m > 0$ under H_1. Then the joint probability density functions of the data are

$$p_0(v) = (2\pi\sigma^2)^{-n/2} \exp\left[-\sum_{k=1}^{n} \frac{v_k^2}{2\sigma^2} \right],$$

$$p_1(v) = (2\pi\sigma^2)^{-n/2} \exp\left[-\sum_{k=1}^{n} \frac{(v_k - m)^2}{2\sigma^2} \right].$$

Of the parameter m all that is known is that it is positive. For any fixed value of m, according to Example 1-2 in Sec. 1.2.4, the optimum decision strategy is equivalent to comparing the sample mean

$$V = \frac{1}{n} \sum_{k=1}^{n} v_k$$

with a fixed critical value M and picking hypothesis H_0 when $V < M$ and H_1 when $V \geq M$. This decision level M is completely determined by the probability Q_0 of an error of the first kind,

$$Q_0 = \int_{M}^{\infty} P_0(V) \, dV,$$

where $P_0(V)$ is a Gaussian density function with expected value zero and variance σ^2/n. The decision surface D is now the hyperplane

$$\sum_{k=1}^{n} v_k = nM,$$

and it is independent of the true expected value m under hypothesis H_1. The test is therefore uniformly most powerful when only positive values of m are possible; it can be used in ignorance of the actual value of m.

It is the exception rather than the rule for the decision surface D maximizing $Q_d(\boldsymbol{\theta})$ of (3-95) for fixed Q_0 to be independent of the parameters $\boldsymbol{\theta}$. In the vast majority of problems the same surface D will not be optimum for all values of $\boldsymbol{\theta}$, and a uniformly most powerful test does not exist. This is so, for instance, when the true expected value m in our example can be either positive or negative.

If the prior probability density function $z(\boldsymbol{\theta})$ of the parameters $\boldsymbol{\theta}$ is known, the *extended* Neyman–Pearson criterion requires that, for preassigned false-alarm probability Q_0, the average probability of detection

$$\overline{Q}_d = \int_{\Theta} z(\boldsymbol{\theta}) Q_d(\boldsymbol{\theta}) \, d^m \boldsymbol{\theta} \qquad (3\text{-}96)$$

be maximum. With this prior density function, the overall probability density function of the data v is

$$p_1(v) = \int_{\Theta} z(\boldsymbol{\theta}) p_1(v|\,\boldsymbol{\theta}) \, d^m \boldsymbol{\theta}, \qquad (3\text{-}97)$$

and the average probability of detection is

$$\overline{Q}_d = \int_{R_1} p_1(v) \, d^n v. \qquad (3\text{-}98)$$

The test in effect chooses either $p_0(v)$ or $p_1(v)$ as better representing the data at hand, and as in Sec. 1.2 the optimum strategy forms the likelihood ratio

$$\Lambda(v) = \frac{p_1(v)}{p_0(v)} \qquad (3\text{-}99)$$

and decides for H_1 if $\Lambda(v) \geq \Lambda_0$ and for H_0 if $\Lambda(v) < \Lambda_0$, with the value of Λ_0 fixed by the preassigned false-alarm probability

$$Q_0 = \Pr[\Lambda(v) \geq \Lambda_0|\,H_0] = \int_{\Lambda_0}^{\infty} P_0(\Lambda) \, d\Lambda.$$

The average probability of detection is then

$$\overline{Q}_d = \Pr[\Lambda(v) \geq \Lambda_0|\,H_1] = \int_{\Lambda_0}^{\infty} P_1(\Lambda) \, d\Lambda.$$

As in Chapter 1, $P_0(\Lambda)$ and $P_1(\Lambda)$ are the probability density functions of the statistic $\Lambda(v)$ under the two hypotheses. Just as in Sec. 3.6.1, $\Lambda(v)$ of (3-99) is an average likelihood ratio and can be expressed as in (3-89). If, as with the detection of a known signal in Gaussian noise, we can take the likelihood ratio $\Lambda(v|\,\boldsymbol{\theta})$ to the limit $n \to \infty$ to obtain the likelihood functional $\Lambda[v(t)|\,\boldsymbol{\theta}]$ for the detection of the signal $s(t; \boldsymbol{\theta})$, the decision that is optimum under this broadened Neyman–Pearson criterion can be based on the average likelihood functional

$$\overline{\Lambda}[v(t)] = \int_{\Theta} z(\boldsymbol{\theta}) \Lambda[v(t)|\,\boldsymbol{\theta}] \, d^m \boldsymbol{\theta}. \qquad (3\text{-}100)$$

In the problem of detecting a narrowband signal of unknown phase ψ in Gaussian noise, treated in Sec. 3.3.2, the unknown parameter is $\boldsymbol{\theta} = \psi$, and there we accepted the prior density function $z(\psi) \equiv (2\pi)^{-1}$ for it. The average likelihood functional (3-100) is then given by (3-60). In Sec. 3.4 we calculated the probability $Q_d(\psi)$ of detecting a signal with a particular phase ψ and found it given by (3-75) independently of ψ.

What is to be done when past experience provides no guidance to selecting the prior probability density function $z(\theta)$? The most prudent course would seem to be to adopt that prior density function $\tilde{z}(\theta)$ for which the maximum average probability of detection

$$\overline{Q}_d[z] = \int_{R_1[z]} p_1(v) \, d^n v = \int_{R_1[z]} d^n v \int_{\Theta} z(\theta) p_1(v|\theta) \, d^m \theta \qquad (3\text{-}101)$$

is least:

$$\overline{Q}_d[\tilde{z}] \le \overline{Q}_d[z], \qquad \forall z(\theta). \qquad (3\text{-}102)$$

The decision region $R_1[z]$ in (3-101) contains those points v for which

$$\Lambda(v) = \frac{p_1(v)}{p_0(v)} = \int_{\Theta} z(\theta) \frac{p_1(v|\theta)}{p_0(v)} \, d^m \theta \ge \Lambda_0$$

with Λ_0 such that the false-alarm probability

$$Q_0[z] = \int_{R_1[z]} p_0(v) \, d^n v \qquad (3\text{-}103)$$

takes on the preassigned value. The prior probability density function $\tilde{z}(\theta)$ defined by (3-102) is termed the *least favorable distribution* of the parameters θ with respect to the Neyman–Pearson criterion. In Sec. 7.6 we shall develop a general criterion that determines whether a given prior probability density function $z(\theta)$ is least favorable in this sense.

3.6.3 Detection of Signals of Unknown Amplitude

Let us denote the amplitude of the signal by A and assume it positive, $A > 0$; the remaining unknown parameters of the signal, including possibly its sign, are designated by θ and lie in a parameter space Θ. The amplitude A will be taken to be statistically independent of those other parameters θ, as will be the case in most detection problems; the joint prior probability density function of the signal parameters then has the form $z_A(A)z(\theta)$. We consider various ways of coping with ignorance of the amplitude A of the signal to be detected. We adopt the extended Neyman–Pearson criterion that for a certain preassigned false-alarm probability the probability of detection, averaged with respect to the prior distribution $z(\theta)$ of the parameters other than amplitude, shall be maximum. That prior distribution $z(\theta)$ may have been selected on the basis of past experience, or it may be the least favorable distribution defined at the end of Sec. 3.6.2.

It may be that the detection strategy that is optimum in this sense for a signal with a given amplitude A turns out not to depend on the value of A. For example, the strategy determined in Sec. 3.3.2 for signals with a uniformly distributed random phase ψ can be put into a form independent of A. Writing the signal as

$$s(t; \psi) = A \operatorname{Re} F(t) \, e^{i\Omega t + i\psi},$$

we easily see that the strategy is equivalent to comparing the sufficient statistic

$$r = \left| \int_0^T G^*(t) V(t)\, dt \right|$$

with a decision level r_0; here $V(t)$ is the complex envelope of the input and $G(t)$ is the solution of the integral equation

$$F(t) = \int_0^T \tilde{\phi}(t, s) G(s)\, ds, \qquad 0 \le t \le T, \qquad (3\text{-}104)$$

with $\tilde{\phi}(t, s)$ the complex autocovariance function of the noise. A signal $s(t; \psi)$ is declared present if $r \ge r_0$, and r_0 is determined by the false-alarm probability, which as in (3-70) is

$$Q_0 = \exp\left[-\frac{r_0^2}{2d'^2}\right], \qquad d'^2 = \int_0^T G^*(t) F(t)\, dt.$$

The probability $Q_d(A, \psi)$ of detecting a signal with amplitude A and phase ψ is by (3-75)

$$Q_d(A, \psi) = Q\left(d, \frac{r_0}{d'}\right), \qquad d = Ad',$$

in terms of Marcum's Q function (3-76) and the signal-to-noise ratio d^2 defined as in (3-53). A test that, like this one, does not require knowing the amplitude A of the signal, yet is optimum under the extended Neyman–Pearson criterion for a signal with arbitrary amplitude A, can be said to be uniformly most powerful with respect to amplitude.

When a test that is uniformly most powerful with respect to amplitude does not exist with the prior probability density function $z(\boldsymbol{\theta})$ adopted for the remaining parameters, the receiver may be designed to be optimum for a particular value A_s of the signal amplitude lying somewhere in the range of expected amplitudes A. We call A_s the *standard amplitude*. For signals of other amplitudes $A \ne A_s$ the receiver will not be optimum, but the loss of signal detectability will seldom be serious.

Alternatively, a prior probability density function $z_A(A)$ may be adopted on the basis of some physical model of how a radar signal is reflected from its target or how a communication signal propagates from transmitter to receiver, as when the channel fades in some manner. The parameters of the prior density function $z_A(A)$ may themselves be only roughly known, and standard values of them must be accepted for receiver design.

When, for instance, the signal strength undergoes *Rayleigh fading*, its amplitude A is a random variable governed by a Rayleigh distribution,

$$z(A) = \frac{A}{s^2} \exp\left(-\frac{A^2}{2s^2}\right) U(A). \qquad (3\text{-}105)$$

Such a signal can often be conveniently considered as having the form

$$s(t) = \mathrm{Re}[aF(t)\, e^{i\Omega t}], \qquad (3\text{-}106)$$

in which $a = a_x + ia_y$ is a complex signal amplitude with a_x and a_y statistically independent Gaussian random variables with zero expected values and equal variances s^2; $A = |a|$. The phase $\psi = \arg a$ is then uniformly distributed over $(0, 2\pi)$.

The *threshold* or *weak-signal* approximation is based on the viewpoint that in the least favorable situation for detecting a signal, the signal is very weak, and a receiver optimum for the weakest signals will be adequate for stronger ones. The average likelihood ratio is now defined by

$$\overline{\Lambda}(v) = \int_0^\infty z_A(A) \int_\Theta z(\boldsymbol{\theta}) \Lambda(v|\ A,\ \boldsymbol{\theta})\ dA\ d^{m-1}\boldsymbol{\theta}. \tag{3-107}$$

The likelihood ratio

$$\Lambda(v|\ A,\ \boldsymbol{\theta}) = \frac{p_1(v|\ A,\ \boldsymbol{\theta})}{p_0(v)}$$

for detection of a signal $s(t;\ A,\ \boldsymbol{\theta})$ with known parameters is expanded in a power series in the amplitude A:

$$\Lambda(v|\ A,\ \boldsymbol{\theta}) = 1 + A\Lambda_A(v|\ 0,\ \boldsymbol{\theta}) + \tfrac{1}{2}A^2\Lambda_{AA}(v|\ 0,\ \boldsymbol{\theta}) + \cdots,$$

in which

$$\Lambda_A(v|\ 0,\ \boldsymbol{\theta}) = \left.\frac{\partial\Lambda(v|\ A,\ \boldsymbol{\theta})}{\partial A}\right|_{A=0}, \qquad \Lambda_{AA}(v|\ 0,\ \boldsymbol{\theta}) = \left.\frac{\partial^2\Lambda(v|\ A,\ \boldsymbol{\theta})}{\partial A^2}\right|_{A=0},$$

and so on. When this series is substituted into (3-107) and the integrations over the amplitude A are carried out, we find

$$\overline{\Lambda}(v) = 1 + \overline{A}\int_\Theta z(\boldsymbol{\theta})\Lambda_A(v|\ 0,\ \boldsymbol{\theta})\ d^{m-1}\boldsymbol{\theta}$$

$$+ \tfrac{1}{2}\overline{A^2}\int_\Theta z(\boldsymbol{\theta})\Lambda_{AA}(v|\ 0,\ \boldsymbol{\theta})\ d^{m-1}\boldsymbol{\theta} + \cdots \tag{3-108}$$

where

$$\overline{A^k} = \int_0^\infty A^k z_A(A)\ dA$$

is the kth moment of the prior probability density function of the amplitude A. The *threshold statistic* is the nonvanishing coefficient

$$g_{\boldsymbol{\theta}} = \int_\Theta z(\boldsymbol{\theta})\left.\frac{\partial^k}{\partial A^k}\ \Lambda(v|\ A,\ \boldsymbol{\theta})\right|_{A=0} d^{m-1}\boldsymbol{\theta} \tag{3-109}$$

of lowest order in this series. In most problems, with proper definition of the amplitude A, this will be the term with $k = 2$. When appropriate, the likelihood functional $\Lambda[v(t)|\ A,\ \boldsymbol{\theta}]$ figures in the definition of the threshold statistic, the limit $n \to \infty$ of an infinite number of samples of the input $v(t)$ having been taken. The threshold statistic $g_{\boldsymbol{\theta}}$ is compared with a decision level set to yield a preassigned false-alarm probability [Mid60a, Sec. 19.4], [Rud61], [Mid66]. Several examples of the threshold statistic will be derived and analyzed in the sequel. We shall find it appropriate mainly when the receiver can base its decision on a large number of statistically independent inputs, whereupon the input signal-to-noise ratio required for it to attain practical values of the false-alarm and detection probabilities is indeed small.

3.6.4 Maximum-likelihood Detection

In some situations the threshold approximation may emphasize signals so weak that there is no hope of detecting them anyhow, and the designer should look for superior detection strategies for signals strong enough to be detected with a useful probability. When the signal is strong, the dominant contribution to the likelihood ratio (3-89), averaged with respect to all the parameters $\boldsymbol{\theta}$, comes from the neighborhood of that point $\boldsymbol{\theta}_m$ of the space Θ for which $\Lambda(v|\,\boldsymbol{\theta})$ is maximum,

$$\Lambda(v|\,\boldsymbol{\theta}_m) \geq \Lambda(v|\,\boldsymbol{\theta}), \qquad \forall \boldsymbol{\theta} \neq \boldsymbol{\theta}_m,$$

and the average likelihood ratio $\overline{\Lambda}(v)$ is then approximately proportional to that maximum value. The *maximum-likelihood* strategy compares the maximum value with a decision level Λ_1 chosen to yield the preassigned false-alarm probability. This strategy will be studied in connection with detecting a signal of unknown time of arrival, as in radar, and it will be found superior to the threshold detector for signals with useful detection probabilities.

As an example of maximum-likelihood detection, let us consider a *spread-spectrum* communication system. In order to deceive an enemy, a transmitter of messages coded into binary digits 0 and 1 utilizes one of M different carrier frequencies Ω_m for sending the pulses representing the 1's in a message; it sends no signal at all for the 0's. The receiver knows in advance the sequence with which the several carrier frequencies Ω_m will be selected. This is a simple version of a spread-spectrum system.

If we are the enemy, however, and ignorant of that sequence, we are confronted with the problem of detecting a signal of the form

$$s_j(t) = \text{Re } S_j(t)\exp(i\Omega_j t + i\psi), \qquad 1 \leq j \leq M, \qquad (3\text{-}110)$$

in which the carrier frequency Ω_j might take on any one of M possible values. These frequencies we assume to be so far apart that the signals $s_j(t)$ are for all practical purposes orthogonal. The phase ψ is unknown and distributed uniformly over $(0, 2\pi)$. The frequency Ω_j can be thought of as an unknown parameter taking on only one of a finite number M of discrete values. Suppose that we have observed that when 1's are transmitted, frequency Ω_j is used with relative frequency—or prior probability—z_j, $1 \leq j \leq M$. These prior probabilities sum to 1:

$$\sum_{j=1}^{M} z_j = 1.$$

The receiver must thus choose between two hypotheses,

$$(H_0): \quad v(t) = n(t),$$
$$(H_1): \quad v(t) = n(t) + s_j(t),$$

where the index j may take any value from 1 to M with conditional probabilities z_j. Let us assume that $n(t)$ is white Gaussian noise with unilateral spectral density N and that the input is as usual observed during the interval $(0, T)$.

If v denotes a set of n samples of our input $v(t)$, the receiver must form an average likelihood ratio

$$\overline{\Lambda}(v) = \sum_{j=1}^{M} z_j \Lambda_j(v), \qquad \Lambda_j(v) = \frac{p_j(v)}{p_0(v)}, \tag{3-111}$$

where $p_0(v)$ is the joint probability density function of the samples v under hypothesis H_0 that only random noise is present. If we pass to the limit of an infinite number of samples as in Sec. 3.3, we see that the receiver must generate the average likelihood functional

$$\overline{\Lambda}[v(t)] = \sum_{j=1}^{M} z_j \Lambda_j[v(t)], \tag{3-112}$$

where $\Lambda_j[v(t)]$ is the likelihood functional for detecting the jth signal $s_j(t)$ among those given in (3-110).

Let us adopt a reference carrier frequency Ω_r in the neighborhood of the carrier frequencies of the signals. With respect to this reference frequency, the complex envelope of the jth signal is $S_j(t) \exp[i(\Omega_j - \Omega_r)t + i\psi]$. According to (3-54), as written for a signal having this complex envelope and received in white noise, the jth likelihood functional is given by

$$\Lambda_j[v(t); \psi] = \exp\left[\text{Re } \frac{1}{N} e^{-i\psi} \int_0^T S_j^*(t) \exp[-i(\Omega_j - \Omega_r)t]V(t)\, dt - \frac{1}{2N} \int_0^T |S_j(t)|^2\, dt\right]$$

before averaging over the phase ψ. If we now average over $0 \le \psi < 2\pi$, we find as in (3-60) the average likelihood functional

$$\overline{\Lambda}_j[v(t)] = \exp(-\tfrac{1}{2}d_j^2)I_0(R_j), \tag{3-113}$$

where

$$d_j^2 = \frac{1}{N} \int_0^T |S_j(t)|^2\, dt \tag{3-114}$$

is the signal-to-noise ratio of the jth signal, and as in (3-80)

$$R_j = \frac{1}{N}\left|\int_0^T S_j^*(t) \exp[-i(\Omega_j - \Omega_r)t]V(t)\, dt\right|.$$

The statistic R_j is the rectified output, sampled at the time $t = T$, of a filter matched to the jth signal $s_j(t)$ of (3-110); the pass frequency of this narrowband filter equals Ω_j. The functional in (3-113) is substituted into (3-112) to determine the average likelihood functional, which the receiver must compare with a decision level

$$\Lambda_0 = \frac{\zeta_0}{\zeta_1};$$

ζ_0 is the relative frequency of 0's, and ζ_1 is the relative frequency of 1's in the messages being transmitted. This receiver requires us to know both the amplitudes of all the signals being utilized and their relative frequencies. Calculating the probability of error it incurs would be extremely difficult because of the nonlinear manner in which the input $v(t)$ is processed.

If the various signals are utilized equally often, $z_j \equiv M^{-1}$, if the signal-to-noise ratios of the signals at the input are the same, $d_j^2 \equiv d^2$, and if the received signals are strong enough to be detected with a low probability of error, there will be one term in the average likelihood functional

$$\overline{\Lambda}[v(t)] = \frac{1}{M} e^{-d^2/2} \sum_{j=1}^{M} I_0(R_j) \tag{3-115}$$

that under hypothesis H_1 is much larger than the rest, and this is most likely to be the term corresponding to the signal actually transmitted. This term will dominate the sum in (3-115) because of the rapidly accelerating increase of the modified Bessel function $I_0(r)$ with increasing r.

A receiver that is nearly as effective as the optimum receiver is therefore one that bases its decision on the largest of the M rectified outputs at time $t = T$. That is, the receiver decides that a 1 was transmitted if

$$\max_j r_j > r_0,$$

where

$$r_j = \frac{R_j}{d_j} = \frac{1}{\sqrt{2NE_j}} \left| \int_0^T S_j^*(t) \exp[-i(\Omega_j - \Omega_r)t] V(t)\, dt \right|, \qquad 1 \le j \le M,$$

is proportional to the rectified output of the jth matched filter, and r_0 is the decision level. (We have normalized the output in this way for convenience in subsequent calculations.) If all the data r_j lie below r_0, the receiver decides that no signal was transmitted at any of the M frequencies, that is, that the message digit is 0. If any datum r_j exceeds r_0, the receiver chooses hypothesis H_1. We can call this a *maximum-likelihood receiver*. The maximum-likelihood receiver is appropriate whenever under hypothesis H_1 one of M orthogonal signals of equal energy may appear, but it is unknown a priori which one it will be.

By the same analysis as in Sec. 3.4 we find that the density functions of each datum r_j under the two hypotheses are

$$p_0(r_j) = r_j \exp(-\tfrac{1}{2} r_j^2) U(r_j) \qquad (H_0) \tag{3-116}$$

when no signal is present and, in terms of the Rayleigh–Rice density function in (3-73),

$$p_1(r_j) = q(d_j, r_j) = r_j \exp[-\tfrac{1}{2}(r_j^2 + d_j^2)] I_0(d_j r_j) U(r_j) \qquad (H_1) \tag{3-117}$$

when a signal $s_j(t)$ with effective signal-to-noise ratio d_j^2, defined as in (3-114), is present.

The false-alarm probability of the maximum-likelihood receiver is 1 minus the probability under hypothesis H_0 that all the rectified outputs r_j fall below the decision level r_0:

$$Q_0 = 1 - [1 - \exp(-\tfrac{1}{2} r_0^2)]^M. \tag{3-118}$$

Because all signals are being treated alike, and all have the same input signal-to-noise ratio $d_j^2 \equiv d^2$, we can calculate the probability $Q_1 = 1 - Q_d$ of missing the signal under the assumption that under hypothesis H_1 the first signal $s_1(t)$ is present. By (3-75)

$$\Pr(r_1 < r_0|\, H_1) = 1 - Q(d, r_0),$$

where again $Q(\cdot, \cdot)$ is Marcum's Q function (3-76). Furthermore, because the signals are assumed orthogonal, the distributions of the rectified outputs of the filters matched to the other $M - 1$ signals are the same as in (3-116), and

$$\Pr(r_j < r_0|\, H_1, \; j > 1) = 1 - \exp(-\tfrac{1}{2}r_0^2),$$

and therefore

$$Q_1 = 1 - Q_d = [1 - Q(d, r_0)][1 - \exp(-\tfrac{1}{2}r_0^2)]^{M-1} \qquad (3\text{-}119)$$

is the probability of false dismissal. The value of the decision level r_0 for minimum error probability is calculated by forming the error probability

$$P_e = \zeta_0 Q_0 + \zeta_1 Q_1,$$

differentiating with respect to r_0, and setting the result equal to zero. The ensuing equation, which we leave for the reader to write out, must be solved by trial and error.

For a receiver based on the Neyman–Pearson criterion, the decision level r_0 must be determined from (3-118) for the preassigned false-alarm probability Q_0:

$$\exp(-\tfrac{1}{2}r_0^2) = 1 - (1 - Q_0)^{1/M}.$$

The signal-to-noise ratio D_M required in order to attain a detection probability Q_d is then found by solving

$$Q(D_M, r_0) = 1 - (1 - Q_d)[1 - \exp(-\tfrac{1}{2}r_0^2)]^{1-M}.$$

The loss in signal detectability caused by the uncertainty about which of the M signals might be present can be measured by the ratio $D_M^2 : D_1^2$, where D_1^2 is the required signal-to-noise ratio when it is known which signal will be present under hypothesis H_1:

$$\exp(-\tfrac{1}{2}\bar r_0^2) = Q_0, \qquad Q(D_1, \bar r_0) = Q_d.$$

In Fig. 3-7 this ratio, expressed in decibels as $10\log_{10}(D_M^2/D_1^2)$, has been plotted versus the number M for a false-alarm probability $Q_0 = 10^{-4}$ and for three values of the probability Q_d of detection. (M is of course an integer, but the computed points have been connected by a continuous curve for the sake of clarity.) The loss is seen to be well under 2 dB even for M as large as 1000.

This receiver, as we shall see in Sec. 7.2, furnishes an approximate model of a maximum-likelihood receiver for detecting a signal of unknown arrival time, as in radar. The unknown parameter θ will initially take on a finite set of values corresponding to the centers of brief intervals into which the entire observation interval $(0, T)$ will have been divided. The signal, when present, may appear during any one of these subintervals, but in which is unknown a priori. When this receiver is treated from the standpoint of the extended Neyman–Pearson criterion, it requires no knowledge of the amplitude of the signal, the decision level r_0 being determined

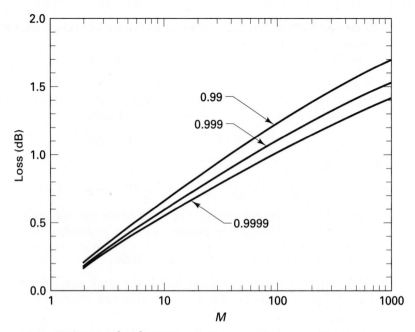

Figure 3-7. Loss D_M^2/D_1^2 in decibels when any of M orthogonal signals with equal signal-to-noise ratio might be present under hypothesis H_1. $Q_0 = 10^{-4}$. The curves are indexed with the probability Q_d of detection.

by the false-alarm probability, as in (3-118). It will be found superior, under ordinary conditions, to the threshold receiver based on the concepts described in Sec. 3.6.3.

Problems

3-1. Consider a memoryless nonlinear device whose output $v_0(t)$ at any time t is a function $g(\cdot)$ of its input $v(t)$ at the same time t:

$$v_0(t) = g(v(t)).$$

If the input is an amplitude-modulated signal $s(t) = S(t) \cos \Omega t$, the output will be a sum of harmonics of the carrier frequency Ω:

$$v_0(t) = \sum_{k=0}^{\infty} G_k[S(t)] \cos k\Omega t.$$

Show how to calculate the functionals $G_k[S(t)]$ by means of the Fourier series for the periodic function $g(S \cos x)$, $-\pi < x \leq \pi$. Evaluate them for the linear full-wave rectifier, $g(v) = |v|$, and for the quadratic rectifier, $g(v) = v^2$.

3-2. Let a rectangular narrowband signal

$$s(t) = \begin{cases} A \sin \Omega t, & 0 \leq t \leq T, \\ 0, & t < 0, \quad t > T, \end{cases}$$

be impressed on the simply resonant circuit of Fig. 3-3. Calculate the complex envelope of the output by means of (3-16). Take the carrier frequency of the signal different from the resonant frequency of the filter by an amount that is arbitrary, but on the order of the bandwidth of the filter.

3-3. Let $\tilde{\phi}(\tau)$ be the complex autocovariance function of a stationary narrowband Gaussian random process $x(t)$ of expected value 0. In terms of $\tilde{\phi}(\tau)$ find the autocovariance function of the output of a quadratic rectifier to which $x(t)$ is applied. The rectifier forms $[x(t)]^2$ and completely attenuates the components of twice the carrier frequency. *Hint:* Use (3-44).

3-4. Let $\tilde{\phi}(\tau)$ be the complex autocovariance function of narrowband Gaussian noise. For two different times t_1 and t_2 let the complex envelope of the noise be $R_j \exp i\theta_j = X(t_j) + iY(t_j), j = 1, 2$. Show that the joint probability density function of the amplitudes R_1 and R_2 is

$$p(R_1, R_2) = \frac{R_1 R_2}{\sigma^4(1 - r^2)} \exp\left[-\frac{R_1^2 + R_2^2}{2\sigma^2(1 - r^2)}\right] I_0\left[\frac{r R_1 R_2}{\sigma^2(1 - r^2)}\right],$$

$$\sigma^2 = \tilde{\phi}(0), \qquad r = \frac{|\tilde{\phi}(t_2 - t_1)|}{\tilde{\phi}(0)}.$$

Hint: Convert (3-40) for $n = 2$ to the joint probability density function of R_1, R_2, θ_1, and θ_2 and integrate out the phases θ_1, θ_2 [Mid48].

3-5. For noise of the same kind as in Problem 3-4 find the probability density function of the phase difference $\psi = \theta_2 - \theta_1$ by integrating the joint probability density function found in that problem over R_1 and R_2 instead of over the phases [Mid48]. *Hint:* Change variables to (z, t), where $R_1 = \sigma z \cos t$ and $R_2 = \sigma z \sin t$, and integrate first over $0 < z < \infty$ and second over $0 < t < \pi/2$. *Answer:* With $a = |r| \cos(\psi - \beta)$ and $\beta = \arg r$,

$$p(\psi) = (2\pi)^{-1}(1 - |r|^2)(1 - a^2)^{-3/2}\left[\sqrt{1 - a^2} + a(\tfrac{1}{2}\pi + \sin^{-1} a)\right].$$

3-6. A sinusoidal signal of amplitude A is added to Gaussian narrowband noise of mean-square amplitude $\sigma^2 = \tilde{\phi}(0)$. Show that the phase θ of the sum, measured with respect to that of the sinusoid, has the probability density function

$$p(\theta) = \frac{1}{2\pi}e^{-a^2/2} + \frac{1}{\sqrt{2\pi}}a \cos \theta \ e^{-\frac{1}{2}a^2 \sin^2 \theta}[1 - \text{erfc}(a \cos \theta)]$$

with $a = A/\sigma$. Work out a Gaussian approximation to this probability density function when $a \gg 1$ [Mid48].

3-7. Let $z_1 = x_1 + iy_1$ and $z_2 = x_2 + iy_2$ have circular Gaussian distributions as in (3-39) with $\tilde{\phi}_{kk} = 1, k = 1, 2$. These random variables are all statistically independent and have expected values zero. Find the probability that

$$|z_1| + |z_2| > a, \qquad a > 0.$$

3-8. Show that the likelihood ratio in (3-60) is equal to dQ_d/dQ_0, where Q_0 and Q_d are the false-alarm and detection probabilities calculated in Sec. 3.4.

3-9. Carry out the derivation of (3-84) from the previous expression for the error probability.

3-10. In an M-ary balanced channel one of M orthogonal signals is transmitted every T seconds with relative frequency $1/M$. The signals are received with energy E in white Gaussian noise of spectral density N, but with an unknown phase that can be taken as uniformly distributed over $(0, 2\pi)$. There is a possibility that fading might destroy the signals, and to indicate this a null zone is provided in the decision mechanism [Blo57]. The rectified outputs of filters matched to each of the signals are compared at the end of each interval $(0, T)$, and the receiver sends to the decoder the symbol corresponding to the filter whose rectified output is the greatest, except that if all the outputs fall below a certain amplitude a, the receiver indicates an "erasure." Calculate the probabilities that the transmitted signal is correctly received, that an erasure is indicated, and that an incorrect symbol is sent on to the decoder.

3-11. The signal $s(t; A) = A \operatorname{Re} F(t) \exp(i\Omega t + i\psi)$ is to be detected in narrowband Gaussian noise having a complex autocovariance function $\tilde{\phi}(t, u)$. The input is $v(t) = \operatorname{Re} V(t) \exp i\Omega t$, and the observation interval is $(0, T)$. The signal is subject to Rayleigh fading, so that its complex amplitude

$$a = A \, e^{i\psi} = a_x + ia_y,$$

as in (3-106), is a circular complex Gaussian random variable with

$$\operatorname{Var} a_x = \operatorname{Var} a_y = s^2;$$

$E(a) = 0$. Write the conditional likelihood functional $\Lambda[v(t)| \, a_x, a_y]$ for detecting this signal in the Gaussian noise in terms of the circular complex Gaussian random variable

$$z = \int_0^T G^*(t)V(t) \, dt,$$

where $G(t)$ is the solution of the integral equation (3-104). Now average the likelihood functional with respect to the random variables a_x and a_y to determine the average likelihood functional $\Lambda[v(t)]$ defined by (3-100), in which $\theta = (a_x, a_y)$. Show how this detection problem is equivalent to that in Example 1-3 with $n = 2$ and

$$N_0 = d_0^2 = \int_0^T G^*(t)F(t) \, dt, \qquad N_1 = N_0 + d_0^4 s^2.$$

One can always normalize the signal amplitude so that $d_0^2 = 1$.

3-12. Let the amplitude A of a narrowband signal $A \operatorname{Re} F(t) \exp(i\Omega t + i\psi)$ be distributed according to the Rayleigh distribution in (3-105), and let the phase ψ be uniformly distributed over $(0, 2\pi)$. It is to be detected in the presence of white Gaussian noise of unilateral spectral density N. Find the optimum detection statistic and relate its decision level to the critical value Λ_0 associated with the Bayes criterion (1-17). Calculate the minimum Bayes cost $\overline{C}_{\min}(s)$ as a function of s and investigate its behavior as s approaches 0. Show that as s vanishes, for $\Lambda_0 > 1$, $\overline{C}_{\min}(0) - \overline{C}_{\min}(s)$ approaches 0 faster than any power of the average signal-to-noise ratio. Take

$$\int_0^T |F(t)|^2 \, dt = 1$$

without loss of generality.

3-13. In the incoherent M-ary channel treated in Sec. 3.5.4, suppose that the amplitude of the signal is subject to Rayleigh fading, so that the complex amplitude of the jth signal has the form

$$S_j(t) = AF_j(t), \qquad 0 \le t \le T,$$

where the complex envelopes $F_j(t)$ are orthonormal over the interval $(0, T)$. The amplitudes A have a Rayleigh distribution as in Problem 3-11, with the variances s^2 the same for all signals. Assume that the signals are received in white Gaussian noise with unilateral spectral density N. Find the optimum receiver for deciding which of the M equally likely signals is present in its input, and calculate its probability of error. *Hint:* One way to do this is to express the likelihood functionals in (3-85) in terms of A and average with respect to the prior probability density function $z(A)$, using the normalization integral for the Rayleigh–Rice distribution in (3-73). It is simpler, however, to use the result of Problem 3-11.

3-14. Evaluate the performance of the maximum-likelihood receiver treated in Sec. 3.6.4 when the signals $s_k(t)$ of (3-110) are subject to Rayleigh fading, their complex amplitudes $a_x + ia_y$ having the circular Gaussian distribution in Problem 3-11 with Var a_x = Var $a_y = s^2$.

3-15. Carry through the analysis of the system described in Problem 3-10 under the same assumption as in Problem 3-13 that the signals are subject to Rayleigh fading.

4

Detection in Multiple Observations

4.1 OPTIMUM DETECTOR

Radar detection of a fixed target at a given range was treated in Chapters 2 and 3 as a matter of deciding whether an echo signal of specified form is present in the input $v(t)$ to a receiver during a certain interval $(0, T)$ after an electromagnetic pulse has been transmitted toward the location in question. How the receiver should process its input $v(t)$ in order to make that decision most efficiently was described there. Usually, however, a radar sends more than one pulse in the direction of a possible target, and the presence of a target is indicated not by only one signal, but by a train of echo signals appearing at the input to the receiver. If there is no target, the input contains only noise.

Denote by $v_k(t)$ the input to the receiver during the T-second interval following transmission of the kth pulse; the time t will be counted in each interval from its beginning. Suppose that the decision about the presence or absence of a target is to be based on the returns from M transmitted pulses, $k = 1, 2, \dots, M$. Denote the echo signal in the kth interval by $s_k(t)$. The receiver must then choose between two hypotheses,

$$
\begin{aligned}
H_0: \quad & v_k(t) = n_k(t), \\
H_1: \quad & v_k(t) = s_k(t) + n_k(t), \qquad 0 \le t \le T, \qquad k = 1, 2, \dots, M,
\end{aligned}
\tag{4-1}
$$

where $n_k(t)$ is the noise during the kth interval. We shall assume that this noise is white and Gaussian, with unilateral spectral density N_k in the kth interval. Our analysis could easily be extended to colored noise, provided that the intervals are separated sufficiently so that the noise in one interval is independent of that in any other, but to do so would not be particularly instructive. The reader will perceive how our equations need to be modified in order to account for noise that is not white.

Hypothesis H_1 will in general be composite; the signals $s_k(t)$ may depend on parameters such as amplitude and phase that are known only imprecisely. These parameters may even vary randomly from one input to another. Suitable assumptions about their prior probability density functions will have to be made. Hypothesis H_0, on the other hand, we shall assume at present to be simple. Although the noise spectral densities N_k may differ from one input to another, we presume they are known.

In a binary communication system transmitting information coded into 0's and 1's, a 1 may be dispatched by sending a signal in each of M successive intervals; these signals are received as $s_k(t)$, $k = 1, 2, \ldots, M$. For the 0's nothing is sent in any interval. The receiver will then be confronted with a hypothesis-testing problem of the type of (4-1). Alternatively, the transmitter might, for each 1, send a quasi-harmonic signal in each of M well separated frequency bands. These signals would be received as

$$s_k(t) = \operatorname{Re} S_k(t) \exp(i\Omega_k t + i\psi_k), \qquad (4\text{-}2)$$

where Ω_k is the carrier frequency at the center of the kth band. The receiver must again choose between hypotheses H_0 and H_1 as in (4-1), but now the inputs $v_k(t)$ on which its decision is based are observed simultaneously,

$$v_k(t) = \operatorname{Re} V_k(t) \exp i\Omega_k t, \qquad 0 \le t \le T;$$

$V_k(t)$ is the complex envelope of the input in the kth frequency band about carrier frequency Ω_k. When the inputs are simultaneous in this way, we assume them far enough apart in frequency so that their noise components $n_k(t)$ are all statistically independent. In *diversity* communications, signals are thus sent at different frequencies simultaneously in order to combat fading that may cause the transmissivity of the medium to vary randomly, but independently, in different frequency bands. Diversity techniques have been described and analyzed by Stein [Ste66] and Kennedy [Ken69].

In sonar detection acoustic echoes are picked up by an array of transducers that convert them to electrical signals. If $v_k(t)$ represents the output of the kth transducer of an M-element array and if $s_k(t)$ is the component of the signal induced in that transducer by the acoustic echo, the receiver must carry out a hypothesis test of the same type as in (4-1). The method by which it combines the inputs $v_k(t)$ in order most effectively to detect echoes from a target in a given direction is called *beamforming*. We shall treat it in an elementary way in Sec. 4.3. In Sec. 4.4 we shall describe a method for comparing the performances of two different ways of processing M independent inputs to the receiver when M is very large. Section 4.5 introduces the subject of distributed-detection systems in which a number of independent sensors search for a common signal and transmit their binary decisions to a central processor that makes the final decision as to its presence or absence.

In what follows we shall analyze the hypothesis test of (4-1) under various assumptions about the signals $s_k(t)$. We shall begin by assuming them completely known, and later we shall permit them to have phases ψ_k that are either identical, but random, or independently random from one signal to another. Both the design of the receiver and—insofar as possible—the calculation of its performance will be presented under these various conditions.

4.1.1 Complete Coherence

When the signals $s_k(t)$ are known in all respects, the optimum detector can be specified by a straightforward extension of the results of Chapter 2. Samples of any one input $v_k(t)$ are independent of those of any other input because we assume statistical independence of all the noise components $n_k(t)$. The joint probability density functions $p_0(v)$ and $p_1(v)$ of all the samples therefore factor into products

$$p_j(v) = \prod_{m=1}^{M} p_j^{(m)}(v), \qquad j = 0, 1,$$

of the probability density functions $p_j^{(m)}(v)$ of samples taken in the several inputs $v_m(t)$, $m = 1, 2, \ldots, M$. The likelihood ratio therefore also factors

$$\Lambda(v) = \prod_{m=1}^{M} \Lambda^{(m)}(v), \qquad \Lambda^{(m)}(v) = \frac{p_1^{(m)}(v)}{p_0^{(m)}(v)},$$

into a product of the likelihood ratios of the M individual inputs. When we pass to the limit of an infinite number of samples of each input, we find that the likelihood functional for the set $\{v_k(t)\}$ of inputs factors as a product

$$\Lambda[\{v_k(t)\}] = \prod_{m=1}^{M} \Lambda^{(m)}[v_m(t)]$$

of the likelihood functionals of each input. These are given by (2-74), and putting them together we write the overall likelihood functional as

$$\Lambda[\{v_k(t)\}] = \exp\left[2\sum_{m=1}^{M} \frac{1}{N_m} \int_0^T s_m(t) v_m(t)\, dt - \sum_{m=1}^{M} \frac{1}{N_m} \int_0^T [s_m(t)]^2\, dt \right].$$

The decision between hypotheses H_0 and H_1 is therefore optimally based on the sufficient statistic

$$g = \sum_{m=1}^{M} \frac{1}{N_m} \int_0^T s_m(t) v_m(t)\, dt. \tag{4-3}$$

The mth term of this sum is the output, at the end of the observation interval $(0, T)$, of a filter matched to the mth signal $s_m(t)$, weighted inversely with the strength N_m of the noise at its input.

As in Chapter 2, the statistic g is a Gaussian random variable under each hypothesis, and the reader can easily show that the false-alarm and detection probabilities are

$$Q_0 = \text{erfc } x, \qquad Q_d = \text{erfc}(x - D),$$

where the effective signal-to-noise ratio D^2 is given by

$$D^2 = \sum_{m=1}^{M} \frac{2E_m}{N_m}, \qquad E_m = \int_0^T [s_m(t)]^2 \, dt. \tag{4-4}$$

This quantity D^2, furthermore, is the maximum signal-to-noise ratio attained by any linear combination

$$g' = \sum_{m=1}^{M} a_m \int_0^T s_m(t) v_m(t) \, dt \tag{4-5}$$

of the outputs of the matched filters. For this reason, the weighting used in (4-3) is called *maximal-ratio combining* in studies of diversity communication systems.

When the signals are quasiharmonic with carrier frequencies Ω_m and phases ψ_m,

$$s_m(t) = \mathrm{Re}\, S_m(t) \exp(i\Omega_m t + i\psi_m), \qquad 1 \leq m \leq M, \tag{4-6}$$

we can write down the likelihood functional by putting $Q(t) = N_m^{-1} S_m(t) \exp i\psi_m$ into (3-54), and we obtain the functional

$$\Lambda[\{v_k(t)\}] = \exp\left[\mathrm{Re} \sum_{m=1}^{M} \exp(-i\psi_m) \frac{1}{N_m} \int_0^T S_m^*(t) V_m(t) \, dt \right.$$
$$\left. - \tfrac{1}{2} \sum_{m=1}^{M} \frac{1}{N_m} \int_0^T |S_m(t)|^2 \, dt \right] \tag{4-7}$$

of the complex envelopes $V_m(t)$ of the M inputs:

$$v_m(t) = \mathrm{Re}\, V_m(t) \exp i\Omega_m t, \qquad 1 \leq m \leq M.$$

When the signals $s_k(t)$ contain only a single common, but unknown phase $\psi_k \equiv \psi$, and the phase of the carrier of one signal relative to that of any other is known to the receiver, the signals are said to be *completely coherent*. If we assign to that common unknown phase ψ the uniform distribution over $(0, 2\pi)$, the average likelihood functional becomes

$$\overline{\Lambda}[\{v_k(t)\}] = e^{-D^2/2} I_0(R),$$

with

$$R = \left| \sum_{m=1}^{M} \frac{1}{N_m} \int_0^T S_m^*(t) V_m(t) \, dt \right|. \tag{4-8}$$

In order to combine inputs associated with different carrier frequencies Ω_m, they must first be brought to a common intermediate frequency (i–f) Ω_0. This can be done by mixing the mth input with a locally generated signal $2\cos(\Omega_m - \Omega_0)t$ by means of a multiplier followed by a filter that discards the components in the product having carrier frequency $2\Omega_m - \Omega_0$:

$$2v_m(t) \cos(\Omega_m - \Omega_0)t$$
$$= \tfrac{1}{2}[V_m(t)e^{i\Omega_m t} + V_m^*(t)e^{-i\Omega_m t}] \cdot [e^{i(\Omega_m-\Omega_0)t} + e^{-i(\Omega_m-\Omega_0)t}]$$
$$= \tfrac{1}{2}[V_m(t)e^{i\Omega_0 t} + V_m^*(t)e^{-i\Omega_0 t} + V_m(t)e^{i(2\Omega_m-\Omega_0)t} + V_m^*(t)e^{-i(2\Omega_m-\Omega_0)t}]$$
$$\rightarrow \text{Re } V_m(t)e^{i\Omega_0 t}.$$

The process is the same as that in a heterodyne receiver.

These inputs, now "at i–f," are passed through filters matched to the signals Re $S_m(t) \exp i\Omega_0 t$. The output of the mth filter is weighted by N_m^{-1}, and the outputs are added, after being brought into simultaneity by appropriate delay lines if necessary. The weighted sum is passed through a linear rectifier, and its output is sampled at what corresponds to the time $t = T$ to produce the statistic R. It is essential that the relative phases of the signals one to another be known precisely if the inputs are to be combined in this manner before rectification; the signals must be completely coherent.

By the same analysis as in Sec. 3.4, the false-alarm and detection probabilities for this receiver are

$$Q_0 = e^{-b^2/2}, \qquad Q_d = Q(D, b) \tag{4-9}$$

in terms of Marcum's Q function (3-76), with the signal-to-noise ratio D^2 given by (4-4), in which the energy of the mth signal is now

$$E_m = \tfrac{1}{2}\int_0^T |S_m(t)|^2 \, dt.$$

4.1.2 Incoherent Signals

A radar receiver is to decide whether a target is present at a given range on the basis of the returns from M transmitted pulses. By

$$v_k(t) = \text{Re } V_k(t) \, e^{i\Omega t}, \qquad 1 \le k \le M,$$

we denote the input to the receiver following the kth transmitted pulse, gated in such a way that the interval $(0, T)$ just encompasses the arrival of an echo from the range in question. The echo signal in the kth interval will be

$$s_k(t) = \text{Re } S_k(t) \exp(i\Omega t + i\psi_k). \tag{4-10}$$

If the receiver fails to synchronize its local oscillator accurately with the transmitted pulses or if the target moves erratically over distances on the order of or greater than a wavelength of the radiation between one observation interval and the next, the phases ψ_k of these echoes will be independently random. We assume them uniformly distributed over $(0, 2\pi)$. The signals are then said to be completely *incoherent*.

Under hypothesis H_0 no target is present, and the M inputs $v_k(t)$ contain only white Gaussian noise of unilateral spectral density N. Under hypothesis H_1 they contain in addition the quasiharmonic signals in (4-10) with independently random phases ψ_k. The likelihood functional for deciding between H_0 and H_1 is then given by (4-7) after it is averaged over the phases ψ_m. If for convenience we utilize the normalized variable z_j defined by

$$z_j = x_j + iy_j = \frac{1}{\sqrt{2N_j E_j}} \int_0^T S_j^*(t) V_j(t)\, dt, \tag{4-11}$$

the average likelihood functional becomes

$$\overline{\Lambda}[\{v_k(t)\}] = \prod_{m=1}^M \int_0^T \frac{d\psi_m}{2\pi} \exp[-\tfrac{1}{2}d_m^2 + d_m \operatorname{Re}[z_m \exp(-i\psi_m)]]$$

$$= \prod_{m=1}^M \exp(-\tfrac{1}{2}d_m^2) I_0(d_m r_m), \qquad r_m = |z_m|, \qquad d_m^2 = \frac{2E_m}{N_m}. \tag{4-12}$$

This average likelihood functional must be compared with a decision level Λ_0, which under the Neyman–Pearson criterion is chosen to yield a preassigned false-alarm probability. Alternatively, the receiver forms the sufficient statistic

$$\Gamma = \sum_{m=1}^M \ln I_0(d_m r_m) \tag{4-13}$$

and compares it with

$$\Gamma_0 = \ln \Lambda_0 + \tfrac{1}{2}D^2, \qquad D^2 = \sum_{m=1}^M d_m^2.$$

This detection statistic Γ can be generated by amplifying the output of the matched filter during the mth interval by a factor proportional to the mth signal strength d_m and applying it to a rectifier whose characteristic is $\ln I_0(x)$, by which we mean that its output is $\ln I_0(|W(t)|)$ when its input is $\operatorname{Re} W(t) \exp i\Omega t$. The outputs of the rectifier are sampled at the end of each interval, the samples are added, and their sum is compared with the decision level Γ_0. If $\Gamma \geq \Gamma_0$, the receiver decides that a train of echoes has arrived.

The form of the rectifier characteristic $y = \ln I_0(x)$ is shown in Fig. 4-1. For small values of x,

$$\ln I_0(x) = \frac{1}{4}x^2 - \frac{1}{64}x^4 + O(x^6), \tag{4-14}$$

and for large values, by (3-61),

$$\ln I_0(x) = x - \frac{1}{2}\ln(2\pi x) + \frac{1}{8x} + O(x^{-2}); \tag{4-15}$$

see (C-16). For small values of $d_m r_m$ the optimum detector uses a rectifier that is nearly quadratic; for large values it is nearly linear. This type of detector was derived by Marcum [Mar48], and its use has been discussed by Woodward and Davies [Woo50], Middleton [Mid53], Fleishman [Fle57], and others. Because of the nonlinearity of the function $\ln I_0(x)$ it is very difficult to determine the false-alarm and detection probabilities for the receiver utilizing the statistic Γ of (4-13).

This receiver depends rigidly on the values of the signal-to-noise ratios d_m^2, $1 \leq m \leq M$. Even if they are all equal, it is necessary to know their common value. The system cannot provide a test for hypothesis H_0 versus hypothesis H_1 that is uniformly most powerful with respect to the amplitudes of the signals.

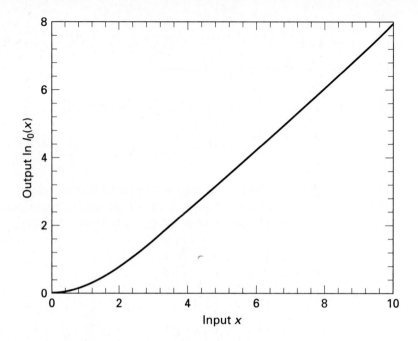

Figure 4-1. Optimum rectifier characteristic, $y = \ln I_0(x)$.

4.2 THE THRESHOLD DETECTOR

4.2.1 The Weak-signal Approximation

If in the detection of M incoherent signals, treated in Sec. 4.1.2, the signal amplitudes are unknown, there are three courses the designer can follow. The first is to pick a typical set of signal-to-noise ratios d_m^2 as specifying a standard set of signals for which the detector is to be optimum. The form of the detector is then given by (4-13). The probability of detecting signals with some other set of parameters d_m^2 will be less than it might have been, had the M values of d_m^2 been known in advance, but the loss of detectability will not in most situations be serious.

A second course is to choose a joint probability density function $z(d_1, d_2, \dots, d_m)$ for the signal-strength parameters d_m and to have the receiver base its decisions on an average likelihood functional

$$\overline{\Lambda} = \int_0^\infty dd_1 \int_0^\infty dd_2 \dots \int_0^\infty dd_M z(d_1, d_2, \dots, d_M) \prod_{k=1}^{M} I_0(d_k r_k) \exp(-\tfrac{1}{2} d_k^2). \qquad (4\text{-}16)$$

If the signals are communication signals that have passed through a fading channel in which their amplitudes fluctuate randomly, such a joint prior density function may be derivable from the nature of the fluctuations.

Suppose, for instance, that the channel suffers independent Rayleigh fading from one input to the next and that the expected signal-to-noise ratio is the same in each input:

$$z(d_1, d_2, \ldots, d_M) = \prod_{k=1}^{M} z(d_k),$$

$$z(d) = \frac{d}{s^2} \exp\left(-\frac{d^2}{2s^2}\right) U(d). \tag{4-17}$$

The integration in (4-16) can be carried out by means of the normalization integral for the Rayleigh–Rice distribution (3-73), or we can more easily use (3-106) and the method of Problem 3-11 to show that the average likelihood functional is

$$\overline{\Lambda} = (1 + s^2)^{-M} \exp\left[\frac{s^2}{2(1 + s^2)} \sum_{k=1}^{M} r_k^2\right], \tag{4-18}$$

where the r_k are, as in (4-12),

$$r_k = \frac{1}{\sqrt{2N_k E_k}} \left|\int_0^T S_k^*(t) V_k(t)\, dt\right|$$

in terms of the complex envelopes $S_k(t)$ of the signals, their energies E_k, and the unilateral spectral densities N_k of the noise in the several inputs. These statistics r_k are independent of the amplitudes of the signals. A sufficient statistic is now

$$U = \tfrac{1}{2} \sum_{k=1}^{M} r_k^2. \tag{4-19}$$

It can be formed by passing each input $\mathrm{Re}\, V_k(t) \exp i\Omega t$ through a filter matched to the signal $\mathrm{Re}\, S_k(t) \exp i\Omega t$, which is followed by a quadratic rectifier whose output is sampled at the end of the observation interval $(0, T)$; the M samples are then appropriately weighted and summed.

A third approach for the designer is to assume the least favorable situation of very weak signals and adopt the threshold statistic defined in (3-109). The product of exponential and Bessel functions in (4-16) is expanded in a power series in which only terms of first order in the signal-to-noise ratios are retained,

$$I_0(d_k r_k) \exp(-\tfrac{1}{2}d_k^2) \approx 1 + \tfrac{1}{4}(r_k^2 - 2)d_k^2,$$

by (3-61) and the series for the exponential function, and

$$\prod_{k=1}^{M} I_0(d_k r_k) \exp(-\tfrac{1}{2}d_k^2) \approx 1 + \tfrac{1}{4} \sum_{k=1}^{M} d_k^2(r_k^2 - 2).$$

Putting this into (4-16) and denoting by $\langle d_k^2 \rangle$ the expected value of the kth signal-to-noise ratio, we find for the average likelihood functional the approximate form

$$\overline{\Lambda} \approx 1 + \tfrac{1}{4} \sum_{k=1}^{M} \langle d_k^2 \rangle (r_k^2 - 2), \qquad \langle d_k^2 \rangle = \int_0^{\infty} d_k^2\, z(d_k)\, dd_k.$$

This is called the *weak-signal* or the *threshold* approximation.

If the signals to be detected are the M repetitions of a pulse conveying the symbol 1 of a binary message and if these signals all arrive with the same average strength in noise of the same spectral density N, the average signal-to-noise ratios

$\langle d_k^2 \rangle$ will be the same, and the average likelihood functional depends on the inputs only through a threshold statistic U that has the same form as that in (4-19). This also serves as the threshold statistic in radar when the target and the antenna are stationary, so that all echoes provide the same average signal-to-noise ratio. The receiver is said to *integrate* M filtered and quadratically rectified inputs.

If, on the other hand, the signals are echoes from a fixed target of constant reflectivity past which the antenna is rotating, the average signal-to-noise ratios $\langle d_k^2 \rangle$ will be proportional to a function $f(\theta)$ representing the combined gains of the radar antenna on transmission and reception:

$$\langle d_k^2 \rangle = \langle d^2 \rangle f(\theta_k - \theta_0), \qquad (4\text{-}20)$$

where θ_k is the azimuth of the antenna at the instant the kth echo is received, and θ_0 is the azimuth of the target. The values of the beam factor $f(\theta)$, if known, could be used to calculate an improved detection statistic of the form

$$U' = \tfrac{1}{2} \sum_{k=1}^{M} f(\theta_k - \theta_0) r_k^2, \qquad (4\text{-}21)$$

which requires knowing more about the signals than does the sum of squares in (4-19). If the energies of the received echoes are truly proportional to $f(\theta_k - \theta_0)$ as in (4-20), the statistic U' will detect the train of signals more reliably than the statistic U of (4-19).

4.2.2 Performance of the Quadratic Threshold Detector

Not only the threshold detector for quasiharmonic signals with independently random phases, but also the optimum detector for such signals when subject to independent Rayleigh fading is based on the sum U of the sampled outputs of quadratic rectifiers, as in (4-19). We now calculate the false-alarm probability Q_0 and the probability Q_d of detecting a train of M signals with given phases and amplitudes:

$$Q_0 = \Pr(U \geq U_0 | H_0), \qquad Q_d = \Pr(U \geq U_0 | H_1),$$

where U_0 is the decision level with which U is compared. The latter probability will be worked out first; the former follows directly from it when we set the signal strengths equal to zero.

The statistic U in (4-19) is the sum of M random variables $\tfrac{1}{2} r_k^2$ that are independent because the noise in one input is statistically independent of that in any other. The probability density function of the sum U of a number of independent random variables is most easily determined from its characteristic function $E(\exp i\omega U | H_1)$ or its moment-generating function

$$h_1(z) = E(e^{-zU} | H_1) = \int_{-\infty}^{\infty} p_1(U) \, e^{-zU} \, dU,$$

which is the Laplace transform of the probability density function $p_1(U)$. For the statistic U in (4-19)

$$h_1(z) = E \exp\left[-\tfrac{1}{2}z \sum_{j=1}^{M} r_j^2 \middle| H_1\right],$$

and because the random variables

$$r_j^2 = x_j^2 + y_j^2 = |z_j|^2$$

are statistically independent, we find

$$h_1(z) = \prod_{j=1}^{M} E\left[\exp[-\tfrac{1}{2}z(x_j^2 + y_j^2)] \middle| H_1\right]. \tag{4-22}$$

Here x_j and y_j are the real and imaginary parts of the circular complex Gaussian random variable z_j defined in (4-11).

The random variables x_j and y_j are independent and Gaussian, with unit variances and expected values given by

$$E(z_j| H_1) = E(x_j + iy_j| H_1) = d_j \exp i\psi_j;$$

that is,

$$E(x_j| H_1) = d_j \cos \psi_j, \qquad E(y_j| H_1) = d_j \sin \psi_j.$$

Their joint probability density function has the circular Gaussian form

$$p_1(x_j, y_j) = \tilde{p}_1(z_j) = \frac{1}{2\pi} \exp[-\tfrac{1}{2}|z_j - d_j \exp i\psi_j|^2]$$

$$= \frac{1}{2\pi} \exp\{-\tfrac{1}{2}[(x_j - d_j \cos \psi_j)^2 + (y_j - d_j \sin \psi_j)^2]\}. \tag{4-23}$$

The moment-generating function of $\tfrac{1}{2}x_j^2$ is therefore, as in (4-22),

$$E[\exp(-\tfrac{1}{2}zx_j^2)] = \frac{1}{\sqrt{2\pi}} \int_{-\infty}^{\infty} \exp[-\tfrac{1}{2}(x_j - d_j \cos \psi_j)^2 - \tfrac{1}{2}zx_j^2] \, dx_j$$

$$= \frac{1}{\sqrt{1 + z}} \exp\left[-\frac{zd_j^2 \cos^2 \psi_j}{2(1 + z)}\right],$$

and the moment-generating function of $\tfrac{1}{2}y_j^2$ has the same form, but with $\cos^2 \psi_j$ replaced by $\sin^2 \psi_j$. Putting these into (4-22), we find for the moment-generating function of the statistic U

$$h_1(z) = \frac{1}{(1 + z)^M} \exp\left[-\frac{D^2 z}{2(1 + z)}\right], \tag{4-24}$$

where D^2 is the total signal-to-noise ratio as in (4-4).

The probability density function of U is the inverse Laplace transform of (4-24) and can be found from a table to have the form

$$p_1(U) = \left(\frac{U}{S}\right)^{(M-1)/2} e^{-U-S} I_{M-1}(2\sqrt{SU})U(U), \qquad S = \tfrac{1}{2}D^2, \tag{4-25}$$

[Erd54, vol. 1, p. 197, eq. (18)], where $I_{M-1}(x)$ is the modified Bessel function of order $M - 1$. This density function depends only on the total energy-to-noise ratio $S = D^2/2$ and not on how the energy in the received signals is divided among them.

The probability Q_d of detecting a set of signals of the form (4-6) with independently random phases ψ_m and effective signal-to-noise ratios d_m^2 is then

$$Q_d = \int_{U_0}^{\infty} p_1(U)\, dU = Q_m(D, \sqrt{2U_0}) \tag{4-26}$$

in terms of the Mth order Q function

$$Q_M(\alpha, \beta) = \int_{\beta}^{\infty} x \left(\frac{x}{\alpha}\right)^{M-1} \exp[-\tfrac{1}{2}(x^2 + \alpha^2)] I_{M-1}(\alpha x)\, dx, \tag{4-27}$$

with D^2 the total signal-to-noise ratio defined in (4-4). The Mth-order Q function generalizes Marcum's Q function defined in (3-76) and is related to it through

$$Q_M(\alpha, \beta) = Q(\alpha, \beta) + e^{-(\alpha^2 + \beta^2)/2} \sum_{r=1}^{M-1} \left(\frac{\beta}{\alpha}\right)^r I_r(\alpha\beta).$$

Recursive methods for computing the detection probability Q_d are outlined in the Appendix, Sec. C.3. These recursive calculations are laborious, however, when M is large. Approximations and alternative methods suitable for $M \gg 1$ will be developed in Chapter 5.

That the probability Q_d in (4-26) is independent of the set of phases ψ_k of the signals indicates—according to a criterion to be developed in Sec. 7.6—that the uniform distribution we assumed for them is least favorable. The probability of detection is also independent of how the total signal-to-noise ratio D^2 is divided among the several signals $s_k(t)$, a confirmation of the natural expectation that the least favorable distribution of the average signal-to-noise ratios is the uniform one $\langle d_k^2 \rangle \equiv \langle d^2 \rangle = D^2/M$ that we used in deriving the threshold statistic (4-19).

The probability density function of the statistic U under hypothesis H_0 can be found by taking the inverse Laplace transform of the moment-generating function obtained by setting $D = 0$ in (4-24). Alternatively we can substitute into (4-25) the power series for the modified Bessel function

$$I_{M-1}(x) = \sum_{k=0}^{\infty} \frac{(x/2)^{M-1+2k}}{k!(M-1+k)!} \tag{4-28}$$

with $x = 2\sqrt{SU}$, after which we let S go to zero. Only the first term of the series remains, and we find

$$p_0(U) = \frac{U^{M-1}}{(M-1)!} e^{-U} U(U), \tag{4-29}$$

which is known as the *gamma* distribution. Integrating (4-29), we find for the false-alarm probability

$$Q_0 = \int_{U_0}^{\infty} p_0(U)\, dU = \sum_{k=0}^{M-1} \frac{U_0^k}{k!} \exp(-U_0), \tag{4-30}$$

which is easily programmed for a calculator. Values can also be found from tables of the incomplete gamma function,

$$Q_0 = 1 - I(M^{-1/2}U_0, M - 1), \qquad I(u, p) = \frac{1}{p!} \int_0^{u\sqrt{p+1}} x^p \, e^{-x} \, dx$$

[Pea34]. For the false-alarm probability $Q_0 = 10^{-p}$, p an integer from 1 to 12 and $1 \leq M \leq 150$, the decision level U_0 can be determined from tables published by Pachares [Pac58]. It is not difficult to program the solution of (4-30) for U_0 by Newton's method:

$$U_0 \leftarrow U_0 + \frac{q(U_0)[\ln q(U_0) - \ln Q_0]}{p_0(U_0)},$$

where $q(U_0)$ is the function on the right side of (4-30). One can start with the trial value $U_0 = -\ln Q_0 + M - 1$. An alternative method, useful when $M \gg 1$, will be presented in Sec. 5.3.3.

Figure 4-2 exhibits the detection probability Q_d versus the quantity $D = \sqrt{2S}$ for a false-alarm probability $Q_0 = 10^{-6}$ and various numbers M of signals. (When the spectral density of the noise equals N in all inputs, $S = E_T/N$, where E_T is the total received signal energy.) Additional graphs can be found in [Mar48] and in [DiF68]. The total energy required to attain a given probability of detection increases with the number M of observations: the greater the number M of incoherent pulses among which its total energy is divided, the more difficult the signal is to detect. For a given energy $E_j \equiv E$ per pulse, on the other hand, the probability of detection of course increases with the number M of signals.

Figure 4-3 shows the loss incurred when the total signal energy E_T is divided among M incoherent signals. Let D_M^2 be the signal-to-noise ratio required to attain a probability Q_d of detection for a false-alarm probability Q_0 given by (4-30). The loss is then defined as $10 \log_{10}(D_M^2/D_1^2)$. For the curves in Fig. 4-3 the false-alarm probability was set at 10^{-6}. For $M = 1000$ the loss is on the order of 9 dB.

In Fig. 4-4 we compare four different situations: (a) The M signals are coherent, and the probability Q_d of detection is given by (4-9). (b) The signal has a fixed amplitude, but appears in only one input, and the decision is made as in the maximum-likelihood receiver described in Sec. 3.6.4, whereupon Q_d is given by (3-119) and the false-alarm probability Q_0 by (3-118). (c) The M signals are incoherent; the total received signal energy is fixed, but arbitrarily divided among all M inputs; and the threshold detector of (4-19) is used, so that Q_d is given by (4-26) and Q_0 by (4-30). (d) The maximum-likelihood receiver is utilized, but the energy of the signal is divided equally among all M inputs, whereupon

$$Q_d = 1 - [1 - Q(M^{-1/2}D, r_0)]^M,$$

with the false-alarm probability again given by (3-118). These detection probabilities have been calculated for $M = 20$ and $Q_0 = 10^{-6}$ and plotted in Fig. 4-4.

4.2.3 Detection Probability for Rayleigh Fading

In Sec. 4.2.1 we found that the optimum statistic for detecting the set of quasi-harmonic signals $s_k(t)$ when they are subject to independent Rayleigh fading is the sum of squares U in (4-19). The average probability $\overline{Q}_d$ of detection with respect to the class of signals whose strength parameters d_k have the Rayleigh distribution

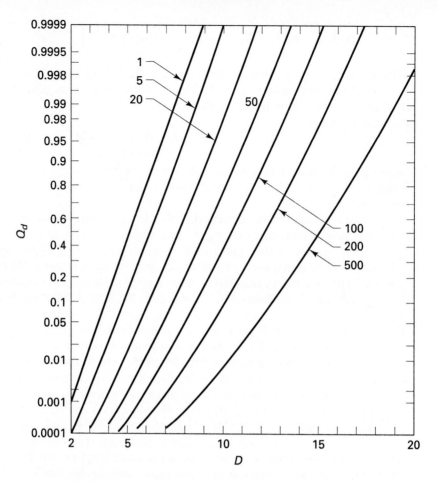

Figure 4-2. Detection probability for incoherent signals of fixed amplitude. $Q_0 = 10^{-6}$. Curves are indexed by the number M of inputs.

can be found by averaging the detection probability Q_d in (4-26) with respect to the joint probability density function of those parameters given in (4-17). It is simpler, however, to proceed as follows.

The probability density function $p_0(U)$ of the statistic U under hypothesis H_0 was shown in Sec. 4.2.2 to be given by (4-29). The likelihood ratio for U is

$$\frac{p_1(U)}{p_0(U)} = (1 + s^2)^{-M} \exp\left(\frac{s^2 U}{1 + s^2}\right)$$

from (4-18) and (4-19). Hence as in (1-30) the probability density function of U under hypothesis H_1 must be

$$p_1(U) = \frac{1}{1 + s^2} \frac{1}{(M - 1)!} \left(\frac{U}{1 + s^2}\right)^{M-1} \exp\left(-\frac{U}{1 + s^2}\right) U(U).$$

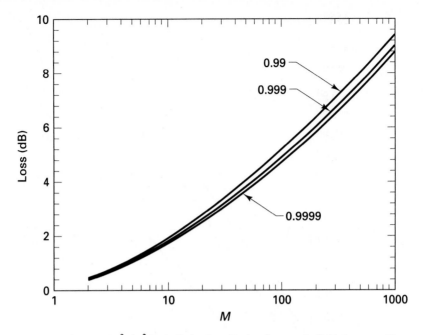

Figure 4-3. Loss D_M^2/D_1^2 in decibels when the signal energy is divided among M incoherent pulses; $Q_0 = 10^{-6}$. The curves are indexed with the probability Q_d of detection.

The random variable $U/(1 + s^2)$ has the gamma distribution. The average probability of detecting these Rayleigh-fading signals is therefore

$$\overline{Q}_d = \sum_{k=0}^{M-1} \frac{U'^k}{k!}\, e^{-U'}, \qquad U' = \frac{U_0}{1 + s^2}, \tag{4-31}$$

and can be calculated by the same algorithm as the false-alarm probability Q_0, which is still given by (4-30).

This average probability of detection is plotted in Fig. 4-5 versus the parameter $(2Ms^2)^{1/2} = \langle D^2 \rangle^{1/2}$ for a false-alarm probability $Q_0 = 10^{-6}$ and various numbers M of signals; $\langle D^2 \rangle$ is the average total signal-to-noise ratio. When the number M of fading signals is small, the average total energy required to attain a detection probability on the order of 0.9 or more is much greater than the total energy required for signals of fixed amplitude. The larger M, however, the more closely do the curves in Fig. 4-5 approach those in Fig. 4-2, and the less deleterious is the effect of the fading on the probability of detection.

If all the signal energy is concentrated in a single Rayleigh-fading signal, the average received energy-to-noise ratio s^2 necessary in order to attain a reasonable average probability Q_d of detection is very large, as can be seen from Fig. 4-5. For instance, for $Q_0 = 10^{-6}$ and $Q_d = 0.999$, $S = s^2 = 13,807.6$. The more independently fading signals among which the total energy is divided—up to a point—, the lower is the required average total received energy-to-noise ratio $S = Ms^2$. In Fig. 4-6 we have plotted this energy-to-noise ratio $S = \langle E_T \rangle/N$ versus the number

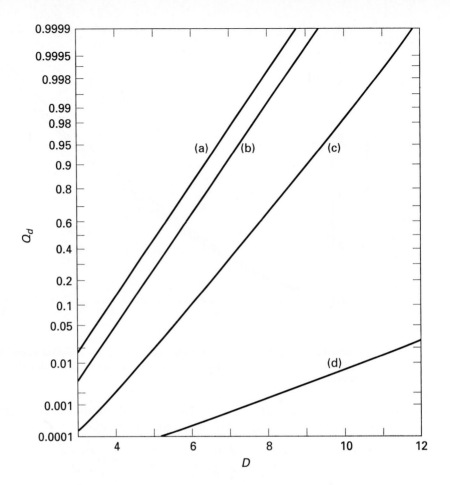

Figure 4-4. Detection probabilities for cases (a) through (d) described in the text; $M = 20$, $Q_0 = 10^{-6}$.

M of inputs for $Q_0 = 10^{-6}$ and three values of the average detection probability $\overline{Q}_d$. The energy-to-noise ratio is smallest when the energy is shared among from fifteen to thirty signals.

Suppose that as in Sec. 3.6.4 it is known that the signal will appear in only one of the M inputs, but in which one is unknown, and suppose that once again the decision that it is present is made whenever any of the M quantities r_j defined in (4-11) exceeds a decision level r_0. This we called the maximum-likelihood receiver. If that signal is subject to Rayleigh fading, the false-alarm probability is again given by (3-118), but the average probability $\overline{Q}_d$ of detecting the signal is now

$$\overline{Q}_d = 1 - \left[1 - \exp\left(-\frac{U_0}{1 + \overline{S}}\right)\right][1 - \exp(-U_0)]^{M-1}, \qquad U_0 = \tfrac{1}{2}r_0^2, \qquad (4\text{-}32)$$

where $\overline{S} = \tfrac{1}{2}\langle D^2 \rangle$ is the average energy-to-noise ratio in that input containing the signal.

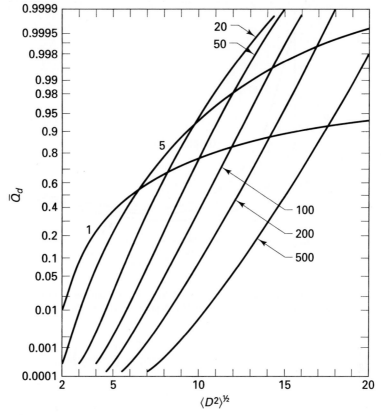

Figure 4-5. Detection probability for Rayleigh-fading incoherent signals versus $\langle D^2 \rangle^{1/2}$; $\langle D^2 \rangle$ is the average signal-to-noise ratio. $Q_0 = 10^{-6}$. Curves are indexed by the number M of inputs.

4.2.4 Other Types of Fading Signals

The decision about the presence or absence of a train of signal pulses is often based on the sum U of the outputs of a quadratic detector (4-19), whatever the distribution of the fading signal amplitudes, although U is in general not the optimum statistic. The average probability $\overline{Q}_d$ of detection can then be obtained by averaging (4-26) with respect to the resultant probability distribution of the quantity D^2 defined in (4-4). This is $D^2 = 2E_T/N$ when the noise spectral densities N_j in all the inputs are assumed equal to N, with E_T the total received signal energy. The decision level U_0 with which U is compared is still determined as in (4-30).

The probability density function $p_1(U)$ of the statistic U under hypothesis H_1 is not usually simple to calculate for an arbitrary type of fading. The moment-generating function

$$\overline{h}(z) = E(e^{-zU} \mid H_1)$$

can easily be written down, however, in terms of that of the total energy-to-noise ratio $S = \frac{1}{2}D^2 = E_T/N$. In (4-24) we replace $\frac{1}{2}D^2$ by S and average with respect to the prior probability density function $z(S)$ of S to obtain

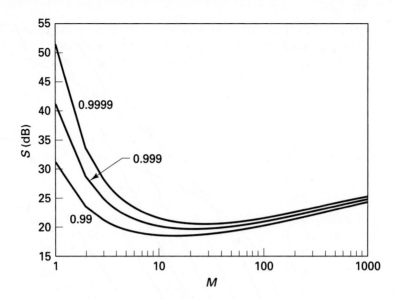

Figure 4-6. Energy-to-noise ratio $10 \log_{10} S$ when the energy of the Rayleigh-fading signals is divided among M independent signals. $Q_0 = 10^{-6}$. The curves are indexed with the average detection probability $\overline{Q}_d$.

$$\overline{h}(z) = \frac{1}{(1 + z)^M} h_S \left(\frac{z}{1 + z} \right) \tag{4-33}$$

with

$$h_S(z) = E(e^{-zS}) = \int_0^\infty z(S) \, e^{-Sz} \, dS$$

the moment-generating function of the total energy-to-noise ratio S.

Swerling [Swe60] described four common types of signal fading, which are known in the radar literature as the Swerling cases. We summarize them as follows.

Case 1. The M signal amplitudes fade together, their common amplitude having a Rayleigh distribution. A distribution of this kind arises when a radar pulse impinges on a complicated target and is reflected from many points thereon. The great number of reflected waves combine with random amplitudes and phases, and the complex amplitude of their sum has real and imaginary parts that are approximately Gaussian random variables by virtue of the central limit theorem. The amplitude of the resultant echo then has a Rayleigh distribution as in (4-17), and the total energy-to-noise ratio S has the exponential distribution

$$z(S) = \frac{1}{\overline{S}} \, e^{-S/\overline{S}} U(S), \tag{4-34}$$

where $\overline{S} = Ms^2$ is the average total energy-to-noise ratio.

Case 2. The M signal amplitudes fade independently and again have a Rayleigh distribution; this is the phenomenon analyzed in Sec. 4.2.3. In contrast to case 1, the radar target changes its position and its orientation significantly and erratically during the intervals between irradiation by the transmitted pulses, and the complex amplitudes of successive echoes can be assumed to be statistically independent. Now the total energy-to-noise ratio S has a scaled gamma distribution:

$$z(S) = \frac{1}{(M-1)! \, \overline{S}'} \left(\frac{S}{\overline{S}'}\right)^{M-1} e^{-S/\overline{S}'} U(S), \qquad \overline{S}' = \frac{\overline{S}}{M}. \qquad (4\text{-}35)$$

Case 3. The M signal amplitudes fade together, and the density function of the common individual signal-to-noise ratio $d_j \equiv d$ has the form

$$z(d) = \frac{d^3}{2s^4} e^{-d^2/2s^2} U(d). \qquad (4\text{-}36)$$

This density function has a narrower peak than the Rayleigh distribution in case 1, as will happen when a few points on the radar target scatter the incident radiation much more strongly than the others. The probability density function of the total energy-to-noise ratio is then

$$z(S) = \frac{4S}{\overline{S}^2} e^{-2S/\overline{S}} U(S). \qquad (4\text{-}37)$$

Case 4. The M signal amplitudes fade independently, the individual signal-to-noise ratios having the same density function as in (4-36). The density function of the total energy-to-noise ratio is then

$$z(S) = \frac{1}{(2M-1)! \, \overline{S}'} \left(\frac{S}{\overline{S}'}\right)^{2M-1} e^{-S/\overline{S}'} U(S), \qquad \overline{S}' = \frac{\overline{S}}{2M}. \qquad (4\text{-}38)$$

In all four of these Swerling cases the total energy-to-noise ratio S has a gamma distribution with density function

$$z(S) = \frac{1}{\Gamma(k)\overline{S}'} \left(\frac{S}{\overline{S}'}\right)^{k-1} e^{-S/\overline{S}'} U(S), \qquad \overline{S}' = \frac{\overline{S}}{k}, \qquad \overline{S} = E(S), \qquad (4\text{-}39)$$

for which the moment-generating function is

$$h_S(z) = (1 + \overline{S}'z)^{-k}. \qquad (4\text{-}40)$$

For the four Swerling cases, $k = 1, M, 2,$ and $2M$, respectively. By (4-33) the quadratic statistic U of (4-19) has the moment-generating function

$$\overline{h}(z) = (1 + z)^{k-M}(1 + bz)^{-k}, \qquad b = 1 + \frac{\overline{S}}{k}. \qquad (4\text{-}41)$$

The probability density functions of the threshold statistic U and the resulting detection probabilities $\overline{Q}_d = \Pr(U \geq U_0 | H_1)$ for the four types of fading represented by the Swerling cases are to be found in the book by DiFranco and Rubin [DiF68], which also presents graphs of $\overline{Q}_d$ versus the average input energy-to-noise ratio for a number of values of M and Q_0. Appendix E modifies the recurrence methods of Appendix C for such fading signals. Closed-form expressions for these detection probabilities have been given by Hou et al. [Hou87]. When $M \gg 1$, however,

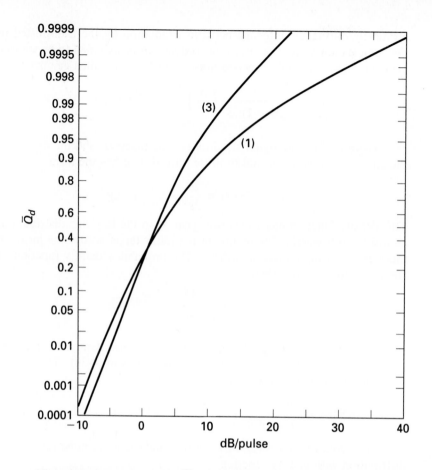

Figure 4-7. Average probability $\overline{Q}_d$ of detection versus average energy-to-noise ratio $10 \log_{10}(\overline{S}/M)$ for Swerling cases 1 and 3; $M = 20$, $Q_0 = 10^{-6}$.

computing $\overline{Q}_d$ by those methods is a lengthy process. Broadly applicable alternative methods will be described in Chapter 5.

In Figs. 4-7 and 4-8 we show the average probability of detection as a function of $\overline{S}/M$ in decibels for "integration" of $M = 20$ pulses suffering fading according to Swerling's four distributions. The curve on Fig. 4-8 marked ∞ refers to unfading signals. When as in cases 1 and 3 the entire pulse train fades together, a much greater average energy-to-noise ratio is required than when the signals fade independently.

When in radar studies such as [Mar48] and [DiF68] the performances of receivers integrating different numbers M of signals are compared, it is often not the false-alarm probability Q_0 that is held fixed, but the *false-alarm number* N_{fa}, defined as

$$N_{\text{fa}} = \frac{M \ln 2}{Q_0}. \tag{4-42}$$

This parameter arises through the following argument.

Detection in Multiple Observations   Chap. 4

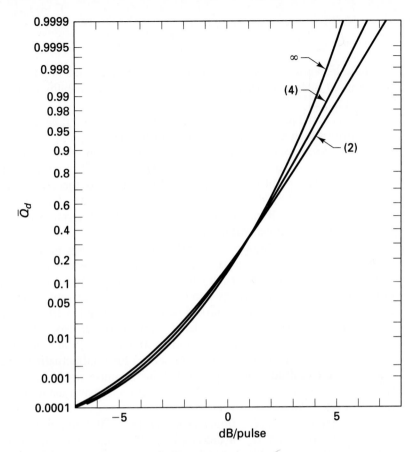

Figure 4-8. Average probability $\overline{Q}_d$ of detection versus average energy-to-noise ratio $10 \log_{10}(\overline{S}/M)$ for Swerling cases 2 and 4; $M = 20$, $Q_0 = 10^{-6}$. The curve marked ∞ represents Q_d for unfading signals.

Consider detection of a target at a particular range by a radar that sends out a pulse every τ_r seconds; $1/\tau_r$ is called the *pulse-repetition rate*. In a total observation time $T_{\text{obs}} \gg \tau_r$, the number of decisions made equals

$$n = \frac{T_{\text{obs}}}{M\tau_r},$$

for when M pulses are integrated, a decision is made only every $M\tau_r$ seconds. The probability that no false alarm occurs in that time T_{obs} will be $(1 - Q_0)^n$. Now define the false-alarm time τ_{fa} as the observation time T_{obs} within which this probability equals $\frac{1}{2}$:

$$(1 - Q_0)^{n'} = \frac{1}{2}, \qquad n' = \frac{\tau_{\text{fa}}}{M\tau_r}. \tag{4-43}$$

Thus in a time τ_{fa} the probability of having at least one false alarm equals $\frac{1}{2}$. With $Q_0 \ll 1$, this equation becomes

$$-n' \ln(1 - Q_0) \approx n'Q_0 = \ln 2$$

or

$$\frac{\tau_{fa}}{\tau_r} = \frac{M \ln 2}{Q_0} = N_{fa}.$$

The false-alarm number can therefore be interpreted as the number of pulses transmitted in an interval of such a duration τ_{fa} that the probability of at least one false alarm equals $\frac{1}{2}$.

An actual radar searches for targets not at a single range, but at all possible ranges from which echoes might return during the interpulse interval τ_r. It is often considered, for reasons we shall encounter later, as in effect dividing the interval $(0, \tau_r)$ into a number $L = W\tau_r$ of *range bins*, where W is the bandwidth of the radar echo. The numbers n and n' of decisions must then be multiplied by this factor L, so that n' in (4-43) is replaced by $n' = W\tau_{fa}/M$. The false-alarm number then becomes $N_{fa} = W\tau_{fa}$ and can be considered as the number of range bins of duration $1/W$ contained within the false-alarm time τ_{fa}.

In Fig. 4-9 we compare the average total signal energy needed to attain a fixed false-alarm number N_{fa} and a fixed probability $\overline{Q}_d$ of detection for signals fluctuating as specified by the four Swerling cases. The average total energy-to-noise ratio $\overline{S}$ in decibels is plotted versus the number M of incoherent pulses among which the energy is divided. The curve marked ∞ refers to signals with a fixed nonrandom total energy; for these, $\overline{Q}_d$ is calculated from (4-26). In cases 2 and 4, the signal amplitudes fluctuate independently, and as M increases, the required average total energy-to-noise ratio $\overline{S}$ approaches that needed for nonfluctuating signals. In cases 1 and 3, the amplitudes of all the signals in the pulse train vary together, and to ensure a given average probability $\overline{Q}_d$ of detection, the average total energy-to-noise ratio $\overline{S}$ must be large for all values of M.

The handbook *Radar Target Detection* [Mey73] contains extensive graphs of the average probability $\overline{Q}_d$ of detection as a function of the number M of pulses integrated. Incoherent detection of a nonfluctuating target and detection of targets fluctuating according to all four Swerling cases are included. The receiver is of the kind treated in this chapter; it sums the quadratically rectified outputs of the matched filter, as in (4-19). Each figure in that book refers to a particular value of the parameter $(\ln 2)/Q_0 = N_{fa}/M$. The curves are indexed with the value of the energy-to-noise ratio per pulse S/M in decibels. The application to radar of the theory we have presented in this section is discussed at length, and an appendix lists the formulas used for computing the detection probabilities.

4.3 BEAMFORMING

4.3.1 Detection by an Array of Transducers

A transducer is a small receptor that responds to an incident electromagnetic or acoustic wave field by generating a current or voltage directly proportional to it. An array of such transducers whose outputs are appropriately weighted and added forms a kind of antenna, which might be used for picking up radar or sonar echoes from a distant target. The transducers, although placed close together, are assumed

Detection in Multiple Observations   Chap. 4

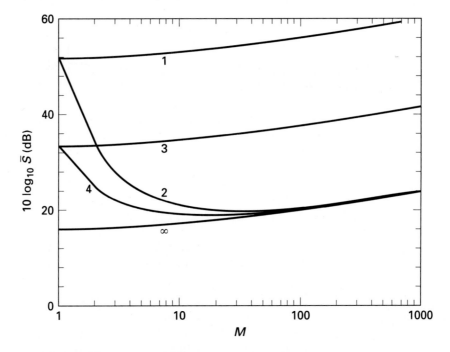

Figure 4-9. Average total energy-to-noise ratio in decibels required to attain false-alarm number $N_{fa} = 10^6$ and probability $\overline{Q}_d = 0.9999$ of detection versus the number M of inputs. Curves are indexed by the number of the Swerling case; ∞ refers to nonfluctuating signal amplitudes.

not to interact or influence each other in any way that cannot be compensated for by the proper circuitry.

Let the jth of M transducers in an array be located at a point $\boldsymbol{\xi}_j = (x_j, y_j)$ in a plane, of which a one-dimensional cross section is shown in Fig. 4-10. (In this section we shall indicate two-component vectors by boldface type.) The output of that transducer during an observation interval $(0, T)$ will be denoted by $v_j(t)$. When no target is present (hypothesis H_0),

$$v_j(t) = n_j(t), \qquad 1 \leq j \leq M,$$

where $n_j(t)$ is white Gaussian noise of unilateral spectral density N. It arises both from the input load resistor of the transducer and from broadband fluctuations in the surrounding medium. The resistor noise will be independent from one transducer to another, and the external noise is assumed to come from so broad a range of directions that it creates uncorrelated and hence statistically independent outputs from the several transducers.

Pulses of the quasiharmonic form Re $F(t) \exp i\Omega t$ and carrier frequency Ω are transmitted toward the anticipated target. The receiver is to process the outputs $v_j(t)$ in such a way as most effectively to decide whether a target is indeed present at a point $(u_x, u_y, R) = (\mathbf{u}, R)$ in a plane parallel to the transducer and a long distance R away. The signal induced in the jth transducer by the echo from the target is

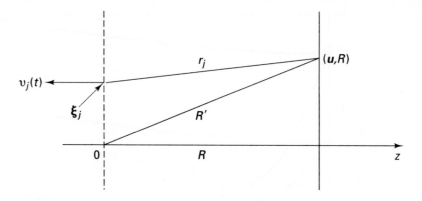

Figure 4-10. Transducer array and signal source.

$$s_j(t) = A \operatorname{Re} F\left(t - \frac{r_j}{c}\right) \exp\left[i\Omega\left(t - \frac{r_j}{c}\right) + i\psi'\right],$$

where r_j is the distance from the target to the jth transducer and c the velocity of the signal—electromagnetic or acoustic, as the case may be. We assume that the complex envelope $F(t - r_j/c)$ varies so slowly that it is approximately the same in all the transducers; differences in the times at which the echo arrives at the various elements of the array are negligible in comparison with the duration of $F(t - r_j/c)$:

$$F\left(t - \frac{r_j}{c}\right) \approx F\left(t - \frac{R'}{c}\right),$$

where R' is the distance from the target to the center of the array.

The distance r_j is given by

$$
\begin{aligned}
r_j &= [R^2 + (x_j - u_x)^2 + (y_j - u_y)^2]^{1/2} = [R^2 + (\boldsymbol{\xi}_j - \mathbf{u})^2]^{1/2} \\
&= [R^2 + \mathbf{u}^2 - 2\boldsymbol{\xi}_j \cdot \mathbf{u} + \boldsymbol{\xi}_j^2]^{1/2} = [R'^2 - 2\boldsymbol{\xi}_j \cdot \mathbf{u} + \boldsymbol{\xi}_j^2]^{1/2} \\
&= R'\left[1 - \frac{2\boldsymbol{\xi}_j \cdot \mathbf{u} - \boldsymbol{\xi}_j^2}{R'^2}\right]^{1/2},
\end{aligned}
$$

where

$$R'^2 = R^2 + \mathbf{u}^2.$$

We expand the bracket in a power series and assume that the source is so remote that we can neglect terms involving $\boldsymbol{\xi}_j/R'$ to powers greater than the first, and we obtain

$$r_j \approx R' - \frac{\boldsymbol{\xi}_j \cdot \mathbf{u}}{R'} = R' - \boldsymbol{\xi}_j \cdot \boldsymbol{\theta}, \tag{4-44}$$

where

$$\boldsymbol{\theta} = \left(\frac{u_x}{R'}, \frac{u_y}{R'}\right)$$

is a 2-vector specifying the direction of the target. After shifting the time variable $t - R'/c \to t$ and absorbing all the neglected terms into the phase ψ, which is unknown and assumed uniformly distributed over $(0, 2\pi)$, we write the quasiharmonic signal from the jth transducer as

Detection in Multiple Observations   Chap. 4

$$s_j(t) = A \operatorname{Re} F(t) \exp(i\Omega t + i\psi + ik\boldsymbol{\xi}_j \cdot \boldsymbol{\theta}), \tag{4-45}$$

where $k = \Omega/c = 2\pi/\lambda$ is the propagation constant of the radiation and λ is its wavelength. This is known as the *paraxial* approximation.

Comparing (4-45) with (4-2), we see that

$$s_j(t) = AF(t) \exp ik\boldsymbol{\xi}_j \cdot \boldsymbol{\theta}, \tag{4-46}$$

is the complex envelope of the signal from the jth transducer, and (4-8) prescribes that the receiver form the statistic

$$r' = \left| \sum_{m=1}^{M} \exp(-ik\boldsymbol{\xi}_m \cdot \boldsymbol{\theta}) \int_0^T F^*(t) V_m(t)\, dt \right| \tag{4-47}$$

and compare it with a decision level r_0', deciding that a target is present when $r' \geq r_0'$. Under the Neyman–Pearson criterion this test will be uniformly most powerful with respect to amplitude. For simplicity of analysis we adopt the proportional statistic $r = |Z|$, where Z is a circular complex Gaussian random variable given by

$$Z = X + iY = M^{-1/2} \sum_{m=1}^{M} z_m, \tag{4-48}$$

$$z_m = C \exp(-ik\boldsymbol{\xi}_m \cdot \boldsymbol{\theta}) \int_0^T F^*(t) V_m(t)\, dt,$$

$$C = N^{-1/2} \left| \int_0^T |F(t)|^2\, dt \right|^{-1/2}. \tag{4-49}$$

The receiver combines the outputs of the transducers after introducing phase shifts $-k\boldsymbol{\xi}_m \cdot \boldsymbol{\theta}$ to compensate for the different phase delays in the paths from the target to the transducers. The combined output is then passed through a filter matched to the signal $\operatorname{Re} F(t) \exp i\Omega t$, rectified, and sampled at the end of the observation interval to produce the statistic r, which is compared with a decision level r_0 chosen so that the false-alarm probability takes on the preassigned value. When the outputs of the transducers are shifted in phase in that way, the antenna is called a *phased array*. The phase shifts $-k\boldsymbol{\xi}_m \cdot \boldsymbol{\theta}$ are just what are required to produce maximum output signal-to-noise ratio by linear combination of echoes coming from the direction $\boldsymbol{\theta}$.

As with (4-11), the real and imaginary parts of z_m have unit variances under both hypotheses—see (4-23)—, and because the z_m's from different transducers are statistically independent, the real and imaginary parts X and Y of Z in (4-48) are also independent Gaussian random variables with unit variance. The statistic $r = |Z|$ therefore has a Rayleigh distribution

$$p_0(r) = r\, e^{-r^2/2} U(r)$$

under hypothesis H_0.

When echoes from a target are present, the statistic r will have the Rayleigh–Rice distribution of (3-73). Let us evaluate the signal-to-noise ratio d^2 appearing in that density function under the supposition that the target lies in a direction $\boldsymbol{\theta}' = \mathbf{u}'/R'$ that may differ from the direction from which the echoes are expected to come. In the signal $S_j(t)$ in (4-46) we now replace $\boldsymbol{\theta}$ by $\boldsymbol{\theta}'$, whereupon the expected value of the circular complex Gaussian random variable Z is

$$E(Z \mid H_1, \boldsymbol{\theta}') = \frac{AC}{\sqrt{M}} e^{i\psi} \sum_{m=1}^{M} \exp[ik\boldsymbol{\xi}_m \cdot (\boldsymbol{\theta}' - \boldsymbol{\theta})] \int_0^T |F(t)|^2 dt = DG(\boldsymbol{\theta}' - \boldsymbol{\theta}) e^{i\psi},$$

where

$$D^2 = M\frac{A^2}{N} \int_0^T |F(t)|^2 dt = \frac{2E_T}{N} \qquad (4\text{-}50)$$

is the total signal-to-noise ratio for a source at $\boldsymbol{\theta}' = \boldsymbol{\theta}$; E_T is the total energy received from the target.

The function

$$G(\boldsymbol{\theta}) = \frac{1}{M} \sum_{m=1}^{M} \exp ik\boldsymbol{\xi}_m \cdot \boldsymbol{\theta} \qquad (4\text{-}51)$$

is the amplitude *gain pattern* of the array. The false-alarm and detection probabilities for a source at $\boldsymbol{\theta}'$ are now, as in Sec. 3.4,

$$Q_0 = e^{-b^2/2}, \qquad Q_d = Q(D_{\text{eff}}, b) \qquad (4\text{-}52)$$

in terms of Marcum's Q function (3-76), with

$$D_{\text{eff}}^2 = D^2 |G(\boldsymbol{\theta}' - \boldsymbol{\theta})|^2$$

the effective signal-to-noise ratio.

The factor $|G(\boldsymbol{\theta}' - \boldsymbol{\theta})|^2$ is often called the *beam pattern* of the array. It is maximum in the direction $\boldsymbol{\theta}' = \boldsymbol{\theta}$ for which the outputs of the transducers have been phased. When the transducers are very small and very close together, the sum in (4-51) can be approximated by an integral,

$$G(\boldsymbol{\theta}) = \frac{1}{A} \int_{-\infty}^{\infty} \int_{-\infty}^{\infty} I(\boldsymbol{\xi}) e^{2\pi i \boldsymbol{\xi} \cdot \boldsymbol{\theta}/\lambda} d^2\boldsymbol{\xi}, \qquad d^2\boldsymbol{\xi} = dx\, dy,$$

where $I(\boldsymbol{\xi})$, the indicator function of the array, equals 1 for points $\boldsymbol{\xi}$ inside the array and 0 for points $\boldsymbol{\xi}$ outside it, and A is the area of the array. The gain pattern is the two-dimensional spatial Fourier transform of the indicator function of the array, a relation familiar from antenna theory.

For a circular array, for instance, of radius a,

$$G(\boldsymbol{\theta}) = \frac{1}{\pi a^2} \int_0^a r\, dr \int_0^{2\pi} \exp\left(\frac{2\pi i r |\boldsymbol{\theta}|}{\lambda} \cos\phi\right) d\phi$$
$$= \frac{2}{a^2} \int_0^a r J_0\left(\frac{2\pi r |\boldsymbol{\theta}|}{\lambda}\right) dr = \frac{2 J_1(y)}{y}, \qquad y = \frac{2\pi a |\boldsymbol{\theta}|}{\lambda}, \qquad (4\text{-}53)$$

in terms of the Bessel function of first order. This is known as the *Airy pattern*. Because the first zero of $J_1(y)$ is at $y = 3.83171$, the angular width of this pattern between its first nulls on each side of the axis is $1.220(\lambda/a)$ radians or $69.88(\lambda/a)°$. The maxima of the first sidelobes of this pattern lie about 17.6 dB below that of the main lobe. The broader an array, in general, the narrower its effective beam pattern for reception of signals from a distant source.

The process of picking the right phases and other weighting factors, if required, for the outputs of an array of transducers in order to maximize the output signal-to-noise ratio and, possibly, to meet other specifications is called *beamforming*. A review, with references to the earlier literature, is to be found in [Cox73].

4.3.2 Elimination of Noise from a Point Source

In Sec. 4.3.1 we assumed that the external noise comes in from so broad a cone of directions that the responses $n_j(t)$ to it in the several transducers are uncorrelated and hence statistically independent. In sonar surveillance the incoming echoes may also be corrupted by acoustic noise from a localized source such as a passing ship. Let us see how and to what extent that can be overcome.

For simplicity we assume that the noise source is small enough to be considered a point; roughly speaking, this entails that the cone of directions from which the noise comes be much narrower than the gain pattern in (4-51). Let the direction of this noise source by given by the 2-vector $\mathbf{w}$. It is far enough away that the approximation (4-44) can be applied to determining the differences in the phases of noise waves reaching the several transducers of the array. Like the other ambient noise, this additional, localized noise has a spectral width, we assume, much broader than that of the echo signals, so that it can be treated as white.

The complex envelope of the noise component of the output $v_j(t)$ from the jth transducer has the form

$$N_j(t) = N_{j0}(t) + N_1(t) \exp ik\boldsymbol{\xi}_j \cdot \mathbf{w},$$

where $N_{j0}(t)$ is the complex envelope of white noise of spectral density N, independent from one transducer to another,

$$\tfrac{1}{2}E[N_{j0}(t_1)N_{m0}^*(t_2)] = N\delta_{jm}\delta(t_1 - t_2),$$

and $N_1(t)$ is the complex envelope of the noise from the point source, whose spectral density is N_s; $N_{j0}(t)$ and $N_1(t)$ are of course statistically independent. Then the complex covariance function of the noise from the jth and the mth transducers is

$$\tfrac{1}{2}E[N_j(t_1)N_m^*(t_2)] = [N\delta_{jm} + N_s \exp ik(\boldsymbol{\xi}_j - \boldsymbol{\xi}_m)\cdot\mathbf{w}]\delta(t_1 - t_2). \qquad (4\text{-}54)$$

The signals $s_j(t)$ arising from the echoes our receiver is trying to detect will again have complex envelopes of the form in (4-46). Because the noise is white, the optimum receiver will as before pass the output $v_j(t)$ of each transducer through a filter matched to the signal Re $F(t) \exp i\Omega t$, where $F(t)$ is the form of the complex envelope of the arriving echo. We can assume, therefore, that its decisions will be based on the M circular complex Gaussian random variables

$$z_j = C\int_0^T F^*(t)V_j(t)\, dt, \qquad (4\text{-}55)$$

where C is as given in (4-49).

How these data z_j shall be combined is determined by their likelihood ratio $\tilde{p}_1(\{z_j\})/\tilde{p}_0(\{z_j\})$, where $\tilde{p}_i(\{z_j\})$ is the joint probability density function under hypothesis H_i of the real and imaginary parts of the M circular complex random variables z_j, $i = 0, 1$.

In order to write down these circular complex Gaussian density functions, which have the forms (3-40) and (3-41), respectively, we need the complex covariance matrix $\boldsymbol{\phi} = \boldsymbol{\mu}^{-1}$ and the expected values $S_j = E(z_j | H_1)$ of the random variables z_j. When the signals in (4-46) are present, by (4-55),

$$S_j = AC \exp(i\psi + ik\boldsymbol{\xi}_j \cdot \boldsymbol{\theta}) \int_0^T |F(t)|^2 dt = M^{-1/2}D \; e^{i\psi} \exp ik\boldsymbol{\xi}_j \cdot \boldsymbol{\theta}; \qquad (4\text{-}56)$$

D^2, defined as in (4-50), is the effective signal-to-noise ratio when no point source of noise is operating and $N_s = 0$.

The elements of the complex covariance matrix $\boldsymbol{\phi}$ of the z_j's are, by (4-54) and (4-49),

$$\phi_{jm} = \tfrac{1}{2}E(z_j z_m^* | H_0) = \tfrac{1}{2}C^2 E \int_0^T \int_0^T F^*(t_1)F(t_2)N_j(t_1)N_m^*(t_2) \; dt_1 \; dt_2$$

$$= C^2 \int_0^T |F(t)|^2 dt \; [N\delta_{jm} + N_s \exp ik(\boldsymbol{\xi}_j - \boldsymbol{\xi}_m) \cdot \mathbf{w}] \qquad (4\text{-}57)$$

$$= \delta_{jm} + h \exp ik(\boldsymbol{\xi}_j - \boldsymbol{\xi}_m) \cdot \mathbf{w}, \qquad h = \frac{N_s}{N};$$

h is the ratio of the spectral density of the noise from the point source to that of the original ambient noise.

The elements of the inverse covariance matrix $\boldsymbol{\mu} = \boldsymbol{\phi}^{-1}$ are

$$\mu_{mn} = \delta_{mn} - \eta \exp ik(\boldsymbol{\xi}_m - \boldsymbol{\xi}_n) \cdot \mathbf{w}, \qquad \eta = \frac{h}{1 + Mh}. \qquad (4\text{-}58)$$

To prove this, we substitute (4-57) and (4-58) into the usual definition of an inverse matrix:

$$\sum_{m=1}^{M} \phi_{jm}\mu_{mn} = \sum_{m=1}^{M} [\delta_{jm} + h \exp ik(\boldsymbol{\xi}_j - \boldsymbol{\xi}_m) \cdot \mathbf{w}][\delta_{mn} - \eta \exp ik(\boldsymbol{\xi}_m - \boldsymbol{\xi}_n) \cdot \mathbf{w}]$$

$$= \delta_{jn} + h \exp ik(\boldsymbol{\xi}_j - \boldsymbol{\xi}_n) \cdot \mathbf{w} - \eta \exp ik(\boldsymbol{\xi}_j - \boldsymbol{\xi}_n) \cdot \mathbf{w}$$

$$\qquad\qquad\qquad\qquad -Mh\eta \exp ik(\boldsymbol{\xi}_j - \boldsymbol{\xi}_n) \cdot \mathbf{w}$$

$$= \delta_{jn}$$

because $h - \eta - Mh\eta = 0$.

Dividing (3-41) by (3-40), we find the likelihood ratio

$$\frac{\tilde{p}_1(\{z_j\})}{\tilde{p}_0(\{z_j\})} = \exp\left[\tfrac{1}{2}\sum_{j=1}^{M}\sum_{m=1}^{M} \mu_{jm}(S_j^* z_m + z_j^* S_m - S_j^* S_m)\right],$$

which we see depends on the data $\{z_j\}$ only through the sufficient statistic $\mathrm{Re}\, Z$, where, by (4-56),

$$Z = \frac{\sqrt{M}}{D} \sum_{j=1}^{M}\sum_{m=1}^{M} S_j^* \mu_{jm} z_m = e^{-i\psi} \sum_{j=1}^{M}\sum_{m=1}^{M} \mu_{jm} z_m \exp(-ik\boldsymbol{\xi}_j \cdot \boldsymbol{\theta})$$

$$= e^{-i\psi} \sum_{m=1}^{M} g_m z_m,$$

with the weighting factors g_m defined by

$$
g_m = \sum_{j=1}^{M} \mu_{jm} \exp(-ik\boldsymbol{\xi}_j \cdot \boldsymbol{\theta}) \tag{4-59}
$$

$$
= \exp(-ik\boldsymbol{\xi}_m \cdot \boldsymbol{\theta}) - \beta G(\mathbf{w} - \boldsymbol{\theta}) \exp(-ik\boldsymbol{\xi}_m \cdot \mathbf{w}), \qquad \beta = \frac{Mh}{1 + Mh};
$$

$G(\cdot)$ is the amplitude gain pattern of the array defined in (4-51). Because the phase ψ of the carrier of the echo pulses is unknown and can be assumed uniformly distributed over $(0, 2\pi)$, the optimum detector will base its decisions on the statistic

$$
r = |Z|, \qquad Z = X + iY = \sum_{m=1}^{M} g_m z_m
$$

with the weighting coefficients g_m given by (4-59).

Let us once again assess the effectiveness of our receiver for detecting a target located not in direction $\boldsymbol{\theta}$, but in direction $\boldsymbol{\theta}'$. Then from (4-55) and (4-46)

$$
E(z_m | H_1, \boldsymbol{\theta}') = M^{-1/2} D\, e^{i\psi} \exp ik\boldsymbol{\xi}_m \cdot \boldsymbol{\theta}'.
$$

Hence the expected value of Z is

$$
E(Z | H_1, \boldsymbol{\theta}') = \frac{D}{\sqrt{M}}\, e^{i\psi} \sum_{m=1}^{M} g_m \exp ik\boldsymbol{\xi}_m \cdot \boldsymbol{\theta}'
$$

$$
= \frac{D}{\sqrt{M}}\, e^{i\psi} \left[\sum_{m=1}^{M} \exp ik\boldsymbol{\xi}_m \cdot (\boldsymbol{\theta}' - \boldsymbol{\theta}) - \beta G(\mathbf{w} - \boldsymbol{\theta}) \sum_{m=1}^{M} \exp ik\boldsymbol{\xi}_m \cdot (\boldsymbol{\theta}' - \mathbf{w}) \right]
$$

$$
= D\sqrt{M}\, e^{i\psi}[G(\boldsymbol{\theta}' - \boldsymbol{\theta}) - \beta G(\mathbf{w} - \boldsymbol{\theta})G(\boldsymbol{\theta}' - \mathbf{w})].
$$

The variances of the real and imaginary parts of Z are

$$
\sigma^2 = \operatorname{Var} X = \operatorname{Var} Y = \tfrac{1}{2}E(ZZ^* | H_0) = \tfrac{1}{2}\sum_{m=1}^{M}\sum_{n=1}^{M} g_m g_n^* E(z_m z_n^* | H_0)
$$

$$
= \sum_{m=1}^{M}\sum_{n=1}^{M} g_m \phi_{mn} g_n^* = \sum_{m=1}^{M}\sum_{n=1}^{M}\sum_{j=1}^{M} g_m \phi_{mn} \mu_{nj} \exp ik\boldsymbol{\xi}_j \cdot \boldsymbol{\theta}
$$

$$
= \sum_{m=1}^{M}\sum_{j=1}^{M} g_m \delta_{mj} \exp ik\boldsymbol{\xi}_j \cdot \boldsymbol{\theta} = \sum_{m=1}^{M} g_m \exp ik\boldsymbol{\xi}_m \cdot \boldsymbol{\theta}
$$

$$
= M[1 - \beta|G(\mathbf{w} - \boldsymbol{\theta})|^2].
$$

The false-alarm and detection probabilities are again given by (4-52), but the effective signal-to-noise ratio is now

$$
D_{\text{eff}}^2 = \frac{|E(Z | H_1, \boldsymbol{\theta}')|^2}{\sigma^2} = D^2 |G'(\boldsymbol{\theta}' - \boldsymbol{\theta};\ \mathbf{w} - \boldsymbol{\theta})|^2,
$$

$$
|G'(\boldsymbol{\theta}';\mathbf{w})|^2 = \frac{|G(\boldsymbol{\theta}') - \beta G(\mathbf{w})G(\boldsymbol{\theta}' - \mathbf{w})|^2}{1 - \beta|G(\mathbf{w})|^2}.
$$

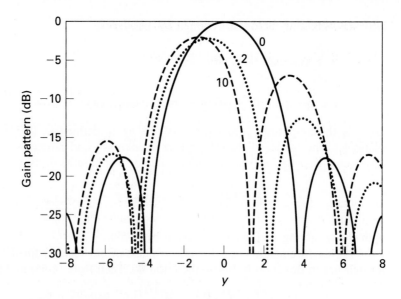

Figure 4-11. Gain pattern $|G'(\boldsymbol{\theta}'; \mathbf{w})|^2$ (dB) in the plane of the noise source and the perpendicular to the circular transducer array versus $y = 2\pi a |\boldsymbol{\theta}'|/\lambda$; $2\pi a |\mathbf{w}|/\lambda = 1$. Solid curve $Mh = 0$; dotted curve: $Mh = 2$; dashed curve: $Mh = 10$; $Mh = MN_s/N$.

When the beam is "on target", $\boldsymbol{\theta} = \boldsymbol{\theta}'$, and the effective signal-to-noise ratio is

$$D_{\text{eff}}^2 = D^2[1 - \beta |G(\mathbf{w} - \boldsymbol{\theta})|^2].$$

If target and noise source are far apart, D_{eff}^2 nearly equals D^2; if, on the other hand, they lie in the same direction, $\boldsymbol{\theta} = \mathbf{w}$, the effective signal-to-noise ratio is reduced by the factor $(1 + Mh)^{-1}$.

Figure 4-11 illustrates how the strength N_s of a point noise source affects the shape of the optimum beam pattern. We take $\boldsymbol{\theta} = (0, 0)$, and we assume a circular array of radius a, so that the beam pattern when $N_s = 0$ is the Airy pattern given in (4-53). It is plotted in decibels as the solid curve marked 0; $y = 2\pi a |\boldsymbol{\theta}'|/\lambda$. The point source of noise is located at an angle $\mathbf{w}$ corresponding to a value of y equal to 1. Thus with $a = 10\lambda$, the first nulls of the Airy pattern lie at $\pm 7°$, and the direction of the noise source makes an angle of $1.8°$ with the perpendicular to the array. The figure shows the function $|G'(\boldsymbol{\theta}'; \mathbf{w})|^2$ for directions $\boldsymbol{\theta}'$ lying in the plane determined by the direction $\mathbf{w}$ of the noise source and that perpendicular. The function is represented in decibels for values of Mh equal to 0, 2, and 10. As $Mh = MN_s/N$ increases, the first right-hand null in the beam pattern moves from $y = 3.83171$ toward the direction ($y = 1$) of the noise source, and the relative strengths of the sidelobes increase. The optimum beam pattern reduces the gain in the direction of the noise source as much as it can without unduly decreasing the gain in the direction $\boldsymbol{\theta} = (0, 0)$ where a target is expected to lie.

4.4 COMPARISON OF RECEIVER PERFORMANCE

4.4.1 Asymptotic Relative Efficiency

A set of signals is either present in or absent from a sequence of M inputs $v_j(t)$ to the receiver, $0 \leq t \leq T$, $1 \leq j \leq M$. In order to establish a simple way of assessing how effectively the receiver detects these signals, we shall suppose that it processes inputs $v_j(t)$ that are statistically homogeneous. We first describe conditions under which statistical homogeneity occurs.

Denote the signal in the jth input by $s(t; a_j, \boldsymbol{\theta}'_j, \boldsymbol{\theta}'')$, where a_j is a parameter representing the strength of the signal, and $\boldsymbol{\theta}'_j$ and $\boldsymbol{\theta}''$ are sets of other parameters characterizing it. The parameters designated collectively by $\boldsymbol{\theta}'_j$ are independently random from one signal to another, and we assume that for all $j \in (1, M)$ their prior probability density functions $z(\boldsymbol{\theta}')$ are the same. The phase ψ_j of a quasiharmonic signal is a typical element of $\boldsymbol{\theta}'_j$. The parameters $\boldsymbol{\theta}''$ are the same in all the signals; we call them the *invariable* parameters. The arrival time τ of a radar echo from a fixed target, for instance, would be an element of $\boldsymbol{\theta}''$. If the target is moving at constant velocity, the Doppler shift w of the carrier and the arrival time τ_1 of, say, the first signal make up $\boldsymbol{\theta}'' = (\tau_1, w)$; the arrival times of the other signals can be expressed in terms of τ_1 and w. In detecting a moving target the receiver can compensate for the changing arrival times τ_j of the echoes from one input to another if the Doppler shift w is known, for the Doppler shift is proportional to the velocity of the target in the direction of the radar antenna. The signal strengths a_j are assumed either to be the same in all inputs or to be independently random from one input to another and identically distributed in each.

The jth input is

$$v_j(t) = n_j(t), \qquad j = 1, 2, \ldots, M, \qquad 0 \leq t \leq T,$$

under hypothesis H_0 and

$$v_j(t) = s(t; a_j, \boldsymbol{\theta}'_j, \boldsymbol{\theta}'') + n_j(t)$$

under hypothesis H_1. The noise $n_j(t)$ is a stochastic process assumed to have identical statistical descriptions in all the inputs, and the noise in one input is statistically independent of that in another. The inputs may be received simultaneously in disjoint frequency bands, or they may be received one after another in the course of time. The receiver processes each input in such a way as to produce a single datum

$$g_j = \mathfrak{f}[v_j(t)], \qquad j = 1, 2, \ldots, M,$$

which is a specified functional $\mathfrak{f}[\cdot]$ of that input. The functional $\mathfrak{f}[\cdot]$ may depend on certain standard values of the invariable parameters $\boldsymbol{\theta}''$. A radar receiver for detecting a target at a fixed location, for instance, will have its output sampled at the time when an echo from the target is expected to appear. The datum g_j might then be

$$g_j = \left| \int_0^T F^*(t; \boldsymbol{\theta}'') V_j(t) \, dt \right|^2,$$

where $a_j F(t; \boldsymbol{\theta}'')$ is the complex envelope of the signal and $V_j(t)$ is the complex envelope of the jth input $v_j(t) = \mathrm{Re}\, V_j(t) \exp i\Omega t$. As a consequence of our assumptions, the random variables g_j under both hypotheses H_0 and H_1 are statistically independent and identically distributed for all $j \in (1, M)$. Their joint probability density functions have the form

$$p_0(\mathbf{g}) = p_0(g_1, g_2, \dots, g_M) = \prod_{j=1}^{M} p_0(g_j),$$

$$p_1(\mathbf{g}) = p_1(g_1, g_2, \dots, g_M) = \prod_{j=1}^{M} p_1(g_j),$$

$\mathbf{g}$ standing for the set of all M statistics g_j. The density function $p_1(g)$ is assumed independent of the values of the random parameters $\boldsymbol{\theta}_j'$ in the signals actually present under hypothesis H_1.

The receiver bases its choice between hypotheses H_0 and H_1 on the M data $\mathbf{g} = (g_1, g_2, \dots, g_M)$. In statistical terminology these are often called *samples*, but they must not be confused with the samples of an input $v(t)$ that we utilized in Chapter 2 to treat the elementary problem of detecting a known signal in Gaussian noise. The datum g_j would ideally represent the result of the optimum processing of the jth input; that is,

$$g_j = \ln \overline{\Lambda}[v_j(t); a_j, \boldsymbol{\theta}''],$$

where

$$\overline{\Lambda}[v_j(t); a_j, \boldsymbol{\theta}''] = \int_{\Theta'} z(\boldsymbol{\theta}_j') \Lambda[v_j(t); a_j, \boldsymbol{\theta}_j', \boldsymbol{\theta}''] \, d^{m'} \boldsymbol{\theta}_j' \qquad (4\text{-}60)$$

is the likelihood functional for the jth input averaged with respect to the prior probability density function $z(\boldsymbol{\theta}_j')$ of any random parameters. This optimum processing would require knowing the values of the invariable parameters $\boldsymbol{\theta}''$ or the assumption that they take certain standard values. Most often g_j will simply be the result of some convenient filtering and rectification of $v_j(t)$.

However the inputs $v_j(t)$ may have been processed to create the "samples" $\{g_j\}$, the optimum way of utilizing them is to form their likelihood ratio

$$\Lambda(\mathbf{g}) = \frac{p_1(\mathbf{g})}{p_0(\mathbf{g})},$$

which is compared with a decision level Λ_0 determined by one's criterion of optimality. Equivalently

$$U = \ln \Lambda(\mathbf{g}) = \sum_{j=1}^{M} \ln \left[\frac{p_1(g_j)}{p_0(g_j)} \right]$$

is compared with the decision level $U_0 = \ln \Lambda_0$. Other schemes are conceivable, however; the g_j's may simply be added to form

$$G = \sum_{j=1}^{M} g_j, \qquad (4\text{-}61)$$

or, as we shall see when we treat nonparametric detection in Chapter 8, they may be ranked in order of magnitude and a decision statistic derived from their places in that order.

Determining the false-alarm and detection probabilities for receivers that base their decisions on statistics such as U and G formed from a large number M of samples may involve lengthy computations. In Chapter 5 we shall describe approximating techniques, but even these may not always be expeditious. One would like to have at least a crude way of assessing and comparing the performance of receivers based on various functionals $\mathfrak{f}[\cdot]$ and various schemes for processing the samples $\{g_j\}$. The simplest way seems to be to compare receivers on the basis of their asymptotic relative efficiency, which we shall now define.

A fixed pair (Q_0, Q_d) of false-alarm and detection probabilities is adopted as the standard of performance, with $Q_0 \ll \frac{1}{2}$, $1 - Q_d \ll \frac{1}{2}$. A typical pair would be $(10^{-6}, 0.99)$. We call the pair (Q_0, Q_d) the *reliability* of the receiver. The inputs to a receiver are taken as statistically identical and homogeneous in the sense just described. Two receivers processing the same kind of inputs $\{v_j(t)\}$ and attaining the same reliability (Q_0, Q_d) are said to be *equipollent*.

Let M_1 be the number of independent inputs required by receiver 1 and let M_2 be the number required by receiver 2 in order for each to attain the reliability (Q_0, Q_d). Then the asymptotic relative efficiency (a.r.e.) of receiver 2 with respect to receiver 1 is defined as

$$\text{a.r.e.}_{2:1} = \lim_{a \to 0} \frac{M_1}{M_2}, \tag{4-62}$$

where a is a parameter specifying the strengths or the average strengths of the signals under hypothesis H_1. In this limit $a \to 0$ the numbers M_1 and M_2 go to infinity; it is the limiting value of their ratio that matters.

Let us suppose that each receiver bases its decisions on the sum

$$G_k = \sum_{j=1}^{M} g_j^{(k)}, \qquad g_j^{(k)} = \mathfrak{f}_k[v_j(t)], \qquad k = 1, 2, \tag{4-63}$$

of the values of a certain functional $\mathfrak{f}_k[\cdot]$ of its input. When the signal strength a is very small, M_1 and M_2 are very large, and by the central limit theorem the statistics G_1 and G_2 have approximately Gaussian distributions. With G_{k0} the decision level on statistic G_k,

$$G_k \geq G_{k0} \Rightarrow \to H_1, \qquad k = 1, 2,$$

the false-alarm and detection probabilities are approximately

$$Q_0 \approx \text{erfc } x, \qquad x = \frac{G_{k0} - E(G_k \mid H_0)}{\sigma_{k0}},$$

$$\sigma_{k0}^2 = \text{Var}(G_k \mid H_0) = \text{Var}_0 \; G_k,$$

$$Q_d \approx \text{erfc}\left[\frac{G_{k0} - E(G_k \mid H_1)}{\sigma_{k1}}\right] = \text{erfc}\left[\frac{\sigma_{k0}}{\sigma_{k1}}(x - \Delta_k)\right],$$

$$\sigma_{k1}^2 = \text{Var}(G_k \mid H_1),$$

where

$$\Delta_k^2 = \frac{[E(G_k \mid H_1) - E(G_k \mid H_0)]^2}{\text{Var}_0 \ G_k} \tag{4-64}$$

is the effective signal-to-noise ratio at the output of receiver k.

When each statistic G_k is the sum of independent and identically distributed random variables as in (4-63),

$$E(G_k \mid H_i) = M_k E(g^{(k)} \mid H_i),$$
$$\text{Var}(G_k \mid H_i) = M_k \ \text{Var}(g^{(k)} \mid H_i), \qquad k = 1, 2; \ i = 0, 1,$$

where $g^{(k)} = f_k[v(t)]$, we can write for the reliability

$$Q_0 \approx \text{erfc} \ x, \qquad Q_d \approx \text{erfc}\left[\frac{\sigma_{k0}}{\sigma_{k1}}(x - M_k^{1/2} D_k)\right]. \tag{4-65}$$

Here

$$D_k^2 = \frac{[E(g^{(k)} \mid H_1) - E(g^{(k)} \mid H_0)]^2}{\text{Var}_0 \ g^{(k)}} \tag{4-66}$$

is the effective signal-to-noise ratio of the output of processor k, $k = 1, 2$; D_k^2 is sometimes called the *deflection* and often is quoted in decibels (dB). Neither adding an arbitrary constant to the statistic $g^{(k)}$ nor multiplying it by an arbitrary constant alters the deflection D_k^2.

If $g_j^{(k)}$ is a linear functional of the input $v_j(t)$, $\sigma_{k0}^2 = \sigma_{k1}^2$. Usually, however, it is nonlinear, and the variances σ_{k0}^2 and σ_{k1}^2 differ because of the interaction between the signal and the noise under hypothesis H_1; but in the limit $a \to 0$ they become equal. Equating the reliabilities (Q_0, Q_d) of the two receivers is then equivalent to putting

$$M_1^{1/2} D_1 = M_2^{1/2} D_2,$$

whereupon the asymptotic relative efficiency of the two equipollent receivers is the ratio of their deflections,

$$\text{a.r.e.}_{2:1} = \lim_{a \to 0} \frac{D_2^2}{D_1^2} \tag{4-67}$$

in the limit of vanishing signal strength a. The concept of asymptotic relative efficiency, attributed to Pitman [Pit49], has been applied to signal detection by Middleton [Mid60a, Ch. 20], Capon [Cap61], and others.

When the signal strength a is very small, the deflection for a statistic g can be written

$$\begin{aligned} D^2 &= \frac{[E(g \mid H_1, a) - E(g \mid H_0)]^2}{\text{Var}_0 \ g} \\ &\approx \frac{a^2}{\text{Var}_0 \ g}\left[\frac{\partial}{\partial a}E(g \mid H_1, a)\Big|_{a=0}\right]^2 = a^2 \eta, \end{aligned} \tag{4-68}$$

where

$$\eta = \frac{1}{\text{Var}_0 \ g}\left[\frac{\partial}{\partial a}E(g \mid H_1, a)\Big|_{a=0}\right]^2 \tag{4-69}$$

is called the *efficacy* of the detector $g = f[v(t)]$. The signal strength a is so defined that the first derivative of $E(g \mid H_1, a)$ is the one of lowest order that does not vanish at $a = 0$. The asymptotic relative efficiency of two equipollent detectors is then the ratio of their efficacies,

$$\text{a.r.e.}_{2:1} = \frac{\eta_2}{\eta_1}.$$

The asymptotic relative efficiency is particularly useful for comparing receivers when the noise is non-Gaussian [Kas88] or the mode of processing the inputs is too complicated for the probability of detection to be calculated exactly. Examples will appear in Chapter 8.

4.4.2 Threshold Detection

The receiver that shows up best in comparisons based on asymptotic relative efficiency is the threshold detector, which as in Sec. 3.6.3 is defined by

$$g_a = g_a[v(t); \boldsymbol{\theta}''] = \left. \frac{\partial \overline{\Lambda}[v(t); a, \boldsymbol{\theta}'']}{\partial a} \right|_{a=0}, \tag{4-70}$$

where $\overline{\Lambda}[v(t); a, \boldsymbol{\theta}'']$ is the likelihood functional averaged over the random parameters $\boldsymbol{\theta}'$ as in (4-60). Again a is a parameter measuring the strength of the input signal and defined in such a way that the derivative $\partial^k \Lambda / \partial a^k$ of lowest order k that does not vanish at $a = 0$ is the first [Rud62]. In the considerations of Sec. 4.2.1, for instance, $a = \langle d_k^2 \rangle$. The threshold detector maximizes the average detection probability in the limit $a \to 0$; in that limit

$$\overline{\Lambda}[v(t); a, \boldsymbol{\theta}''] \approx 1 + a g_a[v(t); \boldsymbol{\theta}'']. \tag{4-71}$$

In order to show that the receiver based on the likelihood ratio is optimum in the sense that it has maximum effective signal-to-noise ratio, we consider an arbitrary functional $G = \mathfrak{f}[\{v_j(t)\}]$ of the set of M inputs $v_j(t)$, $j = 1, 2, \ldots, M$. Its effective signal-to-noise ratio is, as in (4-64),

$$\Delta^2 = \frac{[E(G \mid H_1) - E(G \mid H_0)]^2}{\text{Var}_0 \ G}. \tag{4-72}$$

Let us sample the set of inputs $\{v_j(t)\}$ by some appropriate means, as was done for instance in Chapter 2, obtaining a set v of samples that we eventually take to be infinitely numerous. Then $G = G(v)$ is a function of the set of samples, and when we limit ourselves to only n of them, its expected value under hypothesis H_1 is

$$E(G \mid H_1, n) = \int_{R_n} G(v) p_1(v) d^n v,$$

where $p_1(v)$ is the joint probability density function of $v = (v_1, \ldots, v_n)$. We can then write

$$E(G \mid H_1, n) = \int_{R_n} G(v) \Lambda(v) p_0(v) d^n v = E(G \Lambda(v) \mid H_0, n),$$

where

$$\Lambda(v) = \frac{p_1(v)}{p_0(v)}$$

is the likelihood ratio. When we pass to the limit $n \to \infty$ of an infinite number of samples, this likelihood ratio becomes the likelihood functional

$$\Lambda[\{v_j(t)\}; a, \boldsymbol{\theta}''],$$

which we abbreviate as $\Lambda(v)$, and we find

$$E(G|H_1) = E[G\Lambda(v)|H_0]. \tag{4-73}$$

Putting $G = 1$, we find, furthermore, that $E[\Lambda(v)|H_0] = 1$.

We can then write for the numerator in (4-72)

$$[E(G|H_1) - E(G|H_0)]^2 = \left[E\{G[\Lambda(v) - 1]|H_0\}\right]^2$$
$$= \left[E\{(G - \overline{G}_0)[\Lambda(v) - 1]|H_0\}\right]^2,$$

where $\overline{G}_0 = E(G|H_0)$ is nonrandom. The Schwarz inequality for expectations states that for two random variables A and B,

$$[E(AB)]^2 \le E(A^2)E(B^2), \tag{4-74}$$

with equality when $A = cB$, c any nonrandom constant [Hel91, p. 186], [Pap91, p. 154]. Applying this with $A = G - \overline{G}_0$ and $B = \Lambda(v) - 1$, we find

$$[E(G|H_1) - E(G|H_0)]^2 \le E[(G - \overline{G}_0)^2|H_0]E\{[\Lambda(v) - 1]^2|H_0\}$$
$$= \text{Var}_0\ G\ \text{Var}_0[\Lambda(v) - 1],$$

and by (4-72) we obtain an inequality on the effective signal-to-noise ratio:

$$\Delta^2 \le \text{Var}_0[\Lambda(v) - 1].$$

Equality obtains when

$$G = c[\Lambda(v) - 1] = c\left[\Lambda[\{v_j(t)\}; a, \boldsymbol{\theta}''] - 1\right].$$

Thus for a fixed number M of inputs $v_j(t)$, the likelihood-ratio receiver will attain the largest effective signal-to-noise ratio among all ways of processing them. In the limit $a \to 0$, $\Lambda(v) - 1$ becomes proportional to the sum of threshold statistics $g_a[v_j(t); \boldsymbol{\theta}'']$ as defined in (4-70).

Putting $G = \Lambda(v) - 1$ into (4-73), we find

$$E[\Lambda(v) - 1|H_1] = E\{\Lambda(v)[\Lambda(v) - 1]|H_0\}$$
$$= E[[\Lambda(v) - 1]^2|H_0] = \text{Var}_0\ [\Lambda(v) - 1].$$

The effective signal-to-noise ratio for the likelihood-ratio detector is therefore

$$\Delta^2 = E\{[\Lambda(v) - 1]|H_1\},$$

and by (4-71) that of the threshold detector is

$$\Delta_\theta^2 = MaE\{g_a[v(t); \boldsymbol{\theta}'']|H_1\} \tag{4-75}$$

with $g_a[v(t); \boldsymbol{\theta}'']$ defined in (4-70). This often provides the simplest way of calculating the effective signal-to-noise ratio of the threshold detector.

4.4.3 Comparison of the Linear and the Quadratic Detector

As an example, consider the detection of a known quasiharmonic signal in white Gaussian noise. As we learned in Sec. 4.2.1, the threshold detector utilizes the functional

$$g^{(2)} = \left| \int_0^T F^*(t) V(t) \, dt \right|^2 ,$$

which is the output, sampled at time $t = T$, of a quadratic rectifier following a filter matched to the signal $\mathrm{Re}\, F(t) \exp i\Omega t$. Compare that receiver with one utilizing an mth-law rectifier instead,

$$g^{(m)} = \left| \int_0^T F^*(t) V(t) \, dt \right|^m .$$

In Problem 4-7 you are asked to calculate the asymptotic relative efficiency of these two receivers. The asymptotic relative efficiency a.r.e.$_{2:m}$ is always greater than 1 for $m \neq 2$. For a linear rectifier ($m = 1$)

$$\text{a.r.e.}_{2:1} = \frac{\eta_2}{\eta_1} = \frac{4(4 - \pi)}{\pi} = 1.09296.$$

For these receivers the signal-strength parameter a is proportional to the squared amplitude A^2 of the input signal, that is, to the input energy-to-noise ratio $S = E/N$; E is the input signal energy and N the unilateral spectral density of the input noise. Indicating the receivers by the subscript $k = 1, 2$, and taking $M_1 = M_2 = M \gg 1$, we find from (4-66) and (4-68)

$$D_1^2 = S_1^2 \eta_1 = D_2^2 = S_2^2 \eta_2,$$

whence the ratio of *input* energy-to-noise ratios needed for the two receivers to attain the same reliability must be

$$\frac{S_1}{S_2} = \left(\frac{\eta_2}{\eta_1} \right)^{1/2} = 1.04545 \equiv 0.193 \text{ dB}.$$

That is, the receiver with a linear rectifier requires 0.193 dB more input signal energy than one with a quadratic rectifier when $M \gg 1$.

The relative standings of these receivers are reversed when M is less than about 100, and the receiver with a quadratic rectifier is superior to one with a linear rectifier only when the number M of inputs is so large that the first term of (4-14) is an adequate approximation in (4-13) for signals and noise of such strengths that the receiver attains an acceptable reliability (Q_0, Q_d). As we shall now demonstrate, however, for small numbers M and for reasonable values of the probabilities Q_0 and Q_d, the average value of $d_m r_m$ in (4-13) is so large that it is, for most of the observations, well beyond the quadratic part of the curve in Fig. 4-1, even when no signal is present.

From (4-29) with $U = \frac{1}{2}r^2$, $M = 1$, it follows that with no signal present the random variable r has the Rayleigh distribution

$$p_0(r) = r \, e^{-r^2/2} U(r),$$

and its expected value under hypothesis H_0 is

$$E(r \mid H_0) = \bar{r} = \int_0^\infty r \, p_0(r) \, dr = \left(\frac{\pi}{2} \right)^{1/2}.$$

Hence, if all pulses have equal amplitudes, the average value of the argument $d_m r_m$ of (4-13) is

$$d_m \bar{r}_m = d_m \left(\frac{\pi}{2} \right)^{1/2} = \left(\frac{\pi S}{2M} \right)^{1/2}$$

when no signal is present; S is the total energy-to-noise ratio. If we are concerned with signals yielding, say, a detection probability $Q_d = 0.90$ for a false-alarm probability $Q_0 = 10^{-6}$, we can utilize a routine for calculating the inverse Marcum Q function to determine the following average values of the arguments $d_m r_m$ under hypothesis H_0:

M	1	20	50	100	200
$d_m \bar{r}_m$	5.72	1.81	1.34	1.08	0.89

Referring to Fig. 4-1, we see that all but the last two of these are located on a part of the curve of $\ln I_0(x)$ that is nearly linear. When a signal is present, the average value $d_m \bar{r}_m$ is even larger. Because the values of the random variables $d_m r_m$ tend to cluster about their averages $d_m \bar{r}_m$, most of them will lie on the linear part of the curve of Fig. 4-1 for values of M up to about 50 or 100.

In Fig. 4-12 we have plotted the ratio S_1/S_2 of input energy-to-noise ratios required for equipollence of the linear and quadratic detectors with finite values of M, under the assumption that the false-alarm number N_{fa}, as defined in (4-42), equals 10^6. Curves are presented for four values of the probability Q_d of detection. These curves were computed as shown in [Hel90] by methods to be described in Secs. 5.2 and 5.3.

This ratio approaches 0.193 dB as M increases, but only slowly; and the closer Q_d lies to 1, the slower does the ratio S_1/S_2 converge toward that asymptotic value. Comparing the performances of two receivers through their asymptotic relative efficiency depends, as in (4-65), on the assumption that their decision statistics have Gaussian distributions, but the farther the decision level lies in the tails of the distributions under hypotheses H_0 and H_1, the poorer that Gaussian approximation is. We see, furthermore, that for M less than about 100 the linear detector is the better of the two, in accordance with our remarks at the beginning of this part.

4.4.4 Invariable Parameters Unknown

The threshold detector in (4-70) entails adopting some standard values of the invariable parameters θ''; these are by definition the same in all M inputs $v_j(t)$. If the values of θ'' are unknown a priori and the receiver is to be designed for a broad range of those values, some prior probability density function $z(\theta'')$ must be adopted

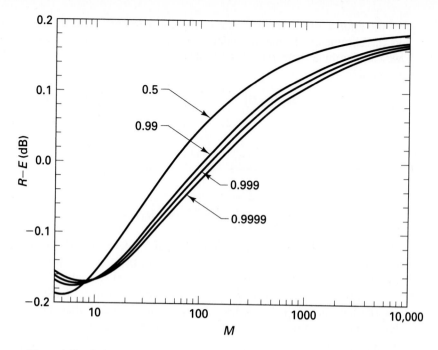

Figure 4-12. Relative efficiency $R - E = 10 \log_{10}(S_1/S_2)$ of linear and quadratic detectors as function of the number M of inputs summed; $N_{\text{fa}} = 10^6$. Curves are indexed with the probability Q_d of detection. [Reprinted from C.W. Helstrom, "Performance of receivers with linear rectifiers," *IEEE Transactions on Aerospace & Electronic Systems*, vol. AES-26 (Mar. 1990), 210–7, ©1990 IEEE.]

for them, and the overall likelihood functional must be averaged with respect to it to form

$$\Lambda[\{v_j(t)\}; a] = \int_{\Theta''} z(\theta'') \Lambda[\{v_j(t)\}; a, \theta''] \, d^{m''}\theta''$$

$$= \int_{\Theta''} z(\theta'') \prod_{k=1}^{M} \Lambda[v_k(t); a, \theta''] \, d^{m''}\theta'',$$

where m'' is the number of invariable parameters θ'' and Θ'' is the space in which they take their values. In the weak-signal limit $a \to 0$, by (4-71),

$$\prod_{k=1}^{M} \Lambda[v_k(t); a, \theta''] \approx \prod_{k=1}^{M} \{1 + a g_a[v_k(t); \theta'']\}$$

$$\approx 1 + a \sum_{k=1}^{M} g_a[v_k(t); \theta'']$$

to first order in a, and

$$\Lambda[\{v_j(t)\}; a] \approx 1 + a \sum_{k=1}^{M} \int_{\Theta''} z(\theta'') g_a[v_k(t); \theta''] \, d^{m''}\theta''. \tag{4-76}$$

In the limit $a \to 0$, therefore, the invariable parameters are treated in the same way as the parameters $\boldsymbol{\theta}'$ that are independently random from one input to another, and we can combine them into $\boldsymbol{\theta} = (\boldsymbol{\theta}', \boldsymbol{\theta}'')$ to write the threshold statistic derived from (4-76) as

$$G_a[\{v_j(t)\}] = \frac{\partial \Lambda[\{v_j(t)\}; a]}{\partial a}\bigg|_{a=0} = \sum_{k=1}^{M} g_a[v_k(t)], \qquad (4\text{-}77)$$

where for all $k \in (1, M)$

$$g_a[v_k(t)] = \frac{\partial \Lambda[v_k(t); a]}{\partial a}\bigg|_{a=0} = \lim_{a \to 0}\left[a^{-1}\{\Lambda[v_k(t); a] - 1\}\right] \qquad (4\text{-}78)$$

with

$$\Lambda[v(t); a] = \int_{\Theta} z(\boldsymbol{\theta})\Lambda[v(t); a, \boldsymbol{\theta}]\, d^m\boldsymbol{\theta}. \qquad (4\text{-}79)$$

Here $\Lambda[v(t); a, \boldsymbol{\theta}]$ is the likelihood functional for detecting the signal $s(t; a, \boldsymbol{\theta})$ in any one of the M inputs $v_j(t)$, and $z(\boldsymbol{\theta})$ is the joint prior probability density function of the m parameters $\boldsymbol{\theta} = (\boldsymbol{\theta}', \boldsymbol{\theta}'')$ other than the signal strength a.

In Sec. 7.6 we shall show how to determine a prior probability density function for the parameters $\boldsymbol{\theta}$ that is least favorable in the sense that the threshold receiver based on it has minimum efficacy among all possible prior density functions $z(\boldsymbol{\theta})$.

4.5 DISTRIBUTED DETECTION

A distributed-detection system consists of a number M of detectors or sensors intended to pick up a common electromagnetic or acoustic disturbance of some prescribed form, the "signal." Placed in separate locations, they transmit information about their ambient fields to a central processor, or *fusion center*, which makes the final decision about the presence or absence of the disturbance. The sensors do not communicate among themselves. In order to reduce to a bare minimum the communication links to the fusion center, each sensor transmits only 0's or 1's: if it in some way decides that the signal is present, it sends a 1, otherwise a 0.

The kth sensor processes its input $v_k(t)$ during an observation interval $(0, T)$ to produce a datum g_k indicative of its estimate of the strength of the signal component of its input. If g_k exceeds a certain decision level g_{k0}, the sensor transmits a 1 to the fusion center, otherwise a 0. Every T seconds the center receives a set of M 0's and 1's, and on the basis of these it decides between the null hypothesis H_0 that the disturbance was absent, the inputs $v_k(t)$ consisting only of random noise, and the alternative hypothesis H_1 that the disturbance was present in the fields incident on the M sensors.

How can each sensor most effectively process its input, and how can the fusion center optimally make its decisions? The latter has a great variety of possibilities. Depending on what it knows about the probability distributions of the inputs $v_k(t)$ under hypotheses H_0 and H_1, it may assign different weights to the 0's and 1's received from different sensors, and the environment of some of the sensors may be so noisy that it is best to disregard their responses altogether.

       Detection in Multiple Observations   Chap. 4

The datum g_k produced in the kth sensor can be represented as a functional $g_k = \mathfrak{f}_k[v_k(t)]$ of its input. How that functional is best designed and how the decision level g_{k0} on g_k is set will depend in general on the joint probability distributions of the inputs to all the sensors and on the rule by which the fusion center makes its decisions. In principle the optimum system will minimize the average cost of its operation, or it will maximize the probability Q_d of deciding that the disturbance is present (hypothesis H_1) for a preassigned value Q_0 of the probability of declaring for H_1 when H_0 is true; a Bayes or a Neyman–Pearson criterion may be adopted. Needless to say, determining the optimum system is, except under special conditions, quite complex.

4.5.1 Identical Independent Sensors

For simplicity let us consider a system in which the M sensors are identical and in which their inputs $\{v_k(t)\}$ are independent and identically distributed. The functionals $\mathfrak{f}_k[\cdot]$ will then likewise be identical. Under hypothesis H_0 the kth input is

$$v_k(t) = n_k(t);$$

the $n_k(t)$ are independent Gaussian random processes with identical autocovariance functions. Under H_1 the kth input is

$$v_k(t) = n_k(t) + s(t; \boldsymbol{\theta}_k),$$

in which the signal components $s(t; \boldsymbol{\theta}_k)$ have the same known form, but may depend on certain unknown parameters $\boldsymbol{\theta}_k$, assumed statistically independent from sensor to sensor. Typically $\boldsymbol{\theta}_k$ may be a random phase ψ_k,

$$s(t; \psi_k) = A \operatorname{Re} F(t) \exp(i\Omega t + i\psi_k),$$

with each ψ_k uniformly distributed over $(0, 2\pi)$. Alternatively, the signals may fade independently, $\boldsymbol{\theta}_k = (A_k, \psi_k)$,

$$s(t; A_k, \psi_k) = A_k \operatorname{Re} F(t) \exp(i\Omega t + i\psi_k),$$

and the random amplitudes A_k may, for example, have a common Rayleigh distribution as in (4-17). Each sensor will then optimally pass its input through a filter matched to the signal $\operatorname{Re} F(t) \exp i\Omega t$; the output, rectified and sampled at the end of each observation interval, produces a datum

$$g_k = \left| \int_0^T F^*(t) V_k(t)\, dt \right|^2,$$

where $V_k(t)$ is the complex envelope of the kth input.

If, on the other hand, the signals, when present, are expected to be periodically repeated, but with independently random phases and possibly independently random amplitudes, the datum g_k might as in (4-19) be the sum of quadratically rectified and sampled outputs of a matched filter. In any case, we assume that the probability density functions $p_0(g_k)$ and $p_1(g_k)$ of each datum under the two hypotheses are known and identical from sensor to sensor. The kth sensor will then transmit a 1 to the fusion center if $g_k \geq g_0$ and a 0 if $g_k < g_0$; the decision level g_0 is the same in all sensors.

The probability of each 1 is

$$q_i(g_0) = \int_{g_0}^{\infty} p_i(g)\, dg, \qquad i = 0, 1, \tag{4-80}$$

under hypothesis H_i, and the probability of a succession of m 1's and $M - m$ 0's received at the fusion center in a given order is

$$\Pr(M - m \text{ 0's, } m \text{ 1's}| H_i) = (1 - q_i)^{M-m} q_i^m, \qquad 0 \leq m \leq M.$$

The likelihood ratio for this observation is a function of m,

$$\Lambda(m) = \frac{(1 - q_1)^{M-m} q_1^m}{(1 - q_0)^{M-m} q_0^m},$$

and we see that the number m of 1's is a sufficient statistic for deciding between the two hypotheses. Ideally the fusion center applies a randomized strategy as described in Sec. 1.2.5, selecting hypothesis H_1 whenever the number m exceeds a decision level m_0; if $m = m_0$, it selects H_1 with probability f. It chooses H_0 whenever $m < m_0$. The probability under hypothesis H_i of observing m 1's and $M - m$ 0's is

$$P_i(m, g_0) = \binom{M}{m} q_i^m (1 - q_i)^{M-m}, \qquad q_i = q_i(g_0), \qquad i = 0, 1. \tag{4-81}$$

The conversion of each datum g_k into 0 or 1 is called *quantization*, and g_0 is the *quantization level*. Under the Neyman–Pearson criterion g_0 should be chosen so that for a preassigned false-alarm probability Q_0 the probability Q_d of detection is as large as possible. Instead of being generated in M distributed sensors, the data $\{g_k\}$ might simply be the outputs of a processing $f[v_k(t)]$ of inputs to a single receiver during a succession of M intervals of duration T, as in Sec. 4.1.1. The receiver then counts the number of times its M outputs g_k exceed the level g_0 and makes its decision about the presence or absence of a signal in the manner just described. In this context, the procedure is called *binary integration* [Sch56], [Sch75, pp. 248–51].

Through (1-57), (1-58), and (4-81) the false-alarm and detection probabilities are functions of the quantization level g_0:

$$Q_0(g_0) = fP_0(m_0, g_0) + \sum_{m=m_0+1}^{M} P_0(m, g_0), \tag{4-82}$$

$$Q_d(g_0) = fP_1(m_0, g_0) + \sum_{m=m_0+1}^{M} P_1(m, g_0). \tag{4-83}$$

Given a value of the quantization level g_0, we determine the integer m_0 and the probability f by the method outlined in connection with (1-57), so that $Q_0(g_0)$ equals the preassigned false-alarm probability. The probability $Q_d(g_0)$ of detection is then given by (4-83).

For Rayleigh-fading signals, for example, we can take, as in Sec. 4.2.3,

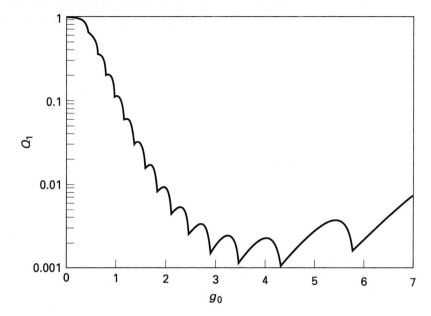

Figure 4-13. False-dismissal probability $Q_1 = 1 - Q_d$ for Rayleigh-fading signals in a distributed-detection system with $M = 15$ sensors as a function of the quantization level g_0; $S = 6.2827$, $Q_0 = 0.001$.

$$q_0 = \exp(-g_0), \qquad g_0 > 0,$$
$$q_1 = \exp\left[-\frac{g_0}{1 + S}\right], \qquad (4\text{-}84)$$

where S is the average energy-to-noise ratio at the input of each sensor. In Fig. 4-13 we show how the false-dismissal probability $Q_1(g_0) = 1 - Q_d(g_0)$ depends on the quantization level g_0 for $M = 15$, $Q_0 = 0.001$, and $S = 6.2827$. Each cusp represents a pure strategy ($f = 0$). As g_0 increases, the probability f increases from 0 to 1 on each convex branch of the curve, and the decision level m_0 on the number m of 1's decreases from one cusp to the next.

To determine the optimum value of the quantization level g_0, it suffices to set $f = 0$ in (4-82) and (4-83). We then equate the right side of (4-82) to the preassigned false-alarm probability, and for a succession of values of m_0 starting at $m_0 = 0$, we solve it for g_0 by Newton's method, substitute g_0 into (4-83) to determine the probability of detection, and stop when this reaches its maximum value [Wor68].

For detection of a narrowband signal of random phase in Gaussian noise, with suitable normalization of the statistic g_k,

$$q_0 = \exp(-g_0), \qquad q_1 = Q(d, \sqrt{2g_0}), \qquad (4\text{-}85)$$

as in (3-70) and (3-75), where $Q(\cdot, \cdot)$ is Marcum's Q function (3-76). In order to show how much is lost by quantizing the data $\{g_k\}$ instead of transmitting them unchanged to the fusion center and having it utilize the threshold statistic U in (4-19), we calculated the ratio of the total energy-to-noise ratio $S_T = \frac{1}{2}Md^2$ required with quantization to the total energy-to-noise ratio S_U required when the receiver bases

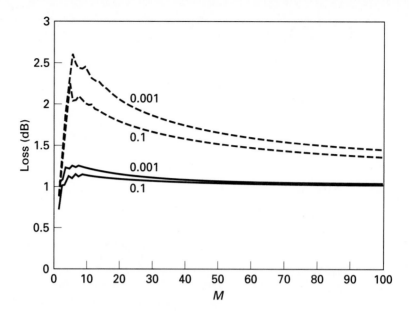

Figure 4-14. Loss (dB) in input energy-to-noise ratio when quantization is used instead of the statistic U in (4-19) versus the number M of samples. Solid curves: fixed signal amplitude; dashed curves: Rayleigh-fading signal. Curves are indexed with the false-dismissal probability $Q_1 = 1 - Q_d$. $Q_0 = 10^{-6}$.

its decisions on the threshold statistic U, both systems attaining the same reliability (Q_0, Q_d). For the latter, Q_0 is given by (4-30) and Q_d by (4-26), with $S_U = \frac{1}{2}D^2$. The loss in input energy-to-noise ratio, defined as S_T/S_U, is plotted in decibels versus the number M of data as the solid curves in Fig. 4-14; two values 0.1 and 0.001 of the false-dismissal probability $Q_1 = 1 - Q_d$ were assumed, and $Q_0 = 10^{-6}$. For $M = 1$, of course, $S_T/S_U = 1$. The ripples in the curves result from integral jumps in the optimum decision level m_0 as M increases.

When the signal suffers Rayleigh fading as in Sec. 4.2.3, the statistic $g = U$ in (4-19) is the optimum detection statistic, and its false-alarm and detection probabilities are given by (4-30) and (4-31); in the latter, s^2 is the average energy-to-noise ratio S_U' per pulse. The probabilities $q_0(g_0)$ and $q_1(g_0)$ are given by (4-84). Again we calculate the values of the energy-to-noise ratios S and S_U' required to attain the same reliability (Q_0, Q_d), and we define the loss as the ratio S/S_U'. It is plotted as the dashed curves in Fig. 4-14. Quantization introduces a somewhat greater loss with Rayleigh-fading signals than with nonfading signals.

It is natural that the optimum strategy will be a pure one ($f = 0$ or $f = 1$), for the fundamental data are the M g_k's, and they are continuous random variables. Randomization at the fusion center would introduce a kind of noise into the system, and we would expect the decision levels g_0 to adjust themselves so as to eliminate it. In order to prove this in detail, we begin by differentiating (4-82) and (4-83) with respect to g_0, using (4-81). All but one of the terms in each summation cancel, and we are left with

$$\frac{dQ_0}{dg_0} = P_0(m_0, g_0) \left\{ \frac{df}{dg_0} + \frac{dq_0}{dg_0} \left[\frac{m_0 f}{q_0} + \frac{(M - m_0)(1 - f)}{1 - q_0} \right] \right\},$$

$$\frac{dQ_d}{dg_0} = P_1(m_0, g_0) \left\{ \frac{df}{dg_0} + \frac{dq_1}{dg_0} \left[\frac{m_0 f}{q_1} + \frac{(M - m_0)(1 - f)}{1 - q_1} \right] \right\}.$$

Because Q_0 is fixed, we can set the first of these derivatives equal to zero, solve it for df/dg_0, and substitute that into the second, obtaining after a bit of algebra

$$\frac{dQ_d}{dg_0}$$

$$= P_1(m_0, g_0) \left[m_0 f \frac{d}{dg_0} \ln\left(\frac{q_1}{q_0} \right) - (M - m_0)(1 - f) \frac{d}{dg_0} \ln\left(\frac{1 - q_1}{1 - q_0} \right) \right]. \tag{4-86}$$

Both ratios

$$\frac{q_1(g_0)}{q_0(g_0)} \quad \text{and} \quad \frac{1 - q_1(g_0)}{1 - q_0(g_0)}$$

appearing in (4-86) are monotonely increasing functions of the quantization level g_0. To see this, consider the operating characteristic of the sensor, which will resemble that in Fig. 1-2, q_1 corresponding to Q_d and q_0 to Q_0. The ratio q_1/q_0 is the slope of the straight chord from the origin $(0, 0)$ to the point on the curve determined by the value of g_0. As g_0 goes from $-\infty$ to ∞—or as in our examples, from 0 to ∞—, that point moves from $(1, 1)$ to $(0, 0)$, and the slope of the chord increases monotonely. Likewise the ratio $(1 - q_1)/(1 - q_0)$ is the slope of a chord from the point on the curve determined by g_0 to the point $(1, 1)$, and this slope also increases monotonely as the point g_0 moves downward along the operating characteristic. Both the logarithmic derivatives multiplying $m_0 f$ and $(M - m_0)(1 - f)$ in (4-86) are therefore positive.

The graph of $Q_d(g_0)$ versus g_0 looks like that in Fig. 4-13, but upside down. As g_0 approaches a cusp from the left, the value of the probability f is approaching 1, and the first term on the right of (4-86) shows that the slope dQ_d/dg_0 of that branch of the curve is positive. To the right of a cusp, f is increasing from 0, and the second term of (4-86) shows that the slope is negative. Each cusp must therefore represent a maximum of the detection probability $Q_d(g_0)$ as a function of the quantization level g_0, and a pure strategy maximizes the detection probability for a fixed false-alarm probability.

A distributed-detection system may be set up as just described, with quantization levels g_{k0} equal to a common value g_0, and with a fixed decision level m_0 on the number m of 1's reaching the fusion center ($f = 0$), but the signal strengths at the several sensors may actually be unequal. The probability

$$q_{1k} = q_{1k}(g_0) = \Pr(g_k \geq g_0 | H_1)$$

that the kth sensor transmits a 1 to the fusion center will then differ from one sensor to another. Let u_k represent the quantized output of the kth sensor:

$$\Pr(u_k = 0 | H_1) = 1 - q_{1k}, \qquad \Pr(u_k = 1 | H_1) = q_{1k}.$$

Then the probability of detection is

$$Q_d(m_0) = \sum_{(\mu)} \prod_{k=1}^{M} [(1 - u_k)(1 - q_{1k}) + u_k q_{1k}], \tag{4-87}$$

in which the summation is over the set (μ) consisting of all sequences $(u_1, u_2, \ldots, u_M)$ having more than m_0 1's. Algorithms for computing this detection probability as a function of m_0 have been outlined by Sarwate [Sar91].

A greater probability of detection can be attained if instead of using only a single quantization level g_0 and sending binary digits, each sensor incorporates a number $L - 1$ of quantization levels $a_1, a_2, \ldots, a_{L-1}$, transmitting a digit r if the datum $g = \mathsf{f}[v(t)]$ lies in the rth interval $a_{r-1} < g \le a_r$, $1 \le r \le L$, $a_0 = -\infty$, and $a_L = \infty$. The receiver bases its decision on a likelihood ratio formed by dividing the probability under hypothesis H_1 of receiving the observed sequence of digits by the probability of receiving it under hypothesis H_0. Determining the optimum set of quantization levels a_r is now a rather more difficult problem than for $L = 2$. Quantization levels maximizing the efficacy of the overall receiver were worked out by Kassam [Kas77] and Cimini and Kassam [Cim83]. An approach based on an approximation method to be introduced in Sec. 5.3.2 is described in [Hel88b].

4.5.2 Nonidentical Independent Sensors

We assume that the kth of M sensors processes its input $v_k(t)$ to produce a datum $g_k = \mathsf{f}_k[v_k(t)]$ in a way that may be optimum for detecting the signal $s_k(t)$ expected to appear at that sensor, or it may be only an approximation to the optimum processing, as when a threshold statistic is generated. As before, the kth sensor transmits a digit $u_k = 0$ to the fusion center if g_k is less than some quantization level g_{k0} and a digit $u_k = 1$ if it is greater. Again we assume that the inputs to the several sensors are statistically independent, but their probability density functions may now differ from one to another. In our effort to find the optimum quantization levels in the M sensors, we begin with the Bayes criterion.

As in Sec. 1.2.1, we define prior probabilities ζ_0 and ζ_1 of hypotheses H_0 and H_1, respectively, and costs C_{ij} attending the final decision for hypothesis H_i when H_j is true; $\zeta_0 + \zeta_1 = 1$. The average cost of operating the distributed-detection system is

$$\overline{C} = \zeta_0[C_{00} \Pr(\to H_0| H_0) + C_{10} \Pr(\to H_1| H_0)]$$
$$+ \zeta_1[C_{01} \Pr(\to H_0| H_1) + C_{11} \Pr(\to H_1| H_1)]$$

as in (1-14), where $\Pr(\to H_i| H_j)$ is the probability that the fusion center decides for hypothesis H_i when H_j is true. Because

$$\Pr(\to H_0| H_j) = 1 - \Pr(\to H_1| H_j),$$

this average cost can be written as

$$\overline{C} = \zeta_0 C_{00} + \zeta_1 C_{01} + \zeta_0(C_{10} - C_{00}) \Pr(\to H_1| H_0) + \zeta_1(C_{11} - C_{01}) \Pr(\to H_1| H_1)$$
$$= \zeta_0 C_{00} + \zeta_1 C_{01} - \zeta_1(C_{01} - C_{11})[\Pr(\to H_1| H_1) - \Lambda_0 \Pr(\to H_1| H_0)],$$

where as in (1-17)

$$\Lambda_0 = \frac{\zeta_0(C_{10} - C_{00})}{\zeta_1(C_{01} - C_{11})}. \tag{4-88}$$

Because $C_{01} > C_{11}$ and $C_{10} > C_{00}$, $\Lambda_0 > 0$; and minimizing the average cost is equivalent to maximizing the function

$$F = \Pr(\rightarrow H_1 | H_1) - \Lambda_0 \Pr(\rightarrow H_1 | H_0). \tag{4-89}$$

We strive to do so by appropriately designing the strategy by which the fusion center processes its input $\mathbf{u} = (u_1, u_2, \dots, u_M)$ and by optimally setting the quantization levels g_{k0} at each sensor.

From what we learned in Sec. 1.2.5 we can state immediately that the optimum strategy for the fusion center is to form from its input the likelihood ratio

$$\Lambda(\mathbf{u}) = \frac{\Pr(\mathbf{u} | H_1)}{\Pr(\mathbf{u} | H_0)},$$

where $\Pr(\mathbf{u} | H_i) = \Pr(u_1, u_2, \dots, u_M | H_i)$ is the probability under hypothesis H_i of its receiving from the sensors the sequence $\mathbf{u}$ of 0's and 1's, $i = 0, 1$. The fusion center decides for hypothesis H_1 whenever $\Lambda(\mathbf{u}) \geq \Lambda_0$; otherwise it chooses hypothesis H_0 that no signal is present. Because the inputs to the sensors are statistically independent, this likelihood ratio and the decision rule can be written as

$$\Lambda(\mathbf{u}) = \prod_{k=1}^{M} \frac{\Pr(u_k | H_1)}{\Pr(u_k | H_0)} \begin{array}{l} \geq \Lambda_0 \Rightarrow \rightarrow H_1, \\ < \Lambda_0 \Rightarrow \rightarrow H_0. \end{array} \tag{4-90}$$

We determine the optimum strategy for each sensor under the assumption that the fusion center and the rest of the sensors adopt their optimum strategies. The result will be a necessary condition for the overall system to be optimum, but the average cost so attained may be only a local and not a global minimum [Rad62]. Under hypothesis H_i, $i = 0, 1$, the probability density function of the datum g_k in the kth sensor is $p_{ik}(g_k)$, and the probability that it sends a 1 to the fusion center is

$$q_{ik}(g_{k0}) = \Pr(u_k = 1 | H_i) = \int_{g_{k0}}^{\infty} p_{ik}(g_k) \, dg_k, \qquad i = 0, 1. \tag{4-91}$$

Let (μ) again denote the set of sequences $\mathbf{u}$ for which the fusion center decides for hypothesis H_1. Its probability of doing so under hypothesis H_i can be written much as in (4-87):

$$G_i(\mathbf{g}_0) = \Pr(\rightarrow H_1 | H_i) = \sum_{(\mu)} \prod_{k=1}^{M} \{(1 - u_k)[1 - q_{ik}(g_{k0})] + u_k q_{ik}(g_{k0})\}, \tag{4-92}$$

where $\mathbf{g}_0$ denotes the set of quantization levels $\{g_{k0}\}$, $1 \leq k \leq M$.

Fixing the decision rule at the fusion center as in (4-90), we maximize

$$F = G_1(\mathbf{g}_0) - \Lambda_0 G_0(\mathbf{g}_0)$$

by varying each quantization level g_{k0}, setting the M partial derivatives of F with respect to the quantization levels equal to zero:

$$\frac{\partial F}{\partial g_{k0}} = \frac{\partial G_1(\mathbf{g}_0)}{\partial g_{k0}} - \Lambda_0 \frac{\partial G_0(\mathbf{g}_0)}{\partial g_{k0}} = 0, \qquad 1 \leq k \leq M. \tag{4-93}$$

By (4-91) and (4-92)

$$\frac{\partial G_i(\mathbf{g}_0)}{\partial g_{k0}} = \sum_{(\mu)} (1 - 2u_k)p_{ik}(g_{k0}) \prod_{m \neq k} \{(1 - u_m)[1 - q_{im}(g_{m0})] + u_m q_{im}(g_{m0})\}$$

$$= \sum_{(\mu)} (1 - 2u_k)p_{ik}(g_{k0})\, \mathrm{Pr}(\mathbf{u}^k \mid H_i), \qquad i = 0, 1,$$

(4-94)

where $\mathbf{u}^k$ denotes the set $(u_1, \ldots, u_{k-1}, u_{k+1}, \ldots, u_M)$ of all the 0's and 1's received by the fusion center in a given trial, omitting the one from the kth sensor. The probabilities figuring in (4-94) are

$$\mathrm{Pr}(\mathbf{u}^k \mid H_i) = \prod_{m \neq k} \mathrm{Pr}(u_m \mid H_i), \qquad i = 0, 1,$$

and

$$\mathrm{Pr}(u_m \mid H_i) = (1 - u_m)\,\mathrm{Pr}(g_m < g_{m0} \mid H_i) + u_m\,\mathrm{Pr}(g_m \geq g_{m0} \mid H_i)$$

$$= (1 - u_m)[1 - q_{im}(g_{m0})] + u_m q_{im}(g_{m0}), \qquad i = 0, 1.$$

We furthermore define the two sequences

$$\mathbf{u}_j^k = (u_1, \ldots, u_{k-1}, u_k = j, u_{k+1}, \ldots, u_M), \qquad j = 0, 1.$$

In (4-94), $1 - 2u_k = 1$ for $u_k = 0$, and $1 - 2u_k = -1$ for $u_k = 1$. If both sequences $\mathbf{u}_0^k$ and $\mathbf{u}_1^k$ lie in the set (μ) of sequences causing the fusion center to decide for hypothesis H_1, therefore, the terms corresponding to these sequences will cancel from the sum over (μ) in (4-94). The only terms remaining are for sequences $\mathbf{u}$ in which $\mathbf{u}_1^k$ lies in (μ) and $\mathbf{u}_0^k$ does not. We designate that set of sequences by (μ_k). These are the sequences $\mathbf{u}$ in which the kth digit u_k is decisive in the sense that if $u_k = 1$ the fusion center decides for hypothesis H_1, and if $u_k = 0$ it decides for hypothesis H_0. The quantization level g_{k0} in the kth sensor does not depend on the probabilities of sequences $\mathbf{u}$ in which the fusion center disregards digit u_k.

In this way (4-93), with (4-94), becomes

$$p_{1k}(g_{k0}) \sum_{(\mu_k)} \mathrm{Pr}(\mathbf{u}^k \mid H_1) - \Lambda_0 p_{0k}(g_{k0}) \sum_{(\mu_k)} \mathrm{Pr}(\mathbf{u}^k \mid H_0) = 0,$$

and the quantization level g_{k0} in the kth sensor is determined by the equation

$$\Lambda_k(g_{k0}) = \Lambda_0 \frac{\displaystyle\sum_{(\mu_k)} \mathrm{Pr}(\mathbf{u}^k \mid H_0)}{\displaystyle\sum_{(\mu_k)} \mathrm{Pr}(\mathbf{u}^k \mid H_1)}, \qquad 1 \leq k \leq M,$$

(4-95)

where

$$\Lambda_k(g_k) = \frac{p_{1k}(g_k)}{p_{0k}(g_k)}$$

is the likelihood ratio for the kth sensor. The M equations (4-95), along with (4-90), determine the quantization levels g_{k0} in each sensor. Alternative derivations of these equations are to be found in [Sri86] and [Hob89].

In order to interpret these equations, let us describe an iterative procedure for calculating the quantization levels leading to minimum Bayes cost for a given set of costs and prior probabilities, that is, for a given ratio Λ_0 (4-88). Given a trial set

$\mathbf{g}_0 = \{g_{k0}\}$ of quantization levels in the M sensors, one's computer runs through all 2^M possible sequences $\mathbf{u} = (u_1, u_2, \ldots, u_M)$ of 0's and 1's and assigns them to H_0 or H_1 according to whether the likelihood ratio $\Lambda(\mathbf{u})$ in (4-90) is less than or greater than the ratio Λ_0 in (4-88). It then uses (4-95) to evaluate a new quantization level g_{k0} for each sensor.

Denote the right side of (4-95) by $R_k(\mathbf{g}_0)$. A stable way of determining the new set of quantization levels, we have found, is to solve the M equations

$$\Lambda_k(g'_{k0}) = R_k(\mathbf{g}_0), \qquad k = 1, 2, \ldots, M,$$

for the M quantities g'_{k0}, and then to take the kth new quantization level as the average of g'_{k0} and the previous level g_{k0},

$$g_{k0} \leftarrow \tfrac{1}{2}(g_{k0} + g'_{k0}).$$

With the new set of quantization levels, the computer reclassifies all 2^M sequences in accordance with (4-90), recalculates the quantization levels by (4-95), and continues thus until the latter cease changing significantly. It can, for instance, calculate the quantity F in (4-89) at each stage and stop when its increase becomes inconsiderable. That quantity is

$$F = Q_d(\Lambda_0) - \Lambda_0 Q_0(\Lambda_0),$$

and the false-alarm and detection probabilities involved are $Q_0(\Lambda_0) = \Pr(\rightarrow H_1 | H_0)$ and $Q_d(\Lambda_0) = \Pr(\rightarrow H_1 | H_1)$, where as in (4-92)

$$\Pr(\rightarrow H_1 | H_i) = \sum_{(\mu)} \prod_{k=1}^{M} \Pr(u_k | H_i), \qquad i = 0, 1, \tag{4-96}$$

with the summation over the set (μ) of all sequences $\mathbf{u}$ leading the fusion center to decide for hypothesis H_1.

If one has adopted the Neyman–Pearson criterion that the probability Q_d of detection shall be maximum for a preassigned false-alarm probability Q_0, one can search for the value of the parameter Λ_0 for which the false-alarm probability $Q_0(\Lambda_0) = \Pr(\rightarrow H_1 | H_0)$ takes on that preassigned value Q_0. For each value of Λ_0 during the search, the computer must work out the optimum set of quantization levels g_{k0}. We now consider how to initiate this search.

Let us suppose that the inputs to all the sensors have the same statistical character, but differ only in their input energy-to-noise ratios $S_1, S_2, \ldots, S_M$. In order to determine an initial set $\mathbf{g}_0$ of quantization levels and an initial value of Λ_0, one pretends that all these input ratios are equal to a common value $\overline{S}$ and that all the data g_k are identically distributed. Then all the quantization levels will initially be equal to a value g_0 that can be calculated by the method described in Sec. 4.5.1. The fusion center will in that situation decide for hypothesis H_1 if it receives more than some number m_0 of 1's and for H_0 if it receives m_0 or fewer. The subsequences $\mathbf{u}^k$ in (4-94) for the sequences $\mathbf{u}$ lying in the sets (μ_k) have exactly m_0 1's and $M - 1 - m_0$ 0's. In the notation of Sec. 4.5.1, (4-95) yields for the initial value of Λ_0

$$\Lambda_0 = \frac{p_1(g_0)[q_1(g_0)]^{m_0}[1 - q_1(g_0)]^{M-1-m_0}}{p_0(g_0)[q_0(g_0)]^{m_0}[1 - q_0(g_0)]^{M-1-m_0}}$$

with $q_i(\cdot)$ given in (4-80); $p_i(g)$ is the probability density function of the datum g in each sensor, which is temporarily assumed to be the same in all ($i = 0, 1$). The subsequent search seems to converge most rapidly if the common input energy-to-noise ratio $\overline{S}$ lies roughly in the middle of the set $S_1, S_2, \ldots, S_M$.

With these initial values of Λ_0 and $g_{k0} \equiv g_0$, $1 \leq k \leq M$, one restores the original distributions of the M data g_k and uses the procedure based on (4-90) and (4-95) to calculate a new set of quantization levels g_{k0}, and one evaluates its false-alarm probability $Q_0(\Lambda_0)$. Next one increases Λ_0 by some small amount ε and by the same method computes $Q_0(\Lambda_0 + \varepsilon)$. The secant method then determines a new value of Λ_0 through

$$\Lambda_0 \leftarrow \Lambda_0 + \varepsilon \frac{Q_0(\Lambda_0) - Q_0}{Q_0(\Lambda_0) - Q_0(\Lambda_0 + \varepsilon)} \tag{4-97}$$

[Pre86, pp. 248–51]. This process is repeated until the values of Λ_0 cease changing significantly. The probability of detection is the final value of $\Pr(\rightarrow H_1 | H_1)$ as in (4-96).

The writer has tried this method for signals suffering Rayleigh fading, for which the datum g_k has density functions and complementary cumulative distributions given by

$$p_{0k}(g) = e^{-g} U(g), \qquad q_{0k}(g_0) = e^{-g_0} U(g_0),$$

$$p_{1k}(g) = a_k e^{-a_k g} U(g), \qquad q_{1k}(g_0) = e^{-a_k g_0} U(g_0), \qquad a_k = \frac{1}{1 + S_k}. \tag{4-98}$$

As many as nine sensors were included, and a small variety of input energy-to-noise ratios S_k were tested. It was found more efficient to work with the variables $\ln Q_0$ and $\ln \Lambda_0$ in the secant method (4-97) than with Q_0 and Λ_0. The method converged in all cases, but no guarantee can be furnished that it will always do so. If it goes awry, some other method of solving the equations (4-95) must be sought. The number 2^M of sequences that must be taken into account, and thus also the required storage and the computation time, increase exponentially with the number M of sensors.

This method produces only a local maximum of the detection probability Q_d for a given false-alarm probability Q_0. To what set of quantization levels it converges may depend on the initial values chosen for them. In Table 4-1 we list three sets that satisfy the optimization equations (4-95) for $Q_0 = 0.001$, along with the resulting false-dismissal probabilities $Q_1 = 1 - Q_d$. Under each set is a list of the sequences of digits that cause the fusion center to decide for hypothesis H_0. The digits are in the same order as the sensor input energy-to-noise ratios in the top half of the table. The first solution resulted from taking all input energy-to-noise ratios equal initially to $S_4 = 34$, the second from taking them equal to 20, and the third from taking them equal to the largest energy-to-noise ratio $S_1 = 40$. The solution in the first column yields the highest probability Q_d of detection. In this example three of the seven sensors have much weaker inputs than the rest.

Equations corresponding to (4-95) when the inputs to the sensors are not statistically independent are to be found in [Hob89], but no procedure for solving them has been recommended. The weak-signal approximation has been utilized in [Blu92]

Table 4-1 Distributed Detection ($Q_0 = 0.001$)

S_k	Quantization levels		
40	8.7296	5.0098	8.6046
38	8.7360	4.9934	8.6110
36	8.7433	4.9776	8.6181
34	8.7513	4.9624	8.6261
14	8.9608	4.9075	4.8105
12	9.0209	4.9312	4.6621
10	9.1056	4.9726	4.4784
Q_1:	2.2704 (−4)	2.6544 (−4)	3.7437 (−4)

Sequences leading to H_0		
0000000	0000000	0000000
	1000000	0000100
	0100000	0000010
	0010000	0000001
	0001000	
	0000100	
	0000010	
	0000001	

to obtain a strategy for distributed-detection systems with a small number of dependent inputs.

Problems

4-1. A binary communication system transmits 0's and 1's by sending a pulse at carrier frequency Ω_1 for each 0 and a pulse at carrier frequency Ω_2 for each 1. The receiver has two narrowband inputs

$$v_1(t) = \text{Re } V_1(t) \exp i\Omega_1 t \quad \text{and} \quad v_2(t) = \text{Re } V_2(t) \exp i\Omega_2 t,$$

both observed during an interval $(0, T)$ and corrupted by statistically independent white noise processes

$$n_1(t) = \text{Re } N_1(t) \exp i\Omega_1 t \quad \text{and} \quad n_2(t) = \text{Re } N_2(t) \exp i\Omega_2 t$$

with unilateral spectral density N.

Under hypothesis H_0

$$v_1(t) = n_1(t) + A \text{ Re } F(t) \exp(i\Omega_1 t + i\psi), \qquad v_2(t) = n_2(t);$$

under hypothesis H_1

$$v_1(t) = n_1(t), \qquad v_2(t) = n_2(t) + A \text{ Re } F(t) \exp(i\Omega_2 t + i\psi).$$

The complex envelope $F(t)$ is known to the receiver, but the phase ψ is unknown and uniformly distributed over $(0, 2\pi)$. The signals are subject to Rayleigh fading; that is, the signal amplitude is a random variable with a Rayleigh distribution

$$z(A) = \frac{A}{\sigma^2} e^{-A^2/2\sigma^2} U(A).$$

How should the receiver process its two inputs in order to decide between hypotheses H_0 and H_1 with minimum probability P_e of error? Assume that 0's and 1's occur with equal relative frequencies. In terms of a suitably defined average energy-to-noise ratio, calculate that minimum probability of error.

4-2. Use (3-106) and the method of Problem 3-11 to derive (4-18).

4-3. Define an effective signal-to-noise ratio for the statistic U by

$$D_U^2 = \frac{[E(U\mid H_1) - E(U\mid H_0)]^2}{\text{Var}_0\ U}$$

as in (4-72), where $\text{Var}_0\ U$ is the variance of U under hypothesis H_0. Determine the effective signal-to-noise ratios D_U^2 and $D_{U'}^2$ for the statistics U and U' of (4-19) and (4-21), respectively, when the signal-to-noise ratio of the radar echo signal actually present in the kth interval is

$$d_k^2 = d^2 f(\theta_k - \theta_0),$$

and $f(\cdot)$ is the combination of the antenna gain patterns on transmission and reception. Use Schwarz's inequality to show that $D_{U'}^2 > D_U^2$.

4-4. Evaluate the effective signal-to-noise ratio $D_{g'}^2$ defined as in Problem 4-3 for the statistic g' of (4-5) and determine the coefficients a_m for which it is maximum.

4-5. As in Example 1-3, a receiver must decide whether a noise source is present on the basis of n independent measurements of a datum v that has a Gaussian distribution with expected value 0. In the absence of the source (hypothesis H_0) its variance is N_0, and in its presence (hypothesis H_1) the variance of v is $N_1 = N_0 + S$, $S > 0$. The receiver bases its decisions on the statistic

$$G_k = \sum_{j=1}^{n} v_j^{2k}.$$

The optimum statistic is G_1. Calculate the asymptotic relative efficiency of a receiver utilizing G_k versus a receiver utilizing G_1 for integral values of $k > 1$. Evaluate it for $k = 2, 3, 4$.

4-6. Under hypothesis H_0 the datum g has the probability density function $p_0(g) = \frac{1}{2} \exp(-|g|)$; under H_1 it is $p_1(g) = \frac{1}{2} \exp(-|g - a|)$, where $a > 0$ represents a signal. On the basis of M independent data $g_1, g_2, \ldots, g_M$ of this kind, a receiver is to decide between these two hypotheses in accordance with the Neyman–Pearson criterion. Find the optimum statistic for this decision as a combination of the data, and calculate its effective signal-to-noise ratio as defined in (4-66). Compare this receiver with one basing its decisions on the statistic

$$G' = \sum_{k=1}^{M} g_k.$$

Calculate the asymptotic relative efficiency of this receiver with respect to the optimum receiver. Be careful in passing to the limit $a \to 0$.

4-7. Determine the efficacy, as defined in (4-69), for a detector that utilizes the statistic

$$g^{(m)} = \left| \int_0^T F^*(t) V(t)\, dt \right|^m, \qquad m > 0,$$

where $v(t) = \text{Re}\, V(t) \exp i\Omega t$ is the input to the receiver and $s(t) = \text{Re}\, S(t) \exp (i\Omega t + i\psi)$ is the signal to be detected. The noise is white and Gaussian. For the necessary moments use (C-7) of Appendix C. Find the asymptotic relative efficiency a.r.e.$_{2:m}$ of a receiver utilizing a quadratic detector ($m = 2$) with respect to a receiver using an mth-law detector ($m \neq 2$). Calculate it for $m = 1, 3, 4,$ and 5 and observe that it is greater than 1.

4-8. In a certain diversity communication system there are M receivers that simultaneously pick up the output of a single transmitter sending messages coded into equally probable 0's and 1's. For a 0 the transmitter sends nothing; for a 1 it sends a narrowband pulse with complex envelope $F(t)$. The digits appear every T seconds, and the signals are confined to intervals of duration T. Each receiver picks up a common "specular" component of the signal, which has a known amplitude and is the same in all the receivers, and an independently randomly scattered component, whose amplitude has a Rayleigh distribution with parameter σ^2 as in Problem 4-1. Thus the input to the kth receiver when a 1 was sent is

$$v_k(t) = n_k(t) + B \operatorname{Re} F(t) \exp(i\Omega t + i\psi) + C_k \operatorname{Re} F(t) \exp(i\Omega t + i\phi_k),$$
$$k = 1, 2, \ldots, M, \qquad 0 \leq t \leq T,$$

and the probability density function of C_k has the Rayleigh form in Problem 4-1. The phases ϕ_k, $1 \leq k \leq M$, and ψ are all independent and uniformly distributed over $(0, 2\pi)$. The random amplitudes $C_1, C_2, \ldots, C_M$ are independent of each other and of the phases. The noise inputs $n_k(t)$ are independent, white, and Gaussian with unilateral spectral density N.

(a) Work out the optimum way of processing and combining the M inputs $v_1(t)$, $v_2(t), \ldots, v_M(t)$ in order to decide with minimum probability of error whether a 0 or a 1 was sent. Draw a block diagram of your system. *Hint:* It is simpler to work with the real and imaginary parts of $C_k \exp i\phi_k$, $k = 1, 2, \ldots, M$. What is their joint probability density function?

(b) Determine the threshold statistic G for this decision problem under the assumption that the signal amplitudes are very small. Assume that the ratio σ^2/B^2 of the scattered component to the specular component is known and fixed. Calculate the effective signal-to-noise ratio, defined as in (4-72), for the statistic G. Here $E(G| H_1)$ includes an average over the distributions of the amplitudes and the phases of the received signals. *Hint:* Use (4-75).

4-9. For all four Swerling cases, derive the probability density functions $z(S)$ of the total energy-to-noise ratio, as given in Sec. 4.2.4, from those given for the individual signal-to-noise ratios d. Calculate the expected values and the variances of the statistic U in (4-19) in all four cases.

5

Evaluating Signal Detectability

5.1 THE EDGEWORTH SERIES

5.1.1 The Moment-generating Function

Radar receivers, as we have seen in Chapter 4, often base their decisions about the presence or absence of a target on multiple inputs, the noise in which is statistically independent from one to another. Most commonly the detection statistic, which we now denote by G', is the sum of statistics g_j formed from each input,

$$G' = g_1 + g_2 + \cdots + g_j + \cdots + g_M; \qquad (5\text{-}1)$$

the statistic g_j is a functional

$$g_j = \mathfrak{f}[v_j(t)], \qquad 1 \leq j \leq M,$$

of the jth input, and M is the number of inputs. Indeed, if signal parameters such as carrier phase and amplitude are independently random from one input to another, the logarithm of the average likelihood ratio has the form (5-1), and in Sec. 4.2.1 we saw that a particular threshold statistic also takes this form. A diversity communication system in which a symbol 1 invokes the simultaneous transmission of signals in disjoint frequency bands likewise decides about the transmitted message digits on the basis of a statistic such as G'. The components g_j of G' are often independent random variables because of the independence of the noise in the several inputs. They may lose that statistical independence, however, if the signal components depend on a common, but random parameter such as the amplitude A in Swerling cases 1 and 3 (Sec. 4.2.4).

The larger the number M of inputs, the more tedious it becomes to calculate the false-alarm and detection probabilities

$$Q_0 = \int_G^\infty P_0(g) \, dg, \qquad Q_d = \int_G^\infty P_1(g) \, dg$$

for a receiver that decides that a signal is present when the statistic G' exceeds a decision level G. (We can use any letter we like to designate our variable of integration, and we shall often use g in this context.) The algorithms in the Appendix, Sec. C.3, for calculating the Mth-order Q function, for instance, take a long time when M is large and may require complicated stratagems in order to avoid overflow and underflow in the process. For some detectors it may not even be possible to determine the density functions $P_0(\cdot)$ and $P_1(\cdot)$ of the statistic G' under hypotheses H_0 and H_1 in closed form.

Methods for calculating *tail probabilities* such as Q_0 and Q_d will now be presented that are generally the more accurate, the larger the number M of independent random variables g_k, yet whose computation times are roughly independent of M. Reference to a particular hypothesis will be omitted. We shall assume that the moment-generating function

$$h(z) = E(e^{-G'z}) = \int_{-\infty}^\infty P(g) \, e^{-gz} \, dg \qquad (5\text{-}2)$$

of the statistic G' is an analytic function in the complex z-plane. The moment-generating function is the Laplace transform of the probability density function $P(\cdot)$ of G'. It was given this name because the coefficients of its Taylor expansion about the origin,

$$h(z) = \sum_{k=0}^\infty E(G'^k) \frac{(-z)^k}{k!} \qquad (5\text{-}3)$$

are proportional to the moments of the random variable G'. The moment-generating function is assumed to be regular in a vertical strip that contains the imaginary axis and has finite or infinite width; we deal only with random variables all of whose moments are finite. This strip, which is called the *regularity domain*, can be represented by

$$c_1 < \text{Re } z < c_2, \qquad c_1 < 0, \qquad c_2 > 0.$$

Along the imaginary axis, $\text{Re } z \equiv 0$, $z = -i\omega$; and $h(-i\omega)$ is the familiar characteristic function, that is, the Fourier transform of the probability density function $P(\cdot)$ [Hel91, pp. 121–3], [Pap91, pp. 115–20]. When, as often, G' is a positive random variable, the regularity domain includes the entire right half-plane, for then

$$|h(z)| = \left| \int_0^\infty P(g) \, e^{-gz} \, dg \right| \leq \int_0^\infty P(g) \, e^{-gx} \, dg \leq \int_0^\infty P(g) \, dg = 1, \qquad x = \text{Re } z > 0,$$

and singularities occur only in the left half-plane.

When the statistic G' is a sum of M independent random variables, the moment-generating function of G' is the product

$$h(z) = \prod_{k=1}^M \eta_k(z)$$

of the moment-generating functions

$$\eta_k(z) = E[\exp(-g_k z)] = \int_{-\infty}^{\infty} p_k(g)\, e^{-gz}\, dg$$

of the statistics $g_k = \mathfrak{f}[v_k(t)]$; $p_k(\cdot)$ is the probability density function of g_k.

5.1.2 The Gram–Charlier and Edgeworth Series

When, as we shall now assume, the random variables g_j in (5-1) are independent and identically distributed, their sum G' possesses a distribution that by virtue of the central limit theorem is the more nearly Gaussian, the larger the number M [Hel91, pp. 260–5], [Pap91, pp. 214–21]. We describe a method that takes advantage of that asymptotic behavior of the probability density function of G'. It requires knowing only the moments of the random variable G', or of its components g_j, and it does not require the characteristic function or the moment-generating function to be available in closed form.

The probability density function of a random variable such as G' is the inverse Laplace transform of its moment-generating function,

$$P(g) = \int_{c-i\infty}^{c+i\infty} e^{gz} h(z) \frac{dz}{2\pi i}. \tag{5-4}$$

The vertical contour of integration can lie anywhere within the regularity domain of the Laplace transform (5-2): $c_1 < c < c_2$.

Into (5-4) we shall introduce the logarithm of the moment-generating function $h(z)$ expanded in a series of powers of $(-z)$,

$$\ln h(z) = -\overline{G}z + \tfrac{1}{2}\sigma^2 z^2 + \sum_{k=3}^{\infty} \kappa_k \frac{(-z)^k}{k!}, \tag{5-5}$$

where $\overline{G} = E(G')$ is the expected value of the statistic G' and $\sigma^2 = \mathrm{Var}\, G'$ is its variance. The quantities κ_k are called the *cumulants* or *semiinvariants* of G', and $\ln h(z)$ is called the *cumulant-generating function*. The reason for the name "cumulant" is that κ_k for a sum G' of independent random variables equals the sum of the cumulants $\kappa_k^{(j)}$ for the components g_j of the sum. The first cumulant κ_1 is the expected value; the second cumulant κ_2 is the variance. Later we shall see how the coefficients of the expansion in (5-5) can be calculated from the moments $E(G'^k)$ of G', $k = 1, 2, \ldots$.

The moment-generating function $h(z)$ of G' is written as the exponential function of the right side of (5-5):

$$h(z) = \exp(-\overline{G}z + \tfrac{1}{2}\sigma^2 z^2)\exp r(z),$$
$$r(z) = \sum_{k=3}^{\infty} \kappa_k \frac{(-z)^k}{k!}. \tag{5-6}$$

When we write $\exp r(z)$ as a power series in $(-z)$ by using the Taylor expansion of the exponential function, and when we then collect terms with like powers of $(-z)$, we obtain the power series

$$e^{r(z)} = 1 + \sum_{k=3}^{\infty} c_k(-z)^k. \tag{5-7}$$

The first few coefficients c_k are given by

$$c_3 = \frac{\kappa_3}{3!}, \qquad c_4 = \frac{\kappa_4}{4!}, \qquad c_5 = \frac{\kappa_5}{5!}, \qquad c_6 = \frac{\kappa_6 + 10\kappa_3^2}{6!},$$

and so on. An algorithm for computing these from the cumulants will be presented later.

After $h(z)$ in this form is substituted into the contour integral in (5-4), we obtain for the probability density function of G'

$$P(g) = \int_{c-i\infty}^{c+i\infty} \exp[(g - \overline{G})z + \tfrac{1}{2}\sigma^2 z^2] \left[1 + \sum_{k=3}^{\infty} c_k(-z)^k \right] \frac{dz}{2\pi i}. \tag{5-8}$$

We shall now evaluate this inverse transform term by term.

The moment-generating function of a Gaussian random variable with expected value $\overline{G}$ and variance σ^2,

$$P_\Gamma(g) = \frac{1}{\sqrt{2\pi\sigma^2}} \exp\left[-\frac{(g - \overline{G})^2}{2\sigma^2} \right],$$

is $\exp(-\overline{G}z + \tfrac{1}{2}\sigma^2 z^2)$, and we can therefore write, by (5-4),

$$P_\Gamma(g) = \int_{c-i\infty}^{c+i\infty} \exp[(g - \overline{G})z + \tfrac{1}{2}\sigma^2 z^2] \frac{dz}{2\pi i} = \frac{1}{\sigma}\phi^{(0)}\left(\frac{g - \overline{G}}{\sigma} \right),$$

where

$$\phi^{(0)}(y) = \frac{1}{\sqrt{2\pi}} e^{-y^2/2} \tag{5-9}$$

is called the *error function*. Differentiating k times with respect to $\overline{G}$, we find

$$\int_{c-i\infty}^{c+i\infty} (-z)^k \exp[(g - \overline{G})z + \tfrac{1}{2}\sigma^2 z^2] \frac{dz}{2\pi i} = \frac{1}{\sigma^{k+1}}\phi^{(k)}\left(\frac{g - \overline{G}}{\sigma} \right),$$

where

$$\phi^{(k)}(y) = (-1)^k \frac{d^k}{dy^k} \phi^{(0)}(y). \tag{5-10}$$

Thus we can write (5-8) for the probability density function of G' as

$$P(g) = \frac{1}{\sigma}\phi^{(0)}\left(\frac{g - \overline{G}}{\sigma} \right) + \frac{1}{\sigma} \sum_{k=3}^{\infty} \frac{c_k}{\sigma^k}\phi^{(k)}\left(\frac{g - \overline{G}}{\sigma} \right), \tag{5-11}$$

which is known as the *Gram–Charlier* series. It represents the density function as a Gaussian density function plus a sequence of correction terms.

The functions $\phi^{(k)}(y)$ are often written as

$$\phi^{(k)}(y) = \phi^{(0)}(y)h_k(y),$$

where $h_k(y)$ is the kth *Hermite polynomial*,

$$h_0(y) = 1, \qquad h_1(y) = y, \qquad h_2(y) = y^2 - 1,$$

and so on. The functions $\phi^{(k)}(y)$, like the Hermite polynomials, are subject to the recurrent relation

$$\phi^{(k+1)}(y) = y\phi^{(k)}(y) - k\phi^{(k-1)}(y), \qquad (5\text{-}12)$$

which follows easily from Leibnitz's rule for differentiating a product:

$$\frac{d^{k+1}}{dy^{k+1}} (e^{-y^2/2}) = \frac{d^k}{dy^k}(-y\,e^{-y^2/2}) = -\sum_{r=0}^{k} \binom{k}{r} \frac{d^r}{dy^r}(y) \frac{d^{k-r}}{dy^{k-r}}(e^{-y^2/2})$$

$$= -y\frac{d^k}{dy^k}(e^{-y^2/2}) - k\frac{d^{k-1}}{dy^{k-1}}(e^{-y^2/2}).$$

In analyzing receivers we most often need not the probability density function of the decision statistic G', but its complementary cumulative distribution

$$q_+(G) = \Pr(G' \ge G) = \int_G^{\infty} P(g)\, dg, \qquad (5\text{-}13)$$

which furnishes us with the false-alarm or the detection probability, as the case may be. When we substitute the series (5-11) into (5-13), the first term yields an error-function integral as defined in (1-11). Integrating the remaining terms reduces the order of the derivative $\phi^{(k)}$ by 1, as can be seen from (5-10), and we obtain

$$q_+(G) = \operatorname{erfc}\left(\frac{G - \overline{G}}{\sigma}\right) + \sum_{k=3}^{\infty} \frac{c_k}{\sigma^k} \phi^{(k-1)}\left(\frac{G - \overline{G}}{\sigma}\right) \qquad (5\text{-}14)$$

for the complementary cumulative distribution of the statistic G'. The cumulative distribution is similarly

$$q_-(G) = \int_{-\infty}^{G} P(g)\, dg = 1 - q_+(G)$$

$$= 1 - \operatorname{erfc}\left(\frac{G - \overline{G}}{\sigma}\right) - \sum_{k=3}^{\infty} \frac{c_k}{\sigma^k} \phi^{(k-1)}\left(\frac{G - \overline{G}}{\sigma}\right). \qquad (5\text{-}15)$$

The functions $\frac{1}{2} - \operatorname{erfc} y$ and $(-1)^k \phi^{(k)}(y)$ have been tabulated for use in (5-11) through (5-15), [Har52], [Abr70, pp. 966–74]. The recurrent relation (5-12) enables us to calculate the functions $\phi^{(k)}(y)$, $k > 1$, one after another, starting with $\phi^{(0)}(y)$ and $\phi^{(1)}(y)$; a calculator can easily be programmed to do so, and the tables are unnecessary. For computation of erfc y see [Hel91, pp. 592–6] or [Abr70, pp. 932–3].

When the moments of the components g_j of the sum G' in (5-1) can be calculated, but not their moment-generating function, the coefficients of the series for $\ln h(z)$ in (5-5) can be determined in the following way. Denote the moment-generating function of each component g_j by $\eta(z)$. Then

$$\eta(z) = 1 + \sum_{k=1}^{\infty} a_k(-z)^k, \qquad a_k = \frac{E(g_j^k)}{k!},$$

and we need the coefficients $b_m = \kappa_m^0/m!$ in the power series

$$\sum_{m=1}^{\infty} b_m(-z)^m = \ln\left[1 + \sum_{k=1}^{\infty} a_k(-z)^k\right].$$

They can be calculated recursively by the algorithm

$$b_k = a_k - \frac{1}{k} \sum_{m=1}^{k-1} m b_m a_{k-m}, \tag{5-16}$$

which is easily programmed for a calculator or a computer. In particular,

$$b_1 = a_1,$$
$$b_2 = a_2 - \tfrac{1}{2} b_1 a_1 = a_2 - \tfrac{1}{2} a_1^2,$$
$$b_3 = a_3 - \tfrac{1}{3}(b_1 a_2 + 2 b_2 a_1) = a_3 - a_1 a_2 + \tfrac{1}{3} a_1^3,$$

and so on. To derive (5-16), differentiate

$$1 + \sum_{k=1}^{\infty} a_k x^k = \exp\left[\sum_{m=1}^{\infty} b_m x^m \right]$$

with respect to x,

$$\sum_{k=1}^{\infty} k a_k x^{k-1} = \sum_{m=1}^{\infty} m b_m x^{m-1} \left[1 + \sum_{k=1}^{\infty} a_k x^k \right]$$
$$= \sum_{k=1}^{\infty} k b_k x^{k-1} + \sum_{k=2}^{\infty} x^{k-1} \sum_{m=1}^{k-1} m b_m a_{k-m},$$

and equate the coefficients of x^{k-1} on both sides of the equation.

The coefficients b_m are then multiplied by M to obtain the coefficients of the series in (5-5); we denote them by $b_k' = \kappa_k / k!$. In particular,

$$E(G') = b_1' = M a_1, \qquad \text{Var } G' = 2 b_2' = 2 M b_2, \qquad b_k' = M b_k.$$

To find the coefficients c_k in (5-11) and subsequent equations, one forms the exponential function of that series, but with the first two terms set to zero:

$$1 + \sum_{k=1}^{\infty} c_k (-z)^k = \exp\left[\sum_{m=1}^{\infty} b_m''(-z)^m \right].$$

Here $b_1'' = b_2'' = c_1 = c_2 = 0$ and $b_m'' = b_m' = M b_m$, $m \geq 3$. The recurrent relation in (5-16) again applies, but in the form

$$c_k = b_k'' + \frac{1}{k} \sum_{m=1}^{k-1} m b_m'' c_{k-m}. \tag{5-17}$$

Because the summations in (5-16) and (5-17) are the same, they can be programmed as a subroutine to be employed in both algorithms. Taking these with the recurrent relation (5-12) for the derivatives of the error function, one can set up a program for computing the Gram–Charlier series in a computer or in a programmable calculator with sufficient program and data memory.

Taking the terms of (5-14) in their natural order is not the most accurate way of summing that series. When as here the statistic G' is the sum of M independent and identically distributed random variables g_j, the cumulants κ_k in (5-5) have a common factor M:

$$\kappa_k = M\kappa_k^0, \qquad \sigma^2 = M\sigma_0^2,$$

where σ_0^2 is the variance of g_j and κ_k^0 is its kth cumulant. We then find upon working out the coefficients $c_k' = c_k/\sigma^k$ in (5-14) that c_3' is of order $M^{-1/2}$, c_4' and c_6' are of order M^{-1}, c_5', c_7', and c_9' of order $M^{-3/2}$, and so on. Keeping together terms whose coefficients are of the same order of magnitude in M amounts to rearranging the series into the form

$$q_+(G) = \text{erfc } y + c_3'\phi^{(2)}(y) + [c_4'\phi^{(3)}(y) + c_6'\phi^{(5)}(y)]$$

$$+ [c_5'\phi^{(4)}(y) + c_7'\phi^{(6)}(y) + c_9'\phi^{(8)}(y)] + \cdots, \qquad y = \frac{G - \overline{G}}{\sigma}. \qquad (5\text{-}18)$$

The subscripts on the coefficients c_k' in the remaining groups follow the pattern $(8, 10, 12)$, $(11, 13, 15)$, ... , $(3k - 1, 3k + 1, 3k + 3)$, ... Rearranged in this way the series (5-18) is known as the *Edgeworth* series. In evaluating it one should include all terms of a given order in M if one includes any of them. The series in (5-11) and (5-15) are similarly rearranged. Further details about the Edgeworth series can be found in [Fry65, pp. 257–64].

Although the coefficients $c_k' = c_k/\sigma^k$ generally decrease in absolute value from term to term, the functions $\phi^{(k)}(y)$ after a certain value of k begin to increase rapidly in magnitude. This behavior renders the computation of the Edgeworth series somewhat tricky. It is really divergent and behaves like what is called an "asymptotic" series, converging up to a certain point and then going awry. The safest way to evaluate (5-18) seems to be to print out the sum after the triplet of terms in each bracket has been added in. These sums will at first stabilize, but later begin to diverge. Take as $q_+(G)$ the value at which the sum stabilizes. The magnitude of the last triplet included indicates the order of magnitude of the error. The Edgeworth series is the more reliable, the larger M is and the smaller the value of $|y|$ is, that is, the closer the value of G lies to the expected value $E(G')$ of the decision statistic.

Let us see how to apply the Edgeworth series to evaluating the false-alarm and detection probabilities for the statistic $G' = U$ in (4-19) characterizing the quadratic threshold detector of Sec. 4.2.1. From (4-24) we find for the logarithm of the moment-generating function

$$\ln h(z) = -M \ln(1 + z) - \frac{Sz}{1 + z} = M \sum_{k=1}^{\infty} \frac{(-z)^k}{k} + S \sum_{k=1}^{\infty} (-z)^k, \qquad S = \tfrac{1}{2}D^2,$$

and by comparison with (5-5) we find under hypothesis H_1 that

$$E(G') = \overline{G} = M + S,$$

$$\text{Var } G' = \sigma^2 = M + 2S,$$

$$b_k' = \frac{\kappa_k}{k!} = S + \frac{M}{k},$$

$$y = \frac{G - M - S}{\sqrt{M + 2S}},$$

Evaluating Signal Detectability   Chap. 5

where $G = U_0$ is the decision level on the threshold statistic U. Here the step represented by the recursion (5-16) is unnecessary; we already have the cumulants in simple form.

In Table 5-1 we show the results of a computation with the Edgeworth series for $M = 50$ and $G = U_0 = 91.0634$. The columns are headed with the subscripts on the coefficients c_k^l in the last group of terms added in. The values calculated by the algorithms of the Appendix, Sec. C.3, are listed in the column headed "Exact." Each value in the table is to be multiplied by 10 raised to the power listed in the second column. Thus the last column of the row marked $D = 0$ reports the false-alarm probability as $9.99995 \cdot 10^{-7}$. Comparison shows that only for values of the decision level $G = U_0$ near the expected value of the random variable $G' = U$ can one rely on the accuracy of the Edgeworth series. Because in assessing the performance of receivers we are usually concerned with false-alarm probabilities Q_0 and false-dismissal probabilities $Q_1 = 1 - Q_d$ that are small, we turn in the next section to a method that is most accurate and most expeditious for values of the decision level G far in one tail or the other of the distribution of the decision statistic.

5.2 NUMERICAL LAPLACE INVERSION

In Sec. 5.1 we developed expressions (5-14) and (5-15) for the tail probabilities of the decision statistic G' that consist of error-function integrals plus correction terms involving the derivatives of the error function. We found these to be unreliable and awkward for decision levels G far from the expected value $E(G')$.

One might also compute the probability density function of the statistic G' numerically by taking the inverse Fourier transform

$$P(g) = \int_{-\infty}^{\infty} h(-i\omega)\, e^{-i\omega g}\, \frac{d\omega}{2\pi}$$

of its characteristic function by means of the fast Fourier transform algorithm. Tail probabilities could then be computed by integrating $P(g)$ numerically. When the decision level G on G' is far from $E(G')$, however, determining tail probabilities accurately from the result of a numerical Fourier transformation would be difficult, for one would have to evaluate the characteristic function $h(-i\omega)$ at a large number of closely spaced sample values of ω in order to avoid aliasing, which is most deleterious in the tails of the transform $P(g)$. The computation would be lengthy. We therefore concentrate here on methods that are most accurate in the far tails of the distribution and do not require determining the entire probability density function of the statistic.

5.2.1 Integration through a Saddlepoint

The probability density function of a random variable such as G' is the inverse Laplace transform of its moment-generating function as in (5-4). The vertical contour of integration in that expression can lie anywhere within the regularity domain

Table 5-1 Quadratic Rectifier (M=50, G=91.0634)

D	Power of ten	0	3	4, 6	5, 7, 9	8, 10, 12	11, 13, 15	14, 16, 18	Exact
					$q_+(G)$				
0	−7	0.0317552	0.324228	1.55031	4.68901	9.18196	10.9378	9.53601	9.99995
2	−6	0.0530787	0.426905	1.58597	3.62488	5.23143	5.04091	4.88535	4.97052
4	−4	0.235244	0.958182	1.78734	2.11375	2.03782	2.04187	2.04483	2.04410
6	−2	0.644144	1.05223	1.07321	1.06354	1.06479	1.06516	1.06528	1.06496
8	−1	1.97978	1.94869	1.94649	1.94618	1.94552	1.94647	1.94639	1.94614
					$q_-(G)$				
10	−1	2.32796	2.37983	2.37955	2.37975	2.37920	2.37878	2.37894	2.37915
12	−3	13.1715	8.80163	8.71511	8.73916	8.74848	8.75511	8.76065	8.76320
14	−5	14.1631	−5.69614	3.09017	2.45375	2.15916	2.09462	2.08248	2.06603
16	−8	33.505	−77.1716	70.906	−27.0661	−2.57219	1.83911	1.43125	0.21812

of the Laplace transform (5-2): $c_1 < c < c_2$. If we put it to the right of the imaginary axis, $0 < c < c_2$, we can integrate (5-4) from $-\infty$ to G, whereupon we obtain the left-hand tail probability

$$q_-(G) = \int_{-\infty}^{G} P(g)\, dg = \int_{c-i\infty}^{c+i\infty} \frac{h(z)}{z}\, e^{Gz}\, \frac{dz}{2\pi i}, \qquad 0 < c < c_2, \qquad (5\text{-}19)$$

which is the cumulative distribution function of G'. If, on the other hand, we put the contour to the left of the imaginary axis, but still within the regularity domain, $c_1 < c < 0$, we can integrate (5-4) from G to ∞, obtaining the right-hand tail probability, or the complementary cumulative distribution

$$q_+(G) = \int_{G}^{\infty} P(g)\, dg = \int_{c-i\infty}^{c+i\infty} \frac{h(z)}{-z}\, e^{Gz}\, \frac{dz}{2\pi i}, \qquad c_1 < c < 0. \qquad (5\text{-}20)$$

The tail probabilities can often be calculated by integrating (5-19) or (5-20) numerically.

The vertical contour of integration can be placed anywhere in the strip we have called the regularity domain of the moment-generating function $h(z)$. A location favorable for numerical integration can be found by considering the appearance of the integrand

$$H(x) = \frac{h(x)}{x}\, e^{Gx} = e^{\Phi(x)} = E\{\exp[x(G - G') - \ln x]\}$$

of (5-19) for real values x of $z = x + iy$ in $0 < x < c_2$. This function $H(x)$ is a convex U function of x in this range. As an illustration, we have plotted in Fig. 5-1 the function $|H(x)|$ for the moment-generating function for the Mth-order Q function given in (4-24).

A function $f(x; G')$ is said to be convex U in x if for any ε in $(0, 1)$, with $\varepsilon' = 1 - \varepsilon$,

$$f(\varepsilon x_1 + \varepsilon' x_2; G') \leq \varepsilon f(x_1; G') + \varepsilon' f(x_2; G').$$

If twice differentiable, $f(x; G')$ has a nonnegative second derivative $f''(x; G')$ at all values of x in the region. Here G' may be a random variable on which that convex function happens also to depend. If we take the expected value of both sides of that relation with respect to the distribution of G', the sense of the inequality does not change:

$$E[f(\varepsilon x_1 + \varepsilon' x_2; G')] \leq \varepsilon E[f(x_1; G')] + \varepsilon' E[f(x_2; G')].$$

The convexity of $H(x)$ follows from the obvious fact that for any value of a random variable G', $x(G - G') - \ln x$ is convex U, and from the easily demonstrated fact that the exponential function of a convex function is convex U: if $f(x)$ has a positive second derivative, so does $\exp f(x)$. Taking the expected value with respect to the random variable G' preserves this convexity. A similar argument shows that the integrand of (5-20) is convex U in $c_1 < x < 0$. The integrands of (5-19) and (5-20) thus possess a single minimum when the complex variable z moves along the real axis within the convergence strip of the Laplace transform in (5-2). These minima occur at points z_0 on the Re z-axis, where z_0 is a solution of the equation

$$\Phi'(z) = \frac{d\Phi(z)}{dz} = 0, \qquad z = z_0, \qquad c_1 < z_0 < c_2. \qquad (5\text{-}21)$$

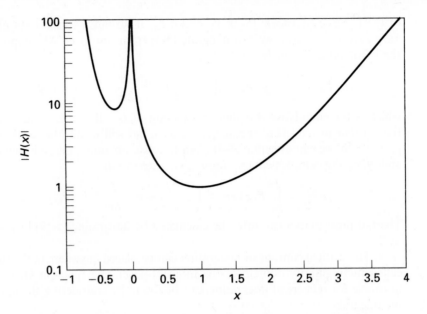

Figure 5-1. $|H(x)|$ for the moment-generating function of the Mth-order Marcum Q function: $M = 5$, $S = \frac{1}{2}D^2 = 6/11$, $G = 40/11$.

It is shown in the theory of complex variables that if an analytic function has a minimum at a point z_0 as z passes through it in one direction—here along the real axis—it has a maximum there as z passes through it in the perpendicular direction. If we run our contour of integration in (5-19) through the point $z = z_0$, the integrand will therefore decrease in absolute value as the point $z = z_0 + iy$ moves up or down the contour away from the real axis. Because of the shape of the surface $|H(z)|$ over the complex z-plane in the neighborhood of $z = z_0$, this point z_0 is called a *saddlepoint* or *col*.

In this region the integrand $H(z)$ can be written in the form

$$
\begin{aligned}
H(z_0 + iy) = \exp[\Phi(z_0) &- \frac{1}{2}\Phi''(z_0)y^2 \\
&+ \frac{1}{3!}\Phi'''(z_0)(iy)^3 + \frac{1}{4!}\Phi^{iv}(z_0)y^4 + \cdots],
\end{aligned}
\tag{5-22}
$$

where $\Phi(z) = \ln H(z)$, $\Phi'(z_0) = 0$, and $\Phi''(z_0) > 0$. The integrand $H(z_0 + iy)$ therefore has a Gaussian behavior as a function of y in the neighborhood of the saddlepoint; we shall take advantage of it in Sec. 5.3.2. Had we placed our contour at any other point z_1 within the strip $0 < \text{Re } z < c_2$, the exponent in (5-22) would have contained a term $\Phi'(z_1)(iy)$, $z = z_1 + iy$, which would cause the integrand to oscillate; and when integrating (5-19) numerically, we should have had to space our samples closely enough to follow the oscillations accurately. By taking the contour through the saddlepoint, oscillations of the integrand are pushed out to the region where its absolute value is small.

For these reasons we set $c = z_0^{(-)}$ in (5-19), where $z_0^{(-)}$ is the solution of (5-21) lying to the right of the origin. Into the integrand of (5-19) we now put $z = z_0^{(-)} + iy$. The imaginary part of the integrand is an odd function of y and integrates to zero. The real part is an even function of y, and we can rewrite (5-19) as

$$q_-(G) = \frac{1}{\pi} \int_0^\infty \mathrm{Re}\left[\frac{h(z_0 + iy)}{z_0 + iy} \exp[(z_0 + iy)G] \right] dy,$$

$$0 < z_0 = z_0^{(-)} < c_2.$$

(5-23)

For the right-hand tail probability a similar argument shows that we should put the contour through the saddlepoint $z_0^{(+)}$ of the function $(-z)^{-1}h(z)\exp Gz$ lying to the left of the origin, $c_1 < c = z_0^{(+)} < 0$; $z_0^{(+)}$ is a second root of (5-21). Then (5-20) becomes

$$q_+(G) = -\frac{1}{\pi} \int_0^\infty \mathrm{Re}\left[\frac{h(z_0 + iy)}{z_0 + iy} \exp[(z_0 + iy)G] \right] dy,$$

$$c_1 < z_0 = z_0^{(+)} < 0.$$

(5-24)

In general it is most efficient to compute $q_+(G)$ for $G > E(G')$ and $q_-(G)$ for $G < E(G')$, although for G near the expected value $E(G')$ it does not much matter which one chooses. The tail probabilities can be determined as accurately as one likes by evaluating (5-23) or (5-24) by a numerical quadrature formula utilizing sufficiently many samples of the integrand spaced closely enough together.

To solve (5-21) for a saddlepoint z_0, Newton's method is generally most expeditious. Starting with a trial value z_0', a new trial value z_0'' is determined from

$$z_0'' = z_0' - \frac{\Phi'(z_0')}{\Phi''(z_0')},$$

(5-25)

where $\Phi''(z)$ is the second derivative of $\Phi(z) = \ln H(z)$. Because the value of this second derivative is not needed to great precision, it suffices, when convenient, to approximate it by calculating $\Phi'(z)$ at nearby values z_0' and $z_0' + \delta z$ and forming

$$\Phi''(z_0') \approx \frac{\Phi'(z_0' + \delta z) - \Phi'(z_0')}{\delta z}.$$

When Newton's method is so modified, it is called the *secant* method [Pre86, pp. 248–51]. Newton's procedure (5-25) is repeated until the value of z_0 ceases changing significantly. Still another way to find the saddlepoint is to use the secant method to search for the root x of

$$\mathrm{Im}\, \Phi(x + i\varepsilon) = \begin{cases} 0, & 0 < x < c_2, \\ -\pi, & c_1 < x < 0, \end{cases}$$

taking ε to be some suitably small number. High accuracy in determining the saddlepoint is unnecessary when (5-23) and (5-24) are to be integrated numerically.

A convenient starting value in the search for the saddlepoint can be derived by approximating the logarithm of the moment-generating function by

$$\ln h(z) \approx -\overline{G}z + \tfrac{1}{2}\sigma_G^2 z^2, \qquad \overline{G} = E(G'), \qquad \sigma_G^2 = \mathrm{Var}\, G'.$$

Then the equation $\Phi'(z) = 0$ for the saddlepoint becomes approximately

$$G - \overline{G} + \sigma_G^2 z - \frac{1}{z} = 0,$$

which can be reduced to a quadratic equation whose roots are

$$z = \frac{\overline{G} - G \pm [(G - \overline{G})^2 + 4\sigma_G^2]^{1/2}}{2\sigma_G^2}.$$

For $G < \overline{G}$ one takes the positive sign, for $G > \overline{G}$ the negative sign. One must check that the starting value lies inside the regularity domain. If not, a value of z on the same side of the origin, but just inside that domain can be chosen to begin the search for the saddlepoint z_0 by (5-25).

Although the integrand $\exp \Phi(z)$ is convex, the phase $\Phi(z)$ is not necessarily so everywhere, and $\Phi''(z)$ may be negative or zero for real values of z in the regularity domain, causing the search for the saddlepoint by Newton's method or the secant method to go awry. In the immediate neighborhood of the saddlepoint z_0, however, $\Phi''(z)$ must be positive.

For integrals of this kind, the trapezoidal rule is both simple and accurate. It approximates a typical semiinfinite integral by

$$I = \int_0^\infty f(y)\, dy \approx \delta y \left[\frac{1}{2} f(0) + \sum_{k=1}^{k_F} f(k\delta y) \right],$$ (5-26)

$$f(y) = \operatorname{Re} H(z_0 + iy).$$

The number k_F of terms is taken large enough that the final value $f(k_F \delta y)$ of the integrand is negligible. One halves the step size δy until the result of the summation stabilizes to the number of significant figures desired. It is unnecessary, after dividing the intervals by 2, to recompute the values of the integrand previously computed, nor even to store them. Before multiplication by δy, one simply adds to the sum previously accumulated the values of the integrand at the new, intermediate points $z = z_0 + iy$. Details of this method of numerical quadrature can be found in *Numerical Recipes* [Pre86, Sec. 4.2, pp. 110–4]. Schwartz [Sch69] and Rice [Ric73] have treated the advantages of this quadrature formula for infinite integrals of analytic functions. The number of reliable significant figures roughly doubles each time one divides the step size δy by 2.

As (5-22) indicates, the width of the integrand of (5-23) and (5-24) is on the order of $[\Phi''(z_0)]^{-1/2}$, and it is convenient to specify the initial interval δy between samples of the integrand as

$$\delta y = \eta[\Phi''(z_0)]^{-1/2},$$ (5-27)

where η is on the order of 1. The summation in (5-26) can be stopped at a point $y = k_F \delta y$ where the ratio of $|H(z_0 + iy)|$ to the absolute value of the sum being accumulated falls below δy times some number ε chosen sufficiently small to ensure the accuracy desired.

As an example, let us consider calculating the Mth-order Q function in the form given in (4-25), (4-26), or as in (C-19) of Appendix C:

$$q_+(G) = Q_M(D, \sqrt{2G}) = Q_M(S, G)$$

$$= \int_G^\infty \left(\frac{U}{S}\right)^{(M-1)/2} e^{-S-U} I_{M-1}(2\sqrt{SU})\, dU, \qquad S = \frac{D^2}{2}.$$

The moment-generating function is

$$h(z) = \frac{1}{(1 + z)^M} \exp\left(-\frac{Sz}{1 + z}\right) \tag{5-28}$$

by (4-24). The "phase" $\Phi(z)$ of the integrand is now

$$\Phi(z) = \ln H(z) = Gz - M \ln(1 + z) - \frac{Sz}{1 + z} - \ln z, \tag{5-29}$$

the saddlepoint z_0 is a root of the equation

$$\Phi'(z) = G - \frac{M}{1 + z} - \frac{S}{(1 + z)^2} - \frac{1}{z} = 0, \tag{5-30}$$

and the second derivative needed in (5-25) and (5-27) is

$$\Phi''(z) = \frac{M}{(1 + z)^2} + \frac{2S}{(1 + z)^3} + \frac{1}{z^2}. \tag{5-31}$$

Equation (5-30) is most easily solved by Newton's method (5-25), although it can be reduced to a cubic equation that can be solved by a computer routine for finding the roots of polynomial equations.

For $M = 10$ and $G = 32.7103$, the second column of Table 5-2 ("Straight Path") shows the values of $q_+(G)$ for $D = 0, 4, 6$ and of $q_-(G)$ for $D = 7, 10, 12$, as calculated by contour integration along a straight vertical contour through the saddlepoint z_0 for $\eta = 1, 0.5, 0.25$. The trapezoidal rule was utilized with $\varepsilon = 10^{-8}$. The number of steps taken in the numerical integration is shown in the third column of the table. For all values of D the results with $\eta = 0.25$ agreed with those calculated by the recurrent algorithm of the Appendix, Sec. C.3. Bounds on the truncation error incurred by cutting off the numerical integration at a finite value of y are to be found in [Hel84d]. The results of that paper justify the simple stopping rule stated under (5-27).

The probability density function $P(G)$ can be calculated by numerical saddlepoint integration of the inverse Laplace transform (5-4). The vertical contour is now taken through a saddlepoint that is the root of $-\psi'(z) = G$, $\psi(z) = \ln h(z)$. Because of the convexity of the integrand $h(z) \exp Gz$ for *real* values of z, a single such saddlepoint exists on the Re z-axis in the regularity domain $c_1 < \text{Re } z < c_2$. Detailed examples were worked out by Rice [Ric80].

5.2.2 Integration on a Curved Path

The magnitude of the integrand in integrals of the types in (5-4), (5-19), and (5-20),

$$q(G) = \int_C e^{\Phi(z)} \frac{dz}{2\pi i}, \tag{5-32}$$

decreases most rapidly along a contour C that is known as the *path of steepest descent* [Car66, pp. 257–66]. The magnitude of the integrand is $\exp[\text{Re } \Phi(z)]$, and along the path of steepest descent Re $\Phi(z)$ drops toward $-\infty$ most precipitously. With $z = x + iy$, the function

$$f_R(x, y) = \text{Re } \Phi(x + iy)$$

Table 5-2 Marcum's Q Function $Q_M(S, G)$ ($M = 10$, $G = 32.7103$)

$D = \sqrt{2S}$	Straight path	Number of steps	Curved path	Number of steps	Saddlepoint approximation
			$Q_M(S, G)$		
0	1.0000709(-6)	23	1.0000428(-6)	10	1.000571(-6)
	1.0000298(-6)	39	1.0000298(-6)	19	
	1.0000298(-6)	78	1.0000298(-6)	37	
4	6.8855663(-3)	14	6.8855271(-3)	8	6.8464(-3)
	6.8855228(-3)	27	6.8855228(-3)	15	
	6.8855228(-3)	49	6.8855228(-3)	30	
6	0.23256222	12	0.23257428	8	0.2196
	0.23255232	23	0.23255233	16	
	0.23255232	37	0.23255232	31	
			$1 - Q_M(S, G)$		
7	0.43102034	14	0.43094908	8	0.4160
	0.43089438	25	0.43089437	15	
	0.43089437	43	0.43089436	30	
10	1.4952758(-3)	9	1.4952678(-3)	6	1.4978(-3)
	1.4952678(-3)	17	1.4952678(-3)	12	
	1.4952678(-3)	31	1.4952678(-3)	23	
12	5.5080475(-7)	9	5.5080370(-7)	6	5.5240(-7)
	5.5080362(-7)	17	5.5080362(-7)	12	
	5.5080362(-7)	31	5.5080362(-7)	23	

changes most rapidly in the direction of its gradient $\nabla f_R(x, y)$, and the gradient is at each point perpendicular to the contours on which $f_R(x, y)$ is constant. In the theory of functions of a complex variable it is shown that the curves on which the function

$$f_I(x, y) = \text{Im } \Phi(x + iy)$$

is constant are orthogonal to those contours. On the path of steepest descent, therefore, Im $\Phi(z)$ is constant.

The behavior of the "phase" $\Phi(z)$ of the integrand in (5-32) can be understood by considering its counterpart in two-dimensional electrostatics, where it plays the role of the complex potential. The lines on which Re $\Phi(x + iy)$ is constant are the equipotentials of the electrostatic field; those on which Im $\Phi(x + iy)$ is constant are the flux lines, or the "lines of force." The flux lines cross the equipotentials at right angles.

Figure 5-2 depicts these curves for the complex potential

$$\Phi(z) = Gz - M \ln(1 + z) + \frac{S}{1 + z} - \ln z - S \tag{5-33}$$

arising in the computation of the Mth-order Marcum Q function. The first term represents a uniform field of strength G, the second term a line charge of $-M$ units at the point $z = -1$, the third term a dipole of strength S at that same point, and the fourth term a unit negative line charge at the origin, $z = 0$. The constant term

Evaluating Signal Detectability   Chap. 5

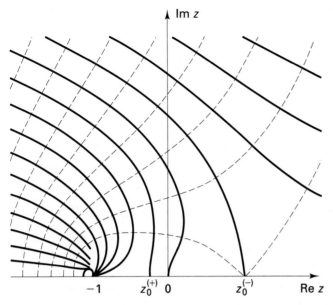

Figure 5-2. Equipotentials (dashed) and flux lines (solid) for $\Phi(z)$ in (5-33): $M = 5$, $S = 6/11$, $G = 40/11$; $Q_M(S, G) = 0.7678$; $z_0^{(-)} = 1$, $z_0^{(+)} = -1/4$. The values of Im $\Phi(z)$ decrease by $\pi/2$ from the upper right to the lower left. On the equipotential emanating from $z_0^{(-)} = 1$, Re $\Phi(z) = -0.1$. On the ones above it, Re $\Phi(z) = -2, \ldots, -14$ from right to left.

$-S$ is inconsequential. The dashed curves in Fig. 5-2 are the equipotentials; the solid curves are the flux lines. Only the upper half of the field (Im $z > 0$) is shown; the lower half is its mirror image. The flux lines of the uniform component of the field, coming from the right, are repelled by the charges at $z = 0$ and $z = -1$.

In this analogy the saddlepoints are neutral points of the field, points where a charge is subject to no force. Figure 5-3 shows a magnified picture of the upper half of the field in the neighborhood of the right saddlepoint $z_0 = z_0^{(-)}$. The positive real axis is a flux line with Im $\Phi(z) \equiv 0$. As a point z moves along it toward the neutral point $z_0^{(-)}$, the value of Re $\Phi(z)$ decreases, as shown in Fig. 5-1. If when the point z reaches $z_0^{(-)}$ it turns upward along the flux line Im $\Phi(z) \equiv 0$ that passes through that point, the value at z of Re $\Phi(z)$ continues to decrease. As z moves along that flux line, the value exp $\Phi(z)$ of the integrand, which remains real, decreases most rapidly toward 0.

If Re $\Phi(z)$ is plotted as a height above some horizontal plane, the dashed lines in Fig. 5-3 form one-half of the contour map of a col whose sides rise to the east and west; the steepest path down to the valley runs north and south.

The second saddlepoint $z_0^{(+)}$ lies on the negative Re z-axis to the left of the charge at $z = 0$. On the segment $-1 < $ Re $z < 0$, Im $\Phi(z) = -\pi$. A flux line on which Im $\Phi(z) = -\pi$ passes vertically through the neutral point $z_0^{(+)}$ and goes off to infinity as shown in Fig. 5-2. The field in the neighborhood of $z_0^{(+)}$ also resembles that shown in Fig. 5-3.

In this example the field possesses a third neutral point to the left of the charge at $z = -1$. It lies outside the regularity domain of the moment-generating function $h(z)$ and need not concern us.

Our vertical path of integration in (5-19) passes through the saddlepoint $z_0^{(-)}$. It can be deformed into the path of steepest descent, Im $\Phi(z) \equiv 0$, through that point without crossing any singularities of the integrand in (5-19). An integration

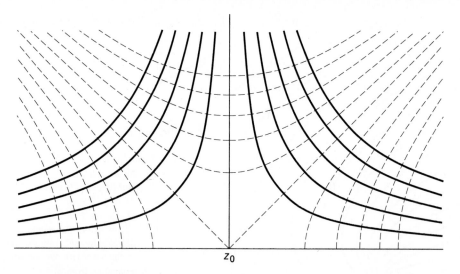

Figure 5-3. Equipotentials (dashed) and flux lines (solid) in the neighborhood of a neutral point (saddlepoint) z_0. These represent a conformal map of the transformation $z - z_0 = w^{1/2}$, $w = 2[\Phi(z) - \Phi(z_0)]/\Phi''(z_0)$.

along that path will therefore yield the same value of $q_-(G)$ as does integration along the vertical path. Along the path of steepest descent the integrand remains real and decreases most rapidly toward zero. Numerical quadrature of the integral along that path would require fewest steps to attain a given precision. The same considerations apply to an evaluation of $q_+(G)$ by (5-20) after deforming the vertical path there into the path of steepest descent, the flux line $\operatorname{Im} \Phi(z) \equiv -\pi$, passing through the saddlepoint $z_0^{(+)}$.

To integrate along the path of steepest descent, however, will in most problems require that path to be computed numerically, and to do so would much protract the numerical integration. Instead we determine a path that, at least as long as the integrand is of significant magnitude, lies close to the path of steepest descent, but is simpler to specify. The simplest path seems in most problems to be a parabola lying symmetrically about the real z-axis and passing through the saddlepoint z_0. The only question is what curvature it should have at the saddlepoint z_0. We want the parabola to fit the path of steepest descent as snugly as possible.

A parabola passing with curvature κ through the saddlepoint z_0 on the Re z-axis is described by the equation

$$z = z_0 + \tfrac{1}{2}\kappa y^2 + iy, \qquad z = x + iy. \tag{5-34}$$

The curvature κ of the parabola that fits the path of steepest descent most tightly is given by

$$\kappa = \frac{\Phi'''(z_0)}{3\Phi''(z_0)}; \tag{5-35}$$

derivatives with respect to z are again indicated by primes. In order to derive this formula, we expand the phase $\Phi(z)$ in the neighborhood of the saddlepoint as

$$\Phi(z) \approx \Phi(z_0) + \tfrac{1}{2}\Phi''(z_0)(z - z_0)^2 + \tfrac{1}{6}\Phi'''(z_0)(z - z_0)^3 + \cdots.$$

Putting $z = z_0 + x + iy$, we find for the imaginary part of the phase on the path of steepest descent

$$\text{Im } \Phi(z) \equiv 0 \approx \Phi''(z_0)xy + \tfrac{1}{6}\Phi'''(z_0)(3x^2y - y^3).$$

Because $x = \tfrac{1}{2}\kappa y^2$, x^2 is on the order of y^4, and keeping only terms through those of third order in y, we find

$$\Phi''(z_0)x = \tfrac{1}{6}\Phi'''(z_0)y^2,$$

whence, by comparison with $x = \tfrac{1}{2}\kappa y^2$, we obtain (5-35). The parabola that fits the path of steepest descent in this manner is called the *osculatory parabola*.

On the path of integration the variable of integration z becomes that given by (5-34), and the element of integration is

$$dz = i(dy - i\,dx) = i(1 - i\kappa y)\,dy.$$

The integral to be evaluated is now, by (5-32),

$$q(G) = \frac{1}{\pi}\int_0^\infty \text{Re}[e^{\Phi(z)}(1 - i\kappa y)]\,dy, \qquad z = z_0 + \tfrac{1}{2}\kappa y^2 + iy. \tag{5-36}$$

Again this is most expeditiously integrated by the trapezoidal rule with the step size chosen as in (5-27).

The phase $\Phi(z)$ contains a term $-\ln z$, which puts a term $-2z_0^{-3}$ into $\Phi'''(z_0)$ in (5-35). When the decision level G is close to the expected value $E(G')$ of the statistic, and when the saddlepoint is $z_0^{(+)} < 0$, this term may dominate that third derivative and cause the curvature κ to be positive. The osculatory parabola is then directed into the right half-plane, and the integral (5-36) along it diverges. In a case such as this, one can use instead the osculatory parabola passing through the right saddlepoint $z_0^{(-)}$, or one can take the curvature κ equal to zero and integrate along a straight vertical path as in Sec. 5.2.1. Alternatively, one can drop the term $-\ln z$ from the phase and utilize the curvature calculated from (5-35) with $\Phi(z)$ replaced by the modified phase $\tilde{\Phi}(z) = \Phi(z) + \ln z$. It is the behavior of the phase $\Phi(z)$ far from the origin that determines the rate of convergence of the integral in (5-36), and on that behavior the term $-\ln z$ has relatively little influence.

In computing the Mth-order Marcum Q function, for which the phase $\Phi(z)$ is given by (5-29), the saddlepoints $z_0^{(-)}$ and $z_0^{(+)}$ lie close to the origin for the large values of the parameters M, S, and G for which this method is most appropriate. It suffices to calculate the curvature κ by evaluating (5-35) for the modified phase $\tilde{\Phi}(z)$ and then setting $z_0 = 0$. We obtain the simple formula

$$\kappa = -\frac{S + (M/3)}{S + (M/2)} \tag{5-37}$$

In the fourth column of Table 5-2 we list the results of computing the Mth-order Q function by numerical integration along a parabola whose curvature is specified by (5-37). The number of steps required in order to attain a certain precision is seen to be reduced by one-third to one-half of the number required when integrating along a straight vertical path. The larger the number M, the less is gained by utilizing a parabolic path, for when M is large, the value of the integrand drops off to zero so rapidly as the point z leaves the saddlepoint z_0 that the curvature of the path is insignificant.

For values of the parameters M, S, and G small enough so that the recursive method of the Appendix, Sec. C.3, does not encounter underflow or overflow in its computation, that method is recommended. Otherwise, the saddlepoint-integration method described here is preferred. For extremely large values of M—in the thousands or larger—, the first three terms on the right side of (5-29) tend to be of the same order of magnitude, and significant figures may be lost in the subtractions. It is best then to write the phase as

$$\Phi(z) = Vz - M[\ln(1 + z) - z] + \frac{Sz^2}{1 + z} - \ln z, \qquad V = G - S - M,$$

into which one puts

$$\ln(1 + z) - z = -\tfrac{1}{2}z^2 F(2, 1; 3; z),$$

and one computes the hypergeometric function $F(2, 1; 3; z)$ either by its power series in z or, as shown in [Hel92d], by its continued fraction. In ordinary practice this complication will be unnecessary.

Saddlepoint integration is no panacea; there are distributions for which it is useless. In considering a new application, it is wise to trace a few paths of steepest descent for typical values of the parameters in the distribution to be computed. Rice [Ric73] has shown how that can quickly, but approximately, be accomplished. If z and $z + \Delta z$ are two close points on the path of steepest descent,

$$\mathrm{Im}[\Phi(z + \Delta z) - \Phi(z)] = 0$$

or, approximately,

$$\mathrm{Im}[\Phi'(z)\Delta z] \approx 0.$$

The complex increment Δz must therefore be proportional to $\Phi'^*(z)$. If the segments on our approximate path are to be of length δ, the point lying next to a point z already computed will be located at

$$z + \Delta z = z - \delta \frac{\Phi'^*(z)}{|\Phi'(z)|}.$$

One starts tracing the path at a point $z_0 + i\varepsilon$ lying just above the saddlepoint z_0.

The smaller one takes δ, the closer the path so traced will lie to the true path of steepest descent. It may be necessary occasionally to correct the approximation by solving the equation $\mathrm{Im}\ \Phi(z) = 0$, $z = x + iy$, using Newton's method or the secant method and varying x or y to bring the point z back to the true path. With Newton's method, for instance, either the x component is corrected by

$$x \leftarrow x - \frac{\mathrm{Im}\ \Phi(z)}{\mathrm{Im}\ \Phi'(z)}, \qquad |\mathrm{Re}\ \Delta z| < |\mathrm{Im}\ \Delta z|,$$

or the y component is corrected by

$$y \leftarrow y - \frac{\mathrm{Im}\ \Phi(z)}{\mathrm{Re}\ \Phi'(z)}, \qquad |\mathrm{Re}\ \Delta z| \geq |\mathrm{Im}\ \Delta z|,$$

until $\mathrm{Im}\ \Phi(z)$ becomes acceptably small. The contours in Fig. 5-2, with appropriate modifications for the equipotentials, were constructed by this method.

In some problems the value of the integrand in (5-19) and (5-20) may decrease to zero very slowly both along the straight vertical path of Sec. 5.2.1 and along whatever path one proposes to use to approximate the path of steepest descent. Changes of the variable of integration that may speed the convergence of the numerical quadrature have been examined by Rice [Ric73]. Applications of saddlepoint integration to computing cumulative distributions of continuous random variables are to be found in [Hel83], [Hel84d], [Hel85a], [Hel86a], [Hel86b], [Hel92a], and [Hel92c]. The same technique can be applied to (5-4) to compute probability density functions if these are needed in order, for instance, to set a decision level under the Bayes criterion.

5.2.3 Integer-valued Random Variables

In the detection of light, as we shall see in Chapter 12, the decision about the presence or absence of a signal is often based on the number n of photoelectrons emitted by a photosensitive surface onto which the signal, accompanied perhaps by random background light, is incident. The number n is a nonnegative integer-valued random variable described by a sequence $p_0, p_1, p_2, \ldots$ of probabilities

$$p_k = \Pr(n = k), \qquad k = 0, 1, 2, \ldots .$$

In certain nonparametric detectors, to be treated in Chapter 8, the decision is also based on an integer-valued random variable. In order to determine false-alarm and detection probabilities, one must be able to calculate the cumulative distribution

$$q_k^{(-)} = \Pr(n < k) = \sum_{r=0}^{k-1} p_r \qquad (5\text{-}38)$$

of the random variable n, or its complement

$$q_k^{(+)} = \Pr(n \geq k) = \sum_{r=k}^{\infty} p_r . \qquad (5\text{-}39)$$

An analysis of the detector often yields the *probability-generating function $h(z)$* of the random variable n, which is defined by

$$h(z) = E(z^n) = \sum_{k=0}^{\infty} p_k z^k \qquad (5\text{-}40)$$

[Hel91, pp. 124–5]. It resembles the z transform of the sequence $\{p_k\}$, except that z^k appears instead of z^{-k}. The probabilities p_k can be recovered from $h(z)$ by differentiation,

$$p_k = \frac{1}{k!} \frac{d^k}{dz^k} [h(z)] \bigg|_{z=0} ,$$

although this is seldom useful for very large values of k unless it can be achieved analytically. Because

$$\sum_{k=0}^{\infty} p_k = 1,$$

$h(1) = 1$. The function $h(z)$ has no singularities within or on the unit circle, for

$$|h(z)| \le \sum_{k=0}^{\infty} p_k |z|^k \le \sum_{k=0}^{\infty} p_k = 1, \qquad |z| \le 1.$$

Derivatives of the probability-generating function $h(z)$ at $z = 1$ provide the *factorial moments*

$$c_m = E[n(n-1)(n-2) \dots (n-m+1)] = \sum_{k=0}^{\infty} k(k-1) \dots (k-m+1)p_k$$

$$= \frac{d^m}{dz^m}[h(z)]\bigg|_{z=1}.$$

In particular,

$$E(n) = c_1 = h'(1) \tag{5-41}$$

and

$$c_2 = h''(1) = E(n^2 - n),$$

with primes denoting differentiation with respect to z. The variance of the random variable n is then

$$\text{Var } n = E(n^2) - [E(n)]^2 = h''(1) + h'(1) - [h'(1)]^2. \tag{5-42}$$

Let us express the probabilities p_r by Cauchy's theorem as

$$p_r = \int_C z^{-(r+1)}h(z)\frac{dz}{2\pi i},$$

where C is a closed curve enclosing the origin of the complex z-plane, but none of the singularities of $h(z)$. Substituting this into (5-38), we find

$$q_k^{(-)} = \int_C \sum_{r=0}^{k-1} z^{-(r+1)}h(z)\frac{dz}{2\pi i} = \int_C \frac{1 - z^{-k}}{z - 1}h(z)\frac{dz}{2\pi i}.$$

If we take the curve C as a closed curve C_- surrounding the origin, but enclosing neither the point $z = 1$ nor any singularities of $h(z)$, then

$$\int_{C_-} \frac{h(z)}{z - 1}\frac{dz}{2\pi i} = 0 \tag{5-43}$$

by Cauchy's theorem, and

$$q_k^{(-)} = \int_{C_-} \frac{z^{-k}h(z)}{1 - z}\frac{dz}{2\pi i}. \tag{5-44}$$

Taking C, on the other hand, as a closed curve C_+ including both $z = 0$ and $z = 1$, but no singularities of $h(z)$, we find that the integral corresponding to (5-43) yields 1, and the complementary cumulative distribution of the number n becomes

$$q_k^{(+)} = 1 - q_k^{(-)} = \sum_{r=k}^{\infty} p_r = \int_{C_+} \frac{z^{-k}h(z)}{z - 1}\frac{dz}{2\pi i}. \tag{5-45}$$

The integrands of (5-44) and (5-45) are convex $\cup$ functions of $x = \operatorname{Re} z$. The integrand of (5-45), for instance, is

$$\frac{x^{-k} h(x)}{x-1} = E\left[\frac{x^{n-k}}{x-1}\right] = E[x^{n-k-1}(1 + x^{-1} + x^{-2} + \cdots)], \qquad x = \operatorname{Re} z > 1,$$

and the bracket is the sum of powers of x, all of which have nonnegative second derivatives and are therefore convex $\cup$. A similar argument holds for the integrand of (5-44). The integrands of (5-44) and (5-45) therefore possess unique saddlepoints in the regions $0 < \operatorname{Re} z < 1$ and $1 < \operatorname{Re} z < \varepsilon_1$, respectively, where ε_1 is the leftmost singularity of $h(z)$, if any, on the positive $\operatorname{Re} z$-axis. Saddlepoint integration should be considered as a way of evaluating the tail probabilities $q_k^{(-)}$ and $q_k^{(+)}$.

The "phase" of the integrand is now

$$\Phi(z) = \ln h(z) - k \ln z - \ln[\pm(z-1)], \tag{5-46}$$

and its derivative is

$$\Phi'(z) = \frac{h'(z)}{h(z)} - \frac{k}{z} - \frac{1}{z-1} = 0 \tag{5-47}$$

at the saddlepoint. If this equation cannot be solved analytically, Newton's method, described in Sec. 5.2.1, is most expeditious. Alternatively one can use the secant method to search for the root of

$$\operatorname{Im} \Phi(x + i\varepsilon) = 0$$

for a suitably small value of ε.

Before attempting to integrate (5-44) or (5-45) numerically, it is wise to plot paths of steepest descent for a few typical values of the parameters in the problem by the technique described at the end of Sec. 5.2.2. For the photoelectron-counting distributions to be studied in Chapter 12, it is found that these paths go off to infinity, and the closed contours in (5-44) and (5-45) can be deformed into osculatory parabolic contours through the saddlepoint without crossing any singularities of the integrand. Their curvatures are calculated by (5-35). The method of Sec. 5.2.2 can then be applied.

For values of k near the expected value $E(n)$, the saddlepoint lies close to the point $z = 1$, and the term $\ln[\pm(z-1)]$ in (5-46) may cause the osculatory parabola to go off in the wrong direction when its curvature is calculated from (5-35), that is, in a direction that leads to divergence of the integral in (5-44) or (5-45). If that happens, one can drop that term in calculating the curvature—as we did in Sec. 5.2.2 in using the modified phase $\tilde{\Phi}(z)$—or one can switch to the saddlepoint lying on the other side of the point $z = 1$, or one can integrate along a straight vertical path through the saddlepoint by setting the curvature κ equal to 0.

As a simple example, we consider the Poisson distribution with expected value μ,

$$p_k = \frac{\mu^k}{k!} e^{-\mu}, \qquad k \geq 0. \tag{5-48}$$

Its probability generating function is

$$h(z) = e^{\mu(z-1)},$$

Table 5-3 Poisson Distribution ($\mu = 20$, $\varepsilon = 10^{-8}$)

k	Numerical integration	Number of steps	Saddlepoint approximation
		$q_k^{(-)}$	
5	0.16944753(−4)	8	0.17145(−4)
	0.16944743(−4)	16	
	0.16944743(−4)	30	
10	0.49954170(−2)	8	0.49684(−2)
	0.49954122(−2)	16	
	0.49954122(−2)	30	
15	0.10486598	10	0.10099
	0.10486428	20	
	0.10486428	38	
		$q_k^{(+)}$	
20	0.52975001	7	0.50880
	0.52974272	14	
	0.52974272	27	
30	0.21818225(−1)	7	0.21764(−1)
	0.21818217(−1)	14	
	0.21818217(−1)	27	
40	0.53202024(−4)	7	0.53268(−4)
	0.53202024(−4)	14	
	0.53202024(−4)	27	
50	0.12458926(−7)	7	0.12477(−7)
	0.12458926(−7)	14	
	0.12458926(−7)	27	

and the "phase" is

$$\Phi(z) = \mu(z - 1) - k \ln z - \ln(z - 1).$$

The saddlepoint z_0 is the root of

$$\Phi'(z) = \mu - \frac{k}{z} - \frac{1}{z - 1} = 0,$$

which leads to a quadratic equation that the reader should write out and solve explicitly.

In Table 5-3 we list the results of the numerical saddlepoint integration of (5-44) and (5-45) along a parabolic contour whose curvature was determined from (5-35). Even for an expected value μ as low as 20, good agreement with the exact cumulative probabilities is obtained with a few steps of numerical integration. Summing Poisson probabilities is a simple computation, of course, and one would turn to saddlepoint integration only when μ is so large that $\exp(-\mu)$ underflows or when n is so large that the number of terms in summing (5-38) is excessive.

If the integer-valued random variable n takes on values only in a finite range $0 \leq n \leq M$, the probability-generating function will be a polynomial of degree M, and the integrands of (5-44) and (5-45) will have M zeros in the complex plane. The path of steepest descent from the saddlepoint will run into one of those zeros, and it is not possible to find a path of integration going out to infinity. The integrand may have numerous other saddlepoints in the complex plane, both above and below the Re z-axis. When $E(n)$ and k are large, however, it often happens that a path can be taken through the real-axis saddlepoint and that along that path the integrand drops to zero so rapidly that by using the stopping rule described in Sec. 5.2.2 one can evaluate the probability $q_k^{(-)}$ or $q_k^{(+)}$ with high accuracy. It is necessary to study each such application carefully, and no general criteria for the applicability of saddlepoint integration can be given. The method has been applied to distributions of nonnegative integer-valued random variables in papers such as [Hel84a], [Hel84b], [Hel84c], [Hel85b], [Hel87], and [Hel88a].

5.3 APPROXIMATIONS

The sorts of computational methods we introduced in Secs. 5.1 and 5.2 may entail rather much programming, particularly when the moment-generating or the probability-generating function is complicated, and for a quick assessment of receiver performance one often resorts to simply calculated approximations to false-alarm and detection probabilities. We shall concern ourselves here with those that are most accurate when the decision statistic G' is the sum of a large number M of terms as in (5-1).

What is usually the simplest approximation consists of taking only the first term in the Gram–Charlier series (5-14),

$$q_+(G) = \Pr(G' \geq G) = \int_G^\infty P(g)\, dg \approx \mathrm{erfc}\left(\frac{G - \overline{G}}{\sigma}\right), \qquad (5\text{-}49)$$

where $\mathrm{erfc}(\cdot)$ is the error-function integral defined in (1-11), $\overline{G} = E(G')$, and $\sigma^2 = \mathrm{Var}\, G'$. This is known as the *Gaussian* approximation. The false-alarm probabilities Q_0 and the false-dismissal probabilities $1 - Q_d$ we are mostly concerned with, however, are quite small, and the decision level G is usually rather far in the right- or left-hand tail of the probability density function of the statistic G', $P_0(\cdot)$ or $P_1(\cdot)$ as the case may be. As we have seen in Sec. 5.1, it is necessary then to take a large number of terms in the Gram–Charlier or Edgeworth series in order to attain acceptable accuracy, and the first term (5-49) is likely to be a very poor approximation.

5.3.1 The Chernoff Bound

At times, and this is particularly true in theoretical studies of communication-system performance, it suffices merely to be sure that the number one has obtained lies above the true value of the tail probability, Q_0 or $Q_1 = 1 - Q_d$. Such assurance is provided by the Chernoff bound. We shall derive it first for the right-hand tail probability $q_+(G)$ defined as in (5-13).

Denote the probability density function of the decision statistic G' by $P(\cdot)$, and let G be the decision level. Then the complementary cumulative distribution function of G' is

$$q_+(G) = \int_G^\infty P(g)\, dg = \int_{-\infty}^\infty U(g - G) P(g)\, dg,$$

where $U(\cdot)$ is the unit step function,

$$U(y) = \begin{cases} 1, & y \geq 0, \\ 0, & y < 0. \end{cases}$$

For any negative real value of x,

$$U(g - G) \leq e^{-(g-G)x}, \qquad x \leq 0,$$

as we can see by plotting both sides as functions of g. Hence

$$q_+(G) \leq \int_{-\infty}^\infty e^{-(g-G)x} P(g)\, dg = E[e^{-(G'-G)x}] = e^{Gx} h(x), \qquad x \leq 0, \qquad (5\text{-}50)$$

where as in (5-2) $h(\cdot)$ is the moment-generating function of the statistic G. We should like to make the bound in (5-50) as tight as possible by choosing a value of x for which the right side is minimum.

The function $e^{Gx} h(x)$ on the right side of (5-50) is convex $\cup$ for values of $x = \operatorname{Re} z$ in the regularity domain $c_1 < x < c_2$ of the Laplace transform in (5-2): The function $\exp[-x(G' - G)]$ is a convex function of x for all values of the random variable G', and as we learned in Sec. 5.2.1, the expected value of a convex function is convex. The expected value in the right side of (5-50) is therefore convex $\cup$. It possesses at most one minimum within the regularity domain. When G is sufficiently large, $G > E(G')$, this minimum will occur at a negative value of x; and this will be the case, for instance, for the small values of the false-alarm probability Q_0 that we are ordinarily concerned with bounding. [If the minimum occurs at a positive value of x, we must put $x = 0$ in (5-50), and we obtain the trivial bound $q_+(G) \leq 1$.]

Minimizing the right side of (5-50) is equivalent to minimizing $\ln h(x) + xG$. With derivatives denoted by primes, the value of x is given by

$$\frac{h'(x)}{h(x)} + G = 0, \qquad x \leq 0. \qquad (5\text{-}51)$$

Call the solution of this equation x_0. Then

$$q_+(G) \leq h(x_0) \exp G x_0. \qquad (5\text{-}52)$$

This result is called the *Chernoff bound* [Che62].

Suppose for example that the random variable G' has a Gaussian distribution with expected value zero and variance 1. Then because the moment-generating function of G' is

$$h(z) = e^{z^2/2},$$

we find from (5-51) that $x_0 = -G$, whereupon we obtain the bound

$$q_+(G) = \operatorname{erfc} G \leq e^{-G^2/2}, \qquad G \geq 0.$$

The left-hand tail probability is

$$q_-(G) = \int_{-\infty}^{G} P(g)\, dg = \int_{-\infty}^{\infty} P(g) U(G - g)\, dg,$$

and by a similar argument we derive the bound

$$q_-(G) \le h(x_1) \exp Gx_1, \qquad x_1 \ge 0,$$

where x_1 is the root of (5-51) lying to the right of the origin. The root x_1 will in general be positive if G is sufficiently smaller than the expected value $E(G')$, as will be the case for false-dismissal probabilities $Q_1 = 1 - Q_d$ of an order of magnitude of interest.

Although the Chernoff bound has proved useful in information theory, it is usually somewhat larger than the exact value of the tail probability $q_+(G)$ or $q_-(G)$ and does not provide an accurate approximation to it. In the next part we shall derive an approximation for the tail probability that is closely related to that bound.

5.3.2 The Saddlepoint Approximation

In most detection problems we are concerned with false-alarm probabilities Q_0 on the order of 10^{-4} or less and with detection probabilities Q_d on the order of 0.99 or more. The decision level G then lies far in either the right or the left tail of the density function of the statistic G', and approximate values of these probabilities with accuracy adequate for most engineering purposes can be determined with much less computation than is involved in the numerical integration described in Sec. 5.2. When the contour of integration passes through the saddlepoint z_0, the integrands of (5-19) and (5-20) have nearly a Gaussian form, and by approximating them as such, the integrals are quickly evaluated.

We write (5-22), which is the integrand of (5-19), as

$$H(z_0 + iy) = \exp\!\left[\Phi(z_0) - \tfrac{1}{2}\Phi''(z_0)y^2\right] \exp r(y),$$

where

$$r(y) = \sum_{k=3}^{\infty} \Phi^{(k)}(z_0)\frac{(iy)^k}{k!}, \qquad \Phi^{(k)}(z) = \frac{d^k \Phi}{dz^k},$$

and we expand the second exponential function as

$$\exp r(y) = 1 + r(y) + \tfrac{1}{2}[r(y)]^2 + \cdots.$$

Putting this into (5-19), collecting terms with identical powers of y, and integrating over $-\infty < y < \infty$, we find that the terms with odd powers of y vanish. When the statistic G' has the form (5-1) with its components g_j independent and identically distributed random variables, the function

$$\Phi(z) = \ln h(z) + Gz - \ln z = M \ln \eta(z) + Gz - \ln z$$

contains a term proportional to the number M of inputs, and $\Phi(z)$ and its derivatives at $z = z_0$ can be said to be of order M.

After carrying out the integration of (5-19) term by term and keeping the terms in the result of order M^{-1}, we obtain for the left-hand tail probability

$$q_-(G) \approx [2\pi\Phi''(z_0)]^{-1/2} \exp \Phi(z_0)[1 + T_1 + \cdots], \qquad 0 < z_0 < c_2, \qquad (5\text{-}53)$$

with

$$T_1 = \frac{\Phi^{iv}(z_0)}{8[\Phi''(z_0)]^2} - \frac{5[\Phi'''(z_0)]^2}{24[\Phi''(z_0)]^3}. \qquad (5\text{-}54)$$

As in Sec. 5.2, z_0 is the solution of

$$\Phi'(z) = \frac{d}{dz} \ln h(z) - \frac{1}{z} + G = 0 \qquad (5\text{-}55)$$

lying to the right of the origin. Treating (5-20) in the same way, we find for the right-hand tail probability the approximation

$$q_+(G) \approx -[2\pi\Phi''(z_0)]^{-1/2} \exp \Phi(z_0)[1 + T_1 + \cdots], \qquad c_1 < z_0 < 0, \qquad (5\text{-}56)$$

with T_1 given again by (5-54), but now the saddlepoint z_0 is the solution of (5-55) lying to the left of the origin. The terms omitted are on the order of M^{-2}. Expressions for these terms are to be found in [Hel78]; they involve derivatives of the phase $\Phi(z)$ of ever higher order. If the moment-generating or probability-generating function is complicated, calculating these derivatives can be troublesome, and one may have to resort to numerical differentiation, which is the less accurate, the higher the order of the derivative. Before attempting to calculate such terms, one should try the method of numerical integration presented in Sec. 5.2. The farther the value of G lies in the tail of the density function of the statistic G' and the larger the number M, the more accurate are the approximations represented by (5-53) and (5-56).

An early application of this technique to finding asymptotic forms of the Hankel functions was made by Peter Debye [Deb09]. He attributed the method to an unpublished paper of Riemann's [Rie53]. Daniels [Dan54] described in some detail the use of a saddlepoint method for approximating probability density functions and probability mass functions, but did not extend the method to cumulative distributions [Hel78]. The method can be applied to tail probabilities of integer-valued random variables as well. One uses the phase $\Phi(z)$ defined in (5-46) in terms of the probability-generating function $h(z)$ (5-40) of the integer-valued random variable in question.

Table 5-4 lists values of the saddlepoint approximation to Marcum's Q function for $M = 50$. Those in the column headed "First Order" were calculated by (5-53) and (5-56); for the column headed "Zero Order," the correction term T_1 was omitted. The phase $\Phi(z)$ is given in (5-29), and the saddlepoint z_0 is determined by solving (5-30). For $D = 8$ and $D = 10$ the decision level G is close to the expected value $E(G') = M + \frac{1}{2}D^2$. In that neighborhood the saddle-point approximation loses accuracy, and the Edgeworth series is more reliable. Values of the saddlepoint approximation in zero order for the Q function with $M = 10$ are listed in the last column of Table 5-2. Again we find that the farther G lies in the tail of the distribution, the more accurate this approximation becomes.

Table 5-4 Saddlepoint Approximation
(Q Function: $M = 50$, $G = 91.0634$)

D	sadpt.	Zero order	First order	Exact
		$q_+(G)$		
0	−0.46364	9.99908 (−7)	1.00019 (−6)	9.99995 (−7)
2	−0.42730	4.96812 (−6)	4.97184 (−6)	4.97052 (−6)
4	−0.33934	2.03945 (−4)	2.04549 (−4)	2.04410 (−4)
6	−0.23064	1.05470 (−2)	1.06804 (−2)	1.06496 (−2)
8	−0.12556	1.86097 (−1)	1.97196 (−1)	1.94614 (−1)
		$q_-(G)$		
10	0.13020	2.31365 (−1)	2.39964 (−1)	2.37915 (−1)
12	0.24202	8.73300 (−3)	8.77383 (−3)	8.76320 (−3)
14	0.37318	2.06676 (−5)	2.06632 (−5)	2.06603 (−5)
16	0.51148	2.18325 (−9)	2.18121 (−9)	2.18116 (−9)

5.3.3 Calculating Approximate Decision Levels for the Neyman–Pearson Criterion

The saddlepoint approximation is particularly useful when seeking the decision level G to attain a prescribed false-alarm probability Q_0 or an energy-to-noise ratio S that yields a prescribed false-dismissal probability $Q_1 = 1 - Q_d$. Let us denote the cumulative distribution and its complement by $q_-(x; S)$ and $q_+(x; S)$, respectively. Because usually $Q_0 \ll 1$ and $Q_1 \ll 1$, these probabilities, $q_+(G; 0)$ and $q_-(G; S)$, respectively, have the decision level G in the far right or the far left tail of the applicable probability density function, where the saddlepoint approximation is most reliable.

We first seek the value of G for which $q_+(G; 0)$ equals the preassigned false-alarm probability Q_0. Putting $\ln h(z) = \psi(z)$ into (5-56), we write the saddlepoint approximation as

$$q_+(G; 0) \approx |z|^{-1}[2\pi\Phi''(z)]^{-1/2} \exp[\psi(z) + Gz], \qquad (5\text{-}57)$$

with the saddlepoint z determined by

$$\Phi'(z) = \psi'(z) + G - z^{-1} = 0.$$

Solving this for G, we replace G in (5-57) by

$$G = z^{-1} - \psi'(z), \qquad (5\text{-}58)$$

writing (5-57) as

$$\ln Q_0 = \psi(z) + 1 - z\psi'(z) - \ln|z| - \tfrac{1}{2}\ln[2\pi\Phi''(z)] \qquad (5\text{-}59)$$

with

$$\Phi''(z) = \psi''(z) + z^{-2}.$$

Table 5-5 Marcum's Q Function ($Q_0 = 10^{-6}$, $Q_d = 0.999$)

	Saddlepoint		Exact	
M	G	S	G	S
10	32.7180	51.3098	32.7103	51.2928
20	48.8296	61.7248	48.8265	61.7187
50	91.0632	82.3819	91.0634	82.3866
100	154.9172	105.5961	154.9190	105.6082
200	274.5543	138.3292	274.5576	138.3492
500	613.5707	203.0855	613.5762	203.1183
1000	1157.5704	275.9176	1157.5779	275.9635

One solves (5-59) for z by the secant method or any other convenient method. Many calculators and mathematical software programs have routines that efficiently solve equations of this kind. One can take the initial value of z just inside the range $c_1 < \text{Re } z < 0$. One then determines the approximate value of the decision level G from (5-58).

A similar method can sometimes be used to determine the total energy-to-noise ratio S required to attain a prescribed probability Q_d of detection. When the statistic G' is governed under hypothesis H_1 by Marcum's Q function, for instance, we can solve (5-30) for S as a function of z, obtaining

$$S(z) = (G - z^{-1})(1 + z)^2 - M(1 + z). \qquad (5\text{-}60)$$

The saddlepoint approximation (5-53) for the false-dismissal probability yields

$$\ln(1 - Q_d) \approx \Phi(z) - \tfrac{1}{2} \ln[2\pi\Phi''(z)],$$

with $\Phi(z)$ and $\Phi''(z)$ given by (5-29) and (5-31), respectively. Into these we substitute $S(z)$ from (5-60), obtaining the equation

$$\ln(1 - Q_d) = Gz - M \ln(1 + z) - \frac{zS(z)}{1 + z} - \ln z - \tfrac{1}{2} \ln(2\pi)$$
$$- \tfrac{1}{2} \ln[M(1 + z)^{-2} + 2S(z)(1 + z)^{-3} + z^{-2}].$$

This equation, with (5-60), is solved for the saddlepoint $z > 0$, either by the secant method or by a computer root-finding algorithm. The result is substituted into (5-60) to obtain an approximation to the energy-to-noise ratio S. Table 5-5 lists values of the decision level G and the energy-to-noise ratio S as calculated by this method, along with their exact values. The saddlepoint approximation yields accurate results even for small values of M.

If greater accuracy is desired, the approximations for G and S as derived here can be used as starting values in the solution of

$$\ln q_+(G; 0) = \ln Q_0,$$
$$\ln q_-(G; S) = \ln(1 - Q_d),$$

by the secant method, in which the quantities on the left side are computed by saddlepoint integration as in Sec. 5.2 or by the recursive algorithms in Appendix C or E. This method was used to compute the curves in Fig. 4-9 and the exact values in Table 5-5.

5.3.4 Calculating Decision Levels for the Bayes Criterion

When the receiver of a binary communication system bases its decisions on the statistic G', it must form the likelihood ratio $\Lambda(G) = P_1(G)/P_0(G)$ for comparison with the decision level Λ_0 of (1-17). As shown by Daniels [Dan54], the density functions of G', if unknown in closed form, can be calculated by a similar saddlepoint approximation. Applying it to the contour integral in (5-4), we find the zero-order approximation

$$P(G) = [2\pi\psi''(\overline{z}_0)]^{-1/2}\psi(\overline{z}_0)$$
$$= [2\pi\psi''(\overline{z}_0)]^{-1/2}h(\overline{z}_0) \exp G\overline{z}_0,$$

where

$$\psi(z) = \ln h(z) + Gz,$$
$$\psi'(\overline{z}_0) = \frac{h'(\overline{z}_0)}{h(\overline{z}_0)} + G = 0. \tag{5-61}$$

When G lies in the right tail, the root of (5-61) must be negative, $\overline{z}_0 < 0$; for G in the left tail, it is $\overline{z}_1 > 0$. The equation to be solved here is the same as that involved in the Chernoff bound—see (5-51).

A decision level G on the statistic G' is normally in the right tail of $P_0(\cdot)$ and in the left tail of $P_1(\cdot)$. Thus the likelihood ratio for G is approximately

$$\left[\frac{\psi_0''(\overline{z}_0)}{\psi_1''(\overline{z}_1)}\right]^{1/2} \frac{h_1(\overline{z}_1)}{h_0(\overline{z}_0)} \exp[(\overline{z}_1 - \overline{z}_0)G] \approx \Lambda_0, \tag{5-62}$$

where $\overline{z}_0 < 0$ is the saddlepoint for $P_0(G)$ and $\overline{z}_1 > 0$ is the saddlepoint for $P_1(G)$; the subscripts now refer to the hypotheses H_0 and H_1. From (5-62) we find an expression for the decision level:

$$G \approx \frac{1}{\overline{z}_1 - \overline{z}_0} \left\{\ln \Lambda_0 - \ln\left[\frac{h_1(\overline{z}_1)}{h_0(\overline{z}_0)}\right] + \tfrac{1}{2} \ln\left[\frac{\psi_1''(\overline{z}_1)}{\psi_0''(\overline{z}_0)}\right]\right\}. \tag{5-63}$$

Starting with a trial value of the decision level G, conveniently taken halfway between the expected values $E(G'| H_0)$ and $E(G'| H_1)$ of G' under the two hypotheses, one calculates $\overline{z}_0 < 0$ and $\overline{z}_1 > 0$ from (5-61), substitutes them into the right side of (5-63) to obtain a new value of G, and repeats the process until the decision level stabilizes. Because the error probability $P_e = \zeta_0 Q_0 + \zeta_1(1 - Q_d)$ is insensitive to the exact location of the decision level G, the approximate value of G resulting from iteration of (5-63) will be adequate for most purposes.

5.3.5 The Uniform Asymptotic Expansion

A saddlepoint approximation for the tail probabilities $q_+(G)$ and $q_-(G)$ that is uniformly accurate over the entire range of values of G, even in the neighborhood of the expected value $E(G')$, has been described by Rice [Ric68] and Lugannani and

Rice [Lug80]. It takes as the saddlepoint the root $\bar{z}_0$ of (5-61), treating the factor z^{-1} in the integrands of (5-19) and (5-20) in a different manner from ours. The terms of lowest order in this uniform asymptotic approximation are, for $G > E(G')$ and $\bar{z}_0 < 0$,

$$q_+(G) \approx \text{erfc}[-2\psi(\bar{z}_0)]^{1/2} + A_0 - B_0, \qquad (5\text{-}64)$$

$$\psi(z) = \ln h(z) + Gz,$$

$$A_0 = |\bar{z}_0|^{-1}[2\pi\psi''(\bar{z}_0)]^{-1/2} \exp \psi(\bar{z}_0),$$

$$B_0 = \tfrac{1}{2}[-\pi\psi(\bar{z}_0)]^{-1/2} \exp \psi(\bar{z}_0).$$

When $G \gg E(G')$, the first term in (5-64) nearly cancels the term B_0, leaving the term A_0, which is close to the zero-order saddlepoint approximation in (5-56). When the decision level G is near to $E(G')$, on the other hand, the terms A_0 and B_0 are nearly equal, and the first term is roughly the same as the Gaussian approximation in (5-49).

For $G < E(G')$ and $\bar{z}_0 > 0$, the left-hand tail probability $q_-(G)$ is approximated by formulas of the same form. The terms of higher order are rather more complicated and can be found in the paper of Lugannani and Rice [Lug80]. Like the terms of higher order in the saddlepoint approximation (5-53) and (5-56), they involve derivatives of the moment-generating function. This paper also examines the question of the existence of the saddlepoint calculated from (5-61) for various types of density function $P(\cdot)$.

Problems

5-1. Use the contour integral for Mth-order Q function $Q_M(S, y)$ defined in (C-19) to show that

$$\frac{\partial Q_M(S, y)}{\partial S} = P_{M+1}(S, y),$$

where $P_{M+1}(S, y) = -\partial Q_{M+1}(S, y)/\partial y$ is the associated probability density function.

5-2. Show by the Chernoff method of Sec. 5.3.1 that

$$\sum_{r=k}^{\infty} \frac{y^r}{r!} e^{-y} \leq \left(\frac{y}{k}\right)^k e^{k-y}, \qquad y < k.$$

By Stirling's formula

$$k! \approx (2\pi k)^{1/2} k^k e^{-k} \left[1 + \frac{1}{12k} + \cdots\right],$$

and we then find

$$\sum_{r=k}^{\infty} \frac{y^r}{r!} e^{-y} \leq (2\pi k)^{1/2} \frac{y^k}{k!} e^{-y}.$$

5-3. Determine, to zero order only, the saddlepoint approximation to erfc G, $G > 0$. Evaluate it for $G = 1$ to $G = 10$ in unit steps, and compare the results with tabulated values. Calculate erfc G for the same values of G by numerical contour integration of (5-24), taking a straight vertical path through the saddlepoint. Why cannot a parabolic contour be used in this computation?

5-4. For each of the Swerling cases of amplitude distributions of fading signals defined in Sec. 4.2.4, determine the saddlepoint approximation for the average probability $\bar{Q}_d$

of detection, working out only the zero-order approximation as in (5-53) and (5-56). Assume that the noise is white and Gaussian and that the receiver sums the outputs of a quadratic rectifier following a filter matched to the signal, as in (4-19).

5-5. Calculate the zero-order saddlepoint approximation for the values of the Poisson probability distribution listed in Table 5-3.

5-6. Calculate zero-order saddlepoint approximations to the cumulative binomial distribution and its complement by applying the technique of Sec. 5.3.2 to the integrals in (5-44) and (5-45).

5-7. Use the saddlepoint approximation to calculate the energy-to-noise ratio plotted in Fig. 4-6 for Rayleigh-fading signals as a function of the number M of inputs. Using the method of Sec. 5.3.3, first determine the decision level U_0 required for a preassigned false-alarm probability by the saddlepoint approximation to (4-30), and then use the same type of approximation to (4-31) to calculate the value of $S = Ms^2$ for a preassigned average detection probability $\overline{Q}_d$. Plot your approximation to the energy-to-noise ratio S versus M for $50 \le M \le 1000$, $Q_0 = 10^{-6}$, and $\overline{Q}_d = 0.9999$, and compare with the curves in Fig. 4-6.

5-8. Work out a saddlepoint approximation to the probability $Q_d(m_0)$ of detection in (4-87), taking $1 - Q_d \ll 1$, $M \gg 1$.

6

Estimation of
Signal Parameters

6.1 THE THEORY OF ESTIMATION

A radar has more to do than detect targets; it must find where they are and how they are moving. For this purpose it must estimate the values of certain parameters of the received echo signals, and because of the noise the estimates will be in error. The theory of estimation shows us how to design the receiver to minimize errors due to noise, and it tells us how large the irreducible residual errors will on the average be.

Locating a target requires specifying its distance and its direction. The distance is proportional to the interval between transmission of a radar pulse and reception of its echo, and to measure it the radar must determine the instant when the echo arrives. It might do so by timing the peak of the received signal, but exactly when this peak occurs is made uncertain by the noise. The azimuth of the target can be estimated by comparing the amplitudes of successive echoes as the radar antenna rotates. By changing those amplitudes in a random, unpredictable way the noise introduces error into that estimate. A measurement of the Doppler-shifted carrier frequency of the echo yields the component of the target velocity in the direction of the radar; this too will be falsified by the noise.

A radar echo can be represented in the form

$$s(t; A, \psi, \tau, \Omega) = A \operatorname{Re} F(t - \tau) \exp(i\Omega t + i\psi),$$

where $F(t)$ is the complex envelope of the pulse, A its amplitude, τ its time of arrival or *epoch*, Ω its carrier frequency, and ψ the phase of its carrier. The input to the receiver is

$$v(t) = s(t; A, \psi, \tau, \Omega) + n(t), \qquad 0 \le t \le T,$$

with $n(t)$ the random noise; and from the input $v(t)$ observed during the interval $(0, T)$ the receiver is to determine the values of the unknown parameters A, τ, and Ω. The joint probability density function of set $v = \{v(t_i)\}$ of samples of the input at times t_i depends on those parameters through the signal,

$$p(v; A, \psi, \tau, \Omega) = p_0(\{v_i - s(t_i; A, \psi, \tau, \Omega)\}),$$

where $p_0(\{n_i\})$ is the joint probability density function of samples $n_i = n(t_i)$ of the random noise at times t_i. In this way the unknowns become parameters of the distribution of the observations v.

The phase ψ carries no useful information about the target when, as often, the phase of the transmitted pulse is uncontrolled. Over such an uninformative parameter the joint probability density function may be averaged with respect to an accepted prior distribution $z(\cdot)$,

$$p(v; A, \tau, \Omega) = \int_0^{2\pi} z(\psi) p(v; A, \psi, \tau, \Omega)\, d\psi,$$

to provide a joint density function of the observations that depends only on the quantities of interest. We seek values of A, τ, and Ω for which the joint probability density function $p(v; A, \tau, \Omega)$ in some sense best describes the observations v. Alternatively, we may estimate the phase and other such "nuisance parameters" and discard the results.

Signal parameters may also need to be estimated in communications. Suppose that we wish to transmit numerical data that can take on arbitrary values within a limited range, as in telemetry when the temperature or pressure at a point is to be conveyed periodically to a distant observer. The amplitudes of a succession of pulses might be set by the data, or their carrier frequencies might be caused to deviate proportionally from a reference frequency stored at transmitter and receiver. The receiver must then estimate the amplitudes or the frequencies as the pulses arrive mixed with random noise, and it should be able to do so as accurately as possible.

The theory of estimation presupposes that one knows the joint probability density function $p(v|\, \theta)$ of the outcomes v of a set of measurements as a function of m unknown parameters $\theta = (\theta_1, \theta_2, \ldots, \theta_m)$. These are called the *estimanda*. They can be represented as a point in an m-dimensional parameter space Θ. If, for instance, θ are the parameters of a signal $s(t; \theta)$ received in the midst of random noise, $p(v|\, \theta)$ will derive from the joint probability density function of samples of the noise.

One seeks a strategy that on the basis of measured values of the n random variables $v_1, v_2, \ldots, v_n$ assigns some value $\hat\theta_k(v)$ as an estimate of the kth parameter θ_k; this function $\hat\theta_k(v)$ of the data v is called an *estimator*. The number that results when a particular set v of data is substituted into this function is called an *estimate* of the parameter θ_k. The collection of m such estimators is designated by the vector

$$\hat{\theta}(v) = (\hat\theta_1(v), \hat\theta_2(v), \ldots, \hat\theta_m(v)).$$

Like the vector θ of true values of the estimanda, it is represented by a point in the parameter space Θ.

Because the data are random variables, no two experiments will yield the same values of the estimates $\hat\theta$, even though the true set of parameters θ is the same for both. The most one can hope for is that the estimates $\hat{\theta}_k$ will be close to the true values θ_k "on the average." Given a set of strategies or estimators $\hat{\theta}(v)$ and the conditional probability density function $p(v|\,\theta)$, one can calculate a conditional probability density function $q(\hat{\theta}|\,\theta)$ of the estimates $\hat{\theta}$. One would like this conditional density function to be sharply peaked in the neighborhood of the true values θ.

If the expected value $\overline{\hat{\theta}}_k = E[\hat{\theta}_k(v)|\,\theta]$ of the estimator $\hat{\theta}_k(v)$ equals the true value θ_k, the estimator is said to be *unbiased*. The difference $\overline{\hat{\theta}}_k - \theta_k$ is called the *bias* of the estimator. The mean-square error incurred by the estimator is defined as

$$E = E\{[\hat{\theta}_k(v) - \theta_k]^2\}$$

and it can be written as

$$E = E[[\hat{\theta}_k(v) - \overline{\hat{\theta}}_k + \overline{\hat{\theta}}_k - \theta_k]^2] = E\{[\hat{\theta}_k(v) - \overline{\hat{\theta}}_k]^2\} + (\overline{\hat{\theta}}_k - \theta_k)^2$$
$$= \text{Var } \hat{\theta}_k(v) + (\overline{\hat{\theta}}_k - \theta_k)^2.$$

The mean-square error therefore equals the sum of the variance $\text{Var } \hat{\theta}_k(v)$ of the estimator and the square of its bias. Both the bias and the variance of an estimator should be small, and it is often necessary to compromise between these desiderata.

6.1.1 Maximum-a-posteriori-probability Estimators

Imagine the space Θ of the estimanda θ divided into a large number M of small regions Δ_j. Denote the center of the jth region Δ_j by the m-vector θ_j, and denote by H_j the proposition "The parameter set θ lies in region Δ_j." Consider a strategy whereby the receiver chooses among the M hypotheses H_j on the basis of the observed data $v = (v_1, v_2, \ldots, v_n)$. When the receiver selects hypothesis H_i, it issues the estimate $\hat{\theta}(v) = \theta_i$. The simplest strategy, as we saw in Sec. 1.1, directs the receiver to choose that hypothesis H_i whose conditional probability $\text{Pr}(H_i|\,v)$ is largest; this is Bayes's rule. Here

$$\text{Pr}(H_i|\,v) = \int_{\Delta_i} p(\theta|\,v) \, d^m\theta, \tag{6-1}$$

where $p(\theta|\,v)$ is the conditional probability density function of the parameter θ, given the data v. By Bayes's theorem for continuous random variables this is

$$p(\theta|\,v) = \frac{z(\theta)p(v|\,\theta)}{p(v)}, \tag{6-2}$$

with

$$p(v) = \int_{\Theta} z(\theta)p(v|\,\theta) \, d^m\theta \tag{6-3}$$

the overall probability density function of the data v [Hel91, p. 157], [Pap91, p. 164].

If now the regions Δ_i become smaller and smaller, the number M of hypotheses increasing, the probability $\text{Pr}(H_i|\,v)$ becomes proportional to the conditional probability density function $p(\theta_i|\,v)$ at the center θ_i of region Δ_i, and from this point of view the best strategy for the receiver is to issue as its estimate that set θ for which the conditional density function $p(\boldsymbol{\theta}|\, v)$ is maximum. That is, the optimum estimator $\hat{\boldsymbol{\theta}}(v)$ is defined by the criterion

$$p(\hat{\boldsymbol{\theta}}(v)|\, v) \geq p(\boldsymbol{\theta}|\, v), \qquad \forall \boldsymbol{\theta} \in \Theta.$$

This is called the *maximum-a-posteriori-probability* (MAP) estimator.

The joint probability density function $q(\hat{\boldsymbol{\theta}}|\, \boldsymbol{\theta})$ of the set $\hat{\boldsymbol{\theta}}(v) = (\hat{\theta}_1(v), ..., \hat{\theta}_m(v))$ of estimators of the m unknown parameters, when their true values are $\boldsymbol{\theta}$, can be expressed as

$$q(\boldsymbol{\theta}'|\, \boldsymbol{\theta}) = \int_{R_n} \delta(\boldsymbol{\theta}' - \hat{\boldsymbol{\theta}}(v))p(v|\, \boldsymbol{\theta})\, d^n v, \qquad (6\text{-}4)$$

where $\delta(\cdot)$ is an m-dimensional delta function. We attach a prime ($'$) to the set of algebraic variables figuring in the density function in order to distinguish them from the set of estimators $\hat{\boldsymbol{\theta}}(v)$ of the m parameters $\boldsymbol{\theta}$. Just as Bayes's rule maximizes the overall probability of correct decision in hypothesis testing, the MAP estimator maximizes the average value

$$\overline{Q} = \int_\Theta z(\boldsymbol{\theta})q(\boldsymbol{\theta}|\, \boldsymbol{\theta})\, d^m\boldsymbol{\theta}.$$

The posterior density function $q(\boldsymbol{\theta}'|\, \boldsymbol{\theta})$ of the estimate is to be heaped up as high as possible at the true value $\boldsymbol{\theta}$, at least on the average. By (6-2) and (6-4) we can write this average as

$$\overline{Q} = \int_{R_n} p(v)p(\hat{\boldsymbol{\theta}}(v)|\, v)\, d^n v,$$

and because $p(v) \geq 0$, this will be largest if to each point v of the data space R_n we assign as $\hat{\boldsymbol{\theta}}(v)$ that set $\boldsymbol{\theta}$ for which $p(\boldsymbol{\theta}|\, v)$ is maximum.

When the estimanda $\boldsymbol{\theta}$ are parameters of a signal $s(t; \boldsymbol{\theta})$, we can introduce the joint probability density function $p_0(v)$ of the data in the absence of any signal, divide it into both numerator and denominator of (6-2),

$$p(\boldsymbol{\theta}|\, v) = \frac{z(\boldsymbol{\theta})p(v|\, \boldsymbol{\theta})/p_0(v)}{p(v)/p_0(v)},$$

and pass to the limit of an infinite number of samples v of the input $v(t)$ of the receiver, whereupon the conditional probability density function of the estimanda becomes the functional

$$p(\boldsymbol{\theta}|\, v(t)) = \frac{z(\boldsymbol{\theta})\Lambda[v(t)|\, \boldsymbol{\theta}]}{\Lambda[v(t)]},$$

$$\Lambda[v(t)] = \int_\Theta z(\boldsymbol{\theta})\Lambda[v(t)|\, \boldsymbol{\theta}]\, d^m\boldsymbol{\theta}, \qquad (6\text{-}5)$$

where $\Lambda[v(t)|\, \boldsymbol{\theta}]$ is the likelihood functional for detecting the signal $s(t; \boldsymbol{\theta})$ in noise. Because the denominator $\Lambda[v(t)]$ does not depend on the parameters $\boldsymbol{\theta}$, it suffices to find those values of the parameters for which the numerator $z(\boldsymbol{\theta})\Lambda[v(t)|\, \boldsymbol{\theta}]$ is maximum.

6.1.2 Maximum-likelihood Estimator

The MAP estimator, by (6-2), is given by the vector $\boldsymbol{\theta}$ for which the product $z(\boldsymbol{\theta})p(v|\, \boldsymbol{\theta})$ is maximum. In many problems it is difficult or impossible to specify a precise prior probability density function $z(\boldsymbol{\theta})$ for the estimanda $\boldsymbol{\theta}$, but usually whatever prior density function $z(\theta)$ is reasonable will be rather broader as a function of θ than the density function $p(v|\theta)$ of the data. The strictly MAP estimate then lies close to the vector θ for which the probability density function $p(v|\theta)$ of the data is maximum. Unless this is the case, one's measurement of the data v is providing little information to improve one's knowledge of θ as embodied in the prior density function $z(\theta)$. It is customary, therefore, to assume that the prior probability density function $z(\theta)$ is so broad that its effect on the maximization of the product $z(\theta)p(v|\theta)$ is negligible. The values of θ for which $p(v|\theta)$ is maximum are termed the *maximum-likelihood* estimates of θ:

$$p(v|\hat{\theta}(v)) \geq p(v|\theta), \qquad \forall\theta \in \Theta.$$

This inequality defines the *maximum-likelihood* estimator. One can say that the maximum-likelihood estimates are those values of the parameters θ for which $p(v|\theta)$ best fits the observed data v. It is this estimator that we shall principally utilize in estimates of the arrival time and carrier frequency of a signal.

In most estimation problems it is desirable that the same value of a parameter θ be obtained whether one estimates θ itself or some monotone function $f(\theta)$ of that parameter. For instance, if one estimates θ^3 and takes the cube root of the result, one would like to find the same value that a direct estimate of θ would yield. Maximum-likelihood estimates possess this property. Maximum-a-posteriori-probability estimates, however, based on some prior density function $z(\theta)$, in general do not, because of the different weightings assigned to corresponding ranges of the parameter. For a discussion of such matters as applied to physical measurements, the article by Annis, Cheston, and Primakoff [Ann53] and the books by Jeffreys [Jef73], [Jef83] and Jánossy [Ján65] may be consulted.

When the estimate of parameters θ of a signal $s(t) = s(t; \theta)$ is to be based on the input $v(t)$ to a receiver, we must pass to the limit of an infinitely dense sampling. This we can again achieve by dividing $p(v|\theta)$ by an appropriate probability density function $p_0(v)$ of the data that is independent of the parameters θ, usually that of samples of pure noise:

$$\lim_{n\to\infty} \frac{p(v|\theta)}{p_0(v)} = \Lambda[v(t)|\theta].$$

Then the maximum-likelihood estimator $\hat{\theta}[v(t)]$ is that set of parameter values for which the likelihood functional $\Lambda[v(t)|\theta]$ for detecting the signal $s(t; \theta)$ is as large as possible. There may be many values of certain of the parameters θ for which this functional $\Lambda[v(t)|\theta]$ possesses local maxima, and the highest of those maxima identifies the maximum-likelihood estimate. This concept will be developed in the following section.

6.1.3 Estimating the Mean of a Gaussian Distribution

As a simple example to illustrate these ideas, let us estimate the mean or expected value m of a Gaussian distribution with known variance δ^2 by n independent observations $v_1, v_2, \ldots, v_n$ of a random variable v. Their joint probability density function is

　　　　　　　　　Estimation of Signal Parameters　　Chap. 6

$$p(v|\,m) = (2\pi\delta^2)^{-n/2} \exp\left[-\sum_{k=1}^{n} \frac{(v_k - m)^2}{2\delta^2}\right]. \tag{6-6}$$

Let us assume that previous experiments have shown the true mean m to be normally distributed with expected value μ and variance β^2,

$$z(m) = \frac{1}{\sqrt{2\pi\beta^2}} \exp\left[-\frac{(m - \mu)^2}{2\beta^2}\right]. \tag{6-7}$$

From the definition (6-2) we can with some labor calculate the posterior probability density function of the true mean m, given the set of outcomes v:

$$p(m|\,v) = \frac{1}{\sqrt{2\pi s^2}} \exp\left\{-\frac{1}{2s^2}\left[m - s^2\left(\frac{nX}{\delta^2} + \frac{\mu}{\beta^2}\right)\right]^2\right\},$$
$$\frac{1}{s^2} = \frac{n}{\delta^2} + \frac{1}{\beta^2}, \tag{6-8}$$

where X is the sample mean of the data,

$$X = \frac{1}{n}\sum_{k=1}^{n} v_k. \tag{6-9}$$

The value of m at which the posterior density function $p(m|\,v)$ is maximum is the MAP estimator

$$\hat{m}(v) = \xi = s^2\left(\frac{nX}{\delta^2} + \frac{\mu}{\beta^2}\right) = \frac{\beta^2 X + \mu\delta^2/n}{\beta^2 + (\delta^2/n)}. \tag{6-10}$$

As always when the distributions of the observations and of the parameters are Gaussian functions of both the observations and the parameters, this estimator is a linear function of the data v. Because this estimator is a function of the data v only through the sample mean X, $\hat{m} = \hat{m}(X)$, X is said to be a *sufficient statistic* for estimating the mean m.

Upon examining the estimate given in (6-10), we see that if the initial uncertainty β in the value of the mean m is very large, $\beta^2 \gg \delta^2/n$, the MAP estimate $\hat{m}$ is nearly equal to the sample mean, $\hat{m} \approx X$. Indeed, the sample mean X is the maximum-likelihood estimator of m, as we can show by differentiating (6-6) with respect to m and setting the result equal to 0. If, on the other hand, the error variance of the measurements is very large, $\delta^2/n \gg \beta^2$, the estimate $\hat{m}$ is close to the prior expected value μ, and the observations contribute little new information about the value of the mean m.

The estimator $\hat{m}$ in (6-10) is biased, for its expected value is

$$E(\hat{m}|\,m) = \langle\hat{m}\rangle = \frac{\beta^2 m + \mu\delta^2/n}{\beta^2 + (\delta^2/n)} \tag{6-11}$$

when the true value of the mean is m. As the number n of measurements increases, this expected value $\langle\hat{m}\rangle$ approaches the true mean m, and we can say that the estimator is asymptotically unbiased.

Because the variance of the sample mean X equals δ^2/n, the variance of the MAP estimator is

$$\text{Var } \hat{m}(v) = \frac{\beta^4}{[\beta^2 + (\delta^2/n)]^2} \frac{\delta^2}{n},$$

and the reader should verify that the mean-square error $E[\hat{m}(v) - m)^2]$ equals the sum of this variance and the square of the bias.

6.1.4 Jointly Gaussian Parameters and Data

We can generalize the problem treated in Sec. 6.1.3 by considering how to estimate m correlated Gaussian random variables $\theta_1, \theta_2, \ldots, \theta_m$ by observing a number n of random variables $v_1, v_2, \ldots, v_n$. These "data" are themselves Gaussian random variables and are correlated with the estimanda $\theta_1, \theta_2, \ldots, \theta_m$ in such a way that the joint probability density function of all $n + m$ random variables has the multivariate Gaussian form. An example is the estimation of a correlated discrete-time Gaussian random process corrupted by additive Gaussian noise.

Without loss of generality we can assume that all these variables have expected values equal to 0, for otherwise we could define new variables by subtracting out their known expected values.

In order to write out the joint probability density function $p(v, \theta)$ of all the variables, we introduce the $(n + m)$-element column vector whose transpose is $(v^T \theta^T)$, where v is the n-element column vector of the data and θ is the m-element column vector of the estimanda. Then

$$p(v, \theta) = p(v_1, \ldots, v_n, \theta_1, \ldots, \theta_m) = C_1 \exp\left[-\tfrac{1}{2}(v^T \theta^T)\mu\binom{v}{\theta}\right], \qquad (6\text{-}12)$$

where C_1 is a normalization constant and the $(n + m) \times (n + m)$ matrix μ is the inverse of the covariance matrix ϕ of data and estimanda,

$$\phi = \begin{bmatrix} \phi_{vv} & \phi_{v\theta} \\ \phi_{\theta v} & \phi_{\theta\theta} \end{bmatrix}. \qquad (6\text{-}13)$$

Here $\phi_{vv} = E(vv^T)$ is the $n \times n$ covariance matrix of the data v, $\phi_{\theta\theta} = E(\theta\theta^T)$ is the $m \times m$ covariance matrix of the estimanda θ, and $\phi_{\theta v} = \phi_{v\theta}^T = E(\theta v^T)$ is the $m \times n$ cross-covariance matrix of the estimanda and the data. The conditional density function of the parameters θ, given the data v, is now

$$p(\theta | v) = \frac{p(v, \theta)}{p(v)} \qquad (6\text{-}14)$$

where

$$p(v) = C_0 \exp(-\tfrac{1}{2}v^T \phi_{vv}^{-1} v), \qquad (6\text{-}15)$$

with C_0 another normalization constant, is the joint probability density function of the data.

The matrix $\mu = \phi^{-1}$ in (6-12), like that in (6-13), can be expressed in block form,

$$\mu = \begin{bmatrix} \mu_{vv} & \mu_{v\theta} \\ \mu_{\theta v} & \mu_{\theta\theta} \end{bmatrix}, \qquad (6\text{-}16)$$

where

$$\boldsymbol{\mu}_{vv} = (\boldsymbol{\phi}_{vv} - \boldsymbol{\phi}_{v\theta}\boldsymbol{\phi}_{\theta\theta}^{-1}\boldsymbol{\phi}_{\theta v})^{-1},$$

$$\boldsymbol{\mu}_{\theta v} = \boldsymbol{\mu}_{v\theta}^{T} = -\boldsymbol{\mu}_{\theta\theta}\boldsymbol{\phi}_{\theta v}\boldsymbol{\phi}_{vv}^{-1}, \qquad (6\text{-}17)$$

$$\boldsymbol{\mu}_{\theta\theta} = (\boldsymbol{\phi}_{\theta\theta} - \boldsymbol{\phi}_{\theta v}\boldsymbol{\phi}_{vv}^{-1}\boldsymbol{\phi}_{v\theta})^{-1},$$

as can be shown by multiplying (6-13) by (6-16) to obtain the $(n + m) \times (n + m)$ identity matrix.

By dividing (6-12) by (6-15), we find for the conditional density function

$$p(\boldsymbol{\theta}|\,v) = (C_1/C_0)\exp[-\tfrac{1}{2}v^T\boldsymbol{\mu}_{vv}v - \tfrac{1}{2}v^T\boldsymbol{\mu}_{v\theta}\boldsymbol{\theta} - \tfrac{1}{2}\boldsymbol{\theta}^T\boldsymbol{\mu}_{\theta v}v - \tfrac{1}{2}\boldsymbol{\theta}^T\boldsymbol{\mu}_{\theta\theta}\boldsymbol{\theta} + \tfrac{1}{2}v^T\boldsymbol{\phi}_{vv}^{-1}v]$$

$$= \frac{C_1}{C_0}\exp[-\tfrac{1}{2}(\boldsymbol{\theta} - \mathbf{M}v)^T\boldsymbol{\mu}_{\theta\theta}(\boldsymbol{\theta} - \mathbf{M}v)], \qquad (6\text{-}18)$$

where the $m \times n$ matrix $\mathbf{M}$ is still to be determined. We find it by writing out the arguments of the two exponential functions and comparing terms, and we obtain for one of those terms

$$\boldsymbol{\theta}^T\boldsymbol{\mu}_{\theta\theta}\mathbf{M}v = -\boldsymbol{\theta}^T\boldsymbol{\mu}_{\theta v}v,$$

whence $\boldsymbol{\mu}_{\theta\theta}\mathbf{M} = -\boldsymbol{\mu}_{\theta v}$, and hence by (6-17) the matrix $\mathbf{M}$ must be

$$\mathbf{M} = -\boldsymbol{\mu}_{\theta\theta}^{-1}\boldsymbol{\mu}_{\theta v} = \boldsymbol{\phi}_{\theta v}\boldsymbol{\phi}_{vv}^{-1}. \qquad (6\text{-}19)$$

The MAP estimators of the parameters $\boldsymbol{\theta}$—those values that maximize the conditional density function $p(\boldsymbol{\theta}|\,v)$ in (6-18)—are *linear* functions of the data v,

$$\hat{\boldsymbol{\theta}}(v) = \mathbf{M}v = \boldsymbol{\phi}_{\theta v}\boldsymbol{\phi}_{vv}^{-1}v. \qquad (6\text{-}20)$$

The estimation matrix $\mathbf{M}$ is the solution of a set of linear equations written in matrix form as

$$\mathbf{M}\boldsymbol{\phi}_{vv} = \boldsymbol{\phi}_{\theta v}. \qquad (6\text{-}21)$$

The $m \times m$ covariance matrix of the errors is furthermore

$$\mathbf{B} = E[(\boldsymbol{\theta} - \mathbf{M}v)(\boldsymbol{\theta} - \mathbf{M}v)^T] = \boldsymbol{\mu}_{\theta\theta}^{-1} = \boldsymbol{\phi}_{\theta\theta} - \boldsymbol{\phi}_{\theta v}\boldsymbol{\phi}_{vv}^{-1}\boldsymbol{\phi}_{\theta v}^T = \boldsymbol{\phi}_{\theta\theta} - \mathbf{M}\boldsymbol{\phi}_{\theta v}^T \qquad (6\text{-}22)$$

by (6-17) and (6-18).

6.1.5 Bayes Estimates

Sometimes it is possible to specify both the cost of an error in estimation and the joint prior probability density function of the estimanda, much as in hypothesis testing the costs of incorrect decisions and the prior probabilities of the hypotheses may be available. The cost to the experimenter of assigning a set of estimates $\hat{\boldsymbol{\theta}}$ to the parameters when their true values are given by $\boldsymbol{\theta} = (\theta_1, \theta_2, \ldots, \theta_m)$ will be a function $C(\hat{\boldsymbol{\theta}}, \boldsymbol{\theta})$ of both the true values and the estimates. As a function of the estimates, it is smallest for $\hat{\boldsymbol{\theta}} = \boldsymbol{\theta}$, often depending only on the differences $(\hat{\theta}_k - \theta_k)$ between the estimates and the true values of the parameters. Given both a prior probability density function $z(\boldsymbol{\theta})$ of the estimanda and a cost function $C(\hat{\boldsymbol{\theta}}, \boldsymbol{\theta})$, the observer is in a position to adopt the Bayes criterion that the estimation strategy should yield the minimum average cost per experiment.

The optimum Bayes strategy for estimation can be derived in much the same way as for hypothesis testing, and as in Sec. 1.1.2 we find that it requires us to determine those values $\hat{\theta}$ of the estimanda θ for which the conditional risk

$$C(\hat{\theta}|\, v) = \int_{\Theta} C(\hat{\theta}, \theta) p(\theta|\, v)\, d^m\theta \qquad (6\text{-}23)$$

is minimum. The Bayes cost associated with any estimator $\hat{\theta}(v)$ is

$$\begin{aligned}
\overline{C} &= \int_{\Theta}\int_{R_n} z(\theta) C(\hat{\theta}(v), \theta) p(v|\theta)\, d^n v\, d^m\theta \\
&= \int_{R_n} C(\hat{\theta}(v)|\, v) p(v)\, d^n v;
\end{aligned} \qquad (6\text{-}24)$$

we have used (6-2). This Bayes cost will be smallest if to each point v in the data space R_n we assign the vector $\hat{\theta}$ that minimizes the conditional risk in (6-23).

The Bayes cost can be written

$$\overline{C} = \int_{\Theta} z(\theta) R[\hat{\theta}(v)|\, \theta]\, d^m\theta, \qquad (6\text{-}25)$$

in terms of the risk associated with an arbitrary estimator $\hat{\theta}(v)$ of the parameters, when their true values are θ:

$$R[\hat{\theta}(v)|\, \theta] = \int_{R_n} C(\hat{\theta}(v), \theta) p(v|\, \theta)\, d^n v. \qquad (6\text{-}26)$$

When the estimanda are parameters θ of a signal $s(t; \theta)$, their Bayes estimator is that set $\hat{\theta} = \hat{\theta}[v(t)]$ for which

$$C[\hat{\theta}|\, v(t)] = \int_{\Theta} C(\hat{\theta}, \theta) p(\theta|\, v(t))\, d^m\theta$$

is minimum, with $p(\theta|\, v(t))$ defined as in (6-5) in terms of the likelihood functional $\Lambda[v(t)|\, \theta]$ for detecting that signal in noise.

6.1.6 The Quadratic Cost Function

Because the mean-square error is one of the principal measures of the quality of an estimator, it is appropriate to consider the quadratic cost function

$$C(\hat{\theta}, \theta) = (\hat{\theta} - \theta)^2, \qquad (6\text{-}27)$$

which because of its mathematical simplicity is indeed the most frequently adopted. The resulting estimator $\hat{\theta}(v)$ is called the *minimum-mean-square-error* (MMSE) estimator.

The conditional risk in (6-23) is then

$$\begin{aligned}
C(\hat{\theta}|\, v) &= \int_{-\infty}^{\infty} (\hat{\theta} - \theta)^2 p(\theta|\, v)\, d\theta \\
&= \hat{\theta}^2 - 2\hat{\theta} \int_{-\infty}^{\infty} \theta p(\theta|\, v)\, d\theta + \int_{-\infty}^{\infty} \theta^2 p(\theta|\, v)\, d\theta,
\end{aligned}$$

and it is minimum when

$$\hat{\theta}(v) = \int_{-\infty}^{\infty} \theta p(\theta | v)\, d\theta = E(\theta | v). \tag{6-28}$$

The Bayes estimator is now the conditional expected value of the parameter θ, given the data v.

The minimum conditional risk equals the conditional variance of the parameter θ, given the data v:

$$\min_{\hat{\theta}(v)} C(\hat{\theta}(v) | v) = \mathrm{Var}(\theta | v).$$

Multiplying by the overall density function $p(v)$ of the data, as defined in (6-3), we find the minimum Bayes cost to be

$$\overline{C}_{\min} = \int_{R_n} \mathrm{Var}(\theta | v) p(v)\, d^n v = E_v[\mathrm{Var}(\theta | v)]. \tag{6-29}$$

When $E(\theta | v)$ is a linear function of the data v, as happens when the joint probability density function $p(\theta, v) = z(\theta) p(v | \theta)$ of the parameter θ and the data v is Gaussian in both θ and v, $\mathrm{Var}(\theta | v)$ is independent of v, and the minimum Bayes cost $\overline{C}_{\min}$ is simply the conditional variance $\mathrm{Var}(\theta | v)$ itself. Although this is now independent of the data v, it is in general smaller than the prior variance $\mathrm{Var}\,\theta$, which is determined entirely by the prior density function $z(\theta)$.

In our example in Sec. 6.1.3 of estimating the mean m of a Gaussian random variable from n observations, we see by (6-8) that the estimator ξ in (6-10) is the conditional expected value of the estimandum m, and it therefore serves not only as the MAP, but also as the MMSE estimator of m. The conditional variance $\mathrm{Var}(m | v)$ of the mean m, given the data v, is by (6-8)

$$\mathrm{Var}(m | v) = s^2 = \frac{\beta^2 \delta^2}{n\beta^2 + \delta^2}, \tag{6-30}$$

and it is independent of v. As in (6-29), this must then equal the minimum Bayes cost,

$$\overline{C}_{\min} = \frac{\delta^2}{n} \frac{\beta^2}{\beta^2 + (\delta^2/n)}.$$

The risk associated with a true mean m is, according to the definition in (6-26),

$$R[\hat{m}(v) | m] = \int_{R_n} [\hat{m}(v) - m]^2 p(v | m)\, d^n v = \int_{-\infty}^{\infty} [\hat{m}(X) - m]^2 p(X | m)\, dX$$

$$= \frac{\delta^2}{n} \left[\frac{\beta^4 + (\mu - m)^2 \delta^2/n}{[\beta^2 + (\delta^2/n)]^2} \right], \tag{6-31}$$

where $p(X | m)$ is the probability density function of the sample mean X when the true mean of the data is m. By using (6-7) and (6-25) the reader should verify the minimum Bayes cost just derived.

When several parameters $\boldsymbol{\theta}$ are to be estimated, what corresponds to the quadratic cost function is the positive-definite quadratic form

$$C(\hat{\boldsymbol{\theta}}, \boldsymbol{\theta}) = \sum_{i=1}^{m} \sum_{j=1}^{m} g_{ij}(\hat{\theta}_i - \theta_i)(\hat{\theta}_j - \theta_j). \tag{6-32}$$

The Bayes estimates are again the conditional expected values of the parameters, given the data v:

$$\hat{\theta}_k(v) = \int_\Theta \theta_k p(\theta|\, v)\, d^m\theta, \qquad k = 1, 2, \ldots, m. \qquad (6\text{-}33)$$

This estimator is the same as though the cost of an error $\hat{\theta}_k(v) - \theta_k$ were simply its square, and the total cost were the sum

$$C(\hat{\theta}, \theta) = \sum_{k=1}^{m} (\hat{\theta}_k - \theta_k)^2. \qquad (6\text{-}34)$$

When as in Sec. 6.1.4 the data v and the estimanda θ are jointly Gaussian random variables, we can see from (6-18) that the conditional expected values of the estimanda form the column vector

$$\hat{\theta}(v) = \mathbf{M}v$$

with the $m \times n$ matrix $\mathbf{M}$ given in (6-19). These therefore serve as the minimum-mean-square-error estimators of the parameters θ in the joint probability density function in (6-12). They constitute linear estimators of the parameters because they depend linearly on the data v.

6.1.7 The Principle of Orthogonality

In many estimation problems one cannot be sure that the data and the estimanda are jointly Gaussian distributed, but one chooses to adopt a linear estimator of the form $\hat{\theta} = \mathbf{M}v$ anyhow. For one thing, linear operations are simplest to carry out by both digital and analog means. One then seeks the estimation matrix $\mathbf{M}$ for which the total mean-square error (6-34) is minimum. It must be the same as the optimum matrix $\mathbf{M}$ when the data and the estimanda are really jointly Gaussian distributed, and it is given by (6-20) and (6-21).

A quick way of deriving the MMSE linear estimator is based on the principle of orthogonality [Hel91, pp. 495–500], [Pap91, p. 204]. Two random variables A and B are said to be *orthogonal* if $E(AB) = 0$. Two n-tuples of random variables $v = (v_1, v_2, \ldots, v_n)^T$ and $\mathbf{w} = (w_1, w_2, \ldots, w_n)^T$ are orthogonal if each element of the $n \times n$ matrix $E(v\mathbf{w}^T)$ equals zero. The "space" of these random variables then takes on the properties of a metric space in which the squared "distance" between two such vectors is specified by the expected value

$$E[(v - \mathbf{w})^2] = E[(v^T - \mathbf{w}^T)(v - \mathbf{w})]$$

of their squared difference.

In our problem of estimating the m random variables $\theta = (\theta_1, \theta_2, \ldots, \theta_m)^T$, the data v are thought of as spanning a subspace V of the metric space. The MMSE linear estimator $\hat{\theta} = \mathbf{M}v$, as a linear combination of the data v, also lies in that subspace V. Because the average cost, measured by the squared distance between $\hat{\theta}$ and the vector θ of true values, is to be minimum, the error vector $\theta - \hat{\theta}$ must be as short as possible and hence perpendicular to the subspace V; that is, it must be orthogonal to all vectors v in V,

$$E[(\boldsymbol{\theta} - \hat{\boldsymbol{\theta}})v^T] = 0.$$

Thus

$$E(\boldsymbol{\theta}v^T) = E(\mathbf{M}vv^T) = \mathbf{M}E(vv^T)$$

or

$$\boldsymbol{\phi}_{\theta v} = \mathbf{M}\boldsymbol{\phi}_{vv},$$

as in (6-21).

The covariance matrix $\mathbf{B}$ of the errors is

$$\mathbf{B} = E[(\boldsymbol{\theta} - \hat{\boldsymbol{\theta}})(\boldsymbol{\theta}^T - \hat{\boldsymbol{\theta}}^T)] = E[(\boldsymbol{\theta} - \hat{\boldsymbol{\theta}})\boldsymbol{\theta}^T] - E[(\boldsymbol{\theta} - \hat{\boldsymbol{\theta}})\hat{\boldsymbol{\theta}}^T].$$

Because the estimator $\hat{\boldsymbol{\theta}}(v)$ lies in the subspace V and the error $\boldsymbol{\theta} - \hat{\boldsymbol{\theta}}$ is orthogonal to it, the second term vanishes, and we find from the first

$$\mathbf{B} = E(\boldsymbol{\theta}\boldsymbol{\theta}^T) - E(\mathbf{M}v\boldsymbol{\theta}^T) = \boldsymbol{\phi}_{\theta\theta} - \mathbf{M}\boldsymbol{\phi}_{\theta v}^T$$

as in (6-22).

The statistical model that most commonly gives rise to an estimation problem of this kind is one in which the data v result from some known linear operation on the unknowns $\theta_1, \theta_2, \dots, \theta_m$, with the addition of independently random errors,

$$v_j = \sum_{i=1}^{m} K_{ji}\theta_i + e_i, \qquad 1 \leq j \leq n,$$

or, in matrix form,

$$v = \mathbf{K}\boldsymbol{\theta} + \mathbf{e}, \tag{6-35}$$

where $\mathbf{e}$ is a column vector of the n errors or "noise" variates, which are independent of the estimanda $\boldsymbol{\theta}$ and usually assumed to have equal variances Var $e_i \equiv \sigma^2$; $\mathbf{K}$ is a known $n \times m$ matrix. Then

$$\boldsymbol{\phi}_{\theta v} = E[\boldsymbol{\theta}(\mathbf{K}\boldsymbol{\theta} + \mathbf{e})^T] = \boldsymbol{\phi}_{\theta\theta}\mathbf{K}^T,$$

$$\boldsymbol{\phi}_{vv} = E[(\mathbf{K}\boldsymbol{\theta} + \mathbf{e})(\boldsymbol{\theta}^T\mathbf{K}^T + \mathbf{e}^T)] = \mathbf{K}\boldsymbol{\phi}_{\theta\theta}\mathbf{K}^T + \sigma^2\mathbf{I},$$

with $\mathbf{I}$ the $n \times n$ identity matrix. The equations to be solved for the $m \times n$ estimation matrix $\mathbf{M}$ are then, from (6-21),

$$\mathbf{M}\mathbf{K}\boldsymbol{\phi}_{\theta\theta}\mathbf{K}^T + \sigma^2\mathbf{M} = \boldsymbol{\phi}_{\theta\theta}\mathbf{K}^T. \tag{6-36}$$

The linear filtering and prediction of discrete-time stochastic processes can be treated in this framework and has given rise to a vast literature. Elementary introductions can be found in such works as [Lar79, vol. 2, pp. 112–31], [Hel91, pp. 507–39], [Pap91, pp. 486–93, 512–28], [Sch91a, pp. 323–33, 423–78], to name only a few. Continuous-time stochastic processes can be linearly filtered and predicted by analogous techniques, which were first developed by Norbert Wiener [Wie60]. The reader can also consult [Mid60a, pp. 697–712], [Van68, Ch. 6], [Hel91, pp. 540–60], and [Pap91, pp. 480–6, 508–11], to cite only a few of the many references. A comprehensive bibliography up to 1974 was listed by Kailath [Kai74]. We shall take up this topic in Chapter 11 when we study the detection of Gaussian stochastic signals in Gaussian noise.

When as in (6-35) the data v are linearly related to the unknown parameters θ, with independently random additive errors, what correspond to the maximum-likelihood estimators introduced in Sec. 6.1.2 are the estimators $\hat{\theta}(v)$ determined by the method of least squares. If the errors e_i are assumed to be independent Gaussian random variables with equal variances σ^2, the conditional probability density function of the data v will be

$$p(v|\,\theta) = C \exp\left[-\frac{1}{2\sigma^2}(v - \mathbf{K}\theta)^T(v - \mathbf{K}\theta)\right],$$

with C a normalization constant, and this is maximum when the values of the estimanda $\theta_1, \theta_2, \ldots, \theta_m$ are such that the sum of squares

$$(v - \mathbf{K}\theta)^T(v - \mathbf{K}\theta) = \sum_{i=1}^{n}\left[v_i - \sum_{j=1}^{m} K_{ij}\theta_j\right]^2$$

is minimum [Bic77, pp. 94–9], [Hel91, pp. 500–6], [Sch91a, pp. 359–415].

6.2 ESTIMATION OF SIGNAL PARAMETERS

6.2.1 Maximum-likelihood Estimators

The maximum-likelihood estimates of a set θ of parameters of a signal $s(t; \theta)$ are defined in Sec. 6.1.2 as those values of θ for which the likelihood functional $\Lambda[v(t)|\,\theta]$ is maximum. When the ambient noise is white and Gaussian, with unilateral spectral density N, this functional is

$$\Lambda[v(t)|\,\theta] = \exp\left[\frac{2}{N}\int_0^T s(t; \theta)v(t)\,dt - \frac{1}{N}\int_0^T [s(t; \theta)]^2\,dt\right] \qquad (6\text{-}37)$$

as in (2-74). Let us distinguish the amplitude A and write the signal as

$$s(t; \theta) = Af(t, \theta'),$$

where θ' is the set of parameters other than the amplitude. Then

$$\ln \Lambda[v(t)|\,A, \theta'] = \frac{2A}{N}\int_0^T f(t; \theta')v(t)\,dt - \frac{A^2}{N}\int_0^T [f(t; \theta')]^2\,dt,$$

and maximizing with respect to the amplitude A we find

$$\max_A \ln \Lambda[v(t)|\,A, \theta'] = \frac{\left[\int_0^T f(t; \theta')v(t)\,dt\right]^2}{N\int_0^T [f(t; \theta')]^2\,dt}, \qquad (6\text{-}38)$$

which remains to be maximized with respect to the other parameters. Denoting the maximum-likelihood estimates of those other parameters by $\hat{\theta}'$, we write the maximum-likelihood estimator of the amplitude A as

$$\hat{A} = \frac{\int_0^T f(t; \hat{\theta}')v(t)\,dt}{\int_0^T [f(t; \hat{\theta}')]^2\,dt}.$$

It is generally difficult to determine the values of the parameters $\boldsymbol{\theta}'$ for which the right side of (6-38) is maximum. One way of doing so approximately is to build a bank of parallel filters, each matched to the signal $f(t; \boldsymbol{\theta}')$ for a different set $\boldsymbol{\theta}'$. These sets are densely spaced over the $(m - 1)$-dimensional region Θ' in which the parameters $\boldsymbol{\theta}'$ are expected to lie. The input $v(t)$ is applied to each of these filters, and the output of each is sampled at the end of the observation interval $(0, T)$, squared, and divided by the denominator of (6-38). The value of $\boldsymbol{\theta}'$ for the filter for which the resulting quantity is largest is then approximately the maximum-likelihood estimate.

If the points $\boldsymbol{\theta}'$ of the space Θ' for which we build filters matched to the signals $f(t; \boldsymbol{\theta}')$ are close enough together, the noise components of the outputs of these filters will be highly correlated. By interpolating the observed values of

$$\Gamma(\boldsymbol{\theta}'_i) = \frac{\left[\int_0^T f(t; \boldsymbol{\theta}'_i)v(t)\, dt\right]^2}{N\int_0^T [f(t; \boldsymbol{\theta}'_i)]^2\, dt}$$

and finding the maximum of the interpolated function $\Gamma(\boldsymbol{\theta}')$, we can then expect to be able to estimate the parameters $\boldsymbol{\theta}'$ nearly as accurately as though—what is impossible—we had filters matched to signals $f(t; \boldsymbol{\theta}')$ for a continuum of values of $\boldsymbol{\theta}'$.

6.2.2 Estimation of Arrival Time

As an example, suppose that the estimandum is the arrival time τ of a signal

$$s(t; A, \tau) = Af(t - \tau).$$

Then by (6-38) we must find the maximum value of

$$\Gamma(\tau) = \max_A \ln \Lambda[v(t)|\, A, \tau] = \frac{\left[\int_0^T f(t - \tau)v(t)\, dt\right]^2}{N\int_0^T [f(t - \tau)]^2\, dt}. \tag{6-39}$$

If we assume the observation interval to be much longer than the duration T' of the signal $f(t)$ $(T \gg T')$ and neglect the possibility that the signal may overlap one end or the other of the interval $(0, T)$, the denominator of (6-39) is independent of τ, and we must maximize

$$G(\tau) = \left|\int_0^T f(t - \tau)v(t)\, dt\right| \approx \left|\int_{-\infty}^{\infty} f(u - \tau)v(u)\, du\right|. \tag{6-40}$$

In order to see how to generate this function $G(\tau)$ in real time, let us consider a filter matched to the signal $f(t)$ over an interval $(0, T')$, outside of which we assume $f(t) \equiv 0$; $T' \ll T$. It will have an impulse response

$$k(s) = \begin{cases} f(T' - s), & 0 \le s \le T', \\ 0, & s < 0, \quad s > T', \end{cases} \tag{6-41}$$

and the output of this filter when the receiver input $v(t)$ is applied to it will be

$$v_0(t) = \int_0^{T'} f(T' - s)v(t - s)\, ds = \int_{-\infty}^{\infty} f(T' - s)v(t - s)\, ds$$
$$= \int_{-\infty}^{\infty} f(T' - t + u)v(u)\, du.$$

By comparison with (6-40) we see that

$$G(\tau) = |v_0(\tau + T')|. \tag{6-42}$$

Hence the maximum-likelihood estimate $\hat\tau$ of the arrival time is given by

$$\hat\tau = t_m - T',$$

where t_m is the time at which the rectified output $|v_0(t)|$ of the matched filter is maximum. That output will have many peaks, and the highest of them identifies the time t_m. It is unnecessary to build a bank of filters matched to signals with different values of τ; a single filter, matched as in (6-41) to $f(t)$, will do.

The question now arises how accurate is this estimate $\hat\tau$ of the arrival time τ. We gauge that by the mean-square error $E[(\hat\tau - \tau_0)^2]$, where τ_0 is the true value of the estimandum τ. In order to evaluate it, we must assume that the signal-to-noise ratio is large. We can then drop the absolute-value signs in (6-42) and maximize only

$$G(\tau) = \int_0^T f(t - \tau)v(t)\, dt;$$

that is, we solve the equation

$$\int_0^T f'(t - \hat\tau)v(t)\, dt = 0, \tag{6-43}$$

the prime indicating differentiation with respect to the argument of the function. When the signal-to-noise ratio is large, the root of this equation corresponding to the largest peak value of $G(\tau)$ will lie close to the true value τ_0 of the arrival time, and we can make the power-series expansion

$$f'(t - \hat\tau) = f'(t - \tau_0) - (\hat\tau - \tau_0)f''(t - \tau_0) + \cdots$$

and neglect terms of higher order in $(\hat\tau - \tau_0)$ than the first. Putting this into (6-43), we find for the error in the estimate

$$\hat\tau - \tau_0 \approx \frac{\int_0^T f'(t - \tau_0)v(t)\, dt}{\int_0^T f''(t - \tau_0)v(t)\, dt}. \tag{6-44}$$

The expected value of the numerator in (6-44) is

$$E\int_0^T f'(t - \tau_0)[Af(t - \tau_0) + n(t)]\, dt = A\int_0^T f'(t - \tau_0)f(t - \tau_0)\, dt$$
$$= \frac{A}{2}\int_0^T \frac{d}{dt}[f(t - \tau_0)]^2\, dt$$
$$= \frac{A}{2}\{[f(T - \tau_0)]^2 - [f(-\tau_0)]^2\},$$

Estimation of Signal Parameters   Chap. 6

and this expected value vanishes because of our assumption that the signal does not overlap either end of the observation interval $(0, T)$. That of the denominator, however, does not:

$$E \int_0^T f''(t - \tau_0)v(t) \, dt = A \int_0^T f''(t - \tau_0)f(t - \tau_0) \, dt$$

$$= -A \int_0^T [f'(t - \tau_0)]^2 \, dt \neq 0.$$

The numerator of (6-44) is therefore of first order in the noise and has no component due to the signal, and within terms of first order in the noise we can replace the denominator by its expected value, obtaining for the error

$$\hat{\tau} - \tau_0 \approx -\frac{\int_0^T f'(t - \tau_0)n(t) \, dt}{A \int_0^T [f'(t - \tau_0)]^2 \, dt}.$$

In this limit of large signal-to-noise ratio, therefore,

$$E(\hat{\tau} - \tau_0) \approx 0,$$

and the estimator is said to be asymptotically unbiased. We shall find this generally true of maximum-likelihood estimators.

The variance of the estimator is

$$\text{Var } \hat{\tau} = \left[A \int_{-\infty}^{\infty} |f'(t)|^2 \, dt \right]^{-2} E \left\{ \left[\int_{-\infty}^{\infty} f'(t - \tau_0)n(t) \, dt \right]^2 \right\}$$

when the observation interval is much longer than the duration of the signal, and by an analysis like that in (2-25) we find

$$E \left\{ \left[\int_{-\infty}^{\infty} f'(t - \tau_0)n(t) \, dt \right]^2 \right\} = \frac{N}{2} \int_{-\infty}^{\infty} [f'(t)]^2 \, dt,$$

whereupon the error variance is

$$\text{Var } \hat{\tau} = \frac{N}{2A^2} \left[\int_{-\infty}^{\infty} [f'(t)]^2 \, dt \right]^{-1} = \frac{N}{2E\beta^2} \qquad (6\text{-}45)$$

with

$$E = A^2 \int_0^T [f(t)]^2 \, dt$$

the energy of the signal and

$$\beta^2 = \frac{\int_{-\infty}^{\infty} [f'(t)]^2 \, dt}{\int_{-\infty}^{\infty} [f(t)]^2 \, dt}. \qquad (6\text{-}46)$$

For a low-pass signal, β is the root-mean-square (rms) bandwidth; in terms of the spectrum $F(\omega)$ of the signal, defined as the Fourier transform of $f(t)$,

$$\beta^2 = \frac{\int_{-\infty}^{\infty} \omega^2 |F(\omega)|^2 \frac{d\omega}{2\pi}}{\int_{-\infty}^{\infty} |F(\omega)|^2 \frac{d\omega}{2\pi}}. \qquad (6\text{-}47)$$

Thus the variance of the error in an estimate of the arrival time τ is inversely proportional to both the signal-to-noise ratio $d^2 = 2E/N$ and the square β^2 of the rms bandwidth of the signal.

6.2.3 Asymptotic Variance of Maximum-likelihood Estimators

Our result for the variance of the maximum-likelihood estimator of the arrival time τ in the limit of large signal-to-noise ratio is a special instance of a general formula due to Fisher [Fis22] for the asymptotic variance of maximum-likelihood estimators. We derive it for a finite set $v = (v_1, v_2, \ldots, v_n)$ of data, passing later to the limit of an infinite number of samples.

The maximum-likelihood estimator of a single parameter θ maximizes the joint probability density function $p(v|\theta)$ of the data or, what is the same thing, its logarithm

$$g(v|\theta) = \ln p(v|\theta). \qquad (6\text{-}48)$$

The estimate $\hat{\theta}$ is a zero of the function

$$\frac{\partial}{\partial \theta} g(v|\theta) = g_\theta(v|\theta), \qquad (6\text{-}49)$$

a subscript θ indicating differentiation with respect to θ. Expanding this function of θ in a Taylor series about the true value θ_0 of the estimandum, we must solve

$$g_\theta(v|\hat{\theta}) = g_\theta(v|\theta_0) + (\hat{\theta} - \theta_0)g_{\theta\theta}(v|\theta_0) + \cdots = 0. \qquad (6\text{-}50)$$

Again assuming that the signal-to-noise ratio is so large, and the estimate so accurate, that terms of higher order in $(\hat{\theta} - \theta_0)$ can be neglected, we find for the error

$$\hat{\theta} - \theta_0 \approx -\frac{g_\theta(v|\theta_0)}{g_{\theta\theta}(v|\theta_0)}. \qquad (6\text{-}51)$$

The numerator of (6-51) has expected value zero, as can be shown by differentiating the normalization condition

$$\int_{R_n} p(v|\theta)\, d^n v = 1$$

with respect to θ:

$$\frac{\partial}{\partial \theta} \int_{R_n} p(v|\theta)\, d^n v = \int_{R_n} p_\theta(v|\theta)\, d^n v$$

$$= \int_{R_n} g_\theta(v|\theta) p(v|\theta)\, d^n v = E[g_\theta(v|\theta)] = 0. \qquad (6\text{-}52)$$

Here E refers to an expectation with respect to the joint probability density function of the data v.

The expected value of the denominator of (6-51), on the other hand, does not in general vanish, for differentiating once more we find

$$\int_{R_n} g_{\theta\theta}(v|\theta) p(v|\theta)\, d^n v + \int_{R_n} g_\theta(v|\theta) p_\theta(v|\theta)\, d^n v$$

$$= \int_{R_n} g_{\theta\theta}(v|\theta) p(v|\theta)\, d^n v + \int_{R_n} [g_\theta(v|\theta)]^2 p(v|\theta)\, d^n v = 0,$$

whereupon

$$E[g_{\theta\theta}(v| \theta)] = -E\{[g_\theta(v| \theta)]^2\},\tag{6-53}$$

and $E\{[g_\theta(v| \theta)]^2\} \neq 0$ unless the probability density function $p(v| \theta)$ does not really depend on the parameter θ at all.

The numerator of (6-51), therefore, is proportional to the noise alone; its signal component must vanish. The denominator is of the form "signal plus noise." To terms of first order in the noise, the denominator of (6-51) can thus be replaced by its expected value,

$$\hat{\theta} - \theta_0 \approx -\frac{g_\theta(v| \theta)}{E[g_{\theta\theta}(v| \theta)]},\tag{6-54}$$

and the maximum-likelihood estimator $\hat{\theta}(v)$ is asymptotically unbiased:

$$E[\hat{\theta}(v)| \theta_0] \approx \theta_0.$$

The variance of this estimator is obtained by squaring (6-54), averaging, and using (6-53),

$$\text{Var } \hat{\theta}(v) \approx \frac{1}{E\{[g_\theta(v| \theta_0)]^2\}} = \frac{1}{\text{Var } g_\theta(v| \theta_0)},\tag{6-55}$$

and this is Fisher's formula.

If in this analysis we divide $p(v| \theta)$ by $p_0(v)$, the joint probability density function of the data when no signal is present, our result will be unchanged, for $p_0(v)$ is independent of the parameter θ. Then because in the limit of infinitely dense sampling of the input $v(t)$

$$\frac{p(v| \theta)}{p_0(v)} \longrightarrow \Lambda[v(t)| \theta],$$

the variance of a maximum-likelihood estimator $\hat{\theta}[v(t)]$ based on the input $v(t)$ to our receiver is

$$\text{Var } \hat{\theta}[v(t)] \approx \frac{1}{\text{Var } g_\theta[v(t)| \theta_0]}.\tag{6-56}$$

Here

$$g[v(t)| \theta] = \ln \Lambda[v(t)| \theta],$$

and $g_\theta[v(t)| \theta]$ is its partial derivative with respect to the estimandum θ. Into this we put the true value $\theta = \theta_0$ of the parameter θ. By virtue of (6-53) this variance can also be written as

$$\text{Var } \hat{\theta}[v(t)] \approx -\frac{1}{E\{g_{\theta\theta}[v(t)| \theta_0]\}},\tag{6-57}$$

which may sometimes be more easily evaluated. Keep in mind that these results are approximations valid when the signal-to-noise ratio is so large that the errors are on the average small.

The reader is invited to show that (6-56) and (6-57) reduce to (6-45) when the parameter θ is the arrival time τ and we assume that the amplitude A of the signal $Af(t - \tau)$ is known and not to be estimated.

When several parameters $\boldsymbol{\theta} = (\theta_1, \theta_2, \ldots, \theta_m)$ are to be estimated, their maximum-likelihood estimators are again asymptotically unbiased in the limit of large signal-to-noise ratio. The estimators may be correlated, and the generalized Fisher formula

$$\text{Cov}(\hat{\theta}_j, \hat{\theta}_k) = E[(\hat{\theta}_j - \theta_{0j})(\hat{\theta}_k - \theta_{0k})] = B_{jk} \approx (\Gamma^{-1})_{jk}, \qquad (6\text{-}58)$$

corresponding to (6-56), approximates their covariances. Here Γ is an $m \times m$ matrix whose elements are

$$\Gamma_{jk} = \text{Cov}\left\{\frac{\partial g}{\partial \theta_j}, \frac{\partial g}{\partial \theta_k}\right\} = -E\left[\frac{\partial^2 g}{\partial \theta_j \partial \theta_k}\right], \qquad g = \ln \Lambda[v(t)| \theta], \qquad (6\text{-}59)$$

into which one substitutes the true values of the estimanda θ. The equality of these two forms will be demonstrated subsequently. The matrix Γ is called the *Fisher information matrix*.

Denoting a partial derivative with respect to the jth parameter θ_j by a subscript θ_j, we write the counterpart of (6-50) as the multivariate Taylor expansion about the point θ_0,

$$g_{\theta_i}(v| \hat{\theta}) \approx g_{\theta_i}(v| \theta_0) + \sum_{k=1}^{m} (\hat{\theta}_k - \theta_{0k})g_{\theta_i \theta_k}(v| \theta_0) \approx 0, \qquad 1 \le i \le m.$$

This provides approximate equations for the estimates $\hat{\theta}_k$. At large signal-to-noise ratio, we can replace the random variables

$$g_{\theta_i \theta_k}(v| \theta_0)$$

by their expected values

$$E[g_{\theta_i \theta_k}(v| \theta_0)] = E\frac{\partial^2 g(v| \theta_0)}{\partial \theta_i \partial \theta_k} = -\Gamma_{ik}$$

as in (6-59). Solving the resulting simultaneous equations for the errors, we find

$$\hat{\theta}_i - \theta_{0i} \approx \sum_{k=1}^{m} (\Gamma^{-1})_{ik} g_{\theta_k}(v| \theta_0), \qquad 1 \le i \le m.$$

If we now introduce a column vector $\mathbf{T}$ whose elements are the errors

$$T_i = \hat{\theta}_i - \theta_{0i}$$

and a column vector $\mathbf{G}$ whose elements are the random variables

$$G_k = g_{\theta_k}(v| \theta_0),$$

we can write this set of equations in matrix form as

$$\mathbf{T} \approx \Gamma^{-1}\mathbf{G}. \qquad (6\text{-}60)$$

The covariance matrix of the errors is then

$$\mathbf{B} = E(\mathbf{TT}^T),$$

where superscript T indicates the transpose of a vector.

By twice differentiating the normalization integral for $p(v|\boldsymbol{\theta})$, we obtain

$$\frac{\partial}{\partial\theta_i}\int_{R_n}\frac{\partial}{\partial\theta_k}p(v|\boldsymbol{\theta})\,d^n v = \frac{\partial}{\partial\theta_i}\int_{R_n}g_{\theta_k}(v|\boldsymbol{\theta})p(v|\boldsymbol{\theta})\,d^n v$$

$$= \int_{R_n}g_{\theta_i\theta_k}(v|\boldsymbol{\theta})p(v|\boldsymbol{\theta})\,d^n v + \int_{R_n}g_{\theta_k}(v|\boldsymbol{\theta})g_{\theta_i}(v|\boldsymbol{\theta})p(v|\boldsymbol{\theta})\,d^n v$$

$$= E[g_{\theta_i\theta_k}(v|\boldsymbol{\theta})] + E[g_{\theta_i}(v|\boldsymbol{\theta})g_{\theta_k}(v|\boldsymbol{\theta})] = 0.$$

Hence

$$\Gamma_{ik} = E[g_{\theta_i}(v|\boldsymbol{\theta})g_{\theta_k}(v|\boldsymbol{\theta})] = \mathrm{Cov}\{g_{\theta_i}, g_{\theta_k}\},$$

as in (6-59), or in matrix notation

$$\boldsymbol{\Gamma} = E(\mathbf{GG}^T).$$

Thus from (6-60) we obtain

$$\mathbf{B} \approx E(\boldsymbol{\Gamma}^{-1}\mathbf{GG}^T\boldsymbol{\Gamma}^{-1}) = \boldsymbol{\Gamma}^{-1}\boldsymbol{\Gamma}\boldsymbol{\Gamma}^{-1} = \boldsymbol{\Gamma}^{-1},$$

which, after we again pass to the limit of infinitely dense sampling of the input $v(t)$, is the matrix form of (6-58).

We shall apply (6-58) in Sec. 6.3 to the simultaneous estimation of the arrival time and carrier frequency of a narrowband signal such as a radar echo.

6.2.4 The Cramér–Rao Inequality

The Fisher formulas (6-56) and (6-58) approximate the mean-square error and the covariance matrix of the maximum-likelihood estimators when the signal-to-noise ratio is large. A bound on the mean-square error of any estimator is provided by the Cramér–Rao inequality [Cra46, Sec. 32.3], [Rao45]. For any estimator $\hat{\theta}(v)$ satisfying certain easy conditions of good behavior, the mean-square error is subject to a lower bound

$$E[(\hat{\theta}(v) - \theta)^2] \geq \frac{(d\bar{\hat{\theta}}/d\theta)^2}{E\{[g_\theta(v|\theta)]^2\}}, \qquad (6\text{-}61)$$

where $g_\theta(v|\theta)$ is defined as in (6-49) and

$$\bar{\hat{\theta}}(\theta) = E[\hat{\theta}(v)|\theta] = \int_{R_n}\hat{\theta}(v)p(v|\theta)\,d^n v. \qquad (6\text{-}62)$$

The numerator in (6-61) equals 1 if $\hat{\theta}(v)$ is an unbiased estimator.

We demonstrate (6-61) as follows. From (6-62) we obtain by differentiation

$$\frac{d\bar{\hat{\theta}}}{d\theta} = \int_{R_n}\hat{\theta}(v)p_\theta(v|\theta)\,d^n v = \int_{R_n}\hat{\theta}(v)g_\theta(v|\theta)p(v|\theta)\,d^n v.$$

Multiplying (6-52) by θ and subtracting from this, we find

$$\frac{d\bar{\hat{\theta}}}{d\theta} = \int_{R_n}[\hat{\theta}(v) - \theta]g_\theta(v|\theta)p(v|\theta)\,d^n v = E\{[\hat{\theta}(v) - \theta]g_\theta(v|\theta)\}.$$

The Schwarz inequality for expectations (4-73) states that for any random variables A and B

$$[E(AB)]^2 \leq E(A^2)E(B^2).$$

Taking $A = \hat{\theta}(v) - \theta$ and $B = g_\theta(v|\theta)$, we obtain

$$\left(\frac{d\bar{\hat{\theta}}}{d\theta}\right)^2 \leq E\{[\hat{\theta}(v) - \theta]^2\}E\{[g_\theta(v|\theta)]^2\},$$

whence (6-61) by division.

As an example consider the estimation of the mean m of n independent Gaussian random variables $v = (v_1, v_2, \ldots, v_n)$, treated in Sec. 6.1.3. From (6-6)

$$g(v|m) = \ln p(v|m) = -\frac{1}{2\delta^2}\sum_{k=1}^{n}(v_k - m)^2 - \frac{n}{2}\ln(2\pi\delta^2).$$

Here the logarithmic derivative is

$$g_m(v|m) = \frac{n(X - m)}{\delta^2}, \tag{6-63}$$

where X is the sample mean defined in (6-9), and the denominator of (6-61) is

$$E\{[g_m(v|m)]^2\} = \text{Var } g_m(v|m) = \left(\frac{n}{\delta^2}\right)^2 \text{Var } X = \frac{n}{\delta^2}.$$

One estimator of the mean m is the biased one in (6-10), and for it, by (6-11),

$$\frac{d\bar{m}}{dm} = \frac{\beta^2}{\beta^2 + \delta^2/n}.$$

The Cramér–Rao inequality (6-61) then states that

$$E\{[\bar{m}(v) - m]^2\} \geq \frac{\delta^2}{n}\left[\frac{\beta^2}{\beta^2 + \delta^2/n}\right]^2.$$

The left side is the risk $R[\hat{m}(v)|m]$ given in (6-31), and it is indeed greater than the right side except when $m = \mu$, whereupon the two sides are equal. For the maximum-likelihood estimator $\hat{m}(v) = X$, on the other hand, both sides of (6-61) are equal to δ^2/n for all values of m, and the Cramér–Rao inequality becomes an equality.

When for a particular estimator $\hat{\theta}(v)$ the two sides of the Cramér–Rao inequality (6-61) are equal, the estimator is said to be *efficient*. The Schwarz inequality (4-73) is an equality if and only if the random variables A and B are proportional, $A = kB$, k nonrandom. The condition for an efficient estimator is therefore that

$$g_\theta(v|\theta) = \frac{\partial}{\partial\theta}[\ln p(v|\theta)] = k(\theta)[\hat{\theta}(v) - \theta], \tag{6-64}$$

in which $k(\theta)$ may depend on the estimandum θ, but not on the data v.

An efficient estimator is unbiased, for upon taking the expected value of (6-64) with respect to the distribution of the data, we find by (6-52) that

$$E[\hat{\theta}(v) - \theta] = 0.$$

For this estimator, when we square (6-64) and average,

$$E\{[g_\theta(v|\theta)]^2\} = [k(\theta)]^2 E\{[\hat{\theta}(v) - \theta]^2\} = \frac{[k(\theta)]^2}{E\{[g_\theta(v|\theta)]^2\}},$$

whence the mean-square error of the efficient estimator is

$$E\{[\hat{\theta}(v) - \theta]^2\} = \frac{1}{k(\theta)}. \tag{6-65}$$

Integrating (6-64) with respect to θ, we find

$$\ln p(v|\,\theta) = \int_{\theta_1}^{\theta} k(\theta')[\hat{\theta}(v) - \theta']\,d\theta' + \ln r(v),$$

in which $\ln r(v)$, independent of θ, serves as a constant of integration. Hence the probability density function of the data must have the form

$$p(v|\,\theta) = r(v)H(\hat{\theta}(v); \theta).$$

Any Bayes estimator of the parameter θ will therefore depend on the data v only through the function $\hat{\theta}(v)$, which is therefore a sufficient statistic for estimating the parameter θ. An efficient estimator is always sufficient, but the opposite may not be true.

For the efficient estimator $\hat{m}(v) = X$ of the mean m of n independent Gaussian random variables, $k(\theta) = n/\delta^2$, as appears by comparing (6-63) and (6-64); and (6-65) then confirms that its variance is δ^2/n. Problem 6-2 directs you to show that the maximum-likelihood estimator of the variance of n independent Gaussian random variables with expected value 0 is also efficient. On the whole, however, efficient estimators are rare.

The Cramér–Rao inequality can be expressed in terms of the input $v(t)$ to the receiver by again dividing the joint probability density function $p(v|\,\theta)$ of the data by the joint density function $p_0(v)$ of the data when noise alone is present, and it becomes

$$E\{[\hat{\theta}[v(t)] - \theta]^2\} \geq \frac{(d\overline{\hat{\theta}}/d\theta)^2}{\mathrm{Var}\, g_\theta[v(t)|\,\theta]} \tag{6-66}$$

with $g_\theta[v(t)|\,\theta] = \frac{\partial}{\partial\theta} \ln \Lambda[v(t)|\,\theta]$ as before.

We learned in Sec. 6.2.3 that when the signal-to-noise ratio is large, the maximum-likelihood estimator is asymptotically unbiased, and the numerator of (6-66) becomes equal to 1. Comparing (6-66) and (6-56) we see that the right sides are then the same, and in this limit the Cramér–Rao inequality (6-66) becomes an equality. We say, therefore, that the maximum-likelihood estimator is *asymptotically efficient*.

For a set of m unbiased estimators $\hat{\theta}_i(v)$, $1 \leq i \leq m$, of the parameters $\boldsymbol{\theta} = (\theta_1, \theta_2, \ldots, \theta_m)$, there exists a matrix counterpart of the Cramér–Rao inequality. The elements of the covariance matrix $\mathbf{B}$ of the errors are

$$B_{ij} = E\{[\hat{\theta}_i(v) - \theta_{0i}][\hat{\theta}_j(v) - \theta_{0j}]\} = \mathrm{Cov}\{\hat{\theta}_i(v), \hat{\theta}_j(v)\}.$$

Then

$$\mathbf{B} \geq \boldsymbol{\Gamma}^{-1}, \tag{6-67}$$

where $\boldsymbol{\Gamma}$ is the Fisher information matrix defined in (6-59). The matrix inequality (6-67) means that the matrix $\mathbf{B} - \boldsymbol{\Gamma}^{-1}$ is nonnegative definite. That is, for any column vector $\mathbf{Y}$ of coefficients

$$(y_1, y_2, \ldots, y_m) = \mathbf{Y}^T,$$

$$\mathbf{Y}^T \mathbf{B} \mathbf{Y} \geq \mathbf{Y}^T \mathbf{\Gamma}^{-1} \mathbf{Y}$$

or

$$\sum_{i=1}^{m} \sum_{j=1}^{m} B_{ij} y_i y_j \geq \sum_{i=1}^{m} \sum_{j=1}^{m} (\mathbf{\Gamma}^{-1})_{ij} y_i y_j.$$

Taking $y_k = 1$, $y_i \equiv 0$, $i \neq k$, for instance, we find that the variance of any unbiased estimator of the parameter θ_k is bounded below by

$$\text{Var } \hat{\theta}_k(v) \geq (\mathbf{\Gamma}^{-1})_{kk}, \qquad 1 \leq k \leq m,$$

where $(\mathbf{\Gamma}^{-1})_{kk}$ is the kth diagonal element of the inverse $\mathbf{\Gamma}^{-1}$ of the Fisher information matrix.

The concentration ellipsoid of a set of unbiased estimators is defined by the equation

$$\mathbf{Z}^T \mathbf{B}^{-1} \mathbf{Z} = \sum_{i=1}^{m} \sum_{j=1}^{m} (\mathbf{B}^{-1})_{ij} z_i z_j \equiv c, \qquad (6\text{-}68)$$

where c is any positive constant. It is stated in [Bel60, p. 59] that (6-67) implies $\mathbf{B}^{-1} \leq \mathbf{\Gamma}$, and this form of the Cramér–Rao inequality tells that the concentration ellipsoid lies outside the ellipsoid

$$\mathbf{Z}^T \mathbf{\Gamma} \mathbf{Z} = \sum_{i=1}^{m} \sum_{j=1}^{m} \Gamma_{ij} z_i z_j \equiv c \qquad (6\text{-}69)$$

for the same value of c. If the errors $z_i = \hat{\theta}_i - \theta_{0i}$ were Gaussian random variables, (6-68) would specify an ellipsoidal surface on which their joint probability density function is constant. The farther from the origin this surface lies, the more broadly dispersed are the estimates; and the ellipsoid in (6-69) sets a bound to how concentrated about the true values that joint density function could be.

6.2.5 Estimation of Signal Parameters in Gaussian Noise

If the signal $s(t; \mathbf{\theta})$ is received in the presence of colored Gaussian noise $n(t)$ with expected value zero and covariance function $\phi(t, s)$, the logarithm of the likelihood functional figuring in all these formulas is given by (2-71) as

$$g[v(t)| \mathbf{\theta}] = \ln \Lambda[v(t)| \mathbf{\theta}]$$

$$= \int_0^T q(t; \mathbf{\theta})v(t)\, dt - \tfrac{1}{2}\int_0^T s(t; \mathbf{\theta})q(t; \mathbf{\theta})\, dt,$$

where $q(t; \mathbf{\theta})$ is the solution of the integral equation

$$s(t; \mathbf{\theta}) = \int_0^T \phi(t, u)q(u; \mathbf{\theta})\, du, \qquad 0 \leq t \leq T, \qquad (6\text{-}70)$$

which corresponds to (2-66). Then

$$g_{\theta_i}[v(t)| \mathbf{\theta}] = \int_0^T \frac{\partial q(t; \mathbf{\theta})}{\partial \theta_i} v(t)\, dt + \text{terms independent of } v(t).$$

The covariance of a linear functional of the input $v(t)$ depends only on the noise component $n(t)$, the signal component eventually cancelling out, and by (6-59) we therefore find for the elements of the Fisher information matrix

$$\Gamma_{ij} = \text{Cov}\{g_{\theta_i}, g_{\theta_j}\} = E \int_0^T \int_0^T \frac{\partial q(u; \boldsymbol{\theta})}{\partial \theta_i} \frac{\partial q(t; \boldsymbol{\theta})}{\partial \theta_j} [n(t)n(u)] \, dt \, du.$$

By the definition of the autocovariance function this is

$$\Gamma_{ij} = \int_0^T \int_0^T \frac{\partial q(u; \boldsymbol{\theta})}{\partial \theta_i} \frac{\partial q(t; \boldsymbol{\theta})}{\partial \theta_j} \phi(t, u) \, dt \, du = \int_0^T \frac{\partial s(t; \boldsymbol{\theta})}{\partial \theta_i} \frac{\partial q(t; \boldsymbol{\theta})}{\partial \theta_j} \, dt,$$

in reducing which we have used the derivative of (6-70) with respect to θ_i.

This matrix element Γ_{ij} is conveniently written as

$$\Gamma_{ij} = \frac{\partial^2 H(\boldsymbol{\theta}_1, \boldsymbol{\theta}_2)}{\partial \theta_{1i} \partial \theta_{2j}} \bigg|_{\boldsymbol{\theta}_1 = \boldsymbol{\theta}_2 = \boldsymbol{\theta}}, \tag{6-71}$$

in which

$$H(\boldsymbol{\theta}_1, \boldsymbol{\theta}_2) = \int_0^T s(t; \boldsymbol{\theta}_1) q(t; \boldsymbol{\theta}_2) \, dt$$

is the *generalized ambiguity function* for the signal $s(t; \boldsymbol{\theta})$. We have introduced two sets of parameters

$$\boldsymbol{\theta}_p = (\theta_{p1}, \theta_{p2}, \ldots, \theta_{pm}), \qquad p = 1, 2,$$

and in (6-71) we differentiate $H(\boldsymbol{\theta}_1, \boldsymbol{\theta}_2)$ with respect to the components of $\boldsymbol{\theta}_1$ and $\boldsymbol{\theta}_2$ separately, afterward setting both sets $\boldsymbol{\theta}_1$ and $\boldsymbol{\theta}_2$ of parameters equal to the true values of the parameters $\boldsymbol{\theta}$. When the noise is white, this ambiguity function is

$$H(\boldsymbol{\theta}_1, \boldsymbol{\theta}_2) = \frac{2}{N} \int_0^T s(t; \boldsymbol{\theta}_1) s(t; \boldsymbol{\theta}_2) \, dt$$

in terms of the unilateral spectral density N of the noise.

When the signals are narrowband modulations of a carrier of angular frequency Ω, we can use the likelihood functional as written in (3-54) to express the generalized ambiguity function as

$$H(\boldsymbol{\theta}_1, \boldsymbol{\theta}_2) = \text{Re} \int_0^T S^*(t; \boldsymbol{\theta}_1) Q(t; \boldsymbol{\theta}_2) \, dt, \tag{6-72}$$

in which $S(t; \boldsymbol{\theta})$ is the complex envelope of the signal and $Q(t; \boldsymbol{\theta})$ is the solution of the integral equation

$$S(t; \boldsymbol{\theta}) = \int_0^T \tilde{\phi}(t, u) Q(u; \boldsymbol{\theta}) \, du, \qquad 0 \leq t \leq T,$$

whose kernel is the complex autocovariance function of the noise. For white noise, by (3-45),

$$H(\boldsymbol{\theta}_1, \boldsymbol{\theta}_2) = \frac{1}{N} \text{Re} \int_0^T S^*(t; \boldsymbol{\theta}_1) S(t; \boldsymbol{\theta}_2) \, dt. \tag{6-73}$$

The elements Γ_{ij} of the Fisher information matrix are again given by (6-71) in terms of this symmetric function $H(\boldsymbol{\theta}_1, \boldsymbol{\theta}_2)$. By utilizing (6-71) the computation of the covariances of maximum-likelihood estimators of arrival time and carrier frequency of a narrowband pulse signal is much simplified, as we shall see in Sec. 6.3.

6.2.6 The Ziv–Zakai Bound

For an unbiased estimator of the arrival time τ of a signal $s(t - \tau)$, the Cramér–Rao inequality (6-66) becomes

$$E[(\hat{\tau} - \tau)^2] \geq \frac{1}{d^2\beta^2}, \tag{6-74}$$

where $d^2 = 2E/N$ is the signal-to-noise ratio and β is the rms bandwidth defined in (6-46) and (6-47). This inequality follows from the identity of the right sides of (6-45), (6-55), and (6-66). For a rectangular signal of duration T',

$$s(t) = \begin{cases} A, & 0 \leq t < T', \\ 0, & t < 0, \quad t \geq T', \end{cases} \tag{6-75}$$

however, $\beta = \infty$, and (6-74) becomes the trivial inequality $E[(\hat{\tau} - \tau)^2] \geq 0$.

A more useful lower bound on the mean-square error of an estimate of signal arrival time τ was discovered by Ziv and Zakai [Ziv69]. We shall briefly sketch their derivation. Consider a communication system sending equally likely binary digits 0 and 1 every T seconds by transmitting either the signal $s(t - \tau_1)$ (hypothesis H_1) or the signal $s(t - \tau_2)$ (hypothesis H_2); $\tau_2 > \tau_1$. Each signal is assumed to arrive well within the observation interval $(0, T)$. The optimum receiver, according to what we learned in Chapter 2—see Problem 2-7—passes its input $v(t)$ through a filter matched to the signal $s(t - \tau_2) - s(t - \tau_1)$ and samples its output at time $t = T$. If that output is positive, it chooses hypothesis H_2; if negative, hypothesis H_1. The solution to Problem 2-8 informs us that the minimum attainable probability of error is

$$P_{e,\min} = \operatorname{erfc} d\sqrt{(1 - \lambda)/2}, \qquad d^2 = \frac{2E}{N}, \tag{6-76}$$

where E is the energy of each signal, N the unilateral spectral density of the noise, and

$$\lambda = \lambda(\tau_2 - \tau_1) = \frac{1}{E} \int_{-\infty}^{\infty} s(t - \tau_1)s(t - \tau_2)\,dt. \tag{6-77}$$

We neglect the possibility that either signal overlaps one end of the observation interval or the other; $T \gg T'$.

An alternative and inferior detector estimates the arrival time τ of the received signal and chooses hypothesis H_1 if $\hat{\tau} < \frac{1}{2}(\tau_1 + \tau_2)$ and hypothesis H_2 if $\hat{\tau} \geq \frac{1}{2}(\tau_1 + \tau_2)$. Its probability of error will be

$$\tfrac{1}{2} \Pr[\hat{\tau} \geq \tfrac{1}{2}(\tau_1 + \tau_2)|\, H_1] + \tfrac{1}{2} \Pr[\hat{\tau} < \tfrac{1}{2}(\tau_1 + \tau_2)|\, H_2],$$

and this must be greater than $P_{e,\min}$. Therefore, with $\delta = \tau_2 - \tau_1 > 0$,

$$\tfrac{1}{2} \Pr(\hat{\tau} - \tau_1 \geq \tfrac{1}{2}\delta|\, H_1) + \tfrac{1}{2} \Pr(\tau_2 - \hat{\tau} > \tfrac{1}{2}\delta|\, H_2) \geq P_{e,\min}.$$

Because

$$\Pr(\hat{\tau} - \tau_1 \geq \tfrac{1}{2}\delta|\, H_1) \leq \Pr(|\hat{\tau} - \tau_1| \geq \tfrac{1}{2}\delta|\, H_1),$$

$$\Pr(\tau_2 - \hat{\tau} > \tfrac{1}{2}\delta|\, H_2) \leq \Pr(|\hat{\tau} - \tau_2| > \tfrac{1}{2}\delta|\, H_2),$$

we can replace that inequality by

$$\tfrac{1}{2} \Pr(|\hat\tau - \tau_1| \geq \tfrac{1}{2}\delta|\ H_1) + \tfrac{1}{2} \Pr(|\hat\tau - \tau_2| > \tfrac{1}{2}\delta|\ H_2) \geq P_{e,\,\min}. \qquad (6\text{-}78)$$

At this point we need the Chebyshev inequality, which for any random variable x having finite variance and for any a and $g > 0$ states that

$$\Pr(|x - a| > g) \leq \frac{E[(x - a)^2]}{g^2}. \qquad (6\text{-}79)$$

To prove it, we write, in terms of the probability density function $p(\cdot)$ of x,

$$
\begin{aligned}
E[(x - a)^2] &= \int_{-\infty}^{\infty} (x - a)^2 p(x)\, dx \\
&\geq \int_{-\infty}^{a-g} (x - a)^2 p(x)\, dx + \int_{a+g}^{\infty} (x - a)^2 p(x)\, dx \\
&\geq g^2 \int_{-\infty}^{a-g} p(x)\, dx + g^2 \int_{a+g}^{\infty} p(x)\, dx = g^2 \Pr(|x - a| > g),
\end{aligned}
$$

and we divide by g^2.

Applying (6-79) to (6-78), putting $g = \tfrac{1}{2}\delta$, and using (6-76), we find the inequality

$$\tfrac{1}{2} E[(\hat\tau - \tau)^2|\ \tau = \tau_1] + \tfrac{1}{2} E[(\hat\tau - \tau)^2|\ \tau = \tau_2] \geq \tfrac{1}{4}\delta^2 \operatorname{erfc} d\sqrt{\frac{1 - \lambda(\delta)}{2}}.$$

If we define ε^2 as the maximum value of $E[(\hat\tau - \tau)^2|\ \tau]$ for any true value of the arrival time in $0 \leq \tau \leq T$, then

$$\varepsilon^2 \geq \tfrac{1}{4}\delta^2 \operatorname{erfc} d\sqrt{\frac{1 - \lambda(\delta)}{2}}. \qquad (6\text{-}80)$$

The tightest lower bound is obtained by maximizing the right side as the separation δ varies over $0 \leq \delta \leq T$.

If the signal vanishes outside an interval $(0, T')$, $\lambda(\delta) \equiv 0$ for $\delta \geq T'$ by (6-77), and the right side of (6-80) is largest for $\delta = T$, whence

$$\varepsilon^2 \geq \tfrac{1}{4} T^2 \operatorname{erfc} \frac{d}{\sqrt{2}}. \qquad (6\text{-}81)$$

In [Ziv69] this is called the *external bound*.

Specializing now to a rectangular signal of duration T' as in (6-75), we find from (6-77) that

$$\lambda(\delta) = \begin{cases} 1 - \dfrac{|\delta|}{T'}, & -T' \leq \delta \leq T' \\ 0, & |\delta| > T', \end{cases}$$

and (6-80) becomes

$$\varepsilon^2 \geq \tfrac{1}{4}\delta^2 \operatorname{erfc}\left[d\left(\frac{|\delta|}{2T'}\right)^{1/2} \right] = \frac{T'^2}{d^4} b^4 \operatorname{erfc} b, \qquad 0 \leq \delta \leq T',$$

with b the argument of erfc $(\cdot)$. The right side is maximum for $b = 1.812$, and we find

$$\varepsilon^2 \geq 0.3772 \frac{T'^2}{d^4} \qquad (6\text{-}82)$$

provided $d^2 = 2E/N \geq 2b^2 = 6.566$. This is called the *internal bound*. For large signal-to-noise ratio d^2, this lower bound decreases more rapidly with increasing signal-to-noise ratio than does the right side of the Cramér–Rao inequality (6-74) for signals with finite bandwidth β. For $d^2 = 2E/N < 6.566$, one uses the external bound (6-81), which depends on the duration T of the observation interval.

The change from one type of bound to the other when the signal-to-noise ratio d^2 passes a certain limit exemplifies what is called the *threshold effect*. At large signal-to-noise ratios the main source of error is the displacement by the noise of the highest peak of the output of the matched filter. When the signal-to-noise ratio is low, one of the many peaks of that output may surpass the height of the peak occurring near the time $\tau_0 + T'$ when the output would be maximum in the absence of noise. The result is an error that may be a considerable fraction of the duration T of the observation interval. This is why the duration T appears in the bound (6-81).

For signals of finite bandwidth β, [Ziv69] presents a lower bound of the same form as in (6-74), except with a different numerical coefficient:

$$\varepsilon^2 \geq \frac{0.16}{d^2 \beta^2}, \qquad d^2 \geq 2.88.$$

Refinements involving the use of other probability bounds than the Chebyshev, but requiring additional assumptions about the probability density function of the estimator $\hat{\tau}[v(t)]$, lead to similar forms for the lower bound on the mean-square estimation error, but with larger multiplicative factors. Remember that (6-81) and (6-82) are only *lower bounds* to the mean-square error $E[(\hat{\tau} - \tau_0)^2]$, and resist the temptation to regard the right side of either as an approximation to the actual mean-square error in an estimate of the arrival time τ.

6.3 ESTIMATION OF PARAMETERS OF A NARROWBAND SIGNAL

6.3.1 Arrival Time

The maximum-likelihood estimation of the arrival time τ of a pulse signal $Af(t - \tau)$ was shown in Sec. 6.2.2 to require timing the peak value of the rectified output $|v_0(t)|$ of a filter matched to the signal $f(t)$. The mean-square error of the resulting estimate was given by (6-45) as

$$\text{Var } \hat{\tau} \approx \frac{1}{d^2 \beta^2}$$

in the limit of large signal-to-noise ratio $d^2 = 2E/N$, with β the rms bandwidth of the signal as defined in (6-46) and (6-47).

In radar the transmitted and received signals are narrowband modulations of a high-frequency carrier, and the delayed echo has the form

$$s(t; \tau) = A \operatorname{Re}[F(t - \tau) \exp i\Omega(t - \tau)],$$

in which $F(t)$ is the complex envelope. For a signal of this kind, as can be seen from (6-47), the bandwidth β is roughly equal to the carrier frequency Ω, which is much larger than the bandwidth $\Delta\omega$ of the envelope $F(t)$. Then (6-45) becomes

$$\operatorname{Var} \hat{\tau} \approx \frac{1}{d^2\Omega^2}, \tag{6-83}$$

which asserts that when the signal-to-noise ratio d^2 is large, the arrival time τ of the radar echo can be measured within a fraction of a period of its carrier. Because the range r of the target equals $2c\tau$, where c is the velocity of electromagnetic radiation, this implies that

$$\operatorname{Var} \hat{r} = 4c^2 \operatorname{Var} \hat{\tau} \approx \frac{4c^2}{d^2\Omega^2} = \frac{\lambda^2}{\pi^2 d^2},$$

where $\lambda = 2\pi c/\Omega$ is the wavelength of the radiation; and when $d^2 \gg 1$, the range can be estimated within a fraction of that wavelength.

To attain such accuracy the oscillations of frequency 2Ω in the squared output $[v_0(t)]^2$ of the matched filter must be observed closely in order to ascertain which is highest. Only when the signal is very strong will this be possible without ambiguity. With weaker signals, even though the output stands well above the noise level, the noise may give a neighboring cycle of those oscillations a greater excursion from the zero line, and the error in the estimate will be some multiple of their period π/Ω. If the measurement is repeated several times, the observed maximum will jump erratically from one cycle to another in the vicinity of the time $\tau_0 + T'$.

When this is happening, knowledge of the phase of the transmitted pulse is no longer of any use; and even if that phase were uncontrolled, the estimate of the arrival time would not be much altered. This estimate and its accuracy will be nearly the same as for the time of arrival of a quasiharmonic signal

$$s(t; A, \psi, \tau) = A \operatorname{Re} F(t - \tau) e^{i\Omega t + i\psi}$$

of unknown phase ψ.

Because the phase of the signal is unknown, the receiver may as well rectify the output of a filter matched to the signal

$$\operatorname{Re} F(t) e^{i\Omega t}$$

and filter off the components of frequency 2Ω in the output of the rectifier, to produce

$$R(t) = \left| \int_0^{T'} F^*(T' - s)V(t - s) \, ds \right|^2,$$

where $V(t)$ is the complex envelope of the input $\operatorname{Re} V(t) \exp i\Omega t$,

$$V(t) = AF(t - \tau) e^{i\psi} + N(t);$$

T' is as before the delay in the matched filter, and $N(t)$ is the complex envelope of the noise.

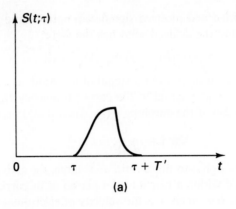

(a)

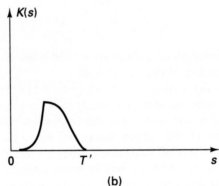

(b)

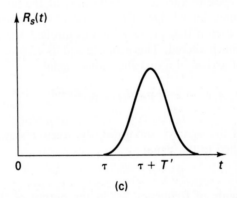

(c)

Figure 6-1. Estimation of target range. (a) Complex envelope of signal arriving at time τ. (b) Complex impulse response of matched filter. (c) Signal component of rectified output of matched filter.

We shall see in the next part that timing the peak value of the function $R(t)$ indeed yields the maximum-likelihood estimate of the arrival time τ of the envelope of this signal. Figure 6-1 illustrates the complex envelope of the delayed signal pulse, the complex impulse response of the matched filter, and the signal component $R_s(t)$ of the output $R(t)$ of the rectifier, that is, the form that output would take, were no noise present. The noise displaces the peak of the rectified output $R(t)$ from that of $R_s(t)$ and thus introduces a random error into the estimate of the arrival time τ.

It is plausible that as will be demonstrated in Sec. 6.3.4, in the limit of large signal-to-noise ratio the variance of the maximum-likelihood estimator of the arrival time τ is expressed as in (6-45), except that the bandwidth β must be replaced by the rms bandwidth $\Delta\omega$ of the complex envelope of the signal,

$$\text{Var } \hat{\tau} \approx \frac{1}{d^2\Delta\omega^2}. \tag{6-84}$$

The mean-square frequency deviation $\Delta\omega^2$ is defined by

$$\Delta\omega^2 = \frac{\int_{-\infty}^{\infty}(\omega - \bar{\omega})^2|f(\omega)|^2\frac{d\omega}{2\pi}}{\int_{-\infty}^{\infty}|f(\omega)|^2\frac{d\omega}{2\pi}} \tag{6-85}$$

with $f(\omega)$ the Fourier transform of the complex envelope $F(t)$ of the signal, as in (3-7), and $\bar{\omega}$ the mean deviation from the carrier frequency,

$$\bar{\omega} = \frac{\int_{-\infty}^{\infty}\omega|f(\omega)|^2\frac{d\omega}{2\pi}}{\int_{-\infty}^{\infty}|f(\omega)|^2\frac{d\omega}{2\pi}}. \tag{6-86}$$

Swerling [Swe59], [Swe64] showed that the arrival time can be determined with a variance given by (6-83) when the signal-to-noise ratio is so large that $d \gg \Omega/\Delta\omega$. When $1 \ll d \ll \Omega/\Delta\omega$, however, the variance is limited by $1/d^2\Delta\omega^2$ as in (6-84). It is not so hard to see why the dividing line between the ranges of validity of the two formulas should occur at a signal-to-noise ratio on the order of $\Omega/\Delta\omega$. In order to measure the time τ within a fraction of a period of the carrier, it must be possible to locate the peak of the envelope of the output of the matched filter with an error somewhat smaller than π/Ω. Hence the signal-to-noise ratio must, by (6-84), be large enough that

$$(\text{Var } \hat{\tau})^{1/2} \approx \frac{1}{d\Delta\omega} \ll \frac{\pi}{\Omega},$$

from which we obtain the condition $d \gg \Omega/(\pi\Delta\omega)$.

The system just derived closely resembles a conventional radar range-measuring device. The intermediate-frequency amplifier of the radar corresponds to the matched filter, although it may not be precisely matched to the signal, but may merely have an approximately equal bandwidth. The target range is measured by timing the peak of the rectifier output as displayed on an A-scope. This output has many peaks due to the noise, but whenever the signal-to-noise ratio suffices for practically certain detection, one peak caused by the signal stands out above the rest. The noise displaces the highest point of this peak by an amount whose mean-square magnitude is given approximately by the variance $\text{Var } \hat{\tau}$ in (6-84). The greater the bandwidth $\Delta\omega$ of the signal, the more accurately the time of arrival τ can be estimated.

6.3.2 Signal Arrival Time and Carrier Frequency

If the radar target is not stationary, as assumed in Sec. 6.3.1, but is moving toward or away from the antenna, the carrier frequency Ω of the echo differs from that of the transmitted pulse because of the Doppler effect. If we pick the transmitted frequency as our reference frequency Ω_r, the echo will have the form

$$s(t; \tau, w) = A \text{ Re } F(t - \tau) \exp[i(w + \Omega_r)(t - \tau) + i\psi], \tag{6-87}$$

where A is its amplitude, ψ its phase, τ its epoch, and $w = \Omega - \Omega_r$ the change in its carrier frequency. This *Doppler shift* w is given by

$$w = \Omega - \Omega_r = \frac{2v}{c}\Omega_r, \qquad (6\text{-}88)$$

where v is the component of target velocity in the direction of the radar antenna and c is the velocity of electromagnetic radiation. A derivation of this formula, which holds when $v \ll c$, can be found in Appendix F.

The pulse envelope $F(t)$ is also compressed by a factor $(1 + 2v/c)^{-1}$, but for targets of ordinary velocities this factor differs negligibly from 1. For a target moving at a rate of 500 mph and for a carrier frequency $\Omega_r = 2\pi \cdot 3 \cdot 10^9$ rad/sec (3000 MHz), the frequency shift is $w = 2\pi \cdot 4.5 \cdot 10^3$ rad/sec (4500 Hz), an appreciable fraction of the 1-MHz bandwidth typical of a radar pulse. For the much larger velocities encountered in tracking missiles and satellites, the Doppler shift will be even greater.

When the echo signal suffers a Doppler shift comparable with its bandwidth, the response to it of a filter matched to the transmitted pulse is much reduced, and the signal may be missed in the noise. Only if the target velocity is known can the receiver filter be properly matched to the echo pulse. If the Doppler shift can be measured, on the other hand, the observer can calculate the component of the target velocity in the direction of the radar antenna, obtaining valuable information for tracking the target efficiently. The possibility thus arises of measuring both the distance of the target and its velocity by estimating the time τ of arrival of the echo signal and the frequency shift w of its carrier.

As the time τ of arrival is unknown, we can combine the term $-\Omega_r\tau$ in (6-87) with the unknown phase ψ and write the signal instead as

$$s(t; A, \psi, \tau, w) = A \operatorname{Re} F(t; \tau, w) \exp[i\Omega_r t + i\psi],$$

with

$$F(t; \tau, w) = F(t - \tau)\, e^{iw(t-\tau)}. \qquad (6\text{-}89)$$

This signal is received in the presence of white, Gaussian noise of unilateral spectral density N, during an observation interval $(0, T)$ that is so long that signals overlapping its ends can be disregarded.

Let us designate by $\boldsymbol{\theta}'$ those parameters other than the amplitude A and the phase ψ; in this problem $\boldsymbol{\theta}' = (\tau, w)$. About the amplitude A we have no prior information, and the phase ψ can be assumed uniformly distributed over $(0, 2\pi)$. Equivalently, with

$$A\, e^{i\psi} = u + iv,$$

the components of the complex amplitude $u + iv$ can be taken as independently random. Although we have no interest in their values, we estimate them anyhow, avoiding the necessity of postulating a prior density function for them.

For the complex envelope of the signal we put

$$S(t; \boldsymbol{\theta}) = (u + iv)\tilde{S}(t; \boldsymbol{\theta}'), \qquad \boldsymbol{\theta}' = (\tau, w).$$

The logarithm of the likelihood functional for detecting this signal in white noise is

$$\ln \Lambda[v(t)|\,\boldsymbol{\theta}] = \text{Re}\left[\frac{u - iv}{N} \int_0^T \tilde{S}^*(t; \boldsymbol{\theta}')V(t)\, dt\right] - \frac{u^2 + v^2}{2N} \int_0^T |\tilde{S}(t; \boldsymbol{\theta}')|^2\, dt \qquad (6\text{-}90)$$

by (3-54) with $Q(t; \boldsymbol{\theta}) = N^{-1}S(t; \boldsymbol{\theta})$ from (3-52) and (3-45). The maximum-likelihood estimates of the parameters u, v, and $\boldsymbol{\theta}' = (\tau, w)$ are those values for which (6-90) is maximum. In terms of the random variables x and y defined by

$$z = x + iy = \frac{1}{N} \int_0^T \tilde{S}^*(t; \boldsymbol{\theta}')V(t)\, dt$$

and the *reduced ambiguity function*

$$H'(\boldsymbol{\theta}_1', \boldsymbol{\theta}_2') = \frac{1}{N}\, \text{Re} \int_0^T \tilde{S}^*(t; \boldsymbol{\theta}_1')\tilde{S}(t; \boldsymbol{\theta}_2')\, dt,$$

we can write (6-90) as

$$\ln \Lambda[v(t)|\,\boldsymbol{\theta}] = ux + vy - \tfrac{1}{2}(u^2 + v^2)H'(\boldsymbol{\theta}', \boldsymbol{\theta}'). \qquad (6\text{-}91)$$

Differentiating first with respect to u and v and setting the results equal to zero, we find for the maximum-likelihood estimates $\hat{u}$ and $\hat{v}$ of these components of the complex amplitude

$$\hat{u} + i\hat{v} = \hat{A}e^{i\hat{\psi}} = \frac{z}{H'(\boldsymbol{\theta}', \boldsymbol{\theta}')} = \frac{1}{NH'(\boldsymbol{\theta}', \boldsymbol{\theta}')} \int_0^T \tilde{S}^*(t; \boldsymbol{\theta}')V(t)\, dt, \qquad (6\text{-}92)$$

into which we must substitute the maximum-likelihood estimates of the parameters $\boldsymbol{\theta}'$ yet to be determined. Putting (6-92) into (6-91), we see that these are the values of $\boldsymbol{\theta}'$ for which

$$\max_{u,v} \ln \Lambda[v(t)|\,\boldsymbol{\theta}] = \frac{1}{2N^2 H'(\boldsymbol{\theta}', \boldsymbol{\theta}')} \left|\int_0^T \tilde{S}^*(t; \boldsymbol{\theta}')V(t)\, dt\right|^2$$

is maximum.

In principle the maximum-likelihood estimates $\hat{\boldsymbol{\theta}}'$ could be obtained by building a bank of parallel filters matched to signals of the form $\text{Re}[F(t; \boldsymbol{\theta}') \exp i\Omega_r t]$ for closely spaced values of $\boldsymbol{\theta}'$ in the space Θ' of these remaining parameters. The outputs of each filter must be passed to a quadratic rectifier, whose output at the end of the interval $(0, T)$ is divided by $[2N^2 H'(\boldsymbol{\theta}', \boldsymbol{\theta}')]$. The parameters of the filter yielding the largest value of the resulting quantity identify the maximum-likelihood estimates of $\boldsymbol{\theta}'$.

When as here $\boldsymbol{\theta}' = (\tau, w)$, and the observation interval $(0, T)$ is so much longer than the duration T' of the signal that we can disregard the possibility that the signal overlaps one end or the other of it, the quantity

$$H'(\boldsymbol{\theta}', \boldsymbol{\theta}') = \frac{1}{N} \int_{-\infty}^{\infty} |F(t - \tau)|^2\, dt$$

is independent of τ and w and can be dropped. One then constructs a bank of parallel filters, each matched to a signal

$$\text{Re}[F(t) \exp i(\Omega_r + w)t]$$

for one of a closely spaced set of values of the Doppler shift w spanning the range of expected Doppler shifts. The complex impulse response of one of these filters can be taken as

$$K_w(s) = \begin{cases} F^*(T' - s)\, e^{-iw(T'-s)}, & 0 \le s \le T', \\ 0, & s < 0, \quad s > T', \end{cases} \tag{6-93}$$

where the interval $(0, T')$ is long enough to contain the entire signal.

The signal component of the output of one of these filters, when the signal

$$s(t; \tau_0, w_0) = A\, \mathrm{Re}\{F(t - \tau_0)\exp[i(\Omega_r + w_0)t + i\psi]\}$$

arrives, will have the complex envelope

$$S_0(t) = A\int_0^T K_w(s)F(t - \tau_0 - s)\exp[iw_0(t - s) + i\psi]\, ds \tag{6-94}$$

$$\approx A\exp(iw_0 t + i\psi')\int_{-\infty}^{\infty} F^*(u)F(t - T' - \tau_0 + u)\exp[-i(w - w_0)u]\, du,$$

with ψ' an inconsequential phase. The output $\mathrm{Re}\, V_0(t; w)\exp i\Omega_r t$ of this filter is quadratically rectified, and the signal component of the output of the rectifier is

$$R_s(t; w) = A^2\left|\int_{-\infty}^{\infty} F^*(u)F(t - T' - \tau_0 + u)\exp[-i(w - w_0)u]\, du\right|^2. \tag{6-95}$$

This signal component reaches its largest peak value in the output of that filter tuned to a signal with Doppler shift $w = w_0$, and that peak value occurs at time $t_0 = \tau_0 + T'$. The noise may cause the largest peak value to appear at the output of a filter tuned for a different value of w and at a different time, introducing errors into the resulting estimates $\hat{\tau}$ and $\hat{w}$ of the time of arrival and the Doppler shift.

If the spacing δw between the values of Doppler shift w for which the matched filters in the bank are tuned is much less than the reciprocal T'^{-1} of the duration of an echo, the noise components of the outputs of adjacent filters will be highly correlated. It will then be possible to estimate τ and w by interpolating among the functions $|V_0(t; w)|^2$ for values of w in the neighborhood of the largest peak output. When the signal-to-noise ratio is large, the rms error in the estimate $\hat{w}$ of the Doppler shift can be made rather smaller than the separation δw of the pass frequencies of the matched filters, and that rms error will be roughly the same as though one had a continuum of matched filters tuned over the entire range of expected carrier frequencies $\Omega_r + w$.

The simplest way to implement this prescription is to construct many copies of a filter matched to the same quasiharmonic signal $\mathrm{Re}\, F(t)\exp i\Omega_I t$, where Ω_I is a suitable intermediate frequency. Preceding each such filter is a mixer in which the input $\mathrm{Re}\, V(t)\exp i\Omega_r t$ is beaten against a wave of angular frequency $\Omega_r + w - \Omega_I$ for one of a uniformly spaced set of values of w. Those waves are generated by local oscillators whose output angular frequencies are displaced from $\Omega_r - \Omega_I$ by signals from a frequency multiplier that produces sinusoids of angular frequencies that are all required multiples of the spacing δw.

If the Doppler shift w is known, only a single matched filter is required, and we can take $w = w_0$ in (6-94) and (6-95). The prescription given in Sec. 6.3.1 for the maximum-likelihood estimator of the arrival time τ then ensues.

6.3.3 The Complex Ambiguity Function

When in Sec. 6.3.4 we calculate the variances and covariance of estimates of the arrival time τ and the Doppler shift w of a narrowband signal and when we interpret the results, we shall need to consider a complex ambiguity function $\lambda(\tau, w)$ defined by

$$\lambda(\tau, w) = \int_{-\infty}^{\infty} F(s - \tfrac{1}{2}\tau)F^*(s + \tfrac{1}{2}\tau)\, e^{-iws}\, ds, \qquad (6\text{-}96)$$

in which we assume the complex envelope so normalized that

$$\int_{-\infty}^{\infty} |F(t)|^2\, dt = 1, \qquad (6\text{-}97)$$

whereupon $\lambda(0, 0) = 1$. In terms of the Fourier transform $f(\omega)$ of the complex envelope $F(t)$—see (3-7)—the ambiguity function can be written

$$\lambda(\tau, w) = \int_{-\infty}^{\infty} f(\omega + \tfrac{1}{2}w)f^*(\omega - \tfrac{1}{2}w)\, e^{-i\omega\tau}\, \frac{d\omega}{2\pi}. \qquad (6\text{-}98)$$

This function will also figure in our analysis of signal resolution in Chapter 10. In order not to digress in the midst of Sec. 6.3.4, we shall now describe some of its properties.

In terms of the ambiguity function, the signal component $R_s(t; w)$ of the rectified output of a filter matched to a signal with frequency shift w is, by (6-95),

$$R_s(t; w) = A^2 |\lambda(t_0 - t, w - w_0)|^2, \qquad t_0 = \tau_0 + T'. \qquad (6\text{-}99)$$

Think now of a bank of filters matched for signals (6-89) with densely spaced carrier frequencies $\Omega_r + w$, and imagine plotting these rectified outputs $R_s(t; w)$ vertically over the (t, w)-plane. The resulting surface will reproduce the absolute square of the ambiguity function, and its peak will lie over the point (t_0, w_0).

Expanding the ambiguity function $\lambda(\tau, w)$ in the neighborhood of the origin, we obtain after some labor

$$\lambda(\tau, w) = 1 - i\overline{\omega}\tau - i\overline{t}w - \tfrac{1}{2}\overline{\omega^2}\tau^2 - \overline{\omega t}w\tau - \tfrac{1}{2}\overline{t^2}w^2 + \cdots. \qquad (6\text{-}100)$$

In this series appear various moments and cross-moments of the complex envelope $F(t)$ and its Fourier transform $f(\omega)$,

$$\overline{t^n} = \int_{-\infty}^{\infty} t^n |F(t)|^2\, dt, \qquad (6\text{-}101)$$

$$\overline{\omega^n} = \int_{-\infty}^{\infty} \omega^n |f(\omega)|^2\, \frac{d\omega}{2\pi}; \qquad (6\text{-}102)$$

there are no denominators because of the normalization (6-97). The quantity $\overline{\omega t}$ is defined by

$$\overline{\omega t} = -\tfrac{1}{2}i - i \int_{-\infty}^{\infty} tF^*(t)F'(t)\, dt, \qquad (6\text{-}103)$$

the prime indicating differentiation. The term $-\tfrac{1}{2}i$ makes this quantity real, as can be shown by integration by parts.

Without loss of generality we can choose the origin of time so that $\bar{\tau} = 0$ and the carrier frequency so that $\bar{\omega} = 0$. Then $\overline{\omega^2}$ reduces to the mean-square frequency deviation $\Delta\omega^2$ defined in (6-85) and $\overline{t^2}$ to the mean-square duration

$$\Delta t^2 = \int_{-\infty}^{\infty} (t - \bar{\tau})^2 |F(t)|^2 \, dt \tag{6-104}$$

of the complex envelope $F(t)$.

If we express the complex envelope as in (3-5),

$$F(t) = M(t) \, e^{i\phi(t)}$$

in terms of the amplitude and phase modulations defined in (3-2), these quantities become

$$\Delta t^2 = \int_{-\infty}^{\infty} t^2 [M(t)]^2 \, dt, \tag{6-105}$$

$$\Delta\omega^2 = \int_{-\infty}^{\infty} [M'(t)]^2 \, dt + \int_{-\infty}^{\infty} [\phi'(t)]^2 [M(t)]^2 \, dt, \tag{6-106}$$

$$\overline{\omega t} = \Delta(\omega t) = \int_{-\infty}^{\infty} t\phi'(t)[M(t)]^2 \, dt, \tag{6-107}$$

primes indicating differentiation with respect to time. Here we have assumed the origins of time and frequency so chosen that $\bar{\tau} = 0$ and $\bar{\omega} = 0$.

The quantity $\Delta(\omega t)$ is thus a weighted average of the product of the time t and the instantaneous frequency modulation $\phi'(t)$; it vanishes for a purely amplitude-modulated signal. The mean-square frequency deviation $\Delta\omega^2$ is composed of a term representing the time dependence of the amplitude modulation and a term corresponding to a weighted squared frequency modulation. The weighting function is always the squared amplitude-modulation $[M(t)]^2$.

The outputs of the bank of matched filters introduced at the beginning of this part are correlated. The complex envelopes of the noise components of the outputs of those filters are

$$V_0(t; w) = \int_{-\infty}^{\infty} K_w(s) N(t - s) \, ds,$$

where $N(t)$ is the complex envelope of the noise. We define their complex cross-covariance function as

$$\phi(t_2 - t_1; w_1, w_2) = \tfrac{1}{2} E[V_0(t_1; w_1) V_0^*(t_2; w_2)| H_0],$$

and we want to express it in terms of the ambiguity function in (6-96).

Using the assumption that the noise is white,

$$\tfrac{1}{2} E[N(t_1) N^*(t_2)] = N\delta(t_1 - t_2),$$

we write it as

$$\phi(t_2 - t_1; w_1, w_2) = \tfrac{1}{2} E \int_{-\infty}^{\infty}\int_{-\infty}^{\infty} K_{w_1}(s_1) K_{w_2}^*(s_2) N(t_1 - s_1) N^*(t_2 - s_2) \, ds_1 \, ds_2$$

$$= N \int_{-\infty}^{\infty}\int_{-\infty}^{\infty} K_{w_1}(s_1) K_{w_2}^*(s_2) \, \delta(t_1 - s_1 - t_2 + s_2) \, ds_1 \, ds_2$$

$$= N \int_{-\infty}^{\infty} K_{w_1}(s_2 - t_2 + t_1) K_{w_2}^*(s_2) \, ds_2.$$

Putting $s = t_2 - t_1$, we find by (6-93) that

$\phi(s; w_1, w_2)$

$$= N \int_{-\infty}^{\infty} F^*(T' - u + s) \exp[-iw_1(T' - u + s)]F(T' - u) \exp[iw_2(T' - u)] \, du$$

$$= N \int_{-\infty}^{\infty} F(T' - u)F^*(T' - u + s) \exp[-i(w_1 - w_2)(T' - u) - iw_1 s] \, du.$$

After comparing this with (6-96) and appropriately changing the integration variable, we find that it can be written as

$$\phi(s; w_1, w_2) = N \exp[-\tfrac{1}{2}i(w_1 + w_2)s]\lambda(s, w_1 - w_2). \tag{6-108}$$

Thus the complex cross-covariance function of the complex envelopes of the outputs of the filters tuned for frequency shifts w_1 and w_2 is proportional to the ambiguity function $\lambda(s, w_1 - w_2)$, where $s = t_2 - t_1$ is the interval between samplings. The dependence of this ambiguity function on the shape of the complex envelope $F(t)$ of the signal will be studied in some detail in Chapter 10, where we treat the resolution of close signals.

From (6-100) the squared ambiguity function $|\lambda(\tau, w)|^2$ in the neighborhood of the origin is

$$|\lambda(\tau, w)|^2 \approx 1 - [\Delta\omega^2 \tau^2 + 2\Delta(\omega t)\tau w + \Delta t^2 w^2],$$
$$\Delta(\omega t) = \overline{\omega t} - \overline{\omega}\,\overline{t}, \tag{6-109}$$

through terms of second order. The width of the ambiguity function $|\lambda(\tau, w)|$ in the w-direction is therefore on the order of $(\Delta t^2)^{-1/2}$, and (6-108) shows that the correlation of the outputs of the matched filters in our filter bank extends over a range of frequencies on the order of the reciprocal of the duration of the signal.

6.3.4 Calculation of the Error Covariances

The covariance matrix of the estimators of the amplitude components u and v, the arrival time τ, and the Doppler shift w is approximately given by (6-58), which involves the inverse of the Fisher information matrix Γ; the input signal-to-noise ratio must be large. For narrowband signals received in white noise the elements of that matrix are in turn given by (6-71) in terms of the generalized ambiguity function $H(\theta_1, \theta_2)$ of (6-73). There $\theta = (u, v, \tau, w)$,

$$S(t; \theta) = (u + iv)F(t - \tau) \exp iw(t - \tau),$$

and

$$H(\theta_1, \theta_2) = N^{-1} \operatorname{Re}[(u_1 - iv_1)(u_2 + iv_2)$$
$$\cdot \int_0^T F^*(t - \tau_1)F(t - \tau_2) \exp[iw_2(t - \tau_2) - iw_1(t - \tau_1)] \, dt].$$

Again invoking the long duration of the observation interval, we can write this as

$$H(\theta_1, \theta_2) = \frac{1}{N} \operatorname{Re}[(u_1 - iv_1)(u_2 + iv_2)\, e^{-i\overline{W}\overline{\tau}}\lambda(\overline{\tau}, \overline{w})] \tag{6-110}$$

in terms of the complex ambiguity function defined in (6-96). Here

$$\bar{\tau} = \tau_2 - \tau_1, \qquad \bar{w} = w_1 - w_2, \qquad W = \tfrac{1}{2}(w_1 + w_2). \qquad (6\text{-}111)$$

In order to find the Fisher information matrix we must differentiate the function in (6-110) with respect to the components of $\boldsymbol{\theta}_1$ and $\boldsymbol{\theta}_2$ separately and then set $\boldsymbol{\theta}_1 = \boldsymbol{\theta}_2 = \boldsymbol{\theta}_0$. When we do so, $\bar{\tau}$ and $\bar{w}$ vanish, and W becomes the true Doppler shift w_0. We normalize the complex envelope $F(t)$ as in (6-97), whereupon $\lambda(0, 0) = 1$, and

$$\frac{u_0^2 + v_0^2}{N} = \frac{A_0^2}{N} = \frac{2E}{N} = d^2$$

is the input signal-to-noise ratio. The rows and columns of the matrix $\boldsymbol{\Gamma}$ will be labeled u, v, τ, and w, in that order.

The differentiations with respect to the elements of the complex amplitude $u + iv$ yield immediately

$$\Gamma_{uu} = \Gamma_{vv} = \frac{1}{N}, \qquad \Gamma_{uv} = 0.$$

We can carry out those with respect to the other parameters most easily by using the expansion (6-100) in the neighborhood of the origin. Substituting it into (6-110), we obtain

$$H(\boldsymbol{\theta}_1, \boldsymbol{\theta}_2) = N^{-1} \, \mathrm{Re}[(u_1 - iv_1)(u_2 + iv_2) \, e^{-iW\bar{\tau}}]$$
$$\cdot \, [1 - \tfrac{1}{2}\Delta\omega^2\bar{\tau}^2 - \Delta(\omega t)\bar{w}\,\bar{\tau} - \tfrac{1}{2}\Delta t^2\bar{w}^2 + \cdots].$$

Now carrying out the differentiations, using (6-111), and at the end setting

$$u_1 + iv_1 = u_2 + iv_2 = u_0 + iv_0, \qquad u_0^2 + v_0^2 = A_0^2,$$
$$\tau_1 = \tau_2 = \tau_0, \qquad w_1 = w_2 = w_0,$$

we find after some labor that the Fisher information matrix is

$$\boldsymbol{\Gamma} = d^2 \begin{bmatrix} A_0^{-2} & 0 & A_0^{-2}w_0v_0 & 0 \\ 0 & A_0^{-2} & -A_0^{-2}w_0u_0 & 0 \\ A_0^{-2}w_0v_0 & -A_0^{-2}w_0u_0 & \Delta\omega^2 + w_0^2 & -\Delta(\omega t) \\ 0 & 0 & -\Delta(\omega t) & \Delta t^2 \end{bmatrix}. \qquad (6\text{-}112)$$

Our concern here is the 2×2 covariance matrix of the estimators $\hat{\tau}[v(t)]$ of the arrival time and $\hat{w}[v(t)]$ of the frequency shift. When we invert the Fisher matrix $\boldsymbol{\Gamma}$, we find that the u and v columns contain terms proportional to the true frequency shift w_0, but the submatrix containing the variances and the covariance of $\hat{\tau}$ and $\hat{w}$ does not. Those terms proportional to w_0 in the u and v columns represent a correlation between the estimators of the phase ψ and of the arrival time τ and frequency shift w, but as we are unconcerned with estimating the phase ψ or the components u and v of the complex amplitude $u + iv = A \exp i\psi$, we shall not bother working them out. It suffices then to invert the Fisher matrix $\boldsymbol{\Gamma}$ after setting w_0 equal to 0, whereupon it becomes

$$\boldsymbol{\Gamma} = d^2 \begin{bmatrix} A_0^{-2} & 0 & 0 & 0 \\ 0 & A_0^{-2} & 0 & 0 \\ 0 & 0 & \Delta\omega^2 & -\Delta(\omega t) \\ 0 & 0 & -\Delta(\omega t) & \Delta t^2 \end{bmatrix}. \qquad (6\text{-}113)$$

Its determinant is

$$\det \Gamma = d^8 A_0^{-4} C$$

with

$$C = \Delta\omega^2 \Delta t^2 - [\Delta(\omega t)]^2, \tag{6-114}$$

and its inverse is the approximate covariance matrix **B** of the errors:

$$\mathbf{B} \approx \Gamma^{-1} = \frac{1}{d^2 C} \begin{bmatrix} A_0^2 C & 0 & 0 & 0 \\ 0 & A_0^2 C & 0 & 0 \\ 0 & 0 & \Delta t^2 & \Delta(\omega t) \\ 0 & 0 & \Delta(\omega t) & \Delta\omega^2 \end{bmatrix}. \tag{6-115}$$

The simplest way to verify this inverse is to multiply the matrices in (6-113) and (6-115) and show that the identity matrix results. The approximate covariance matrix of the estimators of arrival time and Doppler shift is the lower-right block of (6-115), with the value of C given in (6-114). In this way we find that the variances of $\hat{\tau}$ and $\hat{w}$ and their covariance are

$$\text{Var } \hat{\tau} \approx \frac{\Delta t^2}{d^2 C}, \qquad \text{Var } \hat{w} \approx \frac{\Delta\omega^2}{d^2 C}, \qquad \text{Cov}\{\hat{\tau}, \hat{w}\} \approx \frac{\Delta(\omega t)}{d^2 C}.$$

If the echo signal contains no frequency modulation, $\Delta(\omega t) = 0$ and the errors in the estimates of arrival time and Doppler shift are uncorrelated:

$$\text{Var } \hat{\tau} \approx \frac{1}{d^2 \Delta\omega^2}, \qquad \text{Var } \hat{w} \approx \frac{1}{d^2 \Delta t^2}, \qquad \text{Cov}(\hat{\tau}, \hat{w}) \approx 0. \tag{6-116}$$

The variance of the estimator of the frequency shift w is inversely proportional to the mean-square duration Δt^2 of the signal. The variance of the estimator of the arrival time is now the same as that in (6-84), which applies when the carrier frequency $\Omega_r + w_0$ is known and need not be estimated. The same result arises if we discard the fourth column and row of (6-112), set $w_0 = 0$ in the remaining 3×3 matrix, and take its inverse.

For a given signal shape an increase in the bandwidth $\Delta\omega$ entails a decrease in the rms duration Δt, and the variance of the estimate of the arrival time τ cannot be reduced without at the same time accepting less precise measurements of the frequency shift w unless a marked change is made in the shape of the transmitted pulse itself. Still considering signals without frequency modulation, we observe that the product $\Delta\omega^2 \Delta t^2$ can be made large by using signals whose spectra, for a fixed duration Δt, are distributed as much toward high frequencies as possible. This can be achieved by giving the pulses very sharp corners, but these cannot be transmitted unless the antenna and the lines that feed it have large bandwidths themselves. Such signals will yield the most accurate simultaneous estimates of arrival time and frequency. The increased accuracy will not be gained, however, unless the receiver contains filters properly matched to those signals.

The analysis in Sec. 6.2.3 shows that in the limit of large input signal-to-noise ratio the errors in the estimates depend only on the derivatives of $\ln \Lambda[v(t)|\, \boldsymbol{\theta}]$ with respect to the parameters. When the signals are observed in additive Gaussian noise $n(t)$, these derivatives are linear in $n(t)$ and therefore themselves Gaussian random variables. The errors $\hat{\boldsymbol{\theta}}[v(t)] - \boldsymbol{\theta}_0$ are consequently asymptotically Gaussian random variables when the signal-to-noise ratio is large, and their joint probability density function depends on the inverse of the covariance matrix

$$\mathbf{B}' = d^{-2}C^{-1}\begin{bmatrix} \Delta t^2 & \Delta(\omega t) \\ \Delta(\omega t) & \Delta\omega^2 \end{bmatrix},$$

which is

$$\mathbf{B}'^{-1} = d^2\begin{bmatrix} \Delta\omega^2 & -\Delta(\omega t) \\ -\Delta(\omega t) & \Delta t^2 \end{bmatrix}.$$

The joint probability density function of the estimates of arrival time τ and Doppler shift w is thus, asymptotically,

$$p(\hat{\tau}, \hat{w}) \approx \frac{d^2 C^{1/2}}{2\pi}\exp\{-\tfrac{1}{2}d^2[\Delta\omega^2(\hat{\tau} - \tau_0)^2 - 2\Delta(\omega t)(\hat{\tau} - \tau_0)(\hat{w} - w_0) + \Delta t^2(\hat{w} - w_0)^2]\}.$$

The contours of constant probability density are similar to the *uncertainty ellipse*, whose equation is

$$\Delta\omega^2\tau^2 + 2\Delta(\omega t)\tau w + \Delta t^2 w^2 = 1$$

in the (τ, w)-plane. From (6-109) we see that the contours of equal height of the squared ambiguity function in the neighborhood of the origin $(0, 0)$ are also similar to the uncertainty ellipse. The area of this ellipse is

$$A = \frac{\pi}{\sqrt{C}} = \frac{\pi}{\sqrt{\Delta\omega^2\Delta t^2 - [\Delta(\omega t)]^2}}.$$

The larger the quantity C, the smaller the uncertainty ellipse, and the more concentrated is the joint probability density function $p(\hat{\tau}, \hat{w})$ about the true values τ_0 and w_0 of the arrival time and Doppler shift.

In terms of the amplitude and frequency modulations of the signal, the quantity C can be written, from (6-105) through (6-107), as

$$\begin{aligned}
C = &\int_{-\infty}^{\infty} t^2[M(t)]^2\, dt \int_{-\infty}^{\infty} [M'(t)]^2\, dt \\
&+ \int_{-\infty}^{\infty} t^2[M(t)]^2\, dt \int_{-\infty}^{\infty} [\phi'(t)]^2[M(t)]^2\, dt - \left[\int_{-\infty}^{\infty} t\phi'(t)[M(t)]^2\, dt\right]^2.
\end{aligned} \tag{6-117}$$

By the Schwarz inequality the combination of the last two terms is always nonnegative; it vanishes when the frequency modulation is linear in the time,

$$\phi'(t) = at,$$

for some constant a. Signals with this type of frequency modulation are called *chirp pulses*. For fixed amplitude modulation $M(t)$, C can be increased beyond the first term on the right side of (6-117) only by introducing a frequency modulation varying faster than linearly with the time t. That first term is made large by using pulses that rise and fall at beginning and end in as small a fraction of the total duration Δt as possible. When we study the ambiguity function in more detail in Chapter 10 on signal resolution, the effect of the shape of the signal envelope $F(t)$ on errors and ambiguities in the estimates of arrival time τ and the Doppler shift w will be considered at greater length.

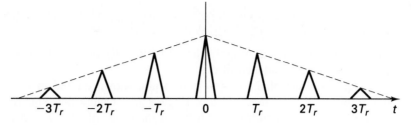

Figure 6-2. Signal component of complex envelope of output of the matched filter for a train of pulses (6-118).

6.3.5 Coherent Pulse Trains

Let us suppose that instead of a single pulse the radar transmits a train of M coherent pulses,

$$F(t) = M^{-1/2} \sum_{k=0}^{M-1} E(t - kT_r). \tag{6-118}$$

Their separation T_r, called the *repetition period*, is assumed to be so long that the component pulses $E(\cdot)$ do not overlap. The mean-square duration Δt^2 of this pulse train will be on the order of $(MT_r)^2$ by (6-104). For amplitude-modulated pulses, which we consider for simplicity, the variance of the maximum-likelihood estimator of the frequency shift w will then be approximately

$$\text{Var } \hat{w} \approx \frac{1}{d^2 M^2 T_r^2} \tag{6-119}$$

by (6-106). The longer the pulse train, it would appear, the more accurately the Doppler shift w of a radar echo can be estimated.

With such a long pulse train, however, one risks ambiguity in estimates of both the arrival time τ and the frequency shift w. Let us examine the former. The complex envelope of the signal component of the output of the filter matched to a signal with the true frequency shift w_0 will be the self-convolution of the pulse train. It is illustrated schematically in Fig. 6-2 for $M = 4$; in practice the repetition period T_r will be much greater, relative to the duration of each component pulse $E(\cdot)$, than depicted there. The total number of output pulses will in general equal $2M - 1$.

Unless the input signal-to-noise ratio d^2 is very high, the noise may cause one of the subsidiary output pulses to surpass the main one, introducing an error into the estimate of the arrival time τ that is some multiple of T_r. An error of this kind is called an *ambiguity*. The longer the pulse train—that is, the larger M—, the more slowly the heights of the subsidiary output pulses decrease on each side, and the greater is the likelihood of an ambiguity.

The approximation to Var $\hat{\tau}$ in (6-116) presumes that the main pulse definitely stands out above the others and that the error $\hat{\tau} - \tau_0$ is due only to a distortion of its peak by the noise. When, as we assume, the pulses $E(t - kT_r)$ do not overlap, the mean-square bandwidth $\Delta\omega^2$ appearing there is the same as for each of the component pulses $E(\cdot)$, as can be seen from an alternative way of writing (6-85):

$$\Delta\omega^2 = \frac{\int_{-\infty}^{\infty}|F'(t)|^2\,dt}{\int_{-\infty}^{\infty}|F(t)|^2\,dt}, \qquad F'(t) = \frac{dF(t)}{dt}, \qquad \overline{\omega} = 0.$$

The reciprocal $1/\Delta\omega$ is roughly the width of any one of the pulses in the train emerging from the matched filter.

The envelope of the output-pulse train, shown dashed in Fig. 6-2, would be the output of a filter matched to a single signal having the same duration MT_r as the pulse train. Its bandwidth will be on the order of $(MT_r)^{-1}$. Ambiguity can be expected with the pulse train (6-118) unless the arrival time of such a single, long signal can be timed within a fraction of T_r, and this requires

$$\frac{(MT_r)^2}{d^2} \ll T_r^2$$

The signal-to-noise ratio d^2 must therefore be large compared with $M^2 \gg 1$ if (6-119) is to hold for a pulse train of length MT_r and ambiguity is to be unlikely.

Ambiguity in measurement of the frequency shift w must also be anticipated when a train of pulses is transmitted. To show that, we consider the signal components of the outputs at time $t_0 = \tau_0 + T'$ of a bank of filters matched to signals with a dense set of frequency shifts. According to (6-99) these will be proportional to $|\lambda(0, w - w_0)|^2$, where w_0 is the true value of the shift in the incoming signal. For $\lambda(\cdot, \cdot)$ we shall use its expression (6-98) in the frequency domain.

The Fourier transform of the pulse train (6-118) is

$$f(\omega) = M^{-1/2} \sum_{k=0}^{M-1} e(\omega)\exp(-ikT_r\omega),$$

where

$$e(\omega) = \int_{-\infty}^{\infty} E(t)\,e^{-i\omega t}\,dt$$

is the Fourier transform of a component pulse. Summing, we write this as

$$f(\omega) = M^{-1/2}e(\omega)\frac{1 - e^{-iMT_r\omega}}{1 - e^{-iT_r\omega}} \tag{6-120}$$

$$= M^{1/2}e(\omega)\exp[-\tfrac{1}{2}i(M-1)T_r\omega]C(\omega),$$

where

$$C(\omega) = \frac{\sin\tfrac{1}{2}MT_r\omega}{M\sin\tfrac{1}{2}T_r\omega}. \tag{6-121}$$

In Fig. 6-3 we have plotted a portion of the function $|C(w)|$ for $M = 20$. It consists of peaks of height equal to 1 that are separated in angular frequency by $2\pi/T_r$, whose widths are on the order of $2\pi/MT_r$, and between which are $M-1$ much lower subsidiary peaks. We can call it a *comb function*. Because the narrowband transfer function of a filter matched to any one of our pulse trains is proportional to $C(\omega - w)$, such a filter has been called a *comb filter*.

The width of the transform $e(\omega)$ of $E(t)$ is on the order of $(\Delta't^2)^{-1/2}$, where $\Delta't^2$ is the mean-square duration of $E(t)$. The factor $e(\omega)$ modulates the comb function in (6-120), and the width of that modulation is much greater than the separation $2\pi/T_r$ between the tines of the comb.

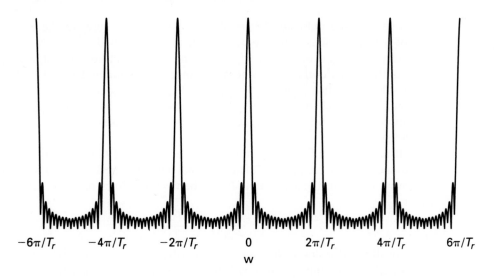

$$-6\pi/T_r \qquad -4\pi/T_r \qquad -2\pi/T_r \qquad 0 \qquad 2\pi/T_r \qquad 4\pi/T_r \qquad 6\pi/T_r$$

w

Figure 6-3. Comb function $|C(\omega)|$ defined in (6-121); $M = 20$.

According to (6-98), the signal components of the outputs of the matched filters at time t_0 are proportional to

$$A^2 |\lambda(0, w - w_0)|^2 = A^2 \left| \int_{-\infty}^{\infty} f(\omega + \tfrac{1}{2}(w - w_0)) f^*(\omega - \tfrac{1}{2}(w - w_0)) \frac{d\omega}{2\pi} \right|^2$$

$$= A^2 \left| \int_{-\infty}^{\infty} f(u - w_0) f^*(u - w) \frac{du}{2\pi} \right|^2$$

$$= A^2 \left| \int_{-\infty}^{\infty} e(u - w_0) e^*(u - w) C(u - w_0) C^*(u - w) \frac{du}{2\pi} \right|^2,$$

as can be seen by a simple change of the variable of integration in the first integral.

The signal component of the output will be largest for the filter tuned for a frequency shift w equal to the true frequency shift w_0, the tines of one comb $C(u - w)$ then having fallen exactly on those of the other. As our imaginative eye moves along the bank and away from that filter, we see the output decrease rapidly, the tines having moved apart; and for $|w - w_0|$ greater than about $2\pi/MT_r$, the output will be very small until we reach the filter tuned for a frequency shift equal to $w_0 + 2\pi/T_r$. There the tines of the function $C(u - w)$ have again coincided with those of $C(u - w_0)$, and a large output will be found. Likewise, the outputs at time t_0 will be large from all filters tuned for frequency shifts equal to $w_0 \pm 2\pi p/T_r$, where p is any integer, provided only that $2\pi p/T_r < (\Delta' t^2)^{-1/2}$.

At low signal-to-noise ratio the noise may cause the output of a filter tuned for one of those other frequency shifts to exceed the output of the "correct" filter tuned for $w = w_0$. The result is an error in the estimate of the angular frequency shift that is a multiple of $2\pi/T_r$, and this represents an ambiguity in the measurement of the frequency of the radar echo.

The origin of this ambiguity is easily understood. Because the transmitted pulses are coherent, the receiver can measure the change in phase of the r–f carrier from one reflected pulse to the next by comparing the phases of the echoes with that of a local oscillator synchronized with the transmitted phase. A target with a velocity v moves a distance $L = vT_r$ in one pulse-repetition period, and the r–f phase of the carrier of the radar echo will change by an amount $\Delta\phi = \Omega_0(2L/c)$. The target velocity is then given by $v = (\Delta\phi/\Omega_0)(c/2T_r)$. Because the receiver cannot distinguish phase changes differing by multiples of 2π, an ambiguity in the true velocity of some multiple of $(2\pi/\Omega_0)(c/2T_r) = \frac{1}{2}\lambda/T_r$ arises; λ is the wavelength of the radiation.

If ambiguity in frequency is to be avoided and (6-119) is to hold, one must be able to measure the frequency shift of a signal having a spectrum $e(\omega)$ with an error somewhat less than the separation $2\pi/T_r$ between the tines of the comb function; that is,

$$\frac{1}{d^2\Delta't^2} \ll \frac{4\pi^2}{T_r^2}.$$

The signal-to-noise ratio d^2 must be on the order of or larger than about $T_r^2/(4\pi^2\Delta't^2)$, and this quantity will be large when, as usually, the pulses of the train are widely separated.

Problems

6-1. Given are n independent measurements $v_1, v_2, \ldots, v_n$ of the noise voltage v at a certain point in a receiver. If the noise v is a Gaussian random variable with expected value 0, what is the maximum-likelihood estimator of its variance? Calculate the expected value and the variance of this estimator as functions of the true variance. Give a sufficient statistic for estimating Var v.

6-2. Show that the estimator of the variance determined in Problem 6-1 is efficient.

6-3. Suppose that in Problem 6-1 both the expected value and the variance of the voltage v are unknown. Work out their maximum-likelihood estimators based on the same n measurements.

6-4. Show that the properties of the MAP estimator in Sec. 6.1.1 follow from the Bayesian estimation theory of Sec. 6.1.4 when the cost function has the bizarre form

$$C(\hat{\boldsymbol{\theta}}, \boldsymbol{\theta}) = A - B\delta(\hat{\boldsymbol{\theta}} - \boldsymbol{\theta}), \qquad B > 0,$$

where $\delta(\cdot)$ is the m-dimensional delta function and m the number of estimanda.

6-5. Analyze the estimation of the mean m of a set of n independent Gaussian random variables, treated in Sec. 6.1.3, in the framework of Sec. 6.1.4. Assume for simplicity that the prior mean μ equals 0. That is, consider the observed data as having the form

$$v_k = m + e_k, \qquad k = 1, 2, \ldots, n,$$

in which m is a Gaussian random variable having expected value 0 and variance β^2, and the errors e_k are independently Gaussian with expected values 0 and variances δ^2. The errors are of course independent of the estimandum m; see (6-35). Use the method of Sec. 6.1.4 to find the MMSE estimator of the mean m.

6-6. Show how to obtain the maximum-likelihood estimator of the azimuth θ of a fixed target on the basis of echoes received by a radar antenna scanning at a uniform rate. The received signals can be taken as

$$f_k(t) = \text{Re } Ar(\theta_0 + k\delta)F(t) \exp(i\Omega t + i\psi_k), \qquad 0 \le t \le T,$$

Estimation of Signal Parameters   Chap. 6

where the amplitude A, the azimuth θ_0 of the target, and the phases ψ_k are independent and unknown, and all are to be estimated. The beam pattern of the antenna is represented by $r(\theta)$. The noise is white and Gaussian with unilateral spectral density N. It is assumed that the transmitting antenna is fixed and that all pulses incident on the target have the same energy. Show from the Fisher formula or otherwise that at large signal-to-noise ratio the variance of that maximum-likelihood estimator of the azimuth is inversely proportional to the total signal-to-noise ratio $d_T^2 = 2E_T/N$ and directly proportional to the mean-square beamwidth Θ^2, defined by

$$\Theta^2 = \frac{\int_{-\pi}^{\pi}[r(\theta)]^2\, d\theta}{\int_{-\pi}^{\pi}[r'(\theta)]^2\, d\theta}.$$

The prime indicates differentiation. Here E_T is the total received energy. Assume that the angle δ through which the antenna turns between reception of one echo and the next is small enough that summations over k can be approximated by integrations over azimuth.

6-7. Let $v(t)$ be a realization of Gaussian random noise of autocovariance $\phi(t, s) = Br(t, s)$, where B is positive, but unknown. If $v(t)$ is given over only a finite interval $(0, T)$, the multiplicative constant B can be estimated as accurately as desired. To show this, consider the estimator

$$b_n = \frac{1}{n}\sum_{k=1}^{n}\frac{v_k^2}{\lambda_k},$$

where

$$v_k = \int_0^T f_k(t)v(t)\, dt,$$

with $\{f_k(t)\}$ the orthonormal eigenfunctions and $\{\lambda_k\}$ the eigenvalues of the integral equation

$$\lambda f(t) = \int_0^T r(t, s)f(s)\, ds, \qquad 0 \le t \le T,$$

$\lambda_1 \ge \lambda_2 \ge \cdots \ge \lambda_n > 0$. Show that b_n is an unbiased estimator of B and that $\operatorname{Var} b_n \to 0$ as $n \to \infty$. Hence the larger the number of terms in b_n, the greater is the accuracy of the estimator.

6-8. The linearly rising signal

$$s(t) = a + bt$$

is observed during an interval $(0, T)$ in the presence of white, Gaussian noise $n(t)$ of unilateral spectral density N. The constants a and b are unknown. Find the maximum-likelihood estimator of the slope b of the signal on the basis of the receiver input $v(t) = s(t) + n(t)$, and calculate the mean-square error of this estimator exactly. Compare your result with that given by the Fisher formula.

6-9. The signal $s(t) = a\cos\Omega t + b\sin\Omega t$ is received in the presence of white, Gaussian noise $n(t)$ of unilateral spectral density N. Find the maximum-likelihood estimators of the parameters a and b based on observation of the input $v(t) = s(t) + n(t)$ during an interval $(0, T)$. Make no approximations, but assume for simplicity that ΩT is an integral multiple of 2π. Calculate the variances and the covariance of the estimators $\hat{a}[v(t)]$ and $\hat{b}[v(t)]$ of a and b.

6-10. Given n independent pairs (x_k, y_k) of correlated Gaussian random variables x and y, known to have expected values zero, determine the maximum-likelihood estimators of their common variance $\sigma^2 = \operatorname{Var} x = \operatorname{Var} y$ and of their covariance $\mu = E(xy)$. The joint probability density function of an individual pair (x, y) is

$$p(x, y) = \frac{1}{2\pi(\sigma^4 - \mu^2)^{1/2}} \exp\left[-\frac{\sigma^2(x^2 + y^2) - 2\mu xy}{2(\sigma^4 - \mu^2)} \right].$$

Hint: First find the maximum-likelihood estimators of two elements of the inverse covariance matrix of x and y. Then solve for the estimators of σ^2 and μ.

6-11. Determine the normalization constants in (6-12) and (6-15) and use (6-18) to show that for a symmetric block matrix such as that in (6-16)

$$\det\begin{bmatrix} \boldsymbol{\mu}_{vv} & \boldsymbol{\mu}_{v\theta} \\ \boldsymbol{\mu}_{\theta v} & \boldsymbol{\mu}_{\theta\theta} \end{bmatrix} = \det \boldsymbol{\mu}_{\theta\theta} \det(\boldsymbol{\mu}_{vv} - \boldsymbol{\mu}_{v\theta}\boldsymbol{\mu}_{\theta\theta}^{-1}\boldsymbol{\mu}_{\theta v}).$$

6-12. A sine wave $A \cos(\Omega t + \theta)$ of known amplitude A is received in the presence of white, Gaussian noise during an interval of duration T. It is desired to estimate the phase θ, which carries information, relative to the phase of a sinusoid $B \cos \Omega t$ available at the receiver. Take the input as

$$v(t) = \operatorname{Re} V(t) e^{i\Omega t}, \qquad V(t) = N(t) + A e^{i\theta}.$$

From the Cramér–Rao inequality determine a lower bound on the mean-square error of an unbiased estimator $\hat{\theta}$ of the phase. Work out the maximum-likelihood estimator of the phase θ and show how it might be realized. Calculate the probability density function of this maximum-likelihood estimator $\hat{\theta}$; it depends on the input signal-to-noise ratio. Derive an approximate form for this probability density function when that signal-to-noise ratio is very large, and from this determine in turn the mean-square error in $\hat{\theta}$ in that limit. Compare with the bound given by the Cramér–Rao inequality.

6-13. Information is sent as the difference ϕ between the phases of two successive narrowband signals of known complex envelope $F(t)$, carrier frequency Ω, and amplitude A. The signals arrive during adjacent intervals, each of duration T, and are corrupted by white, Gaussian noise of unilateral spectral density N. Thus the complex envelopes of successive inputs $v_1(t)$ and $v_2(t)$ to the receiver have the forms

$$V_1(t) = AF(t) e^{i\psi} + N_1(t), \qquad 0 \le t \le T,$$
$$V_2(t) = AF(t) e^{i\psi + i\phi} + N_2(t), \qquad 0 \le t \le T,$$

where the common carrier phase ψ is unknown and uniformly distributed over $(0, 2\pi)$, and $N_1(t)$ and $N_2(t)$ are statistically independent complex envelopes of the noise. Work out the maximum-likelihood estimator of the phase difference ϕ in terms of $V_1(t)$ and $V_2(t)$. Assuming that the input signal-to-noise ratio is large, calculate the mean-square error $\operatorname{Var} \hat{\phi}$ of this estimator $\hat{\phi}$ of ϕ. Make any justifiable approximations that are necessary, and the earlier the better.

6-14. Formulate a Bayesian theory of the combined detection of a signal $s(t; \alpha)$ and the estimation of an unknown parameter α on which the signal depends. Under hypothesis H_0 (noise alone present) the joint probability density function of the data $v = (v_1, v_2, \ldots, v_n)$ is $p_0(v)$; under hypothesis H_1 (signal plus noise present) it is $p_1(v|\alpha)$. The data v are appropriate samples of the input; at the end, let $n \to \infty$. You are to design a receiver that estimates the parameter α each time it decides that a signal is present. If it chooses hypothesis H_0, it issues no estimate.

Define the following symbols:

ζ = prior probability of hypothesis H_0,

$z(\alpha)$ = prior probability density function of the parameter α, given that a signal is present,

C_{00} = cost of choosing hypothesis H_0 when H_0 is true,

$C_{10}(\hat{\alpha})$ = cost of choosing hypothesis H_1 and issuing estimate $\hat{\alpha}$ of the parameter α when H_0 is true,

$C_{01}(\alpha)$ = cost of choosing hypothesis H_0 when H_1 is true and the signal parameter equals α,

$C_{11}(\hat{\alpha}, \alpha)$ = cost of choosing hypothesis H_1 and of issuing the estimate $\hat{\alpha}$ when the signal is present with parameter value α.

Write down the average cost of operation of a strategy for combined detection and estimation, and show how the system should be designed to operate with minimum average cost. Assume that the input to the receiver, observed during the interval $(0, T)$, is

$$v(t) = n(t), \qquad (H_0)$$
$$v(t) = n(t) + s(t; \alpha), \qquad (H_1)$$

under the two hypotheses, and express the prescription for receiver operation in terms of the likelihood functional $\Lambda[v(t)|\,\alpha]$ for detecting a signal $s(t; \alpha)$ in the input $v(t)$.

Taking the cost function $C_{11}(\hat{\alpha}, \alpha) = g(\hat{\alpha} - \alpha)^2$, g a positive constant, with

$$C_{10}(\hat{\alpha}) = C_{11}(\hat{\alpha}, 0), \qquad C_{01}(\alpha) = C_{11}(0, \alpha), \qquad C_{00} \le 0,$$

express the optimum estimator of the parameter α, under this condition of uncertainty as to the presence or absence of the signal, in terms of the optimum estimator of α when it is known that the signal is present. (By optimum we mean minimizing average cost.) Apply the theory to the detection of a signal $Af(t)$ of known form and unknown amplitude $\alpha = A$ in the presence of white, Gaussian noise of unilateral spectral density N. The amplitude A, to be estimated with the quadratic cost function defined above, has a Gaussian prior probability density function with expected value 0 and variance σ^2. Describe the optimum strategy in terms of matched filtering and whatever subsequent processing is necessary.

6-15. Express $\bar{t}$ and $\bar{\omega}$, defined as in (6-101) and (6-102), in terms of the amplitude modulation $M(t)$ and the phase modulation $\phi(t)$ of the narrowband signal as defined in (3-2). Derive (6-105) through (6-107). *Hint:* The Fourier transform of $F'(t)$ is $i\omega f(\omega)$.

7

Detection of Signals with Unknown Parameters

7.1 UNKNOWN ARRIVAL TIME: THE THRESHOLD DETECTOR

Radar is used to detect targets that might be anywhere in a range interval much longer than the electromagnetic pulses it sends out, and the echo signals may arrive at the receiver at any time during the period $(0, T)$ between transmissions. Considering a single such interpulse interval, we treat the receiver as a device for deciding between two hypotheses on the basis of its input $v(t) = \text{Re } V(t) \exp i\Omega t$,

$$H_0: V(t) = N(t),$$
$$H_1: V(t) = N(t) + S(t), \qquad S(t) = AF(t - \tau) e^{i\psi}, \qquad 0 \le t \le T.$$

As we saw at the beginning of Sec. 3.3, the receiver can be presumed to have available the complex envelope $V(t)$ of its input. Here $N(t)$ is the complex envelope of white Gaussian noise of unilateral spectral density N, and $F(t)$ is proportional to the complex envelope of the transmitted radar signal. As usual, Ω is the angular frequency of the carrier of the narrowband signals. The amplitude and phase of the received echo are A and ψ, respectively, and τ is the time at which the radar echo arrives; $\tau = 2R/c$ with R the distance to the target and c the velocity of light. We take the complex envelope $F(t)$ to differ significantly from zero only during an interval $(0, T')$ much shorter than the observation interval $(0, T)$.

We assume that the signal is just as likely to arrive at any time during $(0, T)$, assigning to the epoch τ a uniform prior probability density function

$$z(\tau) \equiv T^{-1}, \qquad 0 \le \tau \le T. \tag{7-1}$$

The phase ψ is as before uniformly distributed over $(0, 2\pi)$, and because the receiver does not keep track of the phase of the transmitted pulse, we take ψ and τ as statistically independent. Because the observation interval $(0, T)$ is so much longer than the duration T' of the pulse $F(t)$, we can disregard the possibility that the echo may overlap one end or the other of that interval. As the size of the target is unknown, the amplitude A conveys no information about how far away it is, and we can take A as independent of ψ and τ.

Were the echo amplitude A known a priori, the optimum receiver would determine the average likelihood functional (3-100), which here becomes

$$\overline{\Lambda}[v(t); A] = \int_0^{2\pi} \frac{d\psi}{2\pi} \int_0^T \Lambda[v(t); A, \psi, \tau] \frac{d\tau}{T},$$

where by (3-54) as written for detection in white noise, with $Q(t) = N^{-1}S(t)$,

$$\Lambda[v(t); A, \psi, \tau] = \exp\left\{ \text{Re}\left[e^{-i\psi} \frac{A}{N} \int_0^T F^*(t - \tau)V(t)\, dt \right] - \frac{A^2}{2N} \int_0^T |F(t - \tau)|^2\, dt \right\}.$$

Upon carrying out the average over the phase ψ, this becomes

$$\overline{\Lambda}[v(t); A] = \frac{1}{T} \int_0^T d\tau \, I_0\left[\frac{A}{N} \left| \int_0^T F^*(t - \tau)V(t)\, dt \right| \right]$$

$$\cdot \exp\left[-\frac{A^2}{2N} \int_0^T |F(t - \tau)|^2\, dt \right], \qquad (7\text{-}2)$$

as in (3-80); $I_0(\cdot)$ is again the modified Bessel function. Because the echo signal is unlikely to overlap either end of the interval $(0, T)$, we can take

$$\int_0^T |F(t - \tau)|^2\, dt \approx \int_{-\infty}^{\infty} |F(t)|^2\, dt$$

to be constant, and we normalize our signal $F(t)$ so that this integral equals 1.

A receiver would need to know or assume some value of the amplitude A in order to implement the average likelihood functional (7-2). It would pass the input $v(t)$ through a filter matched to the signal $F(t)$, which we have taken to differ significantly from zero only over an interval $(0, T')$ for $T' \ll T$. The complex impulse response of this filter is

$$K(s) = \begin{cases} F^*(T' - s), & 0 \le s \le T', \\ 0, & s < 0, \quad s > T', \end{cases}$$

and the complex envelope of the output $v_0(t) = \text{Re } V_0(t) \exp i\Omega t$ of the filter is

$$V_0(t) = \int_0^{T'} F^*(T' - s)V(t - s)\, ds \approx \int_{-\infty}^{\infty} F^*(T' - s)V(t - s)\, ds$$

$$= \int_{-\infty}^{\infty} F^*(u)V(t - T' + u)\, du,$$

as in (3-16). The integral appearing in the argument of the Bessel function in (7-2), on the other hand, is approximately

$$\int_{-\infty}^{\infty} F^*(t - \tau)V(t)\, dt = \int_{-\infty}^{\infty} F^*(u)V(u + \tau)\, du = V_0(\tau + T').$$

Thus the average likelihood functional can be written

$$\overline{\Lambda}[v(t);\, A] = \frac{1}{T}\, \exp\!\left[-\frac{A^2}{2N}\int_{-\infty}^{\infty} |F(t)|^2\, dt\right] \int_{T'}^{T+T'} I_0(AN^{-1}|V_0(t)|)\, dt, \qquad (7\text{-}3)$$

and the receiver would pass the output Re $V_0(t)\exp i\Omega t$ of the matched filter to a rectifier having a characteristic proportional to $I_0(AN^{-1}|V_0|)$, whose output would in turn be integrated over an interval $(T', T + T')$. If the integrated output, proportional to $\overline{\Lambda}[v(t);\, A]$, exceeded a certain decision level, a signal would be deemed to have arrived sometime during the interval $(0, T)$. To evaluate the false-alarm and detection probabilities of this receiver would be extremely difficult.

The threshold detector is based on the assumption that the signals are so weak that both the Bessel function and the exponential function in (7-2) and (7-3) can be approximated by the first two terms of their power series; all powers of the amplitude A higher than the second are neglected. Thus by (3-61) we write (7-3) as

$$\overline{\Lambda}[v(t);\, A] \approx \frac{1}{T}\left[1 - \frac{A^2}{2N}\int_{-\infty}^{\infty} |F(t)|^2\, dt\right]$$

$$\cdot \int_{T'}^{T+T'}\left[1 + \frac{A^2}{4N^2}|V_0(t)|^2\right] dt + O(A^4) \qquad (7\text{-}4)$$

$$\approx 1 + \frac{A^2}{4N^2 T}\int_{T'}^{T+T'} |V_0(t)|^2\, dt - \frac{A^2}{2N}\int_{-\infty}^{\infty} |F(t)|^2\, dt + O(A^4).$$

Multiplying this by an arbitrary prior probability density function $z_A(A)$ of the signal amplitude A, integrating, and comparing the result with (3-108), we find that the term of lowest order in the amplitude is that proportional to A^2, and we obtain the threshold statistic

$$g_A = \frac{1}{2T}\int_{T'}^{T+T'} |V_0(t)|^2\, dt. \qquad (7\text{-}5)$$

The factor $\overline{A^2}/2N^2$ has been absorbed into the decision level.

The threshold receiver, or *weak-signal detector*, therefore passes the output Re $V_0(t)\exp i\Omega t$ of the matched filter through a quadratic rectifier, whose output is integrated during the interval $(T', T + T')$, where T' is the delay in that filter. If the threshold statistic g_A exceeds a decision level, set in such a way that the false-alarm probability equals a preassigned value, the receiver decides that a signal is present. This threshold detector requires no assumption about the actual amplitude A of the signal to be detected. It can be thought of as measuring the total energy picked up by the antenna during the interval $(0, T)$, and it is sometimes called an *energy detector* or *radiometer*.

To compute the false-alarm and detection probabilities exactly for the threshold statistic g_A is most difficult. We resort to an approximation, replacing it by

$$g_A' = \tfrac{1}{2}C \sum_{k=1}^{L} |z_k|^2, \qquad z_k = V_0(t_k). \qquad (7\text{-}6)$$

We are sampling the output of the matched filter at L times uniformly spaced during the interval $(T', T + T')$; the samples are separated by T/L. We want to choose the value of L so that these samples can be considered to be at least approximately independent statistically. The sampling interval T/L should therefore be on the order of the width of the complex autocovariance function $\phi(s)$ of the output of the matched filter. Indeed, if the signal is strictly bandlimited to a frequency range of width W hertz, and $L = WT$, the samples z_k are uncorrelated and, being circular-complex Gaussian random variables, they are statistically independent. By the sampling theorem and with a suitable choice of the constant C, therefore, $g_A = g'_A$ [Bal57], [Pap91, pp. 376–9].

The complex autocovariance function of the output $V_0(t)$ can be read off from (6-108) by setting $w_1 = w_2 = 0$, and using (6-96) and (6-98) we find

$$\phi(s) = \tfrac{1}{2}E[V_0(t)V_0^*(s)|\ H_0] = N\lambda(s, 0)$$
$$= N\int_{-\infty}^{\infty} F(u - \tfrac{1}{2}s)F^*(u + \tfrac{1}{2}s)\ ds \qquad (7\text{-}7)$$
$$= N\int_{-\infty}^{\infty} |f(\omega)|^2 e^{-i\omega s}\ \frac{d\omega}{2\pi},$$

where as before $f(\omega)$ is the Fourier transform of the complex envelope $F(t)$ of the signal.

The random variables $r_k = \tfrac{1}{2}|z_k|^2$ have an exponential distribution under hypothesis H_0,

$$p_0(r) = a\ e^{-ar}\ U(r),$$

and for such a distribution the variance equals the square of the expected value,

$$E(r|\ H_0) = a^{-1}, \qquad \text{Var}(r|\ H_0) = a^{-2} = [E(r|\ H_0)]^2, \qquad (7\text{-}8)$$

as the reader can easily demonstrate.

We want to choose the constants C and L so that the threshold statistic g_A and its approximation g'_A have equal expected values and variances under hypothesis H_0. From (7-5), because under hypothesis H_0 the output of the matched filter is a stationary process,

$$E(g_A|\ H_0) = \tfrac{1}{2}E[|V_0(t)|^2|\ H_0] = E(r|\ H_0) = \phi(0) = N$$

by (7-7), and by (7-6)

$$E(g'_A|\ H_0) = LCE(r|\ H_0) = LCN,$$

whence $C = 1/L$.

The variance of the approximate statistic g'_A under hypothesis H_0 is

$$\text{Var}(g'_A|\ H_0) = LC^2\ \text{Var}(r|\ H_0) = LC^2N^2 = \frac{N^2}{L} \qquad (7\text{-}9)$$

when, as assumed, the terms in (7-6) are statistically independent. To calculate that of the threshold statistic g_A, we write, from (7-5),

$$E(g_A^2|\ H_0) = \frac{1}{4T^2}\int_{T'}^{T+T'}\int_{T'}^{T+T'} E[|V_0(t_1)|^2|V_0(t_2)|^2|\ H_0]\ dt_1\ dt_2.$$

We evaluate the expectation inside the integrand by means of (3-44), in which we put

$$z_1 = z_3 = V_0(t_1), \qquad z_2 = z_4 = V_0(t_2).$$

The first term on the right side of (3-44), substituted into our double integral, yields $[E(g_A| H_0)]^2$, leaving the second term to provide the variance, and we find

$$\text{Var}(g_A| H_0) = \frac{1}{T^2} \int_{T'}^{T+T'} \int_{T'}^{T+T'} |\phi(t_1 - t_2)|^2 \, dt_1 \, dt_2.$$

When, as we assume, $T \gg T'$ and the duration T of the observation interval much exceeds the width of the autocovariance function $\phi(\cdot)$, the integral with respect to t_2 can be extended over $(-\infty, \infty)$, and we obtain

$$\text{Var}(g_A| H_0) = \frac{1}{T} \int_{-\infty}^{\infty} |\phi(s)|^2 \, ds.$$

Equating this to (7-9), we find

$$L = TN^2 \left[\int_{-\infty}^{\infty} |\phi(s)|^2 \, ds \right]^{-1}.$$

By Parseval's theorem and (7-7) this becomes

$$L = T \left[\int_{-\infty}^{\infty} |f(\omega)|^4 \, \frac{d\omega}{2\pi} \right]^{-1} = WT, \qquad (7\text{-}10)$$

when we define the effective bandwidth W of the signal as

$$W = \frac{|\phi(0)|^2}{\int_{-\infty}^{\infty} |\phi(s)|^2 \, ds} = \frac{[\int_{-\infty}^{\infty} |f(\omega)|^2 \, \frac{d\omega}{2\pi}]^2}{\int_{-\infty}^{\infty} |f(\omega)|^4 \, \frac{d\omega}{2\pi}}. \qquad (7\text{-}11)$$

The numerators have been introduced for the sake of generality, eliminating the need for the complex envelope $F(t)$ of the signal to be normalized.

For a strictly bandlimited signal, the quantity W equals its bandwidth in hertz:

$$f(\omega) \equiv \begin{cases} B, & -\pi W \le \omega \le \pi W, \\ 0, & |\omega| > \pi W. \end{cases}$$

For a rectangular signal of duration T', $W = 1.5/T'$. For a Gaussian signal of mean-square bandwidth $\Delta\omega^2$,

$$W = \frac{\sqrt{\Delta\omega^2}}{\sqrt{\pi}}.$$

Except in dealing with the ambiguity function $\lambda(\tau, w)$, the bandwidth as defined in (7-11) is usually more useful than the rms bandwidth $\Delta\omega$.

Assuming now that the terms in our approximation g_A' are roughly independent statistically, we see that it is a detection statistic of the same form as U in (4-19), and we can use the false-alarm and detection probabilities as calculated in Sec. 4.2,

$$Q_0 \approx Q_L(0, b), \qquad Q_d \approx Q_L(D, b),$$

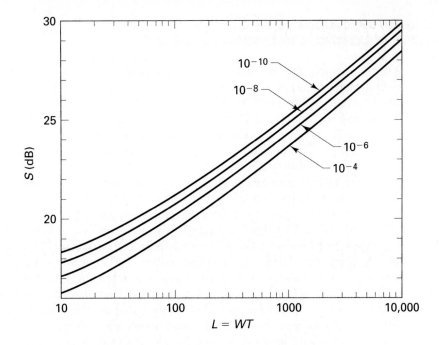

Figure 7-1. Energy-to-noise ratio (dB) for the threshold receiver to attain Q_d = 0.999 versus time–bandwith product WT. Curves are indexed with the false-alarm probability Q_0.

where $Q_L(\alpha, \beta)$ is the generalized Marcum Q function as defined in (4-27) ($L = M$), and D is related to the input energy-to-noise ratio

$$S = \frac{E}{N} = \frac{D^2}{2};$$

E is the energy of the received signal, and N is the unilateral spectral density of the noise. Remember that we showed in Sec. 4.2.2 that the probability of detection depends only on the total energy-to-noise ratio and not on how the signal energy is distributed among the terms of the sum $U = g_A'$. The parameter b, related to the decision level on the threshold statistic g_A and its approximation g_A', is set to provide a preassigned false-alarm probability Q_0.

In Fig. 7-1 we have plotted the energy-to-noise ratio $S = E/N$ (dB) required, in this approximation, to attain a probability Q_d of detection equal to 0.999, for various false-alarm probabilities Q_0, as a function of the time-bandwidth product $L = WT$, $10 \leq L \leq 10^4$. Lengthening the interval during which the signal might arrive by a factor of 1000 increases the required input energy-to-noise ratio by about 12 dB when the threshold detector is employed.

The performance of the radiometer in detecting spread-spectrum signals has been treated at length in [Dil89].

Sec. 7.1   Unknown Arrival Time: The Threshold Detector   **271**

7.2 UNKNOWN ARRIVAL TIME: THE MAXIMUM-LIKELIHOOD DETECTOR

When we sample the rectified output of the matched filter at intervals $\Delta t = W^{-1}$, as in forming the approximate detection statistic g_A' in Sec. 7.1, we in effect reduce our detection problem to that treated in Sec. 3.6.4. There the signal could appear in one and only one of M independent channels or inputs to the receiver. The maximum-likelihood receiver described there decides that a signal is present if the filtered and rectified output of any of the channels exceeds a certain decision level. The comparison illustrated in Fig. 4-4—see curves (b) and (c)—shows that a larger probability of detection can be achieved by that maximum-likelihood receiver than by a receiver that simply adds the quadratically rectified outputs of each channel. The latter corresponds to the energy detector analyzed in Sec. 7.1.

In the context of the sampling approximation of Sec. 7.1, this means that a decision for hypothesis H_1 should be made if any of the samples $|V_0(t_k)|^2$ passes the decision level. Returning to the rectified output $|V_0(t)|^2$ as a continuous function of time, we require a decision for H_1 if $|V_0(t)|^2$ exceeds a decision level a at any time in $T' < t < T + T'$. This is just what an ordinary radar receiver accomplishes. The operator sees the rectified output $|V_0(t)|^2$ displayed on the A-scope [Sko62, pp. 6, 439–40]. It fluctuates owing to the random noise. If any fluctuation is large enough to cross a threshold mentally applied by the operator, he attributes the excursion, or *blip*, to the presence of an echo signal. The time t_m at which the rectified output is maximum provides—as we have seen in Chapter 6—an estimate $\hat{\tau}$ of the arrival time τ of the echo through $\hat{\tau} = t_m - T'$, and that time t_m is close to the instant when the rectified output $|V_0(t)|^2$ crosses the threshold. If $|V_0(t)|^2$ remains below the threshold during the entire interval, $|V_0(t)|^2 < a$, $T' < t < T + T'$, the operator concludes that no signal is present.

In practice the operator sees the output $|V_0(t)|^2$ during intervals $(T', T + T')$ following each of a large number of transmitted pulses, even when the radar is scanning in azimuth. The persistence of a blip at a certain point on the A-scope trace will then enhance the likelihood of perceiving a target. The amount by which this ability increases the probability Q_d of detection is difficult to calculate. We can instead imagine the detection process as carried out automatically by a receiver that sums, or "integrates," the rectified outputs $|V_k(t)|^2$ of the matched filter during a succession of M interpulse intervals, $1 \leq k \leq M$. The crossing of the total output

$$r(t) = \sum_{k=1}^{M} |V_k(t)|^2 \tag{7-12}$$

over a decision level a at any time in $T' < t < T + T'$ is then taken as indicating the presence of a target.

This system is diagrammed in Fig. 7-2. We assume that the target is stationary. The delay line must retard its input by a time accurately equal to the interval T between transmitted pulses. Its output is fed back to its input, where it is added to the output of the quadratic rectifier. During the final observation interval the alarm circuit is activated, and if the output of the adder crosses the decision level built into the alarm, it signals the presence of an echo. At the same time it can turn

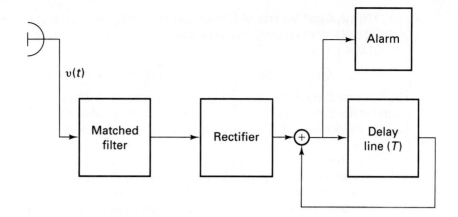

Figure 7-2. Receiver for signals of unknown arrival time.

on another circuit, not shown, to measure the time $t_m = \hat{\tau} + T'$ at which the total output reaches its peak value and thus estimate the range of the target.

The false-alarm and detection probabilities for a receiver that makes its decisions in this manner are

$$
\begin{aligned}
Q_0 &= 1 - \Pr[r(t) < a,\, T' \le t \le T + T' |\, H_0], \\
Q_d &= 1 - \Pr[r(t) < a,\, T' \le t \le T + T' |\, H_1],
\end{aligned}
\tag{7-13}
$$

where for a single interval of observation $r(t) = |V_0(t)|^2$; for a number M of intervals $r(t)$ is given by (7-12). These probabilities are most difficult to calculate exactly. In the next section we shall approximate the false-alarm probability Q_0 by the expected number of times the random process $r(t)$ crosses above the decision level a, an approximation that is valid when—as usually—$Q_0 \ll 1$. In Sec. 7.4 we shall argue that at input energy-to-noise ratios large enough for useful probabilities of detection, Q_d can be approximated by the probability of detection of a signal whose time τ of arrival is precisely known.

7.3 THE FALSE-ALARM RATE

7.3.1 The First-passage Time Problem

The maximum-likelihood receiver for detecting a signal of unknown arrival time in Gaussian noise must determine whether a certain stochastic process crosses a fixed level during the observation interval. For a receiver working with a single input $v(t)$, this process is the rectified output of a filter matched to the expected signal. When several inputs with independent noise components are available, the process is the sum of a number M of such rectified outputs. The false-alarm probability is the probability that the stochastic process crosses the decision level during the interval of observation when the receiver input contains only noise. Now we shall consider the problem of calculating this probability.

The stochastic process of interest will be denoted by $r(t)$; it is stationary. The probability Q_0 that $r(t)$ will exceed a level $r = a$ sometime during an observation interval $0 < t < T$ is given by

$$Q_0 = 1 - P_0(T), \qquad P_0(t) = \Pr[r(t') < a, 0 \le t' \le t].$$

The function $P_0(t)$ is the probability that a process $r(t')$ drawn at random from its ensemble lies below the level $r = a$ throughout the interval $(0, t)$. From the value $P_0(0) = \Pr[r(0) < a]$ at $t = 0$ this function $P_0(t)$ decreases to 0 as the time t goes to infinity.

The negative derivative $q(t) = -dP_0/dt$, $0 < t < \infty$, is the *first-passage-time probability density function*; $q(t)\, dt$ is the probability that the process $r(t)$ crosses the level $r = a$ from below for the first time in the interval $(t, t + dt)$. Calculating the density function $q(t)$ is essentially the same problem as finding the probability $P_0(t)$, and it has been solved for only a few types of stochastic process $r(t)$.

Early work on first-passage-time probabilities was summarized by Siegert [Sie51] in a paper in which he presented a general solution for stochastic processes of the type known as *Markov processes*. For a Markov process $r(t)$ the conditional probability density function of r at time t, given the values $r_k = r(t_k)$ of the process at an arbitrary set of m previous times $t_m < t_{m-1} < \cdots < t_1 < t$, is a function only of r and $r_1 = r(t_1)$:

$$p(r, t \mid r_1, t_1; r_2, t_2; \ldots; r_m, t_m) = p(r, t \mid r_1, t_1)$$

[Hel91, pp. 444–8], [Pap91, pp. 635–7]. The function $p(r, t \mid r_1, t_1)$ is called the *transition probability density function* of the process. The probability density function of the value $r(t)$ of a Markov process at any time t in the future depends only on its probability density function at the present, and not on the past history of the process. Siegert showed how the Laplace transform of the probability density function $q(t)$ could be written in terms of the Laplace transform of the transition probability density function of the process.

The stationary Gaussian Markov process, for example, has an autocovariance function $\phi(\tau) = \phi(0) \exp(-\mu|\tau|)$, and the Laplace transform of its first-passage-time probability function is a quotient of Weber–Hermite functions. When the interval is so long that $\mu T \gg 1$, the probability density function $q(t)$ is governed mainly by the pole of this Laplace transform lying nearest the origin; and if the level $r = a$ is much higher than the rms value $[\phi(0)]^{1/2}$ of the process, the probability Q_0 that it will be crossed at least once during an interval $(0, T)$ is approximately

$$Q_0 \approx 1 - e^{-\eta t}, \quad \eta = \frac{\mu a}{\sqrt{2\pi\phi(0)}} \exp\left[-\frac{a^2}{2\phi(0)}\right], \quad \mu T \gg 1, \quad a^2 \gg \phi(0), \tag{7-14}$$

where η is the reciprocal of the expected value of the first-passage time and has been calculated by Siegert's formulas.

A second Markov process lending itself to this kind of analysis is the squared envelope

$$r(t) = [x(t)]^2 + [y(t)]^2$$

of a narrowband Gaussian process whose complex autocovariance function is the exponential function $\tilde{\phi}(\tau) = \tilde{\phi}(0) \exp(-\mu|\tau|)$. This process, whose quadrature components $x(t)$ and $y(t)$ are independent Gaussian Markov processes, has been treated by Rice [Ric58], Tikhonov [Tik61], and this writer [Hel59]. The Laplace transform of its first-passage-time probability density function can be expressed as a quotient of confluent hypergeometric functions. When $\mu T \gg 1$ and $a \gg \tilde{\phi}(0)$, the probability Q_0 that the level $r = a$ will be crossed at least once in $(0, T)$ is asymptotically

$$Q_0 \approx 1 - e^{-\eta t}, \qquad \eta = \frac{\mu a}{\tilde{\phi}(0)} \exp\left[-\frac{a}{2\tilde{\phi}(0)}\right], \qquad \mu T \gg 1, \quad a \gg \tilde{\phi}(0). \qquad (7\text{-}15)$$

When the expected value of the process $r(t)$ is 0 and $a = 0$, the problem of calculating the probability density function $q(t)$ or the distribution $P_0(t)$ is known as the *zero-crossing problem.* Problems of this type have been extensively studied by Longuet–Higgens [Lon62], McFadden [McF62], Rainal [Rai62], [Rai87], Slepian [Sle62], and others. Besides the first-passage-time probability density function, they have investigated the distribution of the number of times $r(t)$ crosses the level $r = 0$ in a given interval and the distributions of the lengths of the intervals between such crossings.

7.3.2 The Crossing-rate Approximation

In a radar receiver that is to detect signals of unknown arrival time the false-alarm probability must be kept much smaller than 1, simply because the user cannot afford to let it be large. In a defensive system based on radar, for instance, it is so costly to send missiles to attack apparent targets that few sorties can be permitted. It can therefore be assumed that the level a is so much larger than the rms value of the process $r(t)$ that there is only a small probability that the output $r(t)$ of the receiver will exceed it at any time during the observation interval $(0, T)$. In addition, the interval $(0, T)$ is much longer than the correlation time of the stochastic process $r(t)$, which is on the order of the reciprocal of the bandwidth of the signal. Over most of the interval $(0, T)$ the process $r(t)$ has negligible correlation with its initial value $r(0)$ and its initial time derivatives, and the probability $Q_0 = 1 - P_0(T)$ will be almost independent of them as well. Under these conditions it is useful to define an average rate $\eta(a)$ with which the stochastic process $r(t)$ crosses the level $r = a$ from below. For a stationary process this rate η is constant, and in an interval of length T the average number of crossings is ηT. In radar η is called the *false-alarm rate.*

Let $P_n(T)$ be the probability that the rectified process $r(t)$ crosses the decision level a n times during $(0, T)$. Then the average number of crossings is

$$\eta T = P_1(T) + 2P_2(T) + 3P_3(T) + \cdots,$$

and the false-alarm probability is

$$Q_0 = P_1(T) + P_2(T) + P_3(T) + \cdots$$
$$= \eta T - P_2(T) - 2P_3(T) - \cdots.$$

Under the assumption that the decision level a is so high that the probabilities $P_n(T)$, $n \geq 2$, of two or more crossings in $(0, T)$ are negligible, the false-alarm probability is approximately equal to the product of the false-alarm rate η and the duration T of the observation interval,

$$Q_0 \approx \eta T. \tag{7-16}$$

Cramér [Cra66] showed that if $r(t)$ is a Gaussian random process, then in the limit $a^2 \gg \text{Var } r(t)$ the probabilities $P_n(T)$ have approximately the Poisson form

$$P_n(T) \approx \frac{(\eta T)^n}{n!} e^{-\eta T},$$

and again

$$Q_0 \approx 1 - e^{-\eta T} \approx \eta T, \qquad \eta T \ll 1.$$

We now turn to the calculation of the false-alarm rate η.

7.3.3 The Crossing Rate of a Stochastic Process

The false-alarm rate, or the expected number of crossings of $r = a$ per second, is given by Rice's formula

$$\eta(a) = \int_0^\infty r' p(a, r') dr', \tag{7-17}$$

where $r' = dr/dt$ is the rate of change of the stochastic process $r(t)$, which must be differentiable at least once [Ric44, eq. 3.3–5]. (Primes indicate differentiation with respect to the time t.) The joint probability density function of the rectified output r and its rate r' of change is $p(r, r')$. The history of this result has been narrated by Rainal [Rai88].

To derive (7-17) we utilize the *counting functional*

$$g(t) = r'(t) \delta(r(t) - a)$$

introduced by Middleton [Mid60a, p. 426]. If the process $r(t)$ crosses the level $r = a$ in the brief interval $(t - \varepsilon, t + \varepsilon)$ and is then increasing,

$$\int_{t-\varepsilon}^{t+\varepsilon} g(t)\, dt = \int_{t-\varepsilon}^{t+\varepsilon} r'(t) \delta(r(t) - a)\, dt = \int_{r(t-\varepsilon)}^{r(t+\varepsilon)} \delta(r - a)\, dr = 1, \qquad r' > 0.$$

If it crosses $r = a$ while decreasing,

$$\int_{t-\varepsilon}^{t+\varepsilon} g(t)\, dt = \int_{r(t-\varepsilon)}^{r(t+\varepsilon)} \delta(r - a)\, dr = -1, \qquad r' < 0,$$

because then $r(t + \varepsilon) < r(t - \varepsilon)$. If $r(t)$ does not cross $r = a$ during $(t - \varepsilon, t + \varepsilon)$ at all,

$$\int_{t-\varepsilon}^{t+\varepsilon} g(t)\, dt = 0,$$

for then the delta function $\delta(r - a)$ stands outside the interval $(r(t - \varepsilon), r(t + \varepsilon))$. Introducing a factor $U(r')$, with $U(\cdot)$ the unit step function, into the counting functional eliminates the downward crossings from the count. If we break up the interval $(0, T)$ into subintervals $(t - \varepsilon, t + \varepsilon)$, we see that the total number of times that the process $r(t)$ crosses $r = a$ in an upward direction during $(0, T)$ must be the random variable

$$N_+(T) = \int_0^T r'(t) U(r'(t)) \delta(r(t) - a) \, dt.$$

The average number of upward crossings is obtained by multiplying this by the joint probability density function $p(r, r'; t)$ of the random variables r and r' at time t and integrating:

$$E[N_+(T)] = \int_0^T dt \int_{-\infty}^{\infty} dr \int_{-\infty}^{\infty} dr' \, p(r, r'; t) r' U(r') \delta(r - a)$$

$$= \int_0^T dt \int_0^{\infty} r' p(a, r'; t) \, dr'.$$

Because the process $r(t)$ is stationary, $p(r, r'; t) = p(r, r')$, and

$$E[N_+(T)] = \eta T,$$

where η is the constant false-alarm rate given in (7-17).

For a stationary Gaussian random process $r(t)$ with expected value zero and autocovariance function $\phi(\tau)$, the derivative $r'(t)$ has zero expected value and variance

$$\text{Var } r' = -\frac{d^2\phi(\tau)}{d\tau^2}\bigg|_{\tau=0} = |\phi''(0)|,$$

and $r'(t)$ is independent of $r(t)$ [Hel91, p. 412], [Pap91, p. 314]. Hence their joint density function is

$$p(r, r') = \frac{1}{2\pi\sqrt{\phi(0)|\phi''(0)|}} \exp\left[-\frac{r^2}{2\phi(0)} - \frac{r'^2}{2|\phi''(0)|} \right],$$

and from (7-17) the crossing rate is

$$\eta = \frac{1}{2\pi} \left[\frac{|\phi''(0)|}{\phi(0)} \right]^{1/2} \exp\left[-\frac{a^2}{2\phi(0)} \right].$$

It is necessary that $|\phi''(0)|$ be finite,

$$|\phi''(0)| = \int_{-\infty}^{\infty} \omega^2 \Phi(\omega) \, \frac{d\omega}{2\pi} < \infty,$$

where $\Phi(\omega)$ is the spectral density of $r(t)$. The Gauss–Markov process mentioned at the beginning of Sec. 7.3.1 has $|\phi''(0)| = \infty$, but for such a process we can use (7-14) to approximate the false-alarm probability when $\eta T \ll 1$.

When the level a is high, $a^2 \gg \phi(0)$,

$$\text{Pr}(r \ge a) = \text{erfc}\left(\frac{a}{\sqrt{\phi(0)}} \right) \approx \frac{1}{a} \left[\frac{\phi(0)}{2\pi} \right]^{1/2} \exp\left[-\frac{a^2}{2\phi(0)} \right]$$

by (2-135). Then we can write the crossing rate as approximately

$$\eta \approx \left[\frac{2\pi a^2}{\phi(0)} \right]^{1/2} \frac{\beta}{2\pi} \text{Pr}(r > a), \tag{7-18}$$

where

$$\beta^2 = \frac{\int_{-\infty}^{\infty} \omega^2 \Phi(\omega) \, \frac{d\omega}{2\pi}}{\int_{-\infty}^{\infty} \Phi(\omega) \, \frac{d\omega}{2\pi}} \tag{7-19}$$

is the mean-square bandwidth of the process $r(t)$. The first factor on the right of (7-18) is ordinarily on the order of 1. We can interpret the crossing rate η as roughly equal to the number $\beta/2\pi$ of approximately independent samples of $r(t)$ per second, multiplied by the probability $\Pr(r > a)$ that any one of those samples exceeds the level a.

7.3.4 The Crossing Rate of the Rectified Process

In the maximum-likelihood receiver of a signal of unknown arrival time, as described in Sec. 7.2, the process $r(t)$ is the output of a quadratic rectifier following a filter matched to the signal to be detected:

$$r(t) = |V_0(t)|^2 = [x(t)]^2 + [y(t)]^2 = |z(t)|^2, \qquad (7\text{-}20)$$

where $V_0(t) = z(t) = x(t) + iy(t)$ is a circular complex Gaussian random process with expected value zero under hypothesis H_0. We introduce the notation $z(t)$ for convenience. The derivative of this rectified process is

$$r'(t) = 2(xx' + yy') = 2\,\mathrm{Re}[z^*(t)z'(t)], \qquad (7\text{-}21)$$

where

$$z'(t) = \frac{dz}{dt} = x'(t) + iy'(t)$$

is the derivative of the complex envelope $z(t)$.

We need the joint probability density function of r and r', and we obtain it by way of the conditional probability density function $p(r'|z)$. When the complex variable z is fixed, so are x, y, and r. (Unless otherwise noted, all these are samples of random processes taken at the same time t.) Thereupon r' in (7-21) is a linear combination of Gaussian random variables x' and y', and

$$p(r'|z) = p(r'|x, y)$$

must be a Gaussian density function, determined entirely by the conditional expected value $E(r'|x, y)$ and the conditional variance $\mathrm{Var}(r'|x, y)$, which we now calculate.

First we set up the joint circular Gaussian probability density function for the real and imaginary parts of z and $z' = dz/dt$. The complex autocovariance function of the stationary process $z(t)$ is, as in (3-32),

$$\phi(t_2 - t_1) = \tfrac{1}{2}E[z(t_1)z^*(t_2)].$$

Hence

$$\tfrac{1}{2}E[z'(t_1)z^*(t_2)] = \frac{d}{dt_1}\phi(t_2 - t_1), \qquad (7\text{-}22)$$

and setting $t_1 = t_2 = t$, we obtain

$$\tfrac{1}{2}E(z'z^*) = -\frac{d\phi(\tau)}{d\tau}\bigg|_{\tau=0} = -\phi_0',$$

which is purely imaginary. Likewise,

$$\tfrac{1}{2}E(zz'^*) = \phi_0'.$$

Detection of Signals with Unknown Parameters   Chap. 7

Differentiating (7-22) with respect to t_2 and setting $t_1 = t_2 = t$,

$$\tfrac{1}{2} E(z'z'^*) = -\frac{d^2\phi(\tau)}{d\tau^2}\bigg|_{\tau=0} = -\phi_0'' = |\phi_0''|. \tag{7-23}$$

In terms of the narrowband spectral density $\tilde{\Phi}(\omega)$ of the circular-complex process $z(t)$, defined as in (3-19), the complex autocovariance function is

$$\phi(\tau) = 2 \int_{-\infty}^{\infty} \tilde{\Phi}(\omega) \, e^{i\omega\tau} \, \frac{d\omega}{2\pi},$$

and differentiating twice with respect to τ, we find

$$\phi_0' = 2 \int_{-\infty}^{\infty} i\omega\tilde{\Phi}(\omega) \, \frac{d\omega}{2\pi},$$

$$|\phi_0''| = 2 \int_{-\infty}^{\infty} \omega^2\tilde{\Phi}(\omega) \, \frac{d\omega}{2\pi}.$$

By appropriately choosing the carrier frequency Ω, we can make ϕ_0' equal to zero, and we assume that that has been done. Then the complex covariance matrix of z and z' is diagonal, and by (3-40) their joint circular Gaussian density function is

$$\tilde{p}(z, z') = p(x, y, x', y') = \frac{1}{(2\pi)^2\phi_0|\phi_0''|} \exp\left[-\frac{|z|^2}{2\phi_0} - \frac{|z'|^2}{2|\phi_0''|}\right],$$

and x, y, x', and y' are independent.

By dividing by

$$\tilde{p}(z) = \frac{1}{2\pi\phi_0} \exp\left[-\frac{|z|^2}{2\phi_0}\right], \tag{7-24}$$

we find that the conditional probability density function of x' and y', given x and y, is

$$\tilde{p}(z'|z) = p(x', y'|x, y) = \frac{1}{2\pi|\phi_0''|} \exp\left[-\frac{|z'|^2}{2|\phi_0''|}\right].$$

The conditional expected value of the derivative z' is therefore zero, and by (7-21) that of r' is also zero. By (7-21), (7-20), and (7-23),

$$\begin{aligned}
\text{Var}(r'|z) = \text{Var}(r'|r) &= 4(x^2 \,\text{Var}\, x' + y^2 \,\text{Var}\, y') \\
&= 4(x^2 + y^2)|\phi_0''| = 4r|\phi_0''|.
\end{aligned} \tag{7-25}$$

Thus the conditional density function of the conditionally Gaussian random variable r', given x and y and hence r, must be

$$p(r'|r) = \frac{1}{\sqrt{8\pi r|\phi_0''|}} \exp\left[-\frac{r'^2}{8r|\phi_0''|}\right].$$

Now with $r = |z|^2$ we find from the circular Gaussian density function in (7-24) that

$$p(r) = \frac{1}{2\phi_0} \exp\left(-\frac{r}{2\phi_0}\right) U(r), \tag{7-26}$$

and the joint probability density function of r and r' is therefore

$$p(r, r') = \frac{1}{2\phi_0\sqrt{8\pi r |\phi_0''|}} \exp\left[-\frac{r}{2\phi_0} - \frac{r'^2}{8r|\phi_0''|}\right] U(r).$$

Substituting this into Rice's crossing-rate formula (7-17) and integrating, we find that the crossing rate for this quadratically rectified process is

$$\eta(a) = \beta \left[\frac{2\pi a}{\phi_0}\right]^{1/2} \exp\left(-\frac{a}{2\phi_0}\right) \qquad (7\text{-}27)$$

with

$$\beta^2 = \frac{|\phi_0''|}{4\pi^2 \phi_0}.$$

In terms of the narrowband spectral density $\tilde{\Phi}(\omega)$ of the process $V_0(t)$, defined as in (3-19), the parameter β^2 is

$$\beta^2 = \frac{\int_{-\infty}^{\infty} \left(\frac{\omega - \bar{\omega}}{2\pi}\right)^2 \tilde{\Phi}(\omega) \frac{d\omega}{2\pi}}{\int_{-\infty}^{\infty} \tilde{\Phi}(\omega) \frac{d\omega}{2\pi}}$$

with the mean frequency $\bar{\omega}$ of $V_0(t)$ measured from an arbitrary carrier frequency Ω. Thus β is the rms bandwidth (in hertz) of the process $V_0(t)$ at the output of the narrowband matched filter. Because that filter is matched to the signal $F(t)$,

$$\tilde{\Phi}(\omega) = \frac{N}{2}|f(\omega)|^2,$$

where as before $f(\omega)$ is the Fourier transform of the complex envelope $F(t)$ of the signal. Thus $\beta = \Delta\omega/2\pi$, where $\Delta\omega^2$ is the mean-square bandwidth (6-91). For a signal $F(t)$ bandlimited to $-\pi W < \omega < \pi W$, $\beta = W/\sqrt{12}$; and for a Gaussian signal,

$$F(t) = e^{-t^2/2\sigma^2}, \qquad \beta = \frac{1}{2\pi\sqrt{2}\sigma}.$$

If the signal $F(t)$ is rectangular, on the other hand, β is infinite; but a radar signal cannot rise infinitely rapidly, and the crossing rate η is inevitably finite. The bandwidth parameter β is also infinite for a process $r(t) = [x(t)]^2 + [y(t)]^2$ when $x(t)$ and $y(t)$ are independent Gaussian Markov processes, but for such a rectified process we can use (7-15).

We can express the crossing rate η in (7-27) much as in (7-18),

$$\eta = \left[\frac{2\pi a}{\phi_0}\right]^{1/2} \beta \, \Pr(|V_0(t)|^2 > a),$$

and as the first factor is on the order of 1, the rate is nearly the number β of effectively independent samples per second times the probability that any one sample of $|V_0(t)|^2$ exceeds the level a.

If M inputs are filtered, rectified, delayed, and superimposed, as in (7-12), the random process involved is

$$r(t) = \sum_{k=1}^{M} |z_k(t)|^2, \qquad (7\text{-}28)$$

with $z_k(t) = V_k(t) = x_k(t) + iy_k(t)$ the complex envelope of the output of the matched filter during the kth interval. Then as in (7-21)

$$r'(t) = 2 \text{ Re} \sum_{k=1}^{M} z_k^*(t)z_k'(t).$$

By fixing all the z_k's in our condition, we can carry through the same analysis, and (7-25) becomes

$$\text{Var}(r' | \{z_k\}) = 4 \sum_{k=1}^{M} (x_k^2 + y_k^2)|\phi_0''| = 4r|\phi_0''|$$

as before. The only change in our result is to replace $p(r)$ in (7-26) by the gamma density function for M degrees of freedom,

$$p(r) = \frac{1}{2\phi_0(M-1)!} \left[\frac{r}{2\phi_0} \right]^{M-1} \exp\left(-\frac{r}{2\phi_0}\right) U(r), \qquad (7\text{-}29)$$

whereupon the false-alarm rate is

$$\eta(a) = \frac{1}{(M-1)!} \left[\frac{2\pi a}{\phi_0} \right]^{1/2} \left[\frac{a}{2\phi_0} \right]^{M-1} \beta \exp\left(-\frac{a}{2\phi_0}\right), \qquad (7\text{-}30)$$

and the false-alarm probability is approximately $Q_0 = \eta T$.

7.4 UNKNOWN ARRIVAL TIME: THE PROBABILITY OF DETECTION

The probability of detection in a maximum-likelihood receiver, expressed in (7-13), is even more difficult to calculate precisely than the false-alarm probability, for the random process $r(t)$ is no longer stationary; the signal makes its statistical properties, such as its expected value and variance, into functions of time. At large energy-to-noise ratios, however, as we shall now argue, it is a good approximation to set the probability Q_d of detection equal to what it would be if the arrival time τ of the signal were known. In effect we equate the probability that the peak value of the process $r(t)$ exceeds the decision level $r = a$ with the probability that a sample of $r(t)$ taken at the proper instant for detecting a signal of known arrival time will exceed the same level.

The signal component of the rectified, delayed, and summed outputs, that is, of $r(t)$, peaks at the time $t_0 = \tau_0 + T'$, where τ_0 is the exact arrival time of the signal and T' the delay in the matched filter. When the energy-to-noise ratio is large, this time is close to the time $t_m = \hat{\tau} + T'$ when the sum of signal and noise is maximum. The difference $t_m - t_0$ is the error in estimating the arrival time of the radar echo, and as we have seen in Sec. 6.3, its rms value is much less than the reciprocal bandwidth $\Delta\omega^{-1}$ of the signal when the energy-to-noise ratio $S = E/N$ is large. Because of the correlation imposed on the noise by the matched filter, the sum of signal and noise can change only slightly between times t_0 and t_m. Then the probability

$$\Pr[\max r(t) > a, \quad T' \le t \le T + T' | H_1] = \Pr[r(t_m) > a | H_1]$$

that the peak of the output $r(t)$ exceeds the decision level a when H_1 is true must, for $S \gg 1$, be nearly the same as the probability

$$\Pr[r(t_0) > a| H_1].$$

This, however, is just the probability of detecting the signal when its arrival time τ_0 is known, as calculated in Sec. 4.2.2. The detection probability is therefore to good approximation

$$Q_d \approx Q_M(S, y) = \int_y^\infty \left(\frac{r}{S}\right)^{(M-1)/2} e^{-S-r} I_{M-1}(2\sqrt{Sr})\, dr, \qquad y = \frac{a}{2\phi_0}, \qquad (7\text{-}31)$$

M being the number of delayed and rectified outputs of the matched filter that make up the final output $r(t)$. [Here we use the form of the generalized Marcum Q-function in (C-19) for simplicity of writing.]

The false-alarm probability is

$$Q_0 = \frac{1}{(M-1)!} \sqrt{4\pi y}\, y^{M-1}(\beta T)\, e^{-y}, \qquad y = \frac{a}{2\phi_0}, \qquad (7\text{-}32)$$

from (7-30). For a signal bandlimited to W hertz, as we saw in Sec. 7.3, $\beta = W/\sqrt{12}$, and we find

$$Q_0 = \frac{1}{(M-1)!} \sqrt{\frac{\pi y}{3}}\, y^{M-1} L\, e^{-y}, \qquad L = WT.$$

In Fig. 7-3 we have plotted versus $L = WT$ the energy-to-noise ratio $S = E/N$ (dB) required to yield a detection probability $Q_d = 0.999$ for such a bandlimited signal with $M = 1$ and the same false-alarm probabilities as in Fig. 7-1. The maximum-likelihood receiver is seen to be much less sensitive to uncertainty in the arrival time τ than the threshold receiver of Sec. 7.1. Lengthening the observation interval by a factor of 1000 requires an increase in the energy-to-noise ratio S by only 0.7 to 1.3 dB in order to maintain the same reliability (Q_0, Q_d). The factor βT affects the decision level through (7-32) only logarithmically, and the resulting slow rise in the decision level a with increasing βT entails only a slow rise in the requisite energy-to-noise ratio.

7.5 SIGNALS OF UNKNOWN ARRIVAL TIME AND CARRIER FREQUENCY

If a radar target is moving, the carrier frequency of the echo is displaced from that of the transmitted radar pulse by the Doppler shift. As shown in Appendix F, if the target is moving away from the transmitter and receiver with a velocity v, the carrier frequency is altered upon reflection by $w = -(2v/c)\Omega_0$, where c is the velocity of light and Ω_0 the carrier frequency of the transmitted pulse; $|v| \ll c$. The complex envelope of the signal is slightly expanded or contracted, depending on the direction of relative motion, but by a negligible amount when as usual $|v| \ll c$. The receiver must anticipate signals having a carrier frequency Ω lying anywhere in a band $\Omega_0 - \frac{1}{2}W_d < \Omega < \Omega_0 + \frac{1}{2}W_d$, where $\frac{1}{2}W_d$ equals $(2\Omega_0/c)$ times the maximum velocity attainable by a target.

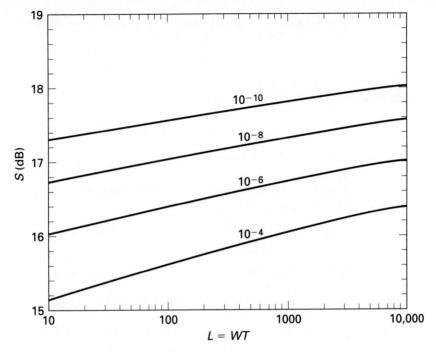

Figure 7-3. Energy-to-noise ratio (dB) for $Q_d = 0.999$, maximum-likelihood detector, versus time–bandwidth product. Curves are indexed with the false-alarm probability Q_0.

A typical echo signal will have the form

$$s(t; A, \psi, \tau, w) = A \operatorname{Re} F(t - \tau) \exp[i(\Omega_0 + w)t + i\psi],$$

where τ is the arrival time of the envelope, w the Doppler shift, A the amplitude, and ψ the phase. The likelihood functional for detecting this signal in white Gaussian noise is

$$\Lambda[v(t); A, \psi, \tau, w] = \exp\left\{ \frac{A}{N} \operatorname{Re} \left[e^{-i\psi} \int_0^T F^*(t - \tau) e^{-iwt} V(t)\, dt \right] - \frac{A^2}{2N} \int_0^T |F(t - \tau)|^2\, dt \right\},$$

as in (6-90), where $V(t)$ is the complex envelope of the input

$$v(t) = \operatorname{Re} V(t) \exp i\Omega_0 t,$$

referred to the carrier frequency Ω_0 of the transmitter. Again the phase ψ can be taken as uniformly distributed over $(0, 2\pi)$, and we assume that the observation interval $(0, T)$ is much longer than the duration T' of the signal.

In the ensemble of signals the arrival time τ and the Doppler shift w can be taken as statistically independent, and if their prior probability density functions are $z_1(\tau)$ and $z_2(w)$, the average likelihood functional for detecting a signal of amplitude A will be

$$\overline{\Lambda}[v(t)] = \int_0^T z_1(\tau)\,d\tau \int_{-W_d/2}^{W_d/2} z_2(w)\,dw\,\exp\left[-\frac{A^2}{2N}\int_0^T |F(t-\tau)|^2\,dt\right.$$
$$\left. \cdot\, I_0\left[\frac{A}{N}\left|\int_0^T F^*(t-\tau)\,e^{-iwt}\,V(t)\,dt\right|\right]\right],$$

much as in (7-2). Again we can put

$$\int_0^T |F(t-\tau)|^2\,dt \approx \int_{-\infty}^{\infty} |F(t)|^2\,dt = 1,$$

and again we can take the prior density function $z_1(\tau)$ of the arrival time as the uniform one in (7-1).

If the input $v(t)$ is passed through a filter matched to a signal with Doppler shift w, whose complex impulse response is

$$K(s; w) = \begin{cases} F^*(T'-s)\,e^{-iw(T'-s)}, & 0 \le s \le T', \\ 0, & s < 0, \quad s > T', \end{cases}$$

as in (6-93), the complex envelope $V_0(t; w)$ of the output will be

$$V_0(t; w) = \int_{-\infty}^{\infty} F^*(u)\,e^{-iwu}\,V(t - T' + u)\,du \qquad (7\text{-}33)$$

as in (3-16), and we can write the average likelihood functional as

$$\overline{\Lambda}[v(t)] = \frac{1}{T}\,\exp\left[-\frac{A^2}{2N}\int_{-\infty}^{\infty} |F(t)|^2\,dt\right]$$
$$\cdot \int_{T'}^{T+T'} d\tau \int_{-W_d/2}^{W_d/2} z_2(w) I_0\left[\frac{A}{N}|V_0(t; w)|\right] dw. \qquad (7\text{-}34)$$

To implement this functional, even after picking some standard amplitude A and making a reasonable assumption about the prior density function $z_2(w)$, would be most difficult, not to speak of trying to evaluate the performance of the resulting receiver.

Because the modified Bessel function $I_0(\cdot)$ is a steeply rising function of its argument, the main contribution to the integral in (7-34) comes from the neighborhood of the maximum value of $|V_0(t; w)|$, provided, as we assume, that the prior density function $z_2(w)$ varies smoothly over the frequency interval $(-\frac{1}{2}W_d < w < \frac{1}{2}W_d)$ and $W_d T' \gg 1$. If that maximum value $|V_0(t_m; w_m)|$ is large, $\overline{\Lambda}[v(t)]$ will be approximately proportional to

$$I_0\left[\frac{A}{N}|V_0(t_m; w_m)|\right]$$

and we can expect $\overline{\Lambda}[v(t)]$ to be large when $|V_0(t_m; w_m)|$ is large. A good approximation to the optimum receiver will again be the maximum-likelihood receiver, which decides that a signal is present (hypothesis H_1) if the maximum value of the function $|V_0(t; w)|^2$ exceeds a decision level a, that is, if $|V_0(t; w)|^2$ crosses a at some time t in $(T', T + T')$ and for some value or values of w.

It is impossible to examine $|V_0(t; w)|^2$ for a continuum of values of w in $(-\frac{1}{2}W_d, \frac{1}{2}W_d)$. Instead, just as in Sec. 6.3.2, one constructs a bank of narrowband filters in parallel, each matched to a signal

$$\text{Re } F(t) \exp i(\Omega_0 + w)t, \qquad 0 \le t \le T',$$

for one of a finite set of values of w uniformly spaced over the range $(-\frac{1}{2}W_d, \frac{1}{2}W_d)$, and one observes all their rectified outputs during the interval $(T', T + T')$. By taking their spacing δw in frequency small enough so that their complex transfer functions $f^*(\omega - w)$ substantially overlap, one loses little of the information contained in $|V_0(t; w)|$ as a continuous function of w. [Here again $f(\omega)$ is the spectrum of the complex envelope of the signal $F(t)$.]

In Secs. 6.3.2 and 6.3.3, we examined the outputs of just such a bank of matched filters. There we showed that the signal component of the quadratically rectified output of the filter $K_w(s)$ matched to a signal with Doppler shift w is

$$R_s(t; w) = A^2 |\lambda(t_0 - t, w - w_0)|^2, \qquad t_0 = \tau_0 + T',$$

where $\lambda(\tau, w)$ is the complex ambiguity function defined in (6-96); $\lambda(0, 0) = 1$. This output is maximum in the filter tuned for $w = w_0$ and at time $t_0 = \tau_0 + T'$. The noise will displace the time at which the peak output occurs and may cause that largest output to take place in a neighboring filter.

The noise components of the filter outputs are, however, correlated. As in (6-108), the complex cross-covariance function of the complex envelopes of the outputs of filters tuned for frequency shifts w_1 and w_2 is

$$\phi(s; w_1, w_2) = \frac{1}{2}E[V_0(t; w_1)V_0^*(t + s; w_2)| H_0]$$
$$= N \exp[-\frac{1}{2}i(w_1 + w_2)s]\lambda(s, w_1 - w_2).$$

The width of this function in the frequency direction is on the order of Δt^{-1}, where again Δt^2 is the mean-square duration of the signal. The rectified outputs of filters having pass frequencies separated by $\delta w \ll \Delta t^{-1}$ will therefore be significantly correlated, and the signal components of those outputs will differ only by a little.

If the rectified output of any of these filters exceeds a decision level a, the receiver decides that a signal is present. The frequency shift w_m associated with the filter having the largest rectified output serves as an estimate of the Doppler shift w of the echo and hence provides an estimate of the component v of the velocity of the target in the direction of the receiver. The time t_m at which the peak output occurs furnishes as before an estimate of the distance of the target through $\hat{\tau} = t_m - T'$.

We can derive a crude approximation to the false-alarm probability by the following reasoning. The factor βT in (7-32) can be thought of as the number of independent opportunities for the rectified output $r(t)$ of the matched filter to cross the decision level a when under hypothesis H_0 it consists of only noise. Looking across the bank of matched filters in the frequency direction, we see outputs that are approximately uncorrelated when separated in angular frequency by roughly Δt^{-1}. The total number of approximately independent outputs in the range of frequencies spanned by the filter bank is about $W_d \Delta t$. The overall false-alarm rate can therefore be approximated by $W_d \Delta t$ times that determined in Sec. 7.3, and for a single input, $M = 1$, by (7-32),

$$Q_0 \approx \sqrt{4\pi y}(\beta T)(W_d \Delta t) e^{-y}, \qquad y = \frac{a}{2\phi_0}, \qquad (7\text{-}35)$$

provided $W_d \Delta t \gg 1$. Here as in Sec. 7.4 y is proportional to the decision level a. If $W_d \Delta t < 1$, the outputs of all the filters will fluctuate more or less together, and in (7-35) we must replace the factor $W_d \Delta t$ by 1.

By an argument similar to that in Sec. 7.4, the probability Q_d of detecting an echo signal with large enough input energy-to-noise ratio S will again be roughly the same as though its arrival time and Doppler shift were known a priori,

$$Q_d \approx Q_1(S, y) = Q(\sqrt{2S}, \sqrt{2y})$$

in terms of Marcum's Q function (3-76). The difference $w_m - w_0$ represents the error in an estimate of the carrier frequency, and we saw in Sec. 6.3.4 that the rms error will be a fraction of Δt^{-1} determined by the signal-to-noise ratio d, and $d \gg 1$. The difference $t_m - t_0$ represents the error in an estimate of the arrival time, and this will be a fraction of $\Delta \omega^{-1}$. If the largest peak output $|V_0(t_m; w_m)|^2$ exceeds the decision level a, therefore, it is nearly certain that the output $|V_0(t_0; w_0)|^2$ will also exceed that decision level. The latter, however, is the decision statistic we should use if we knew the true arrival time and Doppler shift of the signal.

If a number M of successive inputs to the receiver are filtered, rectified, delayed, and summed as in Fig. 7.2, the delay line associated with a given filter must introduce a delay that is compensated for the motion of the target between transmitted pulses. The delay corresponding to a target receding with velocity v must equal $(1 + 2v/c)T = (1 - w/\Omega_0)T$, where w is the resulting Doppler shift. The probability of detection can then as before be approximated by (7-31).

The additional factor $W_d \Delta t$ affects the decision level a only logarithmically, and as with βT, the energy-to-noise ratio required to attain a particular probability Q_d of detection will increase only slowly with an increasing width W_d of the frequency band within which the carrier frequency $\Omega_0 + w$ of the radar echo is uncertain.

When in Chapter 10 we treat the resolution of signals, we shall study the properties of the ambiguity function $\lambda(\tau, w)$ in detail, and we shall consider further how the detection of signals of unknown arrival time and Doppler shift and the estimation of those parameters depend on its form.

7.6 THE LEAST FAVORABLE DISTRIBUTION

7.6.1 The Extended Neyman–Pearson Criterion

In this chapter we have thus far assumed that the unknown arrival time τ and the unknown frequency shift w of the radar echo $s(t; \tau, w)$ are a priori uniformly distributed over their respective ranges $(0, T)$ and $(-\frac{1}{2}W_d, \frac{1}{2}W_d)$. Ordinarily there is no reason to adopt any other prior distributions for such parameters. If the noise were not white, however, but possessed some nonuniform spectral density, or if the noise level varied in some known way during the observation interval, it might be unclear what kind of prior density functions $z_1(\tau)$, $z_2(w)$ would be most reasonable, absent any reliable past experience of the relative frequencies with which various values of the signal parameters τ and w occur.

In order to treat this problem in a general way, we return to the consideration of the extended Neyman–Pearson criterion that we began in Sec. 3.6.2. Let the signal $s(t; \theta)$ depend on m unknown parameters $\theta = (\theta_1, \ldots, \theta_m)$, which lie in a parameter space Θ. As before, we suppose the input $v(t)$ to the receiver to have been appropriately sampled to produce n data $v = (v_1, \ldots, v_n)$ upon which the decision about the presence or the absence of the signal will be based. Under hypothesis H_0—noise alone is present—these are governed by a joint probability density function $p_0(v)$. When under hypothesis H_1 the signal $s(t; \theta)$ is present, their joint probability density function is $p_1(v|\theta)$.

Given the prior probability density function $z(\theta)$ of the unknown parameters,

$$\int_\Theta z(\theta) \, d^m\theta = 1, \tag{7-36}$$

the extended Neyman–Pearson criterion calls for basing the decision between hypotheses H_0 and H_1 on the average likelihood ratio

$$\overline{\Lambda}(v; z) = \frac{p_1(v)}{p_0(v)},$$

where

$$p_1(v) = \int_\Theta z(\theta) p_1(v|\theta) \, d^m\theta \tag{7-37}$$

is the overall probability density function of the data under hypothesis H_1. If $\overline{\Lambda}(v; z)$ exceeds a certain decision level Λ_0, hypothesis H_1 is selected. The decision surface $D[z]$ on which $\overline{\Lambda}(v; z) \equiv \Lambda_0$ divides the space R_n of the data into two regions $R_0[z]$ and $R_1[z]$, and

$$\overline{\Lambda}(v; z) < \Lambda_0, \quad v \in R_0[z]; \qquad \overline{\Lambda}(v; z) \geq \Lambda_0, \quad v \in R_1[z].$$

The value of the decision level Λ_0 is chosen to induce a preassigned value of the false-alarm probability

$$Q_0 = \int_{R_1[z]} p_0(v) \, d^n v. \tag{7-38}$$

The probability of correctly detecting the signal $s(t; \theta)$ with this receiver is

$$Q_d(\theta; z) = \int_{R_1[z]} p_1(v|\theta) \, d^n v, \tag{7-39}$$

and the average probability of detection is

$$\overline{Q}_d[z] = \int_\Theta z(\theta) Q_d(\theta; z) \, d^m\theta. \tag{7-40}$$

Under the extended Neyman–Pearson criterion, as we saw in Sec. 3.6.2, this average detection probability is maximum among all decision strategies satisfying (7-38).

When the prior probability density function $z(\theta)$ is unknown, the most conservative course would seem to be to take it as that density function $\tilde{z}(\theta)$ for which $\overline{Q}_d[z]$ is smallest. As we shall see, with whatever set θ of parameters the signal arrives, the probability $\overline{Q}_d(\theta; \tilde{z})$ of detecting it with the receiver designed for the least favorable distribution $\tilde{z}(\theta)$ will never be less than the average detection probability $\overline{Q}_d[\tilde{z}]$, and a fortiori for any actual prior distribution $z(\theta)$ of the parameters

$$\int_\Theta z(\boldsymbol{\theta}) Q_d(\boldsymbol{\theta}; \tilde{z}) \, d^m\boldsymbol{\theta} \geq \overline{Q}_d[\tilde{z}].$$

If, on the other hand, the receiver were designed for some other prior probability density function $z_1(\boldsymbol{\theta})$, and the true prior probability density function $z(\boldsymbol{\theta})$ differed from it, the average detection probability might be less than $\overline{Q}_d[\tilde{z}]$. We now set out to determine criteria by which the least favorable prior probability density function $\tilde{z}(\boldsymbol{\theta})$ can be identified.

When, as often, the amplitude $A > 0$ of the signal, perhaps among other parameters, is unknown a priori, its least favorable distribution $\tilde{z}(A)$ will obviously be $\delta(A)$; the probability $Q_d(A)$ of detecting a signal of amplitude A is always least when $A = 0$. The amplitude A thus plays a peculiar role among the signal parameters, and it must be treated separately. In what follows we assume that either the amplitude A of the signal is known, or the receiver is designed to be optimum only for signals of some standard amplitude, or the probability density function $p_1(v|\, A, \boldsymbol{\theta})$ has already been averaged with respect to some prior density function $z(A)$ of the amplitude. We turn to the search for a criterion by which the least favorable distribution $\tilde{z}(\boldsymbol{\theta})$ of parameters $\boldsymbol{\theta}$ other than A can be recognized.

For simplicity let us assume at first that the parameter space contains only M discrete points $\boldsymbol{\theta}_1, \boldsymbol{\theta}_2, \dots, \boldsymbol{\theta}_M$. In detection, for instance, it may be that under hypothesis H_1 one and only one of M possible signals $s_j(t) = s(t; \boldsymbol{\theta}_j)$ can be present in a given trial; the observer is not concerned which it is. Let the prior probabilities of these parameter values or signals be $z_1, z_2, \dots, z_M$, with

$$z_1 + z_2 + \cdots + z_M = 1. \tag{7-41}$$

The set of probabilities $\mathbf{z} = (z_1, z_2, \dots, z_M)$ specifies the relative frequencies of the various parameter values $\boldsymbol{\theta}_j$ when hypothesis H_1 is true, that is, when some signal is present. These probabilities can be represented as a point in an M-dimensional Cartesian space, and (7-41) requires the point to lie on a particular hyperplane in that space.

The joint probability density function of the observations v when the jth signal is present is abbreviated as

$$p_1(v|\, \boldsymbol{\theta}_j) = p_j(v);$$

the joint density function of the data v under hypothesis H_0 is still $p_0(v)$. The average probability of detection is then

$$\overline{Q}_d[\mathbf{z}] = \sum_{j=1}^M z_j \int_{R_1[\mathbf{z}]} p_j(v) \, d^n v = \sum_{j=1}^M z_j Q_j[\mathbf{z}], \tag{7-42}$$

where

$$Q_j[\mathbf{z}] = \int_{R_1[\mathbf{z}]} p_j(v) \, d^n v$$

is the probability of detecting the jth signal with the Neyman–Pearson strategy for prior distribution $\mathbf{z}$. Here $R_1[\mathbf{z}]$ is again the region of the space $\boldsymbol{R}_n$ of observations in which H_1 is chosen. The Neyman–Pearson criterion directs us to maximize the average probability $\overline{Q}_d[\mathbf{z}]$ for a fixed false-alarm probability given by (7-38). The regions $R_0[\mathbf{z}]$ and $R_1[\mathbf{z}]$ are separated by the decision surface $D[\mathbf{z}]$, whose position we must determine. In particular, we seek a set of prior probabilities z and a decision surface $D[z]$ such that the maximum value of $\overline{Q}_d[z]$ is as small as possible.

The prior probabilities z are restricted to the range $(0, 1)$,

$$0 \le z_j \le 1, \qquad j = 1, 2, \ldots, M; \tag{7-43}$$

the point z lies in the unit hypercube of the M-dimensional space, a requirement that further constrains the variational problem at hand. With each point z in the intersection of this unit hypercube with the hyperplane of (7-41), there is associated a maximum value of $\overline{Q}_d[z]$ for a given value of Q_0. That maximum value is attained by a receiver that forms the average likelihood ratio

$$\overline{\Lambda}(v) = \sum_{j=1}^{M} z_j \Lambda_j(v), \tag{7-44}$$

where

$$\Lambda_j(v) = \frac{p_j(v)}{p_0(v)}, \qquad 1 \le j \le M,$$

and compares it with a decision level Λ_0, choosing hypothesis H_1 when $\overline{\Lambda}(v) \ge \Lambda_0$ and hypothesis H_0 otherwise. The decision level Λ_0 is set by the requirement that the false-alarm probability equal the preassigned value. This strategy is the same as that described in the beginning, except that we have only a finite number instead of a continuum of possible signals.

Henceforth we define $\overline{Q}_d[z]$ as the average probability of detection when this strategy, optimum under the extended Neyman–Pearson criterion, has been adopted by the receiver. We shall now demonstrate that $\overline{Q}_d[z]$ is a convex U function of the set $z = (z_1, z_2, \ldots, z_M)$ as constrained by (7-41). That is, if z_A and z_B are two such distributions of prior probability among the M signals and if μ_A and μ_B are two constants in $(0, 1)$, with $\mu_A + \mu_B = 1$, then

$$\overline{Q}_d[z] = \overline{Q}_d[\mu_A z_A + \mu_B z_B] \le \mu_A \overline{Q}_d[z_A] + \mu_B \overline{Q}_d[z_B]. \tag{7-45}$$

From this convexity it follows that a least favorable distribution exists.

Consider two scenarios S_A and S_B. Under S_A the transmitter in a binary communication system sends the signals $s_i(t) = s(t; \theta_i)$—when it sends any signal at all—with the relative frequencies

$$z_A = \left(z_1^{(A)}, z_2^{(A)}, \ldots, z_M^{(A)} \right);$$

under S_B it sends them with relative frequencies z_B. It uses scenario S_A with relative frequency μ_A and S_B with relative frequency $\mu_B = 1 - \mu_A$. The actual set of relative frequencies of the signals under hypothesis H_1 is then $\mu_A z_A + \mu_B z_B = \overline{z}$. The transmitter adopts one scenario or the other whether the current message symbol is a 0 or a 1; if it is a 0, of course, no signal is transmitted.

The transmitter informs the receiver which scenario it is using in any given symbol transmission by sending a symbol $u = 0$ or $u = 1$ over a noiseless channel, $u = 0$ indicating it is using scenario S_A and $u = 1$ indicating it is using S_B. The receiver bases its decisions on the data $v_1, v_2, \ldots, v_n$ and on the variable u. If $u = 0$, it forms the average likelihood ratio

$$\overline{\Lambda}_A(v) = \sum_{j=1}^{M} z_j^{(A)} \Lambda_j(v).$$

If $u = 1$, it forms the average likelihood ratio

$$\overline{\Lambda}_B(v) = \sum_{j=1}^{M} z_j^{(B)} \Lambda_j(v).$$

These are compared with decision levels Λ_{A0} and Λ_{B0}, respectively, each chosen to yield the preassigned false-alarm probability Q_0. The average probability of correct decision under hypothesis H_1 is then

$$\overline{Q}_d = \mu_A \overline{Q}_d[\mathbf{z}_A] + \mu_B \overline{Q}_d[\mathbf{z}_B], \tag{7-46}$$

and the false-alarm probability is $\mu_A Q_0 + \mu_B Q_0 = Q_0$.

An alternative strategy for the receiver is to disregard the datum u. The prior probability of the jth signal under hypothesis H_1 is then equal to

$$\overline{z}_j = \mu_A z_j^{(A)} + \mu_B z_j^{(B)},$$

and the receiver must form the average likelihood ratio

$$\overline{\Lambda}(v) = \sum_{j=1}^{M} \overline{z}_j \Lambda_j(v),$$

comparing it with that decision level Λ_0 that yields the false-alarm probability Q_0. This alternative strategy attains an average detection probability $\overline{Q}_d[\overline{\mathbf{z}}]$, with $\overline{\mathbf{z}} = \mu_A \mathbf{z}_A + \mu_B \mathbf{z}_B$. This average detection probability cannot, however, exceed the detection probability in (7-46), for in ignoring the datum u the alternative strategy has discarded part of the available information on which the decisions might be based. The inequality in (7-45) must therefore hold, and the average detection probability $\overline{Q}_d[\mathbf{z}]$ must be a convex U function in the space of the distributions $\mathbf{z}$ of prior probability constrained by (7-41).

Because the average detection probability $\overline{Q}_d[\mathbf{z}]$ is convex U in $\mathbf{z}$, it must possess a unique minimum $\check{Q}_d = Q_d[\check{\mathbf{z}}]$ for some prior distribution $\check{\mathbf{z}} = (\check{z}_1, \dots, \check{z}_M)$ within or on the boundary of the admissible region

$$0 \le \check{z}_j \le 1, \qquad 1 \le j \le M, \qquad \sum_{j=1}^{M} \check{z}_j = 1.$$

We call $\check{\mathbf{z}}$ the *least favorable distribution* of the prior probabilities of the several signals $s_i(t)$. There may be a number of prior probability distributions $\check{\mathbf{z}}$ yielding the same minimum average detection probability $\overline{Q}_d[\check{\mathbf{z}}]$; if so, they form a convex set in the sense that any weighted average of them will attain the same average detection probability. This would be the case, for instance, if certain of the parameters $\boldsymbol{\theta}$ were irrelevant. We disregard this possibility henceforth.

If a uniformly most powerful test of hypothesis H_1 against hypothesis H_0 exists and if it is characterized by the probabilities $Q_i = Q_d(\boldsymbol{\theta}_i)$ of detecting the signals, $1 \le i \le M$, the least favorable distribution will assign probability 1 to the signal for which Q_i is least. (If several signals share a common minimum detection probability, the prior probabilities z_i can be distributed among these signals arbitrarily.) Such is the case, as we have seen, when the signals have the same form, but different positive amplitudes; the weakest signal will be assigned prior probability 1. Henceforth we exclude the existence of a uniformly most powerful test.

When the minimum $\tilde{Q}_d$ occurs at a point $\tilde{\mathbf{z}}$ lying entirely within the admissible region, we can find $\tilde{\mathbf{z}}$ by the technique of Lagrange multipliers. We are seeking a stationary value of (7-42) under the constraints of (7-38), (7-41), and (7-43). We therefore try to find a value of the quantity

$$\Gamma = \sum_{j=1}^{M} z_j \int_{R_1[\mathbf{z}]} p_j(v) d^n v - \lambda \int_{R_1[\mathbf{z}]} p_0(v) d^n v - \mu \sum_{j=1}^{M} z_j \qquad (7\text{-}47)$$

that is maximum for a variation in the decision surface $D[\mathbf{z}]$ and minimum for a variation in the components z_j of $\mathbf{z}$. These variations can now be made independently. The position of the surface $D[\mathbf{z}]$ and the values of the z_j's for which the stationary value is attained are functions of the Lagrange multipliers λ and μ, whose values are later chosen so that the constraints in (7-38) and (7-41) are satisfied.

First taking the z_j's as fixed, we vary the surface $D[\mathbf{z}]$ until the quantity Γ is maximum. As in our analysis in Sec. 1.2, the maximum is attained when the decision surface $D[\mathbf{z}]$ is one of a family of surfaces described by the equation

$$\Lambda(v) = \sum_{j=1}^{M} z_j \frac{p_j(v)}{p_0(v)} = \lambda', \qquad v \in D[\mathbf{z}], \qquad 0 < \lambda' < \infty. \qquad (7\text{-}48)$$

When the decision surface $D[\mathbf{z}]$ is given by (7-48), any small variation in it will produce a decrease in the quantity Γ that is of second order in the magnitude of the change in the position of the surface, as is usually the case with stationary values.

Assuming now that the regions $R_0[\mathbf{z}]$ and $R_1[\mathbf{z}]$ are separated by a surface $D[\mathbf{z}]$ given by (7-48), let us vary each z_j in (7-47) so as to minimize the quantity Γ. This variation will cause a change in the surface $D[\mathbf{z}]$, as well as in the z_j's appearing explicitly in (7-47), but the effect on Γ of the variation of the surface D is of second order because it has been set in accordance with (7-48), and it is only the variations in the explicit z_j's that matter. In this way we obtain the set of M equations

$$\frac{\partial \Gamma}{\partial z_j} = \int_{R_1[\mathbf{z}]} p_j(v) \, d^n v - \mu \equiv 0, \qquad j = 1, 2, \ldots, M.$$

These equations assert that for the set $\mathbf{z} = \{z_j\}$ of prior probabilities that we are seeking, the Neyman–Pearson strategy yields a detection probability for the jth signal (or set of parameter values $\boldsymbol{\theta}_j$) that is the same for all signals, $Q_j \equiv \mu$, $j = 1, 2, \ldots, M$. Along with (7-48) and the constraints of (7-38) and (7-41), these equations suffice to determine the set $\mathbf{z}$ of prior probabilities.

Let us now write (7-47) as

$$\Gamma = \sum_{j=1}^{M} z_j (Q_j[\mathbf{z}] - \mu) - \lambda Q_0[\mathbf{z}] \qquad (7\text{-}49)$$

and assume that for each set $\mathbf{z} = (z_1, \ldots, z_M)$ the Neyman–Pearson detector (7-48) that suffers a false-alarm probability Q_0 is being utilized. Suppose that the minimum value of $\overline{Q}_d[\mathbf{z}]$ occurs for a set $\mathbf{z}$ on the boundary of the admissible region, some of the prior probabilities z_i being zero.

Consider a set of infinitesimal variations δz_j in the prior probabilities z_j; by (7-41),

$$\sum_{j=1}^{M} \delta z_j = 0.$$

The change in Γ is then

$$\delta\Gamma = \sum_{j=1}^{M} \delta z_j (Q_j[\mathbf{z}] - \mu),$$

for as we have seen, the change in

$$\sum_{j=1}^{M} z_j Q_j[\mathbf{z}] - \lambda Q_0[\mathbf{z}]$$

due to the alteration of the decision surface D under the change $\delta\mathbf{z}$ is of second order and negligible. If Γ is to be minimum, this change $\delta\Gamma$ must be nonnegative. If $\tilde{z}_j > 0$, δz_j can be either positive or negative, and $Q_j - \mu$ must vanish. If $\tilde{z}_j = 0$, however, δz_j can only be positive, for the variation must not take the point $\mathbf{z}$ outside the admissible region. Then in order for Γ to be minimum, $Q_j[\tilde{\mathbf{z}}] - \mu$ must be nonnegative, and we obtain the conditions

$$Q_i[\tilde{\mathbf{z}}] \geq \mu, \qquad \tilde{z}_i = 0,$$
$$Q_i[\tilde{\mathbf{z}}] = \mu, \qquad \tilde{z}_i > 0.$$

From these it follows that

$$\tilde{Q}_d = \overline{Q}_d[\tilde{\mathbf{z}}] = \sum_{i=1}^{M} \tilde{z}_i Q_i[\tilde{\mathbf{z}}] = \mu,$$

and we can write the conditions for the least favorable distribution as

$$Q_i[\tilde{\mathbf{z}}] \geq \tilde{Q}_d, \qquad \tilde{z}_i = 0,$$
$$Q_i[\tilde{\mathbf{z}}] \equiv \tilde{Q}_d, \qquad \tilde{z}_i > 0. \qquad (7\text{-}50)$$

If the receiver is designed to meet the extended Neyman–Pearson criterion for the least favorable prior distribution, it attains the same probability $\tilde{Q}_d$ of detecting all signals $s_i(t)$ whose prior probabilities are positive. The other signals can be detected with greater probability $Q_i[\tilde{\mathbf{z}}]$, but the distribution $\tilde{\mathbf{z}}$ assigns them zero prior probability.

When detecting signals in Gaussian noise, the ratio $p_j(v)/p_0(v)$ can be replaced by the likelihood functional

$$\Lambda_j[v(t)] = \Lambda[v(t)|\, \boldsymbol{\theta}_j]$$

by passing to the limit $n \to \infty$ of an infinite number of samples of the input $v(t)$; the likelihood functional is given by (2-71) with $s(t) = s(t;\, \boldsymbol{\theta}_j)$. The decision is based on the average likelihood functional

$$\overline{\Lambda}[v(t);\, \tilde{\mathbf{z}}] = \sum_{j=1}^{M} \tilde{z}_j \Lambda[v(t)|\, \boldsymbol{\theta}_j],$$

which is compared with a decision level Λ_0 whose value is picked so that the false-alarm probability Q_0 equals the preassigned value.

As an example, suppose that it is unknown whether the amplitude of the signal is positive or negative. Denote the sign of the signal amplitude by θ_j; $\theta_1 = 1$, $\theta_2 = -1$. Now $M = 2$. Under hypothesis H_1 the signal is either $s_1(t) = s(t)$ or $s_2(t) = -s(t)$. The signals are to be detected in the presence of white Gaussian noise of unilateral spectral density N. The likelihood functionals are then

$$\Lambda[v(t) | \theta_j] = \exp\left[\frac{2}{N}\theta_j \int_0^T s(t)v(t)\, dt - \frac{1}{N}\int_0^T [s(t)]^2\, dt \right], \qquad j = 1, 2,$$

by (2-74). The symmetry of this problem leads to the conjecture that in the least favorable situation positive and negative signals are equally likely: $\tilde{z}_1 = \tilde{z}_2 = \frac{1}{2}$. The decision is then based on the average likelihood functional

$$\overline{\Lambda}[v(t); \tilde{z}] = \tfrac{1}{2}\Lambda[v(t)|\theta_1] + \tfrac{1}{2}\Lambda[v(t)|\theta_2]$$
$$= e^{-d^2/2}\cosh g,$$

where

$$g = \frac{2}{N}\int_0^T s(t)v(t)\, dt, \qquad d^2 = \frac{2}{N}\int_0^T [s(t)]^2\, dt.$$

This average likelihood functional is a monotone function of the absolute value $|g|$ of the output, at time $t = T$, of a filter matched to the signal $s(t)$. The receiver therefore compares $|g|$ with some decision level $g_1 > 0$, deciding that a signal is present if $|g| > g_1$. The false-alarm probability

$$Q_0 = \Pr(|g| \geq g_1 | H_0) = 2\,\text{erfc}\,x, \qquad x = \frac{g_1}{d},$$

determines the appropriate value of the decision level g_1. The probability of detecting either signal is now

$$Q_d(\theta_j) = \Pr(|g| \geq g_1 | H_1, \theta_j)$$
$$= \text{erfc}(x - d) + \text{erfc}(x + d),$$

and it is the same for both signals, that is, independent of the true value of the parameter θ (the sign of the signal), as required by (7-50). This probability of detection, for a given false-alarm probability Q_0, is smaller than the detection probability determined in Sec. 2.2 for a signal of known sign.

The criterion (7-50) enables us to test whether a conjectured least favorable distribution is correct. Consider a communication system, such as the one we have been discussing, in which the 1's in a message are transmitted by sending either one or the other of two orthogonal signals, $s_1(t)$ or $s_2(t)$. These are received in white, Gaussian noise of unilateral spectral density N, and their input signal-to-noise ratios are

$$d_k^2 = \frac{2}{N}\int_0^T [s_k(t)]^2\, dt, \qquad k = 1, 2.$$

If $d_1^2 = d_2^2$, the least favorable distribution assigns equal probability to each signal: $z_1 = z_2 = 1/2$. The reader can easily verify that the resulting receiver will detect each signal, when it is present, with equal probability: $Q_1 = Q_2$. Suppose, however, that $d_1^2 < d_2^2$. One might think that the least favorable distribution would put all the probability on the first signal: $z_1 = 1$, $z_2 = 0$. The receiver would then consist only of a filter matched to $s_1(t)$, followed by a decision device in which its output is compared with a decision level set to provide the preassigned false-alarm probability Q_0. Such a system would detect the second signal, however, with a probability $Q_2 = Q_0$, and this would be less than the probability Q_1 of detecting the first, in violation of the first line of (7-50). The least favorable distribution must in this case assign some positive probability $z_2 > 0$ to signal $s_2(t)$. Just how much it is difficult to calculate.

Our prescription (7-50) does not depend on the number M of points in the parameter space Θ, and we can allow that space to become a continuum with an infinite number of points. It is only necessary to limit the total range of values of each parameter. The receiver bases its decisions on an average likelihood functional

$$\overline{\Lambda}[v(t); z] = \int_{\Theta} z(\theta)\Lambda[v(t); \theta]\, d^m\theta, \qquad (7\text{-}51)$$

where $\Lambda[v(t); \theta]$ is the likelihood functional for detecting the signal $s(t; \theta)$ in the noise. The probability of detecting this signal depends on the prior probability density function $z(\theta)$ built into the receiver strategy, and we write it

$$Q_d(\theta; z) = \Pr\{\overline{\Lambda}[v(t); z] \geq \Lambda_0 |\, H_1, \theta\}.$$

From (7-50) it follows that the parameter space Θ may be found to be divided into two disjoint regions Θ_+ and Θ_0, which may or may not be simply connected. In the region Θ_+ the least favorable prior probability density function $\tilde{z}(\theta)$ of the parameters is positive, and the probability $Q_d(\theta; \tilde{z})$ of detecting the signal $s(t; \theta)$ is independent of θ:

$$\tilde{z}(\theta) > 0, \qquad Q_d(\theta; \tilde{z}) \equiv \int_{\Theta_+} \tilde{z}(\theta')Q_d(\theta'; \tilde{z})\, d^m\theta' = \overline{Q}_d[\tilde{z}], \qquad \theta \in \Theta_+. \qquad (7\text{-}52a)$$

In the complementary region Θ_0 the least favorable distribution $\tilde{z}(\theta)$ vanishes, and if we were playing a game against a malevolent adversary, we should not expect him to send us signals with parameter values in this region Θ_0; for if such a signal did arrive, its probability $Q_d(\theta; \tilde{z})$ of detection would be greater than the uniformly minimum value for region Θ_+:

$$\tilde{z}(\theta) \equiv 0, \qquad Q_d(\theta; \tilde{z}) > \overline{Q}_d[\tilde{z}], \qquad \theta \in \Theta_0. \qquad (7\text{-}52b)$$

These detection probabilities $Q_d(\theta; \tilde{z})$ are achieved by a receiver that forms from the input $v(t)$ the average likelihood functional

$$\overline{\Lambda}[v(t); \tilde{z}] = \int_{\Theta} \tilde{z}(\theta)\Lambda[v(t)|\, \theta]\, d^m\theta$$

and compares it with a decision level Λ_0 fixed by the preassigned false-alarm probability.

It is usually very difficult to calculate the least favorable prior probability density function $\tilde{z}(\theta)$ if it cannot be discovered immediately through some symmetry, invariance, or natural ordering of the parameters of the signals. In Sec. 3.3 we

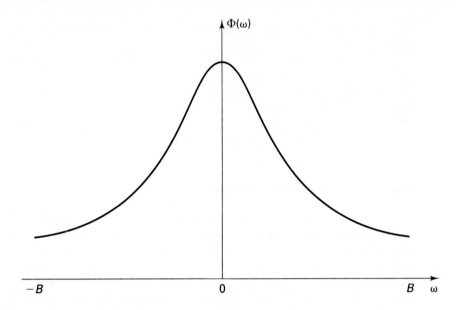

$$\Phi(\omega)$$

$-B \qquad\qquad\qquad 0 \qquad\qquad\qquad\qquad B \quad \omega$

Figure 7-4. Narrowband spectral density of the noise.

analyzed the detection of a narrowband signal of unknown phase ψ under the assumption that the prior probability density function $z(\psi)$ was uniform over $(0, 2\pi)$. The probability $Q_d(\psi)$ of detecting a signal of given phase ψ was shown in Sec. 3.4 to be independent of ψ, and the criterion (7-52) verifies that that uniform prior distribution is indeed the least favorable distribution $\tilde{z}(\psi)$ of the phase ψ of the signal.

In general the criterion (7-52) serves either to confirm or to disprove that a hypothesized prior probability density function $z(\boldsymbol{\theta})$ is least favorable. Consider, for example, the detection of a quasiharmonic signal

$$s(t; \psi, w) = \mathrm{Re}[S(t)\, e^{iwt + i\psi + i\Omega t}]$$

of unknown phase ψ and unknown carrier frequency $\Omega + w$ in the presence of colored Gaussian noise. The parameter w, representing perhaps the Doppler shift of a signal from a transmitter of unknown velocity, is confined to a band $-B < w < B$. The narrowband spectral density $\tilde{\Phi}(\omega)$ of the noise might resemble that in Fig. 7-4. Suppose that the range $2B$ of the Doppler shift w is somewhat greater than the bandwidth of either the signal or the nonwhite component of the noise.

One might think that the signal would least favorably arrive with a Doppler shift $w = 0$, that is, in the midst of the strongest part of the spectral density of the noise, so that the least favorable prior density function of the parameters might be

$$z(\psi, w) = \frac{1}{2\pi}\delta(w), \qquad 0 \le \psi < 2\pi, \qquad -B < w < B. \qquad (7\text{-}53)$$

The optimum detector would then be the same as that worked out in Sec. 3.3 for detecting the signal $s(t; \psi) = s(t; \psi, 0)$, whose carrier frequency is completely known. If we calculated the probability $Q_d(\psi, w)$ of detecting a signal $s(t; \psi, w)$ with a Doppler shift w near the outer limits $\pm B$ of its range, however, we should find it indeed independent of the phase ψ, but not of w, and furthermore, $Q_d(\psi, w)$ would be less than $Q_d(\psi, 0)$, the average detection probability under the conjectured prior density function $z(\psi, w)$ in (7-53). This contradiction of (7-52) establishes that the conjectured least favorable distribution was wrong. Because of the enormous difficulty of calculating the detection probability $Q_d(\psi, w)$ for the optimum detector arising through (7-51) from an arbitrary prior probability density function $z(\psi, w)$, this detection problem remains unsolved. In the next part, however, we shall describe how we can obtain an approximation to the least favorable distribution by considering the threshold detector introduced in Sec. 3.6.3.

7.6.2 The Threshold Detector

A method for determining a prior probability density function $z(\theta)$ of the unknown signal parameters that is at least approximately least favorable rests on the assumption that the signals are so weak that the receiver can use the threshold statistic introduced in Sec. 3.6.3. As we saw in Sec. 4.4.2, this statistic is most appropriate when the receiver can base its choices between hypotheses H_0 and H_1 on a large number M of statistically homogeneous and independent inputs $v_k(t)$, $1 \leq k \leq M$, $0 \leq t \leq T$. As in that section, we designate the input signal strength by a, defining it in such a way that the nonvanishing term of lowest order in an expansion of the likelihood functional in powers of a is proportional to a itself; see (3-108) and (4-71).

We again isolate the signal strength a from the other unknown signal parameters θ, and as at the end of Sec. 4.4.4, we write the likelihood functional for detecting the signal $s(t; a, \theta)$ in any one of the M inputs $v_k(t)$ as $\Lambda[v(t); a, \theta]$. As in (4-77) through (4-79) the threshold statistic is now

$$G_a[\{v_j(t)\}; z] = \sum_{k=1}^{M} g[v_k(t); z], \tag{7-54}$$

where

$$g[v(t); z] = \lim_{a \to 0} [a^{-1}\{\Lambda[v(t); a, z] - 1\}], \tag{7-55}$$

with

$$\Lambda[v(t); a, z] = \int_{\Theta} z(\theta)\Lambda[v(t); a, \theta] \, d^m\theta. \tag{7-56}$$

We are explicitly indicating the dependence of the threshold statistic on the accepted prior probability density function $z(\theta)$ of the signal parameters.

When $M \gg 1$ the probability density functions of the statistic in (7-54) are approximately Gaussian by the central limit theorem, and the false-alarm and detection probabilities are approximately

$$Q_0 \approx \operatorname{erfc} x, \qquad Q_d(\theta; z) \approx \operatorname{erfc}(x - \sqrt{M}D_g(\theta; z)), \tag{7-57}$$

in terms of the error-function integral (1-11), where as in (4-68)

$$[D_g(\theta; z)]^2 = \frac{[E(g \mid H_1, \theta) - E(g \mid H_0)]^2}{\operatorname{Var}_0 g}, \qquad g = g[v(t); z], \tag{7-58}$$

is the effective signal-to-noise ratio for each input $v_k(t)$. As before, Var_0 indicates the variance of a random variable under hypothesis H_0.

The values of the unknown parameters θ appear in (7-57) only through the expected value $E\{g[v(t); z] \mid H_1, \theta\}$ of the statistic defined in (7-55). The criterion (7-52) therefore implies that in the limit $a \to 0$ of weak signals the least favorable distribution $\tilde{z}(\theta)$ in general divides the parameter space Θ into two disjoint regions Θ_+ and Θ_0, such that for some constant h,

$$
\begin{aligned}
E\{g[v(t); \tilde{z}] \mid H_1, \theta\} &\equiv ah, & \tilde{z}(\theta) &> 0, & \theta &\in \Theta_+, \\
E\{g[v(t); \tilde{z}] \mid H_1, \theta\} &\geq ah, & \tilde{z}(\theta) &\equiv 0, & \theta &\in \Theta_0.
\end{aligned}
\tag{7-59}
$$

The constant h is determined by

$$
\begin{aligned}
h &= a^{-1} \int_\Theta \tilde{z}(\theta) E\{g[v(t); \tilde{z}] \mid H_1, \theta\} \, d^m\theta \\
&= a^{-1} E\{g[v(t); \tilde{z}] \mid H_1, \tilde{z}\}.
\end{aligned}
\tag{7-60}
$$

The notation "H_1, z" indicates an expected value with respect to the distribution $z(\theta)$ of the unknown parameters under hypothesis H_1. Henceforth we shall refer to the prior probability density function $\tilde{z}(\theta)$ satisfying (7-59) as the least favorable distribution, suppressing the qualification "in the limit $a \to 0$."

Recalling that the likelihood functional is the limit of the likelihood ratio when the number n of data v grows beyond all bounds,

$$
\Lambda[v(t); a, z] = \lim_{n \to \infty} \frac{p_1(v; a, z)}{p_0(v)},
$$

with

$$
p_1(v; a, z) = \int_\Theta z(\theta) p_1(v \mid a, \theta) \, d^m\theta,
$$

we find

$$
E\{\Lambda[v(t); a, z] \mid H_0\} = \lim_{n \to \infty} \int_{R_n} p_1(v; a, z) d^n v = 1,
$$

so that $E\{g[v(t); z] \mid H_0\} = 0$. Furthermore, with $a \ll 1$,

$$
\begin{aligned}
E\{g[v(t); z] \mid H_1, z\} &= a^{-1} \lim_{n \to \infty} \int_{R_n} \left[\frac{p_1(v; a, z)}{p_0(v)} - 1 \right] p_1(v; a, z) \, d^n v \\
&= a^{-1} E[\{\Lambda[v(t); a, z] - 1\} \Lambda[v(t); a, z] \mid H_0] \\
&= a^{-1} E[\{\Lambda[v(t); a, z] - 1\}^2 \mid H_0] \\
&= a \operatorname{Var}_0 g[v(t); z].
\end{aligned}
$$

[Compare (4-75) and the argument leading up to it.]

For the least favorable distribution $\tilde{z}(\theta)$, therefore, (7-60) yields

$$
h = \operatorname{Var}_0 g[v(t); \tilde{z}],
\tag{7-61}
$$

and by (7-58) the effective signal-to-noise ratio is bounded below by

$$
[D_g(\theta; \tilde{z})]^2 \geq a^2 h = a^2 \operatorname{Var}_0 g[v(t); \tilde{z}],
$$

with equality for parameters θ in the region Θ_+ where $\tilde{z}(\theta) > 0$. Comparison with (4-68) shows that the quantity h represents the efficacy of the threshold detector with respect to a as the measure of input signal strength, the detector having been designed on the basis of the least favorable distribution $\tilde{z}(\theta)$ of the unknown signal parameters.

7.6.3 Narrowband Signals in Gaussian Noise

Let the signals to be detected during $(0, T)$ have the form

$$s(t; A, \theta) = A \, \mathrm{Re} \, F(t; \theta) \, e^{i\Omega t + i\psi},$$

and assume that the phase ψ has its least favorable distribution, the uniform one over $(0, 2\pi)$. Let the narrowband noise $\mathrm{Re} \, N(t) \exp i\Omega t$ have the complex autocovariance function

$$\phi(t, s) = \tfrac{1}{2} E[N(t)N^*(s)]. \tag{7-62}$$

Define $G(t; \theta)$ as the solution of the integral equation

$$F(t; \theta) = \int_0^T \phi(t, s)G(s; \theta) \, ds, \qquad 0 \le t \le T. \tag{7-63}$$

Then as in (3-53) and (3-60), the likelihood functional for detecting the signal $s(t; A, \theta)$ in the input $\mathrm{Re} \, V(t) \exp i\Omega t$ to the receiver is

$$\Lambda[v(t); A, \theta] = \exp[-\tfrac{1}{2} A^2 J(\theta)] I_0(Ar),$$

where

$$r = \left| \int_0^T G^*(t; \theta)V(t) \, dt \right|,$$

$$J(\theta) = \int_0^T G^*(t; \theta)F(t; \theta) \, dt,$$

after averaging over the phase ψ. By using the series expansion (3-61) of the modified Bessel function, we can expand this likelihood functional $\Lambda[v(t); A, \theta]$ in powers of A, and we find

$$\Lambda[v(t); A, \theta] = 1 + \tfrac{1}{4} A^2[r^2 - 2J(\theta)] + O(A^4).$$

Comparison with (7-55) prompts us to take $a = \tfrac{1}{2} A^2$ as our measure of the signal strength, and the threshold statistic for a single input $v(t)$ is then

$$g[v(t); z] = \int_\Theta z(\theta)g[v(t); \theta] \, d^m\theta, \tag{7-64}$$

with

$$g[v(t); \theta] = \tfrac{1}{2}[r^2 - 2J(\theta)]$$
$$= \tfrac{1}{2} \left| \int_0^T G^*(t; \theta)V(t) \, dt \right|^2 - \int_0^T G^*(t; \theta)F(t; \theta) \, dt. \tag{7-65}$$

For this random variable, $E\{g[v(t); \theta] \mid H_0\} = 0$, as can be shown by replacing $V(t)$ by $N(t)$ and taking the expected value with the use of (7-62).

When a signal $s(t; A, \theta)$ is present,

$$V(t) = AF(t; \theta) \, e^{i\Omega t + i\psi} + N(t),$$

and as the reader can easily show by substituting into (7-65),

$$E\{g[v(t); \boldsymbol{\theta}_1]| H_1, \boldsymbol{\theta}\} = a\Gamma(\boldsymbol{\theta}_1, \boldsymbol{\theta}),$$

where

$$\Gamma(\boldsymbol{\theta}_1, \boldsymbol{\theta}_2) = \left| \int_0^T G^*(t; \boldsymbol{\theta}_1)F(t; \boldsymbol{\theta}_2) \, dt \right|^2 \qquad (7\text{-}66)$$

plays a central role in this problem. In terms of it, by (7-64),

$$E\{g[v(t); z]| H_1, \boldsymbol{\theta}\} = a\int_\Theta z(\boldsymbol{\theta}_1)\Gamma(\boldsymbol{\theta}_1, \boldsymbol{\theta}) \, d^m\boldsymbol{\theta}_1.$$

Putting this into (7-59), we find that the criterion for the least favorable distribution $\tilde{z}(\boldsymbol{\theta})$ in the weak-signal limit is

$$
\begin{aligned}
\int_\Theta \tilde{z}(\boldsymbol{\theta}_1)\Gamma(\boldsymbol{\theta}_1, \boldsymbol{\theta}) \, d^m\boldsymbol{\theta}_1 &\equiv h, & \tilde{z}(\boldsymbol{\theta}) &> 0, & \boldsymbol{\theta} &\in \Theta_+, \\
\int_\Theta \tilde{z}(\boldsymbol{\theta}_1)\Gamma(\boldsymbol{\theta}_1, \boldsymbol{\theta}) \, d^m\boldsymbol{\theta}_1 &\geq h, & \tilde{z}(\boldsymbol{\theta}) &\equiv 0, & \boldsymbol{\theta} &\in \Theta_0.
\end{aligned}
\qquad (7\text{-}67)
$$

The solution $\tilde{z}(\cdot)$ of these equations will be proportional to h, and h is determined by the normalization

$$\int_\Theta \tilde{z}(\boldsymbol{\theta}) \, d^m\boldsymbol{\theta} = 1. \qquad (7\text{-}68)$$

Furthermore, multiplying the equations in (7-67) by $\tilde{z}(\boldsymbol{\theta})$ and integrating over the parameter space Θ, we find

$$h = \int_\Theta\int_\Theta \tilde{z}(\boldsymbol{\theta}_1)\Gamma(\boldsymbol{\theta}_1, \boldsymbol{\theta}_2)\tilde{z}(\boldsymbol{\theta}_2) \, d^m\boldsymbol{\theta}_1 \, d^m\boldsymbol{\theta}_2. \qquad (7\text{-}69)$$

The reader should verify by using (3-44) that (7-61) holds for the statistic defined in (7-64) and (7-65).

The kernel $\Gamma(\boldsymbol{\theta}_1, \boldsymbol{\theta}_2)$ is a kind of generalized ambiguity function. Indeed, if the unknown signal parameters are the arrival time τ and the frequency shift w,

$$F(t; \tau, w) = F(t - \tau) \, e^{iwt},$$

and if the noise is white with unilateral spectral density N, so that

$$G(t; \tau, w) = N^{-1}F(t; \tau, w),$$

then as the reader can easily show,

$$\Gamma(\tau_1, w_1; \tau_2, w_2) = N^{-2}|\lambda(\tau_1 - \tau_2, w_1 - w_2)|^2,$$

in terms of the complex ambiguity function defined in (6-96).

The conditions (7-67) are the same as arise when one minimizes the quadratic form on the right side of (7-69) under the constraints (7-68) and $\tilde{z}(\boldsymbol{\theta}) \geq 0$, $\forall\boldsymbol{\theta} \in \Theta$. The only practical way to solve such a problem seems to be to sample the parameter space Θ at a finite number of points and approximate the integral in (7-69) by a double summation and the integrals in (7-67) and (7-68) by single summations over values of $\tilde{z}(\boldsymbol{\theta})$ at those points. The minimization then becomes a problem in quadratic programing [Col71].

As an example, suppose that the only unknown parameter is the frequency shift w of a narrowband signal having the form

$$s(t; w) = A \operatorname{Re} F(t) e^{i(\Omega+w)t+i\psi}.$$

Its complex envelope is taken as

$$F(t; w) = F(t) e^{iwt}.$$

The parameter w is assumed to lie in a finite band $(-B, B)$, and as described at the end of Sec. 7.6.1, the narrowband spectral density $\tilde{\Phi}(\omega)$ of the noise is nonuniformly distributed over that band. In [Hel92b] this problem was treated under the assumptions that the noise is stationary and that the observation interval is so much longer than the duration of the signal that it can be taken as $(-\infty, \infty)$. Then the integral equation (7-63) becomes

$$F(t) e^{iwt} = \int_{-\infty}^{\infty} \tilde{\phi}(t - s)G(s; w) \, ds,$$

where $\tilde{\phi}(\cdot)$ is the complex autocovariance function of the noise. As in Sec. 2.2.2, we can solve this integral equation by Fourier transformation, and the Fourier transform $g(\omega; w)$ of $G(s; w)$ is found to be

$$g(\omega; w) = \frac{f(\omega - w)}{\tilde{\Phi}(\omega)}$$

in terms of the Fourier transform $f(\omega)$ of $F(t)$. The kernel defined in (7-66) becomes

$$\Gamma(w_1, w_2) = \left| \int_{-\infty}^{\infty} \frac{f^*(\omega - w_1) f(\omega - w_2)}{\tilde{\Phi}(\omega)} \frac{d\omega}{2\pi} \right|^2.$$

In [Hel92b] the noise was taken as the sum of white noise and noise with a Lorentz spectral density of bandwidth μ. Without loss of generality the spectral density of the white component was set equal to 1, whereupon

$$\tilde{\Phi}(\omega) = 1 + \frac{2\mu P}{\mu^2 + \omega^2},$$

with P measuring the relative strength of the nonwhite component. The signal was taken to be rectangular.

The frequency band $(-B, B)$ was sampled at a number of uniformly spaced frequencies w_j in order to convert the integrals in (7-67) and (7-68) to summations, and the quadratic programing problem of finding the samples $\tilde{z}(w_j)$ of the least favorable distribution was solved numerically by a method due to Wolfe [Wol59]. It transpired that the least favorable distribution is concentrated in a number M of weighted delta functions:

$$\tilde{z}(w) = \sum_{k=1}^{M} c_k \delta(w - w_k), \qquad w_1 = -B, \ w_M = B. \tag{7-70}$$

The number M and the approximate locations of the delta functions were determined by the quadratic-programing algorithm.

For a prior probability density function of the form in (7-70), the first set of equations in (7-67) becomes

$$\sum_{k=1}^{M} c_k \Gamma(w_j, w_k) \equiv h, \qquad 1 \leq j \leq M, \tag{7-71}$$

when the left side is evaluated at the M points $\theta = w_j$. By (7-69) we must minimize the quadratic form

$$Q(\mathbf{w}, \mathbf{c}) = \sum_{j=1}^{M} \sum_{k=1}^{M} c_j \Gamma(w_j, w_k) c_k, \tag{7-72}$$

$$\mathbf{w} = (w_1, \ldots, w_M), \qquad \mathbf{c} = (c_1, \ldots, c_M),$$

under the constraints

$$c_k > 0, \qquad 1 \leq k \leq M, \qquad \sum_{k=1}^{M} c_k = 1. \tag{7-73}$$

Starting with values of $\mathbf{w}$ provided by the quadratic-programing algorithm, (7-71) was solved with (7-73) to determine the set $\mathbf{c}$ of coefficients more accurately. These were substituted into (7-72), and a gradient algorithm was used to minimize the quadratic form by varying the M frequencies w_k. With the new set $\mathbf{w}$, a new set $\mathbf{c}$ of coefficients was obtained from (7-71) and (7-73). The procedure was repeated until the value of $Q(\mathbf{w}, \mathbf{c})$ ceased changing significantly. Details of the computations can be found in [Hel92b].

The detector optimum in the weak-signal limit now consists of a bank of M narrowband pass filters matched to signals

$$\sigma(t; w_k) = \text{Re } G(t; w_k) e^{i\Omega t}, \qquad 1 \leq k \leq M.$$

After these are quadratic rectifiers whose outputs are weighted with the prior probabilities c_k to form the threshold statistic

$$S = \frac{1}{2} \sum_{k=1}^{M} c_k \left| \int_0^{T'} G^*(T' - s, w_k) V(s) \, ds \right|^2,$$

where T' is a delay long enough to encompass the signals $\sigma(t; w_k)$. The value of S is compared with a decision level set to induce the preassigned false-alarm probability Q_0. When a number of inputs $v_j(t)$ are available one after another, the resulting values of the statistic S are summed before comparison with the decision level. The performance of the system is measured by the minimum value h of the quadratic form $Q(\mathbf{w}, \mathbf{c})$; it decreases with an increasing width $2B$ of the expected range of the frequency shifts w.

Problems

7-1. Compare the rms bandwidth $(\Delta\omega^2)^{1/2}/2\pi$ defined through (6-85) and the bandwidth W defined by (7-11) for strictly bandlimited signals with spectrum $f(\omega)$ uniform in $(-\pi W, \pi W)$, for Gaussian signals, and for rectangular signals.

7-2. Show from (7-7) that for a strictly bandlimited signal Re $F(t)$ exp $i\Omega t$ with a uniform spectrum over the range $-\pi W \leq \omega \leq \pi W$, samples of the complex envelope $V_0(t)$ of the output of the filter matched to the signal are uncorrelated when separated in time by multiples of $1/W$.

7-3. When the time–bandwidth product $L = WT$ is large, the statistic g'_A in (7-6) has approximately a Gaussian distribution, and the false-alarm and detection probabilities are approximately

$$Q_0 \approx \text{erfc } x, \qquad Q_d \approx \text{erfc}(x - D),$$

where D^2 is the effective signal-to-noise ratio. Evaluate D^2 in terms of the signal energy E, the spectral density N of the white noise, and the time–bandwidth product WT. With this Gaussian approximation, calculate the input signal-to-noise ratios S (dB) needed to attain a probability of detection $Q_d = 0.999$ when $WT = 10,000$ and the false-alarm probabilities Q_0 are 10^{-4}, 10^{-6}, 10^{-8}, and 10^{-10}. Compare your results with the values shown in Fig. 7-1.

7-4. Using the counting functional that we employed in deriving Rice's formula (7-17), work out an expression for the variance Var N_+ of the number N_+ of times that a process $r(t)$ crosses the level a in the upward direction during an interval of duration T.

7-5. The signal $s(t) = \theta f(t)$ is to be detected in Gaussian noise having autocovariance function $\phi(t, u)$ by observation of the input $v(t)$ to the receiver during an interval $(0, T)$. The sign θ of the signal is either $+1$ or -1 with equal probabilities. For a false-alarm probability $Q_0 = 10^{-6}$ and for detection probabilities $Q_d = 0.9, 0.99$, and 0.999, calculate the signal-to-noise ratios d^2 required. Compare these with the signal-to-noise ratios needed for the same values of Q_0 and Q_d when the sign θ of the signal is known to be $\theta = +1$. Determine those ratios in decibels and compare with the corresponding "losses" when the phase ψ of a narrowband signal is unknown, as plotted in Fig. 3-5.

7-6. In a binary communication system transmitting 0's and 1's every T seconds, no signal at all is sent when a 0 is the message symbol. When a 1 must be dispatched, the transmitter sends either one, but not both, of two signals, which we label A and B. These suffer Rayleigh fading during propagation to the receiver. Signal A is received as

$$s_A(t) = A \text{ Re } F_1(t) \, e^{i\Omega t + i\psi}$$

and signal B as

$$s_B(t) = A \text{ Re } F_2(t) \, e^{i\Omega t + i\psi},$$

in which the phase ψ is random and uniformly distributed over $(0, 2\pi)$. When signal A is sent, the signal-to-noise ratio d has the Rayleigh distribution

$$z(d) = \frac{d}{s_A^2} \exp\left(-\frac{d^2}{2s_A^2}\right) U(d)$$

as in (4-17); d is proportional to the random amplitude A. When signal B is sent, the signal-to-noise ratio d has the distribution

$$z(d) = \frac{d}{s_B^2} \exp\left(-\frac{d^2}{2s_B^2}\right) U(d)$$

with $s_A^2 \neq s_B^2$ in general. The signal envelopes $F_1(t)$ and $F_2(t)$ are orthogonal. Whenever it needs to transmit a 1, the transmitter sends signal A with probability z_A and signal B with probability z_B; $z_A + z_B = 1$. The receiver does not know which signal, if any, is transmitted. The signals are received in white Gaussian noise with unilateral spectral density N.

(a) Determine the optimum strategy under the Neyman–Pearson criterion by which the receiver should decide whether a 0 or a 1 was sent. The average detection probability $\overline{Q}_d$ is to be maximum when the false-alarm probability has the preassigned value Q_0. Describe how your strategy might be implemented. Show that the decision can be based on the values of just two random variables w_1 and w_2 that have exponential distributions under each hypothesis. [An exponential distribution is one for which the density function has the form $a \exp(-aw)U(w)$ for some positive constant a.] To illustrate the detection strategy, draw a diagram of the (w_1, w_2) plane and its division into the two regions R_0 and R_1 in which hypotheses H_0 and H_1 are respectively chosen.

(b) Show how to calculate for the optimum strategy the false-alarm probability Q_0, the probabilities $\overline{Q}_{dA}$ and $\overline{Q}_{dB}$ of detecting signals A and B, respectively, and the overall average detection probability $\overline{Q}_d$. Reduce each of your expressions for these probabilities until they involve a single integration, but do not attempt to evaluate these single integrals, which would probably have to be computed numerically. One could then search for the values of z_A and z_B for which $\overline{Q}_d$ is minimum by applying the criterion in (7-50).

(c) Now assume that $s_A^2 \ll 1$ and $s_B^2 \ll 1$, and determine the prior probabilities z_A and z_B that are least favorable in the sense of Sec. 7.6.2. The ratio $s_A^2 : s_B^2$ is assumed to stay fixed in the passage to the limit $s_A^2 \to 0$, $s_B^2 \to 0$. Calculate the value of h in (7-60) in terms of s_A^2 and s_B^2.

7-7. Calculate $\mathrm{Var}_0\, g[v(t); z]$ for the threshold statistic $g[v(t); z]$ as given by (7-64) and (7-65) by using (3-44), and show that your result equals the value of h in (7-69).

8

Detection of Signals Under Conditions of Uncertainty

8.1 DETECTION IN NOISE OF UNKNOWN STRENGTH

The detection of signals having unknown parameters was introduced in Sec. 3.6, and our study continued in Chapter 7 with special attention to signals with unknown arrival time and Doppler shift. Now we take up the detection of signals under other kinds of uncertainty, beginning in this section with noise of unknown strength and in the next section treating detection in noise whose probability density functions are only vaguely known.

8.1.1 The CFAR Receiver

When a radar system is being jammed by broadband noise from an enemy transmitter, it is faced with the problem of detecting echo signals in the presence of noise of unknown spectral density. We assume here that that spectral density is so much broader than the spectra of the signals themselves that it can be considered uniform over the frequency band of the echoes, and the noise is effectively white. Were there no other signals about except our echo signals and the enemy's noise and if we knew the shape of its spectral density, but not its total power, we could in principle—as will be seen in Problems 8-1 and 8-2—measure that power exactly and detect the echo signals with the same reliability (Q_0, Q_d) as though the strength of the interfering noise were known a priori. Conditions are in reality never so ideal. The approach usually taken is to estimate the spectral density N of the noise, assumed white, and to substitute that estimate for the true value of N in the specification of the decision level with which the output of the receiver is compared.

The receiver will be assumed to contain a threshold detector; that is, it adds the quadratically rectified outputs of a filter matched to the signal during M successive intervals, as in Secs. 4.2 and 7.2. We now assume, as described in Sec. 7.1, that the rectified output $|V_0(t)|^2$ of the matched filter is sampled at $L = WT$ times during each of the M intervals $(T', T + T')$ in which it is observed; again W is the bandwidth of the signal, defined as in (7-11), and T' is the delay in the matched filter, $T' \ll T$. In radar parlance the interval is divided into L *range bins*, and in effect we assume that an echo will appear in one and only one of these subintervals of duration $W^{-1} = T/L$. We shall deal with a single such sample or range bin in each interpulse period, and our analysis will be the same as though the receiver were deciding about the presence or absence of a target at a fixed, predetermined location.

If the target might be moving, the receiver must again contain a bank of parallel filters matched to signals with a number of discrete carrier frequencies $\Omega_0 + k\delta w$, spaced across the spectral band in which echoes can be expected to appear, as in Sec. 7.5. The instants at which samples are taken during each of the M observation intervals must then be adjusted to compensate for motion of the target between transmitted pulses. Under the presumption that this has been done, we can treat the detection in the same manner as though the target were stationary.

The receiver must choose between the two standard hypotheses

$$v_k(t) = n_k(t), \qquad 1 \le k \le M, \qquad (H_0)$$

and

$$v_k(t) = n_k(t) + s_k(t), \qquad 1 \le k \le M,$$
$$s_k(t) = A \operatorname{Re} F(t) \exp(i\Omega t + i\psi_k), \qquad (H_1)$$

where the phases ψ_k are independently random and uniformly distributed over $(0, 2\pi)$. The noise inputs $n_k(t)$ are white, Gaussian, and independently random from one input to another. The receiver forms the decision statistic

$$U = \tfrac{1}{2} \sum_{k=1}^{M} |z_k|^2, \qquad (8\text{-}1)$$

as in (4-19), with

$$z_k = x_k + iy_k = C \int_0^{T'} F^*(t) V_k(t)\, dt. \qquad (8\text{-}2)$$

As before, $V_k(t)$ is the complex envelope of the kth input, and $(0, T')$ is an interval long enough to contain the entire signal. The components x_k and y_k are independent Gaussian random variables. Under hypothesis H_0 their expected values are zero. Under hypothesis H_1, with the normalization constant C suitably chosen, their expected values are given by

$$E(z_k \mid H_1) = E(x_k \mid H_1) + iE(y_k \mid H_1) = d_k \exp i\psi_k,$$

as in (4-23), where now

$$d_k^2 = \frac{2E_k}{N'}, \qquad 1 \le k \le M, \qquad (8\text{-}3)$$

with E_k the energy of the signal appearing in the kth observation interval. In (8-3) N' is a fiducial noise level introduced for the sake of normalization and taking the place of the true spectral density N, which is now unknown.

Were that spectral density N known, the receiver would compare the statistic U with a decision level U_0 proportional to N. Instead the receiver must estimate that noise level N. To that end it acquires a number M' of auxiliary inputs $v'_j(t)$ containing only white noise that is deemed to be independent of the noise inputs $n_k(t)$ and to possess the same unknown spectral density N. These inputs $v'_j(t)$ will ordinarily be taken in spectral bands close to, but not overlapping that of the signals $s_k(t)$, and care must be taken that no significant vestiges of these signals can appear in them. They are passed through suitable filters whose sampled outputs, much as in (8-2), provide M' complex samples $z'_k = x'_k + iy'_k$, $1 \leq k \leq M'$. In practice there are a number R of such auxiliary inputs and filters, the output of each of which is sampled during each of the M interpulse intervals, whereupon $M' = RM$.

The components x'_k, y'_k are independent Gaussian random variables with expected values zero under both hypotheses H_0 and H_1. These components are furthermore statistically independent of the M components x_k and y_k of the samples $z_k = x_k + iy_k$ in (8-2), and they have the same variances. The statistic

$$U' = \tfrac{1}{2} \sum_{k=1}^{M'} |z'_k|^2 \tag{8-4}$$

then constitutes an estimate of the spectral density N. The expected value of U', as the reader can easily show, is proportional to the actual value N of the spectral density of the noise, and with appropriate scaling U' becomes an unbiased estimator of N.

One replaces the unknown spectral density N, as it figures in the decision level U_0, by the observed quantity U', which is a random variable; and the receiver now decides that a signal is present in the range bin of concern if $U \geq \beta U'$, where β is a constant whose value is specified by the preassigned false-alarm probability Q_0 [Fin68]. Because this receiver suffers the same false-alarm probability no matter what the true value of the spectral density of the total white noise may be, it is called a *constant false-alarm rate* (CFAR) receiver.

8.1.2 False-alarm Probability

The false-alarm probability of this receiver is

$$Q_0 = \Pr(U \geq \beta U' \mid H_0) = \Pr\left(\frac{U}{U'} \geq \beta \mid H_0\right).$$

As shown in elementary texts on probability theory, the probability density function of the random variable $\beta' = U/U'$ is given by

$$p(\beta') = \int_{-\infty}^{\infty} |y| p_U(\beta' y) p_{U'}(y)\, dy,$$

where $p_U(\cdot)$ is the probability density function of the random variable U in (8-1) and $p_{U'}(\cdot)$ is the density function of U' in (8-4) [Hel91, p. 167], [Pap91, p. 138].

With U and U' suitably normalized, these are both gamma distributions with M and $M' = RM$ degrees of freedom, respectively,

$$p_U(y) = \frac{y^{M-1} e^{-y}}{(M-1)!} U(y), \qquad p_{U'}(y) = \frac{y^{M'-1} e^{-y}}{(M'-1)!} U(y), \tag{8-5}$$

under hypothesis H_0, and we find

$$\begin{aligned} p(\beta') &= \frac{\beta'^{M-1}}{(M-1)!(M'-1)!} \int_0^\infty y^{M+M'-1} e^{-(\beta'+1)y} \, dy \\ &= \frac{(M+M'-1)!}{(M-1)!(M'-1)!} \frac{\beta'^{M-1}}{(1+\beta')^{M+M'}} U(\beta'), \end{aligned} \tag{8-6}$$

which is known as the *beta distribution*. Then the false-alarm probability is

$$Q_0 = \frac{(M+M'-1)!}{(M-1)!(M'-1)!} \int_\beta^\infty \frac{y^{M-1}}{(1+y)^{M+M'}} \, dy.$$

This has been tabulated by Pearson [Pea68] as the "incomplete beta distribution." By introducing the variable $x = (1+y)^{-1}$ this is sometimes written as

$$Q_0 = \frac{(M+M'-1)!}{(M-1)!(M'-1)!} \int_0^b x^{M'-1}(1-x)^{M-1} \, dx, \qquad b = \frac{1}{1+\beta}.$$

By expanding $(1-x)^{M-1}$ by the binomial theorem, this can be written in closed form, but as the resulting series contains alternating signs, it is inconvenient for computation, particularly when M and $M' = RM$ are large. Section E.4 of the Appendix presents a series (E-10) due to Robertson [Rob76] that in typical situations converges rapidly and is preferable for computing the false-alarm probability. The relation of the statistic $\beta' = U/U'$ to the F statistic of the analysis of variance is also described there. We turn to the problem of determining the decision level β that yields a preassigned false-alarm probability Q_0.

When M and $M' = RM$ are large, the random variable $V = U - \beta U'$, as the difference of two sums of large numbers of independent random variables, has approximately a Gaussian distribution by virtue of the central limit theorem. Without loss of generality we can take $N' = N = 1$, whereupon

$$E(V) = E(U) - \beta E(U') = M - \beta M'$$

and

$$\sigma_V^2 = \mathrm{Var}(V \mid H_0) = \mathrm{Var}\, U + \beta^2 \mathrm{Var}\, U' = M + \beta^2 M',$$

so that

$$Q_0 = \Pr(V \geq 0 \mid H_0) \approx \mathrm{erfc}\left(-\frac{E(V)}{\sigma_V}\right) = \mathrm{erfc}\, x_0,$$

$$x_0 = \frac{\beta M' - M}{(M + \beta^2 M')^{1/2}}.$$

By solving this equation for β, one can usually determine an initial approximation to the decision level. No real solution exists, however, when $M' < x_0^2$. An adequate starting value in this situation was found to be

$$\beta = M'^{-1}(M + M^{1/2}x_0).$$

Thereafter one can determine the exact value of the constant β by Newton's method, replacing each trial value of β by

$$\beta \leftarrow \beta + \frac{q_0^{(+)}(\beta) - Q_0}{p(\beta)},$$

where $q_0^{(+)}(\beta) = \Pr(\beta' \geq \beta| H_0)$ is computed either by Robertson's series (E-10) or by numerical contour integration as described in the next part; $p(\beta)$ is the probability density function of β given in (8-6). When M or M' or both are large, Stirling's approximation

$$n! \approx (2\pi n)^{1/2} n^n \, e^{-n}$$

can be used for the large factorials.

8.1.3 The Probability of Detection

The probability Q_d that the CFAR receiver correctly reports the presence of a signal,

$$Q_d = \Pr(U \geq \beta U'| H_1) = \Pr\left(\frac{U}{U'} \geq \beta| H_1\right),$$

is given by the complementary cumulative distribution associated with the noncentral beta distribution or with the related noncentral F distribution. References are to be found in Appendix E. Recursive methods are described there for computing the detection probability Q_d for both signals of fixed total energy-to-noise ratio S and signals with fluctuating amplitudes for which S has a gamma distribution as in (4-39). When the numbers M and M' are large, the recursive methods can require lengthy computations liable to round-off error and underflow or overflow.

Because the random variables U and U' are statistically independent, the moment-generating function of the random variable $V = U - \beta U'$ is

$$h(z) = E(e^{-zV}| H_1) = E[e^{-z(U-\beta U')}| H_1]$$
$$= h_U(z)h_{U'}(-\beta z),$$

in which $h_U(z)$ and $h_{U'}(z)$ are the moment-generating functions of U and U', respectively. With U', suitably normalized, governed by the gamma distribution in (8-5) under both hypotheses,

$$h_{U'}(z) = (1 + z)^{-M'}.$$

When the total energy-to-noise ratio S is fixed, the moment-generating function of U is given by (4-24),

$$h_U(z) = \frac{1}{(1 + z)^M} \exp\left(-\frac{Sz}{1 + z}\right), \tag{8-7}$$

and that of the statistic V is therefore

$$h(z) = (1 + z)^{-M}(1 - \beta z)^{-M'} \exp\left(-\frac{Sz}{1 + z}\right). \tag{8-8}$$

If the signal amplitudes are fading, as described in Sec. 4.2.4, the moment-generating function in (8-7) must be averaged with respect to the distribution of the total energy-to-noise ratio S, as in (4-33). If in particular S has the gamma distribution

$$p(S) = \frac{(S/\overline{S}')^{k-1}}{(k-1)!\,\overline{S}'} \, e^{-S/\overline{S}'} U(S), \qquad \overline{S}' = \frac{E(S)}{k}, \qquad k > 0,$$

as in (4-39), we find from (4-41) the moment-generating function

$$h(z) = (1+z)^{k-M}(1+bz)^{-k}(1-\beta z)^{-M'}, \qquad b = 1 + \frac{E(S)}{k}, \qquad (8\text{-}9)$$

for the random variable $V = U - \beta U'$. In any case, the probability density function of V, by the inversion formula for Laplace transforms, is

$$p(V) = \int_{c-i\infty}^{c+i\infty} h(z) \, e^{zV} \, \frac{dz}{2\pi i}; \qquad (8\text{-}10)$$

for (8-8), $-1 < c < \beta^{-1}$; for (8-9), $-b^{-1} < c < \beta^{-1}$.

By taking z with a negative real part, we can obtain the probability of detection by integrating (8-10) over $0 \le V \le \infty$:

$$Q_d = \Pr(V \ge 0 \mid H_1) = -\int_{c-i\infty}^{c+i\infty} z^{-1} h(z) \, \frac{dz}{2\pi i}, \qquad c < 0.$$

Taking z, on the other hand, with a positive real part and integrating over $-\infty < V < 0$, we find

$$1 - Q_d = \Pr(V < 0 \mid H_1) = \int_{c-i\infty}^{c+i\infty} z^{-1} h(z) \, \frac{dz}{2\pi i}, \qquad c > 0.$$

These integrals can be evaluated by numerical integration on a straight vertical contour through the saddlepoint of the integrand lying in the interval specified for the point $z = c$, as described in Sec. 5.2.1. Alternatively, the number of steps in the numerical integration can be reduced by integrating along a parabola as in Sec. 5.2.2. For false-alarm probabilities we set S equal to 0 or b equal to 1. The probabilities of false alarm or detection can be approximated by the saddlepoint method as described in Sec. 5.2.3. The reader can easily work out the details.

In Fig. 8-1 we plot the average probability Q_d of detecting a signal of fixed total energy E_T versus $D = \sqrt{2S}$, where $S = E_T/N$ is the energy-to-noise ratio, with N the unilateral spectral density of the noise actually present. Curves are shown for various numbers M' of independent observations of the noise alone; $M = 20$. Even with ten times as many samples known to be free of any signal component, $M' = 200$, the average probability of detection is still somewhat below the probability of detecting the signal in white noise whose spectral density is known a priori.

In Fig. 8-2 we have plotted for three values of M the loss in decibels incurred when the spectral density N of the white noise is unknown and one must resort to this CFAR receiver. There S is the energy-to-noise ratio the receiver requires to attain a false-alarm probability $Q_0 = 10^{-6}$ and a detection probability $Q_d = 0.999$ when it supplements M inputs with M' signal-free inputs; S_0 is the energy-to-noise ratio required by the threshold receiver when the noise level N is known. The loss is plotted versus $R = M'/M$.

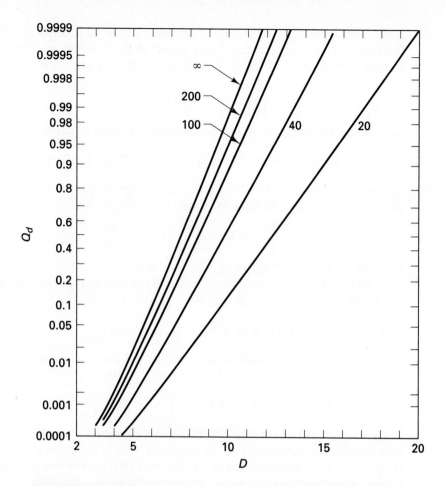

Figure 8-1. Probability Q_d of detection in a CFAR receiver versus $D = \sqrt{2S}$, unfading signals; $Q_0 = 10^{-6}$, $M = 20$. Curves are indexed with the number $M' = RM$ of signal-free observations.

In practice it may be difficult completely to exclude the signals $s_k(t)$ to be detected from the auxiliary inputs $v'_j(t)$, and a certain number of the terms $|z'_k|^2$ in (8-4) may, under hypothesis H_1, contain signal components. The ones most likely to be so corrupted will be the largest, and by raising the value of $\beta U'$ they reduce the probability of correctly choosing H_1. It has been proposed, therefore, to eliminate some fixed number of the largest of the rectified outputs $|z'_k|^2$ from (8-4) and to base the estimate U' on only the remaining terms. The resulting receiver is called a *censored mean–level detector* [Ric77]. The calculation of the false-alarm and detection probabilities for such a receiver is difficult, for the terms remaining in U' are no longer statistically independent. It has been carried out for Rayleigh-fading signals by Ritcey, whose paper [Rit86] furnishes references to other work on this type of detector.

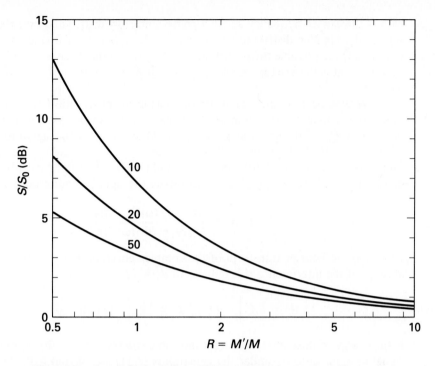

Figure 8-2. Loss S/S_0 in decibels, CFAR receiver; M = number of inputs that may contain a signal, M' the number of signal-free inputs; $Q_0 = 10^{-6}$, $Q_d = 0.999$. Curves are indexed with the value of M.

8.1.4 Colored Interference of Unknown Spectral Density

At the beginning of Sec. 8.1.1 we wrote of an enemy jamming one's radar receiver by transmitting broadband noise of uncertain strength. The receiver was required to estimate the spectral density of the total noise by taking samples of its input in spectral regions where no components of the signals $s_k(t)$ are expected to lie. If, however, the enemy has determined the form of the echo signals, it would be to his advantage to confine his noisy jamming signals to the same spectral band as theirs.

Let us for simplicity consider a single range bin and a single input $v(t)$ to the receiver. Its total noise will then be

$$n(t) = n_w(t) + n_J(t),$$

where $n_w(t)$ is white noise of known unilateral spectral density N, and $n_J(t)$ is colored noise of narrowband spectral density $\Phi_J(\omega)$ coming from the jammer. Its total available power

$$P = 2 \int_{-\infty}^{\infty} \Phi_J(\omega) \, \frac{d\omega}{2\pi} \tag{8-11}$$

is assumed known to the receiver. The signal to be detected is as before

$$s(t) = A \operatorname{Re} F(t) \exp(i\Omega t + i\psi), \qquad 0 \le t \le T,$$

with Ω the carrier frequency and ψ a phase uniformly distributed over $(0, 2\pi)$. How should the jammer distribute its total available power P in frequency in order to reduce as far as possible the probability Q_d with which the receiver detects this signal? That is, what is the least favorable spectral density $\Phi_J(\omega)$ from the standpoint of the receiver?

We assume that the signal arrives during an observation interval that is so much longer than the duration of the signal that it can be taken as $(-\infty, \infty)$. The probability Q_d of detection as given by (3-75) depends on the signal-to-noise ratio d^2 in (3-53). With the interval $(0, T)$ replaced by $(-\infty, \infty)$, we can go through the same analysis as led to (2-80) from (2-66) and (2-67). The total narrowband spectral density is now $\frac{1}{2}N + \Phi_J(\omega)$, and we can write the signal-to-noise ratio as

$$d^2 = A^2 \int_{-\infty}^{\infty} \frac{|f(\omega)|^2}{N + 2\Phi_J(\omega)} \frac{d\omega}{2\pi}, \qquad (8\text{-}12)$$

with $f(\omega)$ the Fourier transform of the complex envelope $F(t)$ of the signal. In the absence of the interference this reduces to $d_0^2 = 2E/N$, where

$$E = \tfrac{1}{2}A^2 \int_{-\infty}^{\infty} |F(t)|^2 \, dt = \tfrac{1}{2}A^2 \int_{-\infty}^{\infty} |f(\omega)|^2 \frac{d\omega}{2\pi} \qquad (8\text{-}13)$$

is the energy of the signal. The least favorable spectral density $\Phi_J(\omega)$ minimizes the signal-to-noise ratio d^2 under the constraints (8-11) and $\Phi_J(\omega) \geq 0$. This problem has been treated by Zetterberg [Zet62].

Introducing the Lagrange multiplier μ^2, we combine (8-12) and (8-11) and minimize

$$\int_{-\infty}^{\infty} \left[\frac{|f(\omega)|^2}{N + 2\Phi_J(\omega)} + 2\mu^2 \Phi_J(\omega) \right] \frac{d\omega}{2\pi}.$$

At frequencies where the spectral density $\Phi_J(\omega)$ is positive, we can find it by differentiating the integrand with respect to $\Phi_J(\omega)$,

$$-\frac{2|f(\omega)|^2}{[N + 2\Phi_J(\omega)]^2} + 2\mu^2 = 0,$$

whence

$$2\Phi_J(\omega) = \frac{|f(\omega)| - \mu N}{\mu} > 0. \qquad (8\text{-}14)$$

This will hold for all frequencies $\omega \in E$ where $\Phi_J(\omega)$ is positive. At all other frequencies $\omega \in E'$, the spectral density $\Phi_J(\omega)$ must vanish. The value of μ is determined by the constraint (8-11):

$$\frac{1}{\mu} \int_E [f(\omega) - \mu N] \frac{d\omega}{2\pi} = P. \qquad (8\text{-}15)$$

The receiver, presuming that the jammer will utilize its least favorable spectral density $\Phi_J(\omega)$, passes its input through a narrowband filter whose narrowband transfer function, as in (2-86), is

$$Y(\omega) = \frac{f^*(\omega)\,e^{-i\omega T'}}{N + 2\Phi_J(\omega)}$$

$$= \mu \frac{f^*(\omega)}{|f(\omega)|}\,e^{-i\omega T'}, \qquad \omega \in E, \tag{8-16}$$

$$= \frac{1}{N} f^*(\omega)\,e^{-i\omega T'}, \qquad \omega \in E'.$$

As before, T' is a delay long enough that $(0, T')$ contains all the signal $s(t)$. The gain characteristic $|Y(\omega)|$ will thus be uniform in the spectral region E; in the complementary region E' it will be the same as for the matched filter designed for detection in white noise.

The minimum value of the signal-to-noise ratio that the jammer can enforce is then, from (8-12),

$$d_{\min}^2 = \mu A^2 \int_E |f(\omega)|\,\frac{d\omega}{2\pi} + \frac{A^2}{N}\int_{E'} |f(\omega)|^2\,\frac{d\omega}{2\pi}$$

$$= d_0^2 - \frac{A^2}{N}\int_E |f(\omega)|\,[|f(\omega)| - \mu N]\,\frac{d\omega}{2\pi}, \tag{8-17}$$

where d_0^2 is the signal-to-noise ratio in the absence of any jamming.

If the jammer does not use its least favorable spectral density, but puts its total available power P into some other spectral density $\Phi_1(\omega)$, while the receiver continues to use the filter prescribed by (8-16), the effective signal-to-noise ratio will be

$$d_{\mathrm{eff}}^2 = d_{\min}^4 \cdot \left[\int_{-\infty}^{\infty} \left| \frac{f(\omega)}{N + 2\Phi_J(\omega)} \right|^2 [N + 2\Phi_1(\omega)]\,\frac{d\omega}{2\pi} \right]^{-1}, \tag{8-18}$$

and it is not difficult to show that $d_{\mathrm{eff}}^2 \geq d_{\min}^2$. The receiver is thus assured that the probability Q_d of detecting the signal, for a preassigned false-alarm probability Q_0, will not be less than

$$Q_d = Q(d_{\min}, b),$$

where $Q(\cdot, \cdot)$ is Marcum's Q function (3-76) and $Q(0, b) = Q_0$. For this reason the matched filter defined by (8-16) might be termed *robust*.

In order to illustrate the effect on signal detectability of a jammer transmitting random noise with the least favorable spectral density $\Phi_J(\omega)$, we assume that the signal has a Gaussian form, with spectrum

$$f(\omega) = A \exp\left(-\frac{\omega^2}{4\Delta\omega^2} \right),$$

where $\Delta\omega^2$ is the mean-square bandwidth. The region E is now defined by $-\omega_0 \leq \omega \leq \omega_0$, where by (8-14)

$$A \exp\left(-\frac{\omega_0^2}{4\Delta\omega^2} \right) = \mu N.$$

In Fig. 8-3 we show as a solid curve the relative signal-to-noise ratio $d_{\min}^2/d_0^2$ in decibels versus the ratio P/NW in decibels; $W = \Delta\omega/\sqrt{\pi}$ is the bandwidth defined

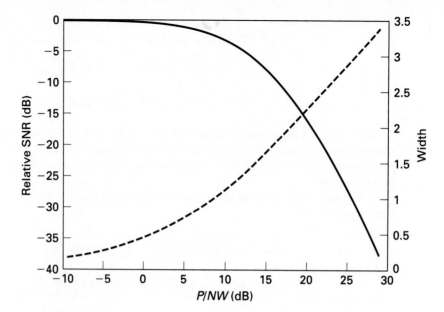

Figure 8-3. Performance of robust matched filter (8-16) for Gaussian signals versus P/NW; P = total jammer power, N = unilateral spectral density of the noise, W = equivalent bandwidth. Solid curve: relative signal-to-noise ratio $d_{\min}^2/d_0^2$ (dB). Dashed curve: relative width $\omega_0/\pi W$.

in (7-11). The dashed curve, referred to the right-hand scale, plots the relative width $\omega_0/\pi W$ of the least favorable spectral density.

The subject of robust matched filtering has been extended well beyond the range of this example. For instance, spectral densities that are unknown, but constrained to lie between upper and lower limits $\Phi_U(\omega)$ and $\Phi_L(\omega)$ and to carry a prescribed total power, have been treated; and the possibility that the signal $F(t)$ may be a distorted version of a given signal $F_0(t)$ under a constraint such as

$$\int_{-\infty}^{\infty} |F(t) - F_0(t)|^2 \, dt \leq \Delta$$

has been taken into account. A broad review of this topic, with numerous references, was published by Kassam and Poor [Kas85].

8.2 NONPARAMETRIC DETECTION

8.2.1 Parametric and Nonparametric Hypotheses

Everything we did in the first seven chapters rested on the postulate that the statistical structure of the noise, embodied in its set of probability density functions, is given. Indeed, in all examples we took the noise to be of the ubiquitous Gaussian type, with expected value zero and a given autocovariance function—the only realistic kind of noise for which the multivariate probability density functions of arbitrary order can

Detection of Signals Under Conditions of Uncertainty   Chap. 8

even be written down. We also assumed that the signals, when present, combine additively with the noise, and that we know everything about them except perhaps the values of a limited number of parameters. When certain statistical properties of the noise are unknown, however, the detection problem becomes one of choosing between composite hypotheses, and a detection theory is more difficult to establish.

There are various levels of ignorance to contend with. If everything is known about the probability density functions of the noise and of the sum of signal and noise except for the values of a finite set of parameters governing them, the two hypotheses H_0 and H_1 are said to be *parametric*. In detecting signals in white Gaussian noise, for instance, the only unknown quantity may be the spectral density of the noise, as in Sec. 8.1. With colored Gaussian noise, the only undetermined parameters may be the variances and the bandwidths of certain components.

If, on the other hand, the very forms of the distributions of the noise are unknown, a finite number of parameters will not suffice to specify them, and the hypotheses are called *nonparametric*. Even Gaussian noise whose autocovariance function or spectral density is unknown gives rise to a nonparametric detection problem. More generally, it may be possible to describe the noise only qualitatively, as by saying that its expected value is 0 or that the probability density function of its amplitudes is an even function, the forms of its density functions remaining otherwise indeterminate. A class of probability density functions, restricted only in such a way and wanting values of an infinitude of parameters for their specification, is termed a *nonparametric* class.

Parametric detection is a matter of choosing between composite hypotheses, under which the probability density functions of the data depend on only a finite number of unknown parameters. It can be handled by the methods introduced in Sec. 3.6. If prior probability density functions of the parameters under the two hypotheses are given, the numerator and the denominator of the likelihood ratio can be averaged with respect to them, as in (3-91), and the decision is based on the value of the resulting ratio. If no prior probability density functions are to be had, least favorable ones can be postulated, or the principle of maximum likelihood can be applied. When only a finite number of parameters are unknown, their values can be estimated separately under each hypothesis, and a likelihood ratio can be determined by substituting the estimates into the density functions.

Detection is particularly simple when the unknown parameters of the noise can be estimated independently of the presence or absence of a signal. Suppose that a coherent communication system transmits pulses of known form that are received in white Gaussian noise whose spectral density N is unknown, as might be the case if a jammer were trying to interfere by emitting broadband noise. When the pulses are orthogonal and are received with equal energies, the optimum receiver does not even depend on the value of N. When on–off pulses $f(t)$ and 0 are used for sending binary messages, the output of a filter matched to the pulse $f(t)$ must be compared with a decision level depending on N, but the minimum attainable probability of error is unaffected by prior ignorance of N, for N can in principle be measured independently. (See Problem 8-1.) In practice N cannot be determined exactly, and in Sec. 8.1 we described how an estimate of N can be utilized instead.

When the signal also depends on unknown parameters or when more than simply the power level of the noise is unknown, the problem of estimating the unknown parameters, designing the detection system, and calculating the false-alarm and detection probabilities is much more complicated. Each such problem must be attacked individually, and little of general applicability can be said.

8.2.2 Nonparametric Receivers

Besides the ever-present Gaussian thermal noise, input processes that cannot easily be described mathematically sometimes perturb a receiver. Randomly occurring impulses due to lightning, sparks in ignition systems, or faulty connections and switches may interfere with communications. A model of such *impulse noise* based on the simplest assumptions provides even first-order probability density functions only at the cost of some difficult calculations [Gil60], [Yue78], and to obtain joint probability density functions of higher order seems hardly possible. Underwater-sound receivers pick up sporadic biological noise, strange croakings and cracklings that are difficult to describe statistically, yet impede the detection of weak signals. Such extraneous noise may not be stationary over a long enough period to allow empirical distributions to be measured with any precision. When neither theory nor experiment is able to furnish detailed statistics, the designer of a detection system must give up characterizing the noise by the usual array of probability density functions and must face the problem of how best to choose between nonparametric hypotheses.

With the distributions of the noise unknown, receivers can no longer be designed to meet a Bayes criterion. Although it may at times be sensible to admit prior probability density functions of a finite number of unknown parameters, it can hardly be meaningful to postulate prior distributions of the probability density functions $p_0(v)$ and $p_1(v)$ themselves. There is no way to determine the average risk of a detection strategy with respect to a nonparametric class of distributions and hence no way of saying that one detector is better than another in the Bayes sense.

Of the Neyman–Pearson criterion all that is left is the directive to attain a specified false-alarm probability, and even this cannot always be achieved. A receiver whose false-alarm probability is the same for all noise distributions in a given nonparametric class is said to be a *nonparametric, distribution–free*, or *constant-false-alarm-rate* (CFAR) receiver. When the alternative hypothesis H_1 is nonparametric, there is no way to specify an average probability of detection and hence no way to maximize one.

If no distributions of the noise are available, nothing can be said about the correlations among values of the input at different times. How they should best be combined becomes uncertain, and matched filtering of the kind we described earlier cannot arise from a theory of nonparametric detection. The input must, however, be filtered in some way that will favor the signals to be detected by, for instance, removing noise of frequencies outside the spectral band that the signals are expected to occupy. We shall suppose this to have been done by prefiltering the input. When the signals to be detected are narrowband signals of unknown phase, the output of the filter may have been rectified as well.

It is customary in studies of nonparametric detection to assume that this pre-filtered and possibly rectified input is sampled at times far enough apart that the samples are at least approximately statistically independent. Alternatively, and more conveniently, one supposes that there are M independent inputs $v_k(t)$, $k = 1, 2, \ldots, M$, each of which is processed in the manner described in Sec. 4.4. The receiver bases its decisions on the M data

$$g_j = \mathfrak{f}[v_j(t)], \qquad j = 1, 2, \ldots, M.$$

These data are statistically independent, and as in Sec. 4.4 they are assumed to be statistically homogeneous: under each hypothesis all M of the data have identical probability density functions. It is on the basis of these M data that the receiver is to decide whether the signal is present or absent.

A restricted approach that has been taken assumes that most of the time the data $g_1, g_2, \ldots, g_M$ have known—or *nominal*—probability density functions $f_0(g)$ and $f_1(g)$ under hypotheses H_0 and H_1, respectively. There are, however, certain known probabilities ε_0 and ε_1 that their probability density functions are instead, say, $h_0(g)$ or $h_1(g)$, so that the overall probability density functions of the data are actually

$$p_i(g) = (1 - \varepsilon_i)f_i(g) + \varepsilon_i h_i(g), \qquad i = 0, 1.$$

One seeks the distributions $h_0(g)$ and $h_1(g)$ that are least favorable in the sense that with the concomitant likelihood-ratio receiver, the Bayes cost is maximum, or in the sense that for a preassigned false-alarm probability, the probability of detection attained by the Neyman–Pearson receiver is minimum. The decision strategy is then said to be *robust*. The difficult problem of applying this concept to signal detection has engendered a considerable literature, which was reviewed by Kassam and Poor [Kas85] and by Kazakos and Papantoni-Kazakos [Kaz90, pp. 154–97]. We deem that topic too specialized for this book, and we turn instead to a brief survey of what has been done when not even nominal density functions for the data can be postulated. It is this subject that properly goes under the name of *nonparametric detection*.

The designer of a nonparametric receiver attempts to exploit the differences, often only qualitative, between the characteristics of the input under the two hypotheses (H_0) "signal absent" and (H_1) "signal present." The values of the input, for instance, might be larger on the average, or more often positive, when a signal is present than when it is not. The designer seeks a receiver whose probability of detecting any of the expected signals is greater than the largest possible value of the false-alarm probability; such a receiver is said to be *unbiased*. Because the data are assumed statistically homogeneous, the receiver strategy should be invariant to permutations of the samples g_k of the input. The already extensive development of nonparametric tests of hypotheses, described in books by Fraser [Fra57], Lehmann [Leh59], and Kendall and Stuart [Ken61, pp. 465–512], has guided the search for receivers with such properties as invariance and absence of bias. Our exposition will follow the surveys by Carlyle and Thomas [Car64] and Carlyle [Car68]. A detailed treatment of this topic is to be found in the book by Gibson and Melsa [Gib75], and a bibliography was published by Kassam [Kas80].

The performance of each nonparametric receiver will be compared with that of a standard receiver, which is optimum when the data $g_1, g_2, \ldots, g_M$ are independently Gaussian random variables. They have expected values zero under hypothesis H_0 that no signal is present. Under hypothesis H_1 they have, by our assumption of statistical homogeneity, a common expected value A. This standard receiver is therefore one that bases its decisions on the sum

$$G = \sum_{i=1}^{M} g_i \tag{8-19}$$

of the data; we call it the *Neyman–Pearson receiver*. Each new receiver will be compared with this one under the condition that both attain the same reliability (Q_0, Q_d), where Q_0 is the false-alarm probability and Q_d the probability of detecting the same signal. When the data g_j are Gaussian random variables, the false-alarm and detection probabilities are as usual given by

$$Q_0 = \text{erfc } x, \qquad Q_d = \text{erfc}(x - D\sqrt{M}),$$

where

$$D^2 = \frac{[E(g|H_1) - E(g|H_0)]^2}{\text{Var}_0\, g} = \frac{A^2}{\sigma^2} \tag{8-20}$$

is the effective signal-to-noise ratio for each datum g_j as defined in (4-66); $\sigma^2 = \text{Var}_0\, g$, and $\text{Var}_0\, g$ is the variance of each datum under hypothesis H_0.

Because of the difficulty of calculating the performance of the Neyman–Pearson receiver for many kinds of non-Gaussian noise and the difficulty of calculating that of certain nonparametric receivers afflicted by any kind of noise for finite M, the analyst must usually resort to assuming $M \gg 1$ and comparing receivers on the basis of their asymptotic relative efficiency. This concept was introduced in Sec. 4.4 and should now be reviewed.

8.2.3 The t-Test

One approach to nonparametric detection is to pretend that the noise has a specific statistical structure, such as the Gaussian, to design the receiver on that basis, and finally to evaluate its performance for noise different from what was assumed. As an illustration we suppose that the signal components of the data $\mathbf{g} = (g_1, g_2, \ldots, g_M)$ are equal to some unknown value A. The expected value of the noise components is taken to be zero, and the signal and noise are assumed to combine additively. Then the expected values of the samples are

$$E(g_j|H_0) \equiv 0, \qquad E(g_j|H_1) \equiv A.$$

As the probability density functions of the g_j's are unavailable, their variances are unknown.

If we pretend that the noise is Gaussian with expected value 0 and unknown variance σ^2, we can derive a detection procedure by the method of maximum likelihood (Sec. 3.6.4). The probability density functions of the samples g_j under the two hypotheses are taken as

$$p_0(\mathbf{g}) = (2\pi\sigma^2)^{-M/2} \exp\left[-\sum_{j=1}^{M} \frac{g_j^2}{2\sigma^2} \right],$$

$$p_1(\mathbf{g}) = (2\pi\sigma^2)^{-M/2} \exp\left[-\sum_{j=1}^{M} \frac{(g_j - A)^2}{2\sigma^2} \right].$$

The value of σ^2 is estimated under hypothesis H_0 by maximizing the density function $p_0(\mathbf{g})$; call this estimate σ_0^2. Estimates of both A and σ^2 under H_1 are found by similarly maximizing $p_1(\mathbf{g})$; call the results $\hat{A}$ and σ_1^2. When they are substituted into $p_0(\mathbf{g})$ and $p_1(\mathbf{g})$, respectively, and the likelihood ratio is formed, it is found to be given simply by $(\sigma_0/\sigma_1)^M$, where

$$\sigma_0^2 = \frac{1}{M} \sum_{j=1}^{M} g_j^2, \qquad \sigma_1^2 = \frac{1}{M} \sum_{j=1}^{M} (g_j - \hat{A})^2, \qquad \hat{A} = \frac{1}{M} \sum_{j=1}^{M} g_j.$$

Comparing this likelihood ratio with a certain decision level is equivalent to comparing the statistic

$$t = \frac{M^{1/2} \hat{A}}{s}, \qquad s^2 = \frac{1}{M-1} \sum_{j=1}^{M} (g_j - \hat{A})^2, \tag{8-21}$$

with some other decision level t_0, as one can show with a little algebra. Here $\hat{A}$ is the sample mean and s the sample standard deviation of the data. The quantity s^2 appearing in (8-21) is an estimate of the variance σ^2 that happens to be unbiased under each hypothesis:

$$E(s^2 | H_0) = E(s^2 | H_1) = \sigma^2.$$

The statistic in (8-21) is known to statisticians as *Student's t statistic*. If the components of the signal are expected to be positive, $A > 0$, hypothesis H_1 is chosen when t exceeds a decision level t_0; this is known as the *one-sided t-test*. If they may be either positive or negative, one requires the absolute value $|t|$ to surpass another level t_0'; this is the *two-sided t-test*. If the noise is known to be Gaussian, the level t_0 or t_0' providing a specified false-alarm probability Q_0 can be obtained from tables of Student's t-distribution, which are available in most statistical handbooks. The probability of detecting a signal by a one-sided t-test can be determined from tables of the noncentral t distribution [Res57]. The false-alarm and detection probabilities for the two-sided t-test can be reduced to those for the central and noncentral F distributions, respectively [Hel85a]. If the noise is not Gaussian or if its first-order probability density function is unknown, the proper setting of the decision level cannot be found from the tables; and with the decision level set as for Gaussian noise, the false-alarm probability may exceed the specified value if the noise actually has some other distribution.

When the number M of samples is very large, the distribution of the statistic t is approximately Gaussian, by virtue of the central limit theorem, for any ordinary distribution of the noise with expected value zero. The variance of t is approximately 1, and the decision levels t_0 and t_0' for a preassigned false-alarm probability are given by the equations

$$Q_0 \approx \text{erfc } t_0, \qquad Q_0 \approx 2 \text{ erfc } t_0'.$$

In this limit $M \gg 1$ the false-alarm probability is the same for all such noise distributions, and the receiver is said to be *asymptotically nonparametric*. The probability of detection can also be estimated in terms of the Gaussian distribution. When M is very large, the sample standard deviation is nearly equal to the true standard deviation of the noise, and the expected value of the statistic t becomes

$$E(t | H_1) \approx \frac{A\sqrt{M}}{\sigma} = D\sqrt{M}$$

when the sample values of the signal are all equal to A; $D^2 = A^2/\sigma^2$ as in (8-20). The variance of t is approximately equal to 1 under hypothesis H_1 as well. The detection probabilities are then

$$Q_d \approx \mathrm{erfc}(t_0 - D\sqrt{M}),$$
$$Q_d \approx \mathrm{erfc}(t_0' - D\sqrt{M}) + \mathrm{erfc}(t_0' + D\sqrt{M})$$

for the one- and two-sided t-tests, respectively.

In this limit of a very large number M of samples, the receiver based on the t-test performs as well as the standard Neyman–Pearson receiver defined in Sec. 8.2.2. The asymptotic relative efficiency of the t-test receiver with respect to that standard receiver is equal to 1. It should be borne in mind, however, that the asymptotic relative efficiency provides a valid basis of comparison only when it is expected that the receivers will actually utilize a very large number of independent samples g_j. If the number M is small, the receivers may behave quite differently from the predictions of the asymptotic theory. As we have seen, the t-test receiver is not nonparametric for finite numbers M of data; its false-alarm probability for fixed decision level t_0 or t_0' depends on the true probability density function of the noise.

8.2.4 The Sign Test

The example of the t-test receiver shows that a receiver utilizing the unmodified amplitudes of the samples g_j is unlikely ever to be nonparametric for finite M. There are two ways to avoid basing the decisions on those amplitudes. One is to discard all but the signs of the samples; the other is to arrange the samples in order of their amplitudes and use their ranks in this arrangement. We shall first describe a receiver that works only with the signs of the samples g_j.

If the noise is as often positive as negative, there will in the absence of a signal usually be nearly as many positive samples as negative. If the signal is known to be positive, therefore, a preponderance of positive samples will lead one to suspect its presence. The receiver can simply count the number n_+ of samples that are positive, choosing hypothesis H_1 when n_+ exceeds a certain decision level n_0. The statistic n_+ is invariant to permutations of the data $\{g_j\}$. Such a receiver is said to carry out the *sign test*.

The sign-test receiver is nonparametric over the class of probability density functions of the noise for which the probability of a positive noise value equals $\frac{1}{2}$, $\Pr(g > 0 | H_0) = \frac{1}{2}$. We are in effect assuming that the median of the noise is known and has been subtracted from all the data. The probability of there being more than n_0 positive samples under hypothesis H_0 is given by the binomial distribution,

$$Q_0 = \Pr(n_+ > n_0 \mid H_0) = 2^{-M} \sum_{i=n_0+1}^{M} \binom{M}{i},$$

independently of the true form of the probability density function of the noise. For a fixed number M, however, only certain values of the false-alarm probability are accessible, and Q_0 cannot be less than 2^{-M} if signals are to be detected at all.

If one wishes the sign-test receiver to achieve an arbitrary false-alarm probability Q_0, it is necessary to introduce randomization as described in Sec. 1.2.5. If the number n_+ of positive signs exceeds n_0, the receiver chooses hypothesis H_1. If $n_+ = n_0$, it chooses H_1 with probability f, and the false-alarm probability is

$$Q_0 = 2^{-M} \left[f \binom{M}{n_0} + \sum_{i=n_0+1}^{M} \binom{M}{i} \right] \qquad (8\text{-}22)$$

as in (1-57). The procedure described there enables one to determine the values of n_0 and f required for a preassigned value of Q_0.

Under our assumption of statistical homogeneity, the probability density function of each sample g_j under hypothesis H_1 will be the same for all; let us denote it by $p_1(g)$. The probability of detecting the signal is then also given in terms of the binomial distribution, and as in (1-58),

$$Q_d = f \binom{M}{n_0} p^{n_0} (1-p)^{M-n_0} + \sum_{j=n_0+1}^{M} \binom{M}{j} p^j (1-p)^{M-j},$$

$$p = \Pr(g > 0 \mid H_1) = \int_0^\infty p_1(g)\, dg. \qquad (8\text{-}23)$$

If $p > \frac{1}{2}$ for all possible signals, the sign-test receiver is unbiased.

The simplest way to compare the performance of the sign-test receiver with that of the Neyman–Pearson receiver is to determine their asymptotic relative efficiency (a.r.e.) for a number of different possible distributions of the noise. We shall show that when signal and noise are additive,

$$p_1(g) = p_0(g - A), \qquad A > 0, \qquad (8\text{-}24)$$

it is given by

$$\text{a.r.e.} = 4\sigma^2 [p_0(0)]^2, \qquad \sigma^2 = \mathrm{Var}_0\, g, \qquad (8\text{-}25)$$

where $p_0(g)$ is the probability density function of the datum g under hypothesis H_0 that no signal is present. Thus when the data are Gaussian random variables,

$$p_0(g) = \frac{1}{\sqrt{2\pi\sigma^2}}\, e^{-g^2/2\sigma^2}, \qquad (8\text{-}26)$$

a.r.e. $= 2/\pi = 0.637$ [Hod56]. Naturally, the Neyman–Pearson receiver is the better one for Gaussian noise. When the noise has a bilateral exponential distribution,

$$p_0(g) = \tfrac{1}{2} e^{-|g|}, \qquad \sigma^2 = 2, \qquad (8\text{-}27)$$

a.r.e. $= 2$. Kanefsky and Thomas [Kan65] calculated the asymptotic relative efficiency of these receivers for a number of other probability density functions $p_0(g)$.

They found that if the density function of the noise is either highly asymmetrical or very much more peaked than the Gaussian, as seems to be the case with certain kinds of impulse noise, the asymptotic relative efficiency exceeds 1 and the sign-test receiver is superior to the Neyman–Pearson receiver, which bases its decisions on the sum G of the data as in (8-19).

The asymptotic relative efficiency can be determined by (4-67),

$$\text{a.r.e.} = \lim_{A \to 0} \frac{D_2^2}{D_1^2}, \tag{8-28}$$

where D_1^2 and D_2^2 are the effective signal-to-noise ratios of the two receivers when the input signal strength equals A. As in (8-20) $D_1^2 = A^2/\sigma^2$ for the Neyman–Pearson receiver.

The sign-test receiver is based on the statistic

$$n_+ = \sum_{j=1}^{M} U(g_j),$$

where $U(\cdot)$ is the unit step function, for which

$$E[U(g)| H_0] = \tfrac{1}{2}, \qquad E[U(g)| H_1] = p = \int_0^\infty p_1(g)\, dg.$$

When we assume as in (8-24) that signal and noise are additive,

$$p = \int_0^\infty p_0(g - A)\, dg = \int_{-A}^\infty p_0(x)\, dx = \tfrac{1}{2} + \int_{-A}^0 p_0(x)\, dx$$
$$\approx \tfrac{1}{2} + A p_0(0),$$

provided that $p_0(0) \neq 0$ and A is small. Now

$$\text{Var}_0\, U(g) = E\{[U(g)]^2| H_0\} - \{E[U(g)| H_0]\}^2 = \tfrac{1}{2} - \tfrac{1}{4} = \tfrac{1}{4},$$

and the effective signal-to-noise ratio of the datum $U(g)$ is

$$D_2^2 = 4A^2 [p_0(0)]^2$$

by the definition in (8-20). Substituting into (8-28), we obtain the asymptotic relative efficiency as given in (8-25).

Unfortunately the asymptotic relative efficiency is an unreliable measure of the performance of the sign-test receiver unless the number M of samples is very large. In Figs. 8-4 and 8-5 we plot in decibels the ratio of the input energy-to-noise ratio required by the Neyman–Pearson receiver (S_1) to that required by the sign-test receiver (S_2) in order to achieve the same reliability (Q_0, Q_d). We set $Q_0 = 10^{-3}$ and 10^{-6} and $Q_d = 0.99$. For Fig. 8-4 it was assumed that the noise is Gaussian, and as $M \to \infty$ the ratio plotted approaches $10 \log_{10}(2/\pi) = -1.9612$, albeit slowly. Figure 8-5 displays the same ratio for noise with the bilateral exponential distribution of (8-27), and as $M \to \infty$ it approaches $10 \log_{10} 2 = 3.0103$. Again the approach is

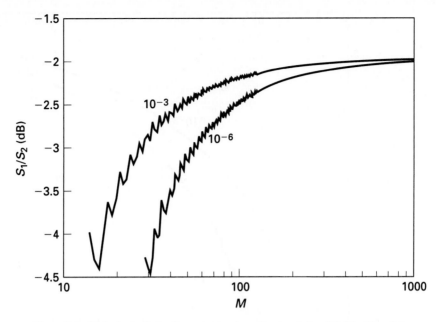

Figure 8-4. Ratio in decibels of energy-to-noise ratios to attain reliability (Q_0, Q_d) by sign-test receiver and Neyman–Pearson receiver; Gaussian noise. $Q_d = 0.99$; curves are indexed with the value of Q_0.

very slow. A calculation by the saddlepoint approximation showed that at $M = 10^5$ the relative efficiency is still only 2.91 dB for $Q_0 = 10^{-6}$, $Q_d = 0.99$.

For numbers M of data less than about 100, however, the ratio is rather less than the limiting value. For noise with the bilateral exponential distribution, Fig. 8-5 shows that the Neyman–Pearson receiver is even superior to the sign-test receiver when $M < 60$ for $Q_0 = 10^{-3}$ and when $M < 120$ for $Q_0 = 10^{-6}$, a conclusion opposite to that indicated by the asymptotic relative efficiency. Comparing receivers on the basis of their asymptotic relative efficiency is unreliable unless the number M of terms summed is very large.

These curves were computed in the following way. For a given number M of samples and a given false-alarm probability Q_0, we first determined n_0 and f from (8-22) and then by Newton's method solved (8-23) for the probability p. From the value of p we determined the input signal amplitude A_2 required to attain the detection probability Q_d with the sign-test receiver. For Gaussian noise of unit variance,

$$p = 1 - \text{erfc } A_2, \tag{8-29}$$

and for the bilateral exponential distribution, by (8-27),

$$p = 1 - \tfrac{1}{2} \exp(-A_2). \tag{8-30}$$

With Gaussian noise the statistic G in (8-19) has expected value MA_1 and variance $M\sigma^2 = M$ under hypothesis H_1, with A_1 determined by

$$Q_0 = \text{erfc } x, \qquad Q_d = \text{erfc}(-y) = \text{erfc}(x - A_1\sqrt{M}),$$

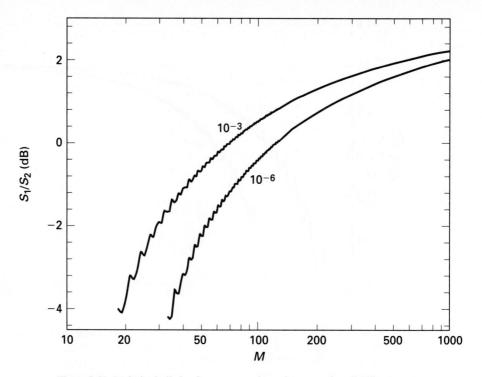

Figure 8-5. Ratio in decibels of energy-to-noise ratios to attain reliability $(Q_0,\ Q_d)$ by sign-test receiver and Neyman–Pearson receiver; bilateral exponential noise distribution. $Q_d = 0.99$; curves are indexed with the value of Q_0.

in which x and y were found from tables of the error-function integral. Then

$$A_1 = \frac{x + y}{\sqrt{M}}.$$

The ratio of the required input energy-to-noise ratios is A_1^2/A_2^2, which is plotted in decibels in Fig. 8-4.

   When the noise distribution is bilateral exponential, it is necessary to calculate the density function $p_0(G)$ of the statistic G in (8-19) by in effect convolving the density function in (8-27) $M - 1$ times. The moment-generating function of the distribution of G is the bilateral Laplace transform

$$h(z) = E(e^{-Gz}\,|\,H_0) = (1 - z^2)^{-M},$$

and the density function $p_0(G)$ was found by evaluating the contour integral

$$p_0(G) = \int_{c-i\infty}^{c+i\infty} (1 - z^2)^{-M} e^{Gz}\,\frac{dz}{2\pi i}, \qquad -1 < c < 1,$$

by the residue theorem. The result was then integrated to determine

$$Q_0 = \int_{G_0}^{\infty} p_0(G)\,dG; \qquad (8\text{-}31)$$

see also [Joh70, Ch. 23]. Furthermore, the probability of detection is

$$Q_d = \int_{G_0}^{\infty} p_0(G - MA_1) \, dG = \int_{G_1}^{\infty} p_0(G) \, dG, \qquad G_1 = G_0 - MA_1. \qquad (8\text{-}32)$$

The values of G_0 and G_1 were obtained by solving (8-31) and (8-32) by Newton's method, and from them the input amplitude A_1 was determined from

$$A_1 = \frac{G_0 - G_1}{M}.$$

The ratio A_1^2/A_2^2 is plotted in decibels in Fig. 8-5.

This method was utilized only for $M \leq 125$, and the jagged curves shown in the figures appeared, the sawteeth resulting from the jumps in the integer decision level n_0 as M increases. For $M > 125$ the serrations of the exact curves became too fine to plot, and the numbers of terms that needed to be summed when computing (8-32) became so large that round-off error began to introduce inaccuracy. The calculation of the probability p was instead based on a contour integral for the cumulative binomial distribution in terms of its probability generating function

$$h(z) = (pz + q)^M, \qquad q = 1 - p,$$

with which we find as in (5-45)

$$q_n^{(+)}(p) = \Pr(k \geq n \mid p) = \int_{C_+} \frac{z^{-n} h(z)}{z - 1} \frac{dz}{2\pi i}, \qquad (8\text{-}33)$$

where k is a binomial random variable; the contour C_+ is a circle centered at the origin and enclosing the point $z = 1$. The integral in (8-33) can be regarded as providing the quantity $q_n^{(+)}(p)$ as a function of a continuous variable n, of which the false-alarm and detection probabilities in (8-22) and (8-23) are polygonal approximations formed by connecting the values of $q_n^{(+)}(p)$ at integral values of n by straight lines. When $M \gg 1$, the polygon lies close to the smooth curve, and we do not make much error if we solve (8-33) with $p = q = \frac{1}{2}$ for the value of n making $q_n^{(+)}(\frac{1}{2}) = Q_0$, and if we then use that value of n, usually nonintegral, to determine the value of p for which $q_n^{(+)}(p) = Q_d$. The latter computation was carried out by the secant method. From the value of p the signal amplitude A_2 was again determined as in (8-30). For bilateral-exponential noise the cumulative distribution of the statistic G needed to determine $Q_d(A_1)$ was computed by the contour-integration method of Sec. 5.2 [Hel89a]. The ratios A_1^2/A_2^2 in decibels plotted in Figs. 8-4 and 8-5 then became smooth curves, which can be considered as approximations to the too finely serrated exact curves for $M > 125$.

From this description it is seen how one can calculate similar performance curves for any kind of noise distribution, provided that one can work out the cumulative distribution of the statistic G in (8-19). Contour-integration methods of the type introduced in Sec. 5.2 will generally be applicable if one can calculate the moment-generating function of the datum g_j under the postulated distribution.

8.2.5 Rank Tests

If one knows more about the noise than what was assumed in Sec. 8.2.4, one can expect to detect the signal more efficiently. The class of even noise density functions

$$p_0(g) = p_0(-g) \tag{8-34}$$

is included in the class for which $\Pr(g > 0| H_0) = \frac{1}{2}$, but is smaller. For this restricted class of noise distributions the *signed rank test* is nonparametric and is superior to the sign test.

When (8-34) holds and when the signal components of the data g_j are positive, not only are there likely to be more positive samples g_j than negative when a signal is present, but the sample values of large absolute value are more likely to be positive than negative. This observation suggests a more elaborate and more efficient test based on ranking the data $\{g_j\}$ in accordance with their absolute values. Again we assume statistical independence and homogeneity of the M data. The signal adds a constant amplitude A to the data, and under hypothesis H_1 the probability density function of each datum g_i has the form in (8-24). The test treats all samples alike and is invariant to their permutation.

Once all M samples have been received, they are arranged according to their absolute values,

$$|g_{i_1}| < |g_{i_2}| < \cdots < |g_{i_M}|,$$

where $i_1, i_2, \ldots, i_M$ is a permutation of all the integers from 1 to M. The kth sample in this arrangement is assigned a "rank" equal to k. The receiver adds the ranks of those samples that are positive, forming what is known as the *Wilcoxon signed rank statistic*

$$r = \sum_{k=1}^{M} kU(g_{i_k}), \tag{8-35}$$

where again $U(\cdot)$ is the unit step function. If the statistic r exceeds a decision level r_0, hypothesis H_1 is chosen. The statistic can also be written as

$$r = \sum_{i=1}^{M} \sum_{j=i}^{M} U(g_i + g_j) = \sum_{j=1}^{M} \sum_{i=1}^{j} U(g_i + g_j), \tag{8-36}$$

which is more easily implemented in an electronic receiver [Car64].

We pause to demonstrate the identity of (8-35) and the rightmost double sum in (8-36). Because both these are invariant to permutations of the data, we can assume that the data are already ranked:

$$|g_1| < |g_2| < \cdots < |g_M|. \tag{8-37}$$

Now consider the jth term in the right-hand sum in (8-36):

$$t_j = \sum_{i=1}^{j} U(g_i + g_j).$$

If g_j is positive, $g_i + g_j$ is positive because $|g_i| < |g_j|$, and $U(g_i + g_j) = 1$ for all i, whereupon $t_j = j$. If $g_j < 0$, on the other hand, $g_i + g_j < 0$ for all i, and $t_j = 0$. Therefore,

$$t_j = jU(g_j),$$

which is the jth term of (8-35) when the data are arranged as in (8-37). Thus both forms of the statistic r yield the same value.

Because the probability density function of the noise is an even function, the probability under hypothesis H_0 that a given datum g_j is positive and its rank will appear in the sum r is equal to $\frac{1}{2}$. The expected value of the rank sum r is then

$$E(r \mid H_0) = \tfrac{1}{2} \sum_{k=1}^{M} k = \tfrac{1}{4} M(M + 1). \tag{8-38}$$

The false-alarm probability is obtained by examining all 2^M partitions of the integers from 1 to M into a primary set with m members ($0 \le m \le M$) and its complementary set with $M - m$ members. Under hypothesis H_0 all primary sets are equally likely to appear as terms in the rank sum (8-35). The false-alarm probability is then 2^{-M} times the number of primary sets whose sum exceeds the decision level r_0.

Again, if an arbitrary false-alarm probability is desired, randomization must be introduced. The receiver decides that a signal is present whenever $r > r_0$; when $r = r_0$, it chooses hypothesis H_1 with probability f. The false-alarm probability is then

$$\begin{aligned} Q_0 &= 2^{-M}\left[f h(r_0; M) + \sum_{r=r_0+1}^{R} h(r; M) \right] \\ &= 2^{-M}\left[\sum_{r=0}^{r_0'-1} h(r; M) + f h(r_0'; M) \right], \qquad r_0' = R - r_0, \end{aligned} \tag{8-39}$$

where $h(r; M)$ is the number of primary sets of integers whose sum equals r, and $R = M(M + 1)/2$ is the maximum possible rank sum. We have used the symmetry $h(r; M) = h(R - r; M)$. The false-alarm probability is independent of the actual probability density function of the samples under hypothesis H_0, provided only that it is an even function, and for this class of noise distributions the rank-test receiver is nonparametric.

The numbers $h(r; M)$ obey the recurrent relations

$$\begin{aligned} h(k; M) &= h(k; M - 1), & 0 \le k < M, \\ h(k; M) &= h(k; M - 1) + h(k - M; M - 1), & M \le k \le R, \end{aligned}$$

$$h(0; 1) = h(1; 1) = 1,$$

which are easily programmed. Tables facilitating the computation of Q_0 and, inversely, the determination of the decision level r_0 are to be found in [Bic77, pp. 479–81] and [Wil73]. For $M \gg 1$ the tables are very lengthy, and the computation of Q_0 takes a long time and requires the storage of many intermediate numbers.

The probabilities

$$p_k^{(M)} = 2^{-M} h(k; M)$$

possess the probability-generating function

$$h_M(z) = \sum_{k=0}^{R} p_k^{(M)} z^k = 2^{-M} \prod_{p=1}^{M} (1 + z^p),$$

and the cumulative probabilities

$$q_k^{(-)} = \sum_{r=0}^{k-1} p_r^{(M)} = q_{R-k+1}^{(+)}$$

can instead be computed by saddlepoint integration (5-44). The "phase" of that integrand is

$$\Phi(z) = \sum_{p=1}^{M} \ln(1 + z^p) - k \ln z - \ln(1 - z) - M \ln 2.$$

There exists a unique saddlepoint z_0 in $0 < \mathrm{Re}\, z < 1$, which is the root of

$$\Phi'(z) = \sum_{p=1}^{M} \frac{pz^{p-1}}{1 + z^p} - \frac{k}{z} - \frac{1}{z-1} = 0. \qquad (8\text{-}40)$$

This can quickly be solved by Newton's method. The curve C_- in (5-44) is conveniently taken as the straight vertical chord of the unit circle passing through the saddlepoint z_0, combined with the portion C' of the unit circle to the left of the chord. When M is on the order of 30 or more, the contribution of the chord dominates. On C' the integrand has many zeros, and that part of the path of integration contributes negligibly to $q_k^{(-)}$. The integration along the chord can be carried out by the trapezoidal rule as described in Sec. 5.2.2, and when $M \geq 30$, the stopping rule given there cuts off the numerical integration before the unit circle is reached. The results of such a numerical saddlepoint integration have been found to have high relative accuracy, and in contrast to the recurrent method, computation time and storage requirements are nearly independent of M and k.

As can be imagined, calculating the probability of detection attained by this receiver is extremely difficult, and one resorts to the asymptotic relative efficiency as a criterion for comparing it with other receivers. It is shown in Appendix G that relative to the Neyman–Pearson receiver the asymptotic relative efficiency of the rank-test receiver is

$$\text{a.r.e.} = 12\sigma^2 \left[\int_{-\infty}^{\infty} [p_0(g)]^2 \, dg \right]^2, \qquad \sigma^2 = \mathrm{Var}_0\, g, \qquad (8\text{-}41)$$

when the signal and noise are additive as in (8-24).

For Gaussian noise the asymptotic relative efficiency equals $3/\pi = 0.955$; for the bilateral exponential distribution in (8-27) it equals $3/2$. If one minimizes

$$\int_{-\infty}^{\infty} [p_0(g)]^2 \, dg$$

under the constraints

$$\int_{-\infty}^{\infty} p_0(g) \, dg = 1, \qquad \int_{-\infty}^{\infty} g p_0(g) \, dg = 0, \qquad \int_{-\infty}^{\infty} g^2 p_0(g) \, dg = \sigma^2,$$

one finds, within an arbitrary scaling factor, that the probability density function for which the asymptotic relative efficiency is minimum has the form

$$p_0(g) = \begin{cases} \frac{3}{4}(1 - g^2), & |g| \leq 1, \\ 0, & |g| > 1, \end{cases}$$

and the attendant asymptotic relative efficiency equals $108/125 = 0.864$.

If the signal may be either positive or negative, but maintains the same sign throughout—as it must under our assumption of statistical homogeneity—the receiver can determine the rank sum r of the positive samples and the rank sum $r' = \frac{1}{2}M(M + 1) - r$ of the negative samples. If the larger of these exceeds a certain decision level r_0', the receiver asserts that a signal is present. With a positive signal the statistic r tends to be large; with a negative signal the statistic r' tends to be large. Tests such as these are known as Wilcoxon tests [Wil45].

The false-alarm probability for this receiver equals 2^{-M} times twice the number of primary sets whose sum exceeds r_0' when no randomization is involved. A randomized test chooses hypothesis H_1 when $\max(r, r') > r_0'$; when $\max(r, r') = r_0'$, it chooses H_1 with probability f. The false-alarm probability is then

$$Q_0 = 2^{-(M-1)} \left[f h(r_0'; M) + \sum_{r=r_0'+1}^{R} h(r; M) \right], \quad r_0' \geq \tfrac{1}{2}R, \quad R = \tfrac{1}{2}M(M + 1).$$

This Wilcoxon test is also nonparametric: its false-alarm probability is independent of the actual density function of the noise.

A radar receiver may be required to detect a train of coherent narrowband pulses having a common phase ψ in each input, and as usual that phase ψ is unknown and may lie anywhere in $(0, 2\pi)$. The input is passed through a narrowband filter matched to the signal $\text{Re}[F(t) \exp i\Omega t]$, and at an appropriate time the complex envelope of its output is sampled to yield, for the jth interpulse interval, a complex sample $v_{cj} + iv_{sj}, 1 \leq j \leq M$. The receiver can then rank the sequences $(v_{c1}, v_{c2}, \ldots, v_{cM})$ and $(v_{s1}, v_{s2}, \ldots, v_{sM})$ individually by their absolute values, so that

$$|v_{c i_1}| < |v_{c i_2}| < \cdots < |v_{c i_M}|, \qquad |v_{s j_1}| < |v_{s j_2}| < \cdots < |v_{s j_M}|,$$

where $(i_1, i_2, \ldots, i_M)$ and $(j_1, j_2, \ldots, j_M)$ are permutations of the integers from 1 to M. The rank sums

$$r_C = \sum_{k=1}^{M} k U(v_{c i_k}), \qquad r_C' = \tfrac{1}{2}M(M + 1) - r_C,$$

$$r_S = \sum_{k=1}^{M} k U(v_{s j_k}), \qquad r_S' = \tfrac{1}{2}M(M + 1) - r_S,$$

are formed. The receiver then compares the statistic $r_C''^2 + r_S''^2$ with a decision level r_0^2, where $r_C'' = \max(r_C, r_C')$, $r_S'' = \max(r_S, r_S')$; if the level is exceeded, a signal is declared present.

If no signal is present, the rank sums will all be roughly equal to their expected value $\frac{1}{4}M(M + 1)$; but if a signal with some arbitrary phase ψ is present, one or the other or both outputs of the sampler will have a preponderance of positive or negative values, and the statistic just defined will, most likely, be larger. If the phases of the successive signals are variable and independently random, however, the test will fail.

8.2.6 Receivers with a Reference Input

Detection is facilitated if the receiver can obtain a separate set of inputs exhibiting the same type of noise as what corrupts the signals, but themselves free of any signals, much as in the CFAR receiver treated in Sec. 8.1. Let us suppose that we have L such reference inputs that provide independent samples $h_1, h_2, \ldots, h_L$ that contain only noise, besides the M independent samples $g_1, g_2, \ldots, g_M$ that may or may not contain a signal.

Adapting from statistics what is variously known as the Wilcoxon two-sample test [Wil45] and the Mann–Whitney test [Man47], Capon [Cap59] analyzed a receiver that detects a constant signal causing the samples $g_1, g_2, \ldots, g_M$ to be generally larger than the reference samples $h_1, h_2, \ldots, h_L$. This signal might be a coherent signal that is always positive or a quasiharmonic signal that has been passed, with its attendant noise, through a rectifier before sampling. The receiver forms all LM possible pairs (g_i, h_j) of samples from the two inputs. It counts the number of pairs in which g_i exceeds h_j to form the statistic

$$V = \sum_{i=1}^{M} \sum_{j=1}^{L} U(g_i - h_j).$$

If V exceeds a decision level V_0, the receiver decides that a signal is present. Alternatively, the receiver can arrange all $(M + L)$ samples in order of their values. The position of a sample in this ordering, starting from the smallest, is its rank. The sum W of the ranks of the M samples g_i is linearly related to the statistic V and can be used instead. This receiver is nonparametric for all noise inputs for which the L samples h_j have the same probability density function as the M samples g_i under hypothesis H_0 and are statistically independent of them. Tables that can be used for setting the decision level V_0 have been published by Fix and Hodges [Fix55] and by Wilcoxon et al. [Wil73].

Under hypothesis H_0 the probability generating function of the rank-sum statistic V is

$$h(z) = c \prod_{j=1}^{M} \frac{z^{L+j} - 1}{z^j - 1}, \qquad c = \binom{L + M}{M}^{-1}$$

[Ken61, p. 494]. The probability distribution of V can be calculated from this probability-generating function by saddlepoint integration as was shown for the signed-rank statistic in (8-39) and (8-40). For M and L greater than about 30, the results are very accurate. For smaller values of M and L the algorithm given by Harding [Har84] is efficient, but the computation time and the storage it requires rise rapidly as M and L increase.

When both numbers L and M of samples are very large, the statistic V is approximately Gaussian. Capon [Cap59] showed that if the ratio L/M remains finite as L and M increase beyond all bounds and the signal strength vanishes, the asymptotic relative efficiency of this receiver relative to the Neyman–Pearson receiver is given by the expression in (8-41). When the noise is Gaussian, the asymptotic relative efficiency is $3/\pi = 0.955$; and whatever the noise distribution, the asymptotic relative efficiency cannot fall below $108/125 = 0.864$. The receiver is thus less affected than the Neyman–Pearson receiver by deviations of the noise distribution from the Gaussian.

8.2.7 Two-input Systems

The nonparametric receivers considered so far have required a certain coherence in the signals, in the sense that the signals must consistently drive the sample values toward more positive or more negative amplitudes. This coherence, however, may not always exist. If the signal is a random process taking on both positive and negative values during the observation or if the signals have random phases, the samples may have various signs, and it will be impossible to distinguish inputs containing signals from pure noise simply by looking at the signs of the samples.

The detection of stochastic signals, which, like random noise, can be described only by means of a collection of probability density functions, will be treated in Chapter 11 by the parametric methods developed heretofore. At present we shall only mention that such signals arise in multipath communications, sonar, and radio astronomy. With these signals the inputs to the receiver may have expected value zero under both hypotheses, the principal differences being a greater power level and, possibly, a different spectral density under hypothesis H_1 from what is observed with a signal absent. If the statistics of the noise are unknown, it is difficult to take advantage of distinctions such as these.

If a reference noise input is available and known to be free of any signals, one might set up a receiver based on the Mann–Whitney–Wilcoxon test just described. The outputs of the prefilter would be applied to a quadratic rectifier before sampling, and one could expect the samples of the input being tested to be mostly larger than those of the reference input when a signal is present.

If the noise arises mainly in the receiver or nearby, it may be simpler to try picking up the signal with two receivers so placed that the signal components of the inputs to each are the same, while the noise components are independently random. The presence of a signal is then indicated by a correlation between the two inputs that is absent when there is no signal. Let the samples of the prefiltered inputs of the two receivers be $v_1, v_2, \ldots, v_M$ and $w_1, w_2, \ldots, w_M$. The samples in each set are supposed to be statistically independent among themselves and statistically homogeneous.

The *sample correlation coefficient*

$$r = \frac{\sum\limits_{i=1}^{M}(w_i - \overline{w})(v_i - \overline{v})}{\left[\sum\limits_{i=1}^{M}(w_i - \overline{w})^2 \sum\limits_{j=1}^{M}(v_j - \overline{v})^2\right]^{1/2}},$$

$$\overline{w} = \frac{1}{M}\sum\limits_{i=1}^{M} w_i, \qquad \overline{v} = \frac{1}{M}\sum\limits_{j=1}^{M} v_j$$

will tend to be small when there is no signal present, and a receiver might compare r with a decision level r_0, choosing hypothesis H_1 when r is the larger. If there is a possibility of a constant, unknown phase shift between the signals at the two receivers, the absolute value $|r|$ should be compared with a level r_0', hypothesis H_0 being chosen when $|r| < r_0'$. These receivers are only asymptotically nonparametric; for a finite number of samples of each input the decision levels r_0 and r_0' for a preassigned false-alarm probability will depend somewhat on the true distribution of the noise.

In order to eliminate this dependence on the distributions, the receiver should work with the signs or the ranks of the samples. The simplest system is the *polarity coincidence correlator*, which has been analyzed by Wolff, Thomas, and Williams [Wol62], Ekre [Ekr63], and Kanefsky and Thomas [Kan65]. It bases its decisions on the signs of the products of the samples,

$$V = \sum_{i=1}^{M} U(v_i w_i),$$

where $U(\cdot)$ is again the unit step function. If this statistic exceeds a decision level, the signs of the two sets of samples have a positive correlation, and the presence of a signal is indicated. This receiver is nonparametric for inputs that under hypothesis H_0 are independent and have even probability density functions $p_0(v) = p_0(-v)$ and $p_0(w) = p_0(-w)$.

If the signals and the noise are independent Gaussian random processes, the optimum receiver based on the Neyman–Pearson criterion simply combines the samples and adds the squares of their pairwise sums, comparing

$$W = \sum_{i=1}^{M}(v_i + w_i)^2 \tag{8-42}$$

with a decision level. With respect to this detector the asymptotic relative efficiency of the polarity coincidence correlator is

$$\text{a.r.e.} = 2(q + \sigma^4)[p_0(0)]^4, \tag{8-43}$$

where σ^2 is the variance and q the fourth central moment of the noise. When the noise is Gaussian, this asymptotic relative efficiency equals $2/\pi^2 = 0.202$, hardly a promising result. Nevertheless, for noise with certain types of cuspidated distribution, Kanefsky and Thomas [Kan65] have found this asymptotic relative efficiency to exceed 1. They were careful to point out, however, that unless the number M of samples is huge and the signal-to-noise ratio infinitesimal, the use of the central limit theorem to evaluate the polarity coincidence correlator with the type of noise probability density functions they considered may seriously overestimate its relative efficiency.

Problems

8-1. A signal $Af(t)$ of known shape, but unknown positive amplitude A, is received in white Gaussian noise of unknown spectral density N. A set of functions $f_1(t), f_2(t), \dots$, orthonormal among themselves and orthogonal to $f(t)$ over the observation interval $(0, T)$, is determined, as by the Gram-Schmidt orthogonalization procedure described in Sec. 2.1.3. The input $v(t)$ is passed through n parallel filters matched to n of these functions $f_j(t)$ to obtain the statistics

$$v_j = \int_0^T f_j(t)v(t)\, dt, \qquad j = 1, 2, \ldots, n.$$

Show that an unbiased estimator of the spectral density N is

$$\hat{N} = \frac{2}{n} \sum_{j=1}^{n} v_j^2,$$

and that by taking n large enough the variance of this estimator can be made as small as desired. Propose a receiver, based on the estimate $\hat{N}$ and on the output of a filter matched to $f(t)$, for detecting the signal $s(t)$, and show that if n is made large enough, this receiver is as reliable as one designed for detection in white noise of known spectral density N.

8-2. A signal of known form $s(t) = Af(t)$, but unknown positive amplitude A, is to be detected in Gaussian noise of autocovariance function $\phi(t, u) = \sigma^2\eta(t, u)$, of which $\eta(t, u)$ is known, but the variance σ^2 is not; $\eta(0, 0) = 1$. As in Sec. 2.1.6, make Karhunen–Loève expansions of the signal, the noise, and the input $v(t)$. Arrange the eigenvalues λ_k of the kernel $\eta(t, u)$ in descending order, and denote the associated eigenfunctions by $f_k(t)$. As data the n quantities

$$v_k = \int_0^T f_k(t)v(t)\, dt, \qquad k = 1, 2, \ldots, n,$$

are to be used. Design a maximum-likelihood receiver to detect the signal on the basis of estimates of the amplitude A and the variance σ^2. Show that if n is taken large enough, the reliability of this receiver will be as great as that of a receiver designed for detection in noise of known variance σ^2 [Sch71].

8-3. Work out the following alternative version of the solution of Problem 8-2. Write the input $v(t)$ as

$$v(t) = v_1(t) + v_2(t), \qquad 0 \leq t \leq T,$$

with $v_1(t)$ defined as

$$v_1(t) = \frac{f(t)}{D^2} \int_0^T g(s)v(s)\, ds, \qquad D^2 = \int_0^T g(u)f(u)\, du,$$

where $g(t)$ is the solution of the integral equation

$$f(t) = \int_0^T \eta(t, s)g(s)\, ds, \qquad 0 \leq t \leq T.$$

Show that $v_2(t)$ is statistically independent of $v_1(t)$ and that its probability density functions are the same whether the signal is present or not. Derive its autocovariance function. Use Problem 6-7 to determine the unknown variance σ^2 from $v_2(t)$, and use this variance in setting the decision level for the detection statistic. What should this detection statistic be?

8-4. Derive the zero-order saddlepoint approximations for the false-alarm and detection probabilities of the CFAR receiver from the moment-generating function $h(z)$ in (8-8). Use the technique described in Sec. 5.3.3 to determine approximately the constant β that yields a preassigned false-alarm probability $Q_0 = \Pr(U \geq \beta U' | H_0)$. Then use your saddlepoint approximation to check the detection probabilities Q_d plotted in Fig. 8-1 in the range $Q_d > 0.9$.

8-5. Suppose that quasiharmonic signals

$$s_k(t) = A \operatorname{Re} F(t) \exp(i\Omega t + i\psi_k), \qquad k = 1, 2, \ldots, n,$$

with independently random phases ψ_k, but with a common, though unknown amplitude A, are all either present or absent in a set of n successive inputs $v_k(t)$ to the receiver. The data on which their detection is to be based are the real and imaginary parts x_k and y_k of the output of a filter matched to the signal,

$$x_k + iy_k = \int_0^T F^*(t)V_k(t)\, dt, \qquad k = 1, 2, \dots , n,$$

where $V_k(t)$ is the complex envelope of the kth input. The noise is white and Gaussian of unknown spectral density. Work out under each hypothesis the maximum-likelihood estimates of the common variance of the n x_k's and the n y_k's, determine the maximum-likelihood estimates of the phases ψ_k and of the common amplitude A, and show that the maximum-likelihood detector is equivalent to one that compares the statistic

$$\frac{\left[\displaystyle\sum_{k=1}^{n}(x_k^2 + y_k^2)^{1/2}\right]^2}{\displaystyle\sum_{k=1}^{n}(x_k^2 + y_k^2)}$$

with a suitable decision level.

8-6. Show how the false-alarm and detection probabilities for the sign test, as given in (8-22) and (8-23), can be calculated by the saddlepoint approximation introduced in Sec. 5.3.2. Assuming that the number n of stages is a continuous variable, as described after (8-23), show how to determine the value of n required for the sign test to attain a preassigned false-alarm probability. By using these saddlepoint approximations, check the efficiencies plotted in Fig. 8-4 at a number of values of $M > 125$. Show how to determine the false-alarm and detection probabilities of the Neyman–Pearson receiver by the saddlepoint approximation when the noise has the bilateral exponential distribution in (8-27), and use your results to check the efficiencies plotted in Fig. 8-5 for several values of $M > 125$.

8-7. The aim of this problem is to work out the asymptotic relative efficiency of the polarity coincidence correlator relative to the detector that is optimum when the data have Gaussian distributions. The inputs to two receivers are corrupted by statistically independent noise processes whose distributions, though unknown, are the same in both receivers. The signals, on the other hand, are identical in the two receivers, but they are random and differ from one observation interval to the next. Denoting the inputs to the two receivers by $x_i(t)$ and $y_i(t)$, we can express them as

$$x_i(t) = s_i(t) + n_i'(t),$$
$$y_i(t) = s_i(t) + n_i''(t), \qquad 1 \le i \le M,$$

under hypothesis H_1. Under hypothesis H_0 the signals $s_i(t)$ are absent. Both receivers process these M inputs in identically linear fashion to remove noise outside the frequency band of the signals, producing a total of $2M$ statistically homogeneous samples:

$$v_i = \mathsf{f}[x_i(t)], \qquad w_i = \mathsf{f}[y_i(t)], \qquad i = 1, 2, \dots , M.$$

Under hypothesis H_0, "noise alone present," the v_i's and w_i's are statistically independent with a common probability density function $p_0(\cdot)$, which is an even function: $p_0(v) = p_0(-v)$ and $p_0(w) = p_0(-w)$. When the signals are present (hypothesis H_1), they add to the noise, so that the conditional probability density functions of the ith samples are $p_0(v_i - s_i)$ and $p_0(w_i - s_i)$, respectively; s_i is the contribution of the signal to each sample. The values s_i of the signal samples are independently random with expected value zero and variance s^2.

(a) Show that if the noise is Gaussian, the optimum receiver forms the statistic W in (8-42) and compares it with a decision level W_0, choosing hypothesis H_1 when $W \geq W_0$. Calculate the effective signal-to-noise ratio

$$D_0^2 = \frac{[E(W|\,H_1) - E(W|\,H_0)]^2}{\mathrm{Var}_0\ W}$$

of this receiver in terms of the signal variance s^2, the noise variance σ^2, and the fourth moment $q = E(n_i'^4) = E(n_i''^4)$ of the noise. Remember that n_i' and n_i'' are statistically independent for all i.

(b) When the noise density function $p_0(\cdot)$ is unknown, but an even function, the receiver will instead base its decision on the output

$$V = \sum_{i=1}^{M} U(v_i w_i)$$

of the polarity coincidence correlator; $U(\cdot)$ is the unit step function. Calculate the effective signal-to-noise ratio D_V^2 of this statistic V in the limit in which $M \gg 1$ and $s^2 \ll 1$. In doing so, assume first that each s_i is a fixed, known quantity, and write an expression for

$$E[U(v_i w_i)|\,H_1] - E[U(v_i w_i)|\,H_0]$$

in terms of $p_0(\cdot)$ and s_i. Then expand this in a power series in s_i, retaining only the term of lowest order, which will turn out to be proportional to s_i^2. Replace s_i^2 by its expected value s^2 before continuing your calculation of D_V^2.

From the results of parts (a) and (b), determine the asymptotic relative efficiency of the polarity coincidence correlator with respect to the Neyman–Pearson receiver. Evaluate it for signal and noise samples that have Gaussian distributions and bilateral exponential distributions (8-27).

9

Sequential Detection

9.1 THE SEQUENTIAL PROBABILITY RATIO TEST

Let us consider a radar that is assigned to deciding about the presence or absence of a target at a specific distance and in a specific direction on the basis of M inputs $v_k(t)$, $1 \leq k \leq M$. These inputs are acquired one after another during successive observation intervals, each initiated by the transmission of an energetic narrowband pulse. If a target is present, each input contains an echo signal having known form, but perhaps randomly varying parameters such as amplitude and phase. The M inputs $v_k(t)$ possess the statistical homogeneity described in Sec. 4.4.1. A continuum of distances is in practice examined simultaneously, but this aspect we shall disregard for the time being. When the M pulses have been transmitted and the decision has been made, the beam of the antenna is shifted to a new azimuth and the procedure is repeated. When the receiver processes each set of M inputs as explained in Sec. 4.4, it is said to be executing a *fixed-sample-size* statistical test. The amount of time needed to search for targets in a certain portion of the sky is proportional to M.

If the antenna is composed of an array of distinct transducers, as described in Sec. 4.3, its beam can be moved electronically by suitably altering the phase shifts in the lines from the transducers to the receiver. In this way the direction of the beam can be placed under the control of the radar observer or his electronic counterpart, and it is unnecessary always to transmit the same number of pulses in each direction. Instead the beam can be shifted to a new direction as soon as the receiver judges that it has information of sufficient quality and quantity to make a reliable decision about the presence or absence of targets at the current azimuth. The number of pulses transmitted in a given direction and of inputs processed before making a decision then depends on the inputs themselves and is a random variable. A detection system that determines the number of its observations according to what it receives is said to operate sequentially. Sequential operation permits searching the sky more rapidly without diminishing the overall reliability of detection.

Sequential detection will be introduced in the context of Sec. 4.4.1, as though a radar were searching for a target at a fixed distance. The inputs to the receiver are denoted as before by $v_j(t)$, $j = 1, 2, \ldots$, and the number of inputs processed before making a decision is no longer fixed, but random. Again the jth input is

$$v_j(t) = n_j(t), \qquad j = 1, 2, \ldots, \qquad 0 \le t \le T,$$

under hypothesis H_0 and

$$v_j(t) = n_j(t) + s(t; a_j, \boldsymbol{\theta}_j', \boldsymbol{\theta}'')$$

under hypothesis H_1. The same assumptions about the noise $n_j(t)$ and about the signals $s(t; a_j, \boldsymbol{\theta}_j', \boldsymbol{\theta}'')$ are being made as in Sec. 4.4. The parameters $\boldsymbol{\theta}_j'$ are independently random from one input to another; the parameters $\boldsymbol{\theta}''$ are invariable. Statistical homogeneity of the inputs is postulated. The receiver processes each input in such a way as to produce a single datum

$$g_j = \mathrm{f}[v_j(t)], \qquad j = 1, 2, \ldots.$$

Its probability density function under hypothesis H_1 is assumed to have been averaged over the values of the random parameters $\boldsymbol{\theta}_j'$ of the signal. By virtue of our assumptions, the data g_j are statistically independent.

When, say, k inputs $v_j(t)$, $1 \le j \le k$, have been received and processed to provide k data, or samples, $g_1, g_2, \ldots, g_k$, the receiver is said to have reached the kth stage of its operation. At each stage the receiver makes one of three decisions: (1) Hypothesis H_0 is true; no signal is present, (2) Hypothesis H_1 is true, and an echo signal of the specified class is present, or (3) Another pulse is to be transmitted, another input $v_{k+1}(t)$ acquired, and a new datum g_{k+1} produced. If one of the first two decisions is made, the sequential test terminates; otherwise it continues through at least one more stage. How shall the decisions be made?

A fully Bayesian approach might be attempted, taking into account the costs entailed by the decisions and the cost of acquiring each new datum. Prior probability density functions of the parameters would be adopted. The data sets $\mathbf{g} = (g_1, g_2, \ldots, g_k, \ldots)$ leading to each possible terminal decision determine a decomposition of the infinite-dimensional space of those sets $\mathbf{g}$ into regions R_0 and R_1, and the boundaries between them must be laid in such a way as to minimize the average cost of operation. Such a minimization would manifestly be an exceedingly difficult mathematical problem.

When the cost of acquiring each new statistically independent input $v_{k+1}(t)$ and generating each new datum g_{k+1} is invariable, and when the costs attending the various decisions are independent of the true values of the signal parameters, the choice among the three decisions (1), (2), and (3) can be based on the posterior probability

$$\Pr(H_1 \mid g_1, g_2, \ldots, g_k) = \Pr(H_1 \mid \mathbf{g}^{(k)})$$

of hypothesis H_1, given the data $\mathbf{g}^{(k)} = (g_1, g_2, \ldots, g_k)$ at hand; $\Pr(H_0|\ \mathbf{g}^{(k)}) = 1 - \Pr(H_1|\ \mathbf{g}^{(k)})$. If that posterior probability is large enough, hypothesis H_1 will be accepted; if small enough, H_0; otherwise the system will proceed to the next stage $k + 1$. Symbolically, there will be two constants α and β, $\alpha > \beta$, such that

$$\Pr(H_1|\ \mathbf{g}^{(k)}) \geq \alpha \Rightarrow \text{choose } H_1,$$

$$\beta \leq \Pr(H_1|\ \mathbf{g}^{(k)}) < \alpha \Rightarrow \text{take another observation,} \qquad (9\text{-}1)$$

$$\Pr(H_1|\ \mathbf{g}^{(k)}) < \beta \Rightarrow \text{choose } H_0.$$

The posterior probability needed is

$$
\begin{aligned}
\Pr(H_1|\ \mathbf{g}^{(k)}) &= \frac{\zeta_1 P_1(\mathbf{g}^{(k)})}{\zeta_0 P_0(\mathbf{g}^{(k)}) + \zeta_1 P_1(\mathbf{g}^{(k)})} \\
&= \frac{\zeta_1 \Lambda(\mathbf{g}^{(k)})}{\zeta_0 + \zeta_1 \Lambda(\mathbf{g}^{(k)})},
\end{aligned}
\qquad (9\text{-}2)
$$

where $P_i(\mathbf{g}^{(k)}) = P_i(g_1, g_2, \ldots, g_k)$ is the joint probability density function of the data $\mathbf{g}^{(k)} = (g_1, g_2, \ldots, g_k)$ under hypothesis H_i, $i = 0, 1$; ζ_0 and ζ_1 are the prior probabilities of hypotheses H_0 and H_1, respectively, and

$$\Lambda(\mathbf{g}^{(k)}) = \frac{P_1(\mathbf{g}^{(k)})}{P_0(\mathbf{g}^{(k)})}$$

is the likelihood ratio at stage k. Because the data are statistically independent,

$$\Lambda(\mathbf{g}^{(k)}) = \prod_{j=1}^{k} \Lambda(g_j)$$

with

$$\Lambda(g) = \frac{P_1(g; S, \boldsymbol{\theta}'')}{P_0(g)}$$

the likelihood ratio for any datum g. It depends on the signal strength S and the set of invariable parameters $\boldsymbol{\theta}''$, but not on the set of random parameters $\boldsymbol{\theta}'$ as these were defined in Sec. 4.4. We suppose that the receiver is set up to detect a standard signal having strength $S = S_s$ and the set $\boldsymbol{\theta}'' = \boldsymbol{\theta}_s''$ of invariable parameters.

Because the data are statistically independent and the cost of making an observation and acquiring a new datum g is constant, the choice the receiver makes depends only on the posterior probability $\Pr(H_1|\ \mathbf{g})$ of hypothesis H_1, $\mathbf{g}$ representing the data so far collected. It does not matter how many data have been collected, for they are all subsumed in that posterior probability. Given $\Pr(H_1|\ \mathbf{g})$, the prospect of the future behavior of the sequential procedure, its eventual outcome, and its eventual total cost appears the same to the receiver whether few or many data g are at hand. The decision levels β and α in (9-1) are therefore independent of the stage k that the test has reached.

Introducing the notation

$$u_j = \ln \Lambda(g_j) = \ln[P_1(g_j; S_s, \boldsymbol{\theta}_s'')/P_0(g_j)], \qquad (9\text{-}3)$$

$$U_k = \ln \Lambda(\mathbf{g}^{(k)}) = \sum_{j=1}^{k} u_j \qquad (9\text{-}4)$$

for the logarithms of the likelihood ratios, we can rewrite the inequalities (9-1) in the form

$$U_k \geq a = \ln A \Rightarrow \text{choose } H_1,$$

$$b = \ln B \leq U_k < a = \ln A \Rightarrow \text{take another observation,} \qquad (9\text{-}5)$$

$$U_k < b = \ln B \Rightarrow \text{choose } H_0,$$

where

$$A = e^a = \frac{\zeta_0 \alpha}{\zeta_1(1 - \alpha)}, \qquad B = e^b = \frac{\zeta_0 \beta}{\zeta_1(1 - \beta)}$$

are the decision levels on the likelihood ratio $\Lambda(\mathbf{g})$. (The notation A for the upper decision level should not be confused with the A we have often used to indicate the amplitude of a signal.) A test basing decisions of this kind on the current value of the likelihood ratio $\Lambda(\mathbf{g}^{(k)})$ or its logarithm U_k is called a *sequential probability ratio test*. Its properties were extensively investigated by Wald [Wal47].

As the test proceeds through its successive stages, the value of the logarithm U of the likelihood ratio changes with the acquisition of each new datum g_k:

$$U_k = U_{k-1} + u_k, \qquad u_k = \ln[P_1(g_k; S_s, \boldsymbol{\theta}_s'')/P_0(g_k)].$$

The point U is said to execute a discrete-time random walk on the real line. Because the data g_j and hence the increments $\ln \Lambda(g_j) = u_j$ of U are independent random variables, the process is said to be an *independent-increment process* [Hel91, p. 321], [Lar79, vol. 1, p. 155]. The false-alarm probability $Q_0(a, b)$ equals the probability under hypothesis H_0 that the random variable U_k crosses the upper "barrier" $a = \ln A$ before passing below the lower barrier $b = \ln B$:

$$Q_0(a, b) = \text{Pr(for some } k: U_k \geq a, b < U_j < a, j = 1, \dots, k - 1 | H_0),$$

and the detection probability $Q_d(a, b)$ is likewise

$$Q_d(a, b) = \text{Pr(for some } k: U_k \geq a, b < U_j < a, j = 1, \dots, k - 1 | H_1).$$

The latter depends on the strength or average strength of the signals, denoted by S, and on the set $\boldsymbol{\theta}''$ of invariable parameters. Wald has shown that this kind of test eventually terminates with probability 1. Calculating the probabilities $Q_0(a, b)$ and $Q_d(a, b)$ in their dependence on the decision levels a and b requires solving an integral equation whose kernel is the probability density function of $u = \ln \Lambda(g)$ under hypothesis H_0 or H_1, as the case may be [Sam48], [Kem50]. To determine the levels a and b that minimize the average cost of operation is a difficult problem [Bla54, Ch. X].

The sequential test can also be set up in such a way as to attain a particular reliability (Q_0, Q_{ds}), where $Q_{ds}(S_s, \boldsymbol{\theta}_s'')$ is the probability of detecting a signal having the standard strength S_s and the standard set $\boldsymbol{\theta}_s''$ of the invariable parameters. One picks the decision levels a and b so that $Q_0(a, b) = Q_0$ and $Q_d(a, b) = Q_{ds}$. This formulation is the counterpart of applying the Neyman–Pearson criterion in a fixed-sample-size test and dispenses with knowledge of the cost matrix $\mathbf{C}$ and the prior probabilities ζ_0 and ζ_1 of the two hypotheses. As aforesaid, computing the functions $Q_0(a, b)$ and $Q_d(a, b)$ and searching for the values of a and b can be expected to be tedious. Wald cut through all these complexities by discovering simple approximations to those functions. They are based on the inequalities

$$A = e^a \leq \frac{Q_{ds}}{Q_0}, \qquad B = e^b \geq \frac{1 - Q_{ds}}{1 - Q_0} \qquad\qquad (9\text{-}6)$$

[Wal47, p. 41], which we shall now derive.

At the nth stage the space R_n of the data

$$\mathbf{g}^{(n)} = (g_1, g_2, \dots, g_n)$$

is divided by the decision rule (9-5) into three regions $R_0^{(n)}$, $R_1^{(n)}$, and $R_2^{(n)}$: $R_0^{(n)}$ is the region of points $\mathbf{g}^{(n)}$ leading to termination of the test with a decision for hypothesis H_0; $R_1^{(n)}$ is the region of points leading to termination with decision for H_1; and $R_2^{(n)}$ is the region of points leading to the decision to take an $(n + 1)$th observation. The probability of choosing hypothesis H_1 at that stage is

$$Q_0^{(n)} = \int_{R_1^{(n)}} P_0(\mathbf{g})\, d^n\mathbf{g}, \qquad d^n\mathbf{g} = dg_1\, dg_2 \dots dg_n,$$

when hypothesis H_0 is true. When hypothesis H_1 is true, this probability is

$$
\begin{aligned}
Q_d^{(n)} &= \int_{R_1^{(n)}} P_1(\mathbf{g}; S_s, \boldsymbol{\theta}_s'')\, d^n\mathbf{g} = \int_{R_1^{(n)}} \Lambda(\mathbf{g}; S_s, \boldsymbol{\theta}_s'') P_0(\mathbf{g})\, d^n\mathbf{g} \\
&\geq A \int_{R_1^{(n)}} P_0(\mathbf{g})\, d^n\mathbf{g} = A Q_0^{(n)}
\end{aligned}
\qquad (9\text{-}7)
$$

because $\Lambda(\mathbf{g}^{(n)}; S_s, \boldsymbol{\theta}_s'') \geq A$ when hypothesis H_1 is chosen. Summing over all stages n, we find

$$Q_{ds} \geq A Q_0, \qquad\qquad (9\text{-}8)$$

whence the first inequality in (9-6). The second follows by considering in a similar manner the probabilities of choosing hypothesis H_0 under the two hypotheses, and one finds

$$1 - Q_{ds} \leq B(1 - Q_0) \qquad\qquad (9\text{-}9)$$

because $\Lambda(\mathbf{g}^{(n)}; S_s, \boldsymbol{\theta}_s'') \leq B$ for $\mathbf{g}^{(n)} \in R_0^{(n)}$.

Wald observed that when the number of stages is on the average very large, which in our context implies that the standard signal strength S_s is very small, the increments u_j in the logarithm U of the likelihood ratio are on the average also very small. When the test terminates with selection of hypothesis H_1, therefore, the value of the likelihood ratio $\Lambda(\mathbf{g}^{(n)}; S_s, \boldsymbol{\theta}_s'')$ in (9-2) will be only slightly greater than A, whereupon (9-7) and hence also (9-8) are nearly equalities. The same argument applies to (9-9), for when H_0 is chosen, $\Lambda(\mathbf{g}^{(n)}; S_s, \boldsymbol{\theta}_s'')$ is very close to B. Thus Wald established the approximate formulas for the decision levels a and b required for attaining the reliability (Q_0, Q_{ds}):

$$a = \ln A \approx \ln\left[\frac{Q_{ds}}{Q_0}\right], \qquad b = \ln B \approx \ln\left[\frac{1 - Q_{ds}}{1 - Q_0}\right] \qquad (9\text{-}10)$$

[Wal47, pp. 44–8].

It is only when the signals are weak and the average number of stages before decision is very large that it is profitable to adopt a sequential test at all. For strong signals the number of stages required will on the average be not much smaller than the number required by a fixed-sample-size test with equal reliability (Q_0, Q_{ds}), and the additional complexity of the sequential test is hardly worth implementing. Wald's approximations are valid under just those conditions when a sequential test is likely to be advantageous. These are the same conditions under which we found in Sec. 4.4 that the asymptotic relative efficiency yields a useful indication of the comparative effectiveness of two receivers or two decision strategies. We now turn to assessing the performance of the sequential probability ratio test under those same circumstances of weak signals and a large average number of observations required to attain a given reliability. The necessary formulas, which were derived by Wald [Wal47], were applied to sequential signal detection by Bussgang and Middleton [Bus55], Blasbalg [Bla57a], [Bla57b], and others. A comprehensive treatment of sequential analysis is to be found in the book by Ghosh [Gho70].

9.2 PERFORMANCE OF THE SEQUENTIAL TEST

When the sequential probability ratio test has been established in the manner just described, it is of interest to assess its performance for signals having possibly different strengths S and possibly different sets θ'' of the invariable parameters from those of the standard signal for which the test has been set up. Because the aim of the sequential test is to reduce the number of samples needed on the average to attain a given reliability, one would like to know not only the probability Q_d of detection, but also the expected number of stages through which the test must pass before making a decision, and these as functions of the actual signal strength S and the actual values of the invariable parameters θ''. To determine the detection probability Q_d and the average number of samples precisely is a difficult problem, and again approximations are sought.

For simplicity of notation we combine the signal strength S and the invariable parameters θ'' into a single vector $\theta = (S, \theta'')$. Any random parameters θ' are as before assumed independently random from one input to another, and we employ the probability density function $P_1(g \mid \theta)$ of the statistic g after averaging over them. The absence of a signal we denote by $\theta = 0$; $P_1(g \mid 0) = P_0(g)$.[1]

Wald's approximations for the detection probability and the average sample number involve the moment-generating function of the logarithm u of the likelihood ratio, defined in (9-3):

$$h(z; \theta) = E(e^{-uz} \mid H_1, \theta) = \int_{-\infty}^{\infty} e^{-uz} P_1(u \mid \theta) \, du$$

$$= \int_{-\infty}^{\infty} \left[\frac{P_0(g)}{P_1(g \mid \theta_s)} \right]^z P_1(g \mid \theta) \, dg. \tag{9-11}$$

As we showed in Sec. 5.2.1, the moment-generating function $h(z; \theta)$ is a convex U function of the parameter z when z is real. The equation

$$h(z; \theta) = 1 \tag{9-12}$$

has in general two roots, one of which is $z = 0$. The second root, which is the one that figures in the subsequent analysis, is a function of the invariable parameters

[1] It is recommended that the reader work Problem 9-1 in the course of reading this section.

$\theta = (S, \theta'')$, and we designate it by $z(\theta)$. It takes on the special values

$$z(0) = -1, \qquad z(\theta_s) = 1,$$

where $\theta_s = (S_s, \theta_s'')$.

When, for instance, the data g_j are Gaussian random variables, as in detecting a signal of known form, but unknown amplitude in Gaussian noise,

$$u = d_s g - \tfrac{1}{2}d_s^2,$$

and

$$h(z; d) = \exp[-(d_s d - \tfrac{1}{2}d_s^2)z + \tfrac{1}{2}d_s^2 z^2], \qquad (9\text{-}13)$$

where $\theta = (d)$; $d_s^2 = 2E_s/N$ is the signal-to-noise ratio of the standard signal, and d^2 is that of the signal whose probability of detection is to be calculated. The root of (9-12) of concern is then

$$z(d) = \frac{2d}{d_s} - 1. \qquad (9\text{-}14)$$

The sequential test, as we have seen, is most appropriate when the input signal-to-noise ratio is small and the average number of stages before termination is large. Then one can in general neglect the cumulants of the statistic u of higher order than the second and approximate the moment-generating function by

$$h(z; \theta) \approx \exp[-E(u|\,\theta)z + \tfrac{1}{2}z^2 \operatorname{Var}(u|\,\theta)], \qquad (9\text{-}15)$$

whereupon the root $z(\theta)$ is approximately

$$z(\theta) \approx \frac{2E(u|\,\theta)}{\operatorname{Var}(u|\,\theta)}. \qquad (9\text{-}16)$$

When $E(u|\,\theta) = 0$, the equation (9-12) has a double root at $z = 0$. This usually occurs for a signal-to-noise ratio on the order of one-half the standard ratio and necessitates special treatment in the derivations to follow.

9.2.1 The Detection Probability

The probability $Q_d(\theta)$ of detection, averaged over the random parameters θ', is given approximately by Wald's formula:

$$Q_d(\theta) \approx \frac{1 - e^{-bz}}{e^{-az} - e^{-bz}}, \qquad z = z(\theta), \qquad (9\text{-}17)$$

where z is the nonzero root of (9-12). For $\theta = 0$ and $\theta = \theta_s$, this reduces to

$$Q_0 \approx \frac{1 - B}{A - B}, \qquad Q_d(\theta_s) \approx \frac{A(1 - B)}{A - B}, \qquad A = e^a, \qquad B = e^b, \qquad (9\text{-}18)$$

which are equivalent to (9-10) and hence to (9-6) treated as an approximation.

Wald derived (9-17) by the following reasoning [Wal47, pp. 48–52]. When the parameter z is a root of (9-12), the function

$$q(g|\,\theta) = \left[\frac{P_0(g)}{P_1(g|\,\theta_s)}\right]^z P_1(g|\,\theta)$$

can be treated as a probability density function, for it is both nonnegative and correctly normalized. Let H be the hypothesis that the datum g has the probability density function $P_1(g|\boldsymbol{\theta})$, and let H^* be the alternative hypothesis that its density function is $q(g|\boldsymbol{\theta})$. We consider a sequential test Σ^* that tests hypothesis H^* against the null hypothesis H. At its kth stage it forms the accumulated likelihood ratio

$$\Lambda_k^* = \prod_{j=1}^{k} \frac{q(g_j|\boldsymbol{\theta})}{P_1(g_j|\boldsymbol{\theta})} = \prod_{j=1}^{k} \left[\frac{P_1(g_j|\boldsymbol{\theta}_s)}{P_0(g_j)}\right]^{-z}$$

and compares it with two decision levels A^* and B^*, deciding for H if $\Lambda_k^* < B^*$, for H^* if $\Lambda_k^* \geq A^*$, and taking another observation if $B^* \leq \Lambda_k^* < A^*$.

Suppose first that $z < 0$, as when the signal-to-noise ratio is small. Then if we take $A^* = A^{-z}$, $B^* = B^{-z}$, the test Σ^* is carrying out the very same operations as our original sequential test, which we denote by Σ. The test Σ compares

$$\Lambda_k = \prod_{j=1}^{k} \frac{P_1(g_j|\boldsymbol{\theta}_s)}{P_0(g_j)} \tag{9-19}$$

with the levels A and B, choosing hypothesis H_1 when Λ_k exceeds A, and so on. When the data are such that test Σ selects hypothesis H_1, the test Σ^* selects H^*. We want the probability $Q_d(\boldsymbol{\theta})$ that Σ selects H_1 when the probability density function of the data is $P_1(g|\boldsymbol{\theta})$, and this equals the probability that Σ^* selects H^* under hypothesis H; that is, it is the false-alarm probability Q_0^* for test Σ^*. That false-alarm probability is approximately, according to (9-18),

$$Q_0^* \approx \frac{1 - B^*}{A^* - B^*} = \frac{1 - B^{-z}}{A^{-z} - B^{-z}} = \frac{1 - e^{-bz}}{e^{-az} - e^{-bz}},$$

which is just the expression (9-17).

For $z > 0$ the passage of Λ_k of (9-19) above the level A, whose overall probability we seek, corresponds to the passage of Λ_k^* below A^{-z}. If we set $B^* = A^{-z}$ and $A^* = B^{-z}$, this event entails the test Σ^* choosing hypothesis H, and now

$$Q_d(\boldsymbol{\theta}) = 1 - Q_0^*.$$

If we apply (9-18) with A and B replaced by B^{-z} and A^{-z}, respectively, this becomes

$$Q_d(\boldsymbol{\theta}) \approx 1 - \frac{1 - B^*}{A^* - B^*} = \frac{A^* - 1}{A^* - B^*} = \frac{B^{-z} - 1}{B^{-z} - A^{-z}},$$

which again reduces to (9-17).

In order to calculate the probability $Q_d(\boldsymbol{\theta})$ of detecting a signal with parameters $\boldsymbol{\theta} = (S, \boldsymbol{\theta}'')$, therefore, one must solve (9-12) for its one root z that differs from 0. That value of $z(\boldsymbol{\theta})$ is then substituted into (9-17). As mentioned before, when $E(u|\boldsymbol{\theta}) = 0$, (9-12) has a double root $z = 0$. Applying L'Hôpital's rule to (9-17) then yields

$$Q_d(\boldsymbol{\theta}) \approx -\frac{b}{a - b}, \qquad E(u|\boldsymbol{\theta}) = 0.$$

When the input signal-to-noise ratio and the standard signal-to-noise ratio are both very small, one can use the approximation (9-16) for $z(\boldsymbol{\theta})$. The resulting probability $Q_d(\boldsymbol{\theta})$ is then the same as that calculated by a different approach from Wald's

[Bar46]. The increments u_j in (9-4) are now very small, on the average, and many stages must usually be traversed before the sum U_k in (9-4) crosses one boundary or the other and the test terminates. A good approximation to the detection probability can therefore be obtained by considering the trajectory of the sum U as a Markov process in continuous time, replacing the stage-number k by a continuous variable t. Were there no barriers at a and b, the variable U would have a Gaussian density function by virtue of the central limit theorem. It would be executing a Brownian motion, with drift velocity $E(u | \theta)$ and diffusion constant $D = \text{Var}(u | \theta)$, and its probability density function would satisfy a Fokker–Planck equation [Hel91, p. 447], [Pap91, p. 652]. By solving that partial differential equation with absorbing barriers at $U = a$ and $U = b$, one can determine the probability that U crosses level a before crossing level b. This is the probability $Q_d(\theta)$ of detection and turns out to be the same as that given by Wald's approximation (9-17) with $u(\theta)$ specified by (9-16) [Hel68, pp. 68–70]. The advantage of this approach is that one does not need to assume that the increments u_j are logarithms of likelihood ratios as in Wald's analysis; u_j may be only an approximation to such, usually the threshold approximation.

In Fig. 9-1 we have plotted the probability $Q_d(d)$ of detection for a sequential receiver of a coherent signal in white Gaussian noise versus the signal-to-noise ratio $d = (2E/N)^{1/2}$. The standard signal has a signal-to-noise ratio $d_s^2 = 1$, and it attains the reliability ($Q_0 = 10^{-6}$, $Q_{ds} = 0.99$). The straight line in the figure represents the detection probability attained by the Neyman–Pearson receiver that sums $M = 50$ data, as in (8-19); it is carrying out a fixed-sample-size test and attains the same reliability for the same standard signal-to-noise ratio.

9.2.2 The Average Sample Number

The number of stages in the sequential test, or the number of inputs $v_j(t)$ used to make the final decision, is a random variable, which will differ from one trial of the test to another, even though the parameters θ are the same. To judge the performance of the sequential receiver it is important to know the average number of stages $\bar{n}(\theta) = E(n_t | H_1, \theta)$ as a function of the invariable parameters θ; here n_t is the number of the stage at which the test terminates. Wald [Wal47, pp. 52–54] showed this "average sample number" to be approximately

$$\bar{n}(\theta) \approx \frac{a Q_d(\theta) + b[1 - Q_d(\theta)]}{E(u | \theta)}, \tag{9-20}$$

in which the denominator is the expected value of the logarithm of the likelihood ratio when a signal with parameters θ is present. The probability $Q_d(\theta)$ is given by (9-17).

This formula is based on the exact relationship

$$E(U_t | \theta) = E(u | \theta) E(n_t | \theta), \tag{9-21}$$

where U_t is the value of the sum U_k in (9-4) at the stage n_t at which the test terminates. This random variable can be written

$$U_t = \sum_{j=1}^{n_t} u_j = \sum_{j=1}^{\infty} y_j u_j,$$

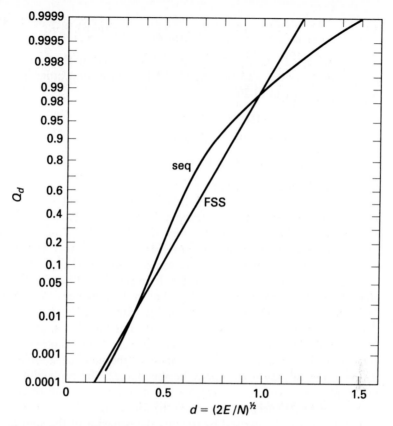

Figure 9-1. Probability $Q_0(d)$ for sequential and fixed-sample-size tests, coherent signal in Gaussain noise, versus signal-to-noise ratio $d = (2E/N)^{1/2}$: $Q_0 = 10^{-6}$, $Q_{ds} = 0.99$, $d_s = 1$, $M = 50$.

in which the y_j are random variables such that $y_j = 1$ if the test passes through stage j ($j \le n_t$) and $y_j = 0$ if it does not ($j > n_t$). Whether the test reaches stage j depends only on the values of the increments u_i for $i < j$, and the random variables y_j and u_j are statistically independent. The expected value of U_t must therefore be

$$E(U_t| \boldsymbol{\theta}) = \sum_{j=1}^{\infty} E(y_j| \boldsymbol{\theta})E(u_j| \boldsymbol{\theta}) = E(u| \boldsymbol{\theta}) \sum_{j=1}^{\infty} E(y_j| \boldsymbol{\theta})$$

$$= E(u| \boldsymbol{\theta})E(n_t| \boldsymbol{\theta}),$$

as in (9-21).

Because the increments u_j are very small, the variable U_t is very close to either a or b when the test terminates, and therefore

$$E(U_t| \boldsymbol{\theta}) \approx aQ_d(\boldsymbol{\theta}) + b[1 - Q_d(\boldsymbol{\theta})],$$

whence (9-20).

When $E(u| \boldsymbol{\theta}) = 0$ and $z(\boldsymbol{\theta}) = 0$, (9-20) must be replaced by

$$\bar{n}(\theta) \approx -\frac{ab}{\mathrm{Var}(u|\,\theta)}, \qquad E(u|\,\theta) = 0. \qquad (9\text{-}22)$$

One obtains this by taking z small in (9-17), expanding the numerator and denominator in powers of z, and keeping terms through second order in z. One then finds for the numerator in (9-20)

$$aQ_d(\theta) + b[1 - Q_d(\theta)] = -\tfrac{1}{2}abz + O(z^2),$$

and as (9-16) is valid in this same limit,

$$E(u|\,\theta) = \tfrac{1}{2}z\,\mathrm{Var}(u|\,\theta) + O(z^2),$$

and (9-20) reduces to (9-22).

As a function of the signal strength, the average number $\bar{n}$ of stages exhibits a peak between zero and the strength S_s of the standard signal. As S increases beyond the standard value, the average sample number $\bar{n}$ decreases toward zero. It is appropriate to compare the value of $\bar{n}(\theta)$ with the number M of stages needed by a fixed-sample-size test that attains the same reliability (Q_0, Q_{ds}) as the sequential test when detecting the same standard signal. In general $\bar{n}(0)$ and $\bar{n}(\theta_s)$ are less than this number M: In the absence of any signal or in the presence of the standard signal, the sequential test attains a specified reliability with a smaller average number of stages than required by the fixed-sample-size test. For signal strengths midway between 0 and the standard, however, the average number $\bar{n}(\theta)$ of stages may exceed M.

A further aspect of the randomness of the number n_t of stages before termination is that it may occasionally be very large. The probability mass function of the number n_t is difficult to calculate, but some indication of the variability of n_t is provided by its variance, for which the following rather complicated and approximate formula can be derived by treating the behavior of the sum U as a continuous Markov process,

$$\mathrm{Var}\,n_t \approx [E(u)]^{-2}\left[1 - \frac{2z(a-b)}{e^{-az} - e^{-bz}}\right]E(n_t)\,\mathrm{Var}\,u - 3(a-b)^2 Q_d(1 - Q_d),$$

$$z = z(\theta), \qquad (9\text{-}23)$$

with all expected values taken for the set of parameters θ. This approximation is valid under the same conditions as the previous ones. For such parameter values that $E(u|\,\theta) = 0$, we must again apply L'Hôpital's rule, and we obtain

$$\mathrm{Var}\,n_t \approx -\frac{ab(a^2 + b^2)}{3[\mathrm{Var}(u|\,\theta)]^2}, \qquad E(u|\,\theta) = 0.$$

At signal strengths intermediate between 0 and the standard one, both $\bar{n}(\theta)$ and $\mathrm{Var}\,n_t$ are relatively large. Signals of these strengths may draw out the sequential test to inordinate lengths before a decision is reached. If such signals are likely to be present, it may be advisable to force the test to yield a decision after a fixed number of stages. Such a truncation of the procedure affects the reliability in a way that is difficult to calculate.

For the fixed-sample-size test (8-19) with Gaussian data, the false-alarm and detection probabilities are

$$Q_0 = \mathrm{erfc}\,\alpha, \qquad Q_d = \mathrm{erfc}\,\beta, \qquad \beta = \alpha - d_s\sqrt{M},$$

where d_s^2 is the standard signal-to-noise ratio, whence

$$M = \frac{(\alpha - \beta)^2}{d_s^2}. \qquad (9\text{-}24)$$

In the sequential test, by (9-13) and (9-14),

$$E(u \mid d) = \tfrac{1}{2}d_s^2 z \,.$$

The ratio of the average sample number to the number M of data needed by the fixed-sample-size test is then, from (9-20) and (9-24),

$$\frac{\bar{n}(d)}{M} \approx \frac{2\{aQ_d(d) + b[1 - Q_d(d)]\}}{(\alpha - \beta)^2 z}, \qquad (9\text{-}25)$$

independently of the standard signal-to-noise ratio. Through (9-17) the right side is a function only of z, which for Gaussian data is given by (9-14). It is conjectured that (9-25) is approximately valid for any sequential probability ratio test in the limit of very small standard signal-to-noise ratio, whereupon $E(n_t) \gg 1$ and $M \gg 1$.

The ratio in (9-25) has been plotted versus z in Fig. 9-2. It exceeds 1 when z is in the neighborhood of 0, that is, for signal strengths roughly equal to one-half the standard signal strength. In that figure the lengths of the error bars are $(\text{Var } n_t)^{1/2}/M$, where Var n_t is given in (9-23). At intermediate signal-to-noise ratios, not only is the expected value of the number of stages before termination very large, but its standard deviation is also large. The expected values of the increments u_j of the logarithmic likelihood ratio U are then small, and U crosses one barrier or the other only after a very large and very variable number of data have been accumulated.

9.3 SEQUENTIAL DETECTION OF SIGNALS OF RANDOM PHASE

When the inputs

$$v_k(t) = n_k(t) + s_k(t; \boldsymbol{\theta}) \qquad (H_1)$$

contain signals $s_k(t; \boldsymbol{\theta})$ that are quasiharmonic pulses

$$s_k(t; \boldsymbol{\theta}) = A \operatorname{Re} F(t; \boldsymbol{\theta}'') \exp(i\Omega t + i\psi_k)$$

with phases ψ_k independently random from one signal to the next, but with a common though unknown amplitude A, not to be confused with the upper threshold in the sequential test, we can use the methods just described to set up and evaluate a sequential test for deciding whether a train of such signals is present in the inputs $v_1(t), v_2(t), \ldots$, successively observed by the receiver. We shall briefly summarize the necessary formulas, leaving their derivation to the interested reader.

As data we take the random variables

$$r_k = C \left| \int_0^T F^*(t) V_k(t)\, dt \right|, \qquad C = \left[N \int_0^T |F(t)|^2\, dt \right]^{-1/2}, \qquad (9\text{-}26)$$

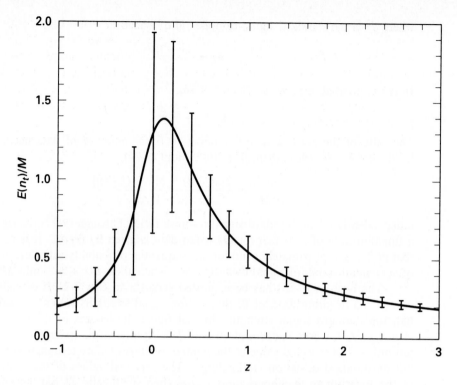

Figure 9-2. Ratio of average sample number $E(n_t) = \bar{n}(d)$ to number M of data in equipollent fixed-sample-size test with Gaussian noise versus parameter z: $Q_0 = 10^{-6}$, $Q_{ds} = 0.99$.

where $V_k(t)$ is the complex envelope of the kth input and $F(t)$ equals $F(t; \boldsymbol{\theta}_s'')$, the complex envelope of a signal with the standard set $\boldsymbol{\theta}_s''$ of invariable parameters; N is the unilateral spectral density of the noise, assumed white and Gaussian. Let

$$D^2 = \frac{2E}{N} = \frac{A^2}{N} \int_0^T |F(t)|^2 \, dt$$

be the signal-to-noise ratio for a signal with amplitude A, and denote the standard signal-to-noise ratio by D_s^2. As in (3-117) the probability density function of each datum r_k is

$$P_1(r; D) = r \, e^{-\frac{1}{2}(r^2+D^2)} I_0(Dr) U(r)$$

when a signal is present with amplitude A; when no signal is present, the density function of the r_k's is

$$P_0(r) = P_1(r; 0) = r e^{-\frac{1}{2}r^2} U(r).$$

The logarithmic likelihood ratio is

$$u = \ln\left[\frac{P_1(r; D_s)}{P_0(r)}\right] = \ln I_0(D_s r) - \tfrac{1}{2}D_s^2 \tag{9-27}$$

and is a monotone function of the datum r. At the end of each interval of duration T the logarithmic likelihood ratio u is determined from the input $v_k(t) =$

Re $V_k(t) \exp i\Omega t$ and added to the sum U carried over from the previous interval. The new value of U is compared with the decision levels $a = \ln A$ and $b = \ln B$ specified by (9-10). If $U < b$, the receiver decides that its inputs have contained only noise; if $U \geq a$, it decides that signals of some positive, but unknown amplitude have also been present. In either event the test ends. If $b \leq U < a$, the receiver causes another pulse to be transmitted and processes the input $v_{k+1}(t)$ during the subsequent interval in the same manner.

To determine the probability $Q_d(D)$ of detection as a function of the signal-to-noise ratio D, one must first calculate the moment-generating function of the logarithmic likelihood ratio u, which is given by (9-11),

$$h(z; D) = \int_0^\infty \left[\frac{P_1(r; D_s)}{P_0(r)} \right]^{-z} P_1(r; D)\, dr$$

$$= \exp(\tfrac{1}{2} D_s^2 z) \int_0^\infty [I_0(D_s r)]^{-z} P_1(r; D)\, dr.$$

(9-28)

This function cannot be obtained in closed form. A power series can be developed by expanding the function $[I_0(x)]^{-z}$ in powers of x, substituting into (9-28), and integrating term by term by means of (C-5). In terms of $S_s = \tfrac{1}{2} D_s^2$ and $S = \tfrac{1}{2} D^2$, the result is

$$h(z; D) = 1 + \tfrac{1}{2} z(1 + z) S_s^2 [1 - \tfrac{2}{3}(2 + z) S_s + \tfrac{1}{4}(11 + 11z + 3z^2) S_s^2]$$

$$- z S S_s [1 - (1 + z) S_s + \tfrac{1}{2}(1 + z)(4 + 3z) S_s^2] + \tfrac{1}{4} z(1 + 2z) S^2 S_s^2 + \cdots.$$

(9-29)

When a signal whose signal-to-noise ratio is D is present, the expected value and the variance of the logarithmic likelihood ratio u are

$$E(u| D) = -\tfrac{1}{2} S_s^2 (1 - \tfrac{4}{3} S_s + \tfrac{11}{4} S_s^2) + S S_s (1 - S_s + 2 S_s^2) - \tfrac{1}{4} S^2 S_s^2 + \cdots,$$

$$\mathrm{Var}(u| D) = S_s^2 (1 + 2S - 2S_s - 6 S_s S) + 0(S_s^4),$$

(9-30)

and from (9-29) the root of the equation $h(z; D) = 1$ is approximately

$$z \approx \frac{2S}{S_s} - 1 = \frac{2D^2}{2D_s^2} - 1.$$

These results can be put into the equations given previously for calculating the probability of detection, the average sample number, and the variance of the number of stages. The approximation is most reliable when both S and S_s are small and the average sample number $\bar{n}(D)$ is very large.

It is a temptation to approximate the logarithmic likelihood ratio in (9-27) by expanding the logarithm as in (4-14) and keeping only the term proportional to r^2. The system would then accumulate the statistics

$$u' = -\tfrac{1}{2} D_s^2 + \tfrac{1}{4} D_s^2 r^2$$

from each observation interval. However, as Bussgang and Mudgett [Bus60] and Blasbalg [Bla61] pointed out, the expected value of u' vanishes when no signal is present, $E(u'| H_0) = 0$, and the test may run through many stages before a decision is reached. They suggested replacing the term of fourth order in the expansion

$$u = -\tfrac{1}{2} D_s^2 + \tfrac{1}{4} D_s^2 r^2 - \tfrac{1}{64} D_s^4 r^4 + 0(r^6)$$

by its expected value under hypothesis H_0 to obtain the approximate test statistic

$$u'' = -\tfrac{1}{2}D_s^2 + \tfrac{1}{4}D_s^2 r^2 - \tfrac{1}{8}D_s^4.$$

By solving the integral equations arising in the exact analysis of the sequential test, Kendall [Ken65] calculated the false-alarm probability Q_0 and the average number $\bar{n}(0)$ of stages under hypothesis H_0 for a sequential test using the statistic

$$u = -\tfrac{1}{2}D_s^2 + \tfrac{1}{4}D_s^2 r^2 + \beta,$$

where β is an arbitrary constant. He found that both Q_0 and $\bar{n}(0)$ are much larger when $\beta = 0$ and $u = u'$ than when $\beta = -D_s^4/8$ and $u = u''$. The principal reason for preferring u' or u'' over u of (9-27) is that a quadratic rectifier is more easily constructed than one having the characteristic $\ln I_0(x)$. The approximation actually used, however, must be carefully selected if the advantages of the sequential test are to be preserved.

9.4 SEQUENTIAL DETECTION OF TARGETS OF UNKNOWN DISTANCE

In searching for targets by means of radar, it is usually necessary to detect those lying anywhere in a range many times longer than the spatial length of a signal. The echoes may arrive at any time during the *interpulse interval* T_p between transmitted pulses. We treated the problem of detecting a signal of unknown arrival time in the input $v(t)$ during a single interpulse interval in Chapter 7, and here we wish to describe some efforts to apply sequential detection to this task.

It is customary to divide the interpulse interval T_p into subintervals of a duration T' on the order of the reciprocal bandwidth of the signals to be detected. The total number of subintervals, or range bins, will be denoted by $L = T_p/T'$. Attention is focused on signals whose leading edges reach the receiver at the beginning of a subinterval; they are substantially past by the end of the same subinterval. In our discussion we suppose the target to be stationary, its echoes appearing at the same relative position in each interpulse interval.

The receiver contains a filter matched to such a signal over an interval of duration T', and the output of this filter is rectified and sampled at the end of each subinterval to provide a statistic r of the form of (9-26). (In practice the matching may be imprecise.) During the jth interpulse interval T_p the receiver thus generates L samples $r_1^{(j)}, r_2^{(j)}, \ldots, r_L^{(j)}$, which are statistically nearly independent. It is on these that its decisions are based.

In the sequential detection system proposed by Marcus and Swerling [Mar62] it is postulated that at most one signal is present and that it may appear with equal probability $1/L$ in any subinterval. At the end of each interpulse interval the receiver forms an average likelihood ratio for the data received since the beginning of the test. Under the approximation that the data $r_k^{(j)}$ are statistically independent, the average likelihood ratio at the end of the mth interpulse interval is

$$\bar{\Lambda}_m = \frac{1}{L}\sum_{s=1}^{L}\prod_{j=1}^{m}\frac{P_1(r_s^{(j)})}{P_0(r_s^{(j)})}, \tag{9-31}$$

where $P_0(r)$ is the probability density function of the datum r when no signal is present, and $P_1(r) = P_1(r; \boldsymbol{\theta}_s)$ is the probability density function of r when a standard signal is present, an average having been taken over any random parameters such as the phase. Because a standard signal can appear in no more than one subinterval, the terms in the summation in (9-31) contained factors $P_0(r_i^{(j)})$, $i \neq s$, which canceled from numerator and denominator.

This average likelihood ratio Λ_m is compared with decision levels A and B determined by Wald's approximations as in (9-10), and the decisions are made as previously described. If $B \leq \Lambda_m < A$, the transmitter is ordered to send out another pulse, and data are collected from the following interval to permit forming a new likelihood ratio Λ_{m+1}. Marcus and Swerling simulated such a sequential test on a digital computer and obtained average sample numbers as functions of the signal-to-noise ratio and the reliability of detection. The test was found to require a somewhat smaller signal-to-noise ratio than a fixed-sample-size test with the same reliability and a total number M of stages equal to the average sample number of the sequential test. The greater the number L of subintervals, the smaller the saving in signal-to-noise ratio. They also observed that for signal strengths intermediate between 0 and the standard strength the average number of stages did not become much larger than the average sample numbers for the zero and the standard strengths. Truncation of such a sequential test to avoid excessive lengths may be unnecessary.

An alternative suggested by Kendall and Reed [Ken63] allows a signal to appear in any subinterval with a probability q; the probability distribution of the total number of targets is then binomial. They presented the form of the likelihood ratio for this test and pointed out that as with the Marcus and Swerling test, simulation would be necessary to evaluate its performance.

In another form of sequential detection a sequential test is applied to the data from each subinterval. Pulses are transmitted in a certain direction until the tests of all L subintervals have terminated. If at the kth stage, for instance, the sequential test for the jth subinterval has not yet reached a final decision, the receiver will use the data $r_j^{(1)}, r_j^{(2)}, \ldots, r_j^{(k)}$ to form a likelihood ratio

$$\Lambda_j^{(k)} = \Lambda(r_j^{(1)}, r_j^{(2)}, \ldots, r_j^{(k)}) = \prod_{s=1}^{k} \frac{P_1(r_j^{(s)})}{P_0(r_j^{(s)})},$$

which is compared with two levels A and B. If $\Lambda_j^{(k)} < B$, the receiver decides that no target is present in the jth subinterval, and if $\Lambda_j^{(k)} > A$, it decides that there is a target there. In either event the test for the jth subinterval terminates, and the data $r_j^{(i)}$ from future interpulse intervals ($i > k$) are disregarded. If, on the other hand, $B \leq \Lambda_j^{(k)} < A$, the jth subinterval will be examined again after the $(k + 1)$th pulse is transmitted.

The average number of pulses needed to attain a specified reliability with such a system was calculated in some representative cases by Reed and Selin [Ree63] and Bussgang and Ehrman [Bus65]. The writer worked out the median number of pulses transmitted when no signal is present, as a function of the number L of subintervals. This number roughly determines the average time needed to scan a fixed portion of the sky that is empty of targets [Hel62].

In that calculation it was assumed that the level $a = \ln A$ is so high and the false-alarm probability Q_0 so small that under hypothesis H_0 the crossings of the upper level a can be neglected. Then one can use a formula given by Wald [Wal47, p. 193] for the probability distribution of the number n_t of stages before a single sequential test terminates:

$$
\begin{aligned}
P_m = \Pr(n_t \le m) &= \left(\frac{\alpha}{2\pi}\right)^{1/2} \int_0^\eta x^{-3/2} \exp\left[-\frac{\alpha}{2x}(x-1)^2\right] dx \\
&= \operatorname{erfc}\left[\left(\frac{\alpha}{\eta}\right)^{1/2}(1-\eta)\right] + e^{2\alpha}\operatorname{erfc}\left[\left(\frac{\alpha}{\eta}\right)^{1/2}(1+\eta)\right],
\end{aligned} \tag{9-32}
$$

where $\eta = m/\bar{n}$, $\bar{n} = \bar{n}(0) = E(n_t|H_0)$, and $\alpha = \bar{n}^2/\operatorname{Var}_0 n_t$. In this situation, with $Q_0 \ll 1$,

$$
\bar{n}(0) \approx \frac{b}{E(u|H_0)} \approx -\frac{2b}{S_s^2}
$$

by (9-20) and (9-30), and

$$
\operatorname{Var}_0 n_t \approx \frac{\bar{n} \operatorname{Var}_0 u}{[E(u|H_0)]^2}
$$

by (9-23), so that

$$
\alpha \approx -\tfrac{1}{2}b = \tfrac{1}{2}|\ln(1 - Q_0)|.
$$

One can also derive (9-32) from the approximate representation of the trajectory of the sum U as a Markov process in continuous time.

If $n_t^{(L)}$ denotes the number of radar pulses that need to be transmitted before the sequential subtests in all L range bins terminate,

$$
\Pr(n_t^{(L)} \le m) = P_m^L,
$$

for $n_t^{(L)} \le m$ if and only if all L of the subtests have terminated with fewer than $m + 1$ stages. The median v of the number of transmitted pulses is therefore given by

$$
\tfrac{1}{2} = P_v^L \qquad \text{or} \qquad P_v = 2^{-1/L}.
$$

The quotient $v/\bar{n}$ can be obtained by setting $P_m = 2^{-1/L}$ in (9-32) and solving for the parameter η. Denote this solution by η_1.

We want to compare this median v with the number M of stages needed by the Neyman–Pearson receiver to attain the same reliability (Q_0, Q_{ds}). That receiver can be regarded as basing its decisions on the sum

$$
G = \sum_{k=1}^M r_k^2
$$

in each range bin, where the r_k are defined in (9-26). This is the threshold receiver, as in (4-19); the input signal-to-noise ratio is assumed so small that the weak-signal approximation can be made, and $M \gg 1$. Then the sum G will have approximately a Gaussian distribution, and the false-alarm and detection probabilities are

$$
Q_0 \approx \operatorname{erfc} g_0, \qquad Q_{ds} \approx \operatorname{erfc} g_1, \qquad g_1 = g_0 - S_s\sqrt{M},
$$

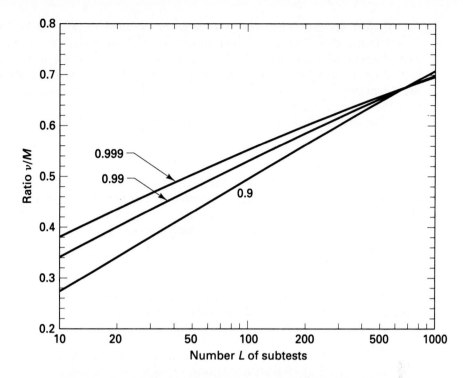

Figure 9-3. Comparison of sequential and conventional detector: v/M versus $L \approx \Delta\omega T$; $Q_0 = 10^{-6}$. Curves are indexed with the probability Q_d of detection.

for $S_s = \frac{1}{2}D_S^2$ is the effective signal-to-noise ratio, defined as in (4-66). Hence the number M of pulses required by the Neyman–Pearson receiver is

$$M \approx \frac{(g_1 - g_0)^2}{S_s^2}.$$

The ratio v/M of the median number of pulses required by the sequential receiver to the number required by the standard receiver is now

$$\frac{v}{M} = \frac{\eta_1 \bar{n}}{M} \approx -\frac{2b\eta_1}{(g_1 - g_0)^2},$$

and it is independent of the standard signal-to-noise ratio, provided that the standard ratio is small. The ratio v/M is plotted in Fig. 9-3 versus the number L of range bins for detection probabilities Q_d of 0.9, 0.99, and 0.999 and an overall false-alarm probability

$$Q_0^T = 1 - (1 - Q_0)^L \approx LQ_0 = 10^{-6}$$

for all L subintervals. The ratio v/M ranges from about 0.3 to about 0.6 as L increases from 10 to 1000. Thus the sequential receiver makes it possible to scan a certain part of the sky in about half the time needed by a conventional receiver when no or few targets are present.

If the radar target might be moving rapidly, the receiver must contain a number of *Doppler channels*, that is, a bank of filters matched to the signal, but with pass frequencies distributed over the range of expected Doppler shifts about the carrier frequency of the transmitted pulses. The output of each such filter during each interpulse interval will again be divided into a number of range bins, and a sequential test will be conducted in each as just described. Each Doppler channel must compensate for the changing arrival times of the echoes from a target moving with the corresponding velocity.

The total number of effectively independent tests will be on the order of

$$L = (\Delta\omega T)(\Delta t W_d),$$

where $\Delta\omega$ is the rms bandwidth of the signal and Δt is its rms duration, as defined in (6-105) and (6-106), and W_d is the width of the expected range of Doppler shifts. With this definition of L, Fig. 9-3 will again roughly represent the saving in the average number of transmitted pulses achieved by using a sequential rather than a fixed-sample-size test.

Problems

9-1. The signals $s_k(t)$ are known completely except for a common amplitude A. They arrive at a known time in the midst of white Gaussian noise of unilateral spectral density N, and in the kth observation interval of duration T the input to the receiver is either $(H_0)\, v_k(t) = n_k(t)$ or $(H_1)\, v_k(t) = s_k(t) + n_k(t)$. Take a standard signal of energy E_s and signal-to-noise ratio $d_s = (2E_s/N)^{1/2}$, which is to be detected with probability $Q_d(d_s)$, and set up a sequential probability ratio test of the kind described in Sec. 9.2.

(a) Express the logarithmic likelihood ratio u_j for the jth interval in terms of the input $v_j(t)$ during that interval.

(b) Show that u_j is a Gaussian random variable and derive its expected value and variance when a signal of energy E is present.

(c) Let $d = (2E/N)^{1/2}$. Find the moment-generating function $h(z; d)$ of the statistic u, and prove (9-14).

(d) Write down formulas for the probability $Q_d(d)$ of detection and the average number $\bar{n}(d)$ of stages of the sequential test.

(e) Plot $\bar{n}(d)$ and $Q_d(d)$ as functions of the signal-to-noise ratio d for $0 < d < d_s = 1$, with $Q_d(d_s) = 0.90$, $Q_0 = 10^{-6}$. Use the approximations given in this chapter.

9-2. Work out a sequential system for deciding whether the variance of a Gaussian noise input is N_0 or N_1, $N_1 > N_0$. The second variance N_1 may represent the sum of the variance N_0 of ordinary noise and the variance $N_s = N_1 - N_0$ of an independent noiselike signal added to it. Such a signal could be the output of a radar-jamming transmitter. Available are independent samples $x_1, x_2, \ldots,$ of the input; their expected values are 0.

(a) Find the logarithmic likelihood ratio $u_j = \ln[\, p_1(x_j)/p_0(x_j)]$, where $p_0(x_j)$ and $p_1(x_j)$ are Gaussian probability density functions of expected value 0 and variances N_0, N_1, respectively.

(b) Show that if the true variance of the data is N, $N > N_0$, the moment-generating function of the statistic u is

$$h(z; N) = (N_0/N_1)^{-z/2}\left[1 + \frac{(N_1 - N_0)N}{N_0 N_1} z\right]^{-1/2}.$$

(c) Calculate the expected value and variance of the statistic u when the true noise variance is N.

(d) Show how to determine the probability of detection and the average number $\bar{n}$ of stages of the test as functions of N. Use the approximate formulas given in this chapter.

9-3. Outline a sequential system to detect the 0's and 1's transmitted over the Rayleigh-fading channel of Sec. 4.2.3. Suppose that there is a noiseless channel whereby the receiver can tell the transmitter when to stop repeating one symbol of the message and go on to the next. Assume that the signal amplitudes and phases are independently random, and set the system up for signals arriving with amplitudes drawn from a standard Rayleigh distribution (4-17) with the parameter s equal to s_0. Determine the logarithmic likelihood ratio u, and calculate its moment-generating function, expected value, and variance when the true parameter of the Rayleigh distribution is s [Bas59].

9-4. In a receiver of optical signals, light falls on a photoelectric detector, and during successive intervals of duration T the receiver counts the numbers $n_1, n_2, \ldots$, of photoelectrons ejected by the light. The receiver must decide between hypothesis H_0 that only background light is present at its input and the hypothesis H_1 that an optical signal is also present. The probabilities of counting n_j electrons in the jth interval under each hypothesis are

$$\Pr(n_j = k \mid H_i) = \frac{\lambda_i^k}{k!} \exp(-\lambda_i), \qquad i = 0, 1, \qquad \lambda_1 > \lambda_0.$$

The numbers n_j in successive intervals are statistically independent.

(a) Show how to set up a sequential detector of the optical signal by using Wald's sequential probability ratio test, assuming a standard expected value λ_{1s} for the numbers n_j under hypothesis H_1. Here the data are governed not by probability density functions, but by probability mass functions.

(b) Using Wald's approximations, determine the decision levels.

(c) Calculate the moment-generating function of the statistic u that is accumulated at each stage of the test, and show how to determine the probability of detection and the average number of stages when the expected number of electrons under hypothesis H_1 is not λ_{1s}, but some other value $\lambda_1 > \lambda_0$.

(d) Show how to compare the performance of this test with that of a fixed-sample-size test that decides between hypothesis H_0 and hypothesis H_1 on the basis of the total number $n_1 + n_2 + \cdots + n_M$ of electrons counted in M successive intervals.

10

Signal Resolution

10.1 SPECIFICATION OF RECEIVERS

10.1.1 Varieties of Resolution

Up to now we have studied the detection only of individually identifiable signals. Radar targets, however, may be so close together that their echoes appear indistinguishable, and the receiver must decide whether to attribute its input to one signal, to several, or to none. A signal overlapping a weaker one of the same kind may well conceal it, as when the radar echo from a large bomber hides one from a small fighter plane nearby. To identify a ballistic missile among a cloud of decoys requires sorting out a multitude of echo signals. The echo from a low-flying aircraft arrives in the midst of a throng of weak, random signals reflected from the ground; these create the interference known as *clutter*. In a ground-mapping radar it is the detailed structure of the reflections themselves that is of interest.

The process of deciding whether the input to a receiver contains one signal or a number of adjacent signals is called *resolution*. The term is borrowed from optics, which has long been concerned with the efficient resolution of close images. One speaks of resolving the echoes of the fighter plane and the bomber or of the decoys and the missile. A ground-mapping radar that accurately reproduces details of the terrain it scans is said to provide good resolution. The quality of signals determining whether they can be easily resolved is called their *resolvability*. Echoes of the same transmitted radar pulse from diverse targets may differ in several respects: in time of arrival τ because of a difference of target distances, in carrier frequency Ω by virtue of a difference of target velocities, and in the antenna azimuth θ for

356

maximum amplitude, owing to a difference of target bearings. Any combination of such parameters may be utilized to resolve the signals.

The study of radar resolution falls into two parts, signal design and receiver design. Most attention has been given to the former, which seeks transmitted pulses whose echoes are most easily resolved. The latter, concerned with how the receiver should be modified to accommodate signals that may overlap, has been less widely developed, possibly because it is so complex.

There are various aspects under which the resolution of signals can be considered, and they are exemplified by the situations mentioned in the first paragraph. One can suppose that at most two signals might be present and that the receiver must either detect a weak signal in the presence of a strong one or decide whether two signals, one, or none is present in its input. Alternatively, the possibility that an arbitrary number of signals are present can be confronted, and one can ask the receiver to estimate the number of signals as well as their arrival times, frequencies, or other parameters. A third type of problem is the detection of a given signal against a background of many weak, random echoes such as make up the ground clutter; it can often be viewed as the detection of a signal in colored Gaussian noise. Ground mappers and similar radars can be regarded as measuring the scattering function of a surface or a volume of space, and their operation can be treated as the estimation of a random process. The first three of these aspects of resolution will be discussed in this chapter; the fourth is beyond the scope of this book. Ground-mapping radars are treated in [Rih69, pp. 441–83], [Har70], [Mor88, pp. 157–85], [Bla91, pp. 39–50], and [Cur91].

In what follows we shall assume that the signals are corrupted by white Gaussian noise of unilateral spectral density N. The modifications needed to accommodate colored noise are mostly evident, but to make them is uninstructive. Although signal resolution will be analyzed from the standpoint of hypothesis testing, it will be found that for any useful progress compromises must be made and something less than an optimum system must be accepted.

10.1.2 Resolution of Two Signals

In the most elementary and least common situation, one or the other, or both or neither, of two signals may be present in the input $v(t)$ of the receiver. We take them to be narrowband signals of the form

$$s_a(t) = \text{Re } S_a(t) \, e^{i\Omega t}, \qquad s_b(t) = \text{Re } S_b(t) \, e^{i\Omega t},$$
$$S_a(t) = AF(t) \exp i\phi_1, \tag{10-1}$$
$$S_b(t) = BG(t) \exp i\phi_2,$$

having independent and unknown phases and amplitudes and arriving in the presence of white Gaussian noise. We suppose that the functions $F(t)$ and $G(t)$ are given, $F(t) \neq G(t)$, and that they are normalized so that

$$\int_0^T |F(t)|^2 \, dt = \int_0^T |G(t)|^2 \, dt = 1. \tag{10-2}$$

The former signal will be called signal A, the latter signal B. Signal B may be a copy of signal A arriving earlier or later, $G(t) = F(t - \tau)$, with a given delay τ. On the basis of the input $v(t)$ received during the observation interval $(0, T)$, the observer must choose one of four hypotheses: (H_0) neither signal A nor signal B is present; (H_1) signal A alone is present; (H_2) signal B alone is present; and (H_3) both signals A and B are present. A prescription for making this choice can be termed a *resolution strategy*. We treat this problem because it provides an insight into how the forms of the signals affect their resolvability and because it represents a simple special case of the more general problem to be analyzed in Sec. 10.1.3.

If no noise were present, the observer could decide without error which of the four hypotheses is true by passing the input through properly matched filters and observing the output. It is easy to verify that the following two test statistics, among others, yield the desired information:

$$\hat{A} = \frac{1}{1 - |\lambda|^2} \left| \int_0^T [F^*(t) - \lambda G^*(t)] V(t) \, dt \right|,$$

$$\hat{B} = \frac{1}{1 - |\lambda|^2} \left| \int_0^T [G^*(t) - \lambda^* F^*(t)] V(t) \, dt \right|,$$

(10-3)

where the complex scalar product

$$\lambda = \int_0^T F^*(t) G(t) \, dt$$

(10-4)

of the two signals measures the extent to which they overlap.

If hypothesis H_0 is true, $\hat{A} = \hat{B} = 0$; under hypothesis H_1, $\hat{A} = A$ and $\hat{B} = 0$; under H_2, $\hat{A} = 0$ and $\hat{B} = B$; and under H_3, $\hat{A} = A$ and $\hat{B} = B$. If either of the quantities in (10-3) vanishes, the corresponding signal is absent. From our work in Sec. 3.3.1 we know that the quantity $\hat{A}$ can be generated by passing the input $v(t)$ through a narrowband filter matched to a signal having the complex envelope $[F(t) - \lambda^* G(t)]/(1 - |\lambda|^2)$, that is, through one with a complex impulse response

$$K_a(s) = \begin{cases} (1 - |\lambda|^2)^{-1}[F^*(T - s) - \lambda G^*(T - s)], & 0 \le s < T, \\ 0, & s < 0, \quad s \ge T. \end{cases}$$

A similar filter with complex impulse response $K_b(s)$, matched to a signal whose complex envelope is $[G(t) - \lambda F(t)]/(1 - |\lambda|^2)$, can be used to form the statistic $\hat{B}$. At the end of the observation interval the rectified outputs of these filters are the desired values of $\hat{A}$ and $\hat{B}$.

If the input contains additive random noise $n(t)$, the statistics $\hat{A}$ and $\hat{B}$ do not vanish, even when the signals are absent, and the observer is faced with a statistical problem. He must set up some strategy—utilizing perhaps the test quantities $\hat{A}$ and $\hat{B}$, perhaps some other functionals of the input $v(t)$—by which to make a choice among the four hypotheses that meets some standard of long-run success. To derive this strategy we turn to the theory of the statistical testing of hypotheses.

The Bayes strategy of Sec. 1.1 can be applied to this problem if one is given a matrix of the costs C_{ij} of choosing hypothesis H_i when hypothesis H_j is really true, and if one is also given the prior probabilities of the four hypotheses and the prior probability density functions of the amplitudes A and B. The Bayes strategy involves calculating the average conditional risk $C(\rightarrow H_i | v(t))$ of each hypothesis, defined as in (1-13), and the observer chooses the one whose conditional risk is the smallest.

When some or all of the prior probabilities and costs are unspecified, the same difficulties arise as in the simpler detection problems treated earlier, and the elementary methods of testing hypotheses become inapplicable. Instead we adopt as a guide the method of maximum likelihood, introduced in Sec. 3.6.4 and applied in Sec. 7.2 to deal with the detection of a signal of unknown time of arrival. The resulting resolution strategy and its probabilities of success and failure shed some light on the resolvability of signals of this kind.

The maximum-likelihood detection of a signal is in effect based on an estimate of its amplitude, derived on the assumption that the input $v(t)$ actually contains the signal. If the estimated amplitude is too small, the result is attributed to a noise fluctuation, and the decision is made that no signal is present. Taking the same viewpoint here, we imagine that the input $v(t)$ contains both signals, and we form the maximum-likelihood estimates of their amplitudes A and B. If the estimate $\hat{A}$ of the amplitude of the first signal is too small, the system decides that the first signal is absent, and the second signal is treated in the same way. If both $|\hat{A}|$ and $|\hat{B}|$ exceed certain decision levels, both signals are declared to be present.

By (3-54) with

$$S(t) = S_a(t) + S_b(t), \qquad Q(t) = N^{-1}S(t),$$

the likelihood functional is

$$\Lambda[v(t); a, b] = \exp\left[\frac{1}{N}\,\mathrm{Re}\int_0^T [S_a^*(t) + S_b^*(t)]V(t)\,dt - \frac{1}{2N}\int_0^T |S_a(t) + S_b(t)|^2\,dt\right],$$

where $V(t)$ is the complex envelope of the input. By introducing the complex signal amplitudes

$$a = A \exp i\phi_1, \qquad b = B \exp i\phi_2,$$

and the circular complex Gaussian random variables

$$z_1 = \int_0^T F^*(t)V(t)\,dt,$$

$$z_2 = \int_0^T G^*(t)V(t)\,dt, \qquad (10\text{-}5)$$

we can write it as

$$\Lambda[v(t); a, b] = \exp\{N^{-1}[\mathrm{Re}(a^*z_1 + b^*z_2) - \tfrac{1}{2}(|a|^2 + 2\,\mathrm{Re}\,\lambda a^*b + |b|^2)]\}; \qquad (10\text{-}6)$$

λ is the *overlap integral* introduced in (10-4).

The likelihood functional is to be maximized with respect to the four variables A, ϕ_1, B, and ϕ_2 or equivalently with respect to a, a^*, b, and b^*. By writing out the exponent in terms of these and differentiating with respect to each as though they were independent variables, we find the equations

$$a + b\lambda = z_1, \qquad a\lambda^* + b = z_2$$

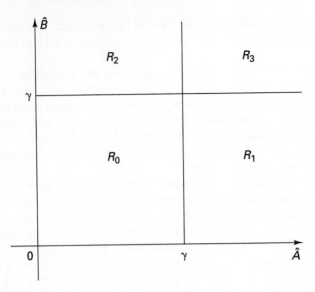

Figure 10-1. Decision regions in choice among four hypotheses.

and their complex conjugates. Solving them, we obtain the maximum-likelihood estimators

$$\hat{a} = \frac{z_1 - \lambda z_2}{1 - |\lambda|^2}, \qquad \hat{b} = \frac{z_2 - \lambda^* z_1}{1 - |\lambda|^2}, \qquad (10\text{-}7)$$

of the complex amplitudes a and b.

The estimators of the amplitudes themselves are then the statistics $\hat{A} = |\hat{a}|$ and $\hat{B} = |\hat{b}|$ given in (10-3). As we said there, these statistics can be generated by passing the input $v(t)$ through appropriate narrowband filters, which are followed by linear rectifiers. The outputs of those rectifiers at the end of the observation interval are compared with a decision level γ, which can be so adjusted that if $\hat{A} < \gamma$, signal A is declared absent; if $\hat{B} < \gamma$, signal B is rejected. Denoting once again the four hypotheses by H_0, H_1, H_2, and H_3 and the decision regions by R_0, R_1, R_2, and R_3, we can represent the decision strategy as in Fig. 10-1.

A false alarm occurs when no signals are present and any of the three hypotheses H_1, H_2, or H_3 is chosen. We shall now describe how to calculate the probability of a false alarm. Under hypothesis H_0 the circular complex Gaussian random variables z_1 and z_2 represent pure noise. By (10-4), (10-2), (10-5), and (3-45) the elements of their complex covariance matrix $\boldsymbol{\phi}_z$ are

$$\tfrac{1}{2}E(z_1 z_1^* \mid H_0) = \tfrac{1}{2}E(z_2 z_2^* \mid H_0) = N,$$
$$\tfrac{1}{2}E(z_1 z_2^* \mid H_0) = N\lambda.$$

Introducing the column vector $\mathbf{z}$ of the z's and its conjugate transposed row vector $\mathbf{z}^+ = (z_1^*, z_2^*)$, we can write this covariance matrix as

$$\boldsymbol{\phi}_z = \tfrac{1}{2}E(\mathbf{z}\mathbf{z}^+ \mid H_0) = N\boldsymbol{\Lambda}, \qquad \boldsymbol{\Lambda} = \begin{bmatrix} 1 & \lambda \\ \lambda^* & 1 \end{bmatrix}. \qquad (10\text{-}8)$$

With **â** the column vector of the estimators $\hat{a}$ and $\hat{b}$,

$$\hat{\mathbf{a}} = \mathbf{Kz},$$

$$\mathbf{K} = \frac{1}{1 - |\lambda|^2} \begin{bmatrix} 1 & -\lambda \\ -\lambda^* & 1 \end{bmatrix} = \mathbf{\Lambda}^{-1},$$

by (10-7). The complex covariance matrix of the estimators is

$$\boldsymbol{\phi}_{ab} = \tfrac{1}{2} E(\hat{\mathbf{a}}\hat{\mathbf{a}}^+|\, H_0) = \mathbf{K}\boldsymbol{\phi}_z \mathbf{K}^+ = N\mathbf{K}\mathbf{\Lambda}\mathbf{K}^+ = N\mathbf{K}^+ = N\mathbf{K},$$

and their inverse covariance matrix is

$$\boldsymbol{\phi}_{ab}^{-1} = N^{-1}\mathbf{\Lambda}.$$

The joint probability density function of the real and imaginary parts of $\hat{a}$ and $\hat{b}$ is therefore

$$\tilde{p}_0(\hat{a}, \hat{b}) = \frac{1 - |\lambda|^2}{(2\pi)^2 N^2} \exp\left[-\frac{|\hat{a}|^2 + \lambda^*\hat{a}\hat{b}^* + \lambda\hat{a}^*\hat{b} + |\hat{b}|^2}{2N} \right]$$

by (3-40).

If we now put

$$\hat{a} = \hat{A} \exp i\theta_A, \qquad \hat{b} = \hat{B} \exp i\theta_B,$$

we can write

$$\lambda^*\hat{a}\hat{b}^* + \lambda\hat{a}^*\hat{b} = 2 \operatorname{Re} \lambda\hat{a}^*\hat{b} = 2|\lambda|\hat{A}\hat{B} \cos(\arg \lambda - \theta_A + \theta_B).$$

Introducing the Jacobian $\hat{A}\hat{B}$ of this transformation to polar coordinates and integrating over $0 \le \theta_A < 2\pi$, $0 \le \theta_B < 2\pi$, we find for the joint probability density function of the estimators $\hat{A}$ and $\hat{B}$ under hypothesis H_0

$$p_0(\hat{A}, \hat{B}) = \frac{1 - |\lambda|^2}{N^2} \hat{A}\hat{B} \exp\left[-\frac{\hat{A}^2 + \hat{B}^2}{2N} \right] I_0\left(\frac{|\lambda|\hat{A}\hat{B}}{N} \right).$$

The false-alarm probability is obtained by integrating this density function over the region R_0 in Fig. 10-1 and subtracting from 1,

$$Q_0 = 1 - (1 - |\lambda|^2) \int_0^c \int_0^c xy\, e^{-\frac{1}{2}(x^2 + y^2)} I_0(|\lambda|xy)\, dx\, dy, \qquad c = \frac{\gamma}{\sqrt{N}},$$

and with some labor this can be reduced to a combination of Q functions,

$$Q_0 = 1 - 2(1 - |\lambda|^2) \int_0^c \int_0^x xy\, e^{-\frac{1}{2}(x^2 + y^2)} I_0(|\lambda|xy)\, dx\, dy$$

$$= 1 - 2(1 - |\lambda|^2) \int_0^c x\, e^{-\frac{1}{2}(1 - |\lambda|^2)x^2}\, dx \int_0^x y\, e^{-\frac{1}{2}(y^2 + |\lambda|^2 x^2)} I_0(|\lambda|xy)\, dy$$

$$= 1 - 2(1 - |\lambda|^2) \int_0^c x\, e^{-\frac{1}{2}(1 - |\lambda|^2)x^2}[1 - Q(|\lambda|x, x)]\, dx$$

$$= e^{-\frac{1}{2}(1 - |\lambda|^2)c^2}[1 - Q(|\lambda|c, c) + Q(c, |\lambda|c)]$$

in terms of Marcum's Q function (3-76). The last step is most easily verified by differentiating the result with respect to c and using (C-9) and (C-10). For $c(1 - |\lambda|) \gg 1$, an adequate approximation is

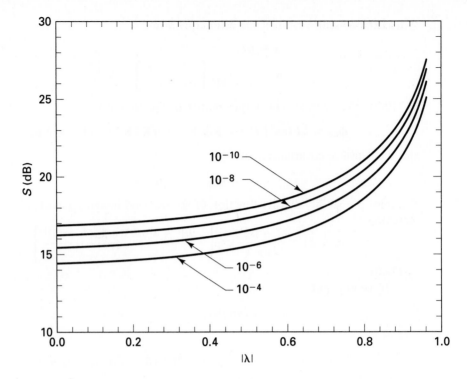

Figure 10-2. Signal-to-noise ratio (dB) required to attain a detection probability $Q_d = 0.999$ for one signal overlapping another, as a function of the overlap parameter $|\lambda|$. The curves are indexed with the false-alarm probability Q_0.

$$Q_0 \approx 2\, e^{-\frac{1}{2}(1-|\lambda|^2)c^2}, \tag{10-9}$$

which corresponds to neglecting the integral over the region R_3 in Fig. 10-1.

When as under hypotheses H_1 and H_3 signal A is present, the probability density function of the real and imaginary parts of the circular complex random variable $\hat{a}$ is

$$\tilde{p}_1(\hat{a}) = \frac{1-|\lambda|^2}{2\pi N}\, \exp\!\left[-\frac{(1-|\lambda|^2)|\hat{a}-a|^2}{2N}\right].$$

As in Sec. 3.4, we find that the probability of correctly deciding that signal A is present is

$$Q_d = \Pr(|\hat{a}| > \gamma \mid H_1 \cup H_3) = Q\!\left[d(1-|\lambda|^2)^{1/2},\ c(1-|\lambda|^2)^{1/2}\right],$$

where $d^2 = 2E/N$ is the input signal-to-noise ratio for signal A; Q_d is the probability of detecting one signal when it is possible that both may be present. In Fig. 10-2 we exhibit the input energy-to-noise ratio $S = \frac{1}{2}d^2$ required to attain a detection probability $Q_d = 0.999$ as a function of the parameter $|\lambda|$, which measures the degree to which the two signals overlap. The closer $|\lambda|$ lies to 1, the more energy is required to attain a specified reliability (Q_0, Q_d) in this decision.

10.1.3 The Resolution of Many Signals

10.1.3.1 Maximum-likelihood detection. A radar may be called on to detect echoes arriving at any time τ within an interval of long duration and having carrier frequencies anywhere within a broad range of values about the frequency Ω of the transmitted pulses. In addition, there may be times when a great many closely spaced targets are to be expected, and it will be necessary to determine both the number of targets and their ranges and range rates. If the echo signals might overlap in time or frequency or both, the problem of resolving them is especially difficult.

One approach is to impose on the space of the invariable parameters $\boldsymbol{\theta}''$ a rectangular grid of more or less uniformly spaced values and to concentrate on the detection of the M signals

$$s_j(t) = A_j \operatorname{Re} F(t; \boldsymbol{\theta}_j'') \exp(i\Omega t + i\phi_j), \qquad j = 1, 2, \ldots, M, \tag{10-10}$$

having parameters $\boldsymbol{\theta}_j''$ at the points of the grid. The signals $F(t; \boldsymbol{\theta}_j'')$ are normalized as in (10-2). The spacings of the points will correspond to the degree of resolution that one wishes to attain with respect to the parameters $\boldsymbol{\theta}''$.

As in Sec. 10.1.2, the detection of these M signals can be treated by the method of maximum likelihood. It is assumed that the input $v(t) = \operatorname{Re} V(t) \exp i\Omega t$ contains all M signals,

$$V(t) = S(t) + N(t), \qquad S(t) = \sum_{j=1}^{M} a_j F(t; \boldsymbol{\theta}_j''), \qquad a_j = A_j \exp i\phi_j, \tag{10-11}$$

where $N(t)$ is the complex envelope of the noise, assumed white. The logarithm of the likelihood functional for detecting the composite narrowband signal $s(t) = \operatorname{Re} S(t) \exp i\Omega t$ in white noise is, by (3-54) with $Q(t) = S(t)/N$,

$$\begin{aligned}
\ln \Lambda[v(t)] &= \frac{1}{N} \operatorname{Re} \int_0^T S^*(t) V(t)\, dt - \frac{1}{2N} \int_0^T |S(t)|^2\, dt \\
&= \frac{1}{2N} \left[\sum_{m=1}^{M} (a_m^* z_m + a_m z_m^*) - \sum_{m=1}^{M} \sum_{n=1}^{M} a_m^* \lambda_{mn} a_n \right],
\end{aligned} \tag{10-12}$$

where

$$z_m = \int_0^T F^*(t; \boldsymbol{\theta}_m'') V(t)\, dt \tag{10-13}$$

is the sample, taken at the end of the observation interval $(0, T)$, of the complex envelope of the output of a filter matched to the signal $s_m(t)$, defined as in (10-10). Furthermore

$$\lambda_{mn} = \int_0^T F^*(t; \boldsymbol{\theta}_m'') F(t; \boldsymbol{\theta}_n'')\, dt, \qquad \lambda_{mm} \equiv 1, \tag{10-14}$$

are elements of the *ambiguity matrix* $\boldsymbol{\Lambda}$ and measure the extent to which signals $s_m(t)$ and $s_n(t)$ overlap.

The maximum-likelihood estimates of the complex amplitudes a_m are obtained by differentiating (10-12) with respect to a_m^* and setting the result equal to zero,

$$z_j = \sum_{n=1}^{M} \lambda_{jn} \hat{a}_n, \qquad 1 \le j \le M, \tag{10-15}$$

and the solutions of these simultaneous equations are

$$\hat{a}_m = \sum_{j=1}^{M} k_{mj} z_j,$$ (10-16)

where the k_{mj} are the elements of the $M \times M$ matrix

$$\mathbf{K} = \mathbf{\Lambda}^{-1}, \qquad \mathbf{\Lambda} = \|\lambda_{mn}\|.$$ (10-17)

The estimated amplitudes $\hat{A}_m = |\hat{a}_m|$ are compared with decision levels γ_m; if $\hat{A}_m > \gamma_m$, the receiver decides that the mth signal, whose parameters are $\boldsymbol{\theta}_m''$, is present.

If there were no noise and if the only signals permitted were those having parameter values $\boldsymbol{\theta}_j''$, this system could correctly identify which signals are present. If a signal might arrive with intermediate values of the parameters, however, two adjacent estimates $\hat{A}_j$ might exceed their decision levels γ_j, indicating the presence of two signals when only one is there. This must be counted as an error of the system. Random noise introduces false alarms that signals are present when they are not. We concentrate now on determining the probability Q_d that the receiver correctly decides that an arbitrary one of the M signals $s_j(t)$ is present.

When the nth signal $s_n(t)$ is present, a situation that we label as hypothesis H_n,

$$E(z_j|\,H_n) = \lambda_{jn} a_n,$$ (10-18)

by (10-13) and (10-14); and by (10-16)

$$E(\hat{a}_m|\,H_n) = a_n \sum_{j=1}^{M} k_{mj} \lambda_{jn} = a_n \delta_{mn} = \begin{cases} a_n, & m = n, \\ 0, & m \neq n, \end{cases}$$ (10-19)

because the matrices $\mathbf{\Lambda}$ and $\mathbf{K}$ are inverses; δ_{mn} is the Kronecker delta.

Let us denote by $\hat{\mathbf{a}}$ the column vector of the estimates $\hat{a}_m$ of the M complex amplitudes and by $\mathbf{z}$ the column vector of the M circular complex Gaussian random variables z_m defined in (10-13). Then by (10-13), (10-14), and (3-45)

$$\tfrac{1}{2} E(z_m z_n^*|\,H_0) = \tfrac{1}{2} \int_0^T \int_0^T F^*(t_1; \boldsymbol{\theta}_m'') F(t_2; \boldsymbol{\theta}_n'') E[V(t_1) V^*(t_2)|\,H_0]\, dt_1\, dt_2$$

$$= N \int_0^T F^*(t; \boldsymbol{\theta}_m'') F(t; \boldsymbol{\theta}_n'')\, dt = N \lambda_{mn},$$ (10-20)

and the complex covariance matrix of the z_m's can be written

$$\boldsymbol{\phi}_z = \tfrac{1}{2} E(\mathbf{zz}^+|\,H_0) = N \mathbf{\Lambda}$$ (10-21)

as in (10-8). Because by (10-16) $\hat{\mathbf{a}} = \mathbf{Kz}$, the complex covariance matrix of the estimates $\hat{a}_m$ is

$$\boldsymbol{\phi}_a = \tfrac{1}{2} E(\hat{\mathbf{a}}\hat{\mathbf{a}}^+|\,H_0) = \tfrac{1}{2} E(\mathbf{Kzz}^+\mathbf{K}^+|\,H_0) = N\mathbf{K}\mathbf{\Lambda}\mathbf{K}^+ = N\mathbf{K}^+ = N\mathbf{K}$$ (10-22)

by (10-17). Thus the variances of the real and imaginary parts of the estimate $\hat{a}_n$ are equal to Nk_{nn}, where k_{nn} is the nth diagonal element of the inverse $\mathbf{K}$ of the ambiguity matrix $\mathbf{\Lambda}$.

If we denote by Q_0 the probability of deciding that the nth signal is present—accepting hypothesis H_n—when it is not, and by Q_d the probability of doing so under hypothesis H_n that the nth signal is present, we find by the methods of Sec. 3.4 that

$$Q_0 = \Pr(\hat{A}_n \geq \gamma_n | H_0) = e^{-\frac{1}{2}c^2},$$
$$Q_d = \Pr(\hat{A}_n \geq \gamma_n | H_n) = Q(d_n, c),$$

in terms of Marcum's Q function. Here c is proportional to the decision level γ_n on the estimate $\hat{A}_n = |\hat{a}_n|$ of the amplitude of the nth signal, and

$$d_n^2 = \frac{2E_n}{Nk_{nn}} \tag{10-23}$$

is the effective signal-to-noise ratio, where $E_n = \frac{1}{2}|a_n|^2$ is the energy of the nth signal.

When we set the decision levels so that all the false-alarm probabilities Q_0 are equal, the overall false-alarm probability

$$Q_0^T = \Pr(\hat{A}_1 \geq \gamma_1 \cup \hat{A}_2 \geq \gamma_2 \cup \cdots \cup \hat{A}_M \geq \gamma_M | H_0)$$

is bounded by

$$Q_0^T \leq MQ_0 \tag{10-24}$$

because for any M events E_i,

$$\Pr\left[\bigcup_{i=1}^{M} E_i\right] \leq \sum_{i=1}^{M} \Pr(E_i); \tag{10-25}$$

this is called the *union bound*. The quantity MQ_0 will be a good approximation to the overall false-alarm probability Q_0^T when $Q_0^T \ll 1$. It corresponds to (10-9) for $M = 2$.

10.1.3.2 Resolution of signals in time. Let us now assume that the signals are uniformly separated in arrival time by δ so that

$$F(t; \boldsymbol{\theta}_m'') = F(t - \tau_m), \qquad \tau_m = m\delta, \qquad 1 \leq m \leq M. \tag{10-26}$$

When as usually the observation interval $(0, T)$ is much longer than the duration T' of each signal, the elements of the ambiguity matrix $\boldsymbol{\Lambda}$ are, by (10-14),

$$\lambda_{mn} = \int_{-\infty}^{\infty} F^*(t - \tau_m)F(t - \tau_n)\, dt = \tilde{\lambda}_{m-n}, \tag{10-27}$$

and $\boldsymbol{\Lambda}$ is a Toeplitz matrix, each row of which is displaced to the right of the row above by one place. The elements of this matrix are

$$\tilde{\lambda}_j = \int_{-\infty}^{\infty} F^*(t - j\delta)F(t)\, dt = \int_{-\infty}^{\infty} |f(\omega)|^2\, e^{ij\delta\omega} \frac{d\omega}{2\pi} \tag{10-28}$$

in terms of the Fourier transform

$$f(\omega) = \int_{-\infty}^{\infty} F(t)\, e^{-i\omega t}\, dt$$

of the complex envelope of the signals. If we now assume that $M \gg 1$ and disregard signals that might arrive near one end or the other of the observation interval,

$T = M\delta \gg T'$, the inverse matrix **K** will also have nearly the Toeplitz form, and its elements will be

$$k_{mn} \approx \tilde{k}_{m-n}$$

with

$$\sum_{j=-\infty}^{\infty} \tilde{k}_{m-j} \tilde{\lambda}_{j-n} = \delta_{mn} = \begin{cases} 1, & m = n, \\ 0, & m \neq n, \end{cases} \qquad (10\text{-}29)$$

by (10-17).

This convolutional equation can be solved by thinking of the matrix elements $\tilde{k}_p$, $\tilde{\lambda}_p$ as Fourier coefficients of periodic functions

$$K(u) = \sum_{p=-\infty}^{\infty} \tilde{k}_p \, e^{-ipu}, \qquad L(u) = \sum_{p=-\infty}^{\infty} \tilde{\lambda}_p \, e^{-ipu}, \qquad -\pi \leq u < \pi, \qquad (10\text{-}30)$$

and by the convolution theorem (10-29) yields

$$K(u) \, L(u) = 1. \qquad (10\text{-}31)$$

By (10-28) the periodic function $L(u)$ is

$$L(u) = \sum_{j=-\infty}^{\infty} \int_{-\infty}^{\infty} |f(\omega)|^2 \, e^{ij(\delta\omega - u)} \frac{d\omega}{2\pi}$$

$$= \int_{-\infty}^{\infty} |f(\omega)|^2 \sum_{k=-\infty}^{\infty} \delta(\delta\omega - u + 2k\pi) \, d\omega \qquad (10\text{-}32)$$

$$= \frac{1}{\delta} \sum_{k=-\infty}^{\infty} \left| f\left(\frac{u - 2k\pi}{\delta}\right) \right|^2 .$$

Here we have used the periodic delta function

$$\sum_{k=-\infty}^{\infty} \delta(x + 2k\pi) = \frac{1}{2\pi} \sum_{j=-\infty}^{\infty} e^{ijx}. \qquad (10\text{-}33)$$

The diagonal element of the inverse matrix **K** is now

$$\tilde{k}_0 = k_{nn} = \int_{-\pi}^{\pi} K(u) \frac{du}{2\pi}$$

$$= \frac{\delta}{2\pi} \int_{-\pi}^{\pi} \frac{du}{\displaystyle\sum_{k=-\infty}^{\infty} \left| f\left(\frac{u-2k\pi}{\delta}\right) \right|^2} \qquad (10\text{-}34)$$

$$= \frac{\delta^2}{2\pi} \int_{-\pi/\delta}^{\pi/\delta} \frac{d\omega}{\displaystyle\sum_{k=-\infty}^{\infty} \left| f\left(\omega - \frac{2k\pi}{\delta}\right) \right|^2} .$$

The denominator of the integrand is sketched in Fig. 10-3, in which we have marked the approximate bandwidth $2\pi W$ of the signals Re $F(t - \tau_n) \exp i\Omega t$ in angular frequency.

When the signals $s_j(t)$ do not overlap significantly, $W\delta \gg 1$, then $\tilde{\lambda}_j \equiv 0$ for $|j| \neq 0$. Because $\tilde{\lambda}_0 = 1$, $\tilde{k}_0 = 1$; and the effective signal-to-noise ratio in (10-23)

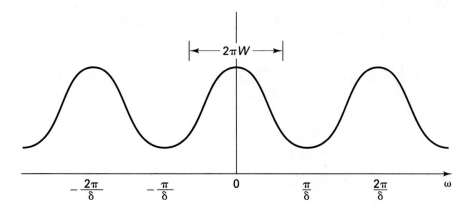

Figure 10-3. Denominator of (10-34).

is equal to the input signal-to-noise ratio. The signals do not interfere and can be detected as effectively as though they arrived alone. When $W\delta \ll 1$, on the other hand, the peaks of the denominator of (10-34) are widely separated, only the central term ($k = 0$) contributes significantly, and the effective output signal-to-noise ratio is approximately

$$d_n^2 \approx \frac{2E_n}{N} \frac{2\pi}{\delta^2} \left[\int_{-\pi/\delta}^{\pi/\delta} \frac{d\omega}{|f(\omega)|^2} \right]^{-1}. \tag{10-35}$$

In order to attain a specified reliability, the input signal-to-noise ratio, which is now

$$\frac{2E_n}{N} \approx d_n^2 \frac{\delta^2}{2\pi} \int_{-\pi/\delta}^{\pi/\delta} \frac{d\omega}{|f(\omega)|^2}, \tag{10-36}$$

must be very large. If the signals $s_j(t)$ were strictly bandlimited to W hertz, the right side of (10-36) would even be infinite when $W\delta < 1$.

Let us now work out the transfer function of a linear, narrowband filter whose rectified output, sampled at the appropriate time, yields the estimate $\hat{A}_n$ of the amplitude of the nth signal. The estimate $\hat{a}_n$ of the complex amplitude of that signal is given approximately as

$$\hat{a}_n = \sum_{j=-\infty}^{\infty} \tilde{k}_{n-j} z_j = \int_{-\infty}^{\infty} \sum_{j=-\infty}^{\infty} \tilde{k}_{n-j} F^*(t - j\delta) V(t)\, dt = \int_{-\infty}^{\infty} H_n^*(t) V(t)\, dt \tag{10-37}$$

by (10-16). Here

$$H_n(t) = \sum_{j=-\infty}^{\infty} \tilde{k}_{n-j}^* F(t - j\delta)$$

has the Fourier transform

$$h_n(\omega) = \sum_{j=-\infty}^{\infty} \tilde{k}_{j-n} f(\omega) e^{-ij\delta\omega} = f(\omega) e^{-in\delta\omega} \sum_{p=-\infty}^{\infty} \tilde{k}_p e^{-ip\delta\omega} = h(\omega) \exp(-i\omega\tau_n),$$

where

$$h(\omega) = K(\delta\omega)f(\omega) = \frac{\delta f(\omega)}{\sum\limits_{k=-\infty}^{\infty} \left| f\left(\omega - \frac{2\pi k}{\delta}\right)\right|^2} \qquad (10\text{-}38)$$

by (10-30) through (10-32). The estimates of the signals can therefore be generated by passing the input $\operatorname{Re} V(t) \exp i\Omega t$ through a narrowband filter whose complex impulse response is

$$K_h(s) = \begin{cases} H^*(T' - s), & 0 \leq s \leq T', \\ 0, & s < 0, \quad s \geq T, \end{cases} \qquad (10\text{-}39)$$

where $H(t)$ is the Fourier transform of $h(\omega)$. Its narrowband transfer function is

$$Y(\omega) = h^*(\omega)e^{-i\omega T'} = \frac{\delta f^*(\omega)\, e^{-i\omega T'}}{\sum\limits_{k=-\infty}^{\infty} \left| f\left(\omega - \frac{2\pi k}{\delta}\right)\right|^2}. \qquad (10\text{-}40)$$

The estimate $\hat{a}_n$ is obtained by sampling the complex envelope of the output of this filter at time $\tau_n + T' = n\delta + T'$. If we place a rectifier after this narrowband filter, the presence of any one of the signals $s_j(t)$ is indicated by the crossing of a decision level γ by the output of this rectifier.

The complex envelope of the signal component $s_n^{(0)}(t)$ of the output of the narrowband filter when the nth signal

$$s_n(t) = \operatorname{Re} a_n F(t - \tau_n)\, e^{i\Omega t}$$

is present has the Fourier transform

$$f_n^{(0)}(\omega) = a_n h^*(\omega)f(\omega) \exp[-i\omega(\tau_n + T')]$$
$$= a_n \frac{\delta |f(\omega)|^2}{\sum\limits_{k=-\infty}^{\infty} \left| f\left(\omega - \frac{2\pi k}{\delta}\right)\right|^2} \exp[-i\omega(\tau_n + T')].$$

When the arrival times τ_n of the signals are closely spaced, $W\delta \ll 1$, this spectrum is nearly uniform over a band of width δ^{-1} hertz, outside of which it nearly vanishes. The signal at the output of the subsequent rectifier therefore has a width on the order of δ, and its peak value occurs at time $\tau_n + T'$. The filter whose narrowband transfer function is given by (10-40) sharpens the incoming signals to such an extent that its outputs due to those signals arriving at times separated by δ do not significantly overlap. The filter also enhances the noise, but maximizes the output signal-to-noise ratio for each of these signals. The shorter the interval δ between the signals that our receiver is required to resolve, the smaller is this maximum output signal-to-noise ratio

$$d^2 = \frac{2E}{N\tilde{k}_0},$$

and the more energy E each signal must carry in order to be detected with the specified reliability (Q_0, Q_d).

10.2 THE DETECTION OF SIGNALS IN CLUTTER

10.2.1 The Spectrum of Clutter Interference

A radar system is often called on to detect a target echo in the presence of a great many other echoes from raindrops or from the surface of the ground or the sea. Interference of this kind is known as *clutter*. In wartime, strips of metal foil known as "chaff" or "window" are dropped from airplanes to confuse enemy radar by creating a similar interference. The reverberation encountered in sonar is another type of clutter. The task of a radar subjected to clutter might be considered as the resolution of a wanted signal from a number of undesirable ones that overlap it in time and frequency.

Because the parameters of the extraneous signals are unpredictable, however, it is more convenient to view the clutter as a type of noise. An apt model pictures it as composed of reflections of the transmitted pulse from a large number of small dispersed scatterers. Because the net voltage they produce at the receiver input is the sum of a large number of weak, random voltages, the clutter can be described by a Gaussian distribution. Methods developed earlier for detecting a signal in colored Gaussian noise can be applied to finding the optimum detection system, which in turn can be analyzed to determine the probability of detection as a function of the strength and distribution of the clutter. One of our purposes in investigating this model is to bring out the similarity of the resultant optimum detector to the maximum-likelihood detector derived in Sec. 10.1.

This clutter noise is not stationary, for the density of scatterers within the transmitted beam varies with the distance, and the total power they reflect is not uniform. Because the density and other characteristics of the scatterers usually change only slightly over a distance on the order of several radar pulse lengths, however, the detectability of an echo signal in the midst of the clutter will be nearly the same as if the clutter had at all times the same statistical properties as in the vicinity of the signal. We can therefore treat the clutter as though it were stationary. As it is Gaussian and has zero expected value, its probability density functions depend only on its spectral density.

If the scatterers are all at rest with respect to the transmitter, the composite echo signal takes the form Re $S(t) \exp i\Omega t$, where

$$S(t) = \sum_n z_n F(t - \tau_n). \tag{10-41}$$

Here $F(t)$ is the complex envelope of the narrowband transmitted pulse, z_n a complex number specifying the amplitude and phase of the nth echo, and τ_n the epoch of the nth echo. We suppose the envelope $F(t)$ normalized as (10-2). The z_n's and the τ_n's are random variables independent from one scattering to another. The infinitesimal echo pulses arrive at a high average rate and produce clutter with a finite average power equal, say, to P_c. Noise of this kind has a spectral density proportional to the absolute square of the spectrum

$$f(\omega) = \int_{-\infty}^{\infty} F(t) e^{-i\omega t} \, dt$$

of each pulse [Hel91, p. 401], [Pap91, p. 360]. Hence the spectral density of the clutter is

$$\Psi(\omega) = \tilde{\Psi}(\omega - \Omega) + \tilde{\Psi}(-\omega - \Omega), \tag{10-42}$$

where $\tilde{\Psi}(\omega)$ is the positive-frequency part of the spectral density as measured from the carrier frequency Ω, and

$$\tilde{\Psi}(\omega) = \tfrac{1}{2}P_c|f(\omega)|^2. \tag{10-43}$$

Usually the scatterers are not all at rest with respect to the transmitter. The transmitter may be on a moving airplane, and the scatterers may themselves be moving erratically, as when trees, bushes, and sea spray are blown about by the wind. In most cases the effect of such motions can be expressed by a convolution,

$$\tilde{\Psi}(\omega) = \tfrac{1}{2}\int_{-\infty}^{\infty} R_c(w)|f(\omega - w)|^2 \frac{dw}{2\pi}, \tag{10-44}$$

where $R_c(w)dw/2\pi$ is the power in the clutter reflected by scatterers inducing a frequency shift in an interval of width $dw/2\pi$ about the frequency $w/2\pi$. The total clutter power is

$$P_c = \int_{-\infty}^{\infty} R_c(w)\frac{dw}{2\pi}.$$

For clutter due to reflections from windblown vegetation, the width of the distribution $R_c(w)$ has been observed to be inversely proportional to the wavelength of the transmitted pulses. The product of this width and the wavelength is on the order of a few centimeters or tens of centimeters per second. Under certain circumstances the distribution of the clutter amplitudes departs from the Gaussian, but for our purposes we must disregard this. Many details of the characteristics of natural and artificial clutter are to be found in [Law50], [Geo52], [McG60], [Sko62, Ch. 12], and [Sch91b, Ch. 4]. Extensive treatments of detection in clutter are to be found in [Rih69, pp. 350–80] and [Sch91b].

The complex autocovariance function $\tilde{\psi}(\tau)$ of the clutter is the Fourier transform of $2\tilde{\Psi}(\omega)$ as in (3-19). Because (10-44) has the form of a convolution, its Fourier transform is simply a product,

$$\tilde{\psi}(\tau) = \lambda(\tau, 0)r_c(\tau),$$

where

$$\lambda(\tau, 0) = \int_{-\infty}^{\infty} F(t - \tfrac{1}{2}\tau)F^*(t + \tfrac{1}{2}\tau)\, dt,$$

$$r_c(\tau) = \int_{-\infty}^{\infty} R_c(\omega)\, e^{i\omega\tau} \frac{d\omega}{2\pi}, \qquad r_c(0) = P_c.$$

Thus the complex autocovariance function of the clutter is the product of the ambiguity function $\lambda(\tau, 0)$ of the transmitted pulse and the Fourier transform of the density $R_c(w)$ of the frequency shifts.

10.2.2 Detection of Single Pulses in Clutter

Let the echo signal be represented by

$$s(t) = A \operatorname{Re} F_s(t - \tau)\, e^{i\Omega t + i\phi},$$

where A and ϕ are its unknown amplitude and phase, τ its epoch, and $F_s(t)$ its complex envelope, normalized as in (10-2). If the signal is a version of the transmitted pulse with the Doppler shift w, as we shall generally presume here, $F_s(t) = F(t) \exp iwt$, $f_s(\omega) = f(\omega - w)$.

We suppose that the receiver is working only with the return from a single transmitted pulse and is detecting echoes individually. The spreading of the spectral density given by (10-44) can then be neglected, for the distribution $R_c(w)$ is much narrower than the spectrum $f(\omega)$ of most radar pulses. By adopting as our reference carrier frequency Ω that of the clutter echoes, we can write the spectral density of the clutter as in (10-42) and (10-43). In addition to the clutter, the input to the receiver will contain white noise of unilateral spectral density N. As we said in Sec. 3.2.4, the white noise can be treated as narrowband noise whose spectral density is uniform over the range of frequencies occupied by the signal. Hence the narrowband spectral density of the total input noise is

$$\tilde{\Phi}(\omega) = \tfrac{N}{2} + \tilde{\Psi}(\omega) = \tfrac{1}{2}[N + P_c|f(\omega)|^2]. \tag{10-45}$$

Because the arrival time of the signal is unknown, we employ the maximum-likelihood detection strategy developed in Sec. 7.2. The input to the receiver is passed through a filter matched to a signal whose complex envelope is

$$Q(t) = \int_{-\infty}^{\infty} q_s(\omega)\, e^{i\omega t}\, \frac{d\omega}{2\pi},$$
$$q_s(\omega) = \frac{f_s(\omega)}{2\tilde{\Phi}(\omega)} = \frac{f_s(\omega)}{N + P_c|f(\omega)|^2}, \tag{10-46}$$

where $f_s(\omega)$ is the Fourier transform of the signal envelope $F_s(t)$; compare (2-77) through (2-79). The narrowband impulse response of the filter is

$$K(\tau) = \begin{cases} Q^*(T' - \tau), & 0 \leq \tau \leq T', \\ 0, & \tau < 0, \quad \tau > T', \end{cases} \tag{10-47}$$

where T' is a delay long enough for the interval $0 < t < T'$ to include most of the reversed signal $Q^*(T' - t)$. The output of this matched filter is rectified and compared with a decision level, which when surpassed indicates the presence of a signal.

It would be difficult in practice to use this maximum-likelihood detector because the clutter power P_c is not constant, but varies in time through the variation with distance of the density of the random scatterers. Because this variation is small in a time on the order of a pulse width, one can envision a system using a time-varying filter so designed that at each instant it is matched to the signal $Q(t)$, (10-46), for the current value of P_c. Such a filter might be difficult to construct. An alternative would be a set of many filters, each matched to $Q(t)$ for a particular value of P_c. The receiver would measure the clutter power independently and by switching arrange to use the filter matched for the nearest value of P_c.

If the delay T' is very long, the complex narrowband transfer function $Y(\omega)$ of the matched filter, defined in Sec. 3.1, is

$$Y(\omega) = q_s^*(\omega)\, e^{-i\omega T'} = \frac{f_s^*(\omega)}{N + P_c|f(\omega)|^2}\, e^{-i\omega T'}. \tag{10-48}$$

The spectrum of the echo pulse as it issues from the matched filter is

$$f_0(\omega) = \frac{A|f_s(\omega)|^2}{N + P_c|f(\omega)|^2} \, e^{-i\omega(T'+\tau)+i\phi}$$

with frequencies still measured from the carrier frequency of the clutter.

In the situation least favorable for detection the target moves with the same relative velocity as the scatterers, there is no relative frequency shift ($w = 0$), and $f_s(\omega) = f(\omega)$. The amplitude $|f_0(\omega)|$ of the output spectrum is then nearly uniform for angular frequencies lying within a band $-\omega_c < \omega < \omega_c$, with the *cutoff frequency* ω_c given roughly by

$$P_c|f(\omega_c)|^2 \approx N.$$

[We suppose the pulse spectrum symmetrical, $|f(\omega)| = |f(-\omega)|$, for simplicity.] At frequencies far from the carrier, for which $|\omega| \gg \omega_c$, the output spectrum drops off to zero. The stronger is the clutter power P_c compared with the product of N and the bandwidth W of the signal, the larger the cutoff frequency ω_c. The width of the output pulse $F_0(t)$ from the matched filter will be on the order of $2\pi/\omega_c$ and much smaller than $2\pi/W$, the approximate width of the echo signal. Thus the effect of the matched filter in (10-47) is to sharpen the returning echo pulses as much as possible without too greatly enhancing the output resulting from the white noise. It is hoped that the signal echo will then stand out over the smaller echoes due to the clutter.

The filter specified by (10-48) for a signal with $w = 0$ is similar to the filters proposed by Urkowitz [Urk53], for which $Y(\omega)$ equals $[f(\omega)]^{-1}$ for $|\omega|$ less than a fixed cutoff frequency, beyond which $Y(\omega)$ vanishes. The cutoff frequency is taken large enough to sharpen the signal pulses as much as possible, yet not so large that the white noise generates an excessive output. The filter in (10-48) also closely resembles the one derived in Sec. 10.1 for distinguishing signals that might arrive at times separated by an interval δ that is much smaller than the reciprocal W^{-1} of their bandwidth. The cutoff frequency ω_c corresponds to the frequency π/δ marked in Fig. 10-3; for $|\omega| < \pi/\delta$, the transfer function $Y(\omega)$ of that filter, as given by (10-40), is roughly proportional to $[f(\omega)]^{-1}$; for $|\omega| \gg \pi/\delta$, $|Y(\omega)| \approx 0$. For $\omega_c \gg W$ and $w = 0$, the spectrum $f_0(\omega)$ of the output of the filter in (10-48) closely resembles that of the filter in (10-40).

The performance of a detection system of this kind is described by its false-alarm and detection probabilities, which can be calculated as in Sec. 7.4. They are approximately

$$
\begin{aligned}
Q_0 &\approx \sqrt{2\pi} \, \beta T b \, e^{-b^2/2}, \\
Q_d &\approx Q(d_{\text{eff}}, b),
\end{aligned}
\tag{10-49}
$$

where $Q(\cdot, \cdot)$ is Marcum's Q function, b is proportional to the decision level, T is the length of the observation interval, d_{eff}^2 is the effective signal-to-noise ratio, given by

$$d_{\text{eff}}^2 = 2E \int_{-\infty}^{\infty} \frac{|f_s(\omega)|^2}{N + P_c|f(\omega)|^2} \, \frac{d\omega}{2\pi}, \tag{10-50}$$

and β is a bandwidth defined by

$$\beta^2 = \frac{\overline{\omega^2} - \overline{\omega}^2}{(2\pi)^2},$$

$$\overline{\omega^n} = \frac{\int_{-\infty}^{\infty} \frac{\omega^n |f_s(\omega)|^2}{\Phi(\omega)} \frac{d\omega}{2\pi}}{\int_{-\infty}^{\infty} \frac{|f_s(\omega)|^2}{\Phi(\omega)} \frac{d\omega}{2\pi}}. \qquad (10\text{-}51)$$

This bandwidth will exist only if $|f_s(\omega)|^2$ decreases to 0 more rapidly than $|\omega|^{-3}$ as $|\omega| \rightarrow \infty$. In (10-50) $E = \frac{1}{2}A^2$ is the signal energy, the normalization in (10-2) being maintained.

For a fixed false-alarm probability, the probability of detection depends most strongly on the effective signal-to-noise ratio d_{eff}^2 given by (10-50). If one alters the shape of the transmitted pulses or even the transmitted power, to which P_c is proportional, the bandwidth β and hence also the quantity b in (10-49) must change. This variation in b, however, has a much smaller effect on the detection probability Q_d than the accompanying change in the effective signal-to-noise ratio d_{eff}^2.

Let us further analyze the detectability of an echo with the same carrier frequency as the clutter ($w = 0$). When $P_c \gg N\delta\omega$, the integrand in (10-50) is nearly uniform for $|\omega| < \omega_c$, where again $P_c |f(\omega_c)|^2 \approx N$. For $|\omega| > \omega_c$ the integrand drops to zero. The effective signal-to-noise ratio d_{eff}^2 is therefore roughly given by

$$d_{\text{eff}}^2 \approx \frac{2E\omega_c}{\pi P_c}.$$

With both the signal energy E and the average clutter power P_c proportional to the power output P_T of the transmitter, the signal-to-noise ratio d_{eff}^2 depends on P_T only through the value of ω_c. If, for instance, $|f(\omega)|$ is proportional to $|\omega|^{-n}$ for large values of $|\omega|$, $|f(\omega)| \approx C_1 |\omega|^{-n}$, we see from the equation

$$|C_1|^2 P_c |\omega_c|^{-2n} \approx N$$

that ω_c is proportional to $P_c^{1/2n}$ and hence to $P_T^{1/2n}$. The effective signal-to-noise ratio d_{eff}^2 determining signal detectability through (10-49) is therefore also proportional to $P_T^{1/2n}$. The larger n is, the smaller the influence of a mere increase of transmitter power on the probability of detection.

In order to study in more detail the effect of increasing the transmitted power, let us write the effective signal-to-noise ratio (10-50) as

$$d_{\text{eff}}^2 = \frac{2EW}{P_c}\gamma, \qquad \gamma = \frac{P_c}{NW} \int_{-\infty}^{\infty} \frac{|f(\omega)|^2}{1 + P_c N^{-1}|f(\omega)|^2} \frac{d\omega}{2\pi}, \qquad (10\text{-}52)$$

for a signal with no Doppler shift ($w = 0$). The factor $(2EW/P_c)$ is independent of the transmitted power. In Fig. 10-4 we have plotted the factor γ versus the clutter-to-noise ratio $r = P_c/NW$ for four types of signal; W is the equivalent bandwidth defined in (7-11). The signals and their equivalent bandwidths W are listed in Table 10-1. The signals have been normalized as in (10-2).

For the bandlimited signal it is simple to show that

$$\gamma = \frac{r}{1 + r}, \qquad r = \frac{P_c}{NW},$$

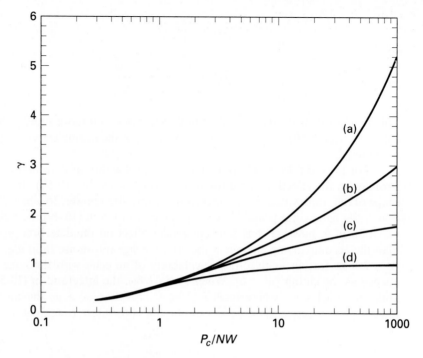

Figure 10-4. Signal-to-noise-ratio factor γ (10-52) as a function of the clutter-to-noise ratio P_c/NW for the signals listed in Table 10-1: (a) rectangular, (b) bilateral exponential, (c) Gaussian, (d) bandlimited pulse.

and increasing the transmitted power, to which r is proportional, leads to saturation. The bandlimited signals Re $F(t)$ exp $i\Omega t$ have long tails and cannot easily be resolved. The more sharply the spectrum $|f(\omega)|$ decreases to zero as $|\omega| \to \infty$, the less effectively the pulses can be resolved or detected in the midst of their own clutter. At the opposite extreme, rectangular pulses, as one might expect, are the most easily resolved and the most reliably detected in clutter noise.

For the bilateral exponential signal, application of [Dwi61, eq. 856.31] or the residue theorem to (10-52) yields

$$\gamma = \frac{r}{(1 + s^2)^{1/2} \, \mathrm{Re}[(1 + is)^{1/2}]}, \qquad s = \sqrt{\frac{8r}{5}}, \qquad r = \frac{P_c}{NW}. \tag{10-53}$$

For the other two types of signal the factor γ was evaluated by numerical integration. For Gaussian signals the integral in (10-52) could be efficiently integrated by the trapezoidal rule. For rectangular signals that integral becomes

$$\gamma = \frac{2r}{\pi} \int_0^\infty \frac{\sin^2 x}{x^2 + C \sin^2 x} \, dx, \qquad C = rWT = \frac{3r}{2}.$$

Because of the slow decline of the integrand to zero as $x \to \infty$, it was necessary to reform the integrand somewhat drastically in order to reduce the time required by the numerical integration. The range from 0 to ∞ was broken into ranges of width π, in each half of the kth of which the substitution

Signal Resolution   Chap. 10

Table 10-1 Signal Parameters

	$F(t)$	$f(\omega)$	W
(a) Rectangular	$\begin{cases} T^{-1/2}, & -\frac{1}{2}T < t < \frac{1}{2}T, \\ 0, & \|t\| \geq \frac{1}{2}T \end{cases}$	$\dfrac{2\sin(\frac{1}{2}\omega T)}{\omega T^{1/2}}$	$\dfrac{3}{2T}$
(b) Bilateral exponential	$\alpha^{1/2}e^{-\alpha\|t\|}$	$\dfrac{2\alpha^{3/2}}{\omega^2 + \alpha^2}$	$\dfrac{2}{5}\alpha$
(c) Gaussian	$(2\sigma^2/\pi)^{1/4}e^{-\sigma^2 t^2}$	$(2\pi/\sigma^2)^{1/4}e^{-\omega^2/4\sigma^2}$	$\pi^{-1/2}\sigma$
(d) Bandlimited	$\dfrac{2\sin(\pi Wt)}{\sqrt{W}t}$	$\begin{cases} W^{-1/2}, & -\pi W < \omega \leq \pi W, \\ 0, & \|\omega\| \geq W \end{cases}$	W

$$x = (k + \tfrac{1}{2})\pi \pm u, \qquad 0 \leq k < \infty,$$

was made, reducing the integral to

$$\gamma = \frac{2r}{\pi}\int_0^{\pi/2}\cos^2 u \sum_{k=0}^{\infty}\left[\frac{1}{[(k + \frac{1}{2})\pi + u]^2 + C\cos^2 u}\right.$$

$$\left. + \frac{1}{[(k + \frac{1}{2})\pi - u]^2 + C\cos^2 u}\right]du.$$

In order further to hasten the convergence, we subtracted the term

$$\gamma_0 = \frac{2r}{\pi}\int_0^{\pi/2}\cos^2 u\, du \sum_{k=0}^{\infty}\frac{2}{(k + \frac{1}{2})^2\pi^2} = \frac{r}{2}$$

[Dwi61, eq. 48.12]. After rather much algebra, we wrote the integral as

$$\gamma = \gamma_0 + \frac{4r}{\pi}\int_0^{\pi/2}\cos^2 u\, G(u)\, du,$$

$$G(u) = \sum_{k=0}^{\infty}\frac{t_k^2(3u^2 - C\cos^2 u) - (u^2 + C\cos^2 u)^2}{t_k^2[(t_k + u)^2 + C\cos^2 u][(t_k - u)^2 + C\cos^2 u]},$$

$$t_k = (k + \tfrac{1}{2})\pi,$$

and this was subjected to a numerical integration routine in the computer. Execution was slow, and the results, plotted as curve (a) in Fig. 10-4, were at last obtained.

When after reflection from a moving target the returning echo has a Doppler shift w with respect to the transmitted carrier frequency Ω, we must replace the numerator of the integrand in (10-52) by $|f(\omega - w)|^2$. For Gaussian pulses the result can be efficiently integrated by the trapezoidal rule, and in Fig. 10-5 we plot the resulting values of the parameter γ versus $r = P_c/NW$ for six values of the Doppler shift parameter $\Delta = w/\sqrt{2\sigma^2}$. The straight line marked ∞ represents $\gamma = r$ as when the echo frequency has been shifted completely outside the spectral band of the clutter. The greater the Doppler shift, the more reliably the signals can be detected in the midst of the clutter produced by the transmitted pulse.

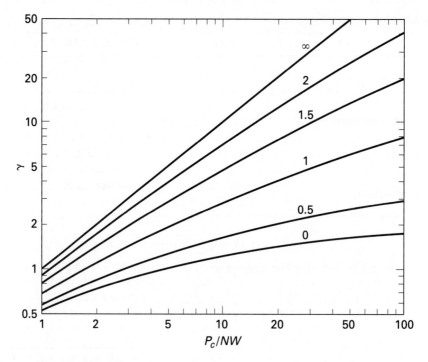

Figure 10-5. Signal-to-noise-ratio factor γ for Doppler-shifted Gaussian signals as a function of the clutter-to-noise ratio P_c/NW. Curves are indexed with the Doppler shift parameter $\Delta = w/\sqrt{2\sigma^2}$. The curve marked ∞ corresponds to $w = \infty$.

When the distribution $R_c(w)$ of Doppler shifts in (10-44) is much narrower than the spectrum $|f(\omega)|$ of the signal, the clutter is said to be *underspread*, and the foregoing analysis applies when the Doppler shift w of the target echo to be detected is zero. If on the other hand the Doppler shift w is large enough so that the signal spectrum $|f_s(\omega)| = |f(\omega - w)|$ lies outside the narrowband spectral density $\tilde{\Psi}(\omega)$ of the clutter, the effective signal-to-noise ratio d_{eff}^2 is the same as the input signal-to-noise ratio $2E/N$.

When the distribution $R_c(w)$ of Doppler shifts of the clutter echoes is much broader than the spectrum $|f(\omega)|$ of the signal, the clutter is said to be *overspread*. If the Doppler-shifted spectrum $|f(\omega - w)|$ of the target echo lies within the region where the clutter spectral density $\tilde{\Psi}(\omega)$ is significantly large, the target echo is being sought in the midst of noise that is effectively white. The optimum filter is then matched to the signal itself.

The similarity of the optimum filter (10-48) for detection of a signal in clutter to that derived in Sec. 10.1 for distinguishing close signals—see (10-40)—makes it plausible that the detection of a signal when other signals of the same form are potentially present can be treated as the detection of the same signal in the presence of Gaussian clutter, which consists of a dense succession of weak signals arriving at random times and with random amplitudes and phases. Resolving the wanted signal in the midst of all those randomly displaced nearby signals is much the same as resolving it from a few strong signals in the same temporal neighborhood. The signals that might arrive close to the one to be detected have the same effect on the structure of the maximum-likelihood receiver as the randomly arriving signals that constitute clutter noise.

If we extend this principle to cover interfering signals that differ in both arrival time τ and Doppler shift w from the one to be detected, with the Doppler shift w anywhere in a band much broader than the signal spectrum $|f(\omega)|$, we are led to the conclusion that the maximum-likelihood receiver for this situation will be much like the maximum-likelihood receiver for detecting the signal in the midst of overspread clutter. The overspread clutter resembles an additional white Gaussian noise, and as we saw in Sec. 7.2, the maximum-likelihood receiver is closely approximated by one that consists of a filter matched to the signal itself,

$$Y(\omega) = f_s^*(\omega)\, e^{-i\omega T'}, \qquad f_s(\omega) = f(\omega - w), \tag{10-54}$$

followed by a rectifier; when the output of the rectifier exceeds a certain decision level, the receiver decides that a signal is present. Because a range of Doppler shifts w is anticipated, the input must be passed through a bank of parallel filters matched to signals as in (10-54) for an array of carrier frequencies $\Omega + w$ uniformly spaced over that range. The resolvability of signals by a receiver of this type is then governed largely by the form of the ambiguity function $\lambda(\tau, w)$ to be studied in the next section, and it is principally by designing the shape of the transmitted signal $\mathrm{Re}\, F(t) \exp i\Omega t$ so that the ambiguity function has the most favorable attainable form that good resolution is to be achieved over a broad range of signal arrival times and Doppler shifts.

10.2.3 Coherent Pulse Trains

If the transmitted signal is a succession of coherent pulses $E(t - kT_r)$ separated by a repetition period T_r,

$$F(t) = M^{-1/2} \sum_{k=0}^{\infty} E(t - kT_r), \tag{10-55}$$

as in (6-118), its spectrum has the form

$$f(\omega) = \sqrt{M}\, e(\omega)\, e^{-i\overline{T}\omega} C(\omega), \qquad \overline{T} = \tfrac{1}{2}(M-1)T_r, \tag{10-56}$$

as in (6-120) and (6-121), where

$$C(\omega) = \frac{\sin \tfrac{1}{2} M T_r \omega}{M \sin \tfrac{1}{2} T_r \omega} \tag{10-57}$$

is the comb function, and $e(\omega)$ is the Fourier transform of the component pulses $E(t)$. The magnitude $|C(\omega)|$ is plotted in Fig. 6-3 for $M = 20$. Then

$$|f(\omega)|^2 = M|e(\omega)|^2|C(\omega)|^2.$$

The narrowband spectral density $\tilde{\Psi}(\omega)$ of the clutter will as in (10-44) be the result of convolving this with the distribution $R_c(w)$ of Doppler shifts induced by the motions of the scatterers. As we saw in Sec. 6.3.5, the width of the main peaks of $|f(\omega)|^2$ is on the order of $2\pi/MT_r$; and when a long pulse train is used, this may be smaller than the width of the scattering distribution $R_c(w)$. The resulting spectral density $\tilde{\Psi}(\omega)$ will be a comb whose tines are separated in angular frequency by $2\pi/T_r$. The width of each tine will be on the order of the sum of $2\pi/MT_r$ and the width of the distribution $R_c(w)$, and in general this will be much smaller than the separation $2\pi/T_r$. The squared pulse spectrum $|e(\omega)|^2$ "modulates" the entire density $\tilde{\Psi}(\omega)$, which spans a range of angular frequencies on the order of Δt^{-1}, where Δt^2 is the mean-square duration of the component pulses $E(t)$.

The transfer function of the optimum filter for detecting a signal having Doppler shift w in the midst of this type of clutter has the form

$$Y(\omega; w) = \frac{f^*(\omega - w)}{\frac{1}{2}N + \tilde{\Psi}(\omega)} e^{-iwT'}, \tag{10-58}$$

an evident modification of (10-46), and the effective signal-to-noise ratio governing the detectability of the signal is

$$d_{\text{eff}}^2 = 2E \int_{-\infty}^{\infty} \frac{|f(\omega - w)|^2}{N + 2\tilde{\Psi}(\omega)} \frac{d\omega}{2\pi}$$

as in (10-50).

The receiver will consist of a bank of filters of the form in (10-58) tuned for a discrete set of Doppler shifts w throughout the expected range, and each is followed by a rectifier whose output is observed during the interval $(T', T + T')$ when target echoes are expected. Each such filter can be considered as a cascade of a filter whose transfer function is

$$Y_0(\omega) = \left[\tfrac{1}{2}N + \tilde{\Psi}(\omega)\right]^{-1}$$

and a filter matched to the Doppler-shifted signal Re $F(t)e^{i(\Omega + w)t}$. Indeed, a single filter $Y_0(\omega)$ could precede a bank of filters matched to those signals. This filter $Y_0(\omega)$ attenuates all frequencies lying within the peaks of the clutter spectral density $\tilde{\Psi}(\omega)$, and signals arriving with Doppler shifts w falling into any of those peaks will be severely attenuated.

When the transmitter is stationary with respect to the bulk of the scatterers, signals arriving with Doppler shifts w that are integral multiples of $2\pi/T_r$ will suffer this attenuation. These Doppler shifts result from targets moving with velocities that are integral multiples of $\frac{1}{2}\lambda/T_r$, where λ is the wavelength of the transmitted radiation. These are called the *blind velocities*. For a transmitted frequency of 3000 MHz, for instance, $\lambda = 0.1$ m; and if the pulse repetition period T_r equals 10^{-3} sec, the blind velocities are multiples of 50 m/sec $\equiv$ 180 km/hr $\equiv$ 112 mph.

Targets moving with such relative velocities that their Doppler shifts w differ from any of those multiples of $2\pi/T_r$ by more than the width of the peaks in the spectral density $\tilde{\Psi}(\omega)$, on the other hand, will be detected with an effective signal-to-noise ratio d_{eff}^2 on the order of $2E/N$. The transmission of coherent pulse trains thus enables the radar to take greater advantage of the motion of its targets relative to that of the clutter scatterers. At the same time, as we discussed in Sec. 6.3.5, it entails the risk of ambiguities in estimates of the arrival times τ and the frequency shifts w of the returning echoes.

10.3 THE SPECIFICATION OF SIGNALS

10.3.1 General Properties of the Ambiguity Function

10.3.1.1 Definitions. Whenever it is necessary to distinguish or resolve two narrowband signals in the presence of white Gaussian noise, the structure of the receiver and its performance depend on their scalar product, or *cross-correlation*. When the complex envelopes of the signals have a common form and the signals differ only in certain nonrandom parameters θ, the cross-correlation is termed the *ambiguity function* of the signals and is written

$$\lambda(\theta_1, \theta_2) = \int_{-\infty}^{\infty} F(t; \theta_1) F^*(t; \theta_2)\, dt. \qquad (10\text{-}59)$$

The normalization

$$\lambda(\theta, \theta) = \int_{-\infty}^{\infty} |F(t; \theta)|^2\, dt = 1 \qquad (10\text{-}60)$$

for all values of the parameters θ is customary.

In radar the parameters chiefly serving to distinguish two echo signals are their arrival times τ and the Doppler shifts w of their carrier frequencies from a common reference value. When the integration in (10-59) is, as usual, carried out over the infinite range, the ambiguity function for these parameters depends only on the differences of the epochs and frequencies of the two signals. If the epochs $-\tau/2$ and $+\tau/2$ and the carrier frequencies $\Omega - w/2$ and $\Omega + w/2$ are assigned to the signals, their complex envelopes can be written as

$$F(t; -\tfrac{1}{2}\tau, -\tfrac{1}{2}w) = F(t - \tfrac{1}{2}\tau)\, e^{-\frac{1}{2}iw(t-\frac{1}{2}\tau)},$$
$$F(t; \tfrac{1}{2}\tau, \tfrac{1}{2}w) = F(t + \tfrac{1}{2}\tau)\, e^{\frac{1}{2}iw(t+\frac{1}{2}\tau)},$$

and the ambiguity function $\lambda(-\tfrac{1}{2}\tau, -\tfrac{1}{2}w; \tfrac{1}{2}\tau, \tfrac{1}{2}w)$ becomes simply

$$\lambda(\tau, w) = \int_{-\infty}^{\infty} F(t - \tfrac{1}{2}\tau) F^*(t + \tfrac{1}{2}\tau) e^{-iwt}\, dt \qquad (10\text{-}61)$$

as in (6-96). Other definitions differing from this by a phase factor have been used, but because of its convenience the form in (10-61) has become standard. The ambiguity function takes on its peak value at the origin, and with the complex envelope $F(t)$ normalized to 1 as in (10-60),

$$|\lambda(\tau, w)| \leq \lambda(0, 0) = 1,$$

as can be shown by means of the Schwarz inequality.

If we introduce the spectrum $f(\omega)$ of the complex envelope $F(t)$,

$$f(\omega) = \int_{-\infty}^{\infty} F(t)\, e^{-i\omega t}\, dt,$$

we find by substituting the inverse Fourier transform into (10-61) that the ambiguity function has much the same form in the frequency domain:

$$\lambda(\tau, w) = \int_{-\infty}^{\infty} f(\omega + \tfrac{1}{2}w) f^*(\omega - \tfrac{1}{2}w) e^{-i\omega\tau} \frac{d\omega}{2\pi}. \qquad (10\text{-}62)$$

10.3.1.2 The ambiguity surface. The ambiguity function $\lambda(\tau, w)$ is in general complex, but the resolvability of two signals with a relative delay τ and a frequency difference w depends only on its magnitude $|\lambda(\tau, w)|$, which it is advantageous to imagine plotted as the height of a surface over the (τ, w)-plane. It can be called the *ambiguity surface*. This quantity $|\lambda(\tau, w)|$ acquires further meaning if one considers a bank of parallel filters used to detect a signal of unknown arrival time and Doppler shift in the presence of white Gaussian noise, as described in Sec. 7.5. We insert a test signal Re $F(t)$ exp $i\Omega t$ at time $t = 0$ and determine the resulting output from the filter matched to a signal Re $F(t)$ exp $i(\Omega + w)t$ with frequency shift w. If $-\tau$ denotes the time measured from the common delay T' of the filters, the signal component of the rectified output of this filter is

$$R_s(T' - \tau; w) = \left| \int_{-\infty}^{\infty} F^*(T' - s) \, e^{-iw(T'-s)} F(T' - \tau - s) \, ds \right|^2$$

$$= \left| \int_{-\infty}^{\infty} F(u) F^*(u + \tau) \, e^{-iwu} \, du \right|^2 = |\lambda(\tau, w)|^2.$$

[Compare (6-99)]. If we suppose that the filters are matched for a dense set of frequencies $\Omega + w$, and if we picture their rectified responses $R_s(T' - \tau; w)$ to the signal Re $F(t)$ exp $i\Omega t$ plotted as a function of time τ and frequency shift w, they will form a surface similar to the ambiguity surface.

For every signal the ambiguity surface is peaked at the origin $(0, 0)$ of the (τ, w)-plane. A second signal arriving with separations τ in time and w in frequency that lie under this central peak will be difficult to distinguish from the first signal. For many types of signal the ambiguity function $|\lambda(\tau, w)|$ exhibits additional peaks elsewhere over the (τ, w)-plane. These sidelobes may conceal weak signals with arrival times and carrier frequencies far from those of the first signal. In a measurement of the arrival time and frequency of a single signal, the noise may cause one of the subsidiary peaks to appear higher than the main one, leading to gross errors in the result. The taller the sidelobes, the greater the probability of such errors, or *ambiguities*, in Doppler shift and signal epoch. It is desirable, therefore, for the central peak of the ambiguity function to be narrow and for there to be as few and as low sidelobes as possible.

10.3.1.3 Restrictions on the ambiguity function. If there existed a signal $F(t)$ whose ambiguity function equaled 1 at $\tau = 0$, $w = 0$ and zero everywhere else, it could be distinguished from another signal having the same form, but separated in time and frequency by displacements however small. The probability of error in resolving two such signals would be no greater than the false-alarm probability for detection. No such signal exists. Indeed, a function $\lambda(\tau, w)$ chosen arbitrarily will not necessarily be the ambiguity function of any signal. Even the magnitude $|\lambda(\tau, w)|$ is not at a designer's disposal, but must satisfy certain conditions.

An example of such a condition is the self-transform property of the squared magnitude $|\lambda(\tau, w)|^2$, due to Siebert [Sie56]:

$$\int_{-\infty}^{\infty} \int_{-\infty}^{\infty} |\lambda(\tau, w)|^2 \, e^{-ix\tau + iyw} \, d\tau \, \frac{dw}{2\pi} = |\lambda(y, x)|^2. \qquad (10\text{-}63)$$

To find a function $|\lambda(\tau, w)|$ possessing this property and having a form assuring good resolution besides is no easy task. Even if it can be found, one must still assign to it such a phase arg $\lambda(\tau, w)$ that the function $|\lambda(\tau, w)|$ arg $\lambda(\tau, w) = \lambda(\tau, w)$ will be the ambiguity function of some signal $F(t)$. Only when the proper phase is known as well can the Fourier transform of $\lambda(\tau, w)$ with respect to w be taken in order to obtain

$$\int_{-\infty}^{\infty} \lambda(\tau, w) \, e^{iwu} \, \frac{dw}{2\pi} = F(u - \tfrac{1}{2}\tau)F^*(u + \tfrac{1}{2}\tau), \tag{10-64}$$

from which the signal envelope $F(t)$ can be found—within an arbitrary constant phase factor—by setting $u = \tfrac{1}{2}t$, $\tau = -t$, and normalizing as in (10-60). Some further restrictions on the amplitude and phase of $\lambda(\tau, w)$ and on its real and imaginary parts have been reported by Stutt [Stu64].

An informative corollary of the self-transform property in (10-63) is derived by setting $x = 0$ and $y = 0$:

$$\int_{-\infty}^{\infty} \int_{-\infty}^{\infty} |\lambda(\tau, w)|^2 \, d\tau \, \frac{dw}{2\pi} = 1. \tag{10-65}$$

The total volume under the surface $|\lambda(\tau, w)|^2$ must be equal to 2π, no matter what the waveform of the signal. This condition prevents our making $|\lambda(\tau, w)|$ small everywhere in the (τ, w)-plane away from the origin. The magnitude $|\lambda(\tau, w)|$ will always have a peak over the point $(0, 0)$, and if we try to make that peak more slender, the values of $|\lambda(\tau, w)|$ elsewhere in the (τ, w)-plane must rise in compensation. Much effort has been expended in searching for signals whose ambiguity function has a magnitude remaining below a specified level over as much of the (τ, w)-plane as possible. Instructive pictures of ambiguity functions of a variety of signals are to be found in the book by Rihaczek [Rih69].

10.3.2 Single Pulses

10.3.2.1 Amplitude-modulated signals. The behavior of the ambiguity function near its peak at the origin can be discovered by expanding the integrand of (10-61) into a double Taylor series in τ and w. Putting

$$F(t - \tfrac{1}{2}\tau) = F(t) - \tfrac{1}{2}F'(t)\tau + \tfrac{1}{8}F''(t)\tau^2 + O(\tau^3),$$
$$e^{-iwt} = 1 - iwt - \tfrac{1}{2}w^2t^2 + O(w^3t^3),$$

substituting into (10-61), evaluating certain of the integrals in t by parts, and using the definitions of the signal moments in Sec. 6.3.3, we obtain finally

$$\lambda(\tau, w) = 1 - i\overline{\omega}\tau - i\overline{t}w - \tfrac{1}{2}\overline{\omega^2}\tau^2 - \overline{\omega t}w\tau - \tfrac{1}{2}\overline{t^2}w^2 + \cdots, \tag{10-66}$$

and through quadratic terms the squared magnitude of the ambiguity function is

$$|\lambda(\tau, w)|^2 \approx 1 - (\Delta\omega^2\tau^2 + 2\Delta(\omega t)w\tau + \Delta t^2 w^2), \tag{10-67}$$

where $\Delta\omega$ and Δt are the rms bandwidth and duration of the signal, and

$$\Delta(\omega t) = \overline{\omega t} - \overline{\omega}\,\overline{t}$$

is its cross-moment of time and frequency. For small values of τ and w the magnitude $|\lambda(\tau, w)|$ is constant along contours similar to the uncertainty ellipse

$$\Delta\omega^2\tau^2 + 2\Delta(\omega t)w\tau + \Delta t^2 w^2 = 1.$$

For the Gaussian pulse this similarity of the contours of constant magnitude holds for all values of τ and w. A signal whose complex envelope is

$$F(t) = \left(\frac{a^2}{\pi}\right)^{1/4} \exp[-\tfrac{1}{2}(a^2 - ib)t^2] \tag{10-68}$$

has an amplitude of the Gaussian shape proportional to $\exp(-\tfrac{1}{2}a^2 t^2)$ and an instantaneous frequency increasing linearly with time: $\phi'(t) = bt$; see (3-2). For this signal the magnitude of the ambiguity function is

$$|\lambda(\tau, w)| = \exp\left[-\frac{(a^4 + b^2)\tau^2 + 2b\tau w + w^2}{4a^2}\right]. \tag{10-69}$$

This function is constant along elliptical contours of the form

$$(a^4 + b^2)\tau^2 + 2b\tau w + w^2 = 4a^2\mu^2,$$

which are similar to the uncertainty ellipse. The area of each contour is equal to $4\pi\mu^2$, which is independent of the rate b of change of the instantaneous frequency. The effect of this linear frequency modulation on the pulse is only to rotate or shear the elliptical contours $|\lambda(\tau, w)|$ = constant without changing their area. An improvement in resolvability in one region due to the frequency modulation is accompanied by a deterioration in some other region.

In applications where negligible Doppler shifts are expected, it is only the behavior of the ambiguity function $\lambda(\tau, 0)$ along the τ-axis that is important, and this can be improved by making the rate b much greater than a^2. An advantage of the Gaussian signal is the absence of subsidiary peaks from its ambiguity function, which much reduces the risk of large errors in measuring its epoch and its frequency. Fowle et al. have extensively treated the generation and detection of the frequency-modulated Gaussian signal [Fow63].

10.3.2.2 Chirp modulation. Linear frequency modulation found one of its first applications to the improvement of range resolution in the design of the "chirp" radar [Kla60a]. This radar transmits a rectangular pulse with a quadratic phase,

$$F(t) = \begin{cases} T^{-1/2}\, e^{\frac{1}{2}ibt^2}, & -\tfrac{1}{2}T \le t \le \tfrac{1}{2}T, \\ 0, & |t| > \tfrac{1}{2}T, \end{cases} \tag{10-70}$$

and the total phase change $\tfrac{1}{4}bT^2$ from beginning to end is very large. The ambiguity function of this signal is

$$\lambda(\tau, w) = \begin{cases} \dfrac{2\sin[\tfrac{1}{2}(b\tau + w)(T - |\tau|)]}{(b\tau + w)T}, & |\tau| \le T, \\ 0, & |\tau| > T. \end{cases} \tag{10-71}$$

Resolution in range only is governed by the values along the τ-axis,

$$\lambda(\tau, 0) = \frac{2 \sin(\frac{1}{2}bT\tau)}{bT\tau}, \qquad |\tau| \ll T,$$

a function that has a narrow peak at the origin $\tau = 0$. The width of this peak, measured between the first zeros, is $(2\pi/bT^2)T$, which when $bT^2 \gg 2\pi$ is much smaller than the duration T of the original signal. For diagrams of this ambiguity function, see [Rih69, pp. 173–5].

This function $\lambda(\tau, 0)$ represents the output of a filter matched to the signal Re $F(t) \exp i\Omega t$ when only that signal is fed into it. Because it is much narrower when $bT^2 \gg 2\pi$ than the envelope of the signal itself, the matched filter is said to compress the pulse. Radars transmitting such frequency-modulated signals and receiving them with matched filters are called *pulse-compression* radars [Mor88, pp. 123–55]. The danger of high-voltage breakdown in the output circuitry of the transmitter limits the peak amplitude of a radar pulse, and the pulse-compression radar can send out signals of much greater total energy than one that simply produces a narrow amplitude-modulated pulse of the same rms bandwidth.

For certain combinations of delay τ and frequency shift w, however, the resolvability of the chirp signal will be no better than that of an unmodulated rectangular pulse of the same duration T and the same energy. Along the line $w + b\tau = 0$, the ambiguity function of the chirp signal is

$$\lambda(\tau, -b\tau) = \begin{cases} \dfrac{T - |\tau|}{T}, & |\tau| \le T, \\ 0, & |\tau| > T, \end{cases}$$

which is the same as the function $\lambda(\tau, 0)$ for the unmodulated square pulse. As with the Gaussian signal, the frequency modulation displaces part of the volume under the ambiguity surface to a different part of the (τ, w)-plane. A Doppler-shifted chirp signal can only with difficulty be distinguished from one that is merely delayed.

Multiplication of the complex envelope of an arbitrary signal by $\exp \frac{1}{2}ibt^2$ shears its ambiguity function. If we define a new signal with complex envelope

$$F_1(t) = F(t) \, e^{\frac{1}{2}ibt^2},$$

its ambiguity function is

$$\lambda_1(\tau, w) = \int_{-\infty}^{\infty} F(t - \tfrac{1}{2}\tau)F^*(t + \tfrac{1}{2}\tau) \exp\left[\tfrac{1}{2}ib(t - \tfrac{1}{2}\tau)^2 - \tfrac{1}{2}ib(t + \tfrac{1}{2}\tau)^2 - iwt \right] dt$$

$$= \int_{-\infty}^{\infty} F(t - \tfrac{1}{2}\tau)F^*(t + \tfrac{1}{2}\tau)e^{-ib\tau t - iwt} \, dt = \lambda(\tau, w + b\tau).$$

If Re $F(t) \exp i\Omega t$ is an amplitude-modulated signal, the principal axes of its ambiguity function stand at right angles to each other; about these axes the ambiguity function is symmetrical. For the "chirp-modulated" signal the axis of symmetry originally along the line $w = 0$ lies instead along the slanting line $w = -b\tau$.

10.3.2.3 Hermite signals. An instructive generalization of the Gaussian signal is the Hermite waveform, whose complex envelope is

$$F(t) = \left[\frac{a}{n!}\right]^{1/2} \pi^{-1/4} e^{-\frac{1}{2}a^2t^2} h_n(at\sqrt{2})$$

for any positive integer n, where $h_n(x)$ is the Hermite polynomial defined in Sec. 5.1.1. These signals have an oscillatory amplitude modulation, which changes sign n times before finally decaying to zero. Klauder [Kla60b] and Wilcox [Wil60] showed that the ambiguity function of this signal is

$$\lambda(\tau, w) = e^{-r^2/4} L_n(\tfrac{1}{2}r^2), \qquad r^2 = a^2\tau^2 + a^{-2}w^2,$$

where $L_n(x)$ is the nth Laguerre polynomial, defined in the Appendix, Sec. C.1. Around the central peak $|\lambda(\tau, w)|$ has elliptical ridges whose heights decrease from the center; between them are elliptical contours on which $\lambda(\tau, w) = 0$. Diagrams are to be found in [Kla60b].

As n increases, the central peak of $|\lambda(\tau, w)|$ becomes narrower and narrower, but at the same time the rms bandwidth and the rms duration of the signals increase. Asymptotically for $n \gg 1$,

$$\lambda(\tau, w) \approx J_0(r\sqrt{2n + 1}), \qquad r = [a^2\tau^2 + a^{-2}w^2]^{1/2}, \qquad 0 \leq r \leq 2\sqrt{2n + 1}$$

[Erd53, vol. 2, p. 199, eq. 10.15(2)]. For large n, $\lambda(\tau, w)$ vanishes for values of τ and w on the ellipses

$$[(n + \tfrac{1}{2})(a^2\tau^2 + a^{-2}w^2)]^{1/2} \approx 1.70, 3.90, 6.12, \dots .$$

The first elliptical ridge surrounding the central peak has a height of about 0.4 and the one next to it a height of about 0.3. Far from the center the function is roughly sinusoidal, but with a slowly decreasing amplitude,

$$\lambda(\tau, w) \approx \left[\frac{2}{(n + \tfrac{1}{2})\pi^2 r^2}\right]^{1/4} \cos(r\sqrt{2n + 1} - \tfrac{1}{4}\pi), \qquad r = [a^2\tau^2 + a^{-2}w^2]^{1/2}.$$

After n zeros there is a final ridge, beyond which $\lambda(\tau, w)$ drops to zero. These ridges in $|\lambda(\tau, w)|$ render the signals liable to ambiguity in time and frequency.

These Hermite signals have mean-square bandwidths and durations given by

$$\Delta t^2 = (n + \tfrac{1}{2})a^{-2}, \qquad \Delta\omega^2 = (n + \tfrac{1}{2})a^2.$$

The central peak of their ambiguity function covers an area of about $9.1/(\Delta\omega\Delta\tau)$ for $n \gg 1$. The area of the (τ, w)-plane occupied by the entire ambiguity function, out to where it begins its final exponential descent to zero, is on the order of the product $\Delta\omega\Delta t$.

10.3.2.4 Moments of the squared ambiguity function. For all amplitude-modulated signals the area covered by the central peak is on the order of $(\Delta\omega\Delta t)^{-1}$, as is evident from (10-67). The part of the ambiguity function significantly greater than zero covers an area on the order of $\Delta\omega\Delta t$, as can be deduced from the relations

$$\frac{1}{2}\int_{-\infty}^{\infty}\int_{-\infty}^{\infty}\tau^2|\lambda(\tau,w)|^2\,d\tau\,\frac{dw}{2\pi} = \Delta t^2,$$

$$\frac{1}{2}\int_{-\infty}^{\infty}\int_{-\infty}^{\infty}\tau w|\lambda(\tau,w)|^2\,d\tau\,\frac{dw}{2\pi} = -\Delta(\omega t), \qquad (10\text{-}72)$$

$$\frac{1}{2}\int_{-\infty}^{\infty}\int_{-\infty}^{\infty}w^2|\lambda(\tau,w)|^2\,d\tau\,\frac{dw}{2\pi} = \Delta\omega^2.$$

These can be derived by making a power-series expansion of the exponential in the integrand of (10-63) and applying (10-67) to the right side. For an amplitude-modulated pulse, for which $\Delta(\omega t) = 0$, the rms widths of the ambiguity function in τ and w are on the same order of magnitude as those of the signal itself in time and frequency. The product of these widths crudely measures the area of the entire ambiguity function.

10.3.2.5 A conjectural ambiguity function.
A function of τ and w that well illustrates these properties was proposed by Charles Persons:

$$L(\tau,w) = \frac{WT}{1+WT}\exp(-\tfrac{1}{2}W^2\tau^2 - \tfrac{1}{2}T^2w^2)$$

$$+ \frac{1}{1+WT}\exp\left[-\frac{\tau^2}{2T^2} - \frac{w^2}{2W^2}\right]. \qquad (10\text{-}73)$$

Whether this is the absolute square of the ambiguity function of any signal, $L(\tau,w) = |\lambda(\tau,w)|^2$, is unknown; it does satisfy the self-transform relation (10-63). In Fig. 10-6 we have sketched the cross sections of $L(\tau,w)$ along the τ- and w-axes for $WT \gg 1$. The mean-square duration and the mean-square bandwidth of whatever signal might possess such an ambiguity function can be shown by (10-67) to be

$$\Delta\omega^2 = \frac{W^3T^3 + 1}{2T^2(1+WT)}, \qquad \Delta t^2 = \frac{W^3T^3 + 1}{2W^2(1+WT)},$$

and one can show that

$$\Delta\omega^2\Delta t^2 = \frac{(W^3T^3 + 1)^2}{4W^2T^2(1+WT)^2}.$$

As is required for any amplitude-modulated signal, $\Delta\omega^2\Delta t^2 \geq \tfrac{1}{4}$. For $WT \gg 1$, $\Delta\omega^2 \approx \tfrac{1}{2}W^2$ and $\Delta t^2 \approx \tfrac{1}{2}T^2$. Near the central peak the first term of (10-73) then dominates; far from it the second dominates. The central peak covers an area on the order of $2\pi/WT$, and the broad skirt represented by the second term in (10-73) covers an area of the order of $2\pi WT$.

In order for amplitude-modulated signals to provide good overall resolution, they must have a large *time–bandwidth product* $\Delta t\Delta\omega$. When $\Delta t\Delta\omega \gg 1$, the central peak of the ambiguity surface will be slender; and as the rest of the function takes up an area on the order of $\Delta t\Delta\omega$, $|\lambda(\tau,w)|$ must attain rather large values outside the center in order to meet the volume constraint given by (10-65). The average level of $|\lambda(\tau,w)|$ in that region will be on the order of $(\Delta t\Delta\omega)^{-1/2}$.

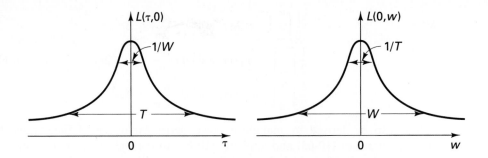

Figure 10-6. Conjectural squared ambiguity function (10-73).

It does not suffice, however, to make $\Delta t \Delta \omega$ large by introducing frequency modulation. As Rihaczek [Rih65] emphasized, the areas in question then involve not $\Delta t \Delta \omega$, but

$$\left\{ \Delta \omega^2 \Delta t^2 - [\Delta(\omega t)]^2 \right\}^{1/2},$$

which also appears in the variances of the errors incurred in simultaneously measuring arrival time and frequency; see (6-114). As we saw in connection with (6-117), linear frequency modulation does not change this quantity at all; its only effect is to rotate or shear the ambiguity surface with respect to the τ- and w-axes. If signals might arrive with time and frequency separations anywhere in a broad area of the (τ, w)-plane, no overall improvement of their resolution can be achieved in this way.

10.3.3 Pulse Trains

10.3.3.1 The ambiguity function of a uniform train. Thus far we have imagined the signal $F(t)$ as consisting of a single pulse. Certain radars, however, transmit a sequence of coherent pulses, constraining their phases to have a definite, known relationship to each other. Let us suppose that a train of M such coherent pulses $E(t)$ of equal amplitudes is received from a target, and let us study the form of its ambiguity function $\lambda(\tau, w)$. The complex envelope of this signal is given in (10-55). We again assume that successive pulses overlap to a negligible degree; T_r is the repetition period between the pulses, and the composite signal is normalized to 1 as in (10-60).

To calculate the ambiguity function of a train of nonoverlapping pulses, we use the form in (10-62), writing the spectrum of the pulse train as

$$f(\omega) = M^{-1/2} e(\omega) \sum_{j=0}^{M-1} e^{-ij\omega T_r};$$

$e(\omega)$ is the Fourier transform of the component pulses $E(t)$. The ambiguity function is then

$$\lambda(\tau, w) = \frac{1}{M} \int_{-\infty}^{\infty} e(\omega + \tfrac{1}{2}w) e^*(\omega - \tfrac{1}{2}w)$$

$$\cdot \sum_{j=0}^{M-1} \sum_{k=0}^{M-1} \exp[-ij(\omega + \tfrac{1}{2}w)T_r + ik(\omega - \tfrac{1}{2}w)T_r - i\omega\tau] \frac{d\omega}{2\pi}$$

$$= \frac{1}{M} \sum_{j=0}^{M-1} \sum_{k=0}^{M-1} e^{-\frac{1}{2}i(j+k)wT_r} \int_{-\infty}^{\infty} e(\omega + \tfrac{1}{2}w) e^*(\omega - \tfrac{1}{2}w) e^{-i\omega[\tau - (j-k)T_r]} \frac{d\omega}{2\pi}$$

$$= \frac{1}{M} \sum_{j=0}^{M-1} \sum_{k=0}^{M-1} e^{-\frac{1}{2}i(j+k)wT_r} \lambda_0(\tau - (j-k)T_r, w),$$

where as in (10-62) $\lambda_0(\tau, w)$ is the ambiguity function of the individual pulses.

Suppose that $(p - \tfrac{1}{2})T_r < \tau < (p + \tfrac{1}{2})T_r$, $p \geq 0$. Then because the pulses do not overlap, the only terms contributing to the double sum are those with $j - k = p$, and

$$\lambda(\tau, w) = \frac{1}{M} \lambda_0(\tau - pT_r, w) \sum_{j=0}^{M-1-p} e^{-\frac{1}{2}i(p+2k)wT_r}, \qquad (p - \tfrac{1}{2})T_r < \tau \leq (p + \tfrac{1}{2})T_r.$$

Now

$$\left| \sum_{k=0}^{M'-1} e^{-ikx} \right| = \left| \frac{1 - e^{-iM'x}}{1 - e^{-ix}} \right| = \left| \frac{\sin \tfrac{1}{2} M'x}{\sin \tfrac{1}{2} x} \right|;$$

for $x = 0$, this sum equals M'. We then obtain

$$|\lambda(\tau, w)| = \sum_{p=-(M-1)}^{M-1} |\lambda_0(\tau - pT_r, w)| \left| \frac{\sin \tfrac{1}{2}(M - |p|)wT_r}{M \sin \tfrac{1}{2} wT_r} \right|, \tag{10-74}$$

$$-(M - \tfrac{1}{2})T_r < \tau \leq (M - \tfrac{1}{2})T_r.$$

For $|\tau| > MT_r$, the shifted pulse trains do not overlap, and $\lambda(\tau, w) \equiv 0$. In particular,

$$|\lambda(pT_r, 0)| = 1 - \frac{|p|}{M}, \qquad -M \leq p \leq M. \tag{10-75}$$

Along the τ-axis ($w = 0$) the ambiguity function consists of repetitions of the function $|\lambda_0(\tau, w)|$ for the component pulses, with $(M - 1)$ peaks on one side of the origin, $(M - 1)$ on the other, and one in the center. The heights of these peaks decrease to each side, as shown by (10-75) and as illustrated in Fig. 6-2. The reason for this behavior is easily seen. Sets of pulses received separated in time by small multiples of the period T_r overlap, except for the pulses at the beginning of one train and at the end of the other. To resolve these composite signals, a receiver must use those pulses that do not overlap; and the more of them there are—the larger the index p—, the more reliable is the resolution of the signal trains.

In the frequency (w) direction the width of the peak of the ambiguity function $|\lambda_0(\tau, w)|$ of the component pulses $E(t)$ is on the order of $(\Delta_0 t)^{-1}$, where $\Delta_0 t$ is the rms duration of $E(t)$. As we have taken $T_r \gg \Delta_0 t$, $(\Delta_0 t)^{-1} \gg T_r^{-1}$. The factor with the sines multiplying $|\lambda_0(\tau - pT_r, w)|$ in (10-74) is the comb function we introduced in (6-121). It breaks up this peak of $|\lambda_0(\tau - pT_r, w)|$ into a succession of narrower peaks having widths on the order of $2\pi/(M - |p|)T_r$ and spaced by $2\pi/T_r$. These peaks resemble the amplitude pattern of light reflected from a diffraction grating.

We fix our attention on the central set of peaks ($p = 0$) of the function $|\lambda(\tau, w)|$, for which the multiplying factor is

$$|C(w)| = \left| \frac{\sin(\frac{1}{2}MwT_r)}{M \sin(\frac{1}{2}wT_r)} \right|.$$

This factor, a portion of which is sketched in Fig. 6-3, reaches a value of 1 whenever $w = 2k\pi/T_r$, k an integer, producing a peak whose width is on the order of $2\pi/MT_r$. Between the tall peaks are a number of ripples whose height is lower by a factor M^{-1}. This "diffraction pattern" is superimposed on the original function $|\lambda_0(\tau, w)|$, breaking it up into many narrow peaks of width $2\pi/MT_r$ and period $2\pi/T_r$, both of which are much smaller than the width $(\Delta_0 t)^{-1}$ of $\lambda_0(\tau, w)$ in the w direction. The rms duration of the signal is now on the order of $\Delta t \approx MT_r$, and the area of the (τ, w)-plane covered by the central peak of the ambiguity function is again on the order of $(\Delta t \Delta\omega)^{-1}$. The entire ambiguity function occupies a total area of about $\Delta t \Delta\omega$.

The breaking up of $|\lambda_0(\tau, w)|$ into peaks and valleys in the w direction indicates that trains of pulses can be more effectively resolved in frequency than single pulses of the same total energy. This can be understood by observing that the coherent repetition M times of the pulse $E(t)$ causes its spectrum $e(\omega)$ to divide into a line spectrum. The lines are separated by $2\pi/T_r$, and their widths are about $2\pi/MT_r$. If the Doppler shifts due to the motions of the radar targets are such that these line spectra for the echoes interlace, filters can be constructed to resolve the signals with high probability. If the relative velocity of the targets is such that the shift w is an integral multiple of the repetition frequency, $w = 2k\pi/T_r$, on the other hand, the line spectra will overlap and resolution will be difficult unless the signals are far enough apart in time.

The measurement of the velocity of an isolated target, by estimating the Doppler shift of its radar echoes as described in Sec. 6.3, can be made more accurately if one utilizes a train of coherent pulses in place of a single pulse. The behavior of the function $|\lambda(\tau, w)|$ indicates, however, that ambiguity will be introduced into the results, frequencies differing by multiples of $2\pi/T_r$ becoming indistinguishable. For a repetition period $T_r = 10^{-3}$ sec and a carrier frequency $\Omega_0 = 2\pi \cdot 3 \cdot 10^9$ sec^{-1} (3000 MHz), the ambiguity in target velocity amounts to $\Delta v = c/[2(\Omega_0/2\pi)T_r] = 180$ km/hr $= 112$ mph; c is the velocity of light.

10.3.3.2 The clear area. If it is only the resolution of very close targets that is of concern, the best signal is one whose ambiguity function $|\lambda(\tau, w)|$ has as slender a central peak as possible, outside of which the function must take on the lowest possible values. We now know that a long train of narrow pulses has these properties. The area A_c occupied by the central peak is inversely proportional to the number M of pulses in the train,

$$A_c \approx \frac{2\pi}{MT_r\Delta_0\omega},$$

where T_r is the pulse repetition period and where $\Delta_0 \omega^{-1}$, the reciprocal of the rms bandwidth of the component pulses, measures the width of their ambiguity function $|\lambda(\tau, w)|$ along the τ-axis. The level of the ambiguity function $|\lambda(\tau, w)|$ between the central peak and the adjacent peaks is on the order of M^{-1}, and by taking M large enough, both this level and the area A_c can be made as small as desired.

The peaks nearest to the central one are separated from it by T_r along the τ-axis and by $2\pi/T_r$ along the w-axis. The area of the (τ, w)-plane over which $|\lambda(\tau, w)|$ can be made arbitrarily small is, therefore, on the order of $(2T_r)(4\pi/T_r) = 8\pi$. The question whether this "clear area" can be made any broader by judicious choice of the signal waveform $F(t)$ has been answered in the negative by Price and Hofstetter [Pri65], who worked out bounds on the size of clear areas of various shapes.

10.3.3.3 Polyphase-coded pulse trains. With a long uniform pulse train the (τ, w)-plane is studded with a great many narrow peaks whose heights near the origin are almost equal to 1, and ambiguities abound. One way to suppress them is to vary the relative phases of the pulses. The signal in (10-55) is replaced by

$$F(t) = \frac{1}{\sqrt{M}} \sum_{k=0}^{M-1} a_k E(t - kT_r), \qquad \sum_{k=0}^{M-1} |a_k|^2 = M, \qquad (10\text{-}76)$$

with the component pulses $E(t)$ normalized as before and assumed not to overlap. The sequence $a_0, a_1, \ldots, a_{M-1}$ is often called a *code*. The amplitude factors a_k are complex; when only the relative phases are being altered, their absolute values $|a_k|$ equal 1. In this way the transmitted power is uniform throughout the sequence, and the pulse train carries the greatest total energy permitted by voltage constraints imposed by the necessity of avoiding breakdown in the antenna feedlines. The sequences $\{a_k\}$ are then called *polyphase* codes.

Instead of (10-74) one now finds for the ambiguity function of the pulse train

$$|\lambda(\tau, w)| = \sum_{p=-(M-1)}^{M-1} |\lambda_0(\tau - pT_r, w)| \, C_p(wT_r),$$

$$C_p(x) = \frac{1}{M} \left| \sum_{k=0}^{M-|p|-1} a_k a_{k+|p|}^* \, e^{-ikx} \right|, \qquad -M < p < M, \qquad C_0(0) = 1. \qquad (10\text{-}77)$$

The function $C_p(x)$ is called the *discrete ambiguity function* [Ger91].

If no Doppler shifts w greater than a fraction of $2\pi/T_r$ are expected, the designer needs to be concerned mainly with the behavior of the M quantities

$$C_p(0) = \frac{1}{M} \left| \sum_{k=0}^{M-|p|-1} a_k a_{k+|p|}^* \right|,$$

which represent the correlation of the sequence a_k with itself, "correlation" being taken, of course, not in the statistical sense. For the sake of comparison, keep in mind that for the uniform pulse train with $a_k \equiv 1$

$$C_p(0) = \begin{cases} 1 - \dfrac{|p|}{M}, & |p| < M, \\ 0, & |p| \geq M. \end{cases}$$

The simplest choices for the a_k's are the numbers $+1$ and -1. In discussing the synchronization of long trains of pulses in a binary communication system, Barker [Bar53] recommended the use of sequences of M positive and negative pulses for whose amplitudes

$$C_p(0) = 1, \qquad C_p(0) \le \frac{1}{M}, \qquad p \ne 0,$$

and he exhibited a number of such sequences. One of length 5 having this property is $+1, +1, +1, -1, +1$, for which

$$C_0(0) = 1, \quad C_1(0) = 0, \quad C_2(0) = \tfrac{1}{5}, \quad C_3(0) = 0, \quad C_4(0) = \tfrac{1}{5}.$$

It is difficult to find long sequences with such an advantageous autocorrelation. In fact, Turyn and Storer [Tur61] showed that there exist no Barker codes of odd length M greater than 13. The Barker code of length 13 is

$$+1, +1, +1, +1, +1, -1, -1, +1, +1, -1, +1, -1, +1. \qquad (10\text{-}78)$$

Golomb and Scholtz [Gol65] studied "generalized Barker sequences" in which the a_k's are the nth roots of 1, that is, powers of $\exp 2\pi i/n$ for integers n. They tabulated a number of these for various values of n and M, and for small values of M they stated for which integers n generalized Barker sequences exist.

Another approach to the design of effective coherent pulse trains considers periodic repetitions of polyphase codes. The periodic correlation of a periodically repeated sequence $a_0, a_1, \ldots, a_{M-1}, a_{k+M} = a_k$, is defined as

$$s_k = \sum_{j=0}^{M-1} a_j a_{j+k}^* = \sum_{j=0}^{M-1-k} a_j a_{j+k}^* + \sum_{j=M-k}^{M-1} a_j a_{j+k-M}^*, \qquad 0 \le k < M. \qquad (10\text{-}79)$$

Techniques have been developed for determining sequences $\{a_k\}$ such that $s_k \equiv 0, 1 \le k < M$—the "perfect periodic codes"—or such that $|s_k|$ is on the order of $1/M$ for $M \gg 1, 1 \le k < M$,—the "asymptotically perfect periodic codes" [Hei61], [Fra62], [Chu72], [Fra80], [Lew82], [Ger91]. Experience has shown that the correlations $C_p(0)$ determining the resolution of signals with zero relative frequency shift ($w = 0$) are then generally small for $1 \le p < M$, so that the resulting signals (10-76) will be effective for radar detection [Fra63]. The practical aspects of utilizing such codes in radar are considered in [Lew81].

If Doppler shifts w much larger than $2\pi/T_r$ may occur, one must investigate the functions $C_p(x)$ for all values of x in the interval $(0, 2\pi)$, which is their basic period. The pattern $C_p(0)$ of peak heights along the τ-axis will be repeated along all lines $w = 2\pi m/T_r$ parallel to the τ-axis for positive and negative integers m. Even with Barker sequences, $C_p(x)$ rises much above the level M^{-1} for values of x between 0 and 2π. Indeed

$$\sum_{p=-(M-1)}^{M-1} \int_0^{2\pi} |C_p(x)|^2 \frac{dx}{2\pi} = 1,$$

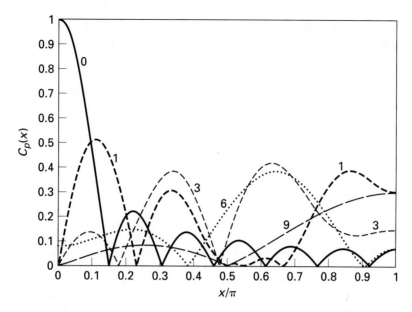

Figure 10-7. Function $C_p(x)$ in (10-77) for the Barker sequence with $M = 13$ for $p = 0, 1, 3, 6, 9$. Curves are indexed with the value of p. Only the segment $0 \leq x \leq \pi$ is shown. The function is symmetrical about $x = 0$ and periodic with period 2π.

which indicates that $C_p(x)$ can be expected to reach heights on the order of $M^{-1/2}$ on the average. In Fig. 10-7 we exhibit the function $C_p(x)$ for the Barker sequence with $M = 13$ for a few values of p. Although small for $x = 0$ and $p > 0$ and along the x-axis away from the origin for $p = 0$, it takes on large values elsewhere. For an illustration of the resulting ambiguity function, see [Rih69, p. 217]. Diagrams of the discrete ambiguity function for polyphase codes with large numbers M of elements are to be found in [Kre83], [Lew86], and [Ger91]; they exhibit similar behavior. Signals with frequency shifts w such that $x = wT_r$ falls into the region where $C_p(x)$ is large will be difficult to detect in the presence of signals with zero frequency shift.

Other sequences of amplitudes a_k investigated include trains of +1's and −1's that can be generated by a binary shift register, particularly sequences of maximal length $2^n - 1$, where n is the number of stages in the shift register [Zie59]. A bibliography of early studies of this problem was drawn up by Lerner [Ler63]. In general, there is no way of finding a sequence $\{a_k\}$ whose functions $C_p(x)$ will be small everywhere in $-\pi < x \leq \pi$, $0 \leq p < M$, except at the origin ($p = 0$, $x = 0$). The most one can usually do is to try a set of a_k's, compute $C_p(x)$ for $p = 0, 1, \ldots, M - 1$ at a number of values of x in $(0, 2\pi)$, and see what it looks like.

The artifice of staggering both the epochs and the frequencies of the component pulses of the train was treated by Rihaczek [Rih64]; see also [Rih69, pp. 308–12]. The usual result is that the clear area becomes filled, and the ambiguity function outside the central peak takes on a jagged structure with an average level on the order of $(\Delta t \Delta \omega)^{-1/2}$. Overall resolution deteriorates. By reducing the heights of the outstanding peaks away from the origin, however, the risk of ambiguities—large errors in the measurement of carrier frequency and arrival time—is diminished.

Problems

10-1. In the context of Sec. 10.1.3, assume that the arrival time of the signals is known, but that they may have any of M carrier frequencies $\Omega + k\Delta\omega$, $0 \leq k < M$, uniformly separated by $\Delta\omega$, which is on the order of the reciprocal $(\Delta t)^{-1}$ of the rms duration of the signal. Then the complex envelope of the kth signal is

$$F(t; \theta_k'') = F(t) \exp ik\Delta\omega t,$$

in which $F(t)$ is given. As in Sec. 10.1.3, the signals may arrive with arbitrary amplitudes and with phases uniformly distributed over $(0, 2\pi)$. Some or all of them may be absent.

  Determine the elements of the ambiguity matrix Λ, and assuming that M is so large that the inverse Λ^{-1} can be taken also to have the Toeplitz form, determine the maximum-likelihood receiver of these signals and evaluate the false-alarm and detection probabilities for the test of hypothesis H_k, "The kth signal is present," versus hypothesis H_k', "The kth signal is absent." Explain how the reliability (Q_0, Q_d) of this test depends on the relation of the frequency interval $\Delta\omega$ to the duration of the signal. Assume that the observation interval $(0, T)$ is long enough to encompass the entire signal and that as usual the noise is white and Gaussian.

10-2. For a radar transmitted pulse of the form

$$F(t) = Ct \, e^{-\mu t} U(t),$$

calculate the signal-to-noise ratio d^2 and the bandwidth β for detection in clutter for $w = 0$; use the formulas in Sec. 10.2.2.

10-3. What is the dependence of the bandwidth β in (10-51) on the transmitted power P_t when $P \gg N\Delta\omega$?

10-4. Calculate the ambiguity function $\lambda(\tau, w)$ for the signal

$$F(t) = \sqrt{2\mu} \, e^{-\mu t} U(t).$$

10-5. Show that for two identical signals of equal energies arriving in white Gaussian noise at known times separated by τ, the signal energy required to decide with given error probabilities which signal is present is proportional to $(\Delta\omega^2 \tau^2)^{-1}$ for small values of τ.

10-6. Verify (10-53).

10-7. Prove Siebert's self-transform property, (10-63).

10-8. Derive (10-72) from the self-transform property and (10-67).

10-9. Derive and sketch the absolute value of the Fourier transform $f(\omega)$ of the chirp signal (10-70) for $bT^2 \ll 1$ and for $bT^2 \gg 1$. Verify (10-71). Why is (10-67) invalid for this signal? *Hint*: Use the Cornu spiral [Jah45, p. 37].

10-10. Define the cross-ambiguity function $\lambda_{12}(\tau, w)$ for signals with complex envelopes $F_1(t)$ and $F_2(t)$ by

$$\lambda_{12}(\tau, w) = \int_{-\infty}^{\infty} F_1(t - \tfrac{1}{2}\tau)F_2^*(t + \tfrac{1}{2}\tau)e^{-iwt} \, dt.$$

Prove the relation

$$\int_{-\infty}^{\infty}\int_{-\infty}^{\infty} \lambda_{12}(\tau - \delta, \phi + \mu)\lambda_{34}^*(\tau + \delta, \phi - \mu) \, e^{i(\tau y - \phi x)} \, d\tau \, \frac{d\phi}{2\pi}$$

$$= \lambda_{13}(x - \delta, y + \mu)\lambda_{24}^*(x + \delta, y - \mu)$$

[Tit66]. Show how to obtain (10-63) from this. What does the above relation become for $F_1(t) = F_2(t) = F_3(t) = F_4(t) = F(t)$ when $x = y = 0$?

10-11. For the 13-element Barker sequence in (10-78) calculate the correlations $C_p(0)$ and $C_p(\pi)$ by (10-77).

10-12. The Frank code has N^2 elements for any positive integer N [Fra62]. One forms an $N \times N$ matrix of which the km-element is $\exp[2\pi i(k - 1)(m - 1)/N]$. The Frank code is composed by writing down the N rows of this matrix, one after another. Write out the code for $N = 4$ and calculate the autocorrelation $C_p(0)$, $-16 < p < 16$, for it. If you have a computer with a fast Fourier transform algorithm in it, evaluate and plot the discrete ambiguity function $C_p(x)$ of (10-77) for a number of values of p between 0 and 15.

11

Stochastic Signals

11.1 STRUCTURE OF THE RECEIVER

11.1.1 Types of Stochastic Signals

Thus far we have presumed that the receiver knows the form of the signals to be detected and may be ignorant only of certain parameters such as amplitude, phase, and time of arrival. It is sometimes impossible, however, to specify the detailed structure of the signal, which may differ from one instance to another. The designer may then have to imagine the signals to have been drawn from an ensemble of random processes with certain statistical properties. Such signals are known as *stochastic signals*. Although their waveforms are usually complicated, it is not their complexity, but the unpredictability of their precise configurations that places them in this category.

Stochastic signals may have been generated in a random manner or, originally possessing a definite form, may have been erratically distorted en route to the receiver. A system for transmitting binary digits, for instance, might send a burst of random noise of fixed duration to represent each 1, with blank intervals standing for the 0's. The signals might have the form

$$s(t) = \text{Re } M(t)Z(t)\, e^{i\Omega t}, \tag{11-1}$$

where $M(t)$ is a fixed modulation, Ω the carrier frequency, and $Z(t)$ the complex envelope of a stationary random process of known complex autocovariance function $\phi(\tau)$:

$$\tfrac{1}{2}E[Z(t_1)Z^*(t_2)] = \phi(t_1 - t_2), \qquad E[Z(t_1)Z(t_2)] \equiv 0.$$

(Henceforth we omit the tildes that in Chapter 3 distinguished complex autocovariance functions and narrowband spectral densities.) The complex autocovariance function of the signal envelope $S(t)$ is then

$$\phi_s(t_1, t_2) = \tfrac{1}{2}E[S(t_1)S^*(t_2)] = M(t_1)M(t_2)\phi(t_1 - t_2). \tag{11-2}$$

The jamming signals transmitted to incommode an enemy radar are sometimes of this nature; they can be generated by amplifying the output of noisy gas-discharge tubes. The signals that radio telescopes pick up from distant parts of the universe are stochastic and usually stationary for relatively long periods of time.

Scatter-multipath communication systems link stations far beyond each other's horizons by emitting signals in such a direction that they will be reflected from the ionosphere. From each determinate transmitted pulse there arrive a large number of weak signals that have traveled paths of slightly different lengths, along which they have suffered a variety of attenuations and distortions. The sum of all these signals strongly resembles a stochastic process [Pri56], [Pri58], [Bel63]. In radar astronomy the signals are reflected from a planet or satellite at a large number of scattering points, and the combination of all the echoes again creates a stochastic signal [Pri60].

When each transmitted pulse Re $F(t) \exp i\Omega t$ is reflected without distortion from a multitude of moving scatterers that introduce Doppler shifts w_m and are so located that the total delays between transmitter and receiver are τ_k, the received signal is Re $S(t) \exp i\Omega t$, and its complex envelope is

$$S(t) = \sum_{k,m} z_{km}F(t - \tau_k)\exp iw_m t.$$

Here z_{km} is a complex number representing the amplitude and phase of the signal with delay τ_k and shift w_m. The complex autocovariance function of the received signal is

$$\phi_s(t_1, t_2) = \tfrac{1}{2}\sum_{k,m}\sum_{k',m'}\mathbf{E}(z_{km}z_{k'm'}^*)F(t_1 - \tau_k)F^*(t_2 - \tau_{k'})\exp(iw_m t_1 - iw_{m'}t_2),$$

where $\mathbf{E}$ denotes an expected value with respect to the ensemble of scatterings. If separate scatterings are assumed statistically independent,

$$\phi_s(t_1, t_2) = \tfrac{1}{2}\sum_{k,m}\mathbf{E}(|z_{km}|^2)F(t_1 - \tau_k)F^*(t_2 - \tau_k)\exp iw_m(t_1 - t_2).$$

When the scatterers are small and dense, this sum can be written as an integral by introducing a function $\sigma(\tau, w)$ defined by

$$\sigma(\tau, w)\,d\tau\,\frac{dw}{2\pi} = \tfrac{1}{2}\sum_{k,m}\mathbf{E}(|z_{km}|^2),$$

in which the summation is taken over those scatterers resulting in a delay between τ and $\tau + d\tau$ and a frequency shift between $w/2\pi$ and $(w + dw)/2\pi$. Then

$$\phi_s(t_1, t_2)$$
$$= \int_{-\infty}^{\infty}\int_{-\infty}^{\infty}\sigma(\tau, w)F(t_1 - \tau)F^*(t_2 - \tau)\exp iw(t_1 - t_2)\,d\tau\,\frac{dw}{2\pi}. \tag{11-3}$$

An example of this type of signal is the clutter noise we discussed in Sec. 10.2.

If we put

$$\psi(\tau, t) = \int \sigma(\tau, w) e^{iwt} \frac{dw}{2\pi},$$

we can write the complex autocovariance function of the received signal as

$$\phi_s(t_1, t_2) = \int_{-\infty}^{\infty} \psi(\tau, t_1 - t_2) F(t_1 - \tau) F^*(t_2 - \tau) \, d\tau. \tag{11-4}$$

Autocovariance functions of this general form were assigned by Price and Green [Pri60] to the echoes expected in radar astronomy. If $\psi(\tau, t)$ as a function of τ is significant over a range of values of τ much longer than the pulse $F(t)$, the target is said to be *deep fluctuating*; if $\psi(\tau, t) = \psi(t)\delta(\tau - \tau_0)$, it is termed a *fluctuating point target*, which is in effect much thinner than the incident signal.

The autocovariance functions in (11-2), (11-3), and (11-4) exemplify those characterizing different kinds of stochastic signals. We shall assume furthermore that the signals are realizations of Gaussian processes of expected value 0, taking the signals and noise to be quasiharmonic and the processes in question to be of the circular Gaussian type described in Sec. 3.2.3. The joint probability density function of any set of samples of their complex envelopes taken at arbitrary times has a circular Gaussian form like that in (3-40).

The stochastic signals, when present, are received in the midst of Gaussian noise, to which we attribute a complex autocovariance function $\phi_0(t_1, t_2)$. For stationary noise $\phi_0(t_1, t_2)$ is a function only of $t_1 - t_2$. If the noise is white with unilateral spectral density N, by (3-45)

$$\phi_0(t_1, t_2) = N\delta(t_1 - t_2). \tag{11-5}$$

Stochastic signals are sometimes picked up not by a single antenna, but by a number of antennas or *sensors* located at different points of space. Many seismometers may be distributed over a broad area for the detection of seismic waves such as might come from an earthquake or nuclear explosion, and arrays of ultrasonic sensors have been constructed for receiving acoustic signals under water, as in sonar. Both seismic and sonar signals can be represented as stochastic processes, and techniques for processing the outputs of such arrays can be derived from the principles of detection theory. Instead of a single input $v(t)$, there are now a number of inputs, the signal and noise components of which are correlated both temporally and spatially. The methods to be described here can be extended to handle multiple inputs, but with some increase of mathematical complexity. For the application to seismology, we refer the reader to the December 1965 issue of the *Proceedings of the IEEE*; for the application to sonar we cite the paper by Middleton and Groginsky [Mid65]. Further references are to be found in both.

The detection of stochastic signals seems to have been treated first by Davis [Dav54] and Youla [You54]. The approach through the theory of hypothesis testing was taken by Middleton [Mid57]. The task of the receiver is viewed as one of choosing between two hypotheses about its input $v(t) = \text{Re } V(t) \exp i\Omega t$. Under hypothesis H_0 $V(t) = N(t)$, where $N(t)$ is the complex envelope of Gaussian narrowband noise of complex autocovariance function $\phi_0(t_1, t_2)$. Under hypothesis H_1 $V(t) = S(t) + N(t)$, where $S(t)$ is a realization of a narrowband Gaussian process of complex autocovariance function $\phi_s(t_1, t_2)$. The signals and noise being independent, the complex autocovariance function of the input $v(t) = \mathrm{Re}\, V(t) \exp i\Omega t$ is

$$\tfrac{1}{2}E[V(t_1)V^*(t_2)| H_1] = \phi_1(t_1, t_2) = \phi_s(t_1, t_2) + \phi_0(t_1, t_2)$$

under hypothesis H_1. The input $v(t)$ is observed during an interval $(0, T)$.

11.1.2 Vector-space Representation

As we learned in Chapter 1, the best strategy for the receiver is to form the likelihood ratio between the joint probability density functions of samples of the input under the two hypotheses. The likelihood ratio is compared with a decision level that depends on the criterion of choice, Bayes or Neyman–Pearson, that the designer has adopted. Before determining the likelihood ratio, however, we shall introduce a convenient notation for handling the signals, the noise, their complex autocovariance functions, and similar functions of one or two time variables that arise in this study. We shall set up a vector-space representation for our input and its constituent signals and noise much like that introduced in Sec. 2.1.

The complex envelope $V(t)$ of the input will be sampled by means of a complete set of functions $f_k(t)$ that are orthonormal over the observation interval $(0, T)$:

$$\int_0^T f_k(t)f_m^*(t)\, dt = \delta_{km} = \begin{cases} 1, & k = m, \\ 0, & k \neq m. \end{cases} \tag{11-6}$$

We can then write the input as

$$V(t) = \sum_k V_k f_k(t), \qquad 0 \leq t \leq T, \tag{11-7}$$

whose complex coefficients V_k are defined by

$$V_k = \int_0^T f_k^*(t)V(t)\, dt. \tag{11-8}$$

[Sums as in (11-7) without indicated limits will be taken to run from $k = 1$ to $k = \infty$.] In this way we set up a correspondence between the temporal function $V(t)$, $0 \leq t \leq T$, and the vector $\mathbf{V} = (V_1, V_2, \ldots, V_k, \ldots)$ of coefficients. In future operations with matrices this vector should be considered as a column vector. Its transposed conjugate row vector

$$\mathbf{V}^+ = (V_1^*, V_2^*, \ldots, V_k^*, \ldots)$$

corresponds to the complex conjugate function $V^*(t)$.

The scalar product of two functions $V(t)$ and $W(t)$, represented respectively by vectors $\mathbf{V}$ and $\mathbf{W}$, is defined as usual by

$$\mathbf{V}^+\mathbf{W} = \sum_k V_k^* W_k = \int_0^T V^*(t)W(t)\, dt, \tag{11-9}$$

the second equality following from the orthonormality relation (11-6).

In like manner we associate with a function $m(t, s)$ the matrix $\mathbf{M} = \|m_{jk}\|$ of coefficients defined by

$$m_{jk} = \int_0^T \int_0^T f_j^*(t) m(t, s) f_k(s) \, dt \, ds,$$

and in terms of these matrix elements m_{jk} the function is

$$m(t, s) = \sum_j \sum_k m_{jk} f_j(t) f_k^*(s). \tag{11-10}$$

When the function $m(t, s)$ has the Hermitian property

$$m(t, s) = m^*(s, t), \tag{11-11}$$

the matrix $\mathbf{M}$ is Hermitian:

$$\mathbf{M} = \mathbf{M}^+, \qquad m_{jk}^* = m_{kj}.$$

To a linear operation of the form

$$W(t) = \int_0^T m(t, s) V(s) \, ds \tag{11-12}$$

corresponds the linear transformation

$$\mathbf{W} = \mathbf{MV} \tag{11-13}$$

of the vector $\mathbf{V}$ into the vector $\mathbf{W}$. Similarly, to the matrix product $\mathbf{LM}$ corresponds the function

$$\int_0^T l(t, u) m(u, s) \, du$$

of t and s, where $l(t, s)$ corresponds to the matrix $\mathbf{L}$ and $m(t, s)$ to the matrix $\mathbf{M}$, as in (11-10). All relations of this kind can be demonstrated by (11-6), and we suggest that the reader do so.

To the matrix $\mathbf{M}^2$ corresponds the function

$$m^{(2)}(t, s) = \int_0^T m(t, u) m(u, s) \, du, \tag{11-14}$$

and higher powers $\mathbf{M}^j$ and the corresponding "iterates" $m^{(j)}(t, s)$ can be defined by continuation of this process. Do not confuse $m^{(j)}(t, s)$ with the jth power of the function $m(t, s)$.

If the functions $f_k(t)$ are eigenfunctions of the operator $m(t, s)$,

$$\mu_k f_k(t) = \int_0^T m(t, s) f_k(s) \, ds, \qquad 0 \le t \le T, \tag{11-15}$$

the μ_k are the eigenvalues of $m(t, s)$. The matrix $\mathbf{M}$ is then diagonal, and the eigenvalues μ_k are its diagonal elements. When $m(t, s)$ is Hermitian as in (11-11), the eigenvalues μ_k are real and the functions $f_k(t)$ orthonormal, as shown in Sec. 2.1.6. If furthermore $m(t, s)$ is positive definite, the eigenvalues μ_k are positive, and we assume them to have been arranged in descending order:

$$\mu_1 \ge \mu_2 \ge \cdots \ge \mu_k \ge \cdots > 0.$$

Mercer's theorem states that the kernel $m(t, s)$ of (11-15) can be expressed in terms of its eigenfunctions as

$$m(t, s) = \sum_k \mu_k f_k(t) f_k^*(s), \qquad (11\text{-}16)$$

and it is a consequence of (11-15) and (11-10); compare (2-42). The iterates of the kernel $m(t, s)$ are similarly

$$m^{(j)}(t, s) = \sum_k \mu_k^j f_k(t) f_k^*(s). \qquad (11\text{-}17)$$

For any function $g(x)$ that possesses a power-series expansion

$$g(x) = \sum_{j=0}^\infty a_j x^j$$

converging in the neighborhood of the origin, we can define the matrix function

$$g(\mathbf{M}) = \sum_{j=0}^\infty a_j \mathbf{M}^j, \qquad (11\text{-}18)$$

and to it will correspond a function of the time variables t and s:

$$G_m(t, s) = \sum_{j=0}^\infty a_j m^{(j)}(t, s), \qquad 0 \le (t, s) \le T.$$

The trace of the matrix $\mathbf{M}$, which is the sum of its diagonal elements or its eigenvalues, is given by

$$\mathrm{Tr}\,\mathbf{M} = \sum_k \mu_k = \int_0^T m(t, t)\, dt, \qquad (11\text{-}19)$$

which we obtain by setting $s = t$ in (11-16) and integrating over the interval $(0, T)$. Alternatively, set $t = s$ in (11-10), integrate over $(0, T)$, and use (11-6).

The function

$$D(z) = \prod_{k=1}^\infty (1 + \mu_k z)$$

is called the *Fredholm determinant* associated with the kernel $m(t, s)$. It will figure prominently in our calculations of the false-alarm and detection probabilities for stochastic signals in white noise. We can write it as

$$D(z) = \det(\mathbf{I} + z\mathbf{M}),$$

where det stands for the determinant of what is here an infinite matrix, and $\mathbf{I}$ is the identity matrix. The determinant of a finite matrix equals the product of its eigenvalues. Taking logarithms turns the product into a sum, and extending this to our infinite matrices we define, for positive-definite $m(t, s)$ and its representative $\mathbf{M}$,

$$\ln \det(\mathbf{I} + z\mathbf{M}) = \mathrm{Tr}\,\ln(\mathbf{I} + z\mathbf{M}). \qquad (11\text{-}20)$$

The eigenvalues of $\mathbf{I} + z\mathbf{M}$ are $1 + z\mu_k$, and provided that the complex number z lies within a circle of radius μ_1^{-1} about the origin—μ_1 is the largest eigenvalue of the kernel $m(t, s)$—the matrix function $\ln (\mathbf{I} + z\mathbf{M})$ is definable by (11-18) because the function $\ln (1 + zx)$ possesses a convergent power series in the neighborhood of the origin. We can express (11-20) as

$$\ln \det(\mathbf{I} + z\mathbf{M}) = \text{Tr} \int_0^z (\mathbf{I} + u\mathbf{M})^{-1}\mathbf{M} \, du = \text{Tr} \int_0^z \mathbf{P}(u) \, du \qquad (11\text{-}21)$$

in terms of a matrix function

$$\mathbf{P}(u) = (\mathbf{I} + u\mathbf{M})^{-1}\mathbf{M}. \qquad (11\text{-}22)$$

The integration in (11-21) can be thought of as carried out by integrating the series expansion of the integrand term by term.

The matrix function $\mathbf{P}(u)$ is the solution of the linear equation

$$\mathbf{P}(u) + u\mathbf{M}\mathbf{P}(u) = \mathbf{M}, \qquad (11\text{-}23)$$

to which corresponds the integral equation

$$P(t, s; u) + u \int_0^T m(t, r)P(r, s; u)\,dr = m(t, s), \qquad 0 \le (t, s) \le T, \qquad (11\text{-}24)$$

for the function $P(t, s; u)$ corresponding to the matrix $\mathbf{P}(u)$. Here u is a parameter, possibly complex. By analytic continuation, we can extend the domains of $\mathbf{P}(u)$ and $P(t, s; u)$ over the entire complex u-plane. In terms of the eigenfunctions $f_k(t)$ of the kernel $m(t, s)$,

$$P(t, s; u) = \sum_k \frac{\mu_k}{1 + u\mu_k} f_k(t)f_k^*(s), \qquad (11\text{-}25)$$

and this function has poles at $u = u_k = -1/\mu_k$ along the negative real axis.

By means of this function $P(t, s; u)$ we can use (11-19) to write (11-21) as

$$\ln D(z) = \ln \det(\mathbf{I} + z\mathbf{M}) = \int_0^z \int_0^T P(t, t; u) \, dt \, du,$$

and the Fredholm determinant becomes

$$D(z) = \exp\left[\int_0^z \int_0^T P(t, t; u) \, dt \, du\right]. \qquad (11\text{-}26)$$

By means of the rules just set forth we shall be able in our subsequent analysis to move freely back and forth between the time domain of functions such as $V(t)$ and $m(t, s)$ and the vector space in which functions $V(t)$ are represented by column vectors $\mathbf{V}$, which are transformed by matrices $\mathbf{M}$ related to $m(t, s)$. We assume throughout that these transitions are legitimate for the kinds of signals and noise we are dealing with, leaving the treatment of exceptional situations to mathematical works on linear operators and functional analysis.

11.1.3 The Likelihood Ratio

The receiver is to decide between two hypotheses

$$V(t) = N(t), \qquad (H_0)$$
$$V(t) = S(t) + N(t), \qquad (H_1)$$

about the complex envelope $V(t)$ of its input Re $V(t) \exp i\Omega t$. Here $N(t)$ and $S(t)$ are circular complex Gaussian random processes with expected values zero and complex autocovariance functions

$$\phi_0(t, s) = \tfrac{1}{2}E[N(t)N^*(s)],$$
$$\phi_s(t, s) = \tfrac{1}{2}E[S(t)S^*(s)]. \tag{11-27}$$

Under hypothesis H_1 the complex autocovariance function of the input is

$$\phi_1(t, s) = \phi_s(t, s) + \phi_0(t, s).$$

The input is observed during an interval $(0, T)$. It is sampled as described in Sec. 11.1.2 in terms of a set of functions orthonormal over $(0, T)$ to produce a vector $\mathbf{V} = (V_1, V_2, \dots)$ of complex samples defined as in (11-8). To the autocovariance functions $\phi_0(t, s)$, $\phi_s(t, s)$, and $\phi_1(t, s)$ now correspond infinite Hermitian matrices $\boldsymbol{\phi}_0$, $\boldsymbol{\phi}_s$, and $\boldsymbol{\phi}_1$, respectively.

Denote by $\mathbf{V}^{(n)}$ the column vector of the first n of the samples:

$$\mathbf{V}^{(n)} = (V_1, V_2, \dots, V_n)^T.$$

The joint probability density function of the real and imaginary parts of these samples has under each hypothesis the circular complex Gaussian form

$$p_j(\mathbf{V}^{(n)}) = (2\pi)^{-n}[\det \boldsymbol{\phi}_j^{(n)}]^{-1} \exp(-\tfrac{1}{2}\mathbf{V}^{(n)+}\boldsymbol{\phi}_j^{(n)-1}\mathbf{V}^{(n)}), \qquad j = 0, 1, \tag{11-28}$$

where $\boldsymbol{\phi}_j^{(n)}$ is the $n \times n$ autocovariance matrix of the n samples V_k under hypothesis H_j:

$$\boldsymbol{\phi}_j^{(n)} = \|\phi_{j,ik}\|, \qquad \phi_{j,ik} = \tfrac{1}{2}E(V_i V_k^* \mid H_j), \qquad 1 \le (i, k) \le n. \tag{11-29}$$

The optimum strategy for deciding between hypotheses H_0 and H_1 on the basis of n samples $(V_1, V_2, \dots, V_n)$ compares the likelihood ratio

$$\Lambda(\mathbf{V}^{(n)}) = \frac{p_1(\mathbf{V}^{(n)})}{p_0(\mathbf{V}^{(n)})}$$
$$= [\det \boldsymbol{\phi}_1^{(n)}\boldsymbol{\phi}_0^{(n)-1}]^{-1} \exp[\tfrac{1}{2}\mathbf{V}^{(n)+}(\boldsymbol{\phi}_0^{(n)-1} - \boldsymbol{\phi}_1^{(n)-1})\mathbf{V}^{(n)}] \tag{11-30}$$

with a suitable decision level. Passing to the limit $n \to \infty$, we find that a sufficient statistic utilizing all the information in the input $v(t) = $ Re $V(t) \exp i\Omega t$ is the logarithmic likelihood ratio

$$\ln \Lambda(\mathbf{V}) = G = \tfrac{1}{2}\mathbf{V}^+\mathbf{H}\mathbf{V} - \ln \det(\boldsymbol{\phi}_1\boldsymbol{\phi}_0^{-1}) = U - B, \tag{11-31}$$

in which the Hermitian matrix

$$\mathbf{H} = \boldsymbol{\phi}_0^{-1} - \boldsymbol{\phi}_1^{-1} \tag{11-32}$$

is the solution of the matrix equation

$$\boldsymbol{\phi}_0\mathbf{H}\boldsymbol{\phi}_1 = \boldsymbol{\phi}_1\mathbf{H}\boldsymbol{\phi}_0 = \boldsymbol{\phi}_1 - \boldsymbol{\phi}_0 = \boldsymbol{\phi}_s, \tag{11-33}$$

and

$$B = \ln \det(\mathbf{I} + \boldsymbol{\phi}_s\boldsymbol{\phi}_0^{-1}). \tag{11-34}$$

Under the Neyman–Pearson criterion the decision about the presence or absence of the signal can be based on the quadratic functional

$$U = \tfrac{1}{2}\mathbf{V}^{\dagger}\mathbf{H}\mathbf{V} = \tfrac{1}{2}\int_0^T \int_0^T V^*(t)h(t, s)V(s)\, dt\, ds \qquad (11\text{-}35)$$

of the input. The kernel of this functional is, by (11-33), the solution of the double integral equation

$$\int_0^T \int_0^T \phi_0(t, u)h(u, v)\phi_1(v, s)\, du\, dv = \phi_s(t, s), \qquad 0 \le (t, s) \le T. \qquad (11\text{-}36)$$

It can be broken into two integral equations

$$\phi_s(t, s) = \int_0^T g(t, u)\phi_1(u, s)\, du, \qquad 0 \le (t, s) \le T, \qquad (11\text{-}37)$$

$$g(t, s) = \int_0^T \phi_0(t, r)h(r, s)\, dr, \qquad 0 \le (t, s) \le T, \qquad (11\text{-}38)$$

which must be solved successively. If the statistic U exceeds a decision level U_0, the receiver decides that a realization of the stochastic process $\operatorname{Re} S(t) \exp i\Omega t$ is present. The decision level is selected to achieve a preassigned false-alarm probability

$$Q_0 = \Pr(U \ge U_0 |\, H_0).$$

11.1.4 Realizations of the Optimum Detector

The test statistic U in (11-35) can be generated by means of a properly matched time-variable or nonstationary linear filter [Pri56]. Because $h(t, s) = h^*(s, t)$, the statistic can be written

$$U = \operatorname{Re} \int_0^T V^*(t)\, dt \int_0^t h(t, s)V(s)\, ds$$

$$= \operatorname{Re} \int_0^T V^*(t)W(t)\, dt, \qquad (11\text{-}39)$$

$$W(t) = \int_0^t h(t, s)V(s)\, ds.$$

For each value of t, the function $W(t)$ is a weighted average of the input that has arrived before time t. It can be generated by passing the input $\operatorname{Re}[V(t)\exp i\Omega t]$ through a time-variable linear filter whose narrowband impulse response is

$$K_t(\tau) = \begin{cases} h(t, t - \tau), & 0 \le \tau \le t, \\ 0, & \tau > t. \end{cases}$$

The output of the filter at time t is $\operatorname{Re} W(t) \exp i\Omega t$, where

$$W(t) = \int_0^t K_t(\tau)V(t - \tau)\, d\tau = \int_0^t h(t, t - \tau)V(t - \tau)\, d\tau,$$

which is the same as (11-39). Because this filter is causal, it can in principle be realized physically; it operates only on the input $v(t)$ previously received.

The output of the time-variable filter is multiplied at each instant by the input $\operatorname{Re} V(t) \exp i\Omega t$, and the high-frequency components of the product are removed by

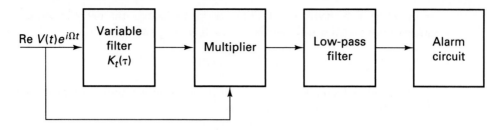

Figure 11-1. Optimum detector of stochastic signals.

filtering. By writing out these factors in terms of $\cos \Omega t$ and $\sin \Omega t$ and multiplying them, the reader can show that their product is

$$\text{Re } V(t) \, e^{i\Omega t} \cdot \text{Re } W(t) \, e^{i\Omega t} = \tfrac{1}{2} \text{ Re } V^*(t)W(t) + \text{ terms of frequency } 2\Omega.$$

The product is integrated by a low-pass filter with impulse response

$$k(\tau) \equiv 1, \qquad 0 \leq \tau < T; \qquad k(\tau) \equiv 0, \qquad \tau \geq T,$$

whose output at time T is

$$\tfrac{1}{2} \text{ Re } \int_0^T V^*(t)W(t) \, dt = \tfrac{1}{2}U.$$

The operation of this receiver is illustrated in Fig. 11-1. The required multiplication can be most easily accomplished by passing the sum $\text{Re}[V(t) + W(t)] \exp i\Omega t$ through a quadratic rectifier, the output of which is

$$|V(t) + W(t)|^2 = |V(t)|^2 + 2 \text{ Re } V^*(t)W(t) + |W(t)|^2.$$

The separately rectified outputs $|W(t)|^2$ and $|V(t)|^2$ are subtracted to leave the desired product.

A second way of generating the test statistic U employs a time-variable linear filter followed by a quadratic rectifier and an integrator [Mid60b]. The output of the filter has the complex envelope

$$X(t) = \int_0^T m(t, s)V(s) \, ds, \tag{11-40}$$

and this output is rectified and integrated to produce

$$U = \tfrac{1}{2}\int_0^T |X(t)|^2 \, dt. \tag{11-41}$$

In order for this quantity to equal the test statistic as given by (11-35), the weighting function $m(t, s)$ must satisfy the nonlinear integral equation

$$h(t, s) = \int_0^T m^*(u, t)m(u, s) \, du, \qquad 0 \leq (t, s) \leq T, \tag{11-42}$$

where $h(t, s)$ is the solution of the integral equation (11-36).

This procedure corresponds to decomposing the matrix $\mathbf{H}$ of (11-32) into a product, $\mathbf{H} = \mathbf{M}^+\mathbf{M}$, and to $X(t)$ corresponds the vector $\mathbf{X} = \mathbf{MV}$, whereupon

$$U = \tfrac{1}{2}\mathbf{X}^+\mathbf{X} = \tfrac{1}{2}\int_0^T |X(t)|^2 \, dt$$

as in (11-41). In order for the filter to be realizable, the decomposition must be carried out in such a way that $m(t, s) \equiv 0$, $s > t$, and then

$$X(t) = \int_0^t m(t, s)V(s)\, ds$$

can be generated simultaneously with the reception of the input $v(t) = \text{Re } V(t) \exp i\Omega t$. To determine a solution of the nonlinear integral equation (11-42) having this causal property is in general most difficult.

11.1.5 Stationarity over a Long Observation Interval

When both signal and noise are realizations of stationary random processes, their autocovariance functions depend on t and s only through $t - s$, and we write them as

$$\phi_0(t, s) \rightarrow \phi_0(t - s), \qquad \phi_s(t, s) \rightarrow \phi_s(t - s),$$

and $\phi_1(t, s)$ is replaced by

$$\phi_1(t - s) = \phi_s(t - s) + \phi_0(t - s).$$

If the observation interval $(0, T)$ is much longer than the correlation times of signal and noise, we can approximate (11-36) by

$$\int_{-\infty}^{\infty} \int_{-\infty}^{\infty} \phi_1(t - u)h(u, v)\phi_0(v - s)\, du\, dv = \phi_s(t - s), \tag{11-43}$$

and $h(t, s)$ must also be a function only of $t - s$. We can then solve (11-43) by applying the convolution theorem for Fourier transforms. Designating these transforms by capital letters, we write it as

$$\Phi_1(\omega)H(\omega)\Phi_0(\omega) = \Phi_s(\omega),$$

which yields

$$H(\omega) = \frac{\Phi_s(\omega)}{\Phi_1(\omega)\Phi_0(\omega)}. \tag{11-44}$$

[The transforms $\Phi_0(\omega)$, $\Phi_s(\omega)$, and $\Phi_1(\omega)$ are, by (3-19), twice the narrowband spectral densities $\tilde{\Phi}_0(\omega)$, $\tilde{\Phi}_s(\omega)$, and $\tilde{\Phi}_1(\omega)$. We adopt this new convention in order to eliminate bothersome factors of 2 from our equations.]

Under these circumstances, the function $m(t, s)$ in (11-42), whose limits are now $(-\infty, \infty)$, also depends on t and s through $t - s$, and in terms of its Fourier transform $M(\omega)$,

$$m(t - s) = \int_{-\infty}^{\infty} M(\omega)\, e^{i\omega(t-s)} \frac{d\omega}{2\pi}.$$

Then we can write (11-42) as $|M(\omega)|^2 = H(\omega)$, and

$$M(\omega) = \left[\frac{\Phi_s(\omega)}{\Phi_1(\omega)\Phi_0(\omega)} \right]^{1/2} e^{i\gamma(\omega)}, \tag{11-45}$$

where $\gamma(\omega)$ is a phase that can be chosen in such a way as to make the filter whose impulse response is $m(\tau)$ physically realizable, whereupon

$$X(t) = \int_0^t m(t - s)V(s)\, ds.$$

The output of this filter is quadratically rectified and integrated over the observation interval $(0, T)$ as in (11-41) to produce an approximation to the optimum statistic U that is the more accurate the longer the interval $(0, T)$.

Suppose, for instance, that the complex autocovariance function of the signal is

$$\phi_s(\tau) = P_s e^{-\mu|\tau|}, \tag{11-46}$$

and that the noise is white, $\phi_0(\tau) = N\delta(\tau)$. Then the signal has a Lorentz spectral density

$$\Phi_s(\omega) = \frac{2P_s\mu}{\omega^2 + \mu^2}, \tag{11-47}$$

and under hypothesis H_1 the spectral density of the input is

$$\Phi_1(\omega) = \frac{2P_s\mu}{\omega^2 + \mu^2} + N = N\left(\frac{\omega^2 + \beta^2}{\omega^2 + \mu^2}\right)$$

with

$$\beta^2 = \mu^2 + \frac{2P_s\mu}{N},$$

whereupon, by (11-44),

$$H(\omega) = \frac{2P_s\mu}{N^2(\omega^2 + \beta^2)}.$$

The transfer function of the causal filter is then

$$M(\omega) = \frac{\sqrt{2P_s\mu}}{N(\beta + i\omega)},$$

and its impulse response is

$$m(\tau) = \frac{\sqrt{2P_s\mu}}{N}e^{-\beta\tau}U(\tau).$$

When $\mu T \gg 1$, $m(t - s)$ will be a close approximation to the optimum time-variable filter $m(t, s)$.

11.1.6 The Threshold Detector

When the signal is much weaker than the noise, we can replace $\phi_1(t, u)$ in (11-36) by $\phi_0(t, u)$, obtaining the integral equation

$$\int_0^T \int_0^T \phi_0(t, u)h_\theta(u, v)\phi_0(v, s)\, du\, dv = \phi_s(t, s), \qquad 0 \le (t, s) \le T, \tag{11-48}$$

for the kernel $h_\theta(t, s)$ of the threshold statistic

$$U_\theta = \tfrac{1}{2}\int_0^T \int_0^T V^*(t)h_\theta(t, s)V(s)\, dt\, ds \tag{11-49}$$

for detection of the stochastic signal Re $S(t) \exp i\Omega t$. This statistic can be implemented by the means outlined in Sec. 11.1.4. When the noise is white, as in (11-5), the kernel of the threshold statistic is simply

$$h_\theta(t, s) = N^{-2}\phi_s(t, s). \tag{11-50}$$

When as in Sec. 11.1.5 the signal and noise are stationary and the observation interval $(0, T)$ is much longer than their correlation times, the threshold statistic can be realized by a time-invariant linear filter followed by a quadratic rectifier. The transfer function of the filter is

$$M_\theta(\omega) = \frac{\sqrt{\Phi_s(\omega)}}{\Phi_0(\omega)} e^{i\gamma(\omega)}, \tag{11-51}$$

as follows from (11-45) by approximating $\Phi_1(\omega)$ by $\Phi_0(\omega)$. Again the phase $\gamma(\omega)$ is chosen so that the filter is realizable. This filter is known as the *Eckart filter* [Eck51], [Mac68].

The signal can now be thought of as a dense succession of pulses whose spectrum is proportional to

$$\sqrt{\Phi_s(\omega)}\, e^{i\gamma(\omega)}$$

and that occur at random times and with independently random amplitudes, as in (10-41). As in (2-86), the Eckart filter is an approximation to the optimum filter for detecting each such pulse in noise of spectral density $\Phi_0(\omega)$, the approximation assuming that the observation interval is much longer than the duration of the pulses. Because they are narrowband signals with random phases and arrive throughout the interval $(0, T)$, one integrates the quadratically rectified output of the filter as in (11-41) to produce the threshold detection statistic U_θ.

11.1.7 The Radiometer

If the signal is an echo from a fluctuating point target, its autocovariance function $\phi_s(t, s)$ can be determined from (11-4) by inserting $\psi(\tau, t) = \psi(t)\delta(\tau - \tau_0)$, and it is

$$\phi_s(t, s) = F(t - \tau_0)\psi(t - s)F^*(s - \tau_0),$$

where τ_0 is the delay to and from the target. Suppose now that the threshold statistic in (11-49) is to be used in a receiver to detect this target in white noise. It is given by the equation

$$\begin{aligned} U_\theta &= \frac{1}{2N^2} \int_0^T \int_0^T V^*(t)F(t - \tau_0)\psi(t - s)F^*(s - \tau_0)V(s)\, dt\, ds \\ &= \frac{1}{2N^2} \int_{-\infty}^{\infty} \int_{-\infty}^{\infty} Y^*(t)\psi(t - s)Y(s)\, dt\, ds, \end{aligned} \tag{11-52}$$

where $Y(t) = F^*(t - \tau_0)V(t)$ can be formed by multiplying the input by a locally generated replica of the transmitted signal with the proper time delay, assumed known. In the second integral the integrations need to be carried out only over intervals during which the signal might arrive and $Y(t) \neq 0$.

The realizations we described in Sec. 11.1.4 might now be applied to U_θ of (11-52), the function $\psi(t - s)$ taking the place of $h(t, s)$ and $Y(t)$ the place of $V(t)$. In particular we might employ the second realization, for which it is necessary to solve an equation similar to (11-42). If the range of integration is extended to span the entire time axis, $-\infty < t < \infty$, the solution becomes a function only of the difference of the two arguments, and the equation itself takes the form

$$\int_{-\infty}^{\infty} g^*(x - t)g(x - s)\, dx = \psi(t - s).$$

Introducing the Fourier transforms

$$G(\omega) = \int_{-\infty}^{\infty} g(t) e^{-i\omega t} dt, \qquad \Psi(\omega) = \int_{-\infty}^{\infty} \psi(t) e^{-i\omega t} dt,$$

we find by the convolution theorem that $|G(\omega)|^2 = \Psi(\omega)$, and $G(\omega)$ can be taken as

$$G(\omega) = \sqrt{|\Psi(\omega)|} \; e^{i\chi(\omega)},$$

the phase $\chi(\omega)$ being chosen so that a filter of impulse response $g(\tau)$ is physically realizable, with $g(\tau) \equiv 0, \tau < 0$. The test statistic then becomes

$$U_\theta = \frac{1}{2N^2} \int_{-\infty}^{\infty} dt \left| \int_{-\infty}^{\infty} g(t-s) F^*(s-\tau_0) V(s) \, ds \right|^2.$$

A device generating the threshold statistic U_θ in this approximate way was proposed by Price and Green [Pri60] for detecting signals in radar astronomy. They called it a *radiometer*. In practice, they pointed out, it will not be necessary to integrate the outputs of the filter or the rectifier over a very long interval in order to achieve a good approximation to the threshold detector.

If the target is deep fluctuating and the autocovariance of the echo signals is given as in (11-4), it is merely necessary to construct a parallel bank of these radiometers, each matched to the transmitted signal with one of a dense set of delays τ_0. The impulse responses $g(\tau)$ may differ from one filter to another. Price and Green [Pri60] termed this more elaborate device a *Rake radiometer*, for it is reminiscent of a similar device used in the detection of signals in a multipath communication system, the Rake receiver [Pri58].

11.1.8 An Example

To illustrate the ideas of this section and the next, it is instructive to have before us the simple example of a stationary stochastic signal with the exponential autocovariance function in (11-46),

$$\phi_s(r, t) \rightarrow \phi_s(r - t) = P_s \, e^{-\mu|r-t|}.$$

The signal is received in the presence of white Gaussian noise of unilateral spectral density N. The modulation $M(t)$, (11-1), is taken as constant, and the signals, which are also Gaussian processes, can be generated by passing white noise through a narrowband simply resonant circuit of bandwidth μ. They are observed during the interval $(0, T)$. This case has been analyzed by Price [Pri56] and others.

For future use it will be convenient to have the solution of the integral equation

$$Nh(r, t; u) + u \int_0^T \phi_s(r, s) h(s, t; u) \, ds = \frac{1}{N} \phi_s(r, t), \qquad 0 \le (r, t) \le T. \qquad (11\text{-}53)$$

The kernel of the test statistic is $h(t, s) = h(t, s; 1)$, by (11-36) with

$$\phi_0(t, s) = N\delta(t - s), \qquad \phi_1(t, s) = N\delta(t - s) + \phi_s(t - s).$$

The integral equation (11-53) takes the form of (2-91) when we identify the kernel as

$$\phi(r - s) = N\delta(r - s) + u\phi_s(r - s),$$

and the method outlined in Sec. 2.3 can be applied. The Fourier transform

$$\Phi(\omega) = \int_{-\infty}^{\infty} \phi(\tau) \, e^{-i\omega\tau} \, d\tau$$

of the kernel is a rational function of frequency:

$$\Phi(\omega) = N + u\Phi_s(\omega) = N + \frac{2\mu P_s u}{\omega^2 + \mu^2} = N\frac{\beta^2 + \omega^2}{\mu^2 + \omega^2}, \qquad \beta^2 = \mu^2 + \frac{2\mu P_s u}{N},$$

where $\Phi_s(\omega)$ is the spectral density of the signal.

The form of the solution of the integral equation is given in (2-100). What corresponds to $q_0(t)$ in (2-100) is the solution $h_0(t - s)$ of (11-53) when the limits of integration are $-\infty$ and $+\infty$ instead of 0 and T, and this solution is the Fourier transform of

$$\int_{-\infty}^{\infty} h_0(t) \, e^{-i\omega t} \, dt = \frac{\Phi_s(\omega)}{N\Phi(\omega)} = \frac{2\mu P_s}{N^2(\beta^2 + \omega^2)},$$

whereupon

$$h_0(t - u) = \frac{\mu P_s}{\beta N^2} e^{-\beta|t-u|}.$$

The terms of (2-100) with delta functions are now absent because the degrees of the numerator and the denominator of $\Phi(\omega)$ are equal. Hence the solution has the form

$$h(r, t; u) = \frac{\mu P_s}{\beta N^2} e^{-\beta|r-t|} + A e^{-\beta r} + B e^{-\beta r}, \tag{11-54}$$

where A and B are functions of t.

To determine the unknown functions $A(t)$ and $B(t)$, we substitute (11-54) into (11-53). When we carry out the integration and use the definition of β, we find that all the terms in $\exp \beta r$ and $\exp(-\beta r)$ cancel, as does the term on the right side of (11-53). We are left only with terms proportional to either $\exp \mu r$ or $\exp(-\mu r)$. Setting the coefficients of each of these functions separately equal to zero, we obtain two simultaneous linear equations for A and B, which are solved in the usual way and yield

$$A = \frac{\mu P_s(\beta - \mu)[(\beta + \mu) \, e^{\beta t} + (\beta - \mu) \, e^{-\beta t}] \, e^{-\beta T}}{N^2 \beta[(\beta + \mu)^2 \, e^{\beta T} - (\beta - \mu)^2 \, e^{-\beta T}]},$$

$$B = \frac{\mu P_s(\beta - \mu)[(\beta + \mu) \, e^{\beta(T-t)} + (\beta - \mu) \, e^{-\beta(T-t)}]}{N^2 \beta[(\beta + \mu)^2 \, e^{\beta T} - (\beta - \mu)^2 \, e^{-\beta T}]}. \tag{11-55}$$

Substituting these into (11-54) and treating the regions $r < t$ and $r > t$ separately, we combine terms to derive the solution

$$h(r, t; u) = \frac{\mu P_s[(\beta + \mu) \, e^{\beta r} + (\beta - \mu) \, e^{-\beta r}][(\beta + \mu) \, e^{\beta(T-t)} + (\beta - \mu) \, e^{-\beta(T-t)}]}{N^2 \beta[(\beta + \mu)^2 \, e^{\beta T} - (\beta - \mu)^2 \, e^{-\beta T}]},$$

$$0 \le r \le t \le T. \tag{11-56}$$

The solution for $0 \le t \le r \le T$ is found by interchanging r and t in this expression. The kernel of the detection statistic in (11-35) is given by (11-56) when one determines β from

$$\beta^2 = \mu^2 + \frac{2\mu P_s}{N}.$$

Substituting into (11-39), we get the test statistic

$$U = \frac{\mu P_s}{N^2\beta[(\beta + \mu)^2 \, e^{\beta T} - (\beta - \mu)^2 \, e^{-\beta T}]} \cdot \text{Re} \int_0^T V^*(s)[(\beta + \mu) \, e^{\beta(T-s)}$$

$$+ (\beta - \mu) \, e^{-\beta(T-s)}] \int_0^s [(\beta + \mu) \, e^{\beta t} + (\beta - \mu) \, e^{-\beta t}] V(t) \, dt \, ds.$$

In the present example the threshold detector for stochastic signals having the autocovariance function in (11-46) is based on the approximation in (11-50) and furnishes the statistic

$$U_\theta = \frac{P_s}{N^2} \, \text{Re} \int_0^T V^*(t) \, dt \int_0^t e^{-\mu(t-s)} V(s) \, ds$$

$$= \frac{P_s}{N^2} \, \text{Re} \int_0^T V^*(t) \, dt \int_0^t e^{-\mu\tau} V(t - \tau) \, d\tau.$$

If the input is turned on at time $t = 0$, the term

$$\int_0^t e^{-\mu\tau} V(t - \tau) \, d\tau$$

is proportional to the envelope of the output of a narrowband simply resonant circuit of bandwidth μ tuned to the carrier frequency Ω. This output is multiplied by the input $v(t) = \text{Re} \, V(t) \exp i\Omega t$ in the manner described in Sec. 11.1.4, and the product is integrated over a period of duration T. Such a threshold detection system can be made independent of the true signal power $\phi_s(0, 0)$, which may not be known in advance; the optimum system, on the other hand, depends on the strengths of both signals and noise.

Unfortunately, there seems to be no such simple approximation in the general case when the signal-to-noise ratio is large. In the present example we see from (11-56) that for large signal-to-noise ratio and long integration time T ($\beta T \gg 1$), the dominant term in the kernel $h(r, t)$ is proportional to $\exp[-\beta|r - t|]$. Hence the optimum detection system is nearly the same as the threshold receiver, except that the bandwidth of the input filter is $(2\mu P_s/N)^{1/2}$ instead of μ. In the next section we shall attack the problem of calculating the false-alarm and detection probabilities for such receivers.

11.1.9 Estimator-correlator Interpretation

In the detection of signals having unknown parameters, the method of maximum likelihood involves our pretending that the signal is present, finding the maximum-likelihood estimates of those parameters, and then constructing—at least conceptually—the optimum or threshold receiver for detecting a signal having parameter values equal to those estimates. When the output of this receiver exceeds a certain decision level, the system decides that a signal is present. When a stochastic signal $s(t) = \text{Re} \, S(t) \exp i\Omega t$ is to be detected, the parameters can be taken as all the sample values

$$S_k = \int_0^T f_k^*(t) S(t) \, dt \tag{11-57}$$

of the signal. When the signal is a Gaussian random process, the real and imaginary parts of any finite number n of these have a joint probability density function of the circular Gaussian form

$$z(\mathbf{S}) = C_1 \exp(-\tfrac{1}{2}\mathbf{S}^{+}\boldsymbol{\phi}_s^{-1}\mathbf{S}),$$

where $\boldsymbol{\phi}_s$ is the corresponding $n \times n$ block of their complex covariance matrix, and C_1 is a normalization constant. [We omit the superscripts (n) used in (11-28)]. The conditional probability density function of the real and imaginary parts of n samples $(V_1, V_2, \ldots, V_n)$ of the input, given the presence of the signal, is

$$p_1(\mathbf{V}|\,\mathbf{S}) = C_2 \exp[-\tfrac{1}{2}(\mathbf{V}^{+} - \mathbf{S}^{+})\boldsymbol{\phi}_0^{-1}(\mathbf{V} - \mathbf{S})],$$

where $\boldsymbol{\phi}_0$ is the corresponding $n \times n$ block of the covariance matrix of the samples of the noise, and C_2 a normalization constant. By the same kind of analysis as in Sec. 6.1.4, we find—see (6-20)—that the vector $\hat{\mathbf{S}}$ of maximum-likelihood estimators of the coefficients S_k is given by

$$\hat{\mathbf{S}} = \boldsymbol{\phi}_s\boldsymbol{\phi}_1^{-1}\mathbf{V}. \tag{11-58}$$

At this point we can let the number n of samples go to infinity so that all the information in the input Re $V(t) \exp i\Omega t$ is utilized in forming the maximum-likelihood estimator.

The maximum-likelihood receiver pretends that it is detecting the signal Re $\hat{S}(t)$ $\exp i\Omega t$, corresponding to the vector $\hat{\mathbf{S}}$, in noise with complex autocovariance function $\phi_0(t, s)$. It must therefore contain a filter matched to the signal Re $\hat{Q}(t) \exp i\Omega t$, where $\hat{Q}(t)$ is the solution of the integral equation

$$\hat{S}(t) = \int_0^T \phi_0(t, u)\hat{Q}(u)\, du, \qquad 0 \le t \le T, \tag{11-59}$$

as in (3-52). By (11-12) and (11-13) the vector representation of this equation is $\hat{\mathbf{S}} = \boldsymbol{\phi}_0\hat{\mathbf{Q}}$, and

$$\hat{\mathbf{Q}} = \boldsymbol{\phi}_0^{-1}\hat{\mathbf{S}} = \boldsymbol{\phi}_0^{-1}\boldsymbol{\phi}_s\boldsymbol{\phi}_1^{-1}\mathbf{V} = \mathbf{HV},$$

for by (11-33) and because of the Hermitian form of the matrix $\mathbf{H}$,

$$\mathbf{H} = \boldsymbol{\phi}_1^{-1}\boldsymbol{\phi}_s\boldsymbol{\phi}_0^{-1} = \boldsymbol{\phi}_0^{-1}\boldsymbol{\phi}_s\boldsymbol{\phi}_1^{-1}.$$

The output of the matched filter, sampled at the end of the observation interval $(0, T)$, is

$$\int_0^T \hat{Q}^{*}(t)V(t)\, dt = \hat{\mathbf{Q}}^{+}\mathbf{V} = \mathbf{V}^{+}\mathbf{HV} = 2U \tag{11-60}$$

by (11-9), and by (11-35) this is twice our detection statistic U. Thus the optimum receiver in effect estimates the complex envelope $S(t)$ of the signal as though the signal were known to be present, and as in (11-60) it "correlates" its input with the solution $\hat{Q}(t)$ of (11-59). This is known as the *estimator-correlator* interpretation of the optimum detector of a Gaussian stochastic signal in Gaussian noise [Kai60]. The maximum-likelihood estimator of the signal is, by (11-58),

$$\hat{S}(t) = \int_0^T g(t, u)V(u)\, du, \qquad 0 \le t \le T,$$

where $g(t, u)$ is the solution of the integral equation (11-37). Because the estimator utilizes the input Re $V(t) \exp i\Omega t$ over the entire interval $(0, T)$, it is not causal, and $\hat{S}(t)$ and $\hat{Q}(t)$ cannot be generated until after the observation interval is past. We shall see later that when the noise is white, it is possible to base the design of the optimum receiver on a causal estimator of the signal, assumed present in the input $v(t)$.

11.1.10 The Question of Singularity

When we studied the detection of a deterministic signal in colored Gaussian noise, we found that under certain circumstances the theory predicted that the signal could be detected with zero probability of error. This situation, known as the "singular case of perfect detection," would arise, for instance, if the spectral density of the noise vanished in a region of frequencies where the spectrum of the signal remained finite. The same possibility of perfect detection must be considered in dealing with stochastic signals.

Usually one's model of the signal and the noise is at least partly conjectural, and its accuracy cannot be completely verified. The model is often one that has been simplified to make it mathematically tractable. If upon an analysis based on it, the singular case turns up and the signals appear to be perfectly detectable, the model must be at fault, for nature never permits complete freedom from the chance of error. Our treatment of the problem of singularity will necessarily be crude. For rigorous proofs the reader must look to the references we shall cite.

When a finite number n of samples are utilized, as (11-30) shows,

$$U_n = \tfrac{1}{2}\mathbf{V}^{(n)+}(\boldsymbol{\phi}_0^{(n)-1} - \boldsymbol{\phi}_1^{(n)-1})\mathbf{V}^{(n)} \tag{11-61}$$

is a sufficient statistic. A stochastic signal will be perfectly detectable only if the probability density functions of that statistic under hypotheses H_0 and H_1 recede so far from each other as n goes to infinity that they no longer overlap. It will then be possible to set the decision level U_0 at such a point between them that $Q_0 = 0$ and $Q_d = 1$. Because the statistic U_n is a quadratic form in Gaussian random variables, however, its probability density function will be finite at all positive values of U_n under both hypotheses. The only way by which the probability density functions $p_0(U_n)$ and $p_1(U_n)$ can cease to overlap as n goes to infinity, therefore, is for the difference of the expected values

$$\Delta U_n = E(U_n| H_1) - E(U_n| H_0)$$

to become ever larger with respect to the standard deviations of U_n under the two hypotheses. Conversely, the probability density functions will continue to overlap if the ratio

$$\frac{(\Delta U_n)^2}{\text{Var}_0\ U_n}$$

remains finite as n grows beyond all bounds. (Var$_j$ indicates the variance under hypothesis H_j.) It is the limiting value of this ratio that settles the question of singularity.

By (11-33) and (11-61)

$$\Delta U_n = \text{Tr } \mathbf{H}\boldsymbol{\phi}_s = \text{Tr}(\boldsymbol{\phi}_s \boldsymbol{\phi}_0^{-1} \boldsymbol{\phi}_s \boldsymbol{\phi}_1^{-1})$$
$$= \text{Tr}(\boldsymbol{\phi}_s^{1/2} \boldsymbol{\phi}_0^{-1} \boldsymbol{\phi}_s^{1/2} \boldsymbol{\phi}_s^{1/2} \boldsymbol{\phi}_1^{-1} \boldsymbol{\phi}_s^{1/2}),$$

where we have used the rule $\text{Tr } \mathbf{AB} = \text{Tr } \mathbf{BA}$ for any two matrices $\mathbf{A}$ and $\mathbf{B}$. Here $\boldsymbol{\phi}_s^{1/2}$ is the square root of the nonnegative-definite matrix $\boldsymbol{\phi}_s$. It can be found if necessary by diagonalizing the matrix $\boldsymbol{\phi}_s$ by a unitary transformation, taking the square roots of the diagonal elements of the transformed matrix (the eigenvalues of $\boldsymbol{\phi}_s$) and performing the inverse unitary transformation. We have dropped the superscripts (n) for simplicity.

By the Cauchy–Schwarz inequality (2-83) the Hermitian matrices $\mathbf{A} = \boldsymbol{\phi}_s^{1/2} \boldsymbol{\phi}_0^{-1} \boldsymbol{\phi}_s^{1/2}$ and $\mathbf{B} = \boldsymbol{\phi}_s^{1/2} \boldsymbol{\phi}_1^{-1} \boldsymbol{\phi}_s^{1/2}$ satisfy the relation

$$|\text{Tr } \mathbf{AB}|^2 = \left| \sum_{i=1}^{n} \sum_{j=1}^{n} A_{ij} B_{ji} \right|^2 \leq \sum_{i=1}^{n} \sum_{j=1}^{n} |A_{ij}|^2 \sum_{i=1}^{n} \sum_{j=1}^{n} |B_{ij}|^2$$

$$= \text{Tr } \mathbf{AA}^{+} \text{ Tr } \mathbf{BB}^{+}.$$

Hence

$$(\Delta U_n)^2 \leq \text{Tr } \boldsymbol{\phi}_s^{1/2} \boldsymbol{\phi}_0^{-1} \boldsymbol{\phi}_s^{1/2} \boldsymbol{\phi}_s^{1/2} \boldsymbol{\phi}_0^{-1} \boldsymbol{\phi}_s^{1/2} \text{ Tr } \boldsymbol{\phi}_s^{1/2} \boldsymbol{\phi}_1^{-1} \boldsymbol{\phi}_s^{1/2} \boldsymbol{\phi}_s^{1/2} \boldsymbol{\phi}_1^{-1} \boldsymbol{\phi}_s^{1/2}$$

$$= \text{Tr } \boldsymbol{\phi}_s \boldsymbol{\phi}_0^{-1} \boldsymbol{\phi}_s \boldsymbol{\phi}_0^{-1} \text{ Tr } \boldsymbol{\phi}_s \boldsymbol{\phi}_1^{-1} \boldsymbol{\phi}_s \boldsymbol{\phi}_1^{-1} = (\text{Var}_1 \ U_n)(\text{Var}_0 \ U_n),$$

and the ratio in question is bounded by

$$\frac{(\Delta U_n)^2}{\text{Var}_0 \ U_n} \leq \text{Var}_1 \ U_n = \text{Tr } \boldsymbol{\phi}_s \boldsymbol{\phi}_0^{-1} \boldsymbol{\phi}_s \boldsymbol{\phi}_0^{-1}.$$

If, therefore, $\text{Var}_1 \ U_n$ stays finite as n goes to infinity, the detection cannot be perfect.

Root [Roo63] studied the singularity of the detection of stochastic signals in terms of the eigenvalues ε_k of the matrix $\boldsymbol{\phi}_0^{-1/2} \boldsymbol{\phi}_1 \boldsymbol{\phi}_0^{-1/2}$, by means of which the variance of the test statistic under hypothesis H_1 can be written

$$\text{Var}_1 \ U_n = \text{Tr } \boldsymbol{\phi}_0^{-1/2} \boldsymbol{\phi}_s \boldsymbol{\phi}_0^{-1/2} \boldsymbol{\phi}_0^{-1/2} \boldsymbol{\phi}_s \boldsymbol{\phi}_0^{-1/2} = \sum_{k=1}^{n} (\varepsilon_k - 1)^2,$$

and he showed that if this sum remains finite as n goes to infinity, the detection process is liable to error. Pitcher [Pit66] has proved that a sufficient condition for this nonsingularity is that the solution $h(t, s)$ of the integral equation (11-36) exist and be continuous in t and s. Further treatments can be found in papers by Hájek [Háj62] and Kadota [Kad64], [Kad65]. Slepian [Sle58b] presented some simple and illuminating examples of singular detection.

It is generally difficult to judge on the basis of the autocovariance functions of signal and noise whether these conditions are fulfilled. Matters are somewhat simpler when both signal and noise are segments of stationary random processes and their autocovariance functions depend on t and s only through $\tau = t - s$. Yaglom [Yag63] showed that when both have rational spectral densities $\Phi_s(\omega)$ and $\Phi_0(\omega)$, detection will be imperfect if and only if

$$\lim_{\omega \to \infty} \frac{\Phi_1(\omega)}{\Phi_0(\omega)} = 1, \qquad \Phi_1(\omega) = \Phi_0(\omega) + \Phi_s(\omega). \qquad (11\text{-}62)$$

This will always be the case when the signal has finite power and the noise contains a component that is white.

We can understand the condition in (11-62) by observing first that the solution of (11-36) will contain a term that is the inverse Fourier transform of $H(\omega)$ in (11-44), along with some delta functions and their derivatives to take care of the end points of the interval $(0, T)$ and, possibly, some exponential functions. When T is large, that term will contribute to $\mathrm{Var}_1\, U_n$ approximately

$$T \int_{-\infty}^{\infty} \left[\frac{\Phi_s(\omega)}{\Phi_0(\omega)} \right]^2 \frac{d\omega}{2\pi}.$$

Middleton [Mid61] showed that the remaining terms are always finite. If the condition (11-62) is satisfied, $\mathrm{Var}_1\, U_n$ will indeed be finite when the spectral densities are rational functions of ω, and the detection will entail a probability of error that vanishes only when the strength of the signal itself grows beyond all bounds.

11.2 THE PERFORMANCE OF THE RECEIVER

11.2.1 The Moment-generating Function of the Test Statistic

Both the optimum detector and the threshold detector of a stochastic signal have the same structure. They determine the value of a quadratic functional

$$U = \tfrac{1}{2} \int_0^T \int_0^T V^*(r) h(r, s) V(s)\, dr\, ds \tag{11-63}$$

and compare it with a decision level U_0. Here $V(t)$ is the complex envelope of the input to the receiver, and $h(t, s)$ is a kernel whose form depends on which detector is adopted. For the optimum detector, $h(t, s)$ is the solution of the integral equation (11-36); for the threshold detector it is the solution of (11-48).

The probability of detecting a particular signal is the probability that U exceeds U_0 when that signal is present. Because what signal might be present is unknown when the signals are stochastic, the only meaningful way to measure the effectiveness of the receiver is to average that probability over all signals of the ensemble.

It is useful to know the average probability of detection not only for the ensemble of signals for which the detector was designed, but also for ensembles of signals of arbitrary average energy. We therefore introduce the hypothesis H_η that a signal is present and that it was drawn from an ensemble in which the autocovariance function is $\eta\phi_s(t, s)$, and we shall attempt to calculate the probability density function $p_\eta(U)$ of the test statistic under that hypothesis. The complex autocovariance function of the process $\mathrm{Re}\, V(t) \exp i\Omega t$ under H_η is

$$\tfrac{1}{2} E[V(t_1) V^*(t_2)| H_\eta] = \phi_\eta(t_1, t_2),$$
$$\phi_\eta(t_1, t_2) = \phi_0(t_1, t_2) + \eta\phi_s(t_1, t_2). \tag{11-64}$$

The probability of detection is

$$Q_d(\eta) = \int_{U_0}^{\infty} p_\eta(U)\, dU$$

for signals of an arbitrary average energy

$$E_\eta = \eta \int_0^T \phi_s(t, t) \, dt = \eta E; \qquad (11\text{-}65)$$

for signals of the designed strength E it is $Q_{ds} = Q_d(1)$. The false-alarm probability is $Q_0 = Q_d(0)$.

The problem of finding the probability density function of a quadratic functional of a random process was first addressed when Kac and Siegert analyzed the filtered output of a quadratic rectifier whose input is Gaussian noise [Kac47]. Let the input to the rectifier be Re $V(t) \exp i\Omega t$. Its filtered output at time t is

$$U(t) = \int_{-\infty}^t k(t - r)|V(r)|^2 \, dr, \qquad (11\text{-}66)$$

where $k(\tau)$ is the impulse response of the filter following the rectifier. If we put

$$h(r, s) = k(t - r)\delta(r - s)$$

into (11-63) and change the limits of integration from 0 and T to $-\infty$ and t, we obtain (11-66). Our subsequent formulas can similarly be modified to apply to the quadratic rectifier by changing the limits of integration in this way.

Among other treatments of the distribution of a quadratic functional of Gaussian noise we cite the work of Siegert [Sie57], Slepian [Sle58a], Grenander, Pollak, and Slepian [Gre59], Turin [Tur60b], and Middleton [Mid60a, Ch. 17]. The usual procedure is to derive first the characteristic function of U and then, when possible, to make a Fourier transformation to obtain the probability density function. This is the course we too shall follow, except that we prefer to work with the equivalent moment-generating function. As we have seen in Chapter 5, it enables us to calculate the cumulative distribution of a statistic such as U by Edgeworth's series or by saddlepoint integration.

We replace the statistic in (11-63) with its matrix form

$$U = \tfrac{1}{2}\mathbf{V}^+\mathbf{H}\mathbf{V}, \qquad (11\text{-}67)$$

where as in Sec. 11.1.2 we represent the complex envelope $V(t)$ by a column vector $\mathbf{V}$ of its samples and $V^*(t)$ by a row vector $\mathbf{V}^+$ of the complex conjugates of those samples. We begin with a restriction to a finite number n of samples, whose joint probability density function under hypothesis H_η is

$$p_\eta(\mathbf{V}^{(n)}) = (2\pi)^{-n}[\det \phi_\eta^{(n)}]^{-1} \exp(-\tfrac{1}{2}\mathbf{V}^{(n)+}\phi_\eta^{(n)-1}\mathbf{V}^{(n)}),$$

where $\phi_\eta^{(n)}$ is the $n \times n$ covariance matrix of the samples $\mathbf{V}^{(n)}$ under hypothesis H_η:

$$(\phi_\eta^{(n)})_{ij} = \tfrac{1}{2}E(V_i V_j^* \mid H_\eta).$$

Denoting by U_n the restriction of (11-67) to the same n samples $\mathbf{V}^{(n)}$, with $\mathbf{H}^{(n)}$ the corresponding $n \times n$ block of the matrix $\mathbf{H}$, we express its moment-generating function under hypothesis H_η as

$$E[\exp(-zU_n)\mid H_\eta] = (2\pi)^{-n}[\det \phi_\eta^{(n)}]^{-1}$$
$$\cdot \int_{R_n}\!\int \exp[-\tfrac{1}{2}\mathbf{V}^{(n)+}\phi_\eta^{(n)-1}\mathbf{V}^{(n)} - \tfrac{1}{2}z\mathbf{V}^{(n)+}\mathbf{H}^{(n)}\mathbf{V}^{(n)}] \, d^n V_x \, d^n V_y,$$

where $d^n V_x d^n V_y$ is the volume element in the space of the real and imaginary parts of the n samples $\mathbf{V}^{(n)}$. By (B-10) this integral equals

$$E[\exp(-zU_n)|\,H_\eta] = [\det \boldsymbol{\phi}_\eta^{(n)}]^{-1} \det[\boldsymbol{\phi}_\eta^{(n)-1} + z\mathbf{H}^{(n)}]^{-1}$$
$$= [\det(\mathbf{I}_n + z\boldsymbol{\phi}_\eta^{(n)}\mathbf{H}^{(n)})]^{-1},$$

where $\mathbf{I}_n$ is the $n \times n$ identity matrix. Passing now to the limit $n \to \infty$, we find for the moment-generating function of the quadratic functional $U = \frac{1}{2}\mathbf{V}^+\mathbf{H}\mathbf{V}$

$$h_\eta(z) = E[e^{-Uz}|\,H_\eta] = [\det(\mathbf{I} + z\boldsymbol{\phi}_\eta\mathbf{H})]^{-1}$$
$$= \exp[-\operatorname{Tr}\ln(\mathbf{I} + z\boldsymbol{\phi}_\eta\mathbf{H})] \tag{11-68}$$

as in (11-20). By the same kind of analysis as in Sec. 11.1.2, we can express this moment-generating function as

$$h_\eta(z) = \exp\left[-\int_0^z du \int_0^T L_\eta(t,\, t;\, u)\, dt\right], \tag{11-69}$$

where the function $L_\eta(t,\, s;\, u)$ corresponds to the matrix

$$\mathbf{L}_\eta = (\mathbf{I} + u\boldsymbol{\phi}_\eta\mathbf{H})^{-1}\boldsymbol{\phi}_\eta\mathbf{H}$$

and is the solution of the integral equation

$$\int_0^T \phi_\eta(t,\, r)h(r,\, s)\, dr = L_\eta(t,\, s;\, u) + u\int_0^T\int_0^T \phi_\eta(t,\, r)h(r,\, v)L_\eta(v,\, s;\, u)\, dr\, dv, \tag{11-70}$$
$$0 \le (t,\, s) \le T,$$

as in (11-24). It would be necessary to solve this integral equation for all values of u in order, through (11-69), to determine the moment-generating function $h_\eta(z)$ of U for all values of z, and this must then be inverted to obtain the probability density function

$$p_\eta(U) = \int_{c-i\infty}^{c+i\infty} h_\eta(z)\, e^{Uz}\, \frac{dz}{2\pi i}, \tag{11-71}$$

where the real quantity c lies to the right of all singularities of $h_\eta(z)$. To carry out this program would be a formidable task.

11.2.2 Detectability in White Noise: The Residue Series

When the noise is white and the quadratic functional U is the optimum statistic for detecting the Gaussian stochastic signal

$$s(t) = \operatorname{Re} S(t)\, e^{i\Omega t}$$

in such noise, the problem of calculating the moment-generating function simplifies to the point where it is feasible to use $h_\eta(z)$ to compute false-alarm and detection probabilities. With white noise the matrix $\boldsymbol{\phi}_0 = N\mathbf{I}$ is diagonal for any set of orthonormal sampling functions $f_k(t)$. The matrix $\mathbf{H}$ figuring in the optimum detection statistic is now, by (11-33),

$$\mathbf{H} = N^{-1}(N\mathbf{I} + \boldsymbol{\phi}_s)^{-1}\boldsymbol{\phi}_s, \tag{11-72}$$

and with
$$\boldsymbol{\phi}_\eta = N\mathbf{I} + \eta\boldsymbol{\phi}_s$$
by (11-64), we find for the moment-generating function in (11-68)

$$h_\eta(z) = \{\det[\mathbf{I} + zN^{-1}(N\mathbf{I} + \eta\boldsymbol{\phi}_s)(N\mathbf{I} + \boldsymbol{\phi}_s)^{-1}\boldsymbol{\phi}_s]\}^{-1}$$

$$= \frac{\det(\mathbf{I} + N^{-1}\boldsymbol{\phi}_s)}{\det[\mathbf{I} + (1+z)N^{-1}\boldsymbol{\phi}_s + z\eta N^{-2}\boldsymbol{\phi}_s^2]}. \tag{11-73}$$

We now seek a residue expansion of this function in order to evaluate the probability density function in (11-71).

Let λ_k be the kth eigenvalue of the matrix $N^{-1}\boldsymbol{\phi}_s$, that is, of the integral equation

$$\lambda_k f_k(t) = \frac{1}{N}\int_0^T \phi_s(t, u)f_k(u)\, du, \qquad 0 \le t \le T. \tag{11-74}$$

Then the denominator of (11-73) can be written as

$$G(z) = \prod_{k=1}^\infty [1 + (1+z)\lambda_k + z\eta\lambda_k^2],$$

and the poles of the moment-generating function $h_\eta(z)$ lie at the points where the factors of $G(z)$ vanish, that is, at

$$z = z_k = -\frac{1 + \lambda_k}{\lambda_k(1 + \eta\lambda_k)}. \tag{11-75}$$

In the neighborhood of its kth zero z_k, the denominator of (11-73) can be written, with some algebra, as

$$G(z) \approx \lambda_k(1 + \eta\lambda_k)\prod_{m \ne k}[1 + \lambda_m + z_k\lambda_m(1 + \eta\lambda_m)](z - z_k)$$

$$= \lambda_k(1 + \eta\lambda_k)\prod_{m \ne k}\left(1 - \frac{\lambda_m}{\lambda_k}\right)\left[1 + \frac{\eta\lambda_m(1 + \lambda_k)}{1 + \eta\lambda_k}\right](z - z_k).$$

As in Sec. 11.1.2 we introduce the Fredholm determinant

$$D(z) = \prod_{k=1}^\infty (1 + \lambda_k z) = \det(\mathbf{I} + zN^{-1}\boldsymbol{\phi}_s), \tag{11-76}$$

in terms of which $G(z)$ can be written

$$G(z) \approx R_k^{-1}\frac{\lambda_k(1 + \eta\lambda_k)^2 D(\alpha_k)}{1 + 2\eta\lambda_k + \eta\lambda_k^2}(z - z_k) = G_k(z - z_k), \tag{11-77}$$

where

$$\alpha_k = \frac{\eta(1 + \lambda_k)}{1 + \eta\lambda_k} \tag{11-78}$$

and

$$R_k = \prod_{m \ne k}\left(1 - \frac{\lambda_m}{\lambda_k}\right)^{-1} = \frac{\lambda_k}{D'(-1/\lambda_k)}, \tag{11-79}$$

the prime indicating differentiation with respect to the argument of the function. The numerator of (11-73) equals $D(1)$. Thus (11-73) can be written

$$h_\eta(z) = D(1) \sum_{k=1}^{\infty} \frac{1}{G_k(z - z_k)}.$$

If we now substitute this into the contour integral (11-71) and complete the contour around the left half-plane, we find for the probability density function of the statistic U under hypothesis H_η

$$p_\eta(U) = \sum_k \frac{R_k(1 + 2\eta\lambda_k + \eta\lambda_k^2)D(1)}{\lambda_k(1 + \eta\lambda_k)^2 D(\alpha_k)} \exp Uz_k,$$

when we apply the residue theorem and use (11-77). The probability $Q_d(\eta)$ of detecting a stochastic signal of relative average strength η is then

$$Q_d(\eta) = \int_{U_0}^{\infty} p_\eta(U) \, dU = \sum_k \frac{R_k(1 + 2\eta\lambda_k + \eta\lambda_k^2)D(1)}{(1 + \lambda_k)(1 + \eta\lambda_k)D(\alpha_k)} \exp U_0 z_k, \qquad (11\text{-}80)$$

with z_k given by (11-75), α_k by (11-78), and R_k by (11-79).

The false-alarm probability is obtained by setting $\eta = 0$,

$$Q_0 = \sum_k \frac{R_k D(1)}{1 + \lambda_k} \exp\left[-\frac{1 + \lambda_k}{\lambda_k} U_0\right], \qquad (11\text{-}81)$$

and the probability of detecting a signal of design strength ($\eta = 1$) is

$$Q_{ds} = Q_d(1) = \sum_k R_k \exp\left(-\frac{U_0}{\lambda_k}\right). \qquad (11\text{-}82)$$

These residue series can be evaluated once the eigenvalues λ_k of (11-74) have been computed, provided that not too many of them have a significant magnitude. As we shall see, when the product of the signal bandwidth W and the observation time T is large, the number of significant eigenvalues is on the order of WT, the eigenvalues lie close together, and the factors R_k, which alternate in sign, become difficult to calculate with sufficient accuracy. These series then become inconvenient for computation and unreliable, and another method must be sought.

When the product of the observation time T by the bandwidth of the stochastic signal, assumed stationary, is much less than 1, only a single eigenvalue λ_1 is significant. The associated eigenfunction $f_1(t)$ is then approximately constant:

$$f_1(t) \approx T^{-1/2}, \qquad 0 \le t \le T.$$

From (11-19) with $m(t, s) = N^{-1}\phi_s(t, s)$ we see that

$$\lambda_1 \approx \frac{1}{N} \int_0^T \phi_s(t, t) \, dt = \frac{E}{N} = S_0,$$

where as in (11-65) $E = E_1$ is the average energy of the signal for which the detector is designed; S_0 is the design energy-to-noise ratio. Now both the optimum and the threshold detectors effectively base their decisions on the squared magnitude $|V_1|^2$ of a single sample

$$V_1 = \frac{1}{\sqrt{T}} \int_0^T V(t)\, dt$$

of the complex envelope of the input; $|V_1|^2$ has an exponential distribution under both hypotheses. The false-alarm and detection probabilities are given by (11-80) and (11-81) as

$$Q_d(\eta) = \exp\left[-\frac{(1 + \lambda_1)U_0}{\lambda_1(1 + \eta\lambda_1)}\right],$$

$$Q_0 = \exp\left[-\frac{(1 + \lambda_1)U_0}{\lambda_1}\right],$$

so that

$$Q_d(\eta) \approx Q_0^{1/(1+\eta S_0)}, \qquad S_0 = \frac{E}{N}. \tag{11-83}$$

11.2.3 The Threshold Detector in White Noise

The threshold detection statistic, according to (11-48) and (11-49), is

$$U_\theta = \tfrac{1}{2} \mathbf{V}^+ \boldsymbol{\phi}_0^{-1} \boldsymbol{\phi}_s \boldsymbol{\phi}_0^{-1} \mathbf{V}$$

in the matrix representation, and when the noise is white, $\boldsymbol{\phi}_0 = N\mathbf{I}$, this becomes

$$U_\theta = \frac{\mathbf{V}^+ \boldsymbol{\phi}_s \mathbf{V}}{2N^2} = \frac{1}{2N^2} \int_0^T \int_0^T V^*(t)\phi_s(t, u)V(u)\, dt\, du. \tag{11-84}$$

This statistic can be realized by the methods described in Sec. 11.1.4, with $h(t, s)$ as used there replaced by $\phi_s(t, s)/N^2$.

Comparison with (11-67) shows that the matrix $\mathbf{H}$ there is now

$$\mathbf{H} = N^{-2}\boldsymbol{\phi}_s,$$

and by (11-68) the moment-generating function of the threshold statistic is

$$h_\eta^\theta(z) = [\det(\mathbf{I} + zN^{-2}\boldsymbol{\phi}_\eta\boldsymbol{\phi}_s)]^{-1}$$
$$= \prod_k [1 + z\lambda_k(1 + \eta\lambda_k)]^{-1}$$

under hypothesis H_η that a signal with complex autocovariance function $\eta\phi_s(t, s)$ is present.

By the same kind of analysis as in Sec. 11.2.2, one finds that the probability of detection attained by the threshold detector is

$$Q_d^\theta(\eta) = \Pr(U_\theta \geq U_0' \mid H_\eta)$$
$$= \sum_k \frac{R_k(1 + 2\eta\lambda_k)}{(1 + \eta\lambda_k)D(\alpha_k')} \exp(z_k' U_0'), \tag{11-85}$$

$$z_k' = -\frac{1}{\lambda_k(1 + \eta\lambda_k)}, \qquad \alpha_k' = \frac{\eta}{1 + \eta\lambda_k}.$$

The false-alarm probability, which determines the decision level U_0', is then

$$Q_0^\theta = \sum_k R_k \exp\left(-\frac{U_0'}{\lambda_k}\right). \tag{11-86}$$

The performance of the threshold detector is independent of the input energy-to-noise ratio adopted in specifying it by (11-84). The same difficulties attend the computation of its false-alarm and detection probabilities as were noted at the end of Sec. 11.2.2.

The optimum statistic U for detection in white Gaussian noise can be written as

$$U = \frac{1}{2N} \sum_k \frac{\lambda_k |V_k|^2}{1 + \lambda_k}, \tag{11-87}$$

and the threshold statistic U_θ as

$$U_\theta = \frac{1}{2N} \sum_k \lambda_k |V_k|^2. \tag{11-88}$$

We expect the threshold statistic to be a good approximation to the optimum one, therefore, when all the eigenvalues λ_k are much less than 1. A crude criterion for this can be derived by the following reasoning, in which it is assumed that the signal is stationary and has an autocovariance function $\phi_s(t - s)$.

From (11-19) applied to $m(t, s) = N^{-1}\phi_s(t - s)$, we find

$$\sum_k \lambda_k = \frac{1}{N} \int_0^T \phi_s(0) \, dt = \frac{T\phi_s(0)}{N} = \frac{E}{N}, \tag{11-89}$$

where E is the average energy in the signal. Similarly,

$$\sum_k \lambda_k^2 = \frac{1}{N^2} \int_0^T \phi_s^{(2)}(t, t) \, dt = \frac{1}{N^2} \int_0^T \int_0^T |\phi_s(t - u)|^2 \, dt \, du,$$

where $\phi_s^{(2)}(t, u)$ is the iterated kernel as in (11-17). This can be written

$$\sum_k \lambda_k^2 = \frac{1}{N^2} \int_{-T}^T (T - |u|)|\phi_s(u)|^2 \, du. \tag{11-90}$$

If now about M of the eigenvalues are nearly equal, $\lambda_k \approx \bar{\lambda}$, and the rest are negligible,

$$M\bar{\lambda} \approx \frac{E}{N}, \qquad M\bar{\lambda}^2 \approx \frac{1}{N^2} \int_{-T}^T (T - |u|)|\phi_s(u)|^2 \, du;$$

and we find for the approximate number of significant eigenvalues

$$M \approx \frac{T^2|\phi_s(0)|^2}{\int_{-T}^T (T - |\tau|)|\phi_s(\tau)|^2 \, d\tau}, \tag{11-91}$$

and

$$\bar{\lambda} = \frac{E}{MN}. \tag{11-92}$$

When the width of the autocovariance function $\phi_s(\tau)$ is much greater than T, we find $M \approx 1$ and $\lambda \approx \lambda_1 \approx E/N$ as before. When, on the other hand, its width is much less than T,

$$M \approx \frac{T|\phi_s(0)|^2}{\int_{-\infty}^\infty |\phi_s(\tau)|^2 \, d\tau} = WT,$$

where W is a signal bandwidth defined by

$$W = \frac{|\phi_s(0)|^2}{\int_{-\infty}^{\infty} |\phi_s(\tau)|^2 \, d\tau} = \frac{\left[\int_{-\infty}^{\infty} \Phi_s(\omega) \frac{d\omega}{2\pi} \right]^2}{\int_{-\infty}^{\infty} [\Phi_s(\omega)]^2 \frac{d\omega}{2\pi}}. \tag{11-93}$$

The threshold approximation can be expected to be valid, therefore, when $\bar{\lambda} = E/(NWT) \ll 1$.

By using the definition in (4-66), the reader can easily show that the deflection of the threshold statistic is

$$\Delta_\theta^2 = \sum_k \lambda_k^2 \approx \frac{E^2}{N^2 WT} = WT \left[\frac{E}{NWT} \right]^2.$$

Thus when $WT \gg 1$, the threshold detector may attain a respectable probability of detection while nevertheless the criterion $E/(NWT) \ll 1$ for its validity is satisfied.

11.2.4 Application of Saddlepoint Integration

When the product WT of the signal bandwidth W and the observation time T is large, the probability of detection by either the optimum or the threshold statistic is most accurately computed by the method of saddlepoint integration described in Sec. 5.2. According to (5-19) and (5-20),

$$1 - Q_d(\eta) = \int_{C_-} z^{-1} h_\eta(z) \exp U_0 z \, \frac{dz}{2\pi i}, \tag{11-94}$$

$$Q_d(\eta) = -\int_{C_+} z^{-1} h_\eta(z) \exp U_0 z \, \frac{dz}{2\pi i}, \tag{11-95}$$

where C_- is a path that passes through the saddlepoint z_0^- of the integrand lying to the right of the origin and C_+ is a path through the saddlepoint z_0^+ to the left of the origin. One uses (11-94) for detection probabilities greater than about 0.5 and (11-95) for those less than about 0.5.

In order to put the moment-generating function $h_\eta(z)$ from (11-73) into a form in which these integrals can be evaluated numerically, we introduce the functions

$$\alpha(z) = \tfrac{1}{2} \{ 1 + z + [(1 + z)^2 - 4\eta z]^{1/2} \},$$
$$\beta(z) = \tfrac{1}{2} \{ 1 + z - [(1 + z)^2 - 4\eta z]^{1/2} \}.$$

For these

$$\alpha(z) + \beta(z) = 1 + z, \qquad \alpha(z)\beta(z) = \eta z,$$

and from (11-73) and (11-76) we find

$$\begin{aligned}
h_\eta(z) &= D(1) \prod_k \frac{1}{1 + (1 + z)\lambda_k + \eta z \lambda_k^2} \\
&= D(1) \prod_k \frac{1}{[1 + \alpha(z)\lambda_k][1 + \beta(z)\lambda_k]} \\
&= \frac{D(1)}{D(\alpha(z))D(\beta(z))}.
\end{aligned} \tag{11-96}$$

If the Fredholm determinant is known in analytic form, therefore, we can use (11-96) in (11-94) and (11-95) and evaluate the integrals numerically as described in Sec. 5.2. It has been found efficient to integrate along a parabolic path as in (5-36) [Hel83].

For the threshold detector one similarly defines

$$\alpha'(z) = \tfrac{1}{2}[z + (z^2 - 4\eta z)^{1/2}], \tag{11-97}$$

$$\beta'(z) = \tfrac{1}{2}[z - (z^2 - 4\eta z)^{1/2}], \tag{11-98}$$

whereupon the moment-generating function $h_\eta^\theta(z)$ of the threshold statistic U_θ is, by (11-84),

$$h_\eta^\theta(z) = \frac{1}{D(\alpha'(z))D(\beta'(z))}. \tag{11-99}$$

When the eigenvalues λ_k of (11-74) are known, one can approximate the Fredholm determinant of (11-76) by taking a finite number of factors and stopping the multiplication when the eigenvalue λ_k is sufficiently small. Otherwise one must seek an alternative method of computing the Fredholm determinant appearing in (11-96) and (11-99) at complex values of z along the path of integration.

If we identify $N^{-1}\phi_s(t, s)$ with the function $m(t, s)$ that figures in Sec. 11.1.2, the eigenvalues μ_k appearing there are identical with the eigenvalues λ_k in (11-74), and the function $P(t, s; u)$ of (11-24) becomes

$$P(t, s; u) = Nh(t, s; u)$$

in terms of the solution of (11-53), whereupon by (11-26) the Fredholm determinant is given by

$$D(z) = \exp\left[N \int_0^z du \int_0^T h(t, t; u)\, dt \right]. \tag{11-100}$$

If one can solve (11-53) analytically, therefore, this equation permits calculating the Fredholm determinant and thence the moment-generating function of the detection statistic. A simpler formula will be derived in Sec. 11.3.

11.2.5 The Toeplitz Approximation

When the stochastic signal is a segment of a stationary process observed in the presence of white noise during an interval $(0, T)$, (11-53) becomes

$$Nh(r, t; u) + u \int_0^T \phi_s(r - s)h(s, t; u)\, ds = \frac{1}{N}\phi_s(r - t), \tag{11-101}$$
$$0 \leq (r, t) \leq T,$$

where $\phi_s(\tau)$ is the complex autocovariance function of the signal. If the product WT of the observation time T and a suitable measure W of the bandwidth of the signal is large, we can approximate (11-101) by extending the limits of integration to $(-\infty, \infty)$, whereupon it is apparent that the function $h(r, t; u)$ depends on r and t only through their difference:

$$h(t, r; u) \approx h_\infty(r - t; u).$$

We can then solve (11-101) approximately by Fourier transformation, much as in (2-77) through (2-79), and setting

$$h_\infty(r - t; u) = \int_{-\infty}^{\infty} H_\infty(\omega; u) \, e^{i\omega(r-t)} \frac{d\omega}{2\pi},$$

we find

$$NH_\infty(\omega; u) + u\Phi_s(\omega)H_\infty(\omega; u) = N^{-1}\Phi_s(\omega)$$

in terms of the spectral density $\Phi_s(\omega)$ of the complex envelope of the signal, that is, the Fourier transform of $\phi_s(\tau)$. Hence

$$H_\infty(\omega; u) = \frac{\Phi_s(\omega)}{N[N + u\Phi_s(\omega)]},$$

and by (11-100) the Fredholm determinant is approximately

$$\begin{aligned}
D(z) \approx D_\infty(z) &= \exp\left[NT \int_0^z h_\infty(0; u) \, du \right] \\
&= \exp\left[T \int_0^z du \int_{-\infty}^{\infty} \frac{\Phi_s(\omega)}{N + u\Phi_s(\omega)} \frac{d\omega}{2\pi} \right] \qquad (11\text{-}102) \\
&= \exp\left[T \int_{-\infty}^{\infty} \ln[1 + zN^{-1}\Phi_s(\omega)] \frac{d\omega}{2\pi} \right].
\end{aligned}$$

If in this limit $WT \gg 1$ one samples all temporal functions at instants uniformly spaced over $(0, T)$, the resulting matrices ϕ_s and $\mathbf{H}$ become Toeplitz matrices in the sense defined in connection with (10-27); their elements depend only on the differences of their indices, and each row is identical to the row above it, but shifted one place to the right. For this reason (11-102) is called the *Toeplitz approximation* [Gre59].

11.2.6 Rational Spectral Densities

The complex autocovariance function of the stationary signal is subject to the Hermitian condition $\phi_s^*(\tau) = \phi_s(-\tau)$. If its spectral density is a real rational function of the frequency ω, we can apply the methods of Appendix A to evaluating the Toeplitz approximation. If $\Phi_s(\omega)$ has $2n$ simple poles, $\phi_s(\tau)$ must have the form (A-3),

$$\begin{aligned}
\phi_s(\tau) &= \sum_{k=1}^{n} f_k \exp(-\mu_k^*\tau), \qquad \tau \geq 0, \\
&= \sum_{k=1}^{n} f_k^* \exp \mu_k\tau, \qquad \tau < 0,
\end{aligned} \qquad (11\text{-}103)$$

for n complex constants μ_k. The term $f_0\delta(\tau)$ in (A-3) is missing because the signal must have finite average power.

The associated spectral density is as given in (A-4),

$$\begin{aligned}
\Phi_s(\omega) &= \sum_{k=1}^{n} \left[\frac{f_k}{\mu_k^* + i\omega} + \frac{f_k^*}{\mu_k - i\omega} \right] \\
&= 2\sum_{k=1}^{n} \frac{f_{kx}\mu_{kx} + f_{ky}(\omega - \mu_{ky})}{(\omega - \mu_{ky})^2 + \mu_{kx}^2}, \qquad \mu_k = \mu_{kx} + i\mu_{ky},
\end{aligned}$$

for real values of ω, and when this is cleared of fractions, it becomes

$$\Phi_s(\omega) = C \prod_{j=1}^{m} |i\omega - h_j|^2 \prod_{k=1}^{n} |i\omega - \mu_k|^{-2}$$

$$= C \frac{\prod_{j=1}^{m} [(\omega - h_{jy})^2 + h_{jx}^2]}{\prod_{k=1}^{n} [(\omega - \mu_{ky})^2 + \mu_{kx}^2]}, \qquad h_j = h_{jx} + ih_{jy}, \tag{11-104}$$

with C a positive constant, and with $m < n$ in order that the signal process have finite average power. As in Appendix A we take the μ_k, $1 \leq k \leq n$, to have positive real parts. This spectral density has simple poles at the n points $-i\mu_k$ in the lower half of the ω-plane and at the n points $i\mu_k^*$ in the upper half.

The argument of the logarithm in (11-102) is then also a rational function of frequency and can be written

$$1 + zN^{-1}\Phi_s(\omega) = \prod_{j=1}^{2n} \frac{i\omega - \beta_j}{i\omega - \mu_j}, \tag{11-105}$$

where the $\beta_j(z)$ are the $2n$ roots of the algebraic equation obtained by clearing the equation

$$1 + zN^{-1}\Phi_s(-ip) = 0 \tag{11-106}$$

of fractions; we have put $\mu_{k+n} = -\mu_k^*$, $1 \leq k \leq n$.

The function in (11-105) can vanish for real values of ω only for z real and negative. As the point z moves from the origin along any path that does not cross the negative real axis in the z-plane, therefore, the trajectories of the roots $\beta_j(z)$ cannot cross the imaginary axis. Provided that z does not lie on the negative real axis, n of the β_j will have positive real parts and n will have negative real parts. We index the roots so that $\beta_1, \ldots, \beta_n, \mu_1, \ldots, \mu_n$ have positive real parts and so that $\beta_{n+1}, \ldots, \beta_{2n}, \mu_{n+1}, \ldots, \mu_{2n}$ have negative real parts.

With this indexing convention, as can be shown from [Gra65, eq. 4.222(1), p. 525], the result of substituting (11-105) into (11-102) and integrating is the Toeplitz approximation

$$D(z) \approx D_\infty(z) = \exp\left[T \sum_{j=1}^{n} (\beta_j - \mu_j)\right]. \tag{11-107}$$

An example of this result will appear in Appendix H.

One uses $D_\infty(z)$ to determine the moment-generating functions $h_\eta(z)$ and $h_\eta^\theta(z)$ of the optimum and threshold detectors for $WT \gg 1$ by (11-96) and (11-99), respectively. It will in general still be necessary to integrate (11-94) and (11-95) numerically along a suitable path, as described in Sec. 5.2, in order to calculate even the approximate false-alarm and detection probabilities by this method.

11.2.7 Detectability of Signals with Lorentz or Rectangular Spectral Densities

Two types of signal spectral densities that manifest opposite extremes of frequency dependence are the Lorentz,

$$\Phi_s(\omega) = \frac{2\mu P_s}{\omega^2 + \mu^2}$$

as in (11-47), and the rectangular,

$$\Phi_s(\omega) = \begin{cases} \dfrac{P_s}{W}, & -\pi W \leq \omega \leq \pi W, \\ 0, & |\omega| > \pi W. \end{cases} \qquad (11\text{-}108)$$

Here $P_s = E/T$ is the average power in the signal.

The Lorentz spectral density has very long tails, and the signal has a substantial fraction of its power at large deviations from the carrier frequency; the rectangular cuts off sharply at each edge of its spectral band. We shall compare the detectabilities of stochastic signals with these spectral densities when either the optimum or the threshold detector is utilized. It will be assumed that both have equal effective bandwidths W as defined by (11-93). For the Lorentz spectral density, the auto-covariance function is given in (11-46), and from (11-93) we find that its effective bandwidth is $W = \mu$. For the rectangular spectral density, substitution of (11-108) into (11-93) shows that W is its effective bandwidth in that same sense.

Figures 11-2 and 11-3 refer to signals having a Lorentz spectral density. In Fig. 11-2 we have plotted the probability $Q_d(1) = Q_{ds}$ that the optimum detector correctly decides that a signal of the standard strength for which it was designed is present versus the square root D of the input energy-to-noise ratio E/N for that standard signal. The curves are indexed with the value of $m = \mu T$, and each point on a curve represents a different detector. In Fig. 11-3 we exhibit as a function of $m = \mu T$ the average energy-to-noise ratio E/N in decibels required to attain three values of the probability Q_{ds} of detecting the standard signal. The solid curves refer to the optimum detector, the dashed curves to the threshold detector of Sec. 11.1.6.

The computations needed to produce these two figures were rather lengthy, and the details have been relegated to Appendix H. The residue series in (11-80) and (11-85) were used for values of $m = \mu T$ less than about 7. Appendix H provides a closed expression for the residue factor R_k in (11-79), circumventing the computation of infinite products. For $7 < m < 12$ saddlepoint integration as in Sec. 11.2.4 was employed. For $m > 12$, as shown in Appendix H, one can apply the Toeplitz approximation of Sec. 11.2.5 and obtain approximate, but accurate false-alarm and detection probabilities in closed form.

For $m \ll 1$ and for $m \gg 1$ the performance of the threshold detector approaches that of the optimum detector. The closer the probability Q_{ds} of detection is to 1, the more slowly that approach takes place and the greater the disparity between the two detectors.

When the stochastic signal has the rectangular spectral density in (11-108), its complex autocovariance function is

$$\phi_s(\tau) = P_s \frac{\sin \pi W \tau}{\pi W \tau}.$$

We can take the observation interval as $-\frac{1}{2}T < t < \frac{1}{2}T$ without changing the eigenvalues λ_k of the integral equation (11-74), which can now be written as

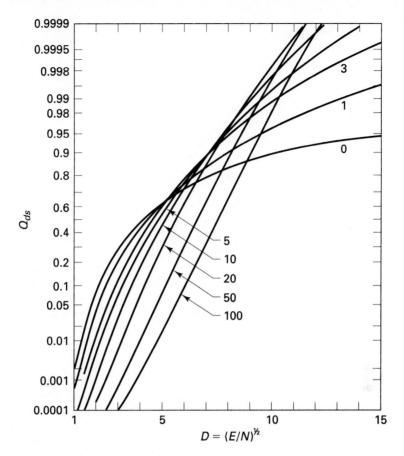

Figure 11-2. Probability Q_{ds} of detection of a Gaussian stochastic signal with a Lorentz spectral density (11-47) versus the square root D of the average energy-to-noise ratio E/N. Curves are indexed with the value of μT; $Q_0 = 10^{-6}$.

$$\lambda_k f_k(t) = \frac{P_s}{NW} \int_{-T/2}^{T/2} \frac{\sin \pi W(t-u)}{\pi(t-u)} f_k(u)\, du, \qquad -\tfrac{1}{2}T \le t \le \tfrac{1}{2}T,$$

and by changing the integration variable we can write this as

$$\lambda_k f_k(\tfrac{1}{2}Tx) = \frac{P_s}{NW} \int_{-1}^{1} \frac{\sin[\tfrac{1}{2}\pi WT(x-y)]}{\pi(x-y)} f_k(\tfrac{1}{2}Ty)\, dy, \qquad -1 \le x \le 1. \qquad (11\text{-}109)$$

Slepian and Sonnenblick [Sle65b] tabulated the eigenvalues $\lambda_k^{(s)}$ of the integral equation

$$\lambda_k^{(s)} F_k(x) = \int_{-1}^{1} \frac{\sin c(x-y)}{\pi(x-y)} F_k(y)\, dy, \qquad -1 \le x \le 1,$$

and comparing this with (11-109) we see that the eigenvalues we have been using are related to theirs by

$$\lambda_k = \frac{P_s}{NW} \lambda_k^{(s)} = \frac{E}{NWT} \lambda_k^{(s)} \qquad (11\text{-}110)$$

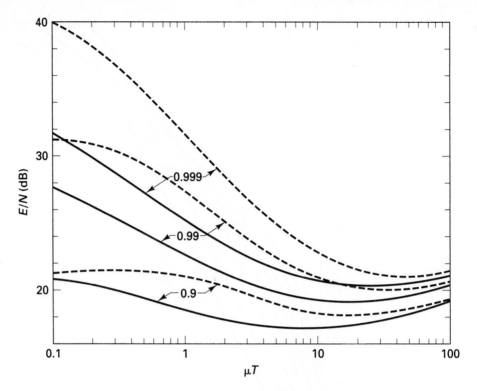

Figure 11-3. Lorentz spectral density: Energy-to-noise ratio (dB) required to attain a probability Q_{ds} of detection for the standard signal as a function of the time–bandwidth product $m = \mu T$; $Q_0 = 10^{-6}$. Solid curves: optimum detector; dashed curves: threshold detector. Curves are indexed with the values of Q_{ds}.

for $c = \frac{1}{2}\pi WT$. For a moderately large time–bandwidth product $WT \gg 1$, there are roughly WT nearly equal eigenvalues

$$\lambda_k \approx \bar{\lambda} = \frac{E}{NWT}, \qquad 0 \le k < WT,$$

and the rest are negligible [Sle65a].

The curves of the detection probability Q_{ds} versus the square root D of the energy-to-noise ratio E/N for the rectangular spectral density look much the same as those for the Lorentz spectral density plotted in Fig. 11-2. The differences between the performances of the optimum and threshold detectors are most clearly evident by comparing Fig. 11-3 with 11-4, which plots the energy-to-noise ratio E/N needed to attain detection probabilities $Q_{ds} = 0.9, 0.99,$ and 0.999 with both detectors when the spectral density is rectangular. The abscissa is the time–bandwidth product WT, which as we have seen corresponds to the parameter μT for the Lorentz spectral density. The disparity between the optimum and the threshold detectors is much smaller for the rectangular than for the Lorentz spectral density.

In calculating these detection probabilities and the false-alarm probabilities needed for setting the decision level U_0, we used the residue series (11-80) and (11-85) for values of WT less than about 7, computing the coefficients R_k by (11-79) car-

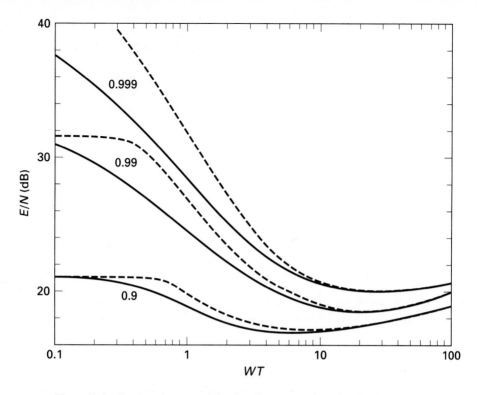

Figure 11-4. Rectangular spectral density: Energy-to-noise ratio (dB) required to attain probability Q_{ds} of detection for the standard signal as a function of the time–bandwidth product $m = WT$; $Q_0 = 10^{-6}$. Solid curves: optimum detector; dashed curves: threshold detector. Curves are indexed with the values of Q_{ds}.

ried to a finite number of factors. For larger values of WT, however, the largest eigenvalues lie too close together to permit accurate computation of the R_k's from the calculated eigenvalues λ_k. It was then necessary to determine Q_0 and Q_{ds} by numerical integration of (11-94) and (11-95). The Fredholm determinant $D(z)$ figuring in the moment-generating functions could be computed accurately enough along the parabolic path of integration by evaluating the product in (11-76) with the eigenvalues λ_k determined from the tabulated eigenvalues $\lambda_k^{(s)}$ as in (11-110).

When $WT \gg 1$, the near equality of the first WT eigenvalues λ_k and the insignificance of the rest permit us to approximate the Fredholm determinant by

$$D(z) = (1 + \bar{\lambda}z)^m, \qquad \bar{\lambda} = \frac{E}{Nm}, \qquad m = WT.$$

As one can see by substituting the rectangular spectral density from (11-108) into (11-102), this corresponds to the Toeplitz approximation.

For integral values of $m = WT$ the false-alarm and detection probabilities are given approximately by scaled cumulative chi-squared distributions. Indeed, our detection statistic is approximately

$$U = \tfrac{1}{2}\mathbf{V}^+\mathbf{H}\mathbf{V} \approx \frac{\bar{\lambda}}{2N(1 + \bar{\lambda})} \sum_{k=1}^{m} |V_k|^2$$

by (11-87), for the matrix $\boldsymbol{\phi}_s$ can now be approximated as an $m \times m$ diagonal matrix with diagonal elements $\bar{\lambda}$. Furthermore, when $m = WT \gg 1$ is so large that we can neglect all but the first m eigenvalues and take these as equal, the threshold statistic (11-88) becomes

$$U_\theta \approx \frac{\bar{\lambda}}{2N} \sum_{k=1}^{m} |V_k|^2,$$

and the optimum and threshold detection statistics U and U_θ differ only by an insignificant constant factor. We expect, therefore, that for $WT \gg 1$ the optimum and the threshold detectors will have nearly the same performance when the spectral density of the signals is rectangular.

The sum

$$\sum_{k=1}^{m} |V_k|^2$$

has a scaled chi-squared distribution under both hypotheses H_0 and H_1, and the false-alarm and detection probabilities for both detectors are approximately

$$Q_0 \approx q(x_0), \qquad Q_{ds} \approx q(x_1), \qquad x_1 = \frac{x_0}{1 + \bar{\lambda}},$$

with

$$q(x) = e^{-x} \sum_{r=0}^{m-1} \frac{x^r}{r!}, \tag{11-111}$$

as in (4-31). Here $x_0 = U_0(1 + \bar{\lambda})/\bar{\lambda}$. For nonintegral values of $m = WT$ it is necessary to compute Q_0 and Q_{ds} by numerical contour integration of (11-94) and (11-95), substituting the approximate forms

$$h_0(z) \approx \left[\frac{1 + \bar{\lambda}}{1 + \bar{\lambda}(1 + z)} \right]^m, \qquad h_1(z) \approx \frac{1}{(1 + \bar{\lambda}z)^m}. \tag{11-112}$$

Indeed, when $m = WT$ is very large, even though an integer, numerical contour integration is more accurate than summing the series in (11-111).

Mismatch between the design signal strength and the strength of the signal actually present only slightly affects the performance of the detector based on the statistic $U = \tfrac{1}{2}\mathbf{V}^+\mathbf{H}\mathbf{V}$ with $\mathbf{H}$ given by (11-72), provided that the design signal strength is equal to or greater than that of the signal actually present. The sensitivity of the performance to the choice of the standard signal strength is greater for the Lorentz than for the rectangular spectral density. This behavior is brought out in Figs. 11-5 and 11-6. In each we plot the input energy-to-noise ratio $S = \eta E/N$ required to attain a certain probability Q_d of detection as a function of the energy-to-noise ratio $S_0 = E/N$ for which the optimum detector is designed via (11-87). The minimum of each curve corresponds to $\eta = 1$. The value $S_0 = 0$ represents the threshold detector.

Crude approximations to the false-alarm and detection probabilities for signals with spectral densities other than the rectangular can be computed by using the approximate moment-generating functions in (11-112). The values of $\bar{\lambda}$ and $m = WT$

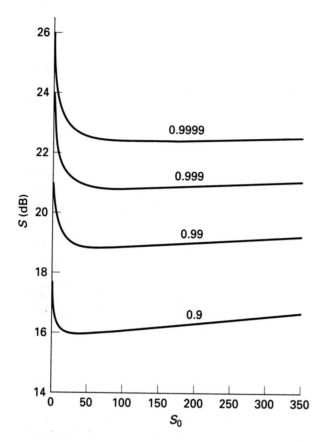

Figure 11-5. Input energy-to-noise ratio $S = \eta E/N$ to yield detection probability Q_d for $Q_0 = 10^{-4}$ versus the design energy-to-noise ratio $S_0 = E/N$; Lorentz spectral density, $\mu T = 14/\pi = 4.4563$. Curves are indexed with the value of Q_d. [Reprinted from C. W. Helstrom, "Evaluating the detectability of Gaussian stochastic signals by steepest-descent integration," *IEEE Transactions on Aerospace & Electronic Systems*, AES-19 (May 1983), 428–37, © 1983 IEEE.]

can be calculated from (11-91) and (11-92). We might call these "equal-eigenvalue" approximations. When m is an integer—and it usually suffices to take it as an integer—, it is simply a matter of summing the series in (11-111). The more the form of the spectral density departs from the rectangular, the less accurate this type of approximation is.

11.3 CAUSAL ESTIMATOR–CORRELATOR REPRESENTATION

11.3.1 Causal Estimator of the Stochastic Signal

In this section we shall derive a simpler way of calculating the Fredholm determinant $D(z)$, defined as in (11-76). As a useful by-product of this analysis we shall formulate the optimum detector of a Gaussian stochastic signal in white noise in such a way as to bring out its resemblance to that for a deterministic narrowband signal. In Sec. 11.4 this detector will be described in terms of a representation of the signal as the output of a linear system driven by white noise, and from a state-space model of that system equations permitting the numerical computation of $D(z)$ and a fortiori of the false-alarm and detection probabilities can be deduced.

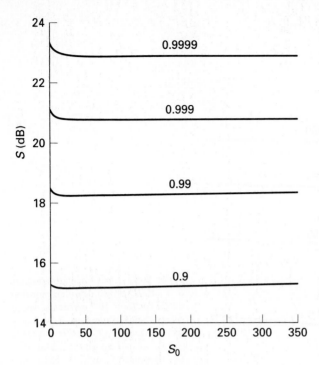

Figure 11-6. Input energy-to-noise ratio $S = \eta E/N$ to yield detection probability Q_d for $Q_0 = 10^{-4}$ versus the design energy-to-noise ratio $S_0 = E/N$; rectangular spectral density, $WT = 14/\pi = 4.4563$. Curves are indexed with the value of Q_d. [Reprinted from C. W. Helstrom, "Evaluating the detectability of Gaussian stochastic signals by steepest-descent integration," *IEEE Transactions on Aerospace & Electronic Systems*, AES-19 (May 1983), 428–37, © 1983 IEEE.]

Let us consider estimating the values of the complex envelope $S(t)$ of the signal during the subinterval $(0, \tau)$ within the observation interval $(0, T)$. The signal is known to be present in the midst of white noise of unilateral spectral density N. For the sake of future calculations, however, we shall assume that the complex autocovariance function of the signal is not $\phi_s(t, u)$, but $z\phi_s(t, u)$, as though the power of the signal were z times as large. We designate the complex envelope of this signal by $S_z(t)$. We seek the maximum-a-posteriori-probability (MAP) estimator of the complex envelope $S_z(\tau)$ at time τ on the basis of the input $v(t) = \mathrm{Re}\, V(t) \exp i\Omega t$ during the preceding interval $(0, \tau)$.

We can again formulate our problem and carry out its analysis in terms of vectors and matrices as in Sec. 11.1, but the functions $f_k(t)$ defining their elements are now orthonormal over the interval $(0, \tau)$. From (11-58) we find that the vector representing the estimator of the complex envelope $S_z(t)$ is

$$\hat{\mathbf{S}}_z = \mathbf{MV}, \qquad \mathbf{M} = z\boldsymbol{\phi}_s\boldsymbol{\phi}_1^{-1}, \qquad \boldsymbol{\phi}_1 = N\mathbf{I} + z\boldsymbol{\phi}_s, \qquad (11\text{-}113)$$

for the covariance matrix of samples of the noise is now $\boldsymbol{\phi}_0 = N\mathbf{I}$. Because the noise is white and $\boldsymbol{\phi}_0 = N\mathbf{I}$, $\boldsymbol{\phi}_1$ and $\boldsymbol{\phi}_s$ commute, $\mathbf{M} = z\boldsymbol{\phi}_1^{-1}\boldsymbol{\phi}_s$, and the matrix $\mathbf{M}$ is the solution of

$$\boldsymbol{\phi}_1\mathbf{M} = (N\mathbf{I} + z\boldsymbol{\phi}_s)\mathbf{M} = z\boldsymbol{\phi}_s. \qquad (11\text{-}114)$$

This matrix equation is equivalent to the integral equation

$$Nm(t, u; z; \tau) + z\int_0^\tau \phi_s(t, r)m(r, u; z; \tau)\, dr = z\phi_s(t, u), \qquad (11\text{-}115)$$

$$0 \le (t, u) \le \tau.$$

The parameter τ in $m(t, u; z; \tau)$ reminds us that this function is defined over $(0, \tau)$ and is the solution of an integral equation (11-115) over that interval.

The estimator of the complex envelope $S_z(t)$ of the signal at time t in $(0, \tau)$ is then

$$\hat{S}_z(t) = \int_0^\tau m(t, u; z; \tau)V(u)\,du, \qquad 0 \le t \le \tau. \qquad (11\text{-}116)$$

In particular, the MAP estimator of $S_z(\tau)$ is

$$\hat{S}_z(\tau) = \int_0^\tau m(\tau, u; z; \tau)V(u)\,du. \qquad (11\text{-}117)$$

This is a *causal* estimator of the signal at time τ because it utilizes only the complex envelope $V(t)$ of the past input during $(0, \tau)$.

11.3.2 Recalculation of the Fredholm Determinant

The Fredholm determinant $D(z; \tau)$ can be defined for an arbitrary interval $(0, \tau)$ as in (11-76), provided that the eigenvalues $\lambda_k = \lambda_k(\tau)$ are those of the integral equation (11-74) with the upper limit T replaced by τ. Denoting the logarithm of $D(z; \tau)$ by $E(z; \tau)$, we find from (11-100)

$$E(z; \tau) = \ln D(z; \tau) = N\int_0^z du \int_0^\tau h(t, t; u; \tau)\,dt, \qquad (11\text{-}118)$$

where by (11-53) the function $h(r, t; u; \tau)$ is the solution of the integral equation

$$Nh(r, t; u; \tau) + u\int_0^\tau \phi_s(r, s)h(s, t; u; \tau)\,ds = N^{-1}\phi_s(r, t),$$
$$0 \le (r, t) \le \tau. \qquad (11\text{-}119)$$

We shall set up a differential equation for $E(z; \tau)$ as a function of τ, and for that we first need an integral equation for the partial derivative of $h(r, t; u; \tau)$ with respect to τ. Following [Sie57] and designating that partial derivative by a subscript τ,

$$\frac{\partial}{\partial \tau}h(r, t; u; \tau) = h_\tau(r, t; u; \tau),$$

we differentiate (11-119) to obtain the integral equation

$$Nh_\tau(r, t; u; \tau) + u\int_0^\tau \phi_s(r, s)h_\tau(s, t; u; \tau)\,ds = -u\phi_s(r, \tau)h(\tau, t; u; \tau). \qquad (11\text{-}120)$$

The solution of this integral equation is

$$h_\tau(r, t; u; \tau) = -Nuh(r, \tau; u; \tau)h(\tau, t; u; \tau), \qquad (11\text{-}121)$$

as we can see by substituting it into (11-120) and using (11-119) with t replaced by τ:

$$-N^2 uh(r, \tau; u; \tau)h(\tau, t; u; \tau) - Nu^2\int_0^\tau \phi_s(r, s)h(s, \tau; u; \tau)h(\tau, t; u; \tau)\,ds$$

$$= -u\phi_s(r, \tau)h(\tau, t; u; \tau).$$

If we now differentiate (11-118) with respect to τ, we find

$$
\begin{aligned}
\frac{\partial E(z;\tau)}{\partial \tau} &= N \int_0^z h(\tau, \tau; u; \tau)\, du + N \int_0^z du \int_0^\tau h_\tau(t, t; u; \tau)\, dt \\
&= N \int_0^z du \left[h(\tau, \tau; u; \tau) - Nu \int_0^\tau h(t, \tau; u; \tau)h(\tau, t; u; \tau)\, dt \right] \quad\quad (11\text{-}122) \\
&= N \int_0^z g(\tau, \tau; u)\, du
\end{aligned}
$$

in terms of the function

$$
g(r, t; u) = h(r, t; u; \tau) - Nu \int_0^\tau h(r, v; u; \tau)h(v, t; u; \tau)\, dv. \quad\quad (11\text{-}123)
$$

We now seek an expression for the last integral in (11-122). As we are through with differentiating with respect to τ, we can return to the matrix domain, where the manipulations are less cumbersome.

Denoting by $\mathbf{H}(u)$ the matrix corresponding to the solution of (11-119), we can express that equation in matrix form as

$$
(N\mathbf{I} + u\boldsymbol{\phi}_s)\mathbf{H}(u) = N^{-1}\boldsymbol{\phi}_s,
$$

so that

$$
\mathbf{H}(u) = N^{-1}(N\mathbf{I} + u\boldsymbol{\phi}_s)^{-1}\boldsymbol{\phi}_s, \quad\quad (11\text{-}124)
$$

and the matrix $\mathbf{G}(u)$ corresponding to the function defined in (11-123) is

$$
\begin{aligned}
\mathbf{G}(u) &= \mathbf{H}(u) - Nu\mathbf{H}(u)\mathbf{H}(u) = \mathbf{H}(u)[\mathbf{I} - Nu\mathbf{H}(u)] \\
&= \mathbf{H}(u)[\mathbf{I} - u(N\mathbf{I} + u\boldsymbol{\phi}_s)^{-1}\boldsymbol{\phi}_s] \\
&= N\mathbf{H}(u)(N\mathbf{I} + u\boldsymbol{\phi}_s)^{-1} = (N\mathbf{I} + u\boldsymbol{\phi}_s)^{-2}\boldsymbol{\phi}_s.
\end{aligned}
$$

Multiplying by du and integrating over $(0, z)$, we obtain

$$
\begin{aligned}
\int_0^z \mathbf{G}(u)\, du &= N^{-1}\mathbf{I} - (N\mathbf{I} + z\boldsymbol{\phi}_s)^{-1} = zN^{-1}(N\mathbf{I} + z\boldsymbol{\phi}_s)^{-1}\boldsymbol{\phi}_s \\
&= z\mathbf{H}(z)
\end{aligned} \quad\quad (11\text{-}125)
$$

by (11-124). Returning to the time domain, we express (11-125) as

$$
\int_0^z g(r, t; u)\, du = zh(r, t; z; \tau),
$$

and (11-122) becomes

$$
\frac{\partial E(z;\tau)}{\partial \tau} = Nzh(\tau, \tau; z; \tau).
$$

Integrating over $(0, T)$, we find

$$
E(z; T) = \ln D(z; T) = Nz \int_0^T h(\tau, \tau; z; \tau)\, d\tau. \quad\quad (11\text{-}126)
$$

This is Siegert's formula for the logarithm of the Fredholm determinant $D(z)$ in terms of the solution of (11-119), with u replaced by z. In contrast to (11-100) it does not require integration over the parameter u.

When the signal to be detected has the Lorentz spectral density (11-47), we find from (11-56), replacing r, t, and T all by τ, and replacing u by z,

$$h(\tau, \tau; z; \tau) = \frac{2\mu P_s}{N^2} \frac{(\beta + \mu)\, e^{\beta\tau} + (\beta - \mu)\, e^{-\beta\tau}}{(\beta + \mu)^2\, e^{\beta\tau} - (\beta - \mu)^2\, e^{-\beta\tau}}$$

$$= \frac{1}{Nz} \frac{\partial}{\partial\tau} \ln[(\beta + \mu)^2\, e^{(\beta-\mu)\tau} - (\beta - \mu)^2\, e^{-(\beta+\mu)\tau}], \qquad (11\text{-}127)$$

$$\beta^2 = \mu^2 + \frac{2\mu P_s z}{N}.$$

Substituting this into (11-126), we obtain

$$\ln D(z; T) = \ln[(\beta + \mu)^2 e^{(\beta-\mu)T} - (\beta - \mu)^2 e^{-(\beta-\mu)T}] - \ln(4\beta\mu)$$

whereupon the Fredholm determinant becomes

$$D(z) = D(z; T) = \frac{(\beta + \mu)^2\, e^{\beta T} - (\beta - \mu)^2\, e^{-\beta T}}{4\beta\mu e^{\mu T}}. \qquad (11\text{-}128)$$

If we compare (11-115) and (11-119), we see that

$$m(t, u; z; \tau) = Nzh(t, u; z; \tau), \qquad (11\text{-}129)$$

and Siegert's formula (11-126) can be written as

$$E(z; T) = \ln D(z) = \int_0^T m(\tau, \tau; z; \tau)\, d\tau \qquad (11\text{-}130)$$

in terms of the solution of the integral equation (11-115). This solution is related to the kernel of the causal estimator of a signal having complex autocovariance function $z\phi_s(t, u)$.

11.3.3 The Mean-square Estimation Error

The total mean-square error in the estimate $\hat{S}_z(t)$ of the complex envelope $S_z(t)$ of the signal $s_z(t) = \operatorname{Re} S_z(t) \exp i\Omega t$ can be defined as

$$\varepsilon_T = \tfrac{1}{2} E \int_0^T |S_z(\tau) - \hat{S}_z(\tau)|^2\, d\tau.$$

The expectation E is taken under hypothesis H_1 that the signal $s_z(t)$ is present. We shall demonstrate that this total error equals $N \ln D(z)$, where $D(z)$ is the Fredholm determinant.

By the principle of orthogonality, which we introduced in Sec. 6.1.4, the error $S_z(\tau) - \hat{S}_z(\tau)$ is orthogonal to the data $V(t)$, $0 \le t \le \tau$, on which the estimator is based through (11-117). (When dealing with circular complex random variables, the transpose used in Sec. 6.1.4 is replaced by the Hermitian conjugate.) It is therefore orthogonal to any linear combination of those data, and in particular to $\hat{S}_z(\tau)$ itself:

$$\tfrac{1}{2} E\{[S_z(\tau) - \hat{S}_z(\tau)]\hat{S}_z^*(\tau)\} = 0.$$

The instantaneous mean-square error is thus, with (11-117),

$$\varepsilon(\tau) = \tfrac{1}{2}E\{[S_z(\tau) - \hat{S}_z(\tau)][S_z^*(\tau) - \hat{S}_z^*(\tau)]\}$$
$$= \tfrac{1}{2}E\{[S_z(\tau) - \hat{S}_z(\tau)]S_z^*(\tau)\}$$
$$= z\phi_s(\tau, \tau) - \tfrac{1}{2}E\int_0^\tau m(\tau, u; z; \tau)V(u)S_z^*(\tau)\,du.$$

Because the signal and the noise are statistically independent,

$$\tfrac{1}{2}E[V(t)S_z^*(\tau)] = \tfrac{1}{2}E[S_z(t)S_z^*(\tau)] = z\phi_s(t, \tau),$$

and the instantaneous mean-square error is

$$\varepsilon(\tau) = z\phi_s(\tau, \tau) - z\int_0^\tau m(\tau, u; z; \tau)\phi_s(u, \tau)\,du. \tag{11-131}$$

To the matrix equation

$$\mathbf{M}\boldsymbol{\phi}_1 = N\mathbf{M} + z\mathbf{M}\boldsymbol{\phi}_s = z\boldsymbol{\phi}_s$$

from (11-113) corresponds the integral equation

$$Nm(r, t; z; \tau) + z\int_0^\tau m(r, u; z; \tau)\phi_s(u, t)\,du = z\phi_s(r, t), \tag{11-132}$$
$$0 \le (r, t) \le \tau;$$

compare (11-115). Putting $r = t = \tau$, we obtain

$$Nm(\tau, \tau; z; \tau) + z\int_0^\tau m(\tau, u; z; \tau)\phi_s(u, \tau)\,du = z\phi_s(\tau, \tau),$$

whence (11-131) becomes

$$\varepsilon(\tau) = Nm(\tau, \tau; z; \tau).$$

The total mean-square error is thus

$$\varepsilon_T = \tfrac{1}{2}\int_0^T |S_z(\tau) - \hat{S}_z(\tau)|^2\,d\tau$$
$$= N\int_0^T m(\tau, \tau; z; \tau)\,d\tau = N\ln D(z) \tag{11-133}$$

by (11-130).

11.3.4 The Fredholm Determinant for a Rational Spectral Density

We shall now use (11-130) to calculate the Fredholm determinant $D(z)$ for detection of a stationary Gaussian stochastic signal having a rational spectral density $\Phi_s(\omega)$ of the form in (11-104). The noise is still white with unilateral spectral density N. The result of our labors will be an expression for $D(z)$ as a quotient of two determinants. When computing false-alarm and detection probabilities by saddlepoint integration of (11-94) and (11-95), these can be evaluated by standard computer routines for evaluating determinants and solving algebraic equations.

The integrand of (11-130) can be obtained from the solution of (11-115) with $u = \tau$. As the signal is stationary, that integral equation becomes

$$NM_z(t) + z \int_0^\tau \phi_s(t - r) M_z(r)\, dr = z\phi_s(t - \tau), \qquad 0 \le t \le \tau, \qquad (11\text{-}134)$$

with $M_z(t) = m(t, \tau; z; \tau)$. Then by (11-130)

$$\ln D(z) = \int_0^T M_z(\tau)\, d\tau. \qquad (11\text{-}135)$$

As in (2-98) and (2-99), $M_z(t)$ must have the form

$$M_z(t) = \sum_{j=1}^{2n} c_j(\tau) \exp \beta_j t, \qquad (11\text{-}136)$$

in which the β_j are the $2n$ roots of the characteristic equation arising from (11-134),

$$N + z\Phi_s(-ip) = 0, \qquad p = \beta_1, \beta_2, \dots, \beta_{2n}; \qquad (11\text{-}137)$$

here $\Phi_s(\omega)$ is the rational spectral density of the signal. Like (11-105), this becomes an algebraic equation of degree $2n$ when we substitute the spectral density from (11-104) and clear fractions. Again, when z does not lie on the negative real axis, there are n roots $\beta_1, \dots, \beta_n$ in the right half of the p-plane and n roots $\beta_{n+1}, \dots, \beta_{2n}$ in the left half-plane. When z goes to 0, these roots go into the $2n$ poles $\mu_1, \dots, \mu_{2n}$ of the spectral density $\Phi_s(-ip)$; $\mu_{k+n} = -\mu_k^*$, $1 \le k \le n$.

When (11-103) and (11-136) are substituted into (11-134) and the integrations are carried out, we obtain—much as in Appendix A—the equation

$$\sum_{j=1}^{2n} c_j \left\{ [N + z\Phi_s(-i\beta_j)]e^{\beta_j t} + z \sum_{k=1}^{n} \left[f_k^* \frac{e^{(\beta_j - \mu_k)\tau}}{\beta_j - \mu_k} e^{\mu_k t} - f_k \frac{1}{\beta_j + \mu_k^*} e^{-\mu_k^* t} \right] \right\}$$

$$= z \sum_{k=1}^{n} f_k^* e^{\mu_k(t-\tau)}, \qquad 0 \le t \le \tau.$$

The first term vanishes because the β_j's are the roots of (11-137). Equating terms proportional to $\exp \mu_k t$ on each side of what is left, we obtain

$$\sum_{k=1}^{2n} c_j \frac{e^{(\beta_j - \mu_k)\tau}}{\beta_j - \mu_k} = e^{-\mu_k \tau}, \qquad 1 \le k \le n, \qquad (11\text{-}138)$$

and equating terms proportional to $\exp(-\mu_k^* t)$, we find

$$\sum_{j=1}^{2n} c_j \frac{1}{\beta_j - \mu_k} = 0, \qquad n + 1 \le k \le 2n. \qquad (11\text{-}139)$$

Denote by $\mathbf{G}'$ the matrix whose elements are

$$G'_{kj} = \begin{cases} \dfrac{e^{(\beta_j - \mu_k)\tau}}{\beta_j - \mu_k}, & 1 \le k \le n, \\[2mm] \dfrac{1}{\beta_j - \mu_k}, & n + 1 \le k \le 2n. \end{cases} \qquad 1 \le j \le 2n, \qquad (11\text{-}140)$$

Then from (11-138) and (11-139)

$$c_j(\tau) = \sum_{k=1}^{n} (\mathbf{G}^{l-1})_{jk}\, e^{-\mu_k \tau},$$

and by (11-136)

$$M_z(\tau) = \sum_{j=1}^{2n} \sum_{k=1}^{n} e^{\beta_j \tau}(\mathbf{G}^{l-1})_{jk}\, e^{-\mu_k \tau}. \qquad (11\text{-}141)$$

Now consider, in view of (11-20),

$$\frac{d}{d\tau}\ln \det \mathbf{G}'(\tau) = \frac{d}{d\tau}\mathrm{Tr}\ln \mathbf{G}'(\tau) = \mathrm{Tr}\left[\mathbf{G}^{l-1}\frac{d\mathbf{G}'}{d\tau}\right] = \sum_{j=1}^{2n}\sum_{k=1}^{2n}(\mathbf{G}^{l-1})_{jk}\frac{dG'_{kj}}{d\tau}.$$

By (11-140)

$$\frac{dG'_{kj}}{d\tau} = \begin{cases} e^{(\beta_j - \mu_k)\tau}, & 1 \le k \le n, \\ 0, & n+1 \le k \le 2n, \end{cases}$$

whereupon by (11-141)

$$M_z(\tau) = \frac{d}{d\tau}\ln \det \mathbf{G}'(\tau).$$

Putting this into (11-135), we find that the Fredholm determinant is

$$D(z) = \frac{\det \mathbf{G}'(T)}{\det \mathbf{G}'(0)}$$

$$= \exp\left[-\sum_{k=1}^{n}\mu_k T\right]\frac{\det \mathbf{G}(T)}{\det \mathbf{G}(0)}, \qquad (11\text{-}142)$$

where the matrix $\mathbf{G}(T)$ is defined by

$$G_{kj}(T) = \begin{cases} \dfrac{e^{\beta_j T}}{\beta_j - \mu_k}, & 1 \le k \le n, \\[4mm] \dfrac{1}{\beta_j - \mu_k}, & n+1 \le k \le 2n, \end{cases} \qquad 1 \le j \le 2n, \qquad (11\text{-}143)$$

as in (A-15), and $\mathbf{G}(0)$ is obtained by putting $T = 0$.

The eigenvalues λ_k of the integral equation

$$\lambda_k f_k(t) = \frac{1}{N}\int_0^T \phi_s(t-r)f_k(r)\,dr, \qquad 0 \le t \le T,$$

are $\lambda_k = -z_k^{-1}$, where by (11-76) $D(z_k) = 0$, and with (11-142) this leads to (A-22).

When T becomes much larger than the reciprocal bandwidth W^{-1} of the signal, those factors $\exp \beta_j T$ in the upper half of the determinant $\det \mathbf{G}(T)$ with $1 \le j \le n$ dominate because $\mathrm{Re}\,\beta_j > 0$ for the first n β_j's. After factoring those exponential factors out of the first n columns of $\mathbf{G}(T)$, we see that $D(z)$ is proportional to $D_\infty(z)$ as given in (11-107). The ratio $D(z)/D_\infty(z) = 1$ at $z = 0$, and when $WT \gg 1$ it is nearly equal to 1 along that part of the path of integration contributing significantly to (11-94) and (11-95).

By the same argument as in Appendix A, we can see that if the spectral density of the signal has a pole μ of order p, both the upper and lower halves of both determinants $\mathbf{G}(T)$ and $\mathbf{G}(0)$ will contain $p - 1$ rows of which each is the derivative with respect to μ of the row above it. In this way the existence of multiple poles of the spectral density $\Phi_s(\omega)$ of the signal can be accommodated.

In order to understand the behavior of the roots of (11-137) as the complex parameter z varies, it is instructive to consider the root locus of the equation

$$N + z\,\Phi_s(\omega) = 0 \qquad (11\text{-}144)$$

in the complex ω-plane. The root locus of (11-137) in the complex p-plane is obtained by rotation counterclockwise through $90°$. The root locus of (11-144) possesses $2n$ branches, and if we think of the complex parameter z as starting out from the origin $z = 0$ of the z-plane, we perceive that each of those $2n$ branches originates at one of the $2n$ poles of the spectral density $\Phi_s(\omega)$. These poles occur in complex-conjugate pairs, $-i\mu_k$ and $i\mu_k^* = -i\mu_{k+n}$. The $2m$ zeros of $\Phi_s(\omega)$ either occur in complex-conjugate pairs or, if real, possess even multiplicity. [If a real zero of $\Phi_s(\omega)$ had odd multiplicity, $\Phi_s(\omega)$ would become negative for a real value of ω in its neighborhood.] Of the $2n$ branches of the root locus, $2(n - m)$ go off to infinity and $2m$ enter the $2m$ zeros of $\Phi_s(\omega)$ when $|z| \longrightarrow \infty$. Of these $2m$ zeros, m lie in the lower half of the ω-plane and m in the upper half-plane, unless there are some zeros on the Re ω-axis. When a branch enters one of these real-axis zeros from the lower half-plane, a mirror image of that branch must enter it from the upper half-plane, for the real-axis zeros have even multiplicity. Thus as $|z| \longrightarrow \infty$, m of the $2n$ branches of the root locus must finish in the lower half of the ω-plane and m must finish in the upper half.

The $2(n - m)$ branches of the root locus heading toward infinity approach asymptotes at angles

$$\arg \omega = \frac{(2k + 1)\pi + \arg z}{2(n - m)}, \qquad 0 \le k < 2(n - m).$$

For $0 \le \arg z < \pi$, half the asymptotes, which are spaced uniformly in angle, will lie in the lower half-plane and half will lie in the upper half-plane. When the ω-plane is rotated counterclockwise to become the p-plane, there will be n branches of the root locus in the right half of that plane and n in the left half. The roots β_k traversing the former set of branches are assigned indices from 1 to n; those traversing the latter set receive indices from $n + 1$ to $2n$.

When z is real and sufficiently negative, at least two of the roots β_k will be purely imaginary. One of each pair of these purely imaginary roots is assigned an index k between 1 and n, the other an index greater by n; it does not matter which is which. These purely imaginary roots of (11-137) arise when calculating the zeros $z_k = -1/\lambda_k$ of the Fredholm determinant in (11-76) and also, possibly, when locating a saddlepoint in Re $z < 0$.

11.3.5 The Likelihood Functional

Our aim is now to express the optimum detection statistic U in terms of the causal estimator $\hat{S}_1(\tau)$ given by (11-117); $z = 1$. If our detector possessed the input $v(t) =$

Re $V(t)$ exp $i\Omega t$ only during the interval $(0, \tau)$, it would optimally base its decision at time τ on the statistic

$$U(\tau) = \tfrac{1}{2} \int_0^\tau \int_0^\tau V^*(r) h(r, t; 1; \tau) V(t) \, dr \, dt$$

$$= \text{Re} \int_0^\tau V^*(r) \int_0^r h(r, t; 1; \tau) V(t) \, dt \, dr \tag{11-145}$$

by (11-35), (11-53), and (11-39). We shall demonstrate that

$$U = U(T) = \frac{1}{N} \text{Re} \int_0^T \hat{S}_1^*(\tau) V(\tau) \, d\tau - \frac{1}{2N} \int_0^T |\hat{S}_1(\tau)|^2 \, d\tau. \tag{11-146}$$

In this analysis we follow [Sch65]. The form (11-146) that results is called the *causal estimator–correlator* representation of the optimum statistic U for detecting the stochastic signal Re $S(t)$ exp $i\Omega t$ in white noise.

Differentiating (11-145) with respect to τ, we obtain

$$\frac{dU}{d\tau} = \tfrac{1}{2} V^*(\tau) \int_0^\tau h(\tau, t; 1; \tau) V(t) \, dt + \text{c.c.} + \tfrac{1}{2} \int_0^\tau \int_0^\tau V^*(r) h_\tau(r, t; 1; \tau) V(t) \, dr \, dt,$$

where c.c. stands for complex conjugate. Now using (11-129), (11-117), and (11-121), we can write this as

$$\frac{dU}{d\tau} = \frac{1}{N} \text{Re}[V^*(\tau)\hat{S}_1(\tau)] - \frac{N}{2} \int_0^\tau \int_0^\tau V^*(r) h(r, \tau; 1; \tau) h(\tau, t; 1; \tau) V(t) \, dr \, dt$$

$$= \frac{1}{N} \text{Re}[V^*(\tau)\hat{S}_1(\tau)] - \frac{1}{2N} |\hat{S}_1(\tau)|^2.$$

Integrating this differential equation over $(0, T)$, we obtain (11-146).

According to (11-31) the likelihood functional for the detection of a narrow-band stochastic signal $s(t) = \text{Re } S(t) \exp i\Omega t$ in an input $v(t) = \text{Re } V(t) \exp i\Omega t$ containing white noise is

$$\Lambda[v(t)] = \exp(U - B)$$

$$= \exp\left[\frac{1}{N} \text{Re} \int_0^T \hat{S}_1^*(t) V(t) \, dt - \frac{1}{2N} \int_0^T |\hat{S}_1(t)|^2 \, dt - B \right] \tag{11-147}$$

by (11-146), where by (11-34)

$$B = \ln \det(\mathbf{I} + N^{-1} \boldsymbol{\phi}_s) = \ln D(1; T) = E(1; T).$$

If we compare this likelihood functional with that in (3-54) as rewritten for detection of a known signal Re $S(t)$ exp $i\Omega t$ in white noise,

$$\Lambda[v(t)] = \exp\left[\frac{1}{N} \text{Re} \int_0^T S^*(t) V(t) \, dt - \frac{1}{2N} \int_0^T |S(t)|^2 \, dt \right],$$

we see that the two are much alike except that the complex envelope $S(t)$ of the signal is replaced by its causal MAP estimator $\hat{S}_1(t)$ and except that (11-147) contains an extraneous term B.

Kailath [Kai69] showed that for the detection of low-pass stochastic signals $s(t)$ in white noise $n(t)$, what corresponds to that term B can be eliminated by altering the definition of the integrals in (11-147) and replacing the ordinary Riemann integral that we have implicitly been utilizing up to now with the *Itô integral*. By doing so, he was able to broaden the scope of the result; the stochastic signal $s(t)$ need not be a sample of a Gaussian random process, but can have any probability measure, provided that by $\hat{s}_1(t)$ one understands that causal estimator of $s(t)$ minimizing the mean-square error $E[\hat{s}_1(t) - s(t)]^2$. According to Sec. 6.1.2, this is the conditional expected value

$$\hat{s}_1(t) = E[s(t)|\, v(t'),\, 0 \leq t' \leq t,\, H_1],$$
$$v(t) = s(t) + n(t).$$

The likelihood functional for detecting this stochastic signal in white noise is written

$$\Lambda[v(t)] = \exp\left[\frac{1}{N} \int_0^T \hat{s}_1(t)v(t)\, dt - \frac{1}{2N} \int_0^T |\hat{s}_1(t)|^2\, dt \right], \tag{11-148}$$

in which $\int$ indicates the Itô integral [Kai69]. For a Gaussian stochastic signal the estimator is the linear MAP estimator defined as in (11-117). For signals with other probability measures it is unknown how to calculate the required conditional expected value $\hat{s}_1(t)$. The considerations underlying this reformulation of the likelihood functional involve the properties of broadband white noise treated as the derivative of the Wiener–Lévy or Brownian stochastic process. In this chapter we are dealing with narrowband signals received in narrowband white noise, that is, noise whose spectral density is flat over a range of frequencies much wider than that occupied by the signals, but not extending beyond a fraction of the carrier frequency Ω of the signals. A formulation of the likelihood functional of the kind in (11-148) is inappropriate, and referring the interested reader to [Kai69], we forbear undertaking its derivation.

11.3.6 The Innovation Process

Let us denote by $\overline{m}(t, s)$ the function

$$\overline{m}(t, s) = \begin{cases} m(t, s; 1; t), & 0 \leq s \leq t \leq T, \\ 0, & 0 < t < s \leq T. \end{cases} \tag{11-149}$$

Then the causal estimator of the envelope $S(t)$ of the signal is

$$\hat{S}_1(t) = \int_0^T \overline{m}(t, s)V(s)\, ds, \qquad 0 \leq t \leq T, \tag{11-150}$$

by (11-117) with $z = 1$, and the integration can be taken over $(0, T)$ by virtue of the second part of (11-149). Substituting this into (11-146), we write the optimum detection statistic U as

$$U = \frac{1}{N} \text{Re} \int_0^T \int_0^T V^*(u)\overline{m}(u, s)V(s)\, du\, ds$$

$$- \frac{1}{2N} \int_0^T \int_0^T \int_0^T V^*(u)\overline{m}^*(t, u)\overline{m}(t, s)V(s)\, dt\, du\, ds$$

$$= \frac{1}{2} \int_0^T \int_0^T V^*(t)h(t, s)V(s)\, dt\, ds,$$

where $h(t, s)$ is the kernel of (11-35). Because this equation must hold for arbitrary $V(t)$, we find the relationship

$$Nh(t, s) = m(t, s; 1; T) = \overline{m}(t, s) + \overline{m}^*(s, t) - \int_0^T \overline{m}^*(u, t)\overline{m}(u, s)\, du,$$

$$0 \leq (t, s) \leq T. \tag{11-151}$$

If we denote by $\overline{\mathbf{m}}$ the matrix corresponding to the function $\overline{m}(t, s)$ and by $N\mathbf{H} = \mathbf{M}$ that corresponding to $m(t, s; 1; T)$, (11-151) becomes

$$\mathbf{M} = \overline{\mathbf{m}} + \overline{\mathbf{m}}^+ - \overline{\mathbf{m}}^+\overline{\mathbf{m}}$$

or

$$\mathbf{I} - \mathbf{M} = \mathbf{I} - \overline{\mathbf{m}} - \overline{\mathbf{m}}^+ + \overline{\mathbf{m}}^+\overline{\mathbf{m}} = (\mathbf{I} - \overline{\mathbf{m}}^+)(\mathbf{I} - \overline{\mathbf{m}}). \tag{11-152}$$

The right side is a factorization of the matrix $\mathbf{I} - \mathbf{M}$ that resembles the decomposition of a matrix into upper-triangular and lower-triangular factors. Because for $t > s$ $\overline{m}^*(s, t) \equiv 0$, (11-151) is equivalent to the nonlinear integral equation

$$m(t, s; 1; T) = \overline{m}(t, s) - \int_t^T \overline{m}^*(u, t)\overline{m}(u, s)\, du, \qquad 0 \leq s \leq t \leq T. \tag{11-153}$$

For signals having a Lorentz spectral density, by (11-56),

$$m(t, s; 1; T) = \frac{\mu P_s}{N\beta} \frac{[(\beta + \mu)e^{\beta s} + (\beta - \mu)e^{-\beta s}][(\beta + \mu)e^{\beta(T-t)} + (\beta - \mu)e^{-\beta(T-t)}]}{(\beta + \mu)^2\, e^{\beta T} - (\beta - \mu)^2\, e^{-\beta T}},$$

$$\beta^2 = \mu^2 + \frac{2\mu P_s}{N}, \qquad 0 \leq s \leq t \leq T,$$

and

$$\overline{m}(u, s) = \frac{2\mu P_s}{N} \frac{(\beta + \mu)\, e^{\beta s} + (\beta - \mu)\, e^{-\beta s}}{(\beta + \mu)^2\, e^{\beta u} - (\beta - \mu)^2\, e^{-\beta u}} U(u - s). \tag{11-154}$$

The reader might amuse himself by showing that these functions indeed satisfy (11-153).

The random process

$$N'(t) = V(t) - \hat{S}_1(t)$$

is called the *innovation process* because it represents the new information at time t about the signal and the noise that is not embodied in the conditional expected value

$$\hat{S}_1(t) = E[S(t)|\, V(t'), 0 \leq t' \leq t, H_1] = \int_0^t \overline{m}(t, u)V(u)\, du.$$

Kailath [Kai68] showed that this process $N'(t)$ is white and possesses the same unilateral spectral density as the input noise $N(t)$. In vector form

$$\mathbf{N}' = \mathbf{V} - \hat{\mathbf{S}}_1 = (\mathbf{I} - \overline{\mathbf{m}})\mathbf{V},$$

and the covariance matrix of the elements of the vector $\mathbf{N}'$ is

$$\boldsymbol{\phi}_n' = \tfrac{1}{2}E(\mathbf{N}'\mathbf{N}'^+|\, H_1) = \tfrac{1}{2}(\mathbf{I} - \overline{\mathbf{m}})E(\mathbf{V}\mathbf{V}^+|\, H_1)(\mathbf{I} - \overline{\mathbf{m}}^+)$$
$$= (\mathbf{I} - \overline{\mathbf{m}})(N\mathbf{I} + \boldsymbol{\phi}_s)(\mathbf{I} - \overline{\mathbf{m}}^+).$$

Now by (11-32),

$$\mathbf{I} - \mathbf{M} = \mathbf{I} - N\mathbf{H} = \mathbf{I} - N(N^{-1}\mathbf{I} - \boldsymbol{\phi}_1^{-1}) = N(N\mathbf{I} + \boldsymbol{\phi}_s)^{-1},$$

whereupon by (11-152)

$$\phi'_n = N(\mathbf{I} - \overline{\mathbf{m}})(\mathbf{I} - \mathbf{M})^{-1}(\mathbf{I} - \overline{\mathbf{m}}^+)$$
$$= N(\mathbf{I} - \overline{\mathbf{m}})(\mathbf{I} - \overline{\mathbf{m}})^{-1}(\mathbf{I} - \overline{\mathbf{m}}^+)^{-1}(\mathbf{I} - \overline{\mathbf{m}}^+) = N\mathbf{I},$$

or in the time domain

$$\tfrac{1}{2}E[N'(t)N'^{*}(s)|\, H_1] = N\delta(t - s),$$

and the innovation process $N'(t)$ is white.

11.4 THE STATE-SPACE FORMULATION

11.4.1 Generation of the Signal by a Linear System

When the narrowband spectral density $\Phi_s(\omega)$ of the signal $s(t) = \mathrm{Re}\ S(t)\exp i\Omega t$ is a rational function of ω, the signal can be thought of as having been generated by a time-invariant linear system whose input is white noise. The output of the system—that is, the signal envelope—is then the solution of a linear differential equation of finite order with constant coefficients. This differential equation can be decomposed into a set of n first-order linear differential equations for a set of random processes $x_k(t)$, which are organized as a column vector having n elements:

$$\mathbf{x}(t) = \begin{bmatrix} x_1(t) \\ x_2(t) \\ \vdots \\ x_n(t) \end{bmatrix}.$$

The integer n is one-half the degree of the denominator of the spectral density $\Phi_s(\omega)$ as a polynomial in ω. The vector $\mathbf{x}(t)$ is called the *state vector* of the linear system, and the set of first-order differential equations, or *state equations*, can be concisely written as

$$\frac{d\mathbf{x}(t)}{dt} = \mathbf{F}\mathbf{x}(t) + \mathbf{G}w(t), \tag{11-155}$$

in which $\mathbf{F}$ is an $n \times n$ matrix, $\mathbf{G}$ is a column vector of n elements, and $w(t)$ is the circular complex white noise driving the system. The n components $x_j(t)$ are considered to be circular complex Gaussian random processes. The elements of $\mathbf{F}$ and $\mathbf{G}$ may be complex. In what follows we refer to the complex envelope $S(t)$ as simply "the signal."

The white noise $\mathrm{Re}\ w(t)\exp i\Omega t$ is statistically independent of the white noise $\mathrm{Re}\ N(t)\exp i\Omega t$ in which the signal $\mathrm{Re}\ S(t)\exp i\Omega t$ is to be detected. Its complex autocovariance function is

$$\tfrac{1}{2}E[w(t_1)w^*(t_2)] = Q\delta(t_1 - t_2), \tag{11-156}$$

and its spectral density Q is proportional to the average power P_s of the stochastic signal $S(t)$. That signal, as a linear combination of the n processes $x_k(t)$, is given by

$$S(t) = \mathbf{C}\mathbf{x}(t), \tag{11-157}$$

in which **C** is a row vector of n elements. A signal generated from white noise in this way might be termed *leucogenic*.

This state-space model of the generation of the stochastic signal by white noise can be generalized by allowing the matrix **F**, the vectors **G** and **C**, and the spectral density Q to be functions of the time t. The signal $S(t)$ is then a nonstationary random process. The outcome of this formulation is a set of nonlinear differential equations from whose solution one can calculate the Fredholm determinant $D(z)$ needed for computing the moment-generating functions of the detection statistic and thence the false-alarm and detection probabilities of the stochastic signal. From related equations the causal estimate $\hat{S}_1(t)$ of the signal and the optimum detection statistic U can be computed in real time. Given an arbitrary autocovariance function $\phi_s(t, s)$ for a nonstationary signal, it is in general difficult to find a state-space model whose output closely approximates the signal process $S(t)$. Methods for doing so are to be found in [Bag71, pp. 292–309], in [Lju83], and in other books on the topic of system identification. In our examples we shall restrict ourselves to stationary stochastic signals with rational spectral densities, taking **F**, **G**, **C**, and Q to be constant.

The first step in the procedure is to set up the state-space model by determining the matrix **F** and the vectors **G** and **C**. It is best illustrated for stationary leucogenic signals by an example, in which the spectral density of the signal will be taken as

$$\Phi_s(\omega) = \frac{Q}{(\omega^2 + a^2)[(\omega - b)^2 + c^2]}, \tag{11-158}$$

and $n = 2$. A certain arbitrariness is involved, and we can choose the vector **C** as we like. Let us take **C** = (1 0) so that the signal is the first component of the state vector. Factoring $\Phi_s(\omega)$, we see that the signal can be generated by passing white noise $w(t)$ through a filter whose narrowband transfer function is

$$Y(\omega) = \frac{1}{(a + i\omega)(c - ib + i\omega)};$$

$\Phi_s(\omega) = Q|Y(\omega)|^2$. Then the output $x_1(t)$ satisfies the differential equation

$$\left[\frac{d}{dt} + a\right]\left[\frac{d}{dt} + c - ib\right] x_1(t) = w(t).$$

A set of state equations is most easily derived by defining

$$x_2(t) = \frac{dx_1}{dt} + (c - ib)x_1,$$

so that

$$\frac{dx_2}{dt} + ax_2 = w(t).$$

Thus the equations (11-155) take the form

$$\frac{dx_1}{dt} = -(c - ib)x_1 + x_2,$$

$$\frac{dx_2}{dt} = -ax_2 + w(t),$$

and we find

$$F = \begin{bmatrix} -(c - ib) & 1 \\ 0 & -a \end{bmatrix}, \qquad G = \begin{bmatrix} 0 \\ 1 \end{bmatrix}.$$

The solution of the state equations (11-155) in terms of the input $w(t)$ and the initial conditions on $x(t)$ at $t = t_0$ can be formally specified in terms of the matrix impulse response $K(t, \tau)$ of the system:

$$\mathbf{x}(t) = \mathbf{K}(t, t_0)\mathbf{x}(t_0) + \int_{t_0}^{t} \mathbf{K}(t, \tau)\mathbf{G}(\tau)w(\tau)\, d\tau. \qquad (11\text{-}159)$$

The $n \times n$ impulse response matrix satisfies the differential equation

$$\frac{\partial}{\partial t}\mathbf{K}(t, \tau) = \mathbf{F}(t)\mathbf{K}(t, \tau), \qquad t \geq \tau, \qquad (11\text{-}160)$$

which must be solved with the initial conditions $\mathbf{K}(\tau, \tau) = \mathbf{I}$, with $\mathbf{I}$ the $n \times n$ identity matrix. For a time-invariant linear system,

$$\mathbf{K}(t, \tau) = e^{\mathbf{F}(t-\tau)}U(t - \tau);$$

the exponential function of a matrix is defined in terms of the Taylor series,

$$e^{\mathbf{F}t} = \mathbf{I} + \mathbf{F}t + \tfrac{1}{2}\mathbf{F}^2 t^2 + \cdots + \frac{1}{k!}\mathbf{F}^k t^k + \cdots.$$

For later use we need the $n \times n$ covariance matrix of the state vector $\mathbf{x}(t)$, defined by

$$\mathbf{L}(t) = \tfrac{1}{2}E[\mathbf{x}(t)\mathbf{x}^+(t)], \qquad (11\text{-}161)$$

in which

$$\mathbf{x}^+(t) = (x_1^*(t), x_2^*(t), \dots, x_n^*(t))$$

is the row vector formed by transposing the state vector and taking the complex conjugates of the elements. Differentiating (11-161) and using the state equations (11-155), we obtain

$$\begin{aligned} \frac{d\mathbf{L}}{dt} &= \tfrac{1}{2}E[\dot{\mathbf{x}}(t)\mathbf{x}^+(t) + \mathbf{x}(t)\,\dot{\mathbf{x}}^+(t)] \\ &= \tfrac{1}{2}E(\mathbf{F}\mathbf{x}\mathbf{x}^+ + \mathbf{G}w\mathbf{x}^+ + \mathbf{x}\mathbf{x}^+\mathbf{F}^+ + \mathbf{x}w^*\mathbf{G}^+), \end{aligned} \qquad (11\text{-}162)$$

the dot standing for a time derivative. Now by (11-159) and (11-156),

$$\begin{aligned} \tfrac{1}{2}E[\mathbf{x}(t)w^*(t)] &= \tfrac{1}{2}E\int_0^t \mathbf{K}(t, \tau)\mathbf{G}(\tau)w(\tau)w^*(t)\, d\tau \\ &= Q\int_0^t \mathbf{K}(t, \tau)\mathbf{G}(\tau)\,\delta(\tau - t)\, d\tau \\ &= \tfrac{1}{2}Q\mathbf{K}(t, t)\mathbf{G}(t) = \tfrac{1}{2}Q\mathbf{G}(t), \end{aligned}$$

the factor of $\tfrac{1}{2}$ entering because the delta function $\delta(\tau - t)$ stands at the end of the interval of integration with only half its "mass" inside. Similarly

$$\tfrac{1}{2}E[w(t)\mathbf{x}^+(t)] = \tfrac{1}{2}Q\mathbf{G}^+(t),$$

and substituting into (11-162) we obtain the differential equation

$$\frac{d\mathbf{L}}{dt} = \mathbf{FL} + \mathbf{LF}^+ + Q\mathbf{GG}^+ \tag{11-163}$$

for the covariance matrix $\mathbf{L}(t)$ of the state vector.

When the input noise is stationary and the linear system is in steady state, $\mathbf{L}$ is a constant matrix $\mathbf{L}_\infty$ and satisfies the steady-state covariance equation

$$\mathbf{FL}_\infty + \mathbf{L}_\infty \mathbf{F}^+ + Q\mathbf{GG}^+ = 0. \tag{11-164}$$

If at time $t = 0$, $\mathbf{L}(0) = \mathbf{L}_\infty$, the output of the system will be a stationary random process. By starting with other initial covariance matrices $\mathbf{L}(0)$, nonstationary outputs with a variety of autocovariance functions can be generated.

For $t_1 > t_2$ the covariance matrix of the state vector is, by (11-161) and (11-159) with $t_1 = t$, $t_2 = t_0$,

$$\begin{aligned}\boldsymbol{\phi}_x(t_1, t_2) &= \tfrac{1}{2}E[\mathbf{x}(t_1)\mathbf{x}^+(t_2)] = \tfrac{1}{2}\mathbf{K}(t_1, t_2)E[\mathbf{x}(t_2)\mathbf{x}^+(t_2)] \\ &= \mathbf{K}(t_1, t_2)\mathbf{L}(t_2), \qquad t_1 \geq t_2,\end{aligned} \tag{11-165}$$

for the white noise $w(\tau)$ for $\tau > t_2$ is uncorrelated with $\mathbf{x}(t_2)$. Similarly

$$\boldsymbol{\phi}_x(t_1, t_2) = \mathbf{L}(t_1)\mathbf{K}^+(t_2, t_1), \qquad t_1 \leq t_2.$$

The autocovariance function of the signal itself is

$$\boldsymbol{\phi}_s(t, u) = \mathbf{C}\boldsymbol{\phi}_x(t, u)\mathbf{C}^+ \tag{11-166}$$

by (11-157).

For a stationary process with $\mathbf{L}$ constant, the covariance matrix of the state vector is

$$\boldsymbol{\phi}_x(\tau) = \tfrac{1}{2}E[\mathbf{x}(t + \tau)\mathbf{x}^+(t)] = \begin{cases} e^{\mathbf{F}\tau}\mathbf{L}_\infty, & \tau \geq 0, \\ \mathbf{L}_\infty e^{-\mathbf{F}^+\tau}, & \tau \leq 0, \end{cases}$$

provided that the linear system is stable, so that stationarity can be attained. For stability the eigenvalues of $\mathbf{F}$ must have negative real parts. The spectral density matrix of the state vector is the Fourier transform

$$\boldsymbol{\Phi}(\omega) = (i\omega\mathbf{I} - \mathbf{F})^{-1}\mathbf{L}_\infty - \mathbf{L}_\infty(i\omega\mathbf{I} + \mathbf{F}^+)^{-1}. \tag{11-167}$$

The spectral density of the signal is then

$$\boldsymbol{\Phi}_s(\omega) = \mathbf{C}\boldsymbol{\Phi}(\omega)\mathbf{C}^+. \tag{11-168}$$

For our example (11-158) the reader should show by substitution into (11-164) that the steady-state covariance matrix is

$$\mathbf{L}_\infty = \frac{Q}{2a} \begin{bmatrix} \dfrac{c + a}{c|c + a - ib|^2} & \dfrac{1}{c + a - ib} \\ \dfrac{1}{c + a + ib} & 1 \end{bmatrix} \tag{11-169}$$

and verify that (11-167) and (11-168) lead back to the spectral density in (11-158).

11.4.2 The Kalman–Bucy Equations

We turn now to determining the causal estimator of the signal $S_z(t)$, whose complex autocovariance function is $z\phi_s(t, u)$, in terms of the state-space model of the linear system that generates $S(t)$. It will then be possible to set up a linear system that generates the optimum statistic U for detecting the signal $s(t) = \operatorname{Re} S(t) \exp i\Omega t$ from the input $v(t) = \operatorname{Re} V(t) \exp i\Omega t$ as it arrives, and we shall be able to compute the Fredholm determinant in terms of the solution of a set of nonlinear differential equations known as the Kalman–Bucy equations. These equations play a central role in the prediction and filtering of random processes.

The signal $S_z(t)$ is given by

$$S_z(t) = \mathbf{C}\mathbf{x}_z(t), \tag{11-170}$$

where the n-vector $\mathbf{x}_z(t)$ obeys the state equations (11-155), except that the strength Q of the driving white noise is replaced by zQ.

Because both the signal $S_z(t)$ and the noise $N(t)$ are circular complex Gaussian random processes, the maximum-a-posteriori-probability (MAP) estimator of this signal at time $t = \tau$ is the causal linear estimator that minimizes the mean-square error

$$\tfrac{1}{2}E|S_z(\tau) - \hat{S}_z(\tau)|^2,$$

and it is the conditional expected value

$$\hat{S}_z(\tau) = E[S_z(\tau)|\ V(t), 0 \le t \le \tau, H_1]$$

of the signal, given the input $V(t) = S_z(t) + N(t)$ to the receiver during the interval $(0, \tau)$. By (11-170)

$$\hat{S}_z(\tau) = \mathbf{C}E[\mathbf{x}_z(\tau)|\ V(t), 0 \le t \le \tau, H_1]$$
$$= \mathbf{C}\hat{\mathbf{x}}_z(\tau),$$

where $\hat{\mathbf{x}}_z(\tau)$ is the minimum-mean-square-error estimator of the state vector $\mathbf{x}_z(\tau)$. The latter estimator will be a linear functional

$$\hat{\mathbf{x}}_z(\tau) = \int_0^\tau \mathbf{k}_z(\tau, t)V(t)\, dt \tag{11-171}$$

of the input $V(t)$ during the interval $(0, \tau)$; here $\mathbf{k}_z(\tau, t)$ is a column vector with n components.

The estimator $\hat{S}_z(\tau)$ is given by (11-117), whose kernel $m(\tau, u; z; \tau)$ is the solution of the integral equation (11-132) with $r = \tau$,

$$Nm(\tau, t; z; \tau) + z\int_0^\tau m(\tau, u; z; \tau)\phi_s(u, t)\, du = z\phi_s(\tau, t), \tag{11-172}$$

$$0 \le t \le \tau,$$

and we can therefore write

$$m(\tau, t; z; \tau) = \mathbf{C}(\tau)\mathbf{k}_z(\tau, t), \tag{11-173}$$

where $\mathbf{k}_z(\tau, t)$ obeys the vector integral equation

$$z\phi_x(\tau, t)\mathbf{C}^+(t) = N\mathbf{k}_z(\tau, t) + z\int_0^\tau \mathbf{k}_z(\tau, u)\phi_s(u, t)\, du. \tag{11-174}$$

Multiplying this equation on the left by the row vector $\mathbf{C}(\tau)$ and using (11-166) yield (11-172).

In Appendix I we show from (11-174) that the column vector $\mathbf{k}_z(\tau, t)$ obeys the vector differential equation

$$\frac{\partial \mathbf{k}_z(\tau, t)}{\partial \tau} = \mathbf{F}(\tau)\mathbf{k}_z(\tau, t) - \mathbf{k}_z(\tau, \tau)\mathbf{C}(\tau)\mathbf{k}_z(\tau, t), \qquad 0 \leq t \leq \tau. \tag{11-175}$$

Differentiating (11-171) with respect to τ and using (11-175), we find that the estimator $\hat{\mathbf{x}}_z(t)$ of the state vector obeys the differential equation

$$\frac{d\hat{\mathbf{x}}_z}{d\tau} = \mathbf{F}\hat{\mathbf{x}}_z(\tau) + \mathbf{k}_z(\tau, \tau)[V(\tau) - \mathbf{C}\hat{\mathbf{x}}_z(\tau)]. \tag{11-176}$$

Because nothing is known about the state vector $\mathbf{x}_z(t)$ at time $t = 0$, the best estimators of its components are zero at $t = 0$, and the initial condition on (11-176) is $\hat{\mathbf{x}}_z(0) = \mathbf{0}$.

The vector differential equation (11-175) must be solved with the initial condition $\mathbf{k}_z(t, t)$ at $\tau = t$. Both the equation itself and its initial condition depend on this vector function $\mathbf{k}_z(\tau, \tau)$. It is given by

$$\mathbf{k}_z(\tau, \tau) = N^{-1}\mathbf{P}_z(\tau)\mathbf{C}^+(\tau) \tag{11-177}$$

in terms of the $n \times n$ matrix function

$$\mathbf{P}_z(\tau) = z\boldsymbol{\phi}_x(\tau, \tau) - z\int_0^\tau \mathbf{k}_z(\tau, u)\mathbf{C}(u)\boldsymbol{\phi}_x(u, \tau)\, du. \tag{11-178}$$

If we multiply this integral equation by N^{-1} and, on the right, by the column vector $\mathbf{C}^+(\tau)$ and recall (11-166), we obtain (11-174) with $t = \tau$.

When the parameter z is real and positive, the matrix $\mathbf{P}_z(\tau)$ is the covariance matrix of the error $\mathbf{x}_z(\tau) - \hat{\mathbf{x}}_z(\tau)$ in the estimator of the state vector $\mathbf{x}_z(\tau)$:

$$\mathbf{P}_z(\tau) = \tfrac{1}{2}E\{[\mathbf{x}_z(\tau) - \hat{\mathbf{x}}_z(\tau)][\mathbf{x}_z^+(\tau) - \hat{\mathbf{x}}_z^+(\tau)]\}.$$

By the principle of orthogonality introduced in Sec. 6.1.4, the error vector $\mathbf{x}_z(\tau) - \hat{\mathbf{x}}_z(\tau)$ is orthogonal to the data $V(t)$ and hence to the estimator $\hat{\mathbf{x}}_z(\tau)$, which is a linear functional of the data:

$$\tfrac{1}{2}E\{[\mathbf{x}_z(\tau) - \hat{\mathbf{x}}_z(\tau)]\hat{\mathbf{x}}_z^+(\tau)\} = \mathbf{0}.$$

Therefore, by (11-171),

$$\mathbf{P}_z(\tau) = \tfrac{1}{2}E\{[\mathbf{x}_z(\tau) - \hat{\mathbf{x}}_z(\tau)]\mathbf{x}_z^+(\tau)\}$$
$$= z\boldsymbol{\phi}_x(\tau, \tau) - \tfrac{1}{2}E\int_0^\tau \mathbf{k}_z(\tau, u)V(u)\mathbf{x}_z^+(\tau)\, du,$$

and if we substitute $V(u) = N(u) + \mathbf{C}\mathbf{x}_z(u)$ and remember that the noise $N(t)$ is independent of the noise $w(t)$ driving the linear system and hence of the state vector $\mathbf{x}_z(t)$, we obtain (11-178).

The integral equation (11-178) is of no use to us at this point because it contains the unknown vector function $\mathbf{k}_z(\tau, u)$. In Appendix I we derive from it the nonlinear matrix differential equation

$$\frac{d\mathbf{P}_z}{dt} = \mathbf{F}\mathbf{P}_z + \mathbf{P}_z\mathbf{F}^+ + zQ\mathbf{G}\mathbf{G}^+ - N^{-1}\mathbf{P}_z\mathbf{C}^+\mathbf{C}\mathbf{P}_z. \tag{11-179}$$

This must be solved with the initial condition $\mathbf{P}_z(0) = z\boldsymbol{\phi}_x(0, 0) = z\mathbf{L}(0)$, which follows from (11-178) with $\tau = 0$. Because of its form it is called a *Riccati* equation. Once we have solved it, we have from (11-177) the vector $\mathbf{k}_z(\tau, \tau)$ needed in (11-175).

Equations (11-175) and (11-179) are the Kalman–Bucy equations [Kal61] for the minimum-mean-square-error estimator (11-171) of the state vector $\mathbf{x}_z(\tau)$, given an input of the form

$$V(t) = N(t) + \mathbf{C}\mathbf{x}_z(t)$$

during the interval $0 \le t \le \tau$. The vector $\mathbf{k}_z(\tau, \tau)$ is called the *Kalman gain*. Most derivations of these equations are statistical in nature, requiring the assumption that the parameter z is real and positive so that $S_z(t)$ can be considered a true random process with a real average power proportional to z. Here we regard these equations as purely a mathematical device for solving the integral equation (11-172) when the underlying signal $S(t)$ can be considered as generated by a linear system described by the state equations (11-155) and driven by circular complex white noise $w(t)$. In this way we are free to take z as a complex variable, as in evaluating saddlepoint integrals.

From the solution of these equations we can calculate the Fredholm determinant $D(z)$, for by (11-173) and (11-177)

$$m(\tau, \tau; z; \tau) = \mathbf{C}(\tau)\mathbf{k}_z(\tau, \tau) = N^{-1}\mathbf{C}(\tau)\mathbf{P}_z(\tau)\mathbf{C}^+(\tau) \qquad (11\text{-}180)$$

and by (11-130)

$$\ln D(z) = \frac{1}{N} \int_0^T \mathbf{C}(\tau)\mathbf{P}_z(\tau)\mathbf{C}^+(\tau) \, d\tau. \qquad (11\text{-}181)$$

This equation holds even for complex values of z and can be used, for instance, to calculate the moment-generating function $h_\eta(z)$ on a path of integration in (11-94) or (11-95) when computing false-alarm and detection probabilities for Gaussian stochastic signals as in Sec. 11.2.4.

The Riccati equation (11-179) is a set of n^2 coupled nonlinear differential equations of first order, and they must in most cases be solved numerically. They can be replaced by twice as many linear equations, embodied in the pair of linear matrix differential equations

$$\frac{d\mathbf{R}_1}{dt} = \mathbf{F}\mathbf{R}_1 + z Q\mathbf{G}\mathbf{G}^+\mathbf{R}_2, \qquad (11\text{-}182)$$

$$\frac{d\mathbf{R}_2}{dt} = N^{-1}\mathbf{C}^+\mathbf{C}\mathbf{R}_1 - \mathbf{F}^+\mathbf{R}_2, \qquad (11\text{-}183)$$

with initial conditions $\mathbf{R}_1(0) = z\mathbf{L}(0)$ and $\mathbf{R}_2(0) = \mathbf{I}$. From their solutions the error-covariance matrix is

$$\mathbf{P}_z(t) = \mathbf{R}_1(t)[\mathbf{R}_2(t)]^{-1},$$

as one can show by differentiating this and using (11-182) and (11-183),

$$\frac{d\mathbf{P}_z}{dt} = \frac{d\mathbf{R}_1}{dt}\mathbf{R}_2^{-1} - \mathbf{R}_1\mathbf{R}_2^{-1}\frac{d\mathbf{R}_2}{dt}\mathbf{R}_2^{-1}$$

$$= \mathbf{F}\mathbf{P}_z + z Q\mathbf{G}\mathbf{G}^+\mathbf{I} - \mathbf{P}_z(N^{-1}\mathbf{C}^+\mathbf{C}\mathbf{P}_z - \mathbf{F}^+\mathbf{I}),$$

which is (11-179). Furthermore,

$$\mathbf{P}_z(0) = \mathbf{R}_1(0) = z\mathbf{L}(0).$$

Now (11-181) becomes

$$\ln D(z) = \frac{1}{N} \int_0^T \mathbf{CR}_1 \mathbf{R}_2^{-1} \mathbf{C}^+ \, dt = \frac{1}{N} \int_0^T \mathrm{Tr}(\mathbf{C}^+ \mathbf{CR}_1 \mathbf{R}_2^{-1}) \, dt$$

by the rule for traces, $\mathrm{Tr}\, \mathbf{AB} = \mathrm{Tr}\, \mathbf{BA}$; as $\mathbf{CR}_1 \mathbf{R}_2^{-1} \mathbf{C}^+$ is a 1×1 matrix, it is equal to its trace. Substituting for $N^{-1} \mathbf{C}^+ \mathbf{CR}_1$ from (11-183), we find

$$\ln D(z) = \int_0^T \mathrm{Tr} \left[\frac{d\mathbf{R}_2}{dt} + \mathbf{F}^+ \mathbf{R}_2 \right] \mathbf{R}_2^{-1} \, dt$$

$$= \mathrm{Tr}\, \ln \mathbf{R}_2(T) + \int_0^T \mathrm{Tr}\, \mathbf{F}^+(t) \, dt$$

$$= \ln \det \mathbf{R}_2(T) + \int_0^T \mathrm{Tr}\, \mathbf{F}^+(t) \, dt,$$

and the Fredholm determinant is

$$D(z) = \det \mathbf{R}_2(T) \exp \left[\int_0^T \mathrm{Tr}\, \mathbf{F}^+(t) \, dt \right]. \tag{11-184}$$

Thus $D(z)$ can be computed directly from the solution of the set (11-182) and (11-183) of linear differential equations without any subsequent integration over $(0, T)$. This result is due to Collins [Col68]. For a time-variable linear system this solution would usually have to be carried out numerically.

As a simple example, suppose that the signal has the Lorentz spectral density (11-47). Then in the state equations (11-155) we can take $F = -\mu$, $G = 1$, $Q = 2\mu P_s$, and $C = 1$, all these now being scalars. The signal is $S(t) = x_1(t)$, which is the single state variable. The steady-state variance equation (11-164) is now

$$2LF = -QG^2 = -2\mu P_s,$$

whereupon $L = P_s$. Now (11-182) and (11-183) become a pair of first-order linear differential equations:

$$\frac{dR_1}{dt} = -\mu R_1 + zQR_2, \tag{11-185}$$

$$\frac{dR_2}{dt} = N^{-1} R_1 + \mu R_2, \tag{11-186}$$

with the initial conditions $R_1(0) = zP_s$, $R_2(0) = 1$. Their solution will have the form

$$\begin{aligned} R_1(t) &= A \exp \beta_1 t + B \exp \beta_2 t, \\ R_2(t) &= C \exp \beta_1 t + D \exp \beta_2 t. \end{aligned} \tag{11-187}$$

When we substitute these into (11-185) and (11-186) and separately equate the coefficients of $\exp \beta_1 t$ and $\exp \beta_2 t$ on both sides, we find

$$\begin{aligned} (\beta_1 + \mu)A &= zQC, & (\beta_2 + \mu)B &= zQD, \\ (\beta_1 - \mu)C &= N^{-1} A, & (\beta_2 - \mu)D &= N^{-1} B, \end{aligned}$$

and these are consistent only if

$$\beta_i^2 - \mu^2 = \frac{zQ}{N} = \frac{2\mu P_s z}{N}, \qquad i = 1, 2.$$

We can therefore set $\beta_1 = \beta$ and $\beta_2 = -\beta$, with β given by (11-127). The initial conditions yield the equations

$$A + B = R_1(0) = P_s z,$$

$$C + D = R_2(0) = 1.$$

Solving these for the coefficients A, B, C, and D, we obtain

$$A = P_s z \frac{\beta + \mu}{2\beta}, \qquad\qquad B = P_s z \frac{\beta - \mu}{2\beta},$$

$$C = P_s z \frac{\beta + \mu}{2N\beta(\beta - \mu)}, \qquad D = -P_s z \frac{\beta - \mu}{2N\beta(\beta + \mu)}.$$

Substituting these into (11-187) and dividing, we find for the error covariance function

$$P(t) = \frac{R_1(t)}{R_2(t)} = 2\mu P_s z \left[\frac{(\beta + \mu)e^{\beta t} + (\beta - \mu)e^{-\beta t}}{(\beta + \mu)^2 e^{\beta t} - (\beta - \mu)^2 e^{-\beta t}} \right].$$

By (11-180) and (11-129) this is $Nm(t, t; z; t) = N^2 zh(t, t; z)$ in agreement with (11-127). Substituting our solution for $R_2(t)$ into (11-184), on the other hand, we obtain the Fredholm determinant (11-128) directly.

A particular example with $n = 2$ has been worked out in detail by Kerr [Ker89].

11.4.3 The Schweppe Likelihood-ratio Receiver

The optimum statistic U for detecting the Gaussian stochastic signal Re $S(t) \exp i\Omega t$ in white noise has been shown to be realizable as in (11-146). When the complex envelope $S(t)$ can be represented as the output $S(t) = \mathbf{C}\mathbf{x}(t)$ of a linear system described as in (11-155) by a set of state equations, the causal estimator needed in (11-146) is

$$\hat{S}_1(t) = \mathbf{C}\hat{\mathbf{x}}(t),$$

where by (11-176) the components of the estimator $\hat{\mathbf{x}}(t)$ of the state vector satisfy the set of differential equations

$$\frac{d\hat{\mathbf{x}}}{dt} = \mathbf{F}\hat{\mathbf{x}}(t) + \mathbf{k}_1(t, t)[V(t) - \hat{S}_1(t)]. \tag{11-188}$$

The Kalman gain vector $\mathbf{k}_1(t, t)$ is determined in advance from the solution of the Riccati equation (11-179) with $z = 1$:

$$\mathbf{k}_1(t, t) = N^{-1}\mathbf{P}_1(t)\mathbf{C}^+(t).$$

The operation of the receiver is shown schematically in Fig. 11-7. The input $V(t)$ is turned on at time $t = 0$, at which time all state variables are zero. From the block marked I_1 the components of the estimate $\hat{\mathbf{x}}(t)$ emerge continually thereafter. (The double lines in the diagram indicate conduction of n vector components.) From that estimate the block "$\mathbf{C}$" generates the causal estimator of the signal by multiplying

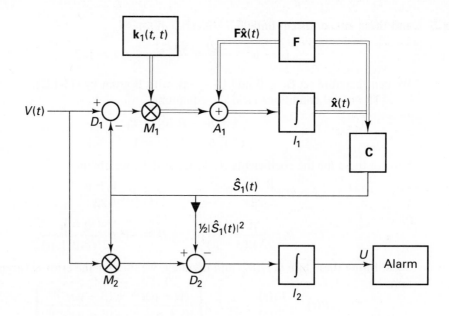

Figure 11-7. Schweppe likelihood-ratio receiver.

the vector $\hat{\mathbf{x}}(t)$ by the row vector $\mathbf{C}$. It is subtracted from the input $V(t)$ at point D_1, and at the multiplier M_1 the difference $V(t) - \hat{S}_1(t)$ multiplies each component of the Kalman gain vector $\mathbf{k}_1(t, t)$. To each of the components of the product the components of the output $\mathbf{F}\hat{\mathbf{x}}(t)$ of the box "$\mathbf{F}$" are added at point A_1. That box carries out the multiplication of the vector $\hat{\mathbf{x}}(t)$ by the $n \times n$ matrix $\mathbf{F}(t)$.

These operations could be carried out at the carrier frequency Ω of the input or, more conveniently, at some lower intermediate frequency down to which the input Re $V(t)$ exp $i\Omega t$ has been beaten. The integrator I_1 can then be a narrowband simply resonant circuit, resonant at the carrier or intermediate frequency and having a bandwidth much smaller than the bandwidth occupied by the input. The elements of the Kalman gain $\mathbf{k}_1(t, t)$ are low-pass functions of time.

At the multiplier M_2 the components of double-frequency 2Ω are filtered off, so that its output is the low-pass signal

$$[\text{Re } V(t) \, e^{i\Omega t}][\text{Re } \hat{S}_1(t) \, e^{i\Omega t}] \rightarrow \text{Re } \hat{S}_1^*(t) V(t).$$

From this the output $\frac{1}{2}|\hat{S}_1(t)|^2$ of the rectifier is subtracted at D_2, and their difference is integrated at I_2. At time $t = T$ the output of this low-pass integrator equals the optimum statistic U. The box marked "Alarm" compares this statistic with the decision level U_0 and reports the presence of a signal if $U \geq U_0$. As the formulation of the optimum detector in (11-146) and its generation by way of the estimator $\hat{\mathbf{x}}(t)$ of a state vector were proposed by Schweppe [Sch65], this system has been termed the Schweppe likelihood-ratio receiver.

11.5 DETECTION OF A COHERENT SIGNAL AND A STOCHASTIC SIGNAL

11.5.1 The Optimum Detector

When a narrowband signal can pass to the receiver both directly and via a scattering medium such as the ionosphere, the receiver is confronted with the choice between two hypotheses

$$V(t) = N(t), \qquad\qquad\qquad (H_0)$$

$$V(t) = N(t) + S(t) + R(t), \qquad (H_1)$$

about the complex envelope of its input $v(t) = \text{Re } V(t) \exp i\Omega t$ during an observation interval $0 \leq t \leq T$. As before in this chapter, $S(t)$ is a realization of a circular complex Gaussian random process with complex autocovariance function $\phi_s(t, u)$. The signal $\text{Re } R(t) \exp i\Omega t$ is deterministic; it is called the *specular component* of the input. We assume that its complex envelope is completely known even unto its r-f phase; the receiver is synchronized with the transmitter. As in most of our work here, we again assume that the noise $\text{Re } N(t) \exp i\Omega t$ is white with unilateral spectral density N.

The optimum detector is most directly derived by expressing the input and its constituents as Karhunen–Loève expansions in terms of the eigenfunctions $f_k(t)$ of the autocovariance function $\phi_s(t, u)$, defined as in (11-74):

$$V(t) = \sum_k V_k f_k(t), \qquad S(t) = \sum_k S_k f_k(t), \qquad R(t) = \sum_k R_k f_k(t).$$

Under hypothesis H_0 the joint probability density function of the real and imaginary parts of the first n of the coefficients, $\mathbf{V}^{(n)} = (V_1, V_2, \ldots, V_n)$, is

$$p_0(\mathbf{V}^{(n)}) = (2\pi N)^{-n} \exp\left[-\sum_{k=1}^{n} \frac{|V_k|^2}{2N}\right]; \qquad (11\text{-}189)$$

compare (3-47). Under hypothesis H_1 the independent circular complex Gaussian random variables V_k have expected values R_k, and the variances of their real and imaginary parts are $N(1 + \lambda_k)$, where the λ_k are the eigenvalues of $N^{-1}\phi_s(t, u)$ as in (11-74). The corresponding joint probability density function is thereupon

$$p_1(\mathbf{V}^{(n)}) = [2\pi N(1 + \lambda_k)]^{-n} \exp\left[-\sum_{k=1}^{n} \frac{|V_k - R_k|^2}{2N(1 + \lambda_k)}\right], \qquad (11\text{-}190)$$

as in (3-49). Dividing and passing to the limit of an infinite number of coefficients, we obtain the likelihood ratio

$$\Lambda(\mathbf{V}) = \frac{1}{D(1)} \exp\left\{\frac{1}{2N} \sum_{k=1}^{\infty} \left[|V_k|^2 - \frac{|V_k - R_k|^2}{1 + \lambda_k}\right]\right\}$$

$$= \frac{1}{D(1)} \exp\left\{\frac{1}{2N} \sum_{k=1}^{\infty} \left[\frac{\lambda_k|V_k|^2}{1 + \lambda_k} + \frac{2 \text{ Re } R_k^* V_k}{1 + \lambda_k} - \frac{|R_k|^2}{1 + \lambda_k}\right]\right\}, \qquad (11\text{-}191)$$

where $D(\cdot)$ is the Fredholm determinant (11-76).

A sufficient statistic for deciding between the hypotheses is therefore $U + W$, where

$$U = \frac{1}{2N} \sum_{k=1}^{\infty} \frac{\lambda_k |V_k|^2}{1 + \lambda_k}$$

is given by (11-35) with the kernel $h(t, s) = h(t, s; 1)$ the solution of (11-53) as before, and

$$W = \frac{1}{N} \, \text{Re} \sum_{k=1}^{\infty} Q_k^* V_k, \qquad Q_k = \frac{R_k}{1 + \lambda_k}. \qquad (11\text{-}192)$$

As in Sec. 3.3.1, the latter can be written

$$W = \frac{1}{N} \, \text{Re} \int_0^T Q^*(t) V(t) \, dt,$$

where $Q(t)$ is the solution of the integral equation

$$R(t) = Q(t) + \frac{1}{N} \int_0^T \phi_s(t, u) Q(u) \, du, \qquad 0 \le t \le T. \qquad (11\text{-}193)$$

The term U can be generated by any of the methods described in Sec. 11.2.4; the term W is the output at time $t = T$ of a narrowband filter matched to the signal $N^{-1} \, \text{Re} \, Q(t) \exp i\Omega t$. The optimum detector is thus a combination of that for the stochastic portion $\text{Re} \, S(t) \exp i\Omega t$ of the input and the optimum detector for the signal $\text{Re} \, R(t) \exp i\Omega t$ as though this were arriving in the presence of both the white noise and the stochastic signal.

11.5.2 The Performance of the Optimum Detector

Calculating the false-alarm and detection probabilities Q_0 and Q_d of the optimum receiver again requires the moment-generating functions $h_0(z)$ and $h_1(z)$ of the optimum statistic under the two hypotheses. The probability of error in the communication system is then, as before,

$$P_e = \zeta_0 Q_0 + \zeta_1 (1 - Q_d),$$

where ζ_0 and ζ_1 are the relative frequencies of 0's and 1's. For simplicity we suppose the decisions to be based on the logarithm $\ln \Lambda[V(t)]$ of the likelihood functional, which equals $U + W$ plus a constant depending on the signal $R(t)$. The statistic

$$U' = \frac{1}{2N} \sum_{k=1}^{\infty} \left[|V_k|^2 - \frac{|V_k - R_k|^2}{1 + \lambda_k} \right] \qquad (11\text{-}194)$$

is compared with a decision level U_0'; if $U' \ge U_0'$, hypothesis H_1 is chosen. Under hypothesis H_0 its moment-generating function is

$$h_0(z) = E(e^{-zU'} | H_0)$$

$$= \prod_{k=1}^{\infty} \frac{1}{2\pi N} \int_{-\infty}^{\infty} \int_{-\infty}^{\infty} \exp \left[-\frac{z}{2N} \left(|V_k|^2 - \frac{|V_k - R_k|^2}{1 + \lambda_k} \right) \right] \exp \left(-\frac{|V_k|^2}{2N} \right) dV_x \, dV_y$$

by (11-189). After combining the terms in the exponents and using (B-10) or, what is the same thing, the normalization integral for the circular complex Gaussian density function, we find

$$h_0(z) = \prod_{k=1}^{\infty} \frac{1 + \lambda_k}{1 + (1 + z)\lambda_k} \exp\left[\frac{z(1 + z)|R_k|^2}{2N[1 + (1 + z)\lambda_k]} \right]$$

$$= \frac{D(1)}{D(1 + z)} e^{z(1+z)J(1+z)},$$

(11-195)

where $D(\cdot)$ is again the Fredholm determinant and

$$J(z) = \frac{1}{2N} \sum_{k=1}^{\infty} \frac{|R_k|^2}{1 + z\lambda_k}.$$

(11-196)

By using the result of Problem 1-12, we find that the moment-generating function of the statistic U' under hypothesis H_1 is

$$h_1(z) = E(e^{-zU'} \mid H_1) = \frac{1}{D(z)} e^{z(z-1)J(z)}.$$

(11-197)

These reduce to the moment-generating functions of U as in (11-73) when $R(t) \equiv 0$.

There are several ways of calculating the term $J(z)$. As in (11-192), we can write it as

$$J(z) = \frac{1}{2N} \int_0^T R^*(t)Q(t; z) \, dt,$$

(11-198)

where $Q(t; z)$ is the solution of the integral equation

$$R(t) = Q(t; z) + \frac{z}{N} \int_0^T \phi_s(t, u)Q(u; z) \, du, \qquad 0 \le t \le T.$$

(11-199)

When the stochastic signal is stationary, $\phi_s(t, u) = \phi_s(t - u)$, and possesses a rational spectral density, the method of Appendix A can be utilized.

If the coherent signal is a sum of sine waves,

$$R(t) = \sum_m r_m \exp iw_m t,$$

(11-200)

and the stochastic signal has a rational spectral density, then as shown in Appendix J, the function $J(z)$ can be expressed as $J(z) = J_1 + J_2$, where

$$J_1 = \frac{T}{2N} \sum_m \sum_n \frac{r_n^* r_m}{H(w_m; z)} e(w_m T - w_n T), \qquad H(w; z) = 1 + \frac{z}{N}\Phi_s(w),$$

$$e(x) = \begin{cases} \dfrac{e^{ix} - 1}{ix}, & x \ne 0, \\ 1, & x = 0, \end{cases}$$

(11-201)

and

$$J_2 = -\frac{1}{2N \det \mathbf{G}} \det\begin{bmatrix} 0 & \mathbf{E} \\ \mathbf{F} & \mathbf{G} \end{bmatrix},$$

(11-202)

with $\mathbf{G}$ the $2n \times 2n$ matrix defined in (11-143). Here $\mathbf{E}$ is a $2n$-element row vector whose elements are

$$E_j = \sum_m r_m^* \left[\frac{\exp(\beta_j - iw_m)T - 1}{\beta_j - iw_m} \right], \qquad 1 \le j \le 2n,$$

(11-203)

and $\mathbf{F}$ is a $2n$-element column vector whose elements are

$$F_k = \sum_m \frac{r_m}{H(w_m; z)} \frac{\varepsilon_{km}}{\mu_k - iw_m}, \qquad 1 \le k \le 2n,$$

$$\varepsilon_{km} = \begin{cases} \exp iw_m T, & 1 \le k \le n, \\ 1, & n+1 \le k \le 2n. \end{cases} \qquad (11\text{-}204)$$

As in Sec. 11.3.4, the β_j are the roots of

$$1 + \frac{z}{N}\Phi_s(-ip) = 0, \qquad p = \beta_1, \beta_2, \dots, \beta_{2n},$$

which can be transformed into an algebraic equation of degree $2n$, and the μ_k are the $2n$ poles of $\Phi_s(-ip)$, arranged so that the first n lie in the right half of the p-plane and the rest lie in the left half; $\mu_{k+n} = -\mu_k^*$. The coefficients r_m might, for instance, be calculated by setting the angular frequencies w_m equal to integral multiples of $2\pi/T$ and taking the Fourier transform of the signal envelope $R(t)$ by means of a fast Fourier transform algorithm. In that case

$$e(w_m T - w_n T) = \delta_{nm}$$

in (11-201).

Another method of calculating the term $J(z)$ in (11-196) arises from the estimator-correlator approach in Sec. 11.3.5. We write it as

$$J(z) = \frac{1}{2N} \sum_{k=1}^{\infty} \left[|R_k|^2 - \frac{z\lambda_k|R_k|^2}{1 + z\lambda_k} \right]$$

$$= \frac{1}{2N} \int_0^T |R(\tau)|^2 \, d\tau - \frac{z}{2} \int_0^T \int_0^T R^*(r)h(r, t; z; T)R(t) \, dr \, dt, \qquad (11\text{-}205)$$

where

$$h(r, t; z; T) = \frac{1}{N} \sum_k \frac{\lambda_k f_k(t)f_k^*(s)}{1 + z\lambda_k} \qquad (11\text{-}206)$$

is the temporal function corresponding as in (11-16) to the matrix

$$\mathbf{H}(z) = N^{-1}(N\mathbf{I} + z\,\boldsymbol{\phi}_s)^{-1}\boldsymbol{\phi}_s$$

in (11-124).

We want to carry through the same type of analysis as led to (11-146), but for the "signal" $S_z(t)$ whose complex autocovariance function is $z\phi_s(t, u)$. We need to adapt (11-121) for complex z. If the $\lambda_k(\tau)$ are the eigenvalues of the integral equation

$$\lambda_k(\tau)f_k(t; \tau) = \frac{1}{N} \int_0^\tau \phi_s(t, u)f_k(u; \tau) \, du, \qquad 1 \le t \le \tau,$$

then as in (11-206)

$$h(r, t; z; \tau) = \frac{1}{N} \sum_k \frac{\lambda_k(\tau)f_k(r; \tau)f_k^*(t; \tau)}{1 + z\lambda_k(\tau)},$$

and interchanging r and t we find

$$h(t, r; z; \tau) = [h(r, t; z^*; \tau)]^* = h^*(r, t; z^*; \tau).$$

Thus Siegert's relation (11-121) for the derivative of $h(r, t; z; \tau)$ with respect to τ can be written

$$h_\tau(r, t; z; \tau) = -Nz h^*(\tau, r; z^*; \tau) h(\tau, t; z; \tau). \qquad (11\text{-}207)$$

Defining

$$U_z(\tau) = \tfrac{1}{2} \int_0^\tau \int_0^\tau R^*(r) h(r, t; z; \tau) R(t)\, dr\, dt$$

and following the procedure that led to (11-146), we find

$$\frac{dU_z}{d\tau} = \tfrac{1}{2} R^*(\tau) \int_0^\tau h(\tau, t; z; \tau) R(t)\, dt + \text{c.c.}$$

$$+ \tfrac{1}{2} \int_0^\tau \int_0^\tau R^*(r) h_\tau(r, t; z; \tau) R(t)\, dr\, dt. \qquad (11\text{-}208)$$

Remembering (11-117) and using (11-129), we define the *causal quasi-estimate*

$$\hat{R}(t; z) = \int_0^\tau m(\tau, u; z; \tau) R(u)\, du = Nz \int_0^\tau h(\tau, u; z; \tau) R(u)\, du, \qquad (11\text{-}209)$$

whereupon, with (11-207), we can write

$$\frac{dU_z}{d\tau} = \frac{1}{2Nz}[R^*(\tau)\hat{R}(\tau; z) + \hat{R}^*(\tau; z^*)R(\tau)]$$

$$- \frac{Nz}{2} \int_0^\tau \int_0^\tau h^*(\tau, r; z^*; \tau) R^*(\tau) h(\tau, t; z; \tau) R(t)\, dr\, dt$$

$$= \frac{1}{2Nz}[R^*(\tau)\hat{R}(\tau; z) + \hat{R}^*(\tau; z^*)R(\tau) - \hat{R}^*(\tau; z^*)\hat{R}(\tau; z)],$$

and integrating over $0 \leq \tau \leq T$, we obtain for (11-205)

$$J(z) = \frac{1}{2N} \int_0^T [R^*(\tau) - \hat{R}^*(\tau; z^*)][R(\tau) - \hat{R}(\tau; z)]\, d\tau. \qquad (11\text{-}210)$$

Whatever means we have for determining the causal estimator $\hat{S}_1(t)$ of the complex envelope $S(t)$ of the signal can be utilized to compute the causal quasi-estimate $\hat{R}(\tau; z)$ needed in evaluating the function $J(z)$ by this expression.

When in particular the stochastic signal $S(t)$ can be generated as the output of a linear system driven by white noise, as in Sec. 11.4, the function $\hat{R}(t; z)$ can be determined in the same way as we there showed one can determine the estimator $\hat{S}_z(t)$. By (11-209) and (11-173),

$$\hat{R}(\tau; z) = \mathbf{C}\hat{\mathbf{r}}_z(\tau)$$

with $\hat{\mathbf{r}}_z(\tau)$ a column vector of n components and $\mathbf{C}$ the row vector in (11-157). As in (11-176), the former obeys the differential equation

$$\frac{d\hat{\mathbf{r}}_z}{d\tau} = \mathbf{F}\hat{\mathbf{r}}_z(\tau) + \mathbf{k}_z(\tau, \tau)[R(\tau) - \mathbf{C}\hat{\mathbf{r}}_z(\tau)],$$

with the initial condition $\hat{\mathbf{r}}_z(0) = \mathbf{0}$. The $n \times n$ matrix $\mathbf{F}$ specifies the dynamics of the linear system as in (11-155), and $\mathbf{k}_z(\tau, \tau)$ is again the Kalman gain in (11-177). The causal quasi-estimate $\hat{R}(\tau; z)$ needed in (11-210) can thus be computed by solving this set of n differential equations, once the Kalman gain has been determined as in Sec. 11.4 by solving the Riccati equation (11-179). This computation can evaluate $J(z)$ and, through (11-181) or (11-184), the Fredholm determinant needed to compute the moment-generating functions of the decision statistic U' at all points of the contours of integration in (11-94) and (11-95) when evaluating the false-alarm and detection probabilities for the receiver of a combination of a coherent and a stochastic signal.

11.6 DISCRETE-TIME PROCESSING[1]

11.6.1 The Optimum Statistic

In order to process the input $v(t)$ digitally, it can be filtered to remove noise with frequencies far from those of the signal to be detected, and the filtered input can be sampled at discrete times uniformly spaced. The decision about the presence or absence of a signal will then be based on a finite number n of samples $V_1, V_2, \ldots, V_n$ of the complex envelope $V(t)$ of the input. We gather these into a column vector $\mathbf{V}$.

When as throughout this chapter the signal and the noise are circular complex Gaussian random processes, the samples V_k are circular complex Gaussian random variables. Under hypotheses H_0 and H_1 they have zero expected values, their $n \times n$ complex covariance matrices are $\boldsymbol{\phi}_0$ and $\boldsymbol{\phi}_1$, respectively, and

$$\boldsymbol{\phi}_1 = \boldsymbol{\phi}_0 + \boldsymbol{\phi}_s,$$

where $\boldsymbol{\phi}_s$ is the $n \times n$ complex covariance matrix of the samples S_k of the complex envelope $S(t)$ of the signal.

By the same procedure as in (11-30) through (11-33) we find that the optimum strategy for deciding between hypotheses H_0 and H_1 is to compare the statistic

$$U = \tfrac{1}{2}\mathbf{V}^+\mathbf{H}\mathbf{V}$$

with a decision level U_0 and to choose hypothesis H_1 if $U \geq U_0$, and otherwise H_0. The level U_0 is set to induce a preassigned false-alarm probability $Q_0 = \Pr(U \geq U_0|\,H_0)$. Now $\mathbf{H}$ is an $n \times n$ matrix given by

$$\mathbf{H} = \boldsymbol{\phi}_0^{-1} - \boldsymbol{\phi}_1^{-1}$$

as in (11-32); it can be computed once for all and stored in the computer that is carrying out this discrete-time processing of the input.

The matrix $\mathbf{H}$ can be decomposed as

$$\mathbf{H} = \mathbf{G}^+\mathbf{G}, \tag{11-211}$$

where $\mathbf{G}$ is a lower-triangular matrix $\|g_{ij}\|$ with $g_{ij} \equiv 0$, $j > i$. This decomposition into the lower-triangular matrix $\mathbf{G}$ and the upper-triangular matrix $\mathbf{G}^+$ can be carried out by standard algorithms [Pre86, pp. 31–8]. The detection statistic can then be written as

$$U = \tfrac{1}{2}\mathbf{W}^+\mathbf{W} = \tfrac{1}{2}\sum_{i=1}^{n} |W_i|^2, \tag{11-212}$$

---

[1]Chapter 12 does not draw upon the material in this section.

where the elements of the column vector $\mathbf{W} = \mathbf{GV}$ are

$$W_i = \sum_{j=1}^{i} g_{ij} V_j. \tag{11-213}$$

Thus W_i depends only on the present sample V_i and the previous samples V_1, V_2, $\ldots$, V_{i-1}, and (11-213) represents a causal discrete-time linear filtering of the input samples V_j. By (11-212) the statistic U can be computed in real time, provided that the computer works fast enough to evaluate (11-213) between samplings.

We assume henceforth that under hypothesis H_0 the real and imaginary parts of the samples $V_k = V_{kx} + iV_{ky}$ are statistically independent and possess variances equal to some number N. For the sake of a less cumbersome notation, we divide all the data—signal and noise—by $\sqrt{N}$, in effect setting N equal to 1. Then under hypothesis H_0 the variances of the real and imaginary parts of the data are

$$\tfrac{1}{2}E(|V_k|^2|\,H_0) = \mathrm{Var}_0\,V_{kx} = \mathrm{Var}_0\,V_{ky} \equiv 1, \tag{11-214}$$

and their covariance matrix $\boldsymbol{\phi}_0$ equals the $n \times n$ identity matrix $\mathbf{I}$. Furthermore $\boldsymbol{\phi}_1 = \mathbf{I} + \boldsymbol{\phi}_s$, and the kernel of the optimum statistic U is

$$\mathbf{H} = \mathbf{I} - (\mathbf{I} + \boldsymbol{\phi}_s)^{-1} = \boldsymbol{\phi}_s(\mathbf{I} + \boldsymbol{\phi}_s)^{-1}. \tag{11-215}$$

When $n \gg 1$, computing the matrices $\mathbf{H}$ and $\mathbf{G}$ and storing both them and the data $\mathbf{V}$ may exceed the capacities of one's computer, and more efficient methods of generating the optimum statistic U must be sought. To that end we define

$$\mathbf{V}^{(k)} = (V_1, V_2, \ldots, V_k)$$

for our temporally ordered samples; $\mathbf{V} = \mathbf{V}^{(n)}$. Under hypothesis H_1 the joint probability density function of the data $\mathbf{V}$ can be expressed as

$$p_1(\mathbf{V}) = p_1(V_n|\,\mathbf{V}^{(n-1)})p_1(\mathbf{V}^{(n-1)}),$$

and continuing thus we write it as

$$p_1(\mathbf{V}) = p_1(V_n|\,\mathbf{V}^{(n-1)})p_1(V_{n-1}|\,\mathbf{V}^{(n-2)}) \ldots$$
$$\cdot\, p_1(V_k|\,\mathbf{V}^{(k-1)}) \ldots p_1(V_2|\,V_1)p_1(V_1). \tag{11-216}$$

Because the data are circular complex Gaussian random variables,

$$p_1(V_k|\,\mathbf{V}^{(k-1)}) = \frac{1}{2\pi W_k}\exp\left[-\frac{1}{2W_k}\left|V_k - E_1(V_k|\,\mathbf{V}^{(k-1)})\right|^2\right] \tag{11-217}$$

where

$$W_k = \tfrac{1}{2}E_1\left[\left|V_k - E_1(V_k|\,\mathbf{V}^{(k-1)})\right|^2\right] \tag{11-218}$$

is the conditional variance of the real and imaginary parts of the datum V_k, given the past input samples V_1, V_2, $\ldots$, V_{k-1}. Here E_1 denotes an expected value under hypothesis H_1. Because $V_k = S_k + N_k$ and the samples N_k of the noise are independent of those of the signal and of each other,

$$E_1(V_k|\,\mathbf{V}^{(k-1)}) = E_1(S_k|\,\mathbf{V}^{(k-1)}) = \hat{S}_k,$$

and $\hat{S}_k$ is the minimum-mean-square-error estimator of the signal sample S_k based on the data collected before "time" k. Furthermore,

$$W_k = \tfrac{1}{2}E_1(|V_k - \hat{S}_k|^2) = 1 + \tfrac{1}{2}E_1(|S_k - \hat{S}_k|^2).$$

Remembering that the joint probability density function of the data $\mathbf{V}$ under hypothesis H_1 is

$$p_1(\mathbf{V}) = (2\pi)^{-n}(\det \boldsymbol{\phi}_1)^{-1} \exp(-\tfrac{1}{2}\mathbf{V}^+\boldsymbol{\phi}_1^{-1}\mathbf{V}),$$

we see from (11-216) and (11-217) that

$$\det \boldsymbol{\phi}_1 = \det(\mathbf{I} + \boldsymbol{\phi}_s) = \prod_{k=1}^{n} W_k. \qquad (11\text{-}219)$$

Now using (11-217) to form the likelihood ratio for our decision between hypotheses H_0 and H_1, we find that the optimum statistic U can be written as

$$U = \tfrac{1}{2}\sum_{k=1}^{n}|V_k|^2 - \tfrac{1}{2}\sum_{k=1}^{n}\frac{|V_k - \hat{S}_k|^2}{W_k}. \qquad (11\text{-}220)$$

Because $\hat{S}_k$ depends only on the past data $\mathbf{V}^{(k-1)}$, this form too can be computed in real time. Our task then is to determine that estimator and the quantities W_k in an efficient manner.

11.6.2 The Kalman Method

The amount of data storage and the duration of computations can be limited when the sequence of signal samples S_k can be modeled as the output of a linear system driven by white noise, the discrete-time counterpart of the dynamical systems considered in Sec. 11.4. The estimates $\hat{S}_k$ can then be calculated by the Kalman equations for the one-step linear predictor of the signal samples on the basis of the data $\{V_k\}$ [Kal60], [Sch65], [Lar79, vol. 2, pp. 120–9], [Hel91, pp. 523–39], [Pap91, pp. 515–24]. The system is described by p circular complex Gaussian state variables, whose values at time k we denote by $y_1[k]$, $y_2[k]$, ... , $y_p[k]$ and collect into a column vector $\mathbf{Y}[k]$. The system evolves in accordance with the dynamical equations

$$\mathbf{Y}[k + 1] = \mathbf{A}(k)\mathbf{Y}[k] + \mathbf{G}(k + 1)\, e[k + 1],$$

where $\mathbf{A}(k)$ is a $p \times p$ matrix, $\mathbf{G}(k + 1)$ is a p-element column vector, and $e[1]$, $e[2]$, ... constitute a sequence of statistically independent "white noise" variates whose real and imaginary parts have zero expected values and variances

$$Q(k) = \tfrac{1}{2}E(|e[k]|^2).$$

The signal sample at time k is a linear combination of the state variables,

$$S_k = \mathbf{D}(k)\mathbf{Y}[k],$$

where $\mathbf{D}(k)$ is a p-element row vector.

When the sequence of signal samples $\{S_k\}$ is a stationary random process, its covariance matrix $\boldsymbol{\phi}_s$ has the Toeplitz form

$$\boldsymbol{\phi}_{s,ij} = \boldsymbol{\phi}_{i-j}, \tag{11-221}$$

and one can define for it the discrete-time spectral density

$$F(w) = \sum_{k=-\infty}^{\infty} \boldsymbol{\phi}_k w^{-k}, \qquad w = e^{i\theta}, \qquad -\pi < \theta \leq \pi,$$

[Hel91, pp. 290–6], [Pap91, pp. 332–6]. If in particular the process $\{S_k\}$ is an autoregressive moving-average (ARMA) process, this spectral density is a rational function of w:

$$F(w) = \frac{A(w)}{B(w)}, \tag{11-222}$$

where $A(w)$ and $B(w)$ are polynomials in w. The degree of $B(w)$ is $2p$; that of $A(w)$ is less than or equal to $2p$. The parameters of the linear system can then be determined much as in Sec. 11.4.1. One factors $F(w)$ into a part having its poles and zeros outside the unit circle $|w| = 1$ in the w-plane and a part having its poles and zeros inside the unit circle; see for instance [Hel91, pp. 325–33]. The method now to be described is more general, however, and it allows the system parameters to be functions of the time k and the sequence $\{S_k\}$ to be nonstationary.

At time k the complex covariance matrix of the state vector $\mathbf{Y}[k]$ is

$$\boldsymbol{\Sigma}(k) = \tfrac{1}{2}E_1(\mathbf{Y}[k]\mathbf{Y}^+[k]),$$

and the specification of the model must include its initial values $\boldsymbol{\Sigma}(1)$. We shall now summarize the equations that define the Kalman method for one-step prediction of a scalar random process—not the most general formulation, but one adequate for our purposes. Their derivation and interpretation can be found in the references just cited and in numerous texts on signal processing and Kalman filtering.

The estimator of the kth signal sample S_k is given by

$$\hat{S}_k = \mathbf{D}(k)\hat{\mathbf{Y}}[k|\, k-1], \tag{11-223}$$

where

$$\hat{\mathbf{Y}}[k|\, k-1] = E_1(\mathbf{Y}[k]|\, \mathbf{V}^{(k-1)})$$

is the one-step predictor of the state vector $\mathbf{Y}[k]$, given the previous data $V_1, V_2, \ldots, V_{k-1}$. The Kalman prediction theory determines it by

$$\hat{\mathbf{Y}}[k|\, k-1] = \mathbf{A}(k-1)\hat{\mathbf{Y}}[k-1],$$

where $\hat{\mathbf{Y}}[k]$ is the estimator of the state vector at time k, $k = 1, 2, \ldots$. It evolves continually through the equation

$$\hat{\mathbf{Y}}[k] = \hat{\mathbf{Y}}[k|\, k-1] + \mathbf{K}(k)(V_k - \hat{S}_k),$$

where $\mathbf{K}(k)$ is a p-element column vector known as the *Kalman gain*. This is determined in advance by the equations

$$\mathbf{K}(k) = W_k^{-1}\mathbf{P}(k|\, k-1)\mathbf{D}^+(k),$$

where

$$W_k = \mathbf{D}(k)\mathbf{P}(k|k-1)\mathbf{D}^+(k) + 1 \qquad (11\text{-}224)$$

is the variance of the innovation $V_k - \hat{S}_k$, and $\mathbf{P}(k|k-1)$ is the $p \times p$ covariance matrix of the error in the predictor $\hat{\mathbf{Y}}[k|k-1]$.

The evolution of the error covariance matrix is described by the equations

$$\mathbf{P}(k|k-1) = \mathbf{A}(k-1)\mathbf{P}(k-1)\mathbf{A}^+(k-1) + Q(k)\mathbf{G}(k)\mathbf{G}^+(k),$$
$$\mathbf{P}(k) = [\mathbf{I} - \mathbf{K}(k)\mathbf{D}(k)]\mathbf{P}(k|k-1). \qquad (11\text{-}225)$$

The initial conditions for this process are

$$\hat{S}_1 = 0, \qquad \hat{\mathbf{Y}}(1|0) = \mathbf{0}, \qquad \hat{\mathbf{Y}}(1) = \mathbf{K}(1)V_1,$$
$$\mathbf{P}(1|0) = \mathbf{\Sigma}(1),$$
$$\mathbf{K}(1) = W_1^{-1}\mathbf{\Sigma}(1)\mathbf{D}^+(1),$$
$$W_1 = \mathbf{D}(1)\mathbf{\Sigma}(1)\mathbf{D}^+(1) + 1,$$
$$\mathbf{P}(1) = [\mathbf{I} - \mathbf{K}(1)\mathbf{D}(1)]\mathbf{\Sigma}(1).$$

At each time k the receiver puts the estimate $\hat{S}_k$ given by (11-223) and the variance W_k given by (11-224) into (11-220) and accumulates the sums in that equation, obtaining the optimum statistic U at the end of the procedure.

11.6.3 Stationary Data

When the sequence $\{V_k\}$ is stationary, the elements of the Hermitian matrix $\boldsymbol{\phi}_1$ have the form

$$\tfrac{1}{2}E_1(V_k V_m^*) = \phi_{1,km} = \delta_{km} + \phi_{k-m} = \delta_{km} + \phi_{m-k}^*. \qquad (11\text{-}226)$$

The conditional expected values appearing in (11-217) are linear combinations of the past data $\mathbf{V}^{(k-1)}$, and we can write them as

$$\hat{V}_k = E_1(V_k | \mathbf{V}^{(k-1)}) = \sum_{m=0}^{k-2} f_m^{(k)} V_{k-1-m}. \qquad (11\text{-}227)$$

The $k-1$ coefficients $f_m^{(k)}$ can be calculated recursively by what is known as the *Durbin algorithm* [Dur60]. We need to modify it slightly in order to accommodate the complex data and the Hermitian covariance matrix in (11-226); in so doing, we follow the exposition in [Bla85, pp. 359–61].

By the principle of orthogonality, which we introduced in Sec. 6.1.7, the errors $V_k - \hat{V}_k$ are orthogonal to the data $V_1, V_2, \ldots, V_{k-1}$; that is,

$$\tfrac{1}{2}E_1[(V_k - \hat{V}_k)V_r^*] \equiv 0, \qquad r = 1, 2, \ldots, k-1. \qquad (11\text{-}228)$$

Putting (11-227) into this and using (11-226), we obtain the set of $k-1$ linear equations

$$\phi_{k-r} = f_{k-1-r}^{(k)} + \sum_{p=0}^{k-2} \phi_{k-1-p-r} f_p^{(k)}, \qquad r = 1, 2, \ldots, k-1,$$

or

$$f_t^{(k)} + \sum_{m=0}^{k-2} \phi_{t-m} f_m^{(k)} = \phi_{t+1}, \qquad 0 \le t \le k - 2. \qquad (11\text{-}229)$$

Initially

$$\hat{V}_1 = 0, \qquad \hat{V}_2 = f_0^{(2)} V_1, \qquad f_0^{(2)} = \frac{\phi_1}{1 + \phi_0}.$$

Given the $(k - 1)$-element column vector $\{f_m^{(k)}\}$ that solves (11-229) at the kth stage of this procedure, $k \ge 2$, one computes the solution for stage $k + 1$ as

$$f_r^{(k+1)} = f_r^{(k)} - \beta_k f_{k-2-r}^{(k)^*}, \qquad 0 \le r \le k - 2,$$
$$f_{k-1}^{(k+1)} = \beta_k, \qquad\qquad\qquad\qquad (11\text{-}230)$$

where

$$\beta_k = \frac{\phi_k - \Gamma_k}{\phi_0 - r_k},$$

with

$$\Gamma_k = \sum_{m=0}^{k-2} \phi_{k-1-m} f_m^{(k)}, \qquad r_k = \sum_{m=0}^{k-2} \phi_{m+1} f_m^{(k)^*}. \qquad (11\text{-}231)$$

The error variance W_k in (11-217) is

$$W_k = \tfrac{1}{2} E_1[(V_k - \hat{V}_k)(V_k^* - \hat{V}_k^*)].$$

Because $\hat{V}_k$ is a linear combination of the data V_r for $r = 1, 2, \ldots, k - 1$ through (11-227), this reduces to

$$W_k = \tfrac{1}{2} E_1[|V_k|^2 - V_k \hat{V}_k^*]$$

by the conjugate of (11-228). Substituting the conjugate of (11-227) and using (11-226), we find that this can be expressed as

$$W_k = 1 + \phi_0 - \sum_{m=0}^{k-2} f_m^{(k)^*} \phi_{m+1} = 1 + \phi_0 - r_k \qquad (11\text{-}232)$$

by (11-231).

The optimum statistic for detecting the stationary signal sequence $\{S_k\}$ in white noise is thus

$$U = \tfrac{1}{2} \sum_{k=1}^{n} |V_k|^2 - \tfrac{1}{2} \sum_{k=1}^{n} \frac{|V_k - \hat{V}_k|^2}{W_k}$$

by (11-217). The $\hat{V}_k$ are computed from (11-227) and the W_k from (11-232) at time k once the solution vector $\{f_m^{(k)}\}$ has been updated from its previous components by (11-230) and (11-231). In contrast to the Kalman method of Sec. 11.6.2, whose storage requirements are independent of the number n of data, the present procedure requires storing both the data vector $\mathbf{V}^{(k)}$ as its components arrive and the k-element vector that solves (11-229), and these vectors eventually grow to length n. The number of computations at each stage is proportional to k and the total number proportional to n^2. The number of computations needed to compute the matrix $\mathbf{H}$ in the original form (11-215), on the other hand, will in general be on the order of n^3.

11.6.4 Performance of the Detector

If, as in Sec. 11.2, we ask for the performance of such a discrete-time receiver when the input signals have a total energy η times that for which the receiver was designed, we must seek the probability $Q_d(\eta)$ that the statistic U exceeds U_0 under a hypothesis H_η that the complex covariance matrix of the input samples is not $\boldsymbol{\phi}_1$, but

$$\boldsymbol{\phi}_\eta = \boldsymbol{\phi}_0 + \eta \boldsymbol{\phi}_s, \qquad (H_\eta). \qquad (11\text{-}233)$$

It can be computed as in (11-94) and (11-95) from the moment-generating function $h_\eta(z)$ of the statistic U under hypothesis H_η, and as in (11-68) this is given by

$$h_\eta(z) = E(e^{-Uz} \mid H_\eta) = [\det(\mathbf{I} + z\,\boldsymbol{\phi}_\eta \mathbf{H})]^{-1}. \qquad (11\text{-}234)$$

Unless the number n is excessive, this moment-generating function can be computed numerically at points z on the contour of integration in (11-94) or (11-95). The false-alarm probability is again $Q_0 = Q_d(0)$, and the probability of detecting the standard signal for which the receiver was designed is $Q_{ds} = Q_d(1)$.

When under hypothesis H_0 the input samples are uncorrelated as in (11-214), the calculation of the probability $Q_d(\eta)$ can proceed as in Sec. 11.2.2, in the equations of which the λ_k are now the eigenvalues of the $n \times n$ Hermitian matrix $\boldsymbol{\phi}_s$; the residue series will have only n terms. Under this circumstance, the moment-generating function of the statistic U is again given by (11-96) in terms of the now finite Fredholm determinant

$$D(z) = \det(\mathbf{I} + z\,\boldsymbol{\phi}_s) = \prod_{k=1}^{n}(1 + z\lambda_k). \qquad (11\text{-}235)$$

When $n \gg 1$, evaluating this $n \times n$ determinant can strain the capacities of all but the largest digital computers. The computation can be expedited if one possesses a simple model of the discrete-time random process $\{S_k\}$. We examine three situations: (1) the signal samples form an ARMA process, (2) they are generated as in Sec. 11.6.2 by a discrete-time linear system, and (3) they constitute as in Sec. 11.6.3 a stationary, but otherwise arbitrary random process.

11.6.4.1 ARMA process. When the spectral density $F(w)$ in (11-222) is a rational function of w, it possesses $2p$ poles $\mu_1, \mu_2, \ldots, \mu_{2p}$, of which $\mu_1, \mu_2, \ldots, \mu_p$—the *exterior* poles—lie outside the unit circle ($|w| = 1$), and the rest, $\mu_{p+1}, \mu_{p+2}, \ldots, \mu_{2p}$—the *interior* poles—lie within it. Here we are confronted with an ARMA process that is the discrete-time counterpart of the random process $S(t)$ treated in Sec. 11.3.4, and it can be shown that for $n > 2p$ the Fredholm determinant can be expressed in a form much like that in (11-142) [Hel89b]. We write the spectral density $F(w)$ as

$$F(w) = R + F'(w),$$

where $F'(w)$—not a derivative!—goes to zero as w goes to 0 or to ∞. The term $R = F(0)$ represents a "white" component of the signal process and is ordinarily zero.

Defining $\beta_1, \beta_2, \ldots, \beta_{2p}$ as the $2p$ roots of the equation

$$1 + zF(w) = 0, \qquad (11\text{-}236)$$

we take $\beta_1, \beta_2, \ldots, \beta_p$ as those roots that go into the p exterior poles and β_{p+1}, $\beta_{p+2}, \ldots, \beta_{2p}$ as those that go into the p interior poles as $z \to 0$. Then the Fredholm determinant is given by

$$D(z) = (1 + Rz)^n \prod_{k=p+1}^{2p} \mu_k^n \frac{\det \mathbf{G}_n(z)}{\det \mathbf{G}_0(z)}, \tag{11-237}$$

where $\mathbf{G}_n(z)$ is a $2p \times 2p$ matrix whose elements are

$$G_{jk}^{(n)} = \begin{cases} \dfrac{\beta_k^n}{\beta_k - \mu_j}, & 1 \le j \le p, & 1 \le k \le 2p, \\[2mm] \dfrac{1}{\beta_k - \mu_j}, & p + 1 \le j \le 2p, & 1 \le k \le 2p. \end{cases} \tag{11-238}$$

The elements of $\mathbf{G}_0(z)$ are obtained by setting $n = 0$. When $n \gg 1$, it is advisable to factor β_j^n out of the jth column of $\det \mathbf{G}_n(z)$ for $1 \le j \le p$ in order to avoid overflow.

As an example, suppose that $\{S_k\}$ is a discrete-time Gaussian Markov process so that its covariances have the form

$$\phi_{s,ij} = C\alpha^{|i-j|}, \qquad |\alpha| < 1. \tag{11-239}$$

Its spectral density is

$$F(w) = \frac{C(1 - \alpha^2)w}{(w - \alpha)(1 - \alpha w)},$$

and $R = 0$, $\mu_1 = \alpha^{-1}$, $\mu_2 = \alpha$. Then (11-236) becomes

$$1 + \frac{Kw}{(w - \alpha)(1 - \alpha w)} = 0, \qquad K = C(1 - \alpha^2)z, \tag{11-240}$$

which reduces to a quadratic equation

$$\alpha w^2 - (1 + \alpha^2 + K)w + \alpha = 0$$

and possesses two roots β_1 and $\beta_2 = \beta_1^{-1}$; as $z \to 0$, $\beta_1 \to \alpha^{-1}$ and $\beta_2 \to \alpha$.

The matrix in (11-238) is now

$$\mathbf{G}_n(z) = \begin{bmatrix} \dfrac{\beta_1^n}{\beta_1 - \alpha^{-1}} & \dfrac{\beta_2^n}{\beta_2 - \alpha^{-1}} \\[3mm] \dfrac{1}{\beta_1 - \alpha} & \dfrac{1}{\beta_2 - \alpha} \end{bmatrix}, \tag{11-241}$$

and after some algebra we find from (11-237) the Fredholm determinant

$$D(z) = \alpha^n \frac{(1 - \alpha\beta_2)^2 \beta_1^{n+1} - (1 - \alpha\beta_1)^2 \beta_2^{n+1}}{(1 - \alpha^2)(\beta_1 - \beta_2)}. \tag{11-242}$$

The roots β_1 and β_2 are functions of z through (11-240).

When z moves away from the origin along the negative Re z-axis, the roots β_1 and β_2 eventually meet the unit circle, and we can put

$$\beta_1 = e^{i\theta}, \qquad \beta_2 = e^{-i\theta}$$

into (11-242). By equating the result to zero, we obtain an equation for θ from whose roots the eigenvalues λ_k of ϕ_s can be calculated: $D(-1/\lambda_k) = 0$.

11.6.4.2 Kalman method. When the signal process $\{S_k\}$ can be modeled as the output of a discrete-time linear system having a finite number of state variables and driven by white noise, the Fredholm determinant can be computed by solving the Kalman equations, much as in Sec. 11.6.2. The ARMA process is a special case. As in Sec. 11.4, we imagine the signal process $\{S_k\}$ replaced by one with a power z times as large, so that throughout ϕ_s is replaced by $z\phi_s$. We can then use the Kalman formulation, except with the $Q(k)$ replaced by $zQ(k)$ and the initial covariance matrix $\Sigma(1)$ replaced by $z\Sigma(1)$. By (11-219) the Fredholm determinant is then

$$D(z) = \det(\mathbf{I} + z\phi_s) = \prod_{k=1}^{n} W_k,$$

where the W_k are again given by (11-224) in the modified system,

$$W_k = 1 + \mathbf{D}(k)\mathbf{P}(k \mid k - 1)\mathbf{D}^+(k),$$

with now

$$\mathbf{P}(k \mid k - 1) = \mathbf{A}(k - 1)\mathbf{P}(k - 1)\mathbf{A}(k - 1)^+ + zQ(k)\mathbf{G}(k)\mathbf{G}^+(k)$$

in place of (11-225). In the initial conditions given at the end of Sec. 11.6.2, $\Sigma(1)$ is replaced by $z\Sigma(1)$. The equations of Sec. 11.6.2 involving the state vector $\mathbf{Y}[k]$ and the data are of course omitted from this computation of the Fredholm determinant.

11.6.4.3 Stationary signal processes If the signal process is merely stationary, its covariance matrix having the Toeplitz form (11-221), the Fredholm determinant can be computed by an extension of the Durbin algorithm of Sec. 11.6.3. Again the signal sequence is replaced by one z times as strong:

$$\phi_k \longrightarrow z\phi_k, \qquad 0 \le k \le n - 1.$$

A complication arises, however, when both z and ϕ_k for $k > 0$ are complex, for the method of Sec. 11.6.3 involves taking complex conjugates of the components of the solution of (11-229). It is now necessary to solve not one set of equations like (11-229), but two, one with the covariances ϕ_k, the other with their complex conjugates ϕ_k^*. These equations are

$$f_t^{(k)} + z \sum_{m=0}^{k-2} \phi_{t-m} f_m^{(k)} = z\phi_{t+1}, \qquad 0 \le t \le k - 2, \qquad \text{(11-243)}$$

$$g_t^{(k)} + z \sum_{m=0}^{k-2} \phi_{t-m}^* g_m^{(k)} = z\phi_{t+1}^*, \qquad 0 \le t \le k - 2. \qquad \text{(11-244)}$$

Initially

$$f_0^{(2)} = \frac{z\phi_1}{1 + z\phi_0}, \qquad g_0^{(2)} = \frac{z\phi_1^*}{1 + z\phi_0}.$$

The two solution vectors are now updated in accordance with the prescription

$$f_r^{(k+1)} = f_r^{(k)} - \beta_k g_{k-2-r}^{(k)}, \qquad 0 \le r \le k - 2,$$

$$g_r^{(k+1)} = g_r^{(k)} - \beta_k' f_{k-2-r}^{(k)},$$

$$f_{k-1}^{(k+1)} = \beta_k, \qquad\qquad\qquad g_{k-1}^{(k+1)} = \beta_k',$$

with

$$\beta_k = \frac{z(\phi_k - \Gamma_k)}{1 + z\phi_0 - zr_k}, \qquad \beta_k' = \frac{z(\phi_k^* - \Gamma_k')}{1 + z\phi_0 - zr_k},$$

where

$$\Gamma_k = \sum_{m=0}^{k-2} \phi_{k-1-m} f_m^{(k)}, \qquad \Gamma_k' = \sum_{m=0}^{k-2} \phi_{k-1-m}^* g_m^{(k)},$$

$$r_k = \sum_{m=0}^{k-2} \phi_{m+1} g_m^{(k)} = \sum_{m=0}^{k-2} \phi_{m+1}^* f_m^{(k)}.$$

(11-245)

The r_k's can be computed recursively as

$$r_{k+1} = r_k + \beta_k(\phi_k^* - \Gamma_k') = r_k + \beta_k'(\phi_k - \Gamma_k), \qquad r_1 = 0.$$

Observe that updating the solution of (11-243) involves the solution of (11-244) and vice versa. In the course of the procedure, one accumulates the Fredholm determinant

$$D(z) = \prod_{k=1}^{n} (1 + z\phi_0 - zr_k)$$

(11-246)

or, if there is danger of overflow, one accumulates its logarithm. If z is real,

$$g_r^{(k)} = f_r^{(k)*},$$

this algorithm reduces to that of Sec. 11.6.3, and only one set of equations needs to be solved. Having an algorithm to compute the Fredholm determinant $D(z)$ for both real and complex values of z, one is in a position to compute the detection probability $Q_d(\eta)$ and its complement as in (11-94) and (11-95).

Problems

11-1. Find the moment-generating function of the quadratic form $U_M = \frac{1}{2}\mathbf{V}^\dagger\mathbf{H}\mathbf{V}$ when the components of the M-dimensional column vector $\mathbf{V}$ are circular Gaussian random variables V_k with expected values S_k and covariances as given in (11-29).

11-2. Using the result of Problem 11-1, show that if $V(t)$ is a circular Gaussian random process of expected value $S(t)$ and complex autocovariance function $\phi(t, u)$, the moment-generating function of the quadratic functional

$$U = \frac{1}{2}\int_0^T \int_0^T V^*(t)h(t, s)V(s)\, dt\, ds$$

is

$$h(z) = h_0(z) \exp\left[-\frac{z}{2}\int_0^T \int_0^T \int_0^T S^*(t)L(t, u; z)h(u, v)S(v)\, dt\, du\, dv\right],$$

where $h_0(z)$ is the moment-generating function for the process of expected value zero, and $L(t, u; z)$ is the solution of an integral equation similar to (11-70).

11-3. For an arbitrary quadratic functional of the type given in (11-63) calculate the effective signal-to-noise ratio defined in (4-66). Show that this effective signal-to-noise ratio is maximum when $h(t, s)$ is equal to the kernel $h_\theta(t, s)$ of the threshold statistic as given by (11-50). Use the Schwarz inequality for traces given in Sec. 11.1.10. This problem is most easily solved in the vector-matrix domain.

11-4. Verify that (11-169) is the solution of the steady-state variance equation (11-164) for the example treated in Sec. 11.4.1, and show that (11-167) and (11-168) yield the spectral density in (11-158).

11-5. For the Lorentz spectral density (11-47) show that (11-142) reduces to the Fredholm determinant in (11-128).

11-6. Set up state equations and determine the matrices $\mathbf{F}$, $\mathbf{G}$, and $\mathbf{L}_\infty$ for a stationary process $x_1(t)$ whose spectral density is

$$\Phi_s(\omega) = \frac{Q(\omega^2 + a^2)}{(\omega^2 + b^2)(\omega^2 + c^2)},$$

taking $\mathbf{C} = (1\ 0)$.

11-7. Let $x(t)$ be a low-pass Gaussian stochastic process with expected value zero and autocovariance function $\phi(t, u) = E[x(t)x(u)]$. Define its average power during the interval $(0, T)$ as the random variable

$$y = \frac{1}{T}\int_0^T [x(t)]^2\,dt.$$

(a) Using the methods of this chapter, derive the moment-generating function $h(z) = E[\exp(-zy)]$ of this average power in terms of the Fredholm determinant associated with the autocovariance function $\phi(t, u)$.

(b) Show how the moment-generating function $h(z)$ could be calculated from a state-variable model of the generation of $x(t)$, as in (11-155) through (11-157).

(c) Calculate the moment-generating function $h(z)$ when $x(t)$ is a Gaussian Markov process with autocovariance function $\phi_0 \exp(-\mu|t - u|)$.

(d) Calculate the conditional moment-generating function

$$h(z) = E(e^{-zy}\,|\,x(0) = 0)$$

when the Gaussian Markov process $x(t)$ is generated as in part (b), but starts at $x(0) = 0$ at time $t = 0$.

     Hint: Use Collins's method described in Sec. 11.4.2 for calculating the Fredholm determinant as in (11-184), taking the initial condition $\mathbf{L}(0) = 0$ on the state-vector covariance matrix $\mathbf{L}(\cdot)$, which, because this is a one-dimensional process, is a scalar and represents the instantaneous variance of $x(t)$.

11-8. A narrowband Gaussian stochastic signal

$$s(t) = \text{Re}\, S(t)\, e^{i\Omega t}$$

is to be detected in white Gaussian noise of spectral density N. The real and imaginary parts of $S(t)$ are independent Wiener–Lévy (or Brownian) processes; that is,

$$S(t) = \int_0^t W(s)\,ds,$$

where $W(t)$ is white Gaussian noise of spectral density R:

$$\tfrac{1}{2}E[W(t)W^*(s)] = R\delta(t - s).$$

Thus

$$\frac{dS}{dt} = W(t), \qquad 0 \le t \le T, \qquad S(0) = 0.$$

Think of this as the dynamical state equation for a linear system with a single state variable $x_1(t) = S(t)$. The input $v(t) = \operatorname{Re} V(t) \exp i\Omega t$ to the receiver is observed during the interval $0 \le t \le T$, and either (H_0) $V(t) = N(t)$ or (H_1) $V(t) = N(t) + S(t)$. The input noise is $\operatorname{Re} N(t) \exp i\Omega t$.

(a) Calculate the complex autocovariance function

$$\phi_s(t, u) = \tfrac{1}{2} E[S(t)S^*(u)]$$

   of the signal.

(b) Use the Kalman–Bucy equations to express the minimum-mean-square-error causal estimator $\hat{S}_1(t)$ of the signal under hypothesis H_1 in terms of the input $V(t'), 0 \le t' \le t \le T$.

(c) Derive and describe the causal estimator–correlator form of the optimum detector for this signal.

(d) Starting with the solution of the Kalman–Bucy equations or otherwise, derive the kernel $h(t, s)$ of the optimum detection statistic U for this signal.

(e) Calculate the moment-generating functions

$$h_i(z) = E(e^{-zU} \mid H_i), \qquad i = 0, 1,$$

   of the optimum detection statistic under both hypotheses and show how to use them to compute the false-alarm probability Q_0 and the probability Q_d of detecting the signal $s(t)$. Use Collins's formula (11-184) to calculate the Fredholm determinant involved.

(f) Determine the eigenvalues of the autocovariance function $\phi_s(t, u)$ of the signal during the interval $(0, T)$, and express the false alarm and detection probabilities Q_0 and Q_d in terms of them and of the decision level U_0 on the statistic U.

11-9. A binary on–off communication system transmits a certain signal for each 1 in the message, and it sends nothing for each 0. The digits appear every T seconds, and 0's and 1's are equally likely. The signal is received as a pulse $s(t)$ of stationary narrowband Gaussian noise of duration T. The narrowband spectral density of this stochastic signal is

$$\Phi_s(\omega) = \frac{C}{(\omega^2 + \mu^2)^2},$$

C a constant. The input to the receiver always contains white Gaussian noise of unilateral spectral density N.

(a) Describe the optimum receiver for deciding every T seconds whether a signal has arrived or not, taking "optimum" in the sense of "attaining minimum probability of error."

(b) Calculate the minimum attainable probability P_e of error in the limit $\mu T \ll 1$, as a function of E/N, where E is the average energy of the received signal $s(t)$. Plot P_e versus E/N for $20 \le E/N \le 1000$.

(c) Calculate the minimum error probability P_e for $\mu T \gg 1$ by making the equal-eigenvalues approximation. That is, as in (11-112) assume that the autocovariance function of the signal has m equal eigenvalues λ_k approximately equal to $\bar{\lambda}$ and that the rest are negligible. As in Sec. 11.2.3, determine $\bar{\lambda}$ and the effective number m of eigenvalues in terms of μT. For $m = 10$ plot the error probability P_e versus the ratio E/N over the same range as in part (b).

(d) Use the Toeplitz approximation in Sec. 11.2.5 to calculate the moment-generating functions $E[\exp(-zU)|\,H_0]$ and $E[\exp(-zU)|\,H_1]$ of the optimum decision statistic U. Then use the saddlepoint approximation to calculate the minimum error probability P_e for the same range of values of E/N as before; $m = 10$. Compare your results with those of part (c).

(e) Describe the threshold detector for these stochastic signals. Under what conditions, in terms of E/N and μT, would you expect it to be nearly as good as the optimum detector, and why?

(f) Describe in detail how to implement the Schweppe likelihood ratio receiver for these signals. Write out in explicit form, component by component, all the Kalman–Bucy equations needed for estimating the incoming signal, and show how their solution is used in the receiver. It is unnecessary to solve any of the differential equations.

11-10. In certain on–off binary communication systems nothing is sent for each 0 in a message; for each 1, signals proportional to $\operatorname{Re} F(t) \exp i\Omega t$ and $\operatorname{Re} G(t) \exp i\Omega t$ are transmitted simultaneously. The input to the receiver when these have been sent (hypothesis H_1) is $v(t) = \operatorname{Re} V(t) \exp i\Omega t$, with a complex envelope

$$V(t) = AF(t)\, e^{i\theta_1} + BG(t)\, e^{i\theta_2} + N(t), \qquad 0 \le t \le T;$$

$N(t)$ is the complex envelope of white Gaussian noise of unilateral spectral density N. Because the channel fades, the amplitudes A and B are independently random and have Rayleigh distributions with the same parameter σ^2:

$$z(A) = \frac{A}{\sigma^2}\, e^{-A^2/2\sigma^2} U(A),$$

$$z(B) = \frac{B}{\sigma^2}\, e^{-B^2/2\sigma^2} U(B).$$

The phases θ_1 and θ_2 are independently random, independent of A and B, and uniformly distributed over $(0, 2\pi)$. Take

$$\int_0^T |F(t)|^2\, dt = \int_0^T |G(t)|^2\, dt = 1, \qquad \int_0^T F(t)G^*(t)\, dt = \lambda \ne 0.$$

Find the optimum receiver for deciding between hypotheses H_0, "0 sent," and H_1, "1 sent," and calculate its probability of error, assuming the symbols 0 and 1 to occur with equal prior probabilities. The matrix methods of this chapter will be found helpful, as will (D-9).

We have simplified this problem by assuming that even though the signal envelopes $AF(t)$ and $BG(t)$ overlap, they fade independently. A more realistic model would assign a correlation coefficient to the received complex amplitudes $A \exp i\theta_1$ and $B \exp i\theta_2$. Describe how that could be done and what modifications correlated fading would require in your analysis.

11-11. A narrowband signal $s(t) = A \operatorname{Re}[F(t - \tau) \exp(i\Omega t + i\phi)]$ of unknown amplitude A, phase ϕ, and arrival time τ is to be detected in the presence of additive white Gaussian noise of unilateral spectral density N. The input $v(t)$ to the receiver is observed during an interval $(0, T)$ that is much longer than the reciprocal of the signal bandwidth W; $WT \gg 1$. Assume that the arrival time τ of the signal is well within the interval so that you can neglect the possibility that the signal overlaps one end of the interval or the other.

The threshold detector for this signal has a filter matched to the signal $\operatorname{Re} F(t) \exp i\Omega t$ over an interval $0 \le t \le T' \ll T$, outside of which we can assume the signal to be negligible. The output Re $V_0(t) \exp i\Omega t$ of the filter is applied to a quadratic rectifier, whose output is in turn integrated during the interval $(T', T + T')$ to generate the threshold statistic

$$U = \tfrac{1}{2} \int_{T'}^{T+T'} |V_0(t)|^2 \, dt, \qquad V_0(t) = \int_{-\infty}^{\infty} F^*(u) V(t - T' + u) \, du,$$

where $V(t)$ is the complex envelope of the input to the receiver.

(a) Using the methods of this chapter and the results of Problem 11-1, determine the moment-generating functions $h_i(z)$ $(i = 0, 1)$ of the statistic U under both hypotheses (H_0) "signal absent" and (H_1) "signal present with a given arrival time τ_0, $0 \ll \tau_0 \ll T$."

(b) Describe how you would use these moment-generating functions to calculate the false-alarm probability Q_0 and the probability Q_d of detecting this signal.

(c) Work out the Toeplitz approximations to the moment-generating functions $h_i(z)$ of U in terms of the spectrum $f(\omega)$ of the complex envelope $F(t)$ of the signal. Then state their forms when the signal is bandlimited to a bandwidth of W hertz,

$$|f(\omega)|^2 \equiv \begin{cases} W^{-1}, & -\pi W \leq \omega \leq \pi W, \\ 0, & |\omega| > \pi W, \end{cases}$$

and determine the false-alarm probability Q_0 and the detection probability Q_d for this special case. Express the latter in terms of the signal-to-noise ratio $D^2 = 2E/N$, E the energy of the received signal, and the time–bandwidth product WT.

11-12. In the system treated in Sec. 11.5, assume that the phase of the coherent signal is unknown; that is, instead of its being Re $R(t) \exp i\Omega t$, it is Re $R(t) \exp(i\Omega t + i\phi)$, with ϕ uniformly distributed over $(0, 2\pi)$. Determine the optimum detector for the combination of this signal and a stochastic signal Re $S(t) \exp i\Omega t$ of the same type as postulated in Sec. 11.5. Assuming that the average total energies of both signal components are proportional to the same positive factor ε, work out the threshold detector for their combination by taking the likelihood functional to the limit $\varepsilon \to 0$. Calculate the effective signal-to-noise ratio—the deflection (4-66)—of this threshold detector. *Hint:* Use (4-75).

11-13. Plot the locus of the roots β_k of (11-137) for the spectral density given in (A-17).

11-14. Show in detail how to calculate the n eigenvalues λ_k of the Toeplitz matrix $\boldsymbol{\phi}_s$, where $\boldsymbol{\phi}_s$ is given in (11-239); use (11-242). Calculate those eigenvalues for $\alpha = 0.5$, $n = 8$, and $C = 1$.

11-15. Write out in detail the steps of the modified Durbin algorithm at the end of Sec. 11.6.4 for $k = 1$ and $k = 2$, and verify that they yield the solutions of the simultaneous equations (11-243) and (11-244) and that (11-246) provides the correct value of the determinant $\det(\mathbf{I} + z\boldsymbol{\phi}_s)$ for $n = 2$.

12

Detection of
Optical Signals

12.1 PHOTOELECTRON COUNTING

12.1.1 Properties of the Light and the Detector

An optical signal is a pulse of electromagnetic waves whose frequencies lie in the infrared or visible range. It may be carrying information, as in an on–off binary communication system in which a pulse of light is transmitted for each 1 in a message, the absence of a pulse indicating a 0. In optical radar the presence of a target is manifested by the reception of a light pulse reflected from it. The simplest method of detecting an optical signal is to focus its source onto a surface that emits photoelectrons and to count the number of electrons emitted during a fixed interval of time $(0, T)$. Sections 12.1 and 12.2 will concentrate on this kind of detection. It is not necessarily the optimum procedure; information contained in the time at which each electron is emitted may enhance the detectability of certain types of optical signal. Such detectors will be briefly considered in Sec. 12.5. In Sec. 12.3 we shall examine receivers utilizing photomultipliers or basing their decisions on a sample of the current at the output of a photosensitive detector, and in Sec. 12.4 the optical counterpart of a heterodyne receiver will be treated.

The light is admitted to an optical receiver through an aperture. The larger the aperture, the more light is collected and the greater the probability of detection. The very best receiver would base its decisions on an observation of the electromagnetic field at its aperture during the interval $(0, T)$. To formulate the optimum processing of that aperture field requires taking account of the laws of quantum mechanics,

470

which govern electromagnetic fields at optical frequencies. A decision theory consistent with quantum mechanics has been formulated; it is outlined in [Hel76]. It shows that in many situations the optimum detector counts photons; in others, it is unknown how physically to realize the optimum processing of the aperture field that the theory prescribes.

The light passing through the aperture of the receiver falls perpendicularly onto the photoelectrically emissive surface, which we shall call simply the *detector* and which has a total area A. The light will be assumed linearly polarized, whereupon its field can be treated as a scalar function of position and time. It will be assumed that the light field is completely coherent over the entire area A of the detector; it consists of plane electromagnetic waves. Any background light will be assumed to have been filtered to remove components far from the frequency of the light to be detected. The incident light can then be represented as a narrowband signal of the form

$$v(t) = \text{Re } V(t) \, e^{i\Omega t}$$

and with complex envelope $V(t)$ and angular carrier frequency Ω; $\Omega = 2\pi f$ and $f = c/\lambda$, where c is the velocity of light and λ is the wavelength of the radiation.

According to the semiclassical theory of light, the field $v(t)$ can be treated as a narrowband Gaussian random process. It may have a nonzero expected value

$$E[v(t)| \, H_1] = \text{Re } S(t) \, e^{i\Omega t}$$

under hypothesis H_1 that a signal is present; $S(t)$ is the complex envelope of a coherent signal that might, for instance, be the output of an ideal pulsed laser. We shall assume that the random portion of the light is a stationary process; indeed, considering how natural light is created, it is difficult to imagine circumstances under which it would not be so. We assign to it the complex autocovariance function

$$\tfrac{1}{2}E\{[V(t_1) - S(t_1)][V^*(t_2) - S^*(t_2)]\} = P\phi(t_1 - t_2). \qquad (12\text{-}1)$$

The Hermitian function $\phi(\tau) = \phi^*(-\tau)$ will be normalized so that $\phi(0) = 1$; P is then the total average power of the random, or *incoherent*, portion of the incident light. The Fourier transform

$$\Phi(\omega) = \int_{-\infty}^{\infty} \phi(\tau) \, e^{-i\omega\tau} \, d\tau \qquad (12\text{-}2)$$

will be called the spectral density of the light, ω representing an angular frequency deviation from the carrier frequency Ω. In the optics literature $\phi(\tau)$ is called the *temporal coherence function* of the light [Man65].

When this light falls on the detector, it ejects photoelectrons. The conditional probability that one electron is ejected in a brief interval Δt is $\lambda(t)\Delta t$, with

$$\lambda(t) = \frac{\eta'}{2hf}|V(t)|^2 \qquad (12\text{-}3)$$

the instantaneous rate of emission; η' is called the *quantum efficiency* of the detector, $0 < \eta' < 1$, $h = 6.626 \cdot 10^{-34}$ J-sec is Planck's constant, and $f = \Omega/2\pi$ is the frequency in hertz; hf is the energy of a single quantum or *photon* of the light. One can think of η' as the probability that an incident photon ejects an electron that is counted in the external circuit of the detector. We abbreviate η'/hf by η. Behind (12-3) lies the assumption that the incident light is not so strong that it significantly depletes the number of electrons available in the material of the detector.

The probability $\lambda \Delta t$ is conditioned on a particular realization of the incident light field, which is a random process. The expected number of photoelectrons counted in $(0, T)$ is

$$E(n) = \bar{n} = \bar{n}_0 + \bar{n}_s,$$

where

$$\bar{n}_0 = \tfrac{1}{2}\eta E \int_0^T |V(t) - S(t)|^2 \, dt = \eta P T \qquad (12\text{-}4)$$

is the expected number due to the random component and

$$\bar{n}_s = \tfrac{1}{2}\eta \int_0^T |S(t)|^2 \, dt \qquad (12\text{-}5)$$

is the expected number due to a coherent component of the light.

Conditionally on a realization $v(t) = \text{Re } V(t) \exp i\Omega t$ of the incident light field, the emission of photoelectrons is assumed to be a Poisson point process. The probability that k electrons are ejected between times t_1 and t_2 has the Poisson form

$$\Pr(n = k \mid V(t), t_1 \le t \le t_2) = \frac{m^k}{k!} e^{-m}, \qquad m = \int_{t_1}^{t_2} \lambda(t) \, dt, \qquad (12\text{-}6)$$

and the numbers ejected in disjoint intervals are statistically independent [Sny75, pp. 38–56], [Hel91, pp. 389–91], [Pap91, pp. 354–8]. We shall be concerned in this section only with the total number n ejected during the observation interval $(0, T)$. The probability that k are ejected is then

$$p_k = \Pr(n = k) = E\left[\frac{m^k}{k!} e^{-m} \right], \qquad m = \tfrac{1}{2}\eta \int_0^T |V(t)|^2 \, dt. \qquad (12\text{-}7)$$

The expectation E is taken with respect to the distribution of the variable m, which is random because $V(t)$ is a circular complex Gaussian random process.

12.1.2 The Probability-generating Function

In detectors of this kind the presence of a signal is indicated when the number n of electrons exceeds a certain decision level n_0. The probabilities of false alarm and detection are then related to the complementary cumulative distribution of the random variable n, and this distribution is in general most directly calculated on the basis of the probability-generating function of the number n of ejected electrons. One often refers to such distributions as *photocount* distributions and speaks of *photon counting*, although it is not photons, but photoelectrons, that are being counted. We shall need only the elementary aspects of this subject. More extensive treatments can be found in [Sny75] and [Sal78]. For descriptions of optical communication and optical radar systems and technical details of their construction, see books such as [Gag76], [Gow84], [Sen85], and [Jel92].

The probability-generating function of a nonnegative integer-valued random variable such as the number n of photocounts is defined as in (5-40),

$$h(z) = E(z^n) = \sum_{n=0}^{\infty} p_k z^k \tag{12-8}$$

with $p_k = \Pr(n = k)$. The probabilities p_k can be recovered from $h(z)$ by

$$p_k = \frac{1}{k!} \frac{d^k}{dz^k} h(z) \Big|_{z=0} . \tag{12-9}$$

The cumulative distribution of the number n is defined as

$$q_k^{(-)} = \Pr(n < k) = \sum_{r=0}^{k-1} p_r , \tag{12-10}$$

and as shown in Sec. 5.2.3, it is given by the contour integral

$$q_k^{(-)} = \int_{C_-} \frac{z^{-k} h(z)}{1 - z} \frac{dz}{2\pi i} , \tag{12-11}$$

where C_- is a closed curve surrounding the origin, but enclosing neither the point $z = 1$ nor any singularities of $h(z)$. Taking C on the other hand as a closed curve C_+ including both $z = 0$ and $z = 1$, but no singularities of $h(z)$, we can write the complementary cumulative distribution of the number n as

$$q_k^{(+)} = 1 - q_k^{(-)} = \sum_{r=k}^{\infty} p_r = \int_{C_+} \frac{z^{-k} h(z)}{z - 1} \frac{dz}{2\pi i} . \tag{12-12}$$

For complicated distributions for which the probabilities p_k cannot easily be calculated or for numbers k of electrons so large that (12-10) cannot be summed accurately enough, the cumulative probabilities $q_k^{(-)}$ and $q_k^{(+)}$ can often be conveniently computed by numerical integration of (12-11) or (12-12) along a suitably chosen contour. In particular, saddlepoint integration similar to that described in Sec. 5.2.3 may well be expeditious. When the number k is far in one tail or the other of the distribution of the number n of counts, the saddlepoint approximation introduced in Sec. 5.3.2 is useful.

12.1.3 Evaluating the Probability-generating Function

The probability-generating function for the Poisson distribution with expected value m is

$$\sum_{k=0}^{\infty} \frac{m^k}{k!} e^{-m} z^k = e^{m(z-1)} \tag{12-13}$$

by (12-8). For the number n of photoelectrons ejected by the type of light treated in Sec. 12.1.1, we find from (12-7) the probability-generating function

$$h(z) = E\left\{ \exp\left[\tfrac{1}{2}\eta(z - 1) \int_0^T |V(t)|^2 \, dt \right] \right\}, \tag{12-14}$$

in which $V(t)$ is the complex envelope of a narrowband Gaussian random process with expected value $\text{Re}[S(t) \exp i\Omega t]$. In order to evaluate $h(z)$, we introduce the eigenvalues λ_k and the eigenfunctions $f_k(t)$ of the temporal coherence function $\phi(\tau)$:

$$\lambda_k f_k(t) = \frac{1}{T} \int_0^T \phi(t - u) f_k(u)\, du, \qquad 0 \leq t \leq T. \tag{12-15}$$

Because $\phi(0) = 1$, the eigenvalues λ_k sum to 1.

Define the Fourier coefficients

$$V_k = \int_0^T f_k^*(t) V(t)\, dt,$$

$$S_k = E(V_k) = \int_0^T f_k^*(t) S(t)\, dt.$$

Then the V_k are independent circular complex Gaussian random variables whose real and imaginary parts have variances $P\lambda_k T$:

$$\tfrac{1}{2} E[|V_k - S_k|^2] = P\lambda_k T.$$

The joint probability density function of these parts has the circular Gaussian form— see (3-49)—

$$\tilde{p}(V_k) = \frac{1}{2\pi P\lambda_k T} \exp\left[-\frac{|V_k - S_k|^2}{2P\lambda_k T} \right].$$

Furthermore in (12-14)

$$\int_0^T |V(t)|^2\, dt = \sum_k |V_k|^2,$$

and the probability-generating function can be written

$$h(z) = \prod_k E\left[\exp\left[\tfrac{1}{2}\eta(z - 1)|V_k|^2\right] \right]$$

with

$$E\left[\exp\left[\tfrac{1}{2}\eta(z - 1)|V_k|^2\right] \right]$$

$$= \frac{1}{2\pi P\lambda_k T} \int_{-\infty}^{\infty} \int_{-\infty}^{\infty} \exp\left[\tfrac{1}{2}\eta(z - 1)|V_k|^2 - \frac{|V_k - S_k|^2}{2P\lambda_k T} \right] dV_{kx}\, dV_{ky}$$

$$= \frac{1}{1 - \bar{n}_0 \lambda_k (z - 1)} \exp\left[\frac{\tfrac{1}{2}\eta(z - 1)|S_k|^2}{1 - \bar{n}_0 \lambda_k (z - 1)} \right].$$

The probability-generating function of the number of photoelectrons is therefore

$$h(z) = \prod_k \frac{1}{1 - \bar{n}_0 \lambda_k (z - 1)} \exp\left[\tfrac{1}{2}\eta(z - 1) \sum_k \frac{|S_k|^2}{1 - \bar{n}_0 \lambda_k (z - 1)} \right]. \tag{12-16}$$

If we define the Fredholm determinant associated with the kernel $\phi(\tau)$ of (12-15) as in (11-76),

$$D(x) = \prod_k (1 + \lambda_k x), \tag{12-17}$$

the first factor in $h(z)$ is $[D(\bar{n}_0(1 - z))]^{-1}$.

Replacing $\bar{n}_0(1-z)$ by x, we write the quadratic form in the exponent of (12-16) as

$$\sum_k \frac{|S_k|^2}{1 + \lambda_k x} = \sum_k S_k^* Q_k, \qquad Q_k = \frac{S_k}{1 + \lambda_k x},$$

and to $(1 + \lambda_k x)Q_k = S_k$ corresponds the integral equation

$$S(t) = Q(t; x) + \frac{x}{T} \int_0^T \phi(t - u)Q(u; x)\, du, \qquad 0 \le t \le T, \qquad (12\text{-}18)$$

as in (11-193). Our probability-generating function thus becomes

$$h(z) = \frac{1}{D(x)} e^{\frac{1}{2}\eta(z-1)J(x)}, \qquad x = \bar{n}_0(1-z), \qquad (12\text{-}19)$$

with

$$J(x) = \int_0^T S^*(t)Q(t; x)\, dt \qquad (12\text{-}20)$$

as in (11-198). When the spectral density $\Phi(\omega)$ is a rational function of the frequency ω, all the methods developed in Secs. 11.3 through 11.5 can be applied to evaluating (12-19). We begin with simple cases.

When the complex envelope $V(t)$ of the incident light is written as a Karhunen–Loève expansion in terms of the eigenfunctions $f_k(t)$ of the autocovariance function $\phi(\tau)$, defined by (12-15), the individual terms are often called *temporal modes* of the light field. If the light, instead of being linearly polarized as we have been assuming, is unpolarized, its incoherent part can be decomposed into two statistically independent components, each of which can in turn be broken into a set of statistically independent temporal modes whose coefficients $V_k^{(1)}$ and $V_k^{(2)}$ have the same statistical properties. The number of modes is in effect doubled, and the factor $[D(x)]^{-1}$ in (12-19) is simply replaced by $[D(x)]^{-2}$. If the incoherent portion of the light is partially polarized, it can again be decomposed into statistically independent components, but these will account for different expected numbers $\bar{n}_0^{(1)}$ and $\bar{n}_0^{(2)}$ of photoelectrons [Hel64]. One then replaces $D(x)$ in (12-19) by

$$D(x_1)D(x_2), \qquad x_i = \bar{n}_0^{(i)}(1-z), \qquad i = 1, 2.$$

12.1.4 A Single Temporal Mode

When the product WT of the bandwidth W of the light and the observation time T is small, $WT \ll 1$, a single eigenvalue λ_1 of (12-15) predominates, $\lambda_1 \approx 1$, and the rest are negligible. The random component of the light is then completely temporally coherent and can be described during $(0, T)$ by a single circular Gaussian random variable V_1. The magnitude $|V_1|$ has a Rayleigh–Rice distribution as in (3-71). This temporal coherence should be distinguished from that characterizing the output of an ideal laser, such as would be represented by the signal $s(t) = \operatorname{Re} S(t) \exp i\Omega t$. For the latter the magnitude of the field is fixed; for the former it is a random variable.

In the absence of a deterministic component $S(t)$, the probability-generating function of the number n of photoelectrons ejected during $(0, T)$ when this light strikes our detector is, by (12-16),

$$h(z) = \frac{1}{1 - \bar{n}_0(z - 1)},$$

where $\bar{n}_0 = \eta P T$ is the expected number of electrons. The probability p_r of counting r electrons is then

$$p_r = (1 - v)v^r, \qquad r = 0, 1, 2, \ldots, \qquad v = \frac{\bar{n}_0}{1 + \bar{n}_0} < 1,$$

and the cumulative distributions are

$$q_k^{(+)} = v^k, \qquad q_k^{(-)} = 1 - v^k, \qquad k \geq 0.$$

This is called the "geometric" or sometimes the "exponential" distribution.

When a signal component is present and ejects an expected number $\bar{n}_s$ of electrons during $(0, T)$, the probability-generating function is, by (12-16),

$$h(z) = \frac{1}{1 - \bar{n}_0(z - 1)} \exp\left[\frac{\bar{n}_s(z - 1)}{1 - \bar{n}_0(z - 1)}\right], \tag{12-21}$$

and one finds by using the generating function of the Laguerre polynomials $L_r(\cdot)$ that

$$p_r = (1 - v)\, e^{-\bar{n}_s(1-v)} v^r L_r\left[-\frac{\bar{n}_s(1 - v)^2}{v}\right] \tag{12-22}$$

[Erd53, vol. 2, p. 189, eq. (17)]. These polynomials obey the recurrent relation in (C-6).

12.1.5 Many Temporal Modes

Comparing (12-21) and (12-16) we see that the kth temporal mode $V_k f_k(t)$ can be thought of as accounting for a random number n_k of photoelectrons; the total number ejected in $(0, T)$ is the sum

$$n = \sum_k n_k,$$

whose terms n_k are independently random. The number n_k has a Laguerre distribution as in (12-22),

$$\Pr(n_k = r) = (1 - v_k) \exp[-R_k(1 - v_k)]v_k^r L_r\left[-\frac{R_k(1 - v_k)^2}{v_k}\right], \tag{12-23}$$

$$v_k = \frac{\bar{n}_0 \lambda_k}{1 + \bar{n}_0 \lambda_k}, \qquad R_k = \tfrac{1}{2}\eta|S_k|^2,$$

with $\bar{n}_0 = \eta P T$ the total expected number of counts arising from the incoherent part of the light and

$$\bar{n}_s = \sum_k R_k$$

the total expected number arising from the coherent part.

　　　　　　　　　　　Detection of Optical Signals　　Chap. 12

The probability distribution $\{p_r\}$ of the total number n of electrons can be computed by successively convolving the distributions in (12-23) until the expected value $\bar{n}_0 \lambda_k + R_k$ associated with the latest convolvendum becomes negligible. The cumulative distribution $q_k^{(-)}$ can then be calculated by summing the p_r's. Alternatively one can put $z = \exp(-i\omega)$ into (12-16) and compute the probabilities p_r by means of the fast Fourier transform of $h[\exp(-i\omega)]$. These methods will be feasible if neither the number of significant eigenvalues λ_k nor the number k of electrons for which $q_k^{(-)}$ is needed is excessive.

When the light has a Lorentz spectral density

$$\Phi(\omega) = \frac{2\mu}{\omega^2 + \mu^2}, \tag{12-24}$$

and there is no coherent component, the probability-generating function of the number of electrons is

$$h(z) = e^m \left\{ \cosh g + \tfrac{1}{2} \left[\frac{g}{m} + \frac{m}{g} \right] \sinh g \right\},$$

$$m = \mu T, \qquad g = \left[m^2 - 2m\bar{n}_0(z - 1) \right]^{1/2},$$

by (12-19) and (11-128). Bédard described a recurrent technique for evaluating the probabilities p_r for this type of incident light [Béd66].

12.1.6 The Toeplitz Approximation

When the time–bandwidth product WT is large, the Fredholm determinant $D(x)$ involved in (12-16) will be

$$D(x) \approx \exp\left\{ T \int_{-\infty}^{\infty} \ln \left[1 + \frac{x}{T}\Phi(\omega) \right] \frac{d\omega}{2\pi} \right\}$$

as in (11-102). The limits of integration in (12-18) can be replaced by $-\infty$ and ∞ and $Q(t; x)$ by $Q_\infty(t; x)$, where

$$S(t) = Q_\infty(t; x) + \frac{x}{T} \int_{-\infty}^{\infty} \phi(t - u) Q_\infty(u; x) \, du.$$

This equation can be solved by Fourier transformation to yield

$$q_\infty(\omega; x) = \int_{-\infty}^{\infty} Q_\infty(t; x) e^{-i\omega t} \, dt = \frac{s(\omega)}{1 + (x/T)\Phi(\omega)},$$

where

$$s(\omega) = \int_{-\infty}^{\infty} S(t) e^{i\omega t} \, dt.$$

Then in (12-19)

$$J(x) \approx \int_{-\infty}^{\infty} s^*(\omega) q_\infty(\omega; x) \frac{d\omega}{2\pi},$$

and we obtain for the probability-generating function of the number of electrons, approximately,

$$h(z) \approx \exp\left\{-T\int_{-\infty}^{\infty} \ln\left[1 + \frac{x}{T}\Phi(\omega)\right]\frac{d\omega}{2\pi}\right.$$

$$\left. + \frac{1}{2}\eta(z-1)\int_{-\infty}^{\infty} \frac{|s(\omega)|^2}{1 + xT^{-1}\Phi(\omega)}\frac{d\omega}{2\pi}\right\}, \qquad x = \bar{n}_0(1-z).$$

(12-25)

Computing even approximate cumulative distributions $q_k^{(-)}$ and $q_k^{(+)}$ from this probability-generating function will usually require numerical contour integration as in (12-11) and (12-12).

12.1.7 Rectangular Spectral Density

When the spectral density of the random component of the light is rectangular with bandwidth W,

$$\Phi(\omega) \equiv \begin{cases} W^{-1}, & -\pi W \leq \omega \leq \pi W, \\ 0, & |\omega| > \pi W, \end{cases}$$

(12-26)

and $WT \gg 1$, the Toeplitz approximation (12-25) yields for the probability-generating function of the number of ejected electrons

$$h(z) = \frac{1}{[1 + N(1-z)]^M} \exp\left[\frac{\bar{n}_s(z-1)}{1 + N(1-z)}\right],$$

$$M = WT, \qquad N = \frac{\bar{n}_0}{WT},$$

(12-27)

with $\bar{n}_0$ and $\bar{n}_s$ the expected numbers of electrons ejected by the random and the coherent components of the light, respectively. The parameter N is the expected number of electrons ejected by each significant temporal mode of the random component of the incident light. Here

$$\bar{n}_s = \frac{1}{2}\eta \int_{-\pi W}^{\pi W} |s(\omega)|^2 \frac{d\omega}{2\pi};$$

it is assumed that all frequency components of both the signal and the noise outside the band $-\pi W \leq \omega \leq \pi W$ have been eliminated by filtering at the input.

We can write (12-27) as

$$h(z) = \left[\frac{1-v}{1-vz}\right]^M \exp\left[\frac{\bar{n}_s(1-v)(z-1)}{1-vz}\right], \qquad v = \frac{N}{1+N}.$$

(12-28)

Then by the generating function of the associated Laguerre polynomials $L_n^{(\alpha)}(\cdot)$ we can write the probability p_r that r electrons are counted as

$$p_r = (1-v)^M \exp[-\bar{n}_s(1-v)]v^r L_r^{(M-1)}\left[-\frac{\bar{n}_s(1-v)^2}{v}\right]$$

(12-29)

[Erd53, vol. 2, p. 189, eq. (17)]. These associated Laguerre polynomials obey the recurrent relation given in (C-22).

In the absence of a coherent component, $\bar{n}_s = 0$, the number of photoelectrons counted has a "hypergeometric" or "Bose–Einstein" or "negative binomial" distribution,

$$p_r = \frac{\Gamma(M+r)}{r!\,\Gamma(M)}(1-v)^M v^r,$$

(12-30)

with $M = WT$ and v as in (12-28). The time–bandwidth product WT need not be an integer. The use of this distribution as an approximation for light with other spectral densities was proposed in [Béd67]; one can take the effective bandwidth W as defined in (11-93). The approximation loses accuracy in the far tails of the distribution when the spectral density $\Phi(\omega)$ much departs from the rectangular form.

When the time–bandwidth product WT is extremely large, whatever the form of the spectral density $\Phi(\omega)$ of the light, the term $\Phi(\omega)$ in (12-25) becomes very small, and we find approximately

$$h(z) = \exp[(\bar{n}_0 + \bar{n}_s)(z - 1)], \qquad \bar{n}_0 = \eta PT, \qquad \bar{n}_s = \tfrac{1}{2}\eta \int_{-\infty}^{\infty} |s(\omega)|^2 \, \frac{d\omega}{2\pi}.$$

The distribution of the number of electrons then takes the Poisson form in (12-6) with $m = \bar{n}_0 + \bar{n}_S$. The integration time T in (12-14) is now so long that

$$\tfrac{1}{2}\int_0^T |V(t)|^2 \, dt$$

equals its expected value PT within a vanishingly small fluctuation, whereupon the probability-generating function in (12-14) takes the form of (12-13), and that leads to the Poisson distribution.

12.1.8 Light with a Rational Spectral Density

When the spectral density $\Phi(\omega)$ of the light is a rational function of the frequency so that the temporal coherence function $\phi(\tau)$ has the form in (11-103), the Fredholm determinant $D(x)$ in (12-19) can be calculated as in (11-142). The matrix $\mathbf{G} = \mathbf{G}(T)$ figuring there is as written in (11-143). As in Sec. 11.3.4 the parameters μ_k, $1 \le k \le 2n$, are the poles of the function $1 + xT^{-1}\Phi(-ip)$, the first n $(\mu_1, \ldots, \mu_n)$ lying in the right half of the p-plane, the rest $(\mu_{n+1}, \ldots, \mu_{2n})$ in the left half. The parameters β_k, $1 \le k \le 2n$, are the zeros of that function.

If the complex envelope $S(t)$ of the coherent component of the light can be written as a sum of sine waves,

$$S(t) = \sum_m \sigma_m \exp iw_m t,$$

then the function $J(x)$ appearing in the probability-generating function (12-19) can be written as in Sec. 11.5.2:

$$J(x) = J_1 + J_2,$$

where by (11-201), *mutatis mutandis*,

$$J_1 = T \sum_n \sum_m \frac{\sigma_n^* \sigma_m}{H(w_m; x)} \, e(w_m T - w_n T),$$

$$H(w; x) = 1 + \frac{x}{T}\Phi(w),$$

$$e(y) = \begin{cases} \dfrac{e^{iy} - 1}{y}, & y \ne 0, \\[2mm] 1, & y = 0, \end{cases}$$

and by (11-202)

$$J_2 = -\frac{1}{\det \mathbf{G}} \det \begin{bmatrix} 0 & \mathbf{E} \\ \mathbf{F} & \mathbf{G} \end{bmatrix}.$$

As in (11-203), the $2n$ elements of the row vector $\mathbf{E}$ are

$$E_j = \sum_m \sigma_m^* \frac{\exp(\beta_j - iw_m)T - 1}{\beta_j - iw_m}, \qquad 1 \le j \le 2n,$$

and as in (11-204) the $2n$ elements of the column vector $\mathbf{F}$ are

$$F_k = \sum_m \frac{\sigma_m}{H(w_m; x)} \frac{\varepsilon_{km}}{\mu_k - iw_m}, \qquad 1 \le k \le 2n,$$

$$\varepsilon_{km} = \begin{cases} \exp iw_m T, & 1 \le k \le n, \\ 1, & n+1 \le k \le 2n. \end{cases}$$

Once the coefficients σ_m of the signal have been calculated by, for instance, taking the w_m as multiples of $2\pi/T$ and evaluating the Fourier transform of $S(t)$ by means of a fast Fourier transform algorithm, the probability-generating function in (12-19) can be computed at points on the contours of integration in (12-11) and (12-12) by standard routines for solving algebraic equations and evaluating determinants.

Alternatively, one can apply (11-130) to calculating the Fredholm determinant by

$$\ln D(x) = \int_0^T m(\tau, \tau; x; \tau) \, d\tau,$$

where $m(\tau, t; x; \tau)$ is the solution of the integral equation

$$m(\tau, t; x; \tau) + \frac{x}{T} \int_0^\tau m(\tau, u; x; \tau)\phi(u - t) \, du = \frac{x}{T}\phi(\tau - t), \qquad 0 \le t \le \tau \le T,$$

corresponding to (11-132) with $r = \tau$. Furthermore, as in (11-210) the function $J(x)$ can be determined by

$$J(x) = \int_0^T [S^*(\tau) - \hat{S}^*(\tau; x^*)][S(\tau) - \hat{S}(\tau; x)] \, d\tau,$$

where

$$\hat{S}(\tau; x) = \int_0^\tau m(\tau, u; x; \tau)S(u) \, du$$

is the causal quasi-estimate of the coherent signal $S(t)$ as in (11-209). Like the Fredholm determinant $D(x)$, that quasi-estimate can be computed by solving Kalman–Bucy equations as described in Secs. 11.4 and 11.5. These methods have been used in [Hel87] to calculate photocount distributions.

12.2 DETECTABILITY OF OPTICAL SIGNALS BY PHOTOCOUNTING

12.2.1 Negligible Background Light

According to the Planck law, thermal background light contributes an average number

$$\overline{N} = \frac{1}{e^{hf/kT} - 1} \tag{12-31}$$

of photons to each temporal mode at frequency f hertz, where $k = 1.38 \cdot 10^{-23}\text{J/K}$ is Boltzmann's constant and T is the absolute temperature of the background radiation; hf is again the energy per photon of the light. At the wavelength $\lambda = 5890$ Å $= 0.589$ μm, which is that of the chief line in the spectrum of sodium, the quantity hf/k equals $2.446 \cdot 10^4$ K. Any background at an effective temperature much less than this will contribute a negligible number of photons and hence eject a negligible average number $\eta'\bar{N}$ of photoelectrons. If our receiver is far out in space, therefore, and not pointed toward the sun or any other intensely radiating body, we can set $\bar{n}_0 = 0$ under hypothesis H_0 that no signal is present.

The receiver will thereupon decide for hypothesis H_1 whenever any photoelectrons at all are counted. In order to attain a preassigned false-alarm probability Q_0, it must utilize a randomized strategy as described in Sec. 1.2.5. When no photoelectrons at all are counted, it chooses hypothesis H_1 with probability Q_0. The probability of detection is then

$$Q_d = Q_0 \Pr(n = 0 \mid H_1) + \Pr(n > 0 \mid H_1)$$
$$= 1 - (1 - Q_0) \Pr(n = 0 \mid H_1). \tag{12-32}$$

If $h_1(z)$ is the probability-generating function of the number n of electrons under hypothesis H_1,

$$Q_d = 1 - (1 - Q_0)h_1(0) \tag{12-33}$$

by (12-8). For false-alarm probabilities Q_0 less than 10^{-4} or so, the term Q_0 in (12-33) can be neglected, and for Q_d greater than about 0.1 the detection probability is nearly independent of the false-alarm probability, which may as well be set equal to zero.

We first consider the detection of incoherent light, $S(t) = 0$, in the absence of thermal background. If the light has a Lorentz spectral density as in (12-24), we find from (11-128) that

$$Q_d = 1 - h_1(0) = 1 - \frac{4g\, e^m}{(g + 1)^2\, e^{mg} - (g - 1)^2\, e^{-mg}}, \qquad g = \left[1 + \frac{2\bar{n}_s}{m}\right]^{1/2},$$

where $m = \mu T$ is the time–bandwidth product and $\bar{n}_s$ is the expected number of photoelectrons ejected by the signal during the observation interval $(0, T)$. In Fig. 12-1 we have plotted this probability of detection for a number of values of $m = \mu T$. The larger the effective number m of temporal modes among which the light is divided, the greater is the probability of detecting it. The curve $m = \infty$ represents the probability

$$Q_d = 1 - \exp(-\bar{n}_s) \tag{12-34}$$

and also represents the probability of detecting a coherent optical signal Re $S(t)$ exp $i\Omega t$ in the absence of background light.

For incoherent signal light with an arbitrary spectral density, the probability of detection is, by (12-16) with $z = 0$, $Q_0 \ll 1$,

$$Q_d = 1 - \prod_k (1 + \bar{n}_s \lambda_k)^{-1},$$

where λ_k are the eigenvalues of (12-15). When the light is incoherent and has the rectangular spectral density (12-26), as in (11-110),

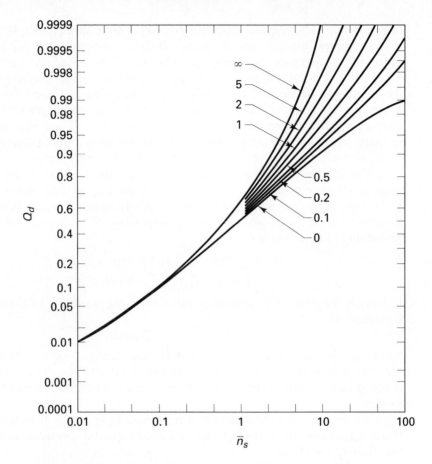

Figure 12-1. Detection probability Q_d of incoherent light with a Lorentz spectral density versus the expected number $\bar{n}_s$ of photoelectrons ejected by the signal. The curves are indexed by the time–bandwidth product $m = \mu T$.

$$\lambda_k = \frac{\lambda_k^{(s)}}{WT} = \frac{\pi}{2c}\lambda_k^{(s)},$$

the $\lambda_k^{(s)}$ being the eigenvalues tabulated in [Sle65b], where the parameter c equals $\pi WT/2$. In Fig. 12-2 we have plotted the probability of detection for various values of WT. As we said at the beginning of Sec. 11.2.7, the bandwidth W as defined by (11-93) equals the parameter μ in the Lorentz spectral density. By comparing Figs. 12-1 and 12-2 we see that for equal values of WT, the signal with a Lorentz spectral density is the more easily detected.

12.2.2 Detection of a Coherent Signal

When a coherent signal $s(t) = \text{Re } S(t) \exp i\Omega t$, such as the output of an ideal laser, is to be detected in the presence of background radiation admitted in a spectral band of width W hertz, the receiver must decide between two hypotheses, (H_0)

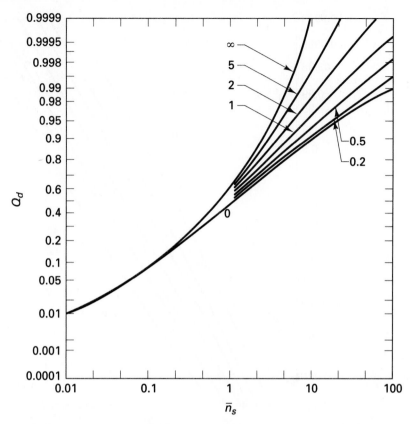

Figure 12-2. Detection probability Q_d of incoherent light with a rectangular spectral density versus the expected number $\bar{n}_s$ of photoelectrons ejected by the signal. The curves are indexed by the time–bandwidth product WT.

the number n of photoelectrons counted during the observation interval $(0, T)$ has the hypergeometric distribution in (12-30), or (H_1) the number n obeys the Laguerre distribution in (12-29). There $M = WT$ is the time–bandwidth product, $N = \bar{n}_0/WT$ is the expected number of photoelectrons ejected by the background light in each temporal mode, and as in (12-5) $\bar{n}_s$ is the expected number ejected by the signal.

Randomization is necessary if the receiver is to attain a preassigned false-alarm probability Q_0. It decides for hypothesis H_1 when the number n exceeds a decision level n_0; when $n = n_0$, it chooses hypothesis H_1 with probability f. These parameters are determined by (1-57) with the probabilities in (12-30), and the probability Q_d of detection is calculated from (1-58) with those in (12-29). In Fig. 12-3 we have plotted as solid lines the detection probability for $N = 0.1$, $Q_0 = 10^{-6}$, versus the quantity

$$D = \left[\frac{2\bar{n}_s}{1 + N} \right]^{1/2},$$

which measures the strength of the signal. The dashed line (H) represents the detection probability attained by the heterodyne receiver to be described in Sec. 12.4.

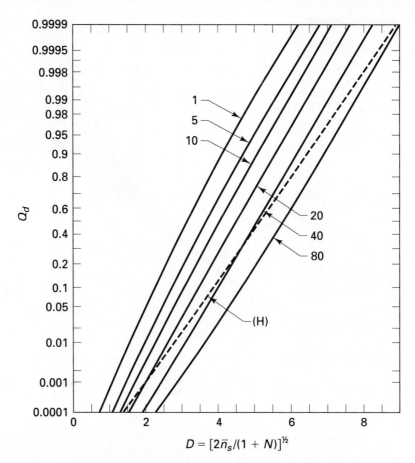

Figure 12-3. Detection probability of a coherent signal in background light that provides N photoelectrons per temporal mode versus the signal-to-noise ratio $D = [2\bar{n}_s/(1 + N)]^{1/2}$, $Q_0 = 10^{-6}$, $N = 0.1$. The curves are indexed by the number $M = WT$ of temporal modes of the background. The dashed line (H) gives the detection probability attained by a heterodyne detector.

12.2.3 The Poisson Limit

When the time–bandwidth product WT is very large, even though the expected number of photoelectrons ejected by the background light in each temporal mode may be very small, the total expected number $\bar{n}_0$ of these may be substantial. Whether the signal is incoherent (Gaussian) light spread over a broad band of frequencies or completely coherent light, like the output of a perfect laser, the receiver must then choose between two hypotheses:

(H_0): Only background light is present, and the probability of k counts is

$$\Pr(n = k\,|\,H_0) = \frac{\bar{n}_0^k}{k!}\,\exp(-\bar{n}_0), \qquad (12\text{-}35)$$

or

(H_1): Background light and a signal are both present, and the probability of k counts is

$$\Pr(n = k \mid H_1) = \frac{m_1^k}{k!} \exp(-m_1), \qquad m_1 = \bar{n}_0 + \bar{n}_s, \qquad (12\text{-}36)$$

where $\bar{n}_s$ is the expected number of photoelectrons ejected by the signal alone. We call this the *Poisson limit*. The number $\bar{n}_0$ may represent the expected number of counts resulting from dark current in the detector, that is, from stray electrons ejected by thermal agitation of the cathode, background radioactivity, or the like.

The optimum treatment of the datum n requires randomization as described in Sec. 1.2.5 in order to attain a preassigned false-alarm probability Q_0. By applying (1-57) and (1-58) one can compute the probability Q_d of detection. One obtains curves like those in Fig. 12-4, for which $Q_0 = 10^{-6}$ and in which the curves are indexed by the expected number $\bar{n}_0$ of electrons under hypothesis H_0. Even a very small average number $\bar{n}_0$ of these significantly increases the expected number $\bar{n}_s$ of electrons that must be ejected by the signal in order to attain a detection probability of the order of 0.99 or greater.

12.3 PHOTOMULTIPLICATION AND SHOT NOISE

12.3.1 Single-stage and Multistage Photomultipliers

In detecting very weak light signals there is a danger that the small number of photoelectrons ejected may be lost amid the noise in the external circuit that is to count them. Figure 12-5 shows the effect of noise added after a photoelectric detector with no background light incident upon it. The Poisson limit $WT \gg 1$ was assumed. The photoelectrons ejected into the external circuit of the detector might, for instance, be stored on a capacitor, to whose voltage a subsequent amplifier adds Gaussian noise. In effect, then, hypothesis H_1 is chosen when the random variable $v = n + x$ exceeds a decision level v_0. Under both hypotheses x is a Gaussian random variable with expected value zero and variance σ^2. Under H_1 n has a Poisson distribution with expected value $\bar{n}_s$; under H_0 $n \equiv 0$. The decision level v_0 was selected to yield a false-alarm probability

$$Q_0 = \mathrm{erfc}\left(\frac{v_0}{\sigma}\right) = 10^{-6}.$$

The curves in Fig. 12-5 are indexed by the standard deviation σ of the noise.

The moment-generating function of the random variable v under hypothesis H_1 is

$$\begin{aligned} h_v(z) &= E(e^{-z(n+x)} \mid H_1) \\ &= h_n(e^{-z})h_x(z), \end{aligned} \qquad (12\text{-}37)$$

where $h_n(\cdot)$ is the *probability*-generating function of the number n of photoelectrons. Here

$$h_v(z) = \exp\left[\bar{n}_s(e^{-z} - 1) + \tfrac{1}{2}\sigma^2 z^2\right]$$

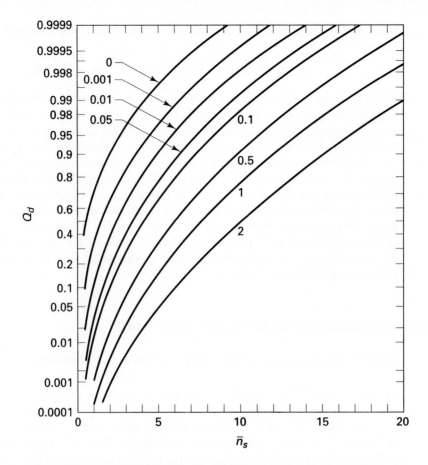

Figure 12-4. Detection probability Q_d of light in the Poisson limit versus the expected number $\bar{n}_s$ of electrons ejected by the signal. The curves are indexed by the expected number $\bar{n}_0$ of electrons under hypothesis H_0 that the signal is absent. $Q_0 = 10^{-6}$.

by (12-13). The probability Q_d of detection can be computed by saddlepoint integration of (5-19) or (5-20), depending on whether the decision level v_0 lies below or above the expected value $\bar{n}_s$ of v. Alternatively, the series to be derived in Problem 12-6 can be utilized.

This noise can be overcome by amplifying the number of photoelectrons by photomultiplication. In a typical photomultiplier primary electrons ejected by the incident light are accelerated by an applied voltage and strike an electrode, or *dynode*, from which they eject secondary electrons. These are accelerated by a further voltage drop and strike a second dynode, ejecting tertiary electrons. The process may continue through some number M of stages. The randomness of the number of secondary electrons per primary at each dynode in effect adds noise to the signal, but if the overall gain of the photomultiplier is large enough, the Gaussian noise represented by x can be overcome.

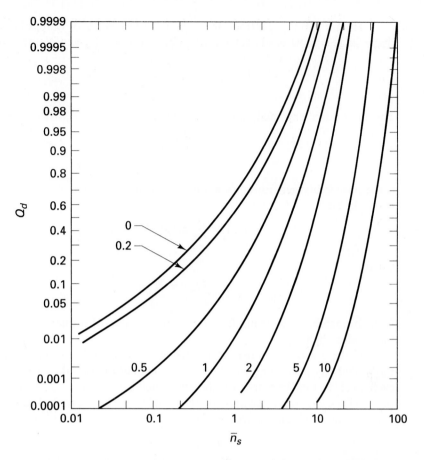

Figure 12-5. Detection probability Q_d of light ejecting electrons with a Poisson distribution in the absence of background light, but with the addition of Gaussian noise in the external circuit, versus the expected number $\bar{n}_s$ of electrons generated by the signal; $Q_0 = 10^{-6}$. The curves are indexed with the standard deviation σ of the additive noise.

We consider first a single stage of multiplication. The number of secondary electrons ejected by each primary electron is a random variable. Denote by $p_j^{(s)}$ the probability that a single primary electron ejects j secondaries from the dynode. The number of primary electrons is also a random variable; let π_r be the probability that r of them strike the first dynode during the observation interval. The total number of output electrons is

$$n = \sum_{k=1}^{r} n_k,$$

where n_k is the random number of secondary electrons ejected by the kth primary electron; the number r of terms in the sum is random. It is assumed that the numbers n_k of secondaries ejected by the several primary electrons are statistically independent.

The probability distribution of the total number n of output electrons from a single stage is most easily calculated from its probability-generating function

$$h(z) = E(z^n) = \sum_{r=0}^{\infty} \pi_r E(z^n \mid r).$$

The conditional expected value $E(z^n \mid r)$ is equal to $[g(z)]^r$, where

$$g(z) = \sum_{j=0}^{\infty} p_j^{(s)} z^j \qquad (12\text{-}38)$$

is the probability-generating function of the number of secondary electrons per primary. Hence

$$h(z) = \sum_{r=0}^{\infty} \pi_r [g(z)]^r = f(g(z)), \qquad (12\text{-}39)$$

where

$$f(z) = \sum_{r=0}^{\infty} \pi_r z^r \qquad (12\text{-}40)$$

is the probability-generating function of the number of primary electrons ejected by the incident light.

If there are a number M of dynodes instead of only one, we let $g_j(z)$ be the probability-generating function of the number of secondary electrons ejected by each electron that impinges on the jth dynode, $j = 1, 2, \ldots, M$. Let $\Gamma_j(z)$ be the probability-generating function of the number of electrons leaving the Mth (last) dynode when a single electron strikes the jth one. Then by (12-39)

$$\Gamma_M(z) = g_M(z),$$
$$\Gamma_{M-1}(z) = g_{M-1}(\Gamma_M(z)),$$
$$\cdots$$
$$\Gamma_j(z) = g_j(\Gamma_{j+1}(z)), \qquad (12\text{-}41)$$
$$\cdots$$
$$h(z) = f(\Gamma_1(z)).$$

The *gain* G_j at the jth dynode is the expected number of electrons ejected from it per electron incident on it,

$$G_j = \frac{d}{dz} g_j(z) \Big|_{z=1},$$

and the overall gain of the device—that is, the expected number of electrons at the output for each primary photoelectron at the input—is

$$G_0 = \prod_{j=1}^{M} G_j. \qquad (12\text{-}42)$$

Recurrent equations for the probabilities $p_n^{(j)}$ that n electrons are ejected from the last dynode when a single electron strikes the jth can be developed by means of

Detection of Optical Signals   Chap. 12

formulas given by Rice for differentiating functions of functions [Ric68, p. 1998]. In (12-41)

$$\Gamma_j(z) = \sum_{n=0}^{\infty} p_n^{(j)} z^n,$$

and by (12-9) and Rice's procedure

$$p_n^{(j)} = \frac{1}{n!} \frac{d^n}{dz^n} \Gamma_j(z) \bigg|_{z=0} = \frac{1}{n!} \frac{d^n}{dz^n} [g_j(\Gamma_{j+1}(z))] \bigg|_{z=0}$$

$$= \sum_{k=1}^{n} g_j^{(k)} b_{n,k}^{(j)}, \qquad n > 0,$$

where

$$g_j^{(k)} = \frac{d^k}{du^k} g_j(u) \bigg|_{u=p_0^{(j+1)}}, \qquad (12\text{-}43)$$

and the array of coefficients $b_{n,k}^{(j)}$ is determined from the probabilities $p_k^{(j+1)}$ by the recurrent relations

$$b_{n+1,1}^{(j)} = p_{n+1}^{(j+1)},$$

$$b_{n+1,k+1}^{(j)} = \frac{1}{n+1} \sum_{m=k}^{n} (n+1-m) p_{n+1-m}^{(j+1)} b_{m,k}^{(j)}.$$

In particular

$$b_{n,n}^{(j)} = \frac{\left[p_1^{(j+1)} \right]^n}{n!}.$$

These relations can be programmed for a computer. (The $b_{n,k}^{(j)}$ are Rice's $c_{n,k}$ divided by $n!$.) Furthermore,

$$p_0^{(j)} = \Gamma_j(0) = g_j(p_0^{(j+1)}), \qquad p_0^{(M)} = g_M(0),$$

by (12-41) with $z = 0$. One starts with $j = M$ and works down to $j = 0$, taking $g_0(z) = f(z)$.

If, for instance, the single-stage distribution is Poisson with gain G at each stage,

$$p_k^{(s)} = \frac{G^k}{k!} e^{-G}, \qquad g_j(z) = e^{G(z-1)}, \qquad (12\text{-}44)$$

and (12-43) becomes

$$g_j^{(k)} = G^k p_0^{(j)} = G^k \exp[G(p_0^{(j+1)} - 1)].$$

Recurrent relations for this Poisson single-stage distribution were given by Lombard and Martin [Lom61]. For the Polyà distribution, or generalized hypergeometric distribution, whose probability-generating function is

$$g(z) = [1 - bG(z - 1)]^{-1/b},$$

the recurrent relations were worked out by Prescott [Pre66].

If the gains or the number of stages or both are large, the expected number of electrons emerging from the photomultiplier will be very large, and these recurrence methods become lengthy and cumbersome, requiring great amounts of computer memory to store the probabilities $p_k^{(j)}$ and the array of coefficients $b_{n,k}^{(j)}$ from stage to stage. Round-off error corrupts the results, particularly in computing the complementary cumulative distribution $q_k^{(+)}$ (12-12) for $k > E(n) = m_1 G_0$, where m_1 is the expected number of primary electrons ejected by the incident light. The method of saddlepoint integration has been found efficacious under these circumstances [Hel84b]. Its application to a single stage of photomultiplication and a treatment of recurrence methods are to be found in [Hel84a].

How a single stage of photomultiplication overcomes the noise in the external circuit is illustrated in Fig. 12-6. Background light is absent; $\bar{n}_0 = 0$. The number of primary photoelectrons under hypothesis H_1 has a Poisson distribution with expected value $\bar{n}_s$, and as in (12-44) the number of secondaries per primary has a Poisson distribution with expected value G, the gain. The number of electrons counted at the output during $(0, T)$ then has a Neyman type A distribution [Tei81]. The probability-generating function of the number n of output electrons is

$$h_n(z) = \exp\{\bar{n}_s[e^{G(z-1)} - 1]\} \tag{12-45}$$

by (12-39), and the probabilities p_k of counting k output electrons obey Neyman's recurrent relations

$$p_0 = \exp[-\bar{n}_s(1 - e^{-G})],$$

$$p_{k+1} = \frac{\bar{n}_s G e^{-G}}{k+1} \sum_{r=0}^{k} \frac{G^r}{r!} p_{k-r}, \qquad k \geq 0,$$

[Ney39], [Hel91, p. 269].

To the number n of output electrons is added Gaussian random noise x with expected value 0 and variance σ^2 as in our example at the beginning of this section. The receiver chooses hypothesis H_1 when the sum $n + x$ exceeds a decision level v_0 set for a preassigned false-alarm probability Q_0; as before, $Q_0 = \mathrm{erfc}(v_0/\sigma)$. The moment-generating function of the sum is determined as in (12-37). In Fig. 12-6 we plot versus the gain G the expected numbers $\bar{n}_s$ of input photoelectrons required to attain two values of the probability Q_d of detection; $Q_0 = 10^{-6}$. The performance of the simple photoelectric counter assumed in Fig. 12-5 is surpassed for gains only a little larger than 1.

As the gain G increases, the expected number $\bar{n}_s$ approaches the value determined by (12-34),

$$\bar{n}_s \to -\ln(1 - Q_d).$$

When $G \gg 1$, a false dismissal occurs only when the number n of output electrons is zero, and

$$p_0 = \Pr(n = 0) \approx \exp(-\bar{n}_s).$$

The rest of the distribution of n is clustered around its expected value $G\bar{n}_s \gg 1$. For $1 \leq k < v_0$, $\Pr(n = k) \ll p_0$, and

$$Q_1 = 1 - Q_d \approx p_0 \Pr(x < v_0) = p_0(1 - Q_0) \approx p_0.$$

The number $\bar{n}_s$ was determined by the secant method applied to the false-dismissal probability

$$Q_1(\bar{n}_s) = \Pr(n + x < v_0 | H_1),$$

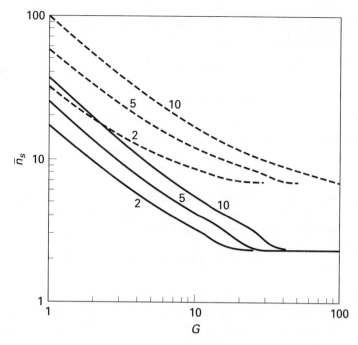

Figure 12-6. Expected number $\bar{n}_s$ of input photoelectrons to attain a fixed probability Q_d of detection with a single stage of photomultiplication versus the gain G of that stage. To the output number Gaussian noise of variance σ^2 is added. Solid curves: $Q_d = 0.9$; dashed curves: $Q_d = 0.999$; $Q_0 = 10^{-6}$. Curves are indexed with the standard deviation σ of the noise.

which during the first part of the search was calculated by the saddlepoint approximation (5-53). After that, Q_1 was computed by numerical integration of (5-19) along a straight vertical path through the saddlepoint.

12.3.2 Avalanche Photodiodes

A somewhat different type of photomultiplier is the avalanche photodiode. Light striking it on one face causes the injection of photoelectrons into the body of the device. Under an applied voltage these are accelerated to the point where they create pairs of holes and electrons by collision with the atoms of the material. These new electrons and holes are themselves accelerated and create further hole–electron pairs. The number n of electrons at the output face of the device during the observation interval $(0, T)$ is counted, and if it exceeds a certain decision level, the receiver decides that a light signal was present at the input.

The number n of electrons produced at the output by a single photoelectron at the input to the device is called the *random avalanche gain*. The distribution of this number depends on the ratio K between the probability that a hole creates a hole–electron pair by collision and the probability that an electron does so; $0 < K < 1$. Under the assumption that this ratio K is independent of the energy of the hole or the electron, Personick [Per71] showed that the probability-generating function $M(z)$

of the number n of output electrons per incident primary electron is given by the implicit equation

$$z = M[1 + a(M - 1)]^{-b}, \qquad b = \frac{1}{1 - K} > 1,$$

$$M(z) = \sum_{n=0}^{\infty} \Pr(n| 1)z^n, \tag{12-46}$$

where the parameter a is related to the gain G of the photodiode by

$$G = \frac{1}{1 - ab}, \qquad 0 < a < 1 - K.$$

At the same time, McIntyre [McI72] had worked out the probabilities $\Pr(n| r)$ that n electrons appear at the output when r primary electrons are injected at the input; the probability-generating function of these is

$$[M(z)]^r = \sum_{n=0}^{\infty} \Pr(n| r)z^n.$$

Only later was it shown that McIntyre's probabilities indeed follow from Personick's probability-generating function [Bal76]. When the probability-generating function of the number of primary electrons is $f(z)$, as in (12-40), that of the total number of output electrons counted in $(0, T)$ is

$$h(z) = \sum_{r=0}^{\infty} p_r z^r = f(M(z)). \tag{12-47}$$

For Poisson input probabilities Conradi [Con72] calculated the output probabilities

$$p_n = \sum_{r=0}^{\infty} \pi_r \Pr(n| r).$$

Because of the complexity of the formulas for $\Pr(n| r)$, this computation and that of the cumulative probabilities $q_n^{(-)}$ and $q_n^{(+)}$ can be cumbersome, particularly for large numbers n of output electrons.

Direct use of the probability-generating function $M(z)$, with (12-47), in numerical integration of (12-11) and (12-12) for these cumulative probabilities is difficult because of the necessity of solving (12-46) for $M(z)$ at each point on the contour of integration. When calculating the cumulative distribution of the number of output electrons, it has been found advantageous instead to use (12-46) to change the variable of integration in the contour integral from z to M. This transformation is single-valued provided one cuts the z-plane along the positive Re z-axis to the right of the point x_* given by

$$x_* = M_*[1 + a(M_* - 1)]^{-b} > 1, \qquad M_* = \frac{1 - a}{a(b - 1)} > 1. \tag{12-48}$$

One can then transform the contours C_- and C_+ of integration into parabolic contours passing through saddlepoints M_0^- and M_0^+ on the Re M-axis:

$$0 < M_0^- < 1, \qquad 1 < M_0^+ < M_*.$$

Detection of Optical Signals   Chap. 12

Details can be found in [Hel84c] and in [Hel88a]. This method has proved quite efficient for calculating the cumulative output distributions even for large numbers of electrons at the output of the photodiode.

In more complicated situations, as when Gaussian noise is added to the output or when, as we shall see in Sec. 12.3.3, it is the output current, rather than the number of output electrons, whose distribution is sought, what one needs is $M(e^{-z})$ as in (12-37). It is necessary to solve

$$-z = \ln M - b \ln [1 + a(M - 1)] \tag{12-49}$$

for M. In the neighborhood of the point $M = 1$, $z = 0$, one can use the power series

$$M(e^{-z}) = 1 + \sum_{k=1}^{\infty} c_k(-z)^k, \qquad |z| < z_* = \ln x_*,$$

whose coefficients are proportional to the moments of the random avalanche gain. They can be computed by the recurrent relations [Hau78]

$$c_{n+1} = \frac{G}{n + 1} \left[(a + 1)c_n + a \sum_{r=1}^{n-1} c_r c_{n-r} + a(b - 1) \sum_{r=0}^{n-1} (r + 1)c_{r+1} c_{n-r} \right],$$

$$n > 1.$$

The recurrences begin with

$$c_1 = G, \qquad c_2 = \tfrac{1}{2} G^3 (1 - a^2 b).$$

A series that is valid everywhere except on the cut in the z-plane extending from $-z_*$ to $-\infty$ is

$$M(e^{-z}) = M_* - \sum_{k=0}^{\infty} e_k t^{k+1}, \qquad t = [\Gamma(z + z_*)]^{1/2}, \qquad \Gamma = \frac{2M_*^2}{K},$$

whose coefficients are generated by the recurrence

$$e_n = \frac{1}{n + 2} \left[\frac{1 - K}{M_*^2} [\delta_{n-2} - (b + 1)M_* e_{n-1}] - \sum_{r=1}^{n-1} (r + 1)e_r e_{n-r} \right],$$

$$\delta_n = \sum_{r=0}^{n} e_r e_{n-r},$$

starting with

$$e_0 = 1, \qquad e_1 = -\frac{2}{3} \frac{b + 1}{b - 1} \frac{M_*}{\Gamma}$$

[Hel92c]. Far from the branch point $-z_*$ the number of terms one needs to include in this series becomes excessive, and one must resort to Newton's method in order to solve (12-49).

12.3.3 Shot Noise

The numbers of electrons at the output of a photodetector are seldom counted directly. In most detectors they induce a current $j(t)$ in an external circuit, and that current, possibly after amplification, is what is measured. In binary detection, hypothesis H_1 that a signal is present is selected when the value of the current exceeds a certain decision level. We turn, therefore, to an introduction to how one can compute the distribution of the current at the output of a photodetector.

The output current has the form

$$j(t) = \sum_{m=-\infty}^{\infty} f(t - \tau_m), \qquad (12\text{-}50)$$

where τ_m is the time at which the mth photoelectron crosses from cathode to anode, and $f(t)$ is the current pulse it induces in the external circuit. The shape and width of $f(t)$ depend on the effective resistance, inductance, and capacitance of that circuit. We shall normalize the output pulses so that

$$\int_{-\infty}^{\infty} f(t)\, dt = 1.$$

The dimension of the current $j(t)$ is then electrons per second. By multiplying by the electronic charge $q = 1.6 \cdot 10^{-19}$ coulombs, one obtains the current in amperes.

The times τ_m form a Poisson point process with rate $\lambda(t)$ given by (12-3). The probability that one electron is emitted during a brief interval δt is $\lambda \delta t$, and the numbers of emissions in disjoint intervals are statistically independent. We seek the conditional moment-generating function

$$h(z) = E \exp[-z\, j(t)] \qquad (12\text{-}51)$$

of the current $j(t)$ at time t, as in (12-50). The expectation E is taken with respect to the random times τ_m, and the rate $\lambda(t)$ is assumed given.

To this end we divide the time into intervals δt so brief that the probability that more than one electron is emitted during any one interval is negligible. We define the random variable ε_m as the number of electrons emitted during the mth interval. Then

$$\Pr(\varepsilon_m = 0) = 1 - \lambda(t_m)\delta t,$$
$$\Pr(\varepsilon_m = 1) = \lambda(t_m)\delta t,$$

with $t_m = m\delta t$; the ε_m's are statistically independent.

We approximate the current by

$$\tilde{j}(t) = \sum_m \varepsilon_m f(t - t_m), \qquad (12\text{-}52)$$

as though the electrons were emitted—if at all—only at the beginning of the intervals. As δt decreases, $\tilde{j}(t)$ approaches the true current $j(t)$. Conditioned on a given emission rate $\lambda(t)$, its moment-generating function is

$$\tilde{h}(z) = E \exp[-z\tilde{j}(t)]$$

$$= E \exp\left[-z\sum_m \varepsilon_m f(t - t_m)\right]$$

$$= \prod_m E \exp[-z\varepsilon_m f(t - t_m)]$$

$$= \prod_m \{1 - \lambda(t_m)\delta t + \lambda(t_m)\delta t \exp[-zf(t - t_m)]\} \qquad (12\text{-}53)$$

$$\approx \prod_m \exp[\lambda(t_m)\delta t \{\exp[-zf(t - t_m)] - 1\}]$$

$$= \exp \sum_m \lambda(m\delta t)\{\exp[-zf(t - m\delta t)] - 1\}\delta t.$$

In the fifth line we used the approximation $1 + x \approx e^x$ for $x \ll 1$ in order to introduce the exponential function. When we pass to the limit $\delta t \rightarrow 0$, this moment-generating function becomes

$$h(z) = \exp\left[\int_{-\infty}^{\infty} \lambda(\tau)\{\exp[-zf(t - \tau)] - 1\} \, d\tau\right]. \qquad (12\text{-}54)$$

Inversion of this moment-generating function to obtain the probability distribution of the current $j(t)$ is a difficult problem when the expected number

$$\bar{n} = \int_{-\infty}^{\infty} \lambda(\tau) \, d\tau$$

of electrons is small; see [Yue78]. For $\bar{n}$ greater than about 10, saddlepoint integration as in (5-19) and (5-20) becomes feasible. In most cases the integral in the exponent of (12-54) must be evaluated numerically at each point on the contour of integration. The path of steepest descent of the integrand has an infinite number of hairpinlike branches, opening to the right and going off to infinity, one above the other. Each passes through a saddlepoint of the integrand. Only when $\bar{n} \gg 1$ does it suffice to calculate the contributions only of the main branch cutting the Re z-axis and one or two adjacent branches. Computations seem to have been limited to shot noise for which the rate $\lambda(\tau)$ is constant and for output pulses $f(t)$ of simple shapes such as a triangle or a half-cycle of a sine wave, restricted to a finite interval outside of which they vanish.

When $f(t)$ is a rectangle,

$$f(t) = \begin{cases} \dfrac{1}{T}, & 0 \le t \le T, \\ 0, & t < 0, \quad t > T, \end{cases}$$

$$h(z) = \exp\left\{\int_{t-T}^{t} \lambda(\tau)\left[\exp\left(-\frac{z}{T}\right) - 1\right] d\tau\right\} = \bar{h}(e^{-z/T}),$$

where $\bar{h}(z)$ is the probability-generating function of the number n of electrons counted during the past T seconds, as defined in (12-14).

When the rate $\lambda(t)$ is itself a random process, as in (12-3) with $V(t)$ a circular-complex Gaussian random process, one must average (12-54) with respect to the distributions of $\lambda(\tau)$. The moment-generating function $h(z)$ then has a form similar to that of the rectified output of a quadratic detector, (11-66), and to calculate it the methods described in Sec. 11.2.1 and in [Hel86b] can be tried. Extensive numerical computation will in any case be required in order to evaluate the probability distribution of the output current $j(t)$.

When the current $j(t)$ results from electrons generated in a photomultiplier, we replace (12-52) by

$$\tilde{j}(t) = \sum_m \varepsilon_m n_m f(t - t_m), \tag{12-55}$$

where n_m is the random number of electrons produced at the output of the device when a single primary electron is injected at its input. It is assumed that these n_m output electrons are produced in a burst that is much shorter than the duration of the output current pulse $f(\cdot)$. The numbers n_m generated by each primary electron are furthermore assumed to be statistically independent. This requires the incident light to be so weak that the secondary electrons hardly perturb the electromagnetic fields within the device and do not inhibit the generation or passage of subsequent secondary electrons.

When the approximate current $\tilde{j}(t)$ is given by (12-55), we replace ε_m by $\varepsilon_m n_m$ in (12-53), and the fourth line of that equation becomes

$$\tilde{h}(z) = \prod_m E\{1 - \lambda(t_m)\delta t + \lambda(t_m)\delta t \exp[-n_m z f(t - t_m)]\},$$

in which the expected value is taken with respect to the distribution of the random numbers n_m of secondaries per primary electron. This averaging results in

$$\tilde{h}(z) = \prod_m \{1 - \lambda(t_m)\delta t + \lambda(t_m)\delta t \overline{M}[zf(t - t_m)]\}, \qquad \overline{M}(z) = M(e^{-z}),$$

where $\overline{M}(z)$ is the moment-generating function and $M(z)$ is the probability-generating function of the number of secondaries per primary. For a single-stage photomultiplier, $M(z)$ equals $g(z)$ in (12-38); for multiple stages it equals $\Gamma_M(z)$ in (12-41); and for an avalanche photodiode it is given by (12-46). Continuing as in (12-53), we find in the limit $\delta t \to 0$ that the conditional moment-generating function of the output current is

$$h(z) = E[\exp(-zj(t))] = \exp\left\{\int_{-\infty}^{\infty} \lambda(\tau)[\overline{M}(zf(t - \tau)) - 1]\, d\tau\right\}. \tag{12-56}$$

Again, if $\lambda(\tau)$ is random, as in (12-3), this moment-generating function must be averaged with respect to the distributions of $\lambda(\tau)$.

When subsequent amplification in effect adds Gaussian noise $n_G(t)$ with expected value zero and equivalent variance σ^2 to the current $j(t)$, the moment-generating function of the sum $j(t) + n_G(t)$ will be

$$h_t(z) = \exp\left\{\int_{-\infty}^{\infty} \lambda(\tau)[\overline{M}(zf(t - \tau)) - 1]\, d\tau + \tfrac{1}{2}\sigma^2 z^2\right\}.$$

The values of this moment-generating function along a path of integration must usually be computed by numerical integration of the exponent in (12-56). When calculating error probabilities of optical receivers by saddlepoint integration, it is important to sketch the path of steepest descent through each saddlepoint in a few cases in order to determine a suitable approximation to it into which to deform the

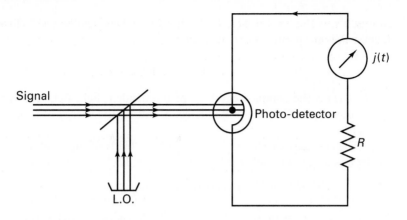

Figure 12-7. Heterodyne receiver.

contour of integration in (5-19) and (5-20). An application to receivers embodying avalanche diodes in a fiber-optical communication system is to be found in [Hel92c].

12.4 OPTICAL HETERODYNE AND HOMODYNE DETECTION

12.4.1 The Heterodyne Receiver

In an optical heterodyne receiver the incoming light, carrying the signal to be detected, is combined at a half-silvered mirror with a strong beam of coherent light from a laser, which might be termed the local oscillator (L.O.). As shown in Fig. 12-7, the sum of the two beams falls on a photoelectric cell, whose emissive surface acts much like a quadratic rectifier. In the interaction of the incident beam with the surface, shot noise is generated as a result of the random emission of photoelectrons. We shall study the effect of this noise on the detectability of the signal.

The field of the local oscillator beam in the heterodyne receiver must precisely match that of the signal over the surface of the detector in both spatial phase and polarization; otherwise the effective signal-to-noise ratio will be diminished. This matching is most easily achieved at infrared frequencies, where the wavelength of the light is conveniently long. Some signal energy will be lost at the mirror. The more transparent this is made for the signal, the more transparent it will be for the local-oscillator beam, and the more powerful this beam must be in order for our subsequent approximations to be valid. If the beam is too strong, however, it may damage the mirror. The technical details of constructing a heterodyne receiver are treated in the literature [Kin78].

Each photoelectron creates a pulse of current in the external circuit of the detector, shown in Fig. 12-7. The form $f(t)$ of this pulse depends on the resistance and the stray inductance and capacitance of the circuit. The total current in the circuit is, as in (12-50),

$$j(t) = \sum_i f(t - \tau_i), \qquad \int_{-\infty}^{\infty} f(t)\, dt = 1,$$

in which τ_i is the time at which the ith electron was emitted. As before, the times τ_i form a Poisson point process with rate

$$\lambda(t) = \tfrac{1}{2}\eta|V(t)|^2, \qquad (12\text{-}57)$$

where $V(t)$ is the complex envelope of the incident light field and $\eta = \eta'/hf$, with η' the quantum efficiency of the detector and hf the energy of a single quantum of the light.

The complex envelope of the sum of the fields of the local-oscillator beam and of the incoming light, labeled "Signal" in Fig. 12-7, can be written

$$V(t) = L_0 \exp(-iw_0 t) + V_i(t), \qquad (12\text{-}58)$$

where L_0 is the constant complex amplitude of the beam coming from the local oscillator,

$$w_0 = \Omega - \Omega_{\text{L.O.}}$$

is the offset frequency, and $V_i(t)$ is the complex envelope of the incoming light. This light has been filtered to remove components of the background light outside the frequency band of the signal being sought. We assume that the local-oscillator beam is so strong that $|L_0| \gg |V_i(t)|$.

We first determine the joint moment-generating function of scaled samples $j(t_m)/|L_0|$ of the current $j(t)$ at times $t_1, t_2, \ldots, t_m, \ldots$:

$$h(z_1, z_2, \ldots, z_m, \ldots) = E\left\{\exp\left[-\frac{1}{|L_0|}\sum_m z_m j(t_m)\right]\right\}$$

$$= E\left\{\exp\left[-\frac{1}{|L_0|}\sum_m \sum_i z_m f(t_m - \tau_i)\right]\right\}.$$

If we compare this with what we obtained when we substituted (12-50) into (12-51), we see that our joint moment-generating function can be written down from $h(z)$ as derived there by replacing $zf(t - \tau)$ everywhere by

$$\frac{1}{|L_0|}\sum_m z_m f(t_m - \tau).$$

Thus from (12-54),

$$h(z_1, z_2, \ldots, z_m, \ldots) = \exp \int_{-\infty}^{\infty} \lambda(\tau)\left\{\exp\left[-\frac{1}{|L_0|}\sum_m z_m f(t_m - \tau)\right] - 1\right\} d\tau$$

$$= \exp \frac{\eta}{2} \int_{-\infty}^{\infty} |V(t)|^2 \left\{\exp\left[-\frac{1}{|L_0|}\sum_m z_m f(t_m - t)\right] - 1\right\} dt$$

by (12-57). This joint moment-generating function is conditioned on a given input process Re $V(t) \exp i\Omega t$.

Under the assumption that $|L_0|$ is large, we make a power-series expansion of the innermost exponential in the moment-generating function, obtaining

$$h(z_1, z_2, \ldots, z_m, \ldots) = \exp\left\{-\frac{\eta}{2|L_0|}\int_{-\infty}^{\infty}|V(t)|^2\sum_m z_m f(t_m - t)\, dt\right.$$

$$+ \frac{\eta}{4|L_0|^2}\int_{-\infty}^{\infty}|V(t)|^2\sum_m\sum_n z_m z_n f(t_m - t)f(t_n - t)\, dt \qquad (12\text{-}59)$$

$$\left. - \frac{\eta}{12|L_0|^3}\int_{-\infty}^{\infty}|V(t)|^2\sum_m\sum_n\sum_p z_m z_n z_p f(t_m - t)f(t_n - t)f(t_p - t)\, dt + \cdots\right\}.$$

The leading term in $|V(t)|^2$, by (12-58), is $|L_0|^2$, and we see that the term in (12-59) with the triple summation is therefore proportional to $|L_0|^{-1}$. In the limit $|L_0| \to \infty$ this term and all those beyond it will vanish. Because only the terms linear and quadratic in the Laplace variables z_m remain, the current $j(t)$ in this limit is a Gaussian stochastic process. We can therefore use the results of Secs. 3.3 and 3.4 to determine the optimum processing of the output of the photodetector and the resultant false-alarm and detection probabilities.

From (12-59) the expected value of the current is

$$E[\,j(t)|\, V(t'), 0 \le t' \le T] = \tfrac{1}{2}\eta\int_{-\infty}^{\infty}|V(s)|^2 f(t - s)\, ds, \qquad (12\text{-}60)$$

conditionally on the complex envelope $V(t)$ of the light field incident on the photocell, and its conditional autocovariance function is

$$\mathrm{Cov}[\,j(t_1),\, j(t_2)|\, V(t'), 0 \le t' \le T] =$$
$$\tfrac{1}{2}\eta\int_{-\infty}^{\infty}|V(s)|^2 f(t_1 - s)f(t_2 - s)\, ds, \qquad (12\text{-}61)$$

which represents *Campbell's theorem*.

We now assume that the bandwidth of the detection circuit is so much greater than that of the fluctuations of the complex envelope $V(t)$ of the light field that we can replace $f(t)$ by a delta function in (12-60). Then the expected value of the current is

$$E[\,j(t)]|V(t'), 0 \le t' \le T] = \tfrac{1}{2}\eta|V(t)|^2. \qquad (12\text{-}62)$$

Because $|L_0| \gg |V_i(t)|$, the dominant term in (12-61) is

$$\mathrm{Cov}[\,j(t_1),\, j(t_2)] = \tfrac{1}{2}\eta|L_0|^2\int_{-\infty}^{\infty} f(t_1 - s)f(t_2 - s)\, ds$$

$$= \tfrac{1}{2}\eta|L_0|^2\int_{-\infty}^{\infty}|F(\omega)|^2 \exp[i\omega(t_1 - t_2)]\,\frac{d\omega}{2\pi},$$

where

$$F(\omega) = \int_{-\infty}^{\infty} f(t)\, e^{-i\omega t}\, dt$$

is the Fourier transform of the current pulse $f(t)$ and is proportional to the transfer function of the external circuit. The spectral density of the current fluctuations is therefore

$$\Phi_{sh}(\omega) = \tfrac{1}{2}\eta|L_0|^2|F(\omega)|^2.$$

Because the external circuit is assumed to have a broad passband, $|F(\omega)| \approx |F(0)| = 1$, and the unilateral spectral density of the shot-noise component of the current $j(t)$ is

$$N_{sh} = 2\Phi_{sh}(\omega) = \eta|L_0|^2. \qquad (12\text{-}63)$$

The shot noise is effectively white.

With $|L_0| \gg |V_i(t)|$, the expected value of the current is, from (12-58) and (12-62),

$$E[j(t)| \, V_i(t'), 0 \le t' \le T] = \tfrac{1}{2}\eta|L_0|^2 + \eta \, \text{Re}[L_0^* V_i(t) \exp iw_0 t]. \qquad (12\text{-}64)$$

The first term is biased out or absorbed into the decision level. In (12-58)

$$V_i(t) = S(t) + V_b(t),$$

where $S(t)$ is the complex envelope of a coherent signal to be detected, and $V_b(t)$ is the complex envelope of the background light field. Then the second term in (12-64) represents the combination of a coherent narrowband signal at the intermediate frequency w_0 with complex envelope

$$S_{if}(t) = \eta L_0^* S(t) \qquad (12\text{-}65)$$

and narrowband Gaussian noise $\text{Re } W(t) \exp iw_0 t$ with complex envelope

$$W(t) = \eta L_0^* V_b(t).$$

The complex autocovariance function of this noise is

$$\tfrac{1}{2}E[W(t_1)W^*(t_2)] = \eta^2|L_0|^2 P\phi(t_1 - t_2)$$

by (12-1). This corresponds to narrowband noise with a narrowband spectral density

$$\tilde{\Phi}(\omega) = \tfrac{1}{2}\eta^2|L_0|^2 P\Phi(\omega)$$

by (3-20) and (12-2), where $\Phi(\omega)$ is the normalized spectral density of the background light. If as in (12-26) that light is assumed to have a uniform spectral density $\Phi(\omega) = W^{-1}$ over a band of angular frequencies, $-\pi W < \omega < \omega W$, much broader than that occupied by the signal, the background contributes an additional white noise having a unilateral spectral density

$$N_b = \eta^2|L_0|^2 \frac{P}{W}. \qquad (12\text{-}66)$$

The signal $\text{Re } S_{if}(t) \exp iw_0 t$ is thus to be detected in the presence of white Gaussian noise of unilateral spectral density $N_{sh} + N_b$. As its phase is in general unknown, we are confronted with the same detection problem as that treated in Sec. 3.3. The voltage developed across the resistor in the external circuit of the photodetector will be applied to a filter matched to that signal, and its output will be passed to a rectifier, whose output at time $t = T$ will be compared with a decision level set to attain a preassigned false-alarm probability Q_0.

The probability of detecting the signal is, as in (3-75),

$$Q_d = Q(D, b) \qquad (12\text{-}67)$$

in terms of Marcum's Q function, where b is determined by the false-alarm probability

$$Q_0 = e^{-b^2/2},$$

and the effective signal-to-noise ratio D^2 is twice the energy of the signal $S_{if}(t)$ divided by the unilateral spectral density $N_{sh} + N_b$ of the noise,

$$D^2 = \frac{1}{N_{sh} + N_b} \int_{-\infty}^{\infty} |S_{if}(t)|^2 \, dt.$$

Here

$$\int_{-\infty}^{\infty} |S_{if}(t)|^2 \, dt = \eta^2 |L_0|^2 \int_{-\infty}^{\infty} |S(t)|^2 \, dt = 2\eta |L_0|^2 \bar{n}_s$$

by (12-65) and (12-5), where $\bar{n}_s$ is the average number of photoelectrons that would be ejected by the coherent component of the incident light in the absence of the local oscillator. By (12-63) and (12-66), furthermore,

$$N_{sh} + N_b = \eta |L_0|^2 + \eta^2 |L_0|^2 \frac{P}{W} = \eta |L_0|^2 \left[1 + \frac{\bar{n}_0}{WT} \right],$$

with $\bar{n}_0 = \eta PT$ as in (12-4). Thus we can write the effective signal-to-noise ratio as

$$D^2 = \frac{2\bar{n}_s}{1 + N}, \qquad N = \frac{\bar{n}_0}{WT}. \tag{12-68}$$

The number N is the expected number of photoelectrons per temporal mode under those conditions, and it is given by $\eta' \overline{N}$, where $\overline{N}$ is specified by the Planck law (12-31) and η' is the quantum efficiency.

In Fig. 12-3 we plotted as a dashed line (H) the probability Q_d of detecting a coherent narrowband light signal by the heterodyne receiver, as given by (12-67), for a false-alarm probability $Q_0 = 10^{-6}$. We took $N = 0.1$, and the abscissa is the effective signal-to-noise ratio D defined by (12-68). The solid lines represent the probability of detecting the same signal by counting the number of photoelectrons it ejects when the local-oscillator beam is turned off; they were calculated as described in Sec. 12.2.2. The curves are indexed by the effective number $M = WT$ of temporal degrees of freedom. Only if the background light is admitted over a broad range of frequencies is a heterodyne receiver superior to one merely counting the number of photoelectrons directly ejected by the light.

Figure 12-8 shows the gain achieved by using a heterodyne detector rather than a photoelectric counter; the gain is defined as the ratio of the expected number $\bar{n}_s$ of photoelectrons needed by the photoelectric counter to attain a given probability Q_d of detection to the expected number $\bar{n}_s^{(H)}$ needed by the heterodyne receiver, for the same false-alarm probability Q_0. Here, by (12-68),

$$\bar{n}_s^{(H)} = \tfrac{1}{2} D^2 (1 + N),$$

where D^2 is the signal-to-noise ratio specified by (12-67). The gain is plotted against the time–bandwidth product $M = WT$ for four values of the expected number N of background photoelectrons per temporal mode. The smaller the expected number

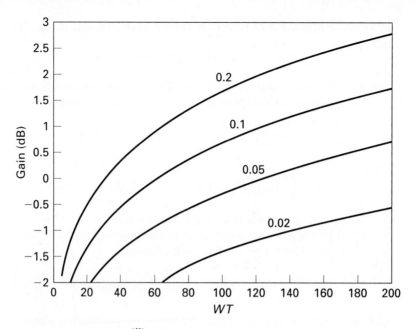

Figure 12-8. Gain $\bar{n}_s/\bar{n}_s^{(H)}$ (dB) of input signal strength when using a heterodyne detector rather than a photoelectric counter versus the time–bandwidth product $M = WT$; $Q_d = 0.99$, $Q_0 = 10^{-6}$. Curves are indexed with the expected number N of photoelectrons per temporal mode.

N is, the greater must the time–bandwidth product WT be before it is advantageous to use a heterodyne detector.

In plotting the curves in Fig. 12-8, we began by determining by (1-57) the combination of the decision level n_0 on the number n of photoelectrons and the randomization fraction f that yields the preassigned false-alarm probability Q_0 for the hypergeometric distribution in (12-30). The number $\bar{n}_s$ required to attain the probability Q_d of detection with the photoelectric counter was first calculated by the saddlepoint approximation to (12-11), in which the probability-generating function $h(z)$ is given by (12-27). By eliminating the parameter $\bar{n}_s$ from the equations for the "phase" and its first two derivatives, it was possible to apply the secant method to search for the saddlepoint $z_0^{(-)}$ yielding the desired value of $Q_1 = 1 - Q_d$ and thence to calculate an approximate value of $\bar{n}_s$. With this as a starting value, the secant method was then used with the exact Laguerre distribution (12-29) in (1-58) to compute the expected number $\bar{n}_s$ yielding the prescribed detection probability Q_d.

In this analysis we have assumed the ideal photoelectric detector shown in Fig. 12-7. In a practical heterodyne receiver it may be necessary to use photosensitive solid-state devices such as avalanche photodiodes. These will contribute additional noise that must be taken into account in assessing the performance of the receiver.

Detection of Optical Signals   Chap. 12

12.4.2 The Homodyne Receiver

If we know the frequency and the phase of the signal precisely—and this knowledge is much more difficult to acquire at optical than at radio frequencies—we can set the local oscillator beam at the same frequency and phase. The receiver is then termed *homodyne*. When L_0 and the complex envelope $S(t)$ of the signal are real, the conditional expected value of the current from the detector will be

$$E[j(t)] = \tfrac{1}{2}\eta|L_0|^2 + \eta L_0[S(t) + \text{Re } V_b(t)]$$
$$= \tfrac{1}{2}\eta|L_0|^2 + s(t) + n(t) \tag{12-69}$$

from (12-64), where $s(t) = \eta L_0 S(t)$. Here $n(t) = \eta L_0 \text{ Re } V_b(t) = \eta L_0 V_{bx}(t)$ is Gaussian noise with autocovariance function

$$\phi_b(t_1 - t_2) = E[n(t_1)n(t_2)]$$
$$= \eta^2 L_0^2 E[V_{bx}(t_1)V_{bx}(t_2)] = \eta^2 L_0^2 P \text{ Re } \phi(t_1 - t_2)$$

by (3-34) and (12-1). Taking $\phi(\tau)$ as real, we find for the spectral density of the noise due to the background light

$$\Phi_b(\omega) = \eta^2 L_0^2 P \Phi(\omega) = \eta^2 L_0^2 \frac{P}{W}$$

when as before the background light has a uniform spectral density over a band of width W hertz. This corresponds to white noise of unilateral spectral density

$$N_b' = 2\eta^2 L_0^2 \frac{PT}{WT} = 2\eta L_0^2 \frac{\bar{n}_0}{WT} = 2\eta L_0^2 N, \tag{12-70}$$

where as before $\bar{n}_0 = \eta PT$ is the total average number of photoelectrons that would be ejected if the background light fell directly onto the detector, and N is as in (12-68). The probabilities of false alarm and detection are now, as in (2-72),

$$Q_0 = \text{erfc } x, \qquad Q_d = \text{erfc}(x - D),$$

with the effective signal-to-noise ratio given by

$$D^2 = \frac{2}{N_{\text{sh}} + N_b'} \int_{-\infty}^{\infty} [s(t)]^2 \, dt = \frac{4\bar{n}_s}{1 + 2N}. \tag{12-71}$$

Here

$$\int_{-\infty}^{\infty} [s(t)]^2 \, dt = \eta^2 L_0^2 \int_{-\infty}^{\infty} |S(t)|^2 \, dt = 2\eta L_0^2 \bar{n}_s$$

by (12-5), and for N_{sh} and N_b' we have substituted (12-63) and (12-70), respectively. Again, $\bar{n}_s$ is the expected number of photoelectrons that would be ejected by the coherent component of the incident light if it fell directly onto the photodetector.

Comparing (12-68) and (12-71), we see that in the classical limit, $N \gg 1$, both signal-to-noise ratios approach the same value $2\bar{n}_s/N$, and by the Planck law (12-31), with $kT \gg hf$, this ratio becomes

$$D^2 \rightarrow \frac{2\bar{n}_s}{N} \rightarrow \frac{2\eta E_s}{\eta hf(kT/hf)} = \frac{2E_s}{kT},$$

where E_s is the energy of the incident signal, k is Boltzmann's constant, and T is the effective absolute temperature of the background radiation. This is the same as the maximum attainable signal-to-noise ratio deduced in Sec. 2.4. In the quantum limit $N \to 0$, on the other hand, the homodyne receiver provides twice as large an effective signal-to-noise ratio as the heterodyne receiver. The l's in the denominators of (12-68) and (12-71) arise from the irreducible quantum fluctuations of the electromagnetic field.

12.5 DETECTION BASED ON COUNTING TIMES

In the previous sections it was assumed that the decision about the presence or absence of a signal is based only on the total number n of photoelectrons counted during the observation interval $(0, T)$. The question arises whether a greater probability of detection could be attained by observing the times at which the individual electrons are ejected by the incident light. In general this must be the case, for if one discarded that information and used only the total number of electrons, one would expect on the basis of our discussion in Chapter 1 to suffer a loss of signal detectability. To calculate just how great the loss is, however, is extremely difficult.

Conditioned on a particular realization of the incident field $v(t) = \operatorname{Re} V(t) \exp i\Omega t$, the probability that one electron is emitted during each of the m subintervals $(t_i, t_i + dt)$, $1 \le i \le m$, and none during the rest of the observation interval $(0, T)$, is

$$\Pr(t_i \le \tau_i < t_i + dt, 1 \le i \le m \mid V(t), 0 \le t \le T)$$

$$= \exp\left[-\int_0^T \lambda(t)\, dt\right] \prod_{i=1}^m [\lambda(t_i)\, dt], \qquad (12\text{-}72)$$

where τ_i is the time at which the ith electron is emitted, and as in (12-3)

$$\lambda(t_i) = \tfrac{1}{2}\eta |V(t_i)|^2$$

with $\eta = \eta'/hf$ as in Sec. 12.1.1 [Bar69].

To show this, we divide the interval $(0, T)$ into a large number of disjoint subintervals Δt. The emission times τ_i form a Poisson point process, and the probabilities that no electron and one electron is emitted during any subinterval are, respectively,

$$\Pr[0 \text{ e's in } (t, t + \Delta t)] = \exp[-\lambda(t)\Delta t], \qquad (12\text{-}73)$$

$$\Pr[1 \text{ e in } (t, t + \Delta t)] = \lambda(t)\Delta t \exp[-\lambda(t)\Delta t]. \qquad (12\text{-}74)$$

Whether an electron is emitted in any subinterval is independent of whether one is emitted in any other subinterval. Hence we can multiply the probabilities (12-73) and (12-74) for the several subintervals, and letting Δt become infinitesimal and replacing it by dt, we obtain (12-72).

This probability must be averaged over the ensemble of the processes $V(t)$. The value of the likelihood ratio when m emissions have been observed to take place at times $\tau_1, \tau_2, \ldots, \tau_m$ and none at any other times is then

$$\Lambda(\tau_1, \tau_2, \ldots, \tau_m) = \frac{E_1\left\{\prod_{j=1}^m \lambda_1(\tau_j) \exp\left[-\int_0^T \lambda_1(t)\, dt\right]\right\}}{E_0\left\{\prod_{j=1}^m \lambda_0(\tau_j) \exp\left[-\int_0^T \lambda_0(t)\, dt\right]\right\}}, \qquad (12\text{-}75)$$

where $\lambda_i(\cdot)$ is the emission rate under hypothesis H_i, and E_i indicates an average with respect to the ensemble of processes $V(t)$ under that hypothesis, $i = 0, 1$. This likelihood ratio is compared with a decision level Λ_0, which if exceeded evokes the decision that H_1 is true and a signal is present. The value of Λ_0 is determined by the false-alarm probability Q_0. In general the probability distributions of the statistic in (12-75) are difficult to compute under either hypothesis.

12.5.1 The Poisson Limit

Suppose that the background light falling on the photoelectrically emissive surface of the detector has a bandwidth W much greater than the reciprocal of the resolving time of the device recording the emission times τ_j. Then under hypothesis H_0 the fluctuations in the rate are negligible, and we can take $\lambda_0(t) \equiv \lambda_0$ as constant. Suppose that the signal to be detected is the output $s(t) = \operatorname{Re} S(t) \exp i\Omega t$ of an ideal laser. Then under hypothesis H_1 we can take

$$\lambda_1(t) = \lambda_0 + \tfrac{1}{2}\eta|S(t)|^2$$

as nonrandom as well. The logarithm of the likelihood ratio in (12-75) is now

$$\ln \Lambda(\tau_1, \tau_2, \ldots, \tau_m) = \sum_{j=1}^{m} \ln \left[\frac{\lambda_1(\tau_j)}{\lambda_0} \right] - \int_0^T [\lambda_1(t) - \lambda_0]\, dt.$$

The statistic

$$g = \sum_{j=1}^{m} \ln \left[\frac{\lambda_1(\tau_j)}{\lambda_0} \right] = \sum_{j=1}^{m} \ln \left[1 + \frac{\eta}{2\lambda_0}|S(\tau_j)|^2 \right] \tag{12-76}$$

is thus sufficient for deciding about the presence or absence of the signal. If the signal envelope is constant, $S(t) \equiv S_0$, the statistic g is simply proportional to the total number m of electrons counted in $(0, T)$, and their emission times τ_j provide no useful additional information. If the signal $S(t)$ is variable, they do provide information, and one can anticipate that utilizing the statistic in (12-76) will enhance signal detectability.

To calculate the performance of a receiver based on g of (12-76) is difficult. The moment-generating function of g under hypothesis H_i, $i = 0, 1$, can be worked out by the following analysis. Divide the interval $(0, T)$ into a large number of brief subintervals (t_{k-1}, t_k), $1 \le k \le M$. Then the statistic g is approximately

$$g = \sum_{k=1}^{M} n_k \ln \left[\frac{\lambda_1(t_k)}{\lambda_0} \right], \qquad t_k = k\,\Delta t, \qquad \Delta t = \frac{T}{M}, \tag{12-77}$$

where n_k is the number of photoelectrons emitted in the kth subinterval. For Δt sufficiently small, n_k is either 0 or 1 with much greater probability than that $n_k > 1$, and (12-77) reduces to (12-76). The numbers n_k are independent Poisson-distributed random variables with expected values

$$E(n_k \mid H_0) = \lambda_0 \Delta t, \qquad E(n_k \mid H_1) = \lambda_1(t_k)\Delta t.$$

The moment-generating function of the statistic g is then approximately

$$h_i(z) = E(e^{-gz} \mid H_i) = \prod_{k=1}^{M} E\left\{\left[\frac{\lambda_1(t_k)}{\lambda_0}\right]^{-n_k z} \middle| H_i\right\}. \qquad (12\text{-}78)$$

For a Poisson random variable with expected value m, as in (12-13),

$$E(z^n) = e^{m(z-1)}.$$

Replacing z by $[\lambda_1(t_k)/\lambda_0]^{-z}$ and m by $\lambda_i(t_k)\Delta t$ in the kth factor in (12-78), we find

$$h_i(z) = \prod_{k=1}^{M} \exp\left\{\lambda_i(t_k)\left[\left[\frac{\lambda_1(t_k)}{\lambda_0}\right]^{-z} - 1\right]\Delta t\right\}$$

$$= \exp\left\{\sum_{k=1}^{M} \lambda_i(t_k)\left[\left[\frac{\lambda_1(t_k)}{\lambda_0}\right]^{-z} - 1\right]\Delta t\right\},$$

and in the limit $M \to \infty$, $\Delta t \to 0$ this becomes

$$h_i(z) = \exp\left\{\int_0^T \lambda_i(t)\left[\left[\frac{\lambda_1(t)}{\lambda_0}\right]^{-z} - 1\right]dt\right\} \qquad (12\text{-}79)$$

with

$$\lambda_0(t) \equiv \lambda_0, \qquad \lambda_1(t) = \lambda_0 + \tfrac{1}{2}\eta|S(t)|^2.$$

These moment-generating functions cannot be inverted analytically to obtain the probability density functions of the statistic g unless $|S(t)|^2$ is constant, whereupon, as we already know, the statistic g is proportional to a Poisson-distributed integer-valued random variable. Otherwise the false-alarm and detection probabilities might be computed by numerical integration of the contour integrals in (12-11) and (12-12), but it would be necessary to evaluate $h_i(z)$ at each point z on the contour of integration by integrating the exponent in (12-79) numerically. These moment-generating functions have the same form as that in (12-54) if there we put $t = 0$ and $f(-\tau) = \ln[\lambda_1(\tau)/\lambda_0]$. The same problems as attend calculating the distribution of shot noise will arise here.

12.5.2 Incoherent Gaussian Light

The probability density functions in the numerator and denominator of (12-75) are

$$p_i(\tau_1, \tau_2, \ldots, \tau_m) = E\left\{\prod_{j=1}^{m} \lambda_i(\tau_j) \exp\left[-\int_0^T \lambda_i(t)\,dt\right] \middle| H_i\right\}, \qquad (12\text{-}80)$$

$i = 0, 1$; E denotes an expectation with respect to the ensemble of stochastic processes $\lambda_i(t)$. Let us now suppose that under both hypotheses the complex envelope $V(t)$ is a stationary circular complex Gaussian random process with expected value zero and autocovariance function $P\phi(t - u)$ as in Sec. 12.1. The rate of emission of photoelectrons is

$$\lambda(t) = \tfrac{1}{2}\eta|V(t)|^2$$

as in (12-3). We drop the subscripts referring to the hypothesis. How to evaluate the expectation in (12-80) has been described by Macchi [Mac71]; see also [Sny75, pp. 303–309]. Here we present a perhaps simpler method.

We denote the product in (12-80) by

$$R[V] = \prod_{j=1}^{m} \lambda(\tau_j) = (\tfrac{1}{2}\eta)^m \prod_{j=1}^{m} |V(\tau_j)|^2.$$

In order to find the expected value

$$
\begin{aligned}
E &= E\left\{ R[V] \exp\left[-\int_0^T \lambda(t)\, dt \right] \right\} \\
&= E\left\{ R[V] \exp\left[-\frac{\eta}{2} \int_0^T |V(t)|^2\, dt \right] \right\}
\end{aligned}
\tag{12-81}
$$

we express $V(t)$ in terms of an orthonormal set of sampling functions $f_k(t)$, as before. Then $R[V]$ depends on the V_k's in a way that we need not specify. We evaluate (12-81) by multiplying the contents of the expectation by the joint probability density function of the real and imaginary parts of a finite number M of samples $V_k = V_{kx} + iV_{ky}$ in circular Gaussian form, and we then integrate over the entire $2M$-dimensional space of V_{kx} and V_{ky},

$$E = (2\pi)^{-M} |\det \mathbf{\Phi}|^{-1} \int_{-\infty}^{\infty} \cdots \int_{-\infty}^{\infty} R(\mathbf{V}) \exp[-\tfrac{1}{2}\eta \mathbf{V}^+\mathbf{V} - \tfrac{1}{2}\mathbf{V}^+\mathbf{\Phi}^{-1}\mathbf{V}]\, d^M V_{kx}\, d^M V_{ky},$$

where $\mathbf{\Phi}$ is the complex autocovariance matrix of the samples V_k, and we have put

$$\int_0^T |V(t)|^2\, dt = \sum_{k=1}^{M} |V_k|^2 = \mathbf{V}^+\mathbf{V}, \qquad R[V] \to R(\mathbf{V}).$$

We write this expected value E as

$$
\begin{aligned}
E &= \frac{1}{(2\pi)^M |\det \mathbf{\Phi}|} \int_{-\infty}^{\infty} \cdots \int_{-\infty}^{\infty} R(\mathbf{V}) \exp[-\tfrac{1}{2}\mathbf{V}^+(\eta\mathbf{I} + \mathbf{\Phi}^{-1})\mathbf{V}]\, d^M V_{kx}\, d^M V_{ky} \\
&= \frac{1}{|\det(\eta\mathbf{I} + \mathbf{\Phi}^{-1})| |\det \mathbf{\Phi}|} \int_{-\infty}^{\infty} \cdots \int_{-\infty}^{\infty} R(\mathbf{V}) q(\{V_k\})\, d^M V_{kx}\, d^M V_{ky},
\end{aligned}
\tag{12-82}
$$

where $q(\{V_k\})$ is a circular Gaussian density function of samples V_k with covariance matrix

$$(\eta\mathbf{I} + \mathbf{\Phi}^{-1})^{-1} = \mathbf{\Phi}(\mathbf{I} + \eta\mathbf{\Phi})^{-1} = \mathbf{M}.$$

This matrix $\mathbf{M}$ is the solution of the matrix equation

$$\mathbf{M} + \eta\mathbf{M}\mathbf{\Phi} = \mathbf{\Phi}.$$

The dimension M of these matrices is now allowed to go to infinity. The auto-covariance function corresponding to the matrix $\mathbf{\Phi}$ is $P\phi(t - u)$, and the function corresponding to the matrix $\mathbf{M}$ is the solution $m(t, u; n)$ of the integral equation

$$m(t, u; n) + \frac{n}{T} \int_0^T m(t, v; n)\phi(v - u)\, dv = P\phi(t - u),$$

$$0 \le (t, u) \le T,$$

(12-83)

with $n = \eta PT$ the expected number of photoelectrons under whichever hypothesis is involved. The factor in front of the second integral in (12-82) becomes

$$\det(\mathbf{I} + \eta\mathbf{\Phi})^{-1} = [D(n)]^{-1}, \qquad n = \eta PT,$$

in terms of the Fredholm determinant $D(x)$ defined in (12-17). Thus we can write the expectation in (12-82), and hence that in (12-80), as

$$p(\tau_1, \tau_2, \ldots, \tau_m) = \frac{1}{D(n)} E_q\left[(\tfrac{1}{2}\eta)^m \prod_{j=1}^{m} |V(\tau_j)|^2 \right],$$

where E_q denotes an expectation with respect to an ensemble of circular complex Gaussian processes $V(t)$ with autocovariance function $m(t, u; n)$. When calculating such a probability density function under hypothesis H_0, one puts n equal to the expected number $\bar{n}_0$ of photoelectrons under that hypothesis; and when calculating it under hypothesis H_1, one puts $n = \bar{n}_1 = \bar{n}_0 + \bar{n}_s$.

In particular

$$p_0 = E_q \exp\left[-\int_0^T \lambda(t)\, dt \right] = \frac{1}{D(n)}$$

is the probability that no photoelectrons are ejected at all under the current hypothesis. This is the same as one obtains from (12-19) with $z = 0$, $S(t) \equiv 0$, and $\bar{n}_0 = n$.

Furthermore

$$p(\tau_1) = \frac{1}{D(n)} \tfrac{1}{2}\eta E_q\big[|V(\tau_1)|^2\big]$$

$$= \frac{\eta}{D(n)} m(\tau_1, \tau_1; n),$$

and by (3-44)

$$p(\tau_1, \tau_2) = \frac{\eta^2}{D(n)} \frac{1}{4} E_q[V(\tau_1)V^*(\tau_1)V(\tau_2)V^*(\tau_2)]$$

$$= \frac{\eta^2}{D(n)} (m_{11}m_{22} + m_{12}m_{21}),$$

with

$$m_{ij} = m(\tau_i, \tau_j; n)$$

in terms of the solution of (12-83). For the density functions of higher order we follow the rule stated after (3-44). For $m = 3$, for instance,

$$p(\tau_1, \tau_2, \tau_3) = \frac{\eta^3}{D(n)} (m_{11}m_{22}m_{33} + m_{13}m_{21}m_{32} + m_{12}m_{23}m_{31}$$

$$+ m_{13}m_{22}m_{31} + m_{11}m_{23}m_{32} + m_{12}m_{21}m_{33})$$

$$= \frac{\eta^3}{D(n)} \overset{+}{\begin{bmatrix} m_{11} & m_{12} & m_{13} \\ m_{21} & m_{22} & m_{23} \\ m_{31} & m_{32} & m_{33} \end{bmatrix}} \overset{+}{} = \frac{\eta^3}{D(n)} P_3(\mathbf{m}).$$

Here $P_3(\mathbf{m})$ is the 3×3 *permanent* of the matrix $\mathbf{m} = \|m_{ij}\|$. It is evaluated like a determinant, but all terms have positive signs. Continuing thus, we see that the general form for the probability density function of the counting times $\tau_1, \tau_2, \ldots, \tau_m$ in (12-80) is

$$p(\tau_1, \tau_2, \ldots, \tau_m) = \frac{\eta^m}{D(n)} P_m(\mathbf{m}),$$

with $P_m(\mathbf{m})$ the $m \times m$ permanent with elements $m_{ij} = m(\tau_i, \tau_j; n)$, $1 \leq (i, j) \leq m$.

With these density functions evaluated under the two hypotheses, one can form the likelihood ratio $\Lambda(\tau_1, \tau_2, \ldots, \tau_m)$ by dividing the density function under hypothesis H_1 by that under hypothesis H_0, where $\tau_1, \tau_2, \ldots, \tau_m$ are the times at which photoelectrons have been observed to have been ejected by the incident light. If this likelihood ratio exceeds a certain decision level Λ_0, the receiver decides for hypothesis H_1; otherwise the null hypothesis H_0 is accepted. The problem of calculating the false-alarm and detection probabilities for this kind of receiver has not, to the writer's knowledge, been solved.

Problems

12-1. In a binary pulse-position modulation system, digits 0 and 1 are transmitted every T seconds. For a 0, a light pulse is received during the first half of the interval $(0, T)$; for a 1, a pulse is received during the second half of the interval. The light signal and background radiation fall onto a photoelectric detector, which counts the numbers n_1 and n_2 of photoelectrons emitted during the intervals $(0, \frac{1}{2}T)$ and $(\frac{1}{2}T, T)$, respectively. Under hypothesis H_0 that a 0 has come in, the numbers n_1 and n_2 have Poisson distributions with expected values m_1 and m_0, respectively; $m_1 > m_0$. Under hypothesis H_1 the expected values are reversed. The numbers n_1 and n_2 are statistically independent, and the 0's and 1's are equally likely. When $n_1 > n_2$, the receiver decides for hypothesis H_0, and when $n_1 < n_2$, for H_1. When $n_1 = n_2$, it chooses one or the other at random and with equal probabilities $\frac{1}{2}$. Calculate the probability of error in this receiver. Express the result in terms of Marcum's Q function. *Hint:* Express the probability $\Pr(n_1 - n_2 = k)$ in terms of the modified Bessel function of order k.

12-2. In a pulse-position modulation (PPM) system transmitting M equally likely symbols every T seconds, the interval $(0, T)$ is divided into M equal parts. For the kth symbol a light pulse is transmitted during the kth subinterval, but nothing in any of the others. The receiver counts the numbers $n_1, n_2, \ldots, n_M$ of photoelectrons emitted from a detector during each of the M subintervals. The receiver decides that the symbol that was transmitted was the one corresponding to the largest of those data. If a certain number r of the data are equal and exceed the rest, the receiver chooses at random and with probability $1/r$ one of the r corresponding symbols. Each of the M random variables n_k has a Poisson distribution with expected value m_0 when no pulse arrives during the kth subinterval and with expected value m_1 when one does arrive; $1 \leq k \leq M$.

Show that the probability of error can be written as

$$P_e = 1 - \frac{1}{M} \sum_{p=0}^{\infty} \frac{m_1^p}{p!} e^{-m_1} \sum_{k=0}^{M-1} S_p^k S_{p-1}^{M-1-k},$$

where

$$S_p = \sum_{j=0}^{p} \frac{m_0^j}{j!} e^{-m_0}, \qquad S_{-1} = 0.$$

This problem is a generalization of Problem 12-1 [Gag76, pp. 261–72].

12-3. Let the number n of photoelectrons have a Laguerre distribution as in (12-29). Set up the computation of the cumulative probability $q_k^{(-)}$ and its complement $q_k^{(+)}$ by numerical integration of (12-11) and (12-12). Show how to determine the saddlepoints through which the contour of integration should pass. Develop saddlepoint approximations for these cumulative probabilities as in Sec. 5.3.2. Compare the results of this approximation with the exact cumulative probabilities for $M = WT = 20$, $\bar{n}_0 = 4$, $\bar{n}_s = 50$, and $r = 10(10)100$.

12-4. Show that the overall gain G_0 in the photomultiplier treated in Sec. 12.3.1 equals the product of the gains G_j at each dynode.

12-5. In the photomultiplier treated in Sec. 12.3.1, assume that the probability that an incident electron ejects k secondary electrons from the jth dynode is

$$p_k^{(s)} = a_j v_j^k, \qquad k > 0, \qquad 0 < v_j < 1,$$

$$p_0^{(s)} = 1 - \frac{a_j v_j}{1 - v_j} \geq 0,$$

for positive constants a_j. Calculate the probability generating function $g_j(z)$ and find the gain G_j of the jth stage in terms of a_j and v_j. Show that when a single primary electron strikes the first dynode, the number n of electrons at the output of the Mth stage has a distribution of the same form as this, but with different parameters. (*Hint*: Look up the *homographic transformation* in a text on complex variables.)

12-6. Consider the sum v of Poisson-distributed photoelectron counts n and a Gaussian random variable x representing the noise in a subsequent amplifier. Assume that n has expected value $\bar{n}_s$, proportional to the strength of an incident light signal, and that x has expected value zero and variance σ^2. Determine an infinite series for the probability $Q = \Pr(n + x > V)$ that the sum v of signal and noise exceeds a decision level V. Calculate the moment-generating function $E[\exp(-zv)]$ of $v = n + x$. Show how to calculate Q and its complement $1 - Q$ by integration along a vertical path in the complex z-plane through a saddlepoint on the Re z-axis, and show how to find the appropriate saddlepoint. Use the saddlepoint approximation to check a few of the values in Fig. 12-5 for $Q_d > 0.95$, $\sigma > 0$.

12-7. Show that the function $M(z)$ defined by (12-46) must have a branch point at $z = x_*$ as given in (12-48). *Hint*: Sketch the curve of z versus M represented by (12-46) for real values of M.

12-8. In (12-80) assume that the rate $\lambda_i(\cdot)$ is nonrandom. By integrating the density function given there over $\tau_1, \tau_2, \ldots, \tau_m$, show that the probability that m electrons are emitted during $(0, T)$ has the Poisson form. Keep in mind the constraint $0 \leq \tau_1 \leq \tau_2 \leq \cdots \leq \tau_m \leq T$. *Hint*: Try mathematical induction.

12-9. Let the distribution of the number of secondary electrons per primary electron in a photomultiplier have the Poisson form in (12-44) with gain G per stage. Show that the probabilities $p_k^{(j)}$ that k electrons are ejected from the last dynode when a single electron strikes the jth dynode can be calculated from the recurrent relations

$$p_k^{(j)} = \frac{G}{k} \sum_{r=0}^{k-1} (r + 1) p_{r+1}^{(j+1)} p_{k-1-r}^{(j)}, \qquad k > 0,$$

$$p_0^{(j)} = \exp[G(p_0^{(j+1)} - 1)],$$

beginning with

$$p_r^{(M)} = \frac{G^r}{r!} e^{-G}, \qquad r \geq 0,$$

for $j = M$ and working down to $j = 1$ [Lom61].

Appendix A

Solution of the Detection Integral Equations

A.1 Inhomogeneous Equations

Our first task is to solve the integral equation

$$\int_0^T \phi(t - u)q(u)\, du = s(t), \qquad 0 \le t \le T, \tag{A-1}$$

in which $\phi(\tau)$ is a kernel that is the Fourier integral of a rational spectral density with simple poles:

$$\Phi(\omega) = \int_{-\infty}^{\infty} \phi(\tau)\, e^{-i\omega\tau}\, d\tau = \frac{N(\omega)}{P(\omega)}$$

$$= C \frac{\displaystyle\prod_{j=1}^{2m} (i\omega - \beta_j)}{\displaystyle\prod_{k=1}^{n} |i\omega - \mu_k|^2} \tag{A-2}$$

with C a positive constant. We take the μ_k's to have positive real parts. The $2n$ poles of the spectral density lie at $i\mu_k^*$ in the upper half-plane and at $-i\mu_k$ in the lower half-plane, $1 \le k \le n$. This more general form of the kernel $\phi(\tau)$ allows for complex autocovariance functions of narrowband noise as in (3-52) and for applications in Chapters 11 and 12 where $\Phi(\omega)$ contains a complex parameter.

Decomposition of $\Phi(\omega)$ into partial fractions and Fourier transformation lead to the expressions

511

$$\phi(t - u) = f_0\delta(t - u) + \sum_{k=1}^{n} f_k \exp[-\mu_k^*(t - u)], \qquad t \geq u,$$

$$= f_0\delta(t - u) + \sum_{k=1}^{n} g_k \exp[\mu_k(t - u)], \qquad t < u, \tag{A-3}$$

for the kernel of (A-1). This partial-fraction decomposition has the form

$$\Phi(\omega) = f_0 + \sum_{k=1}^{n} \left[\frac{f_k}{\mu_k^* + i\omega} + \frac{g_k}{\mu_k - i\omega} \right]. \tag{A-4}$$

When $m < n$, $f_0 = 0$.

As in (2-100) we write the solution of (A-1) in the form

$$q(t) = q_\infty(t) + q_h(t), \tag{A-5}$$

$$q_h(t) = \sum_{j=0}^{n-m-1} \left[a_j\delta^{(j)}(t) + b_j\delta^{(j)}(t - T) \right] + \sum_{j=1}^{2m} c_j \exp \beta_j t. \tag{A-6}$$

Here $\delta^{(j)}(t)$ is the jth derivative of the delta function, defined by

$$\int_{-\infty}^{\infty} f(u)\delta^{(j)}(u - t) \, du = (-1)^j \frac{d^j f(t)}{dt^j}.$$

The delta functions in (A-6) are assumed to stand just inside the interval $(0, T)$, so that they contribute their full weight when substituted into (A-1). In particular,

$$\int_0^T \phi(t - u)\delta^{(j)}(u - a) \, du = (-1)^j \frac{d^j}{da^j} \int_0^T \phi(t - u)\delta(u - a) \, du$$

$$= (-1)^j \frac{d^j}{da^j} \phi(t - a) \tag{A-7}$$

$$= \phi^{(j)}(t - a), \qquad a = 0^+, \, T^-.$$

The particular solution $q_\infty(t)$ in (A-5) is the solution of the integral equation

$$s(t) = \int_{-\infty}^{\infty} \phi(t - u)q_\infty(u) \, du. \tag{A-8}$$

It is the same as $q_\infty(t)$ in (2-77) and can be found as in (2-78) and (2-79) when the signal $s(t)$ is differentiable $2(n - m)$ times, and in this appendix we shall assume that this is the case. When—as is most common—the input contains a white-noise component, this assumption is valid. If there is no white-noise component and the signal is not sufficiently differentiable, one must seek a particular solution by other methods or resort to the more complicated procedure presented by Slepian and Kadota [Sle69].

Substituting (A-8) into the right side of (A-1) and (A-5) into the left side, we obtain

$$\int_0^T \phi(t - u)q_h(u) \, du = \left[\int_{-\infty}^0 + \int_T^{\infty} \right] \phi(t - u)q_\infty(u) \, du.$$

Then substituting from (A-6) and using (A-7) we find

$$\sum_{j=0}^{n-m-1}\left[a_j\phi^{(j)}(t)+b_j\phi^{(j)}(t-T)\right]+\sum_{j=1}^{2m}c_j\int_0^T\phi(t-u)\,e^{\beta_j u}\,du$$

$$=\left[\int_{-\infty}^0+\int_T^\infty\right]\phi(t-u)q_\infty(u)\,du. \tag{A-9}$$

When we introduce (A-3), the first summation on the left side of (A-9) yields the terms

$$\sum_{k=1}^{n}\sum_{j=0}^{n-m-1}\left[f_k a_j(-\mu_k^*)^j\,e^{-\mu_k^* t}+g_k b_j\mu_k^j\,e^{\mu_k(t-T)}\right].$$

These terms arise only if $m < n$, whereupon $f_0 = 0$. The second summation yields terms of the form

$$c_j\left\{f_0 e^{\beta_j t}+\sum_{k=1}^{n}\left[f_k\int_0^t\exp[-\mu_k^*(t-u)+\beta_j u]\,du+g_k\int_t^T\exp[\mu_k(t-u)+\beta_j u]\,du\right]\right\}$$

$$=c_j\left\{f_0+\sum_{k=1}^{n}\left[\frac{f_k}{\mu_k^*+\beta_j}+\frac{g_k}{\mu_k-\beta_j}\right]\right\}e^{\beta_j t}-c_j\sum_{k=1}^{n}\left[f_k\frac{e^{-\mu_k^* t}}{\mu_k^*+\beta_j}+g_k e^{\mu_k t}\frac{e^{(\beta_j-\mu_k)T}}{\mu_k-\beta_j}\right].$$

These are summed over $1 \le j \le 2n$. Putting $\omega = -i\beta_j$ into (A-4), we see that the first term vanishes,

$$c_j\,e^{\beta_j t}\Phi(-i\beta_j)=0,$$

by (A-2).

Substituting (A-3) into the right side of (A-9) produces the terms

$$\sum_{k=1}^{n}\left[f_k\,e^{-\mu_k^* t}L_k+g_k\,e^{\mu_k(t-T)}R_k\right],$$

where

$$L_k=\int_{-\infty}^0 e^{\mu_k^* u}q_\infty(u)\,du,\qquad R_k=\int_T^\infty e^{-\mu_k(u-T)}q_\infty(u)\,du. \tag{A-10}$$

In order for the integral equation (A-1) to be satisfied, the terms proportional to $\exp(-\mu_k^* t)$ and to $\exp[\mu_k(t-T)]$ must vanish individually, and that requirement produces $2n$ linear simultaneous equations

$$\sum_{j=0}^{n-m-1}a_j(-\mu_k^*)^j-\sum_{j=1}^{2m}\frac{c_j}{\beta_j+\mu_k^*}=L_k,\qquad 1\le k\le n, \tag{A-11}$$

$$\sum_{j=0}^{n-m-1}b_j\mu_k^j+\sum_{j=1}^{m}\frac{c_j e^{\beta_j T}}{\beta_j-\mu_k}=R_k,\qquad 1\le k\le n. \tag{A-12}$$

By defining

$$\mu_{k+n}=-\mu_k^*,\qquad 1\le k\le n,$$

we can write (A-11) as

$$-\sum_{j=0}^{n-m-1}a_j\mu_k^j+\sum_{j=1}^{2m}\frac{c_j}{\beta_j-\mu_k}=-L_{k-n},\qquad n+1\le k\le 2n, \tag{A-13}$$

and the set of $2n$ simultaneous linear equations can be concisely written as

$$(\mathbf{B} \quad \mathbf{G})\begin{bmatrix} \mathbf{a} \\ \mathbf{c} \end{bmatrix} = \begin{bmatrix} \mathbf{R} \\ -\mathbf{L} \end{bmatrix}, \qquad \text{(A-14)}$$

where

$$\mathbf{B} = \begin{bmatrix} \mathbf{B}_1 & \mathbf{0} \\ \mathbf{0} & \mathbf{B}_2 \end{bmatrix}$$

is a $2(n-m) \times 2(n-m)$ block matrix in which the (kj)-element of block $\mathbf{B}_1$ is μ_k^{j-1}, $1 \leq k \leq n$, and that of block $\mathbf{B}_2$ is $(-\mu_k^*)^{j-1} = \mu_{k+n}^{j-1}$. The $\mathbf{0}$'s are $(n-m) \times (n-m)$ matrices of zeros. The $2n \times 2m$ matrix $\mathbf{G} = \mathbf{G}(T)$ is

$$G_{kj}(T) = \begin{cases} \dfrac{\exp \beta_j T}{\beta_j - \mu_k}, & 1 \leq k \leq n, \\[4mm] \dfrac{1}{\beta_j - \mu_k}, & n+1 \leq k \leq 2n. \end{cases} \qquad 1 \leq j \leq 2m, \qquad \text{(A-15)}$$

In (A-14) $\mathbf{a}$ is a column vector whose transpose is

$$\mathbf{a}^T = (b_0 \ \dots \ b_{n-m-1} \ -a_0 \ \dots \ -a_{n-m-1}),$$

and $\mathbf{c}$ is the $2m$-element column vector of the coefficients c_j. There $\mathbf{R}$ and $\mathbf{L}$ are the n-element column vectors of the quantities R_k and L_k on the right sides of (A-12) and (A-13), respectively.

Solving these equations requires inverting the $2n \times 2n$ matrix of the coefficients of the $n-m$ a's and b's and the $2m$ c's. It is possible in general to eliminate the $n-m$ a's and b's from (A-12) and (A-13) so as to reduce these to a system of only $2m$ simultaneous equations for the c's [Hel65], but the terms of the new equations are rather more complicated than those in (A-12) and (A-13), and the subsequent calculation of the a's and b's is then cumbersome. Given a digital computer, it is more expeditious to let its program for solving linear simultaneous equations carry the burden of the complexities of solving (A-12) and (A-13), whose coefficients are relatively simple to program. As we saw in Chapter 2, the a's and b's do not appear when $n = m$. When $m = 0$, the c's are absent, and (A-12) and (A-13) can then be solved separately.

If the signal $s(t)$ is a sinusoid, $s(t) = \exp iwt$, the right sides of (A-12) and (A-13) reduce to

$$-L_{k-n} = \frac{1}{(\mu_k - iw)\Phi(w)}, \qquad R_k = \frac{e^{iwT}}{(\mu_k - iw)\Phi(w)}. \qquad \text{(A-16)}$$

Furthermore, by carrying out a discrete Fourier transform of the signal $s(t)$ over the interval $(0, T)$, one can approximate it as

$$s(t) \approx \sum_{j=-M}^{M} s_j \exp iw_j t, \qquad w_j = \frac{2\pi j}{T},$$

and one can compute the n L_k's and R_k's by

$$-L_k = \sum_{j=-M}^{M} \frac{s_j}{(\mu_{k+n} - iw_j)\Phi(w_j)}, \qquad R_k = \sum_{j=-M}^{M} \frac{s_j}{(\mu_k - iw_j)\Phi(w_j)}, \qquad 1 \le k \le n.$$

As an example, let us take the signal to be $\exp iwt$ with angular frequency $w = \pi/2$, let $T = 1$, and let the spectral density of the noise be

$$\Phi(w) = \frac{24w^2 + 16}{(w^4 + 4)(w^2 + 4)}. \qquad (A\text{-}17)$$

Now $m = 1$, $n = 3$, and

$$C = 24, \qquad \beta_1 = (\tfrac{2}{3})^{1/2}, \qquad \beta_2 = -\beta_1,$$

$$\mu_1 = 2, \quad \mu_2 = 1 + i, \quad \mu_3 = 1 - i, \quad \mu_4 = -2, \quad \mu_5 = -1 + i, \quad \mu_6 = -1 - i,$$

and

$$q_\infty(t) = \frac{e^{iwt}}{\Phi(w)} = 0.867398 \, e^{i\pi t/2}.$$

Then we find the coefficients on the right side of (A-13) to have the values

$$-L_1 = -0.2682368 + 0.2106727i,$$
$$-L_2 = -0.6542405 + 0.3734381i,$$
$$-L_3 = -0.1139964 + 0.2930614i,$$

and those on the right side of (A-12) are

$$R_1 = iL_1^*, \qquad R_2 = iL_2^*, \qquad R_3 = iL_3^*,$$

the factor i arising because $\exp iwT = i$ in (A-16). Solving the six simultaneous linear equations (A-12) and (A-13), the computer gives us the coefficients

$$a_0 = 0.473772 - 0.478199i,$$
$$a_1 = 0.0573014 - 0.177352i,$$
$$b_0 = ia_0^*, \qquad b_1 = -ia_1^*,$$
$$c_1 = 0.0861594 + 0.0916176i, \qquad c_2 = ic_1^*.$$

In general, indeed, when $s(t) = \exp iwt$,

$$R_k = L_k^* \, e^{iwT}, \qquad b_j = (-1)^j a_j^* \, e^{iwT}, \qquad c_{j+m} = c_j^* \, e^{iwT}, \qquad 1 \le j \le m,$$

as one can show by substituting these into (A-11) and (A-12) and comparing the results. Taking the imaginary part of our solution yields the result of the example in [Lan56, Sec. 8.4, pp. 309–29].

A.2 Homogeneous Equations and Their Eigenvalues

We seek a method for computing the eigenvalues and eigenfunctions of the integral equation

$$\lambda f(t) = \int_0^T \phi(t - u)f(u)\,du, \qquad 0 \le t \le T, \qquad (A\text{-}18)$$

when $\phi(\tau) = \phi^*(-\tau)$ is an Hermitian kernel of the same type as in (A-3). We assume, without loss of generality, that $m < n$, so that $f_0 = 0$.

The same technique as in Sec. A.1 is employed, but as if the kernel of the integral equation (A-1) were $\lambda\delta(t - u) - \phi(t - u)$ and the right side were zero. The Fourier transform of the kernel, when cleared of fractions, is a rational function with numerator and denominator of the same degree $2n$. Delta functions are then absent from the solution. From (A-6) we see that the eigenfunctions $f(t)$ of (A-18) have the form

$$f(t) = \sum_{j=1}^{2n} c_j \exp \beta_j t, \qquad (A\text{-}19)$$

where by (A-2) the β_j are the $2n$ roots of the algebraic equation that results from clearing the equation

$$\lambda = \Phi(-i\beta), \qquad \beta = \beta_j, \qquad 1 \le j \le 2n, \qquad (A\text{-}20)$$

of fractions. When $\phi(\tau)$ and λ are both real, the roots β_j either are real or occur in complex-conjugate pairs.

Because $s(t)$ is now zero, so is $q_\infty(t)$, and in (A-12) and (A-13) $L_k \equiv R_k \equiv 0$. The terms a_j and b_j have disappeared. Thus (A-14), with $m = n$, becomes simply $\mathbf{G}(T)\mathbf{c} = \mathbf{0}$, where $\mathbf{G}(T)$ is the matrix, now $2n \times 2n$, whose elements are specified by (A-15). In order for nonzero solutions of these $2n$ homogeneous linear equations to exist, the determinant of their coefficients must vanish:

$$\det \mathbf{G}(T) = 0. \qquad (A\text{-}21)$$

This equation is equivalent to

$$\frac{\det \mathbf{G}(T)}{\det \mathbf{G}(0)} = 0, \qquad (A\text{-}22)$$

with the elements of $\mathbf{G}(0)$ given by (A-15) with $T = 0$. For computation (A-22) is preferred, for as will be seen in Chapter 11, the left side of (A-22) is proportional to the Fredholm determinant

$$D(z) = \prod_{j=1}^{\infty} (1 + \lambda_j z), \qquad z = -\frac{1}{\lambda}, $$

and will be real whenever z is real; $\det \mathbf{G}(T)$ by itself may be purely imaginary. As a function of z, $D(z)$ oscillates with ever increasing amplitude as $z \to -\infty$.

The eigenvalues λ of (A-18) are calculated by solving (A-22) by the secant method. The left side of (A-22) is computed as a function of λ by first finding the $2n$ roots β_j of the algebraic equation derived from (A-20),

$$\lambda P(-i\beta) - N(-i\beta) = 0, \qquad \beta = \beta_1, \beta_2, \ldots, \beta_{2n}, \qquad (A\text{-}23)$$

and substituting them into (A-22). Here $N(\omega)$ and $P(\omega)$ are the polynomials of degree $2n$ in the numerator and denominator, respectively, of $\Phi(\omega)$ as in (A-2). The determinants can be evaluated and the algebraic equation (A-23) solved by standard computer routines. The search is conveniently started by taking λ just below the maximum value of the spectral density $\Phi(\omega)$ for ω real.

In order to see how to modify (A-22) to accommodate multiple poles of the spectral density $\Phi(\omega)$, suppose that μ_1 and μ_2 have coalesced to form double poles at $i\mu_1$ and $-i\mu_1$. Let $\mu_2 = \mu_1 + \varepsilon$, $\varepsilon \ll |\mu_1|$. Just before their coalescence, the first and second rows of $\mathbf{G}(T)$ are nearly equal, and so are the $(n + 1)$th and $(n + 2)$th rows. The determinant det $\mathbf{G}(T)$ is unchanged if we subtract the first row from the second, element by element. Each element of the new second row is approximately ε times the derivative with respect to μ_1 of the element above it. Factor ε from the elements of that row. Carry out the same procedure with the first and second rows of det $\mathbf{G}(0)$. The ε's will cancel from (A-22), and we can now let ε go to zero. The first row of both det $\mathbf{G}(T)$ and det $\mathbf{G}(0)$ will be unchanged; the elements of the second row will have been replaced by the first derivatives $\partial/\partial\mu_1$ of those in the first row. The same change must be made in the $(n + 2)$th row of both $\mathbf{G}(T)$ and $\mathbf{G}(0)$.

We thus see that if $\Phi(\omega)$ has poles $\pm i\mu$ of, say, order $p > 1$, the upper halves of det $\mathbf{G}(T)$ and det $\mathbf{G}(0)$ will contain $p - 1$ rows whose elements are the first through the $(p - 1)$th derivatives, with respect to μ, of the elements of the row involving μ, and the same will be true of the lower halves of det $\mathbf{G}(T)$ and det $\mathbf{G}(0)$.

Once the eigenvalues are known, the eigenfunctions of (A-18) are found by solving the homogeneous equations $\mathbf{G}(T)\mathbf{c} = \mathbf{0}$ for each set of $2n$ c_j's. These will contain an arbitrary constant factor. One substitutes them into (A-19) and evaluates the constant factor from the requirement

$$\int_0^T |f(t)|^2 \, dt = 1.$$

The β_j's in (A-19) will have been found from (A-20) in the course of searching for the eigenvalue λ.

For the spectral density in (A-17) with $T = 1$, (A-20) reduces to the algebraic equation

$$\beta^6 - 4\beta^4 + 4(1 + 6z)\beta^2 - 16(1 + z) = 0, \qquad z = -\frac{1}{\lambda}.$$

The roots β_j are the positive and negative square roots of the roots of the cubic equation

$$x^3 - 4x^2 + 4(1 + 6z)x - 16(1 + z) = 0, \qquad \beta = \pm\sqrt{x}.$$

As before,

$$\mu_1 = 2, \qquad \mu_2 = 1 + i, \qquad \mu_3 = 1 - i,$$
$$\mu_4 = -2, \qquad \mu_5 = -1 + i, \qquad \mu_6 = -1 - i,$$

in (A-15). By solving (A-22) by the secant method, the eigenvalues listed in Table A-1 were obtained. The second column of Table A-1 lists the frequencies ω_k such that $\lambda_k = \Phi(\omega_k)$ in order to illustrate the fact that the eigenvalues tend to be samples of the spectral density $\Phi(\omega)$ at angular frequencies separated when $k \gg 1$ by approximately π/T.

The eigenvalues in Table A-1 sum to 0.9998904; the sum of all the eigenvalues must equal

$$\sum_{k=1}^{\infty} \lambda_k = \phi(0) = \int_{-\infty}^{\infty} \Phi(\omega)\frac{d\omega}{2\pi} = 1.$$

The slow decrease of the spectral density, and hence of the eigenvalues, to zero with increasing frequency and index k accounts for the discrepancy.

Table A-1 Eigenvalues and Equivalent Frequencies

λ_k	ω_k/π
0.8181314	0.59627
0.1557310	1.04294
1.939974(−2)	1.84274
4.290960(−3)	2.72199
1.316924(−3)	3.67553
5.265503(−4)	4.63280
2.450937(−4)	5.61580
1.293174(−4)	6.59398
7.396168(−5)	7.58604
4.541252(−5)	8.57257

Problem

A-1. By solving (A-22) calculate the first six eigenvalues of (A-18) for the spectral density

$$\Phi(\omega) = \frac{\omega^2}{1 + \omega^4}$$

and for $T = 1$. For each eigenvalue calculate the two angular frequencies ω such that $\lambda = \Phi(\omega)$. Compare the sum of your eigenvalues with the sum of all eigenvalues λ_k, $1 \leq k < \infty$.

Appendix B

Circular Gaussian Density Functions

To prove that (3-40) indeed represents the joint probability density function of the $2n$ random variables $x_1, x_2, \ldots, x_n, y_1, y_2, \ldots, y_n$, we shall show how it can be reduced to the conventional form (2-1). We take the expected values of all the variables equal to zero; to go from (3-40) to (3-41) is straightforward.

The calculation is simplest in matrix notation. We write the $2n$-element column vector made up of the real random variables x_m and y_m, $1 \leq m \leq n$, arranged vertically, as

$$\begin{bmatrix} \mathbf{x} \\ \mathbf{y} \end{bmatrix},$$

where $\mathbf{x}$ is the n-element column vector of the x_m's and $\mathbf{y}$ the n-element column vector of the y_m's. The $2n \times 2n$ matrix $\tilde{\boldsymbol{\Phi}}$ in (3-38) is similarly written in block form,

$$\tilde{\boldsymbol{\Phi}} = \begin{bmatrix} \tilde{\boldsymbol{\phi}}_x & -\tilde{\boldsymbol{\phi}}_y \\ \tilde{\boldsymbol{\phi}}_y & \tilde{\boldsymbol{\phi}}_x \end{bmatrix}; \tag{B-1}$$

the elements of $\tilde{\boldsymbol{\phi}}_x = \|\tilde{\phi}_{x,mn}\|$ and $\tilde{\boldsymbol{\phi}}_y = \|\tilde{\phi}_{y,mn}\|$ are defined by (3-37).

Let $\mathbf{T}$ be the matrix

$$\mathbf{T} = 2^{-1/2} \begin{bmatrix} \mathbf{I} & i\mathbf{I} \\ \mathbf{I} & -i\mathbf{I} \end{bmatrix} \tag{B-2}$$

with $\mathbf{I}$ the $n \times n$ identity matrix. (A scalar in front of a matrix multiplies each element of the matrix.) The Hermitian conjugate of $\mathbf{T}$ is

$$\mathbf{T}^+ = 2^{-1/2} \begin{bmatrix} \mathbf{I} & \mathbf{I} \\ -i\mathbf{I} & i\mathbf{I} \end{bmatrix}, \tag{B-3}$$

and by the rules of matrix multiplication

$$\mathbf{T}\mathbf{T}^{+} = \mathbf{T}^{+}\mathbf{T} = \begin{bmatrix} \mathbf{I} & \mathbf{0} \\ \mathbf{0} & \mathbf{I} \end{bmatrix},$$

where $\mathbf{0}$ is the $n \times n$ matrix of zeros. Thus the matrix $\mathbf{T}$ is unitary, and in terms of it we can write

$$\begin{bmatrix} \mathbf{z} \\ \mathbf{z}^{*} \end{bmatrix} = 2^{1/2}\,\mathbf{T}\begin{bmatrix} \mathbf{x} \\ \mathbf{y} \end{bmatrix}, \qquad (\mathbf{z}^{*T}\mathbf{z}^{T}) = 2^{1/2}(\mathbf{x}^{T}\mathbf{y}^{T})\mathbf{T}^{+}, \tag{B-4}$$

where $\mathbf{z}$ is the column vector with elements $z_m = x_m + iy_m$, $1 \le m \le n$. The superscript ^{T} indicates a simple interchange of rows and columns, or a *transpose*, without complex conjugation. The row vector $\mathbf{z}^{*T}$ has elements z_m^{*}, $1 \le m \le n$, and $\mathbf{z}^{T} = (z_1, z_2, \dots, z_n)$.

Working backward, we write the quadratic form in (3-40) as

$$\begin{aligned} Q &= \sum_{m=1}^{n}\sum_{k=1}^{n} z_m^{*}\tilde{\mu}_{mk} z_k = \tfrac{1}{2}(\mathbf{z}^{*T}\mathbf{z}^{T})\begin{bmatrix} \tilde{\mu} & \mathbf{0} \\ \mathbf{0} & \tilde{\mu}^{T} \end{bmatrix}\begin{bmatrix} \mathbf{z} \\ \mathbf{z}^{*} \end{bmatrix} \\ &= (\mathbf{x}^{T}\mathbf{y}^{T})\mathbf{T}^{+}\begin{bmatrix} \tilde{\mu} & \mathbf{0} \\ \mathbf{0} & \tilde{\mu}^{T} \end{bmatrix}\mathbf{T}\begin{bmatrix} \mathbf{x} \\ \mathbf{y} \end{bmatrix} = (\mathbf{x}^{T}\mathbf{y}^{T})\mathbf{M}\begin{bmatrix} \mathbf{x} \\ \mathbf{y} \end{bmatrix}, \end{aligned} \tag{B-5}$$

where

$$\mathbf{M} = \mathbf{T}^{+}\begin{bmatrix} \tilde{\mu} & \mathbf{0} \\ \mathbf{0} & \tilde{\mu}^{T} \end{bmatrix}\mathbf{T} = \tfrac{1}{2}\begin{bmatrix} \tilde{\mu} + \tilde{\mu}^{T} & i\,(\tilde{\mu} - \tilde{\mu}^{T}) \\ i\,(\tilde{\mu}^{T} - \tilde{\mu}) & \tilde{\mu} + \tilde{\mu}^{T} \end{bmatrix} \tag{B-6}$$

by (B-2) and (B-3) and the rules of matrix multiplication.

When as here the matrix μ is Hermitian, $\tilde{\mu}^{T} = \tilde{\mu}^{*}$, the matrix $\mathbf{M}$ can be written

$$\mathbf{M} = \begin{bmatrix} \mu_x & -\mu_y \\ \mu_y & \mu_x \end{bmatrix}, \qquad \mu = \mu_x + i\mu_y.$$

Then the quadratic form in (B-5) becomes

$$\begin{aligned} Q &= (\mathbf{x}^{T}\mathbf{y}^{T})\begin{bmatrix} \tilde{\mu}_x & -\tilde{\mu}_y \\ \tilde{\mu}_y & \tilde{\mu}_x \end{bmatrix}\begin{bmatrix} \mathbf{x} \\ \mathbf{y} \end{bmatrix} = \mathbf{x}^{T}\tilde{\mu}_x\mathbf{x} - \mathbf{x}^{T}\tilde{\mu}_y\mathbf{y} + \mathbf{y}^{T}\tilde{\mu}_y\mathbf{x} + \mathbf{y}^{T}\tilde{\mu}_x\mathbf{y} \\ &= \sum_{m=1}^{n}\sum_{k=1}^{n}\left[\tilde{\mu}_{x,mk}(x_m x_k + y_m y_k) + \tilde{\mu}_{y,mk}(y_m x_k - x_m y_k)\right]. \end{aligned} \tag{B-7}$$

With $\tilde{\phi} = \tilde{\mu}^{-1}$ and because $\mathbf{T}^{-1} = \mathbf{T}^{+}$, we find, by (B-1) and as in (B-6),

$$\mathbf{M}^{-1} = \mathbf{T}^{+}\begin{bmatrix} \tilde{\phi} & \mathbf{0} \\ \mathbf{0} & \tilde{\phi}^{T} \end{bmatrix}\mathbf{T} = \begin{bmatrix} \tilde{\phi}_x & -\tilde{\phi}_y \\ \tilde{\phi}_y & \tilde{\phi}_x \end{bmatrix} = \tilde{\Phi}, \qquad \tilde{\phi} = \tilde{\phi}_x + i\tilde{\phi}_y, \tag{B-8}$$

so that the matrix $\mathbf{M}$ of the quadratic form Q is indeed the inverse of the covariance matrix $\tilde{\Phi}$ of the $2n$ random variables $x_1, \dots, x_n, y_1, \dots, y_n$. Furthermore, because $\det(\mathbf{T}\mathbf{T}^{+}) = \det \mathbf{T}^{+} \det \mathbf{T} = 1$,

$$\det \tilde{\Phi} = \det \mathbf{T}^{+} \det \begin{bmatrix} \tilde{\phi} & \mathbf{0} \\ \mathbf{0} & \tilde{\phi}^{T} \end{bmatrix} \det \mathbf{T} = (\det \tilde{\phi})^{2}. \tag{B-9}$$

Putting (B-7) and (B-9) into (3-40), we find the usual form for the joint probability density function of those $2n$ random variables,

$$p(x_1, \dots, x_n, y_1, \dots, y_n) = (2\pi)^{-n}[\det \tilde{\Phi}]^{-1/2}\,e^{-Q/2}.$$

A useful integral is

$$(2\pi)^{-n}\int_{-\infty}^{\infty}\cdots\int_{-\infty}^{\infty}\exp\left[-\frac{1}{2}\sum_{m=1}^{n}\sum_{k=1}^{n}z_m^*\tilde{\mu}_{mk}z_k + \frac{1}{2}\sum_{m=1}^{n}(v_m^*z_m + w_m z_m^*)\right]\prod_{m=1}^{n}dx_m\,dy_m$$

$$= \det\tilde{\phi}\,\exp\left[\frac{1}{2}\sum_{m=1}^{n}\sum_{k=1}^{n}v_m^*\tilde{\phi}_{mk}w_k\right],$$

(B-10)

where $\tilde{\phi} = \tilde{\mu}^{-1}$, and the v_m's and the w_m's are arbitrary complex numbers. Now the matrix $\tilde{\mu}$ is not necessarily Hermitian, and the value of the integral may be complex. It is only necessary that the integral be finite, which requires $\tilde{\mu}$ to be such that $\text{Re}(\mathbf{z}^{*T}\tilde{\mu}\mathbf{z})$ is a positive definite quadratic form in the $2n$ variables $x_j, y_j, 1 \le j \le n$. We shall use (B-10) in Chapters 11 and 12 under circumstances in which $\tilde{\mu}$ has the form $\boldsymbol{\mu}_1 + z\boldsymbol{\mu}_2$, with $\boldsymbol{\mu}_1$ and $\boldsymbol{\mu}_2$ Hermitian, but z is a complex variable, whereupon $\boldsymbol{\mu}_1 + z\boldsymbol{\mu}_2$ is not Hermitian.

To derive (B-10) we introduce the $2n$ complex numbers $(s_1', \ldots, s_n', s_1'', \ldots, s_n'')$, which make up the column vector

$$\mathbf{a} = \begin{bmatrix} \mathbf{s}' \\ \mathbf{s}'' \end{bmatrix}$$

composed of the two n-element column vectors $\mathbf{s}'$ and $\mathbf{s}''$ of the s_j' and the s_j'', respectively. They are defined by the transformation

$$\mathbf{a} = \begin{bmatrix} \mathbf{s}' \\ \mathbf{s}'' \end{bmatrix} = 2^{-1/2}\mathbf{T}^+\begin{bmatrix} \mathbf{w} \\ v^* \end{bmatrix},$$

in terms of the v_m's and the w_m's, whence

$$\begin{bmatrix} \mathbf{w} \\ v^* \end{bmatrix} = 2^{1/2}\mathbf{T}\begin{bmatrix} \mathbf{s}' \\ \mathbf{s}'' \end{bmatrix} = \begin{bmatrix} \mathbf{s}' + i\mathbf{s}'' \\ \mathbf{s}' - i\mathbf{s}'' \end{bmatrix}.$$

(B-11)

In addition, if we again use the superscript ^{T} to denote the transpose of a vector or a matrix—without complex conjugation of the elements—, we can write the row vector

$$(v^{*T}\mathbf{w}^T) = 2^{1/2}(\mathbf{s}'^T\mathbf{s}''^T)\mathbf{T}^+,$$

(B-12)

and by (B-4) the second summation in the exponent in (B-10) is

$$\frac{1}{2}(v^{*T}\mathbf{w}^T)\begin{bmatrix} \mathbf{z} \\ \mathbf{z}^* \end{bmatrix} = (\mathbf{s}'^T\mathbf{s}''^T)\begin{bmatrix} \mathbf{x} \\ \mathbf{y} \end{bmatrix} = \mathbf{s}'^T\mathbf{x} + \mathbf{s}''^T\mathbf{y}.$$

By (B-5) and (B-9) the integral to be evaluated in (B-10) is now

$$I = (2\pi)^{-n}\int_{-\infty}^{\infty}\cdots\int_{-\infty}^{\infty}\exp\left[-\frac{1}{2}Q + \sum_{j=1}^{n}(s_j'x_j + s_j''y_j)\right]\prod_{k=1}^{n}dx_k\,dy_k,$$

with Q given in (B-5) in terms of the matrix $\mathbf{M}$ as defined in (B-6).

Now for an $N \times N$ *symmetric* matrix $\mathbf{C}$ and a row vector $\mathbf{a}^T = (a_1, \ldots, a_N)$,

$$\int_{-\infty}^{\infty} \ldots \int_{-\infty}^{\infty} \exp(-\tfrac{1}{2}\mathbf{t}^T\mathbf{C}\mathbf{t} + \mathbf{a}^T\mathbf{t})\, dt_1 \ldots dt_N$$

$$= \int_{-\infty}^{\infty} \ldots \int_{-\infty}^{\infty} \exp\left[-\frac{1}{2}\sum_{i=1}^{N}\sum_{j=1}^{N} C_{ij}t_i t_j + \sum_{i=1}^{N} a_i t_i\right] dt_1 \ldots dt_N \qquad \text{(B-13)}$$

$$= (2\pi)^{N/2}(\det \mathbf{C})^{-1/2}\, \exp(\tfrac{1}{2}\mathbf{a}^T\mathbf{C}^{-1}\mathbf{a}),$$

and

$$\mathbf{a}^T\mathbf{C}^{-1}\mathbf{a} = \sum_{i=1}^{N}\sum_{j=1}^{N} c_{ij}a_i a_j = \mathbf{a}^T\mathbf{c}\mathbf{a}, \qquad \mathbf{c} = \mathbf{C}^{-1}.$$

This result holds even though the elements of the matrix $\mathbf{C}$ are complex, provided that the integral is finite, and that requires that

$$\sum_{i=1}^{N}\sum_{j=1}^{N} (\operatorname{Re} C_{ij})t_i t_j \geq 0;$$

that is, the matrix $\operatorname{Re} \mathbf{C}$ must be positive definite [Hel91, pp. 237–9]. The proof rests on the ability to transform the quadratic form $\mathbf{t}^T\mathbf{C}\mathbf{t}$ into a sum of squares by a linear transformation $\mathbf{y} = \mathbf{U}\mathbf{t}$, where $\mathbf{U}$ is an orthogonal matrix, and this can be done provided the matrix $\mathbf{C}$ is symmetric, $C_{ij} = C_{ji}$, whether or not its elements are complex.

We now take $N = 2n$, $\mathbf{C} = \mathbf{M} = \tilde{\boldsymbol{\Phi}}^{-1}$, $\mathbf{c} = \tilde{\boldsymbol{\Phi}}$, $\det \mathbf{C} = \det \mathbf{M} = (\det \tilde{\boldsymbol{\phi}})^{-2}$, and $\mathbf{a}^T = (\mathbf{s}'^T\ \mathbf{s}''^T)$; and we observe from (B-6) that the matrix $\mathbf{M}$ is symmetrical: $\mathbf{M} = \mathbf{M}^T$. Thus we obtain from (B-13)

$$I = \det \tilde{\boldsymbol{\phi}}\, \exp(\tfrac{1}{2}\mathbf{a}^T\tilde{\boldsymbol{\Phi}}\mathbf{a}).$$

By (B-8) the quadratic form in the exponent can be written as

$$\mathbf{a}^T\tilde{\boldsymbol{\Phi}}\mathbf{a} = (\mathbf{s}'^T\ \mathbf{s}''^T)\mathbf{T}^+\begin{bmatrix} \tilde{\boldsymbol{\phi}} & \mathbf{0} \\ \mathbf{0} & \tilde{\boldsymbol{\phi}}^T \end{bmatrix}\mathbf{T}\begin{bmatrix} \mathbf{s}' \\ \mathbf{s}'' \end{bmatrix} = \tfrac{1}{2}(v^{*T}\ \mathbf{w}^T)\begin{bmatrix} \tilde{\boldsymbol{\phi}} & \mathbf{0} \\ \mathbf{0} & \tilde{\boldsymbol{\phi}}^T \end{bmatrix}\begin{bmatrix} \mathbf{w} \\ v^* \end{bmatrix}$$

$$= \tfrac{1}{2}(v^{*T}\tilde{\boldsymbol{\phi}}\mathbf{w} + \mathbf{w}^T\tilde{\boldsymbol{\phi}}^T v^*) = v^{*T}\tilde{\boldsymbol{\phi}}\mathbf{w} = \sum_{m=1}^{n}\sum_{k=1}^{n} v_m^* \tilde{\phi}_{mk} w_k,$$

which is the exponent in (B-10); here we have used (B-8), (B-11), and (B-12). The characteristic function in (3-42) can be derived from (B-10) by replacing v_m^* by iw_m^* and w_m by iw_m and by taking $\tilde{\boldsymbol{\phi}}$ as the complex covariance matrix of the n circular complex Gaussian variables $z_1, z_2, \ldots, z_n$.

Appendix C

Q Function

C.1 Properties

The probability density function

$$q(\alpha, x) = x \, \exp[-\tfrac{1}{2}(x^2 + \alpha^2)]I_0(\alpha x)U(x) \tag{C-1}$$

is called the *noncentral Rayleigh*, the *Rayleigh–Rice*, or the *Rice–Nakagami* density function. It describes the distribution of the distance from a point in a plane to the origin when the Cartesian coordinates of the point are independent Gaussian random variables of unit variance and expected values $\alpha \cos \psi$ and $\alpha \sin \psi$, ψ an arbitrary angle. The complementary cumulative distribution

$$Q(\alpha, \beta) = \Pr(x > \beta) = \int_{\beta}^{\infty} q(\alpha, x)\, dx \tag{C-2}$$

is known as *Marcum's Q function*. Some of its properties were given by Rice [Ric44], and it was extensively calculated and utilized by Marcum [Mar48], [Mar50]. In this appendix we shall list some of its properties, with brief derivations of a few of them. In Secs. C.2 and C.3 we treat the generalized Q function $Q_M(\alpha, \beta)$ as defined in (4-27), and we present methods for computing it. These can be applied to computing $Q(\alpha, \beta)$ as well by taking $M = 1$.

Particular values of the Q function are

$$Q(\alpha, 0) = 1, \qquad Q(0, \beta) = e^{-\beta^2/2}. \tag{C-3}$$

The normalization equation $Q(\alpha, 0) = 1$ gives us the useful integral

$$\int_0^\infty x\, e^{-ax^2/2} I_0(bx)\, dx = \frac{1}{a}\, e^{b^2/2a} \qquad\qquad (\text{C-4})$$

by a change of variables.

The generating function of the even moments of the distribution is

$$\int_0^\infty e^{zx^2} q(\alpha, x)\, dx = (1 - 2z)^{-1} \exp\left[\frac{\alpha^2 z}{1 - 2z}\right]$$

by (C-4). The resemblance of this formula to the generating function of the Laguerre polynomials,

$$L_n(x) = \frac{1}{n!}\, e^x \frac{d^n}{dx^n}(x^n e^{-x}),$$

[Erd53, vol. 2, p. 189, eq. (17)], gives us the moments of even order:

$$E(x^{2n}) = 2^n n! L_n\left(-\frac{\alpha^2}{2}\right). \qquad\qquad (\text{C-5})$$

In particular

$$E(x^2) = 2 + \alpha^2, \qquad E(x^4) = 8 + 8\alpha^2 + \alpha^4.$$

Even moments of higher order can be calculated by the recurrent relation for the Laguerre polynomials,

$$(n + 1)L_{n+1}(y) = (2n + 1 - y)L_n(y) - nL_{n-1}(y),$$
$$L_0(y) = 1, \qquad L_1(y) = 1 - y, \qquad\qquad (\text{C-6})$$

[Erd53, vol. 2, p. 190, eq. (23)].

The moments of all orders are most conveniently written in terms of the confluent hypergeometric function,

$$E(x^m) = 2^{m/2}\Gamma(\tfrac{1}{2}m + 1)_1F_1(-\tfrac{1}{2}m; 1; -\tfrac{1}{2}\alpha^2), \qquad\qquad (\text{C-7})$$

which reduces to (C-5) when m equals $2n$ [Ric44, part 2, eq. (3.10–12), p. 107]. In particular,

$$E(x) = \left(\frac{\pi}{2}\right)^{1/2} e^{-\alpha^2/4}\left[\left(1 + \frac{1}{2}\alpha^2\right)I_0\left(\frac{\alpha^2}{4}\right) + \frac{1}{2}\alpha^2 I_1\left(\frac{\alpha^2}{4}\right)\right]$$

and

$$E(x^3) = \left(\frac{\pi}{2}\right)^{1/2} e^{-\alpha^2/4}\left[\left(\frac{1}{2}\alpha^4 + 3\alpha^2 + 3\right)I_0\left(\frac{\alpha^2}{4}\right) + \alpha^2\left(2 + \frac{1}{2}\alpha^2\right)I_1\left(\frac{\alpha^2}{4}\right)\right],$$

which can be obtained by expressing the Bessel functions as confluent hypergeometric functions and applying the recurrent relations for the hypergeometric functions [Ric44, part 2, eq. (4.2–3), p. 119], [Mid60a, p. 1076]. It is shown in [Hel90] that the quantities $a_m = E(x^m)/m!$ obey the recurrent relation

$$a_{m+2} = \frac{(2m + 2 + \alpha^2)(m - 1)a_m - ma_{m-2}}{(m^2 - 1)(m + 2)}. \qquad\qquad (\text{C-8})$$

The partial derivatives of the Q function are

$$\frac{\partial Q(\alpha, \beta)}{\partial \alpha} = \beta\, e^{-\frac{1}{2}(\alpha^2 + \beta^2)} I_1(\alpha\beta), \qquad\qquad (\text{C-9})$$

which is obtained by differentiating the defining equation (C-2) and integrating by parts, and

$$\frac{\partial Q(\alpha, \beta)}{\partial \beta} = -\beta\, e^{-\frac{1}{2}(\alpha^2+\beta^2)} I_0(\alpha\beta). \tag{C-10}$$

The formula

$$Q(\alpha, \beta) + Q(\beta, \alpha) = 1 + e^{-\frac{1}{2}(\alpha^2+\beta^2)} I_0(\alpha\beta) \tag{C-11}$$

can be proved by noting that the first partial derivatives of both sides with respect to α are equal, by (C-9) and (C-10). Because the formula holds at $\alpha = 0$ by (C-3), it must hold for all values of α.

The equation

$$Q(\alpha, \beta) = e^{-\frac{1}{2}(\alpha^2+\beta^2)} \sum_{n=0}^{\infty} \left(\frac{\alpha}{\beta}\right)^n I_n(\alpha\beta) \tag{C-12}$$

holds at $\alpha = 0$ by virtue of (C-3), and its partial derivative with respect to α agrees with (C-9), as can be shown by using

$$\frac{d}{d\alpha}[\alpha^n I_n(\alpha\beta)] = \beta\alpha^n I_{n-1}(\alpha\beta).$$

Hence it is valid for all values of α. Combining (C-12) and (C-11) and interchanging α and β, we get

$$Q(\alpha, \beta) = 1 - e^{-\frac{1}{2}(\alpha^2+\beta^2)} \sum_{n=1}^{\infty} \left(\frac{\beta}{\alpha}\right)^n I_n(\alpha\beta). \tag{C-13}$$

This formula was attributed by Rice [Ric44] to W. R. Bennett.

The following asymptotic formulas are given in [Ric44, pp. 108–9]:

$$Q(\alpha, \beta) \approx \operatorname{erfc}(\beta - \alpha) + \frac{1}{2\alpha\sqrt{2\pi}}\left[1 - \frac{\beta - \alpha}{4\alpha} + \frac{1 + (\beta - \alpha)^2}{8\alpha^2}\right] e^{-\frac{1}{2}(\beta-\alpha)^2},$$

$$\alpha\beta \gg 1, \qquad \alpha \gg \beta - \alpha \gg 1.$$

For $\alpha\beta \gg 1$, $\alpha - \beta \gg 1$,

$$Q(\alpha, \beta) \approx 1 - \frac{1}{\alpha - \beta}\left(\frac{\beta}{2\pi\alpha}\right)^{1/2} e^{-\frac{1}{2}(\alpha-\beta)^2}\left[1 - \frac{3(\alpha + \beta)^2 - 4\beta^2}{8\alpha\beta(\beta - \alpha)^2} + \cdots\right].$$

This can be reduced to

$$Q(\alpha, \beta) \approx 1 - \left(\frac{\beta}{\alpha}\right)^{1/2} \operatorname{erfc}(\alpha - \beta)\left[1 - \frac{\beta + 3\alpha}{8\alpha\beta(\alpha - \beta)} + \cdots\right] \tag{C-14}$$

by using the asymptotic form of the error-function integral,

$$\operatorname{erfc} x \approx \frac{1}{x\sqrt{2\pi}} e^{-\frac{1}{2}x^2}\left[1 - \frac{1}{x^2} + \frac{3}{x^4} - \cdots\right], \qquad x \gg 1, \tag{C-15}$$

[Abr70, p. 932, eq. 26.2.12]. By using (C-11) with the asymptotic form of the Bessel function $I_0(\alpha\beta)$,

$$I_0(\alpha\beta) \approx \frac{e^{\alpha\beta}}{\sqrt{2\pi\alpha\beta}}\left(1 + \frac{1}{8\alpha\beta} + \cdots\right) \tag{C-16}$$

[Abr70, p. 977, eq. 9.7.1], and with (C-15), we obtain

$$Q(\alpha, \beta) \approx \left(\frac{\beta}{\alpha}\right)^{1/2} \text{erfc} \, (\beta - \alpha)\left[1 + \frac{\beta + 3\alpha}{8\alpha\beta(\beta - \alpha)} + \cdots\right],$$

$$\alpha\beta \gg 1, \qquad \beta - \alpha \gg 1. \tag{C-17}$$

C.2 Mth-order Q Function

The Mth-order or *generalized* Q function is defined by

$$Q_M(\alpha, \beta) = \int_\beta^\infty x\left(\frac{x}{\alpha}\right)^{M-1} e^{-\frac{1}{2}(x^2 + \alpha^2)} I_{M-1}(\alpha x) \, dx. \tag{C-18}$$

For computational purposes we shall also consider the equivalent form

$$\begin{aligned}
Q_M(S, y) &= Q_M(\sqrt{2S}, \sqrt{2y}) \\
&= \int_y^\infty \left(\frac{w}{S}\right)^{(M-1)/2} e^{-S-w} I_{M-1}(2\sqrt{Sw}) \, dw,
\end{aligned} \tag{C-19}$$

which represents the probability that a random variable having the probability density function

$$p(w) = \left(\frac{w}{S}\right)^{(M-1)/2} e^{-S-w} I_{M-1}(2\sqrt{Sw})U(w) \tag{C-20}$$

exceeds y; $w = \frac{1}{2}x^2$, $y = \frac{1}{2}\beta^2$, and $S = \frac{1}{2}\alpha^2$.

As in (4-24) the moment-generating function of this random variable is

$$h(z) = E(e^{-zw}) = \frac{e^{-S}}{(1 + z)^M} \exp\left(\frac{S}{1 + z}\right). \tag{C-21}$$

By using the generating function for the associated Laguerre polynomials as given in [Erd53, vol. 2, eq. 10.12(17), p. 189], we can write $h(z)$ as

$$h(z) = \sum_{n=0}^\infty E(w^n)\frac{(-z)^n}{n!} = \sum_{n=0}^\infty L_n^{(M-1)}(-S)(-z)^n,$$

whence the moments of $w = \frac{1}{2}x^2$ are

$$E(w^n) = n! L_n^{(M-1)}(-S),$$

which reduces to (C-5) when $M = 1$. The associated Laguerre polynomials are defined by

$$L_n^{(\alpha)}(x) = \frac{x^{-\alpha} e^x}{n!} \frac{d^n}{dx^n}(x^{n+\alpha} e^{-x}),$$

and they can be calculated by the recurrent relation

$$(n + 1)L_{n+1}^{(\alpha)}(y) = (2n + \alpha + 1 - y)L_n^{(\alpha)}(y) - (n + \alpha)L_{n-1}^{(\alpha)}(y),$$

$$L_0^{(\alpha)}(y) = 1, \qquad L_1^{(\alpha)}(y) = \alpha + 1 - y \tag{C-22}$$

[Erd53, vol. 2, p. 190, eq. (23)].

Corresponding to (C-13) the generalized Q function has an expansion

$$Q_M(\alpha, \beta) = 1 - e^{-\frac{1}{2}(\alpha^2+\beta^2)} \sum_{n=M}^{\infty} \left(\frac{\beta}{\alpha}\right)^n I_n(\alpha\beta), \tag{C-23}$$

as can be shown by differentiating both sides with respect to β and using

$$\frac{d}{d\beta}[\beta^n I_n(\alpha\beta)] = \alpha\beta^n I_{n-1}(\alpha\beta).$$

Hence by (C-13) it is related to the first-order Q function by

$$Q_M(\alpha, \beta) = Q(\alpha, \beta) + e^{-\frac{1}{2}(\alpha^2+\beta^2)} \sum_{n=1}^{M-1} \left(\frac{\beta}{\alpha}\right)^n I_n(\alpha\beta).$$

Because $I_{-n}(\alpha\beta) = I_n(\alpha\beta)$, we can write this as

$$Q_M(\alpha, \beta) = Q(\alpha, \beta) + e^{-\frac{1}{2}(\alpha^2+\beta^2)} \sum_{k=1-M}^{-1} \left(\frac{\alpha}{\beta}\right)^k I_k(\alpha\beta),$$

and by using (C-12) for $Q(\alpha, \beta)$, we find

$$Q_M(\alpha, \beta) = e^{-\frac{1}{2}(\alpha^2+\beta^2)} \sum_{k=1-M}^{\infty} \left(\frac{\alpha}{\beta}\right)^k I_k(\alpha\beta). \tag{C-24}$$

A trigonometrical integral for the generalized Q function can be found by substituting the integral

$$I_k(\alpha\beta) = \int_0^{2\pi} \cos k\theta \, e^{\alpha\beta \cos \theta} \, \frac{d\theta}{2\pi} \tag{C-25}$$

into (C-24) and interchanging summation and integration:

$$Q_M(\alpha, \beta) = e^{-\frac{1}{2}(\alpha^2+\beta^2)} \int_0^{2\pi} e^{\alpha\beta \cos \theta} \sum_{k=1-M}^{\infty} \left(\frac{\alpha}{\beta}\right)^k \cos k\theta \, \frac{d\theta}{2\pi}.$$

Writing $\cos k\theta = \frac{1}{2}(e^{ik\theta} + e^{-ik\theta})$, summing the resulting geometric progressions, and combining, we find

$$Q_M(\alpha, \beta)$$
$$= \left(\frac{\beta}{\alpha}\right)^M \alpha \, e^{-\frac{1}{2}(\beta-\alpha)^2} \int_0^{2\pi} e^{-\alpha\beta(1-\cos \theta)} \frac{\beta \cos (M-1)\theta - \alpha \cos M\theta}{\alpha^2 + \beta^2 - 2\alpha\beta \cos \theta} \, \frac{d\theta}{2\pi}, \tag{C-26}$$
$$\beta > \alpha.$$

For $\alpha > \beta$, on the other hand, we make the same substitution into (C-23), and the same procedure leads to the integral

$$1 - Q_M(\alpha, \beta)$$
$$= \left(\frac{\beta}{\alpha}\right)^M \alpha \, e^{-\frac{1}{2}(\alpha-\beta)^2} \int_0^{2\pi} e^{-\alpha\beta(1-\cos \theta)} \frac{\alpha \cos M\theta - \beta \cos (M-1)\theta}{\alpha^2 + \beta^2 - 2\alpha\beta \cos \theta} \, \frac{d\theta}{2\pi}, \tag{C-27}$$
$$\beta < \alpha.$$

—— If one has a programmable calculator or computer software with a built-in numerical integration routine, the most easily programmed technique to compute the function $Q_M(\alpha, \beta)$ is to integrate the right side of (C-26) numerically. When $\alpha > \beta$, the result is negative and equal to $-[1 - Q_M(\alpha, \beta)]$. It suffices to take the integral from 0 to π and multiply by 2. Because the integrand resembles a semi-Gaussian of width on the order of $(\alpha\beta)^{-1/2}$ peaked at $\theta = 0$, it is advisable to break the integral into two parts, one running from $\theta = 0$ to $\theta = \theta_1$, the other from $\theta = \theta_1$ to π, where θ_1 is on the order of $(10/\alpha\beta)^{1/2}$. Such a program may take longer to execute, however, than the methods to be described in Sec. C.3.

When $\alpha = \beta$, we must proceed as follows. Combining (C-23) and (C-24) yields

$$Q_M(\alpha, \beta) + Q_M(\beta, \alpha) = 1 + e^{-\frac{1}{2}(\alpha^2+\beta^2)} \sum_{n=1-M}^{M-1} \left(\frac{\beta}{\alpha}\right)^n I_n(\alpha\beta),$$

so that

$$Q_M(\alpha, \alpha) = \tfrac{1}{2} + \tfrac{1}{2}e^{-\alpha^2} \sum_{k=1-M}^{M-1} I_k(\alpha^2) = \tfrac{1}{2} + e^{-\alpha^2}\left[\tfrac{1}{2}I_0(\alpha^2) + \sum_{k=1}^{M-1} I_k(\alpha^2)\right], \tag{C-28}$$

which can be evaluated by using tabulated values of the modified Bessel functions and their recurrent relation

$$I_{n+1}(x) = I_{n-1}(x) - \frac{2n}{x}I_n(x).$$

Lacking a table, one can numerically integrate

$$Q_M(\alpha, \alpha) = \tfrac{1}{2} + \tfrac{1}{2}\int_0^{2\pi} e^{-\alpha^2(1-\cos\theta)} \frac{\sin(M - \tfrac{1}{2})\theta}{\sin\tfrac{1}{2}\theta}\frac{d\theta}{2\pi},$$

which is derived by putting (C-25) into (C-28).

C.3 Computation by Recurrence

We evaluate the generalized Q function in the notationally more convenient form (C-19), and we utilize the moment-generating function in (C-21). Applying the power series for e^x, we write this as

$$h(z) = e^{-S} \sum_{r=0}^{\infty} \frac{S^r}{r!}(1 + z)^{-M-r},$$

whose inverse Laplace transform immediately yields for the probability density function

$$p(w) = e^{-S} \sum_{r=0}^{\infty} \frac{S^r}{r!} \frac{w^{M+r-1}\,e^{-w}}{(M + r - 1)!}.$$

This also follows from the power-series expansion of the Bessel function in (C-20). Integrating it term by term, we obtain for (C-19)

$$Q_M(S, y) = \sum_{r=0}^{\infty} b_r g_{M+r-1}, \tag{C-29}$$

where

$$b_r = \frac{S^r e^{-S}}{r!}, \qquad g_k = \frac{1}{k!}\int_y^{\infty} w^k e^{-w}\,dw. \tag{C-30}$$

Integrating the latter by parts, we find the recurrence

$$g_k = \delta_k + g_{k-1}, \qquad \delta_0 = g_0 = e^{-y},$$

$$\delta_k = \frac{y^k e^{-y}}{k!} = \frac{y}{k}\delta_{k-1}, \tag{C-31}$$

which with the recurrence

$$b_r = \frac{S}{r}b_{r-1}, \qquad b_0 = e^{-S}, \tag{C-32}$$

defines an easily programmed algorithm for computing the Mth-order Q function through (C-29). It is most suitable when the limit y of integration lies above the expected value $E(w) = M + S$ of the density function in (C-20).

When y is a decision level chosen for a certain false-alarm probability Q_0,

$$Q_0 = g_{M-1} = \sum_{r=0}^{M-1} \frac{y^r e^{-y}}{r!},$$

whereupon (C-29) becomes

$$Q_M(S, y) = Q_0 e^{-S} + \sum_{r=1}^{\infty} b_r g_{M+r-1}, \qquad y \geq M + S,$$

which is sometimes a useful form. This algorithm was given by Dillard [Dil73] as an extension of McGee's [McG70] for the ordinary Q function $Q(\alpha, \beta)$.

It is shown in [Hel92d] that the error E_k incurred by terminating this series with the kth term is bounded by

$$E_k \leq \sqrt{2\pi k}\, b_k,$$

provided $k > S$ and $k \gg 1$. One can stop the summation in (C-29) when the quantity on the right side, divided by the accumulated sum, falls below a preassigned relative error.

When $y < M + S$, it is preferable for the sake of accuracy to compute $1 - Q_M(S, y)$. From (C-31)

$$g_k = \sum_{m=0}^{k} \delta_m,$$

so that (C-29) can be written

$$Q_M(S, y) = \sum_{r=0}^{\infty} b_r \sum_{m=0}^{M+r-1} \delta_m,$$

and

$$1 - Q_M(S, y) = \sum_{r=0}^{\infty} b_r \sum_{m=M+r}^{\infty} \delta_m.$$

Changing the order of summation yields

$$1 - Q_M(S, y) = \sum_{m=0}^{\infty} \delta_{m+M} \sum_{r=0}^{m} b_r = \sum_{m=0}^{\infty} h_m \delta_{m+M}, \qquad y < M + S, \tag{C-33}$$

where the h_m's are computed by the recursion

$$h_m = h_{m-1} + b_m, \qquad h_0 = b_0 = e^{-S}, \qquad \text{(C-34)}$$

the b_m's being calculated as in (C-32). The δ_{m+M}'s are calculated as in (C-31).

It is shown in [Hel92d] that the error E_k incurred by terminating this series with the kth term is bounded by

$$E_k \lesssim \sqrt{2\pi(M + k)}\, \delta_{M+k},$$

provided that $M + k > y$ and $k \gg 1$. As described above, this bound can be used to tell the computer when to stop summing.

Both these recurrent algorithms permit easy generalization to situations in which either the signal strength S or the decision level y or both are random variables; see Appendix E. Related algorithms were given by Brennan and Reed [Bre65], Robertson [Rob69], and Shnidman [Shn89]. When M is large, all these methods suffer from computer underflow or overflow: for ordinary values of the false-alarm and detection probabilities, S and y are then large and e^{-S} and e^{-y} extremely small. Complicated stratagems are necessary to cope with that problem, and the contour-integration methods described in Sec. 5.2 are preferable. Quite a different method has been described by Parl [Par80], but it too is inefficient for large values of M.

Appendix D

Error Probability for a Channel Carrying Two Nonorthogonal Signals with Random Phases

According to Sec. 3.5.2, we want the probability that $R_0 > R_1$, where

$$R_i = \left| \int_0^T Q_i^*(t)V(t)\,dt \right|, \qquad i = 0, 1,$$

under hypothesis H_1 that $V(t) = F_1(t) + N(t)$. Here $N(t)$ is the complex envelope of Gaussian random noise with autocovariance function

$$\phi(\tau) = \operatorname{Re} \tilde\phi(\tau)e^{i\Omega\tau},$$

and $Q_0(t)$ and $Q_1(t)$ are solutions of the integral equations

$$F_i(t) = \int_0^T \tilde\phi(t - u)Q_i(u)\,du, \qquad 0 \le t \le T, \qquad i = 0, 1.$$

The signals are so chosen that

$$\int_0^T Q_1^*(t)F_1(t)\,dt = \int_0^T Q_0^*(t)F_0(t)\,dt = d^2.$$

For detection in white noise this means that they are received with equal energies E, and $d^2 = 2E/N$. In this analysis we follow [Hel55].

We introduce the circular complex Gaussian random variables

$$z_j = x_j + iy_j = \int_0^T Q_j^*(t)V(t)\,dt, \qquad j = 0, 1.$$

The joint distribution of their real and imaginary parts x_0, y_0, x_1, and y_1 is conveniently written in terms of z_0 and z_1 in the manner introduced in Sec. 3.2.3. Their expected values are

$$E(z_0|\,H_1) = \zeta_0 = e^{i\psi} \int_0^T Q_0^*(t)F_1(t)\,dt = \lambda d^2\,e^{i\psi},$$

$$E(z_1|\,H_1) = \zeta_1 = e^{i\psi} \int_0^T Q_1^*(t)F_1(t)\,dt = d^2\,e^{i\psi},$$

(D-1)

where the complex number λ is defined by

$$\lambda = d^{-2} \int_0^T Q_0^*(t)F_1(t)\,dt = d^{-2} \int_0^T \int_0^T Q_0^*(t)\tilde\phi(t-u)Q_1(u)\,dt\,du.$$

The variances and covariances of x_0, y_0, x_1, and y_1 can be obtained from the relations

$$\{z_0, z_0^*\} = \{z_1, z_1^*\} = \tfrac{1}{2}E \int_0^T \int_0^T Q_0^*(t_1)Q_0(t_2)N(t_1)N^*(t_2)\,dt_1\,dt_2$$

$$= \int_0^T \int_0^T Q_0^*(t_1)Q_0(t_2)\tilde\phi(t_1-t_2)\,dt_1\,dt_2$$

(D-2)

$$= \int_0^T Q_0^*(t)F_0(t)\,dt = d^2,$$

$$\{z_0, z_1^*\} = \int_0^T \int_0^T Q_0^*(t_1)Q_1(t_2)\tilde\phi(t_1-t_2)\,dt_1\,dt_2 = \lambda d^2.$$

(D-3)

Here we use the notation

$$\{w_1, w_2^*\} = \tfrac{1}{2}E\{[w_1 - E(w_1)][w_2^* - E(w_2^*)]\}$$

for the complex covariances of two circular complex variables w_1 and w_2. The complex covariance matrix of the complex variables z_0, z_1, as defined in (3-36), is therefore

$$\phi = \begin{bmatrix} d^2 & \lambda d^2 \\ \lambda^* d^2 & d^2 \end{bmatrix},$$

its determinant is

$$\det \phi = d^4(1 - |\lambda|^2),$$

and its inverse is

$$\phi^{-1} = \mu = \frac{1}{d^2(1 - |\lambda|^2)} \begin{bmatrix} 1 & -\lambda \\ -\lambda^* & 1 \end{bmatrix}.$$

Thus the joint probability density function x_0, y_0, x_1, and y_1 can be written in circular Gaussian form as in (3-41):

$$p(x_0, y_0, x_1, y_1) = \tilde p(z_0, z_1)$$

$$= \frac{1}{(2\pi)^2 d^4(1 - |\lambda|^2)} \exp\left[-\frac{|z_0 - \zeta_0|^2 + |z_1 - \zeta_1|^2 - 2\,\mathrm{Re}\,\lambda(z_0^* - \zeta_0^*)(z_1 - \zeta_1)}{2d^2(1 - |\lambda|^2)} \right].$$

When we substitute from (D-1), we find that a number of terms cancel, and we obtain

$$\tilde{p}(z_0, z_1) = \frac{1}{(2\pi)^2 d^4 (1 - |\lambda|^2)}$$

$$\cdot \exp\left[-\frac{|z_0|^2 + |z_1|^2 + (1 - |\lambda|^2)d^4 - 2d^2(1 - |\lambda|^2)\,\mathrm{Re}(z_1 e^{-i\psi}) - 2\,\mathrm{Re}(\lambda z_0^* z_1)}{2d^2(1 - |\lambda|^2)} \right].$$

Introducing polar coordinates, we write

$$z_0 = R_0\, e^{i\phi_0}, \qquad z_1 = R_1\, e^{i\phi_1}, \qquad \lambda = |\lambda|\, e^{i\beta}.$$

The joint probability density function of the random variables R_0, R_1, ϕ_0, and ϕ_1 is then

$$p(R_0, R_1, \phi_0, \phi_1) = \frac{R_0 R_1}{(2\pi)^2 d^4 (1 - |\lambda|^2)}\, e^{-\frac{1}{2}d^2}$$

$$\cdot \exp\left[-\frac{R_0^2 + R_1^2 - 2d^2(1 - |\lambda|^2)R_1 \cos(\phi_1 - \psi) - 2|\lambda|R_0 R_1 \cos(\phi_1 - \phi_0 + \beta)}{2d^2(1 - |\lambda|^2)} \right]$$

after we multiply by the Jacobian factor $R_0 R_1$. We first integrate over $0 \le \phi_0 < 2\pi$ and then over $0 \le \phi_1 < 2\pi$, and by using (3-61) we obtain the joint density function of R_0 and R_1 under hypothesis H_1:

$$p(R_0, R_1) = \frac{R_0 R_1}{d^4(1 - |\lambda|^2)} \exp\left[-\frac{R_0^2 + R_1^2}{2d^2(1 - |\lambda|^2)} - \frac{d^2}{2} \right] I_0(R_1) I_0\left(\frac{|\lambda| R_0 R_1}{d^2(1 - |\lambda|^2)} \right).$$

The probability of error is

$$P_e = \mathrm{Pr}(R_0 > R_1 \mid H_1) = \int_0^\infty dR_1 \int_{R_1}^\infty dR_0\, p(R_0, R_1).$$

Introducing the new integration variables

$$x = \frac{R_1}{d\sqrt{1 - |\lambda|^2}}, \qquad y = \frac{R_0}{d\sqrt{1 - |\lambda|^2}},$$

we write the error probability as the double integral

$$P_e = (1 - |\lambda|^2) e^{-\frac{1}{2}d^2} \int_0^\infty x I_0(d\sqrt{1 - |\lambda|^2}\, x) e^{-\frac{1}{2}x^2}\, dx \int_x^\infty y\, e^{-\frac{1}{2}y^2} I_0(|\lambda| xy)\, dy$$

$$= (1 - |\lambda|^2)\, e^{-\frac{1}{2}d^2} \int_0^\infty x\, e^{-\frac{1}{2}(1 - |\lambda|^2)x^2} I_0(d\sqrt{1 - |\lambda|^2}\, x)$$

$$\cdot \int_x^\infty y\, e^{-\frac{1}{2}y^2 - \frac{1}{2}|\lambda|^2 x^2} I_0(|\lambda| xy)\, dy\, dx \tag{D-4}$$

$$= (1 - |\lambda|^2)\, e^{-\frac{1}{2}d^2} \int_0^\infty x\, e^{-\frac{1}{2}(1 - |\lambda|^2)x^2} I_0(d\sqrt{1 - |\lambda|^2}\, x) Q(|\lambda| x,\, x)\, dx$$

in terms of the Q function as defined in (3-76).

Let us now write (C-26) for $M = 1$ as

$$Q(\alpha, \beta) = e^{-\frac{1}{2}(\alpha^2 + \beta^2)} \int_0^{2\pi} e^{\alpha\beta \cos \theta} \frac{1 - (\alpha/\beta) \cos \theta}{1 + (\alpha/\beta)^2 - 2(\alpha/\beta) \cos \theta} \frac{d\theta}{2\pi}, \tag{D-5}$$

which yields

$$Q(|\lambda| x,\, x) = e^{-\frac{1}{2}(1 + |\lambda|^2)x^2} \int_0^{2\pi} e^{|\lambda| x^2 \cos \theta} \frac{1 - |\lambda| \cos \theta}{1 + |\lambda|^2 - 2|\lambda| \cos \theta} \frac{d\theta}{2\pi}.$$

Putting this into (D-4), we integrate over $0 < x < \infty$ by means of (C-4),

$$\int_0^\infty x\, e^{-x^2(1-|\lambda|\cos\theta)} I_0(d\sqrt{1-|\lambda|^2}\,x)\, dx = \frac{1}{2(1-|\lambda|\cos\theta)}\exp\left[\frac{d^2(1-|\lambda|^2)}{4(1-|\lambda|\cos\theta)}\right],$$

and we obtain for (D-4) the trigonometric integral

$$P_e = \tfrac{1}{2}(1-|\lambda|^2)e^{-\frac{1}{2}d^2}\int_0^{2\pi}\frac{1}{1-2|\lambda|\cos\theta+|\lambda|^2}\exp\left[\frac{d^2(1-|\lambda|^2)}{4(1-|\lambda|\cos\theta)}\right]\frac{d\theta}{2\pi}. \qquad \text{(D-6)}$$

For simplicity, we henceforward drop the absolute-value signs on $|\lambda|$.

This integral can be simplified by changing the integration variable by

$$\cos\theta = \frac{\lambda+\cos\phi}{1+\lambda\cos\phi},$$

for which a certain amount of algebra produces the relations

$$1-\lambda\cos\theta = \frac{1-\lambda^2}{1+\lambda\cos\phi},$$

$$d\theta = \frac{\sqrt{1-\lambda^2}}{1+\lambda\cos\phi}\,d\phi,$$

$$1+\lambda^2-2\lambda\cos\theta = \frac{(1-\lambda^2)(1-\lambda\cos\phi)}{1+\lambda\cos\phi},$$

and these enable us to write (D-6) as

$$P_e = \tfrac{1}{2}\sqrt{1-\lambda^2}\,e^{-\frac{1}{4}d^2}\int_0^{2\pi}\frac{\exp(\frac{1}{4}\lambda d^2\cos\phi)}{1-\lambda\cos\phi}\frac{d\phi}{2\pi}. \qquad \text{(D-7)}$$

If we now write (C-27) for $M=1$, $\alpha>\beta$, as

$$1-Q(\alpha,\beta) = \beta\, e^{-\frac{1}{2}(\alpha^2+\beta^2)}\int_0^{2\pi} e^{\alpha\beta\cos\phi}\frac{\alpha\cos\phi-\beta}{\alpha^2+\beta^2-2\alpha\beta\cos\phi}\frac{d\phi}{2\pi}$$

and add to it (D-5) after interchanging α and β, we find

$$1-Q(\alpha,\beta)+Q(\beta,\alpha) = e^{-\frac{1}{2}(\alpha^2+\beta^2)}\int_0^{2\pi}\frac{(\alpha^2-\beta^2)e^{\alpha\beta\cos\phi}}{\alpha^2+\beta^2-2\alpha\beta\cos\phi}\frac{d\phi}{2\pi}.$$

Comparing this with (D-7), we see that the integrals have the same form after we put

$$\lambda = \frac{2\alpha\beta}{\alpha^2+\beta^2}, \qquad \tfrac{1}{4}\lambda d^2 = \alpha\beta,$$

that is, if we take α and β as

$$\alpha = \tfrac{1}{2}d[1+\sqrt{1-\lambda^2}]^{1/2},$$

$$\beta = \tfrac{1}{2}d[1-\sqrt{1-\lambda^2}]^{1/2}.$$

The probability of error thus becomes

$$P_e = \tfrac{1}{2}\left[1-Q(\tfrac{1}{2}d\sqrt{1+k},\tfrac{1}{2}d\sqrt{1-k})+Q(\tfrac{1}{2}d\sqrt{1-k},\tfrac{1}{2}d\sqrt{1+k})\right],$$

$$\qquad \text{(D-8)}$$

$$k = \sqrt{1-|\lambda|^2},$$

as in (3-82). The second form given there results from using (C-11) for the first two terms in the brackets. A number of formulas of this type have been given by Price [Pri64].

The probability of error is equal to the probability that the quadratic form

$$U = \tfrac{1}{2}(|z_1|^2 - |z_0|^2)$$

is less than zero. This is a special case of the general Hermitian quadratic form

$$U = \tfrac{1}{2}\mathbf{V}^+\mathbf{H}\mathbf{V}$$

in n circular complex Gaussian random variables $(V_1, V_2, \dots, V_n)$, which we write as a column vector $\mathbf{V}$;

$$\mathbf{V}^+ = (V_1^*, V_2^*, \dots, V_n^*)$$

is the conjugate transposed row vector, and $\mathbf{H}$ is an $n \times n$ Hermitian matrix. We designate the column vector of the expected values of the random variables by $\mathbf{S}$, $S_m = E(V_m)$. The $n \times n$ complex covariance matrix of these variables is denoted by $\boldsymbol{\phi}$:

$$\phi_{jm} = \tfrac{1}{2}E[(V_j - S_j)(V_m^* - S_m^*)],$$
$$\boldsymbol{\phi} = \tfrac{1}{2}E[(\mathbf{V} - \mathbf{S})(\mathbf{V}^+ - \mathbf{S}^+)].$$

The moment-generating function of the random variable U then takes the form

$$h(z) = E(e^{-Uz}) = \frac{1}{\det(\mathbf{I} + z\mathbf{H}\boldsymbol{\phi})} \exp\left[-\tfrac{1}{2}z\mathbf{S}^+(\mathbf{I} + z\mathbf{H}\boldsymbol{\phi})^{-1}\mathbf{H}\mathbf{S}\right], \qquad \text{(D-9)}$$

as can be shown by means of (B-10); $\mathbf{I}$ is the $n \times n$ identity matrix. If furthermore we are given a random variable U that is the sum of, say, M independent random variables U_k of this type,

$$U = \sum_{k=1}^{M} U_k, \qquad U_k = \mathbf{V}_k^+\mathbf{H}\mathbf{V}_k,$$

and if the components of each $\mathbf{V}_k$ have the same covariance matrix $\boldsymbol{\phi}$, but possibly different expected values $\mathbf{S}_k$, it is only necessary to multiply their moment-generating functions to obtain that of U:

$$h(z) = [\det(\mathbf{I} + z\mathbf{H}\boldsymbol{\phi})]^{-M} \exp\left[-\tfrac{1}{2}z \sum_{k=1}^{M} \mathbf{S}_k^+(\mathbf{I} + z\mathbf{H}\boldsymbol{\phi})^{-1}\mathbf{H}\mathbf{S}_k\right]. \qquad \text{(D-10)}$$

For example, Proakis [Pro89, pp. 344–9] utilized a form of this type in order to derive an expression for the probability that $U > 0$ when the matrix $\mathbf{H}$ is a 2×2 nonpositive-definite matrix

$$\mathbf{H} = \begin{bmatrix} A & C \\ C^* & B \end{bmatrix}. \qquad \text{(D-11)}$$

His expression involves the Q function and a series of modified Bessel functions and is a generalization of (D-8). For the latter, $C = 0$, $A = 1$, and $B = -1$.

As shown first by Rice [Ric80], the probability density function of a quadratic form such as U can be computed from the moment-generating function (D-9) or (D-10) by numerical integration of the Laplace inversion integral (5-4), and the numerical methods described in Sec. 5.2 can be applied to evaluating the cumulative distribution of U and its complement. In the case of (D-11), for instance, the matrix $\mathbf{I} + z\,\mathbf{H}\boldsymbol{\phi}$ involved in (D-9) and (D-10) is a 2×2 matrix, and its inverse and its determinant are easily programmed for evaluation of the integrands of (5-4), (5-19), and (5-20) when computing those inversion integrals numerically.

Appendix E

Recursive Methods for Detection Probabilities: Fading Signals and Random Decision Levels

E.1 Fading Signals

When the receiver sums or "integrates" M quadratically rectified outputs of a filter matched to the signal to be detected, as in (4-19), and the input signal-to-noise ratio ratios $d_1^2, d_2^2, \ldots, d_M^2$ are fixed, the probability

$$Q_d = \Pr(U \geq y \mid H_1) = Q_M(S, y) \tag{E-1}$$

of detection can be calculated by either of the two recursive methods in (C-29) through (C-32) and (C-33) and (C-34); here

$$S = \tfrac{1}{2} \sum_{k=1}^{M} d_k^2 = \frac{E_T}{N}$$

is the total energy-to-noise ratio, E_T being the total received energy and N the unilateral spectral density of the noise.

If now the input energy-to-noise ratios are not fixed, but random, the average probability $\overline{Q}_d = \Pr(U > y \mid H_1)$ of choosing hypothesis H_1 when signals are actually present can be calculated by recursive procedures derived from those in Sec. C.3 simply by averaging the terms with respect to the resultant probability distribution of the total energy-to-noise ratio [Mit71]. Thus we find

$$\overline{Q}_d = \sum_{m=0}^{\infty} \langle b_m \rangle g_{M+m-1}, \tag{E-2}$$

where the g_k are given as in (C-31) in terms of the decision level y, and where from (C-30)

$$\langle b_m \rangle = \frac{E(S^m e^{-S})}{m!}$$

is the expected value of the term b_m with respect to the distribution of the total energy-to-noise ratio S. The decision level y in (E-1) is still determined by the preassigned false-alarm probability

$$Q_0 = \Pr(U \geq y \mid H_0) = e^{-y} \sum_{k=0}^{M-1} \frac{y^k}{k!},$$

as in (4-30), where $U_0 = y$. If $h_s(z) = E(e^{-Sz})$ is the moment-generating function of S, then

$$\langle b_m \rangle = \frac{(-1)^m}{m!} \frac{d^m}{dz^m} h_s(z) \bigg|_{z=1}.$$

If, for instance, the total energy-to-noise ratio S has a gamma distribution as in (4-39), $h_s(z)$ is given in (4-40), and we find

$$\langle b_0 \rangle = (1 + \overline{S}')^{-k},$$

$$\langle b_m \rangle = \frac{\Gamma(k + m)}{\Gamma(k)m!} \frac{\overline{S}'^m}{(1 + \overline{S}')^{k+m}}$$

$$= \frac{k + m - 1}{m} \left(\frac{\overline{S}'}{1 + \overline{S}'} \right) \langle b_{m-1} \rangle, \tag{E-3}$$

$$\overline{S}' = \frac{E(S)}{k} = \frac{\overline{S}}{k}.$$

One uses (E-2) and (E-3) when $y > E(U \mid H_1) = \overline{S} + M$. Here $k = 1$, M, 2, and $2M$ for Swerling cases 1, 2, 3, and 4, respectively.

When on the other hand $y < \overline{S} + M$, we modify (C-33) and (C-34) in the same way. Now

$$1 - \overline{Q}_d = \sum_{m=0}^{\infty} \langle h_m \rangle \delta_{m+M},$$

with the δ_k still given by (C-31), but with the coefficients $\langle h_m \rangle$ formed by averaging (C-34),

$$\langle h_0 \rangle = \langle b_0 \rangle = \frac{1}{(1 + \overline{S}')^k},$$

$$\langle h_m \rangle = \langle h_{m-1} \rangle + \langle b_m \rangle, \tag{E-4}$$

and the $\langle b_m \rangle$'s are calculated as in (E-3).

E.2 Random Decision Level, Fixed Signal Strengths

In the constant-false-alarm-rate (CFAR) receiver studied in Sec. 8.1, the decision level $y = U_0$ with which the quadratic statistic U of (4-19) is compared is made proportional to an estimate of the unknown spectral density N of the ambient white noise; that is, $y = \beta x$, where x is a random variable with a gamma distribution

$$z(x) = \frac{x^{M'-1} e^{-x}}{(M' - 1)!} \tag{E-5}$$

when suitably normalized. Here M' is the number of quadratically rectified outputs of a narrowband filter that are integrated to form the estimate of the spectral density N; these outputs are assumed to have no components due to the signal to be detected, and they are supposed to have identical exponential distributions. The constant β is set to provide a preassigned false-alarm probability, which is calculated as in Sec. E.4.

We assume that the M input signal strengths, proportional to $d_1^2, d_2^2, \ldots, d_M^2$, are fixed. The N figuring in the definition of $d_m^2 = 2E_m/N$ can be taken as a fiducial level and cancels out of the final result. The average probability $\overline{Q}_d$ of detecting the signal is obtained by averaging the recurrent relations in Sec. C.3 with respect to the prior probability density function of $y = \beta x$, β a fixed number. Thus when

$$E(y) = \beta E(x) = \beta M' \geq S + M,$$

we compute the average probability of detection from

$$\overline{Q}_d = \sum_{r=0}^{\infty} b_r \langle g_{M+r-1} \rangle, \tag{E-6}$$

where the b_r's are calculated as in (C-32), and where by (C-31)

$$\langle g_s \rangle = \langle \delta_s \rangle + \langle g_{s-1} \rangle$$

and

$$\begin{aligned}\langle \delta_s \rangle &= \frac{E(y^s e^{-y})}{s!} = \int_0^{\infty} \frac{(\beta x)^s e^{-\beta x}}{s!} \frac{x^{M'-1} e^{-x}}{(M' - 1)!} dx \\ &= \frac{\Gamma(s + M')\beta^s}{s!\,\Gamma(M')(1 + \beta)^{s+M'}} = \frac{s + M' - 1}{s} \frac{\beta}{1 + \beta} \langle \delta_{s-1} \rangle,\end{aligned} \tag{E-7}$$

$$\langle g_0 \rangle = \langle \delta_0 \rangle = (1 + \beta)^{-M'}.$$

For $\beta M' < S + M$, on the other hand, we compute $1 - \overline{Q}_d$ from the average of (C-32),

$$1 - \overline{Q}_d = \sum_{m=0}^{\infty} h_m \langle \delta_{m+M} \rangle, \tag{E-8}$$

with h_m calculated as in (C-34) and the $\langle \delta_s \rangle$ as in (E-7).

E.3 Random Signal Strengths and Random Decision Level

When both the signal strengths and the decision level are random, we average the terms in the algorithms of Sec. C.3 with respect to both. Assuming the same gamma distributions as in (4-39) and (E-5), we find as in Secs. E.1 and E.2, from (C-29) through (C-32),

$$\overline{Q}_d = \Pr(U \geq \beta x \mid H_1) = \sum_{m=0}^{\infty} \langle b_m \rangle \langle g_{M+m-1} \rangle, \qquad \text{(E-9)}$$

with the $\langle b_m \rangle$'s calculated as in (E-3) and the $\langle g_s \rangle$'s as in (E-7). One utilizes (E-9) when $\beta M' \geq E(S) + M$. When $\beta M' < E(S) + M$, on the other hand, one utilizes the average of (C-33) and (C-34),

$$1 - \overline{Q}_d = \sum_{m=0}^{\infty} \langle h_m \rangle \langle \delta_{m+M} \rangle$$

with the $\langle \delta_s \rangle$'s calculated as in (E-7) and the $\langle h_m \rangle$'s calculated as in (E-4).

E.4 False-alarm Probability, Random Decision Level

From either (E-6) or (E-9) we find, for $S = 0$ or $E(S) = 0$, the false-alarm probability

$$Q_0 = \Pr(U \geq \beta x \mid H_0) = \langle g_{M-1} \rangle = \sum_{s=0}^{M-1} \langle \delta_s \rangle$$

$$= \frac{1}{(1 + \beta)^{M'}} \sum_{s=0}^{M-1} \frac{\Gamma(s + M')}{s! \Gamma(M')} \left(\frac{\beta}{1 + \beta} \right)^s,$$

in which the $\langle \delta_s \rangle$'s can be computed recursively as in (E-7). For $M \gg 1$ calculating this sum can be tedious, and a more rapidly converging series results from noting that

$$Q_0 = 1 - \Pr(U < \beta x \mid H_0) = 1 - \Pr\left(x \geq \frac{U}{\beta} \mid H_0 \right)$$

and utilizing (E-8) with M and M' interchanged, β replaced by β^{-1}, and $h_m \equiv h_0 = 1$, whereupon

$$Q_0 = \sum_{m=0}^{\infty} \frac{\Gamma(m + M + M')}{(m + M')! \Gamma(M)} \frac{\beta^{-(m+M')}}{(1 + \beta^{-1})^{m+M+M'}}$$

$$= \left(\frac{\beta}{1 + \beta} \right)^M \sum_{m=0}^{\infty} \frac{\Gamma(m + M + M')}{(m + M')! \Gamma(M)} \frac{1}{(1 + \beta)^{m+M'}} \qquad \text{(E-10)}$$

$$= \left(\frac{\beta}{1 + \beta} \right)^M \sum_{k=M'}^{\infty} \frac{\Gamma(M + k)}{k! \Gamma(M)} (1 + \beta)^{-k}$$

[Rob76]. This sum can be computed recursively:

$$Q_0 = \sum_{k=M'}^{\infty} \varepsilon_k, \qquad \varepsilon_0 = \left(\frac{\beta}{1 + \beta} \right)^M, \qquad \varepsilon_k = \frac{M + k - 1}{k(1 + \beta)} \varepsilon_{k-1}.$$

One generates the ε_k's starting from $k = 0$, but does not begin summing them till $k = M'$; and one stops when the last term added in falls below ε times the accumulated sum, with ε small enough to ensure the number of accurate significant figures desired.

The probability Q_0 thus calculated is the probability that the ratio of two chi-squared-distributed random variables η_1 and η_2 with $f_1 = 2M$ and $f_2 = 2M'$ degrees of freedom, respectively, exceeds a level β. Here

$$\eta_1 = \sum_{j=1}^{f_1} x_j^2, \qquad \eta_2 = \sum_{j=1}^{f_2} x_j'^2,$$

where the x_j and x_j' are independent Gaussian random variables with expected values zero and equal variances. Then

$$Q_0 = \Pr(\beta' \geq \beta) = Q(\beta; f_1, f_2)$$

is called the complementary β-distribution and has been tabulated by Pearson [Pea68]. The random variable figures in statistical tests of linear hypotheses as in the analysis of variance, and tables of its percentage points are to be found in many statistical handbooks under the rubric *F distribution*. The probability $\overline{Q}_d$ derived in Sec. E.2 then corresponds to the noncentral *F* distribution, and our parameter S, or sometimes $2S$, is termed the *noncentrality parameter* [Tan38], [Pat49], [Rob76].

Appendix F

Pulse Reflected from a Moving Target

A radar antenna emits a pulse signal $s(t)$, which travels toward a moving target and is reflected from it, and a portion of the reflected pulse is picked up by a receiving antenna in the same vicinity as the transmitter. We want to find the form of the received pulse when the target is moving away from the antenna with velocity V.

Let F be a frame of reference fixed with respect to the antenna, which is at the origin of coordinates. The x-axis points in the direction of the target. In F the field of the transmitted pulse at point x is proportional to

$$s\left(t - \frac{x}{c}\right);$$

c = velocity of light.

Let F' be a frame of reference moving with velocity V relative to F; in F' the target is stationary. The coordinate in F' parallel to the velocity of F' relative to F is designated by x' and the time in F' by t'. Then the Lorentz–Fitzgerald transformation equations relate t and x to t' and x' through

$$t = \Gamma\left(t' + \frac{Vx'}{c^2}\right), \qquad x = \Gamma(x' + Vt'), \tag{F-1}$$

with

$$\Gamma = \left[1 - \frac{V^2}{c^2}\right]^{-1/2} \tag{F-2}$$

[Tol34, pp. 18–21]. At point x' in F' at time t' the transmitted signal field is

$$s\left(t - \frac{x}{c}\right) = s\left[\Gamma\left(1 - \frac{V}{c}\right)\left(t' - \frac{x'}{c}\right)\right], \tag{F-3}$$

as we find by substituting from (F-1).

Let d_0' be the distance from transmitter to target in F'. Then the field at the target when the pulse strikes it is given by (F-3) with $x' = d_0'$. The reflected field is propagating in the opposite direction, but within an inconsequential sign and reflection coefficient it must have the same time dependence as that in (F-3) for $x' = d_0'$. Hence the reflected pulse can be represented by

$$s_r(x', t') = s\left[\Gamma\left(t' + \frac{x' - 2d_0'}{c}\right)\left(1 - \frac{V}{c}\right)\right].$$

We translate this into the coordinates of frame F by using (F-1), from which we obtain

$$t' + \frac{x'}{c} = \Gamma\left(t + \frac{x}{c}\right)\left(1 - \frac{V}{c}\right),$$

whereupon

$$s_r(x, t) = s\left[\Gamma\left(\Gamma\left(t + \frac{x}{c}\right)\left(1 - \frac{V}{c}\right) - \frac{2d_0'}{c}\right)\left(1 - \frac{V}{c}\right)\right]. \qquad \text{(F-4)}$$

Now the coordinate of the target at time $t = 0$ when the pulse was transmitted was $x = d_0$ in frame F and $x' = d_0'$ in frame F'. By the inverse transformation to that in (F-1),

$$x' = \Gamma(x - Vt),$$

these distances are related by

$$d_0' = \Gamma d_0, \qquad d_0 = d_0'\left[1 - \frac{V^2}{c^2}\right]^{1/2}, \qquad \text{(F-5)}$$

which expresses the Lorentz contraction; to someone stationary in the frame F of the antenna, the distance d_0 to the target seems shorter than to someone moving along with the target. Putting (F-5) into (F-4) and setting $x = 0$, we find the received field at the antenna to be

$$s_r(0, t) = s\left[\Gamma^2 t\left(1 - \frac{V}{c}\right)^2 - \frac{2\Gamma^2 d_0}{c}\left(1 - \frac{V}{c}\right)\right] = s\left[\left(\frac{c - V}{c + V}\right)(t - \tau)\right]$$

when we use (F-2). The term

$$\tau = \frac{2d_0}{c - V}$$

represents the interval between transmission and reception of the pulse. The pulse is stretched in time by a factor

$$r^{-1} = \frac{c + V}{c - V},$$

which is greater than 1 when the target is moving away from the antenna.

For a narrowband transmitted pulse

$$s(t) = \text{Re } F(t)\, e^{i\Omega t},$$

the received echo, within a constant of proportionality, is

$$s_r(0, t) = \text{Re}\Big[F(r(t - \tau)) \, e^{i\Omega' t + i\psi} \Big],$$

$$r = \frac{c - V}{c + V}, \qquad \Omega' = r\Omega, \qquad \psi = -\frac{2d_0\Omega}{c + V}.$$

To terms of first order in V/c, the angular carrier frequency of the received signal is

$$\Omega' \approx \left(1 - \frac{2V}{c} \right)\Omega = \Omega + w,$$

where $w = -(2V/c)\Omega$ is the Doppler shift. The stretching of the complex envelope, indicated by the factor r, is usually insignificant.

When a train of pulses separated by T,

$$s(t) = \text{Re} \sum_{k=1}^{M} F[t - (k - 1)T] \, e^{i\Omega[t - (k - 1)T]},$$

is transmitted, the echo signal is

$$s_r(0, t) = \text{Re} \sum_{k=1}^{M} F[r(t - \tau) - (k - 1)T] \, e^{i\Omega[r(t-\tau) - (k-1)T]},$$

and to first order in V/c this is

$$s_r(0, t) \approx \text{Re} \sum_{k=1}^{M} F[r(t - \tau - r^{-1}(k - 1)T)] \, e^{i\Omega'[t - \tau - r^{-1}(k-1)T]}$$

$$= \text{Re} \sum_{k=1}^{M} F\left[r\left(t - \tau - \left(1 + \frac{2V}{c} \right)(k - 1)T \right) \right]$$

$$\cdot \exp\left\{ i\Omega' \left[t - \tau - \left(1 + \frac{2V}{c} \right)(k - 1)T \right] \right\}.$$

The factor r in the argument of the complex envelope can usually be neglected. The factor

$$\left(1 + \frac{2V}{c} \right) T$$

in the complex envelope accounts for the displacement of the target by VT during each interpulse interval.

Appendix G

Asymptotic Relative Efficiency of the Rank Test

In order to calculate the asymptotic relative efficiency of the rank test, we must work out the effective signal-to-noise ratio of the rank statistic r as given in (8-35) and (8-36). We use the latter to determine the expected value of r under hypothesis H_1. In that sum there are M terms with $i = j$ and $\frac{1}{2}M(M-1)$ with $i \neq j$, and its expected value is

$$E(r|\,H_1) = ME[U(g_i)|\,H_1] + \tfrac{1}{2}M(M-1)E[U(g_i + g_j)|\,H_1, i \neq j].$$

The first term is simply

$$E[U(g_i)|\,H_1] = \int_0^\infty p_1(g)\,dg = \int_0^\infty p_0(g - A)\,dg = 1 - P_0(-A) = P_0(A),$$

by (8-24), where

$$P_0(g) = \int_{-\infty}^g p_0(g')\,dg'$$

is the cumulative distribution function of the datum g. Furthermore,

$$E[U(g_i + g_j)|\,H_1, i \neq j]$$

$$= \int_{-\infty}^\infty dg_i \int_{-g_i}^\infty p_1(g_i)p_1(g_j)\,dg_j$$

$$= \int_{-\infty}^\infty p_0(g_1 - A)\,dg_1 \int_{-g_1}^\infty p_0(g_2 - A)\,dg_2 = \int_{-\infty}^\infty p_0(g_1 - A)\,dg_1 \int_{-g_1 - A}^\infty p_0(x)\,dx$$

$$= \int_{-\infty}^\infty p_0(y)\,dy \int_{-y - 2A}^\infty p_0(x)\,dx = \int_{-\infty}^\infty p_0(y)\,dy \int_{-\infty}^{y + 2A} p_0(z)\,dz$$

$$= \int_{-\infty}^\infty p_0(y)P_0(y + 2A)\,dy$$

in terms of the probability density function and the cumulative distribution function of the noise. The numerator of the expression for the effective signal-to-noise ratio of the statistic r is thus

$$E(r \mid H_1) - E(r \mid H_0)$$

$$= \tfrac{1}{2} M(M - 1) \int_{-\infty}^{\infty} p_0(y)[P_0(y + 2A) - P_0(y)] \, dy + M[P_0(A) - P_0(0)],$$

and because for $A \ll \sigma$ we can write

$$P_0(y + 2A) - P_0(y) \approx 2Ap_0(y),$$

this becomes

$$E(r \mid H_1) - E(r \mid H_0) \approx M(M - 1)A \int_{-\infty}^{\infty} [p_0(y)]^2 \, dy + O(M). \qquad \text{(G-1)}$$

Here $O(M)$ is the second term, which is of order M and becomes relatively negligible when we pass to the limit $M \gg 1$.

We also need the variance of the rank statistic r when no signal is present. This we obtain from (8-35). Introducing the random variables $w_k = U(g_k)$, for which

$$\Pr(w_k = 1 \mid H_0) = \Pr(w_k = 0 \mid H_0) = \tfrac{1}{2},$$

and which are independent under hypothesis H_0, we can write

$$r = \sum_{k=1}^{M} k w_k,$$

and its variance is

$$\text{Var}_0 \, r = \sum_{k=1}^{M} k^2 \, \text{Var}_0 \, w_k = \tfrac{1}{4} \sum_{k=1}^{M} k^2 = \tfrac{1}{24} M(M + 1)(2M + 1).$$

Then for $M \gg 1$ the effective signal-to-noise ratio of r is, by (G–1) and (4–64),

$$D_r^2 = 12 \, MA^2 \left[\int_{-\infty}^{\infty} [p_0(y)]^2 \, dy \right]^2.$$

For the statistic G in (8-19) the effective signal-to-noise ratio is

$$D_G^2 = \frac{MA^2}{\sigma^2},$$

and dividing we find the asymptotic relative efficiency as stated in (8-41).

Appendix H

Probability of Detection: Lorentz Spectral Density

H.1 The Residue Series

When the stochastic signals to be detected in white noise have the Lorentz spectral density (11-47), the Fredholm determinant can be written

$$D(z) = \frac{(g+1)^2 e^{mg} - (g-1)^2 e^{-mg}}{4g} e^{-m},$$

$$m = \mu T, \qquad g = \sqrt{1 + Kz}, \qquad K = \frac{2E}{Nm}, \tag{H-1}$$

from (11-128). Here $E = P_s T$ is the average energy of the stochastic signal for which the detector is designed to be optimum [Sie57]. When z is real and less than $-K^{-1}$, the parameter g is imaginary, and putting $g = ig'$, we find

$$D(z) = e^{-m} \frac{\mathrm{Im}[(1 + ig')^2 e^{img'}]}{2g'}$$

$$= e^{-m} \left[\cos mg' + \frac{1 - g'^2}{2g'} \sin mg' \right], \qquad g' = \sqrt{-Kz - 1}. \tag{H-2}$$

By equating the bracket to zero, one finds an equation from which the eigenvalues λ_k can be calculated. (In this appendix eigenvalues are indexed from $k = 0$.) With $z = -1/\lambda_k$,

$$2g'_k \cos mg'_k + (1 - g'^2_k) \sin mg'_k = 0,$$

where

$$g'_k = (K\lambda_k^{-1} - 1)^{1/2}.$$

When as in Example 2-4 of Sec. 2.3.2 we substitute $g'_k = \cot \theta_k$, we obtain

$$m \cot \theta_k = 2\theta_k + k\pi, \qquad k = 0, 1, 2, \ldots, \qquad 0 < \theta_k < \tfrac{1}{2}\pi, \qquad \text{(H-3)}$$

and the eigenvalues are

$$\lambda_k = \frac{K}{1 + g'^2_k} = K \sin^2 \theta_k. \qquad \text{(H-4)}$$

The derivation of an expression for the parameter R_k in (11-79) is quite tedious. For the time being we omit subscripts. In (H-2) we replace mg' by c, writing it as

$$D(z) = e^{-m} \left[\cos c + \frac{1}{2}\left(\frac{m}{c} - \frac{c}{m} \right) \sin c \right],$$

and we observe that at $z = -1/\lambda$

$$\cot c = \frac{c^2 - m^2}{2mc}. \qquad \text{(H-5)}$$

Furthermore, because

$$c = mg' = m(-Kz - 1)^{1/2}$$

and

$$\frac{dc}{dz} = -\frac{1}{2}mK(-Kz - 1)^{-1/2} = -\frac{m^2 K}{2c},$$

we find that

$$\frac{dD}{dz} = D'(z) = -\frac{m^2 K}{2c} \left[-\sin c + \frac{1}{2}\left(\frac{m}{c} - \frac{c}{m} \right) \cos c - \frac{1}{2}\left(\frac{m}{c^2} + \frac{1}{m} \right) \sin c \right] e^{-m}$$

$$= \frac{m^2 K}{2c} \left[\left(1 + \frac{m^2 + c^2}{2mc^2} \right) \sin c + \left(\frac{c^2 - m^2}{2mc} \right) \cos c \right] e^{-m}.$$

Using (H-5) we write this as

$$D'(z) = \frac{m^2 K}{2c} \left[\left(1 + \frac{m^2 + c^2}{2mc^2} \right) \sin c + \frac{\cos^2 c}{\sin c} \right] e^{-m}$$

$$= \frac{m^2 K e^{-m}}{2c \sin c} \left[1 + \left(\frac{m^2 + c^2}{2mc^2} \right) \sin^2 c \right] \qquad \text{(H-6)}$$

$$= \frac{m^2 K e^{-m}}{2c \csc c} \left[\csc^2 c + \frac{m^2 + c^2}{2mc^2} \right].$$

Now with $g' = c/m$ we find from (H-4) that

$$\lambda = \frac{m^2 K}{m^2 + c^2} \qquad \text{(H-7)}$$

or

$$c^2 + m^2 = \frac{m^2 K}{\lambda}, \qquad c^2 = m^2 \frac{K - \lambda}{\lambda}, \qquad \text{(H-8)}$$

and

$$\csc^2 c = \cot^2 c + 1 = \left(\frac{c^2 - m^2}{2mc} \right)^2 + 1 = \left(\frac{c^2 + m^2}{2mc} \right)^2,$$

from (H-5), so that by (H-8)

$$c \csc c = \pm \frac{mK}{2\lambda}.$$ (H-9)

Putting this into (H-6), we obtain

$$D'\left(-\frac{1}{\lambda}\right) = \pm m\lambda \frac{mK}{2\lambda}\left(\frac{mK}{2\lambda} + 1\right)\frac{\lambda}{m^2(K-\lambda)}e^{-m}$$

$$= \pm \frac{K(mK + 2\lambda)}{4(K-\lambda)}e^{-m},$$

and from (11-79), at last, with $k = 0, 1, 2, \ldots$,

$$R_k = (-1)^k \frac{4\lambda_k (K-\lambda_k)}{K(mK + 2\lambda_k)}e^m.$$

The factor $(-1)^k$ arises because by (H-3) $c = mg_k^l$ lies between $k\pi$ and $(k+1)\pi$ and $\csc c$ in (H-9) is therefore positive for k even and negative for k odd. This formula was derived by S. O. Rice. If we use (H-4), we can write it

$$R_k = (-1)^k \frac{e^m \sin^2 2\theta_k}{m + 2\sin^2 \theta_k}$$ (H-10)

as in [Hel83].

From (H-7)

$$\lambda_k = \frac{m^2 K}{m^2 + c_k^2} = \frac{1}{N}\Phi_s(\omega_k), \qquad \frac{k\pi}{T} \leq \omega_k = \frac{c_k}{T} \leq \frac{(k+1)\pi}{T}.$$

Because the Lorentz spectral density drops off to zero with increasing frequency ω very slowly, the eigenvalues λ_k decrease slowly to zero, and it is necessary to include a great many terms in (11-80) in order to compute the detection probability $Q_d(\eta)$ accurately. The greater the time–bandwidth product $m = \mu T$ and the greater the input energy-to-noise ratio $\eta E/N$, the more terms must be taken into the summation. The coefficients R_k, given by (H-10), alternate in sign, and when $m \gg 1$ their absolute values are large and lie close together for small indices k. The corresponding terms of (11-80) are then large and not much different in magnitude and must each be computed to high precision if their sum is to be reliable. For these reasons we found it necessary to turn to the method of saddlepoint integration for values of $m = \mu T$ greater than about 7.

H.2 The Saddlepoint Method

As described in Sec. 5.2, it is first necessary to find the saddlepoint of the integrand of (11-94) or (11-95). Because the Fredholm determinant as given by (H-1) is rather complicated, the most expeditious method of computing the saddlepoint seems to be by numerically minimizing the phase

$$\Phi(z) = \ln h_\eta(z) + U_0 z - \ln |z|$$ (H-11)

for real values of z. For (11-94) the saddlepoint lies to the right of the origin; for (11-95) it lies between the origin and the rightmost pole of $h_\eta(z)$, which by (11-75) is located at

$$z = -\frac{1 + \lambda_0}{\lambda_0(1 + \eta\lambda_0)},$$

where λ_0 is the largest eigenvalue of (11-74). This eigenvalue can be rather quickly calculated by first finding the root θ_0 of (H-3) lying between 0 and $\frac{1}{2}\pi$ and then determining λ_0 from (H-4). It is wise to do so in order to determine the region in which to search for the left-hand saddlepoint of the integrand of (11-95). This saddlepoint need not be located with high precision, and even a crude minimization of $\Phi(z)$ in (H-11) will usually suffice.

In accordance with the Neyman–Pearson criterion, the decision level U_0 is chosen to yield a preassigned false-alarm probability

$$Q_0 = -\int_{C_+} z^{-1} h_0(z) \exp(U_0 z) \frac{dz}{2\pi i}, \qquad h_0(z) = \frac{D(1)}{D(1 + z)}. \tag{H-12}$$

The quickest way to compute the decision level U_0 seems to be first to approximate (H-12) by the saddlepoint approximation (5-56), neglecting the correction term T_1:

$$Q_0 \approx [2\pi\Phi''(z_0)]^{-1/2} \exp\Phi(z_0),$$
$$\Phi(z) = \ln D(1) - \ln D(1 + z) + U_0 z - \ln(-z). \tag{H-13}$$

At the saddlepoint $z = z_0$,

$$\Phi'(z) = -\frac{D'(1 + z)}{D(1 + z)} + U_0 - \frac{1}{z} = 0. \tag{H-14}$$

Solving this for U_0 and substituting it into (H-13), we obtain

$$\ln Q_0 \approx \ln D(1) - \ln D(1 + z)$$
$$+ \frac{zD'(1 + z)}{D(1 + z)} + 1 - \ln(-z) - \frac{1}{2}\ln[2\pi\Phi''(z)] \tag{H-15}$$

with

$$\Phi''(z) = \left[\frac{D'(1 + z)}{D(1 + z)}\right]^2 - \frac{D''(1 + z)}{D(1 + z)} + \frac{1}{z^2}.$$

The derivatives are evaluated numerically. One uses the secant method to find the value of $z = z_0$ satisfying (H-15), with Q_0 the preassigned false-alarm probability, and then one substitutes into (H-14) and solves for the value of the decision level U_0. One then turns to numerical integration of (11-95) with $\eta = 0$, again applying the secant method, but varying the level U_0 until the preassigned false-alarm probability Q_0 is reached. At this stage it is unnecessary to alter the intersection z_0 of the path of integration with the real z-axis. We used a parabolic path of integration, determining its curvature as described in Sec. 5.2.2.

H.3 The Toeplitz Approximation

For time–bandwidth products $m = \mu T$ greater than about 12, it sufficed for the sake of plotting the curves in Fig. 11-2 to use the Toeplitz approximation described in Secs. 11.2.5 and 11.2.6. Substituting (11-47) into (11-105), we find

$$1 + zN^{-1}\Phi_s(\omega) = \frac{\beta^2 + \omega^2}{\mu^2 + \omega^2}$$

with

$$\beta T = \mu T \left(1 + \frac{2P_s z}{\mu N}\right)^{1/2} = m\sqrt{1 + Kz},$$

and by (11-107) the Fredholm determinant is approximately

$$D(z) \approx \exp m[\sqrt{1 + Kz} - 1], \tag{H-16}$$

which corresponds to the limiting form of (H-1) when $m \gg 1$ and we approximate g by 1 everywhere except in $\exp mg$. Now the moment-generating function of the optimum detection statistic under hypothesis H_1 is, by (11-73) or (11-96), with $\eta = 1$,

$$h_1(z) = \mathcal{L}[p_1(U)] = \frac{1}{D(z)}$$
$$= \exp m[1 - \sqrt{1 + Kz}].$$

Here $\mathcal{L}$ stands for the Laplace transform.

The random variable $V = U/K$ then has the probability density function

$$\overline{P}_1(V) \approx \mathcal{L}^{-1}\left[\exp m(1 - \sqrt{1 + z})\right].$$

Recognizing that for any function $F(z)$,

$$\mathcal{L}^{-1}F(1 + z) = e^{-V}\mathcal{L}^{-1}F(z),$$

and using [Erd54, eq. 5.6(1), p. 245], we find for the probability density function of V under hypothesis H_1

$$\overline{P}_1(V) \approx \frac{1}{2}\pi^{-1/2}mV^{-3/2}\exp\left(m - V - \frac{m^2}{4V}\right),$$

and the random variable

$$x = \frac{U}{S} = \frac{2V}{m}, \qquad S = \frac{E}{N},$$

has the probability density function

$$P_1(x) \approx \sqrt{\frac{m}{2\pi}}\, x^{-3/2}\exp\left[-\frac{m}{2x}(x - 1)^2\right] \tag{H-17}$$

in this Toeplitz approximation.

From (11-31) through (11-34) we find the likelihood ratio

$$\Lambda(U) = \frac{p_1(U)}{p_0(U)} = \frac{P_1(x)}{P_0(x)} = \frac{e^U}{D(1)},$$

so that

$$P_0(x) = D(1)\,e^{-Sx}P_1(x).$$

Now $D(1) = \exp m[(1 + K)^{1/2} - 1]$ in this approximation, by (H-16), and substituting (H-17) we find after a little algebra that under hypothesis H_0 the probability density function of $x = U/S$ is

$$P_0(x) \approx \sqrt{\frac{m}{2\pi}}\, x^{-3/2}\exp\left[-\frac{m}{2x}(x\sqrt{1 + K} - 1)^2\right]. \tag{H-18}$$

Both (H-17) and (H-18) can be integrated in closed form to obtain the false-alarm and detection probabilities

$$Q_0 \approx \text{erfc}\left[\left(\frac{M}{x_0'}\right)^{1/2} (x_0' - 1)\right] - e^{2M} \text{erfc}\left[\left(\frac{M}{x_0'}\right)^{1/2} (x_0' + 1)\right], \qquad \text{(H-19)}$$

$$M = m\sqrt{1 + K}, \qquad x_0' = x_0\sqrt{1 + K}, \qquad K = \frac{2E}{Nm},$$

$$Q_{ds} = \text{erfc}\left[\left(\frac{m}{x_0}\right)^{1/2} (x_0 - 1)\right] - e^{2m} \text{erfc}\left[\left(\frac{m}{x_0}\right)^{1/2} (x_0 + 1)\right], \qquad \text{(H-20)}$$

in which $x_0 = U_0/S$.

Given the false-alarm probability Q_0, one calculates x_0' from (H-19) by Newton's method and then substitutes $x_0 = x_0'(1 + K)^{-1/2}$ into (H-20) to determine the probability $Q_{ds} = Q_d(1)$ of detecting the standard signal.

Appendix I

Derivation of the Kalman–Bucy Equations

In order to derive (11-175) we write (11-174) as

$$z\,\boldsymbol{\phi}_x(\tau, t)\mathbf{C}^+(t) = \int_0^\tau \mathbf{k}_z(\tau, u)\boldsymbol{\phi}_z(u, t)\,du, \tag{I-1}$$

where by (11-166)

$$
\begin{aligned}
\boldsymbol{\phi}_z(t, u) &= N\delta(t - u) + z\,\boldsymbol{\phi}_s(t, u) \\
&= N\delta(t - u) + z\,\mathbf{C}\boldsymbol{\phi}_x(t, u)\mathbf{C}^+.
\end{aligned} \tag{I-2}
$$

Differentiating (11-165) and using (11-160), we obtain

$$\frac{\partial \boldsymbol{\phi}_x(t, u)}{\partial t} = \mathbf{F}(t)\boldsymbol{\phi}_x(t, u), \qquad u < t. \tag{I-3}$$

Differentiating (I-1) with respect to τ and using (I-3), we find

$$
\begin{aligned}
z\frac{\partial \boldsymbol{\phi}_x(\tau, t)}{\partial \tau}\mathbf{C}^+(t) &= z\,\mathbf{F}(\tau)\boldsymbol{\phi}_x(\tau, t)\mathbf{C}^+(t) \\
&= \int_0^\tau \frac{\partial \mathbf{k}_z(\tau, u)}{\partial \tau}\boldsymbol{\phi}_z(u, t)\,du + \mathbf{k}_z(\tau, \tau)\boldsymbol{\phi}_z(\tau, t) \\
&= \int_0^\tau \frac{\partial \mathbf{k}_z(\tau, u)}{\partial \tau}\boldsymbol{\phi}_z(u, t)\,du + z\,\mathbf{k}_z(\tau, \tau)\mathbf{C}\boldsymbol{\phi}_x(\tau, t)\mathbf{C}^+, \qquad \tau > t,
\end{aligned}
$$

by (I-2). Hence, using (I-1) again,

$$
\begin{aligned}
\int_0^\tau \frac{\partial \mathbf{k}_z(\tau, u)}{\partial \tau}\boldsymbol{\phi}_z(u, t)\,du &= z\,[\mathbf{F}(\tau) - \mathbf{k}_z(\tau, \tau)\mathbf{C}]\,\boldsymbol{\phi}_x(\tau, t)\mathbf{C}^+ \\
&= \int_0^\tau [\mathbf{F}(\tau) - \mathbf{k}_z(\tau, \tau)\mathbf{C}]\,\mathbf{k}_z(\tau, u)\boldsymbol{\phi}_z(u, t)\,du,
\end{aligned}
$$

or

$$\int_0^\tau \left[\frac{\partial \mathbf{k}_z(\tau, u)}{\partial \tau} - \mathbf{F}(\tau)\mathbf{k}_z(\tau, u) + \mathbf{k}_z(\tau, \tau)\mathbf{C}\mathbf{k}_z(\tau, u) \right] \phi_z(u, t) \, du = 0. \qquad \text{(I-4)}$$

This equation implies that the contents of the brackets must vanish for all u, that is,

$$\frac{\partial \mathbf{k}_z(\tau, u)}{\partial \tau} = \mathbf{F}(\tau)\mathbf{k}_z(\tau, u) - \mathbf{k}_z(\tau, \tau)\mathbf{C}\mathbf{k}_z(\tau, u), \qquad \tau > u, \qquad \text{(I-5)}$$

which corresponds to (11-175).

In order to justify this step, replace any component of the vector in the brackets in (I-4) by $f^*(u)$ and use (I-2), whereupon (I-4) becomes

$$\int_0^\tau f^*(u)[N\delta(u - t) + z\phi_s(u, t)] \, du = 0,$$

or

$$z \int_0^\tau f^*(u)\phi_s(u, t) \, du = -Nf^*(t).$$

Multiply by $f(t) \, dt$ and integrate. Then

$$z \int_0^\tau \int_0^\tau f^*(u)\phi_s(u, t)f(t) \, du \, dt = -N \int_0^\tau |f(t)|^2 \, dt.$$

Because $\phi_s(u, t)$ is a positive-definite kernel, this equation can hold only if z is real and negative or if $f^*(u) \equiv 0$. We exclude the former possibility. Equation (I-5) will then be valid everywhere in the complex z-plane except possibly for points on the negative Re z-axis. By analytical continuation, however, our final result will hold there as well, except at the points $z = -1/\lambda_k$, where the functions involved are singular, the λ_k being the eigenvalues of (11-74).

Now we seek the differential equation satisfied by $\mathbf{P}_z(t)$ as defined by (11-178). Recall that by (11-161) and (11-165) $\phi_x(t, t) = \mathbf{L}(t)$, which obeys the differential equation (11-163). Differentiating (11-178) with respect to time τ and using (I-3), (I-5), and (11-163), we obtain

$$\frac{d\mathbf{P}_z}{d\tau} = z\frac{d\mathbf{L}}{d\tau} - z\mathbf{k}_z(\tau, \tau)\mathbf{C}\mathbf{L}(\tau) - z\int_0^\tau \frac{\partial \mathbf{k}_z(\tau, u)}{\partial \tau}\mathbf{C}\phi_x(u, \tau) \, du$$

$$- z\int_0^\tau \mathbf{k}_z(\tau, u)\mathbf{C}\frac{\partial \phi_x(u, \tau)}{\partial \tau} \, du$$

$$= z(\mathbf{FL} + \mathbf{LF}^+ + Q\mathbf{GG}^+) - z\mathbf{k}_z(\tau, \tau)\mathbf{C}\mathbf{L}(\tau) - z\int_0^\tau \left[\mathbf{Fk}_z(\tau, u)\right.$$

$$\left. - \mathbf{k}_z(\tau, \tau)\mathbf{C}\mathbf{k}_z(\tau, u)\right]\mathbf{C}\phi_x(u, \tau) \, du - z\int_0^\tau \mathbf{k}_z(\tau, u)\mathbf{C}\phi_x(u, \tau)\mathbf{F}^+(\tau) \, du.$$

Substituting from (11-178) and (11-177), we obtain

$$\frac{d\mathbf{P}_z}{d\tau} = \mathbf{FP}_z + \mathbf{P}_z\mathbf{F}^+ + zQ\mathbf{GG}^+ - z\mathbf{k}_z(\tau, \tau)\mathbf{C}\left[\mathbf{L}(\tau) - \int_0^\tau \mathbf{k}_z(\tau, u)\mathbf{C}\phi_x(u, \tau) \, du\right]$$

$$= \mathbf{FP}_z + \mathbf{P}_z\mathbf{F}^+ + zQ\mathbf{GG}^+ - N^{-1}\mathbf{P}_z\mathbf{C}^+\mathbf{CP}_z,$$

which is the Riccati equation (11-179).

Appendix J

Moment-generating Function: Coherent Plus Stochastic Signal

When the stochastic signal in Sec. 11.5 is stationary, the equation (11-199) becomes

$$R(t) = Q(t; z) + \frac{z}{N} \int_0^T \phi_s(t - u)Q(u; z) \, du, \qquad 0 \le t \le T, \qquad \text{(J-1)}$$

and it can be solved by the techniques of Appendix A. We identify $s(t)$ of (A-1) with $R(t)$ and $q(u)$ with $Q(u; z)$, and we take as the kernel of that integral equation

$$\phi(t - u) = \delta(t - u) + \frac{z}{N}\phi_s(t - u).$$

The Fourier transform of this kernel is

$$\Phi(\omega) = 1 + \frac{z}{N}\Phi_s(\omega),$$

and it can be expressed as in (A-2) with $m = n$. The β_j's are the roots of

$$1 + \frac{z}{N}\phi_s(-ip) = 0, \qquad p = \beta_1, \beta_2, \dots, \beta_{2n},$$

as in (11-106). The $2n$ linear simultaneous equations (A-14) now become

$$Gc = \begin{bmatrix} R \\ -L \end{bmatrix},$$

where by (A-15) the $2n \times 2n$ matrix $G = G(T)$ is the same as in (11-143). The n elements of the column vectors R and L are given as in (A-10) by

$$L_k = \int_{-\infty}^0 e^{\mu_k^* u} Q_\infty(u; z) \, du, \qquad R_k = \int_T^\infty e^{\mu_k(T-u)} Q_\infty(u; z) \, du \qquad \text{(J-2)}$$

in terms of the solution of (A-8), which we write as

$$R(t) = Q_\infty(t; z) + \frac{z}{N} \int_{-\infty}^{\infty} \phi_s(t - u) Q_\infty(u; z) \, du, \qquad 0 \le t \le T. \qquad \text{(J-3)}$$

When as in (11-200) the signal envelope $R(t)$ is a sum of sine waves, the solution $Q_\infty(t; z)$ is

$$Q_\infty(t; z) = \sum_m \frac{r_m}{H(w_m; z)} \exp i w_m t,$$

$$H(w; z) = 1 + \frac{z}{N} \Phi_s(w), \qquad \text{(J-4)}$$

as can be seen by substitution. Then we find from (J-2)

$$R_k = \sum_m \frac{r_m}{H(w_m; z)} \int_T^{\infty} \exp[\mu_k(T - u) + i w_m u] \, du$$

$$= \sum_m \frac{r_m}{H(w_m; z)(\mu_k - i w_m)} \exp i w_m T, \qquad \text{(J-5)}$$

$$L_k = \sum_m \frac{r_m}{H(w_m; z)} \int_{-\infty}^0 \exp(\mu_k^* u + i w_m u) \, du$$

$$= -\sum_m \frac{r_m}{H(w_m; z)(\mu_{k+n} - i w_m)}, \qquad \text{(J-6)}$$

when we put $\mu_{k+n} = -\mu_k^*$, $1 \le k \le n$, as before.
   With

$$Q(t; z) = Q_\infty(t; z) + \sum_{j=1}^{2n} c_j \exp \beta_j t$$

as in (A-5) and (A-6), we can write $J(z)$ in (11-198) as

$$J(z) = J_1 + J_2,$$

with

$$J_1 = \frac{1}{2N} \int_0^T R^*(t) Q_\infty(t; z) \, dt, \qquad \text{(J-7)}$$

$$J_2 = \frac{1}{2N} \sum_{j=1}^{2n} c_j \int_0^T R^*(t) \exp \beta_j t \, dt. \qquad \text{(J-8)}$$

From (11-200) and (J-4) we find

$$J_1 = \frac{1}{2N} \sum_n \sum_m \frac{r_n^* r_m}{H(w_m; z)} \int_0^T \exp i(w_m - w_n) t \, dt,$$

which yields (11-201) immediately.

Now with (11-200), (J-8) becomes

$$J_2 = \frac{1}{2N} \sum_m \sum_{j=1}^{2n} r_m^* c_j \int_0^T \exp(\beta_j - iw_m)t \, dt$$

$$= \frac{1}{2N} \sum_m \sum_{j=1}^{2n} r_m^* c_j \left[\frac{\exp(\beta_j - iw_m)T - 1}{\beta_j - iw_m} \right]$$

$$= \frac{1}{2N} \sum_{j=1}^{2n} E_j c_j,$$

with the E_j as in (11-203). Now defining

$$F_k = R_k, \qquad 1 \leq k \leq n, \qquad F_k = -L_{k-n}, \qquad n + 1 \leq k \leq 2n,$$

as in (11-204), we must by (A-14) solve the $2n$ equations

$$\mathbf{Gc} = \mathbf{F},$$

or $\mathbf{c} = \mathbf{G}^{-1}\mathbf{F}$, and we can write

$$J_2 = \frac{1}{2N} \mathbf{E}\mathbf{G}^{-1}\mathbf{F},$$

taking $\mathbf{E}$ as a row vector of the E_j's and $\mathbf{F}$ as a column vector of the F_k's. By the rule for the determinant of a block matrix,

$$\det \begin{bmatrix} \mathbf{A} & \mathbf{B} \\ \mathbf{C} & \mathbf{D} \end{bmatrix} = \det \mathbf{D} \det(\mathbf{A} - \mathbf{B}\mathbf{D}^{-1}\mathbf{C}),$$

with $\mathbf{A} = 0$, $\mathbf{B} = \mathbf{E}$, $\mathbf{D} = \mathbf{G}$, $\mathbf{C} = \mathbf{F}$, and using the fact that $\mathbf{E}\mathbf{G}^{-1}\mathbf{F}$ is a matrix with a single element and hence equal to its determinant, we obtain (11-202).

Bibliography

[Abr70] Abramowitz, M., and I. A. Stegun, *Handbook of Mathematical Functions with Formulas, Graphs, and Mathematical Tables.* Washington, D.C.: U.S. Government Printing Office, 1970.

[Ann53] Annis, M., W. Cheston, and H. Primakoff, "On statistical estimation in physics," *Reviews of Modern Physics*, 25 (Oct. 1953), 818–30.

[Are57] Arens, R., "Complex processes for envelopes of normal noise," *IRE Transactions on Information Theory*, IT-3 (Sept. 1957), 204–7.

[Bag69] Baggeroer, A. B., "A state-variable approach to the solution of Fredholm integral equations," *IEEE Transactions on Information Theory*, IT-15 (Sept. 1969), 557–70.

[Bag71] Baggeroer, A. B., "State variable analysis procedures," Appendix A in H. L. Van Trees, *Detection, Estimation, and Modulation Theory*, vol. II. (New York: John Wiley & Sons, Inc., 1971), 286–327.

[Bak77] Baker, C. T. H., *The Numerical Treatment of Integral Equations.* New York: Oxford University Press, 1977.

[Bal57] Balakrishnan, A. V., "A note on the sampling principle for continuous signals," *IRE Transactions on Information Theory*, IT-3 (June 1957), 143–6.

[Bal76] Balaban, P., P. E. Fleisher, and H. Zucker, "The probability distribution of gains in avalanche photodiodes," *IEEE Transactions on Electron Devices*, ED-23 (Oct. 1976), 1189–90.

[Bar46] Bartlett, M. S., "The large-sample theory of sequential tests," *Proceedings of the Cambridge Philosophical Society*, 42 (1946), 239–44.

[Bar53] Barker, R. H., "Group synchronizing of binary digital systems." in *Communication Theory*, ed. W. Jackson, (New York: Academic Press, Inc., 1953), 273–87.

558

[Bar69] Bar-David, I., "Communication under the Poisson regime," *IEEE Transactions on Information Theory*, IT-15 (Jan. 1969), 31–7.

[Bas59] Basharian, A. E., and B. S. Fleishman, "The application of sequential analysis to binary communication systems with a Rayleigh distribution of signal intensity fluctuations." *Radio Engineering & Electronic Physics*, 4, 2 (1959), 1–9.

[Bay63] Bayes, T., "An essay toward solving a problem in the doctrine of chances," *Philosophical Transactions of the Royal Society*, 53 (1763), 370–418. Reprinted in *Biometrika*, 45 (1958), 296–315.

[Béd66] Bédard, G., "Photon counting statistics of Gaussian light," *Physical Review*, 151 (Nov. 25, 1966), 1038–9.

[Béd67] Bédard, G., J. C. Chang, and L. Mandel, "Approximate formulas for photoelectric counting distributions," *Physical Review*, 160 (Aug. 25, 1967), 1496–1500.

[Bel60] Bellman, R., *Introduction to Matrix Analysis*. New York: McGraw-Hill Book Co., 1960.

[Bel63] Bello, P. A., "Characterization of randomly time-variant linear channels," *IEEE Transactions on Communication Systems*, CS-11 (Dec. 1963), 360–93.

[Bic77] Bickel, P. J., and K. A. Doksum, *Mathematical Statistics: Basic Ideas and Selected Topics*. San Francisco: Holden-Day, Inc., 1977.

[Bla54] Blackwell, D., and M. A. Girshick, *Theory of Games and Statistical Decisions*. New York: John Wiley & Sons, Inc., 1954.

[Bla57a] Blasbalg, H., "The relationship of sequential filter theory to information theory and its application to the detection of signals in noise by Bernoulli trials," *IRE Transactions on Information Theory*, IT-3 (June 1957), 122–31.

[Bla57b] ———, "The sequential detection of a sine-wave carrier of arbitrary duty ratio in Gaussian noise." *IRE Transactions on Information Theory*, IT-3 (Dec. 1957), 248–56.

[Bla61] ———, "On the approximation to likelihood ratio detectors [sic] laws (The threshold case)," *IRE Transactions on Information Theory*, IT-7 (July 1961), 194–5.

[Bla85] Blahut, R. E., *Fast Algorithms for Digital Signal Processing*. Reading, MA: Addison-Wesley Publishing Co., 1985.

[Bla91] Blahut, R. E., W. Miller, Jr., and C. H. Wilcox, *Radar and Sonar, Part I*. New York: Springer-Verlag, 1991.

[Blo57] Bloom, F. J., S. S. S. L. Chang, B. Harris, A. Hauptschien, and K. C. Morgan, "Improvement of binary transmission by null-zone reception," *Proceedings of the IRE*, 45 (July 1957), 963–75.

[Blu92] Blum, R. S., and S. A. Kassam, "Optimum distributed detection of weak signals in dependent sensors." *IEEE Transactions on Information Theory*, IT-38 (May 1992), 1066–79.

[Bod50] Bode, H. W., and C. E. Shannon, "A simplified derivation of linear least-square smoothing and prediction theory," *Proceedings of the IRE*, 38 (Apr. 1950), 417–25.

[Bre65] Brennan, L. E., and I. S. Reed, "A recursive method of computing the Q-function," *IEEE Transactions on Information Theory*, IT-11 (Apr. 1965), 312–3.

[Bun49] Bunimovich, V. I., "The fluctuation process as a vibration with random amplitude and phase," *Journal of Technical Physics, USSR*, 19 (Nov. 1949), 1231–59.

[Bus55] Bussgang, J. J., and D. Middleton, "Optimum sequential detection of signals in noise," *IRE Transactions on Information Theory*, IT-1 (Dec. 1955), 5–18.

[Bus60] ———, and W. L. Mudgett, "A note of caution on the square-law approximation to an optimum detector," *IRE Transactions on Information Theory*, IT-6 (Sept. 1960), 504–5.

[Bus65] ——, and L. Ehrman, "A sequential test for a target in one of k range positions," *Proceedings of the IEEE*, 53 (May 1965), 495–6.

[Cap59] Capon, J., "A nonparametric technique for the detection of a constant signal in additive noise," *IRE Wescon Convention Record*, pt. 4 (1959), 92–103.

[Cap61] ——, "On the asymptotic efficiency of locally optimum detectors," *IRE Transactions on Information Theory*, IT-7 (Apr. 1961), 67–71.

[Car64] Carlyle, J. W., and J. B. Thomas, "On nonparametric signal detectors," *IEEE Transactions on Information Theory*, IT-10 (Apr. 1964), 146–52.

[Car66] Carrier, G. F., M. Krook, and C. E. Pearson, *Functions of a Complex Variable*. New York: McGraw-Hill Book Co., 1966.

[Car68] Carlyle, J. W., "Nonparametric methods in detection theory," in *Communication Theory*, ed. A. V. Balakrishnan (New York: McGraw-Hill Book Co., 1968), 293–319.

[Che59] Chernoff, H., and L. E. Moses, *Elementary Decision Theory*. New York: John Wiley & Sons, Inc., 1959.

[Che62] ——, "A measure of asymptotic efficiency for tests of a hypothesis based on the sum of observations," *Annals of Mathematical Statistics*, 23 (1952), 493–507.

[Chu72] Chu, D. C., "Polyphase codes with good periodic correlation properties," *IEEE Transactions on Information Theory*, IT-18 (July 1972), 531–2.

[Cim83] Cimini, L. J., Jr., and S. A. Kassam, "Data quantization for narrowband signal detection," *IEEE Transactions on Aerospace & Electronic Systems*, AES-19 (Nov. 1983), 848–58.

[Col68] Collins, L. D., "Closed-form expressions for the Fredholm determinant for state-variable covariance functions," *Proceedings of the IEEE*, 56 (Mar. 1968), 350–1.

[Col71] Collatz, L., and W. Wetterling, *Optimierungsaufgaben*. Berlin: Springer-Verlag, 1971.

[Con72] Conradi, J., "The distribution of gains in uniformly multiplying avalanche photodiodes: experimental," *IEEE Transactions on Electron Devices*, ED-19 (June 1972), 713–8.

[Cou31] Courant, R., and D. Hilbert, *Methoden der mathematischen Physik*. Berlin: J. Springer Verlag, 1931. English translation, *Methods of Mathematical Physics*. New York: Interscience Publishers, 1953.

[Cox73] Cox, H., "Resolving power and sensitivity to mismatch of optimum array processors," *Journal of the Acoustical Society of America*, 54 (Sept. 1973), 771–85.

[Cra46] Cramér, H. *Mathematical Methods of Statistics*. Princeton, NJ: Princeton University Press, 1946.

[Cra66] ——, "On the intersections between the trajectories of a normal stationary stochastic process and a high level," *Arkiv för Matematik*, 6, 20 (1966), 337–49.

[Cur91] Curlander, J. C., and R. N. McDonough, *Synthetic Aperture Radar: Systems and Signal Processing*. New York: John Wiley & Sons, Inc., 1991.

[Dan54] Daniels, H. G., "Saddle point approximations in statistics," *Annals of Mathematical Statistics*, 25 (1954), 631–50.

[Dav54] Davis, R. C., "The detectability of random signals in the presence of noise," *IRE Transactions on Information Theory*, PGIT-3 (Mar. 1954), 52–62.

[Dav58] Davenport, W. B., Jr., and W. L. Root, *An Introduction to the Theory of Random Signals and Noise*. New York: McGraw-Hill Book Co., 1958.

[Deb09] Debye, P., "Näherungsformeln für die Zylinderfunktionen für grosse Werte des Arguments und unbeschränkt veränderliche Werte des Index." *Mathematische Annalen*, 67 (1909), 535–58.

[DiF68] DiFranco, J. V., and W. L. Rubin, *Radar Detection*. Englewood Cliffs, NJ: Prentice Hall, 1968.

[Dil73] Dillard, G. M., "Recursive computation of the generalized Q-function," *IEEE Transactions on Aerospace & Electronic Systems*, AES-9 (July 1973), 614–5.

[Dil89] Dillard, R. A., and G. M. Dillard, *Detectability of Spread-spectrum Signals*. Norwood, MA: Artech House, Inc., 1989.

[Dug58] Dugundji, J., "Envelopes and pre-envelopes of real waveforms," *IRE Transactions on Information Theory*, IT-4 (Mar. 1958), 53–7.

[Dur60] Durbin, J., "The fitting of time-series models," *Review of the International Statistical Institute*, 23 (1960), 233–44.

[Dwi61] Dwight, H. B., *Tables of Integrals and Other Mathematical Data*. New York: Macmillan, Inc., 1961.

[Dwo50] Dwork, B. M., "The detection of a pulse superimposed on fluctuation noise," *Proceedings of the IRE*, 38 (July 1950), 771–4.

[Eck51] Eckart, C., "The theory of noise suppression by linear filters," La Jolla, CA: Scripps Institution of Oceanography, SIO 51–44 (Oct. 8, 1951).

[Edw73] Edwards, C. H., *Advanced Calculus of Several Variables*. New York: Academic Press, Inc., 1973.

[Ekr63] Ekre, H., "Polarity coincidence correlation of a weak noise source," *IEEE Transactions on Information Theory*, IT-9 (Jan. 1963), 18–23.

[Erd53] Erdélyi, A., et al., *Higher Transcendental Functions*, 3 vols. New York: McGraw-Hill Book Co., 1953.

[Erd54] Erdélyi, A., ed., *Tables of Integral Transforms*, 2 vols. New York: McGraw-Hill Book Co., 1954.

[Fin68] Finn, H. M., and R. S. Johnson, "Adaptive detection mode with threshold control as a function of spatially sampled clutter-level estimates," *RCA Review*, 29 (Sept. 1968), 414–64.

[Fis22] Fisher, R. A., "On the mathematical foundations of theoretical statistics," *Philosophical Transactions of the Royal Society*, A222 (1922), 309–68. Reprinted as #10 in R. A. Fisher, *Contributions to Mathematical Statistics*. New York: John Wiley & Sons, Inc., 1950.

[Fix55] Fix, E., and J. L. Hodges, Jr., "Significance probabilities of the Wilcoxon test," *Annals of Mathematical Statistics*, 26 (June 1955), 301–12.

[Fle57] Fleishman, B. S., "On the optimal detector with a log I_0 characteristic for the detection of a weak signal in the presence of noise," *Radio Engineering and Electronic Physics*, 2, 6 (1957), 74–85.

[Fow63] Fowle, E. N., D. R. Carey, R. E. Vander Schuur, and R. C. Yost, "A pulse-compression system employing a linear FM Gaussian signal," *Proceedings of the IEEE*, 51 (Feb. 1963), 304–12.

[Fra57] Fraser, D. A. S., *Nonparametric Methods in Statistics*, New York: John Wiley & Sons, Inc., 1957.

[Fra62] Frank, R. L., S. A. Zadoff, and R. C. Heimuller, "Phase shift pulse codes with good periodic correlation properties." *IRE Transactions on Information Theory*, IT-8 (Oct. 1962), 381–2.

[Fra63] Frank, R. L., "Polyphase codes with good nonperiodic correlation properties," *IEEE Transactions on Information Theory*, IT-9 (Jan. 1963), 43–5.

[Fra80] ———, "Polyphase complementary codes," *IEEE Transactions on Information Theory*, IT-26 (Nov. 1980), 641–7.

[Fry65] Fry, Thornton C., *Probability and Its Engineering Uses*, 2nd ed. Princeton, NJ: D. Van Nostrand Company, 1965.

[Gab46] Gabor, D., "Theory of Communication," *Journal of the IEE*, 93 (1946), 429–57.

[Gag76] Gagliardi, R. M., and S. Karp, *Optical Communications*. New York: John Wiley & Sons, Inc., 1976.

[Geo52] George, T. S., "Fluctuations of ground clutter return in airborne radar equipment," *Proceedings of the IEE*, 99 (IV) (Apr. 1952), 92–9.

[Ger91] Gerlach, K., and F. E. Kretschmer, "Reciprocal radar waveforms," *IEEE Transactions on Aerospace & Electronic Systems*, AES-27 (July 1991), 646–54.

[Gho70] Ghosh, B. K., *Sequential Tests of Statistical Hypotheses*. Reading, MA: Addison-Wesley Publishing Co., 1970.

[Gib75] Gibson, J. D., and J. L. Melsa, *Introduction to Nonparametric Detection with Applications*. New York: Academic Press, Inc., 1975.

[Gil60] Gilbert, E. N., and H. O. Pollak, "Amplitude distribution of shot noise," *Bell System Technical Journal*, 39 (Mar. 1960), 333–50.

[Gol65] Golomb, S. W., and R. A. Scholtz, "Generalized Barker sequences," *IEEE Transactions on Information Theory*, IT-11 (Oct. 1965), 533–7.

[Gow84] Gowar, J., *Optical Communication Systems*. Englewood Cliffs, NJ: Prentice Hall, 1984.

[Gra65] Gradshteyn, I. S., and I. M. Ryzhik, *Tables of Integrals, Series, and Products*, 4th ed. New York: Academic Press, Inc., 1965.

[Gre50] Grenander, Ulf, "Stochastic processes and statistical inference," *Arkiv för Matematik*, 1, 17 (1950), 195–277.

[Gre59] Grenander, U., H. O. Pollak, and D. Slepian, "The distribution of quadratic forms in normal variates: A small sample theory with applications to spectral analysis," *Journal of the Society for Industrial and Applied Mathematics (SIAM)*, 7 (1959), 374–401.

[Gui57] Guillemin, E. A., *Synthesis of Passive Networks*. New York: John Wiley & Sons, Inc., 1957.

[Gui63] ———, *Theory of Linear Physical Systems*. New York: John Wiley & Sons, Inc., 1963.

[Háj62] Hájek, J., "On linear statistical problems in stochastic processes," *Czechoslovak Mathematical Journal*, 12, 87 (1962), 404–44.

[Har52] Harvard Computation Laboratory, *Tables of the Error Function and of Its First Twenty Derivatives*. Cambridge, MA: Harvard University Press, 1952.

[Har70] Harger, R. O., *Synthetic Aperture Radar Systems: Theory and Design*. New York: Academic Press, Inc., 1970.

[Har84] Harding, E. F., "An efficient minimal-storage procedure for calculating the Mann-Whitney U, generalized U and similar distributions," *Applied Statistics*, 33 (1984), 1–6.

[Hau58] Haus, H. A., and R. B. Adler, "Optimum noise performance of linear amplifiers," *Proceedings of the IRE*, 46 (Aug. 1958), 1517–33.

[Hau78] Hauk, W., F. Bross, and M. Ottka, "The calculation of error rates for optical fiber systems," *IEEE Transactions on Communications*, COM-26 (July 1978), 1119–26.

[Hei61] Heimiller, R. C., "Phase shift pulse codes with good periodic correlation properties," *IRE Transactions on Information Theory*, IT-7 (Oct. 1961), 254–7.

[Hel55] Helstrom, C. W., "The resolution of signals in white, Gaussian noise," *Proceedings of the IRE*, 43 (Sept. 1955), 1111–8.

[Hel59] ———, "A note on a Markov envelope process," *IRE Transactions on Information Theory*, IT-5 (Sept. 1959), 139–40.

[Hel62] ———, "A range-sampled sequential detection system," *IRE Transactions on Information Theory*, IT-8 (Jan. 1962), 43–7.

[Hel64] ———, "The distribution of photoelectric counts from partially polarized Gaussian light," *Proceedings of the Physical Society of London*, 83 (1964), 777–82.

[Hel65] ———, "Solution of the detection integral equation for stationary filtered white noise," *IEEE Transactions on Information Theory*, IT-11 (July 1965), 335–9.

[Hel68] ———, "Markov processes and applications," in *Communication Theory*, ed. A. V. Balakrishnan (New York: McGraw-Hill Book Co., 1968), 26–87.

[Hel76] ———, *Quantum Detection and Estimation Theory*. New York: Academic Press, Inc., 1976.

[Hel78] ———, "Approximate evaluation of detection probabilities in radar and optical communications," *IEEE Transactions on Aerospace & Electronic Systems*, AES-14 (July 1978), 630–40.

[Hel83] ———, "Evaluating the detectability of Gaussian stochastic signals by steepest-descent integration," *IEEE Transactions on Aerospace & Electronic Systems*, AES-19 (May 1983), 428–37.

[Hel84a] ———, and S. O. Rice, "Computation of counting distributions arising from a single-stage multiplicative process," *Journal of Computational Physics*, 54 (May 1984), 289–324.

[Hel84b] ———, "Output distributions of electrons in a photomultiplier,"*Journal of Applied Physics*, 55 (Apr. 1, 1984), 2786–92.

[Hel84c] ———, "Computation of output electron distributions in avalanche photodiodes," *IEEE Transactions on Electron Devices*, ED-31 (July 1984), 955–8.

[Hel84d] ———, and J. A. Ritcey, "Evaluating radar detection probabilities by steepest descent integration," *IEEE Transactions on Aerospace & Electronic Systems*, AES-20 (Sept. 1984), 624–34.

[Hel85a] ———, and J. A. Ritcey, "Evaluation of the noncentral *F*-distribution by numerical contour integration," *SIAM Journal of Scientific and Statistical Computing*, 6 (July 1985), 505–14.

[Hel85b] ———, "Computation of photoelectron counting distributions by numerical contour integration," *Journal of the Optical Society of America*, A2 (May 1985), 674–82.

[Hel86a] ———, "Calculating error probabilities for intersymbol and cochannel interference," *IEEE Transactions on Communications*, COM-34 (May 1986), 430–5.

[Hel86b] ———, "Distribution of the filtered output of a quadratic rectifier computed by numerical contour integration," *IEEE Transactions on Information Theory*, IT-32 (July 1986), 450-63.

[Hel87] ———, "Computation of photoelectron counting distributions by numerical contour integration. II. Light with rational spectral density," *Journal of the Optical Society of America*, A4 (July 1987), 1245–55.

[Hel88a] ———, "Computing the performance of optical receivers with avalanche diode detectors," *IEEE Transactions on Communications*, COM-36 (Jan. 1988), 61–6.

[Hel88b] ———, "Improved multilevel quantization for detection of narrowband signals," *IEEE Transactions on Aerospace & Electronic Systems*, AES-24 (Mar. 1988), 141–7.

[Hel89a] ———, "Detectability of signals in Laplace noise," *IEEE Transactions on Aerospace & Electronic Systems*, AES-25 (Mar. 1989), 190–6.

[Hel89b] ———, "Distribution of the average power of a normal time series," *SIAM Journal of Scientific and Statistical Computing*, 10 (May 1989), 432–46.

[Hel90] ———, "Performance of receivers with linear rectifiers," *IEEE Transactions on Aerospace & Electronic Systems*, AES-26 (Mar. 1990), 210–7.

[Hel91] ———, *Probability and Stochastic Processes for Engineers*, 2nd ed. New York: Macmillan, Inc., 1991.

[Hel92a] ———, "Detection probabilities for correlated Rayleigh fading signals," *IEEE Transactions on Aerospace & Electronic Systems*, AES-28 (Jan. 1992), 259–67.

[Hel92b] ———, "Minimax detection of signals with unknown parameters," *Signal Processing*, 27 (May 1992), 145–59.

[Hel92c] ———, and C. L. Ho, "Analysis of avalanche diode receivers by saddlepoint integration," *IEEE Transactions on Communications*, COM-40 (Aug. 1992), 1327–38.

[Hel92d] ———, "Computing the generalized Marcum Q-function," *IEEE Transactions on Information Theory*, IT-38 (July 1992), 1422–28.

[Hil53] Hilbert, D., *Grundzüge einer allgemeinen Theorie der linearen Integral-gleichungen*. New York: Chelsea Publishing Co., 1953.

[Hob89] Hoballah, I. Y., and P. K. Varshney, "Distributed Bayesian signal detection," *IEEE Transactions on Information Theory*, 35 (Sept. 1989), 995–1000.

[Hod56] Hodges, J. L., Jr., and E. L. Lehmann, "The efficiency of some nonparametric competitors of the *t*-test," *Annals of Mathematical Statistics*, 27 (June 1956), 324–35.

[Hol82] Holevo, A. S., *Probabilistic and Statistical Aspects of Quantum Theory*. New York: Elsevier-North Holland Publishing Co., 1982.

[Hou87] Hou, X. Y., N. Morinaga, and T. Namekawa, "Direct evaluation of radar detection probabilities," *IEEE Transactions on Aerospace & Electronic Systems*, AES-23 (July 1987), 418–24.

[Jah45] Jahnke, E., and F. Emde, *Tables of Functions with Formulae and Curves*. New York: Dover Publications, 1945.

[Ján65] Jánossy, L., *Theory and Practice of the Evaluation of Measurements*. New York: Oxford University Press, 1965.

[Jef73] Jeffreys, Sir Harold, *Scientific Inference*, 3rd ed. New York: Cambridge University Press, 1973.

[Jef83] Jeffreys, Sir Harold, *Theory of Probability*, 3rd ed. New York: Oxford University Press, 1983.

[Jel92] Jelalian, A. V., *Laser Radar Systems*. Boston: Artech House, 1992.

[Joh70] Johnson, N. L., and S. Kotz, *Continuous Univariate Distributions*, vol. 2. Boston: Houghton-Mifflin, 1970.

[Kac47] Kac, M., and A. J. F. Siegert, "On the theory of noise in radio receivers with square law detectors," *Journal of Applied Physics*, 18 (Apr. 1947), 383–97.

[Kad64] Kadota, T., "Optimum reception of binary Gaussian signals," *Bell System Technical Journal*, 43 (Nov. 1964), 2767–810.

[Kad65] ———, "Optimum reception of binary sure and Gaussian signals," *Bell System Technical Journal*, 44 (Oct. 1965), 1621–58.

[Kad67] ———, "Differentiation of Karhunen–Loève expansion and application to optimum reception of sure signals in noise," *IEEE Transactions on Information Theory*, IT-13 (Apr. 1967), 255–60.

[Kai60] Kailath, T., "Correlation detection of signals perturbed by a random channel," *IRE Transactions on Information Theory*, IT-6 (June 1960), 361–6.

[Kai66a] ———, "Some results on singular detection," *Information & Control*, 9 (Apr. 1966), 130–52.

[Kai66b] ———, "Some integral equations with 'nonrational' kernels," *IEEE Transactions on Information Theory*, IT-12 (Oct. 1966), 442–7.

[Kai67] ———, "A projection method for signal detection in colored Gaussian noise," *IEEE Transactions on Information Theory*, IT-13 (July 1967), 441–7.

[Kai68] ———, "An innovations approach to least-squares estimation, I: Linear filtering in additive white noise," *IEEE Transactions on Automatic Control*, AC-13 (Dec. 1968), 646–55.

[Kai69] ———, "A general likelihood-ratio formula for random signals in Gaussian noise," *IEEE Transactions on Information Theory*, IT-15 (May 1969), 350–61.

[Kai71] ———, "RKHS approach to detection and estimation problems—Part I: Deterministic signals in Gaussian noise," *IEEE Transactions on Information Theory*, IT-17 (Sept. 1971), 530–49.

[Kai72] ———, R. T. Geesey, and H. L. Weinert, "Some relations among RKHS norms, Fredholm equations, and innovations representations," *IEEE Transactions on Information Theory*, IT-18 (May 1972), 341–8.

[Kai74] ———, "A view of three decades of linear filtering theory," *IEEE Transactions on Information Theory*, IT-20 (Mar. 1974), 146–81.

[Kal60] Kalman, R. E., "A new approach to linear filtering and prediction problems," *Transactions of the ASME, Series D, Journal of Basic Engineering*, 82 (Mar. 1960), 34–45.

[Kal61] Kalman, R. E., and R. S. Bucy, "New results in linear filtering and prediction theory," *Transactions of the ASME, Series D, Journal of Basic Engineering*, 83 (Dec. 1961), 95–107.

[Kan65] Kanefsky, M., and J. B. Thomas, "On polarity detection schemes with non-Gaussian inputs," *Journal of the Franklin Institute*, 280 (Aug. 1965), 120–38.

[Kar47] Karhunen, K., "Über linearen Methoden in der Wahrscheinlichkeitsrechnung," *Annales Academiae Scientiarum Fennicae, Series A1, Mathematica-Physica*, 37 (1947), 1–79.

[Kas77] Kassam, S. A., "Optimum quantization for signal detection," *IEEE Transactions on Communications*, COM-25 (May 1977), 479–84.

[Kas80] ———, "A bibliography on nonparametric detection," *IEEE Transactions on Information Theory*, IT-26 (Sept. 1980), 595–602.

[Kas85] ———, and H. V. Poor, "Robust techniques for signal processing: a survey," *Proceedings of the IEEE*, 73 (Mar. 1985), 433–81.

[Kas88] ———, *Signal Detection in Non-Gaussian Noise*. New York: Springer-Verlag, 1988.

[Kaz90] Kazakos, D., and P. Papantoni-Kazakos, *Detection and Estimation*. New York: Computer Science Press, 1990.

[Kem50] Kemperman, H. H. B., "The General One-dimensional Random Walk with Absorbing Barriers with Applications to Sequential Analysis," M. A. Thesis, University of Amsterdam, 1950.

[Ken61] Kendall, M. G., and A. Stuart, *The Advanced Theory of Statistics*, vol. 2. New York: Hafner Publishing Co., 1961; 2nd ed., 1967.

[Ken63] Kendall, W. B., and I. S. Reed, "A sequential test for radar detection of multiple targets," *IEEE Transactions on Information Theory*, IT-9 (Jan. 1963), 51–3.

[Ken65] Kendall, W. B., "Performance of the biased square-law sequential detector in the absence of signal," *IEEE Transactions on Information Theory*, IT-11 (Jan. 1965), 83–90.

[Ken68] Kendall, M. G., and A. Stuart, *The Advanced Theory of Statistics*, vol. 3. London: Charles Griffin & Company Ltd., 1966; New York: Hafner Publishing Company, 1968; 3rd ed., London: Charles Griffin & Company, 1976.

[Ken69] Kennedy, R. S., *Fading Dispersive Communication Channels*. New York: Wiley-Interscience, 1969.

[Ker89] Kerr, T. H., "Analytic example of a Schweppe likelihood-ratio detector," *IEEE Transactions on Aerospace & Electronic Systems*, AES-25 (July 1989), 545–58.

[Kin78] Kingston, R. H., *Detection of Optical and Infrared Radiation*. New York: Springer-Verlag, 1978.

[Kla60a] Klauder, J. R., A. C. Price, S. Darlington, and W. J. Albersheim, "The theory and design of chirp radars," *Bell System Technical Journal*, 39 (July 1960), 745–808.

[Kla60b] Klauder, J. R., "The design of radar signals having both high range resolution and high velocity resolution," *Bell System Technical Journal*, 39 (July 1960), 809–20.

[Kot59] Kotel'nikov, V. A., *The Theory of Optimum Noise Immunity*, trans. R. A. Silverman. New York: McGraw-Hill Book Co., 1959. Reprinted by Dover Publications, New York; 1960, 1968.

[Kre83] Kretschmer, F. F., Jr., and B. L. Lewis, "Doppler properties of polyphase coded pulse compression waveforms," *IEEE Transactions on Aerospace & Electronic Systems*, AES-19 (July 1983), 521–31. Reprinted in [Lew86], 78–88.

[Lan56] Laning, J. H., and R. H. Battin, *Random Processes in Automatic Control*. New York: McGraw-Hill Book Co., 1956.

[Lan61] Landau, H. J., and H. O. Pollak, "Prolate spheroidal wave functions, Fourier analysis and uncertainty, II," *Bell System Technical Journal*, 40 (Jan. 1961), 65–84.

[Lan62] ———, "Prolate spheroidal wave functions, Fourier analysis and uncertainty, III," *Bell System Technical Journal*, 41 (July 1962), 1295–336.

[Lar79] Larson, H. J., and B. O. Shubert, *Probabilistic Models in Engineering Sciences*; vol. 1, *Random Variables and Stochastic Processes*; vol. 2, *Random Noise, Signals, and Dynamic Systems*. New York: John Wiley & Sons, Inc., 1979.

[Law50] Lawson, J. L., and G. E. Uhlenbeck, *Threshold Signals*. New York: McGraw-Hill Book Co., 1950.

[Leh59] Lehmann, E. L., *Testing Statistical Hypotheses*. New York: John Wiley & Sons, Inc., 1959; 2nd ed., 1986.

[Ler63] Lerner, R. M., "Report on progress in information theory in the U.S.A., Communication and Radar—Section B, Radar waveform selection," *IEEE Transactions on Information Theory*, IT-9 (Oct. 1963), 246–8.

[Lew81] Lewis, B. L., and F. E. Kretschmer, Jr., "A new class of polyphase compression codes and techniques," *IEEE Transactions on Aerospace & Electronic Systems*, AES-17 (May 1981), 364–72. Reprinted in [Lew86], 29–37.

[Lew82] ———, and F. E. Kretschmer, Jr., "Linear frequency modulation derived polyphase pulse compression codes," *IEEE Transactions on Aerospace & Electronic Systems*, AES-18 (Sept. 1982), 637–41. Reprinted in [Lew86], 89–93.

[Lew86] ———, F. E. Kretschmer, Jr., and W. W. Shelton, *Aspects of Radar Signal Processing.* Norwood, MA: Artech House, 1986.

[Lju83] Ljung, L., and T. Söderström, *Theory and Practice of Recursive Estimation.* Cambridge, MA: MIT Press, 1983.

[Loè45] Loève, M., "Sur les fonctions aléatoires stationnaires du second ordre," *Revue Scientifique*, 83 (1945), 297–303.

[Loè46] ———, "Fonctions aléatoires du second ordre," *Revue Scientifique*, 84 (1946), 195–206.

[Lom61] Lombard, F. J., and F. Martin, "Statistics of electron multiplication," *Review of Scientific Instruments*, 32 (Feb. 1961), 200–1.

[Lon62] Longuet-Higgins, M. S., "The distribution of intervals between zeros of a stationary random function," *Philosophical Transactions of the Royal Society of London*, A254 (May 24, 1962), 557–99.

[Lov24] Lovitt, W. V., *Linear Integral Equations.* New York: McGraw-Hill Book Co., 1924. Reprinted by Dover Publications, New York; 1950.

[Luc57] Luce, R. D., and H. Raiffa, *Games and Decisions.* New York: John Wiley & Sons, Inc., 1957.

[Lug80] Lugannani, R., and S. O. Rice, "Saddlepoint approximation for the distribution of the sum of independent random variables," *Advances in Applied Probability*, 12 (1980), 475–90.

[Mac68] MacDonald, V. H., and P. M. Schultheiss, "Optimum passive bearing estimation in a spatially incoherent noise environment," *Journal of the Acoustical Society of America*, 46 (Jan. 1968), 37–43.

[Mac71] Macchi, O., "Distribution statistique des instants d'émission des photoélectrons d'une lumière thermique," *Comptes Rendus*, 272 (Feb. 8, 1971), 437–40.

[Man47] Mann, H. B., and D. R. Whitney, "On a test of whether one of two random variables is stochastically larger than the other," *Annals of Mathematical Statistics*, 18 (1947), 50–60.

[Man65] Mandel, L., and E. Wolf, "Coherence properties of optical fields," *Reviews of Modern Physics*, 37 (Apr. 1965), 231–87.

[Mar48] Marcum, J. I., *A Statistical Theory of Target Detection by Pulsed Radar*, Rand Corp. Report RM-753, July 1, 1948. Reprinted in *IRE Transactions on Information Theory*, IT-6 (Apr. 1960), 59–267.

[Mar50] ———, "Table of Q-functions," Rand Corp. Report RM-339 (Jan. 1, 1950).

[Mar62] Marcus, M. B., and P. Swerling, "Sequential detection in radar with multiple resolution elements," *IRE Transactions on Information Theory*, IT-8 (Apr. 1962), 237–45.

[McD56] McDonald, J. R., and M. K. Brachman, "Linear-system integral transform relations," *Reviews of Modern Physics*, 28 (Oct. 1956), 393–422.

[McF62] McFadden, J. A., "On the lengths of intervals in a stationary point process," *Journal of the Royal Statistical Society of London* (B), 24, 2 (1962), 364–82.

[McG60] McGinn, J. W., Jr., and E. W. Pike, "A study of sea clutter spectra," in *Statistical Methods of Radio Wave Propagation*, ed. W. C. Hoffman (London: Pergamon Press Ltd., 1960), 49–92.

[McG70] McGee, W. F., "Another recursive method of computing the Q function," *IEEE Transactions on Information Theory*, IT-16 (July 1970), 500–1.

[McI72] McIntyre, R. J., "The distribution of gains in uniformly multiplying avalanche photodiodes: theory," *IEEE Transactions on Electron Devices*, ED-19 (June 1972), 703–13.

[Mey73] Meyer, D. P., and H. A. Mayer, *Radar Target Detection*. New York: Academic Press, Inc., 1973.

[Mid48] Middleton, D., "Some general results in the theory of noise through nonlinear devices," *Quarterly of Applied Mathematics*, 5, 4 (1948), 445–98.

[Mid53] ———, "Statistical criteria for the detection of pulsed carriers in noise," *Journal of Applied Physics*, 24 (Apr. 1953), 371–8, 379–91.

[Mid55] ———, and D. Van Meter, "On optimum multiple-alternative detection of signals in noise," *IRE Transactions on Information Theory*, IT-1, 2 (Sept. 1955), 1–9.

[Mid57] ———, "On the detection of stochastic signals in additive normal noise—Part I," *IEEE Transactions on Information Theory*, IT-3 (June 1957), 86–121.

[Mid60a] ———, *An Introduction to Statistical Communication Theory*. New York: McGraw-Hill Book Co., 1960.

[Mid60b] ———, "On new classes of matched filters and generalizations of the matched filter concept," *IRE Transactions on Information Theory*, IT-6 (June 1960), 349–60.

[Mid61] ———, "On singular and nonsingular optimum (Bayes) tests for the detection of normal stochastic signals in normal noise," *IRE Transactions on Information Theory*, IT-7 (Apr. 1961), 105–113.

[Mid65] ———, and H. L. Groginsky, "Detection of random acoustic signals by receivers with distributed elements: Optimum receiver structures for normal signal and noise fields," *Journal of the Acoustical Society of America*, 38 (Nov. 1965), 727–37.

[Mid66] ———, "Canonically optimum threshold detection," *IEEE Transactions on Information Theory*, IT-12 (Apr. 1966), 230–43.

[Mil56] Miller, K. S., and L. A. Zadeh, "Solution of an integral equation occurring in the theories of prediction and detection," *IRE Transactions on Information Theory*, IT-2 (June 1956), 72–5.

[Mit71] Mitchell, R. L., and J. F. Walker, "Recursive methods for computing detection probabilities," *IEEE Transactions on Aerospace & Electronic Systems*, AES-7 (July 1971), 671–6.

[Mor53] Morse, P. M., and H. Feshbach, *Methods of Theoretical Physics*. New York: McGraw-Hill Book Co., 1953.

[Mor88] Morris, G. V., *Airborne Pulsed Doppler Radar*. Norwood, MA: Artech House, 1988.

[Ney33a] Neyman, J., and E. Pearson, "On the problem of the most efficient tests of statistical hypotheses," *Philosophical Transactions of the Royal Society of London*, A231, 9 (Feb. 16, 1933), 289–337.

[Ney33b] ———, and E. Pearson, "The testing of statistical hypotheses in relation to probability *a priori*," *Proceedings of the Cambridge Philosophical Society*, 29, 4 (1933), 492–510.

[Ney39] ———, "On a new class of 'contagious' distributions, applicable in entomology and bacteriology," *Annals of Mathematical Statistics*, 10 (1939), 35–57.

[Nor43] North, D. O., "An analysis of the factors which determine signal-noise discrimination in pulsed carrier systems," RCA Laboratory Report, PTR-6C (1943); reprinted in *Proceedings of the IEEE*, 51 (July 1963), 1016–27.

[Nyq28] Nyquist, H., "Thermal agitation of electric charge in conductors," *Physical Review*, 32 (July 1928), 110–3.

[Pac58] Pachares, J., "A table of bias levels useful in radar detection problems," *IRE Transactions on Information Theory*, IT-4 (Mar. 1958), 38–45.

[Pap91] Papoulis, A., *Probability, Random Variables, and Stochastic Processes*, 3rd ed. New York: McGraw-Hill Book Co., 1991.

[Par80] Parl, S., "A new method of calculating the generalized Q function," *IEEE Transactions on Information Theory*, IT-26 (Jan. 1980), 121–4.

[Pat49] Patnaik, P. B., "The non-central χ^2 and F-distributions and their applications," *Biometrika*, 36 (1949), 202–32.

[Pea34] Pearson, K., *Tables of the Incomplete Gamma-Function*. New York: Cambridge University Press, 1934; reissued, 1951.

[Pea68] ———, *Tables of the Incomplete Beta Function*, 2nd ed. New York: Cambridge University Press, 1968.

[Per71] Personick, S., "Statistics of a general class of avalanche detectors with applications to optical communication," *Bell System Technical Journal*, 50 (Dec. 1971), 3075–95.

[Pet54] Peterson, W. W., T. G. Birdsall, and W. C. Fox, "The theory of signal detectability," *IRE Transactions on Information Theory*, PGIT-4 (Sept. 1954), 171–212.

[Pit49] Pitman, E. J. G, "Lecture notes on nonparametric statistical inference," Columbia University, New York, N.Y. (1949).

[Pit66] Pitcher, R. S., "An integral expression for the log likelihood ratio of two Gaussian processes," *SIAM Journal on Applied Mathematics*, 14 (Mar. 1966), 228–33.

[Poo88] Poor, H. Vincent, *An Introduction to Signal Detection and Estimation*. New York: Springer-Verlag, 1988.

[Pre66] Prescott, J. R., "A statistical model for photomultiplier single-electron statistics," *Nuclear Instruments & Methods*, 39 (Jan. 1966), 173–79.

[Pre86] Press, W. H., B. P. Flannery, S. A. Teukolsky, and W. T. Vetterling, *Numerical Recipes*. New York: Cambridge University Press, 1986.

[Pri56] Price, R., "Optimum detection of random signals in noise, with application to scatter-multipath communication," *IRE Transactions on Information Theory*, IT-2 (Dec. 1956), 125–35.

[Pri58] ———, and P. E. Green, Jr., "A communication technique for multipath channels," *Proceedings of the IRE*, 46 (Mar. 1958), 555–70.

[Pri60] ———, and P. E. Green, Jr., *Signal Processing in Radar Astronomy--Communication via Fluctuating Multipath Media*, Lincoln Laboratory, M.I.T., Lexington, MA, Tech. Report 234 (Oct. 6, 1960).

[Pri64] ———, "Some non-central F-distributions expressed in closed form," *Biometrika*, 51, 1 and 2 (1964), 107–22.

[Pri65] ———, and E. M. Hofstetter, "Bounds on the volume and height distributions of the ambiguity function," *IEEE Transactions on Information Theory*, IT-11 (Apr. 1965), 207–14.

[Pro89] Proakis, J. G., *Digital Communications*, 2nd ed. New York: McGraw-Hill Book Co., 1989.

[Rad62] Radner, R., "Team decision problems," *Annals of Mathematical Statistics*, 33 (Sept. 1962), 857–81.

[Rai62] Rainal, A. J., "Zero-crossing intervals of Gaussian processes," *IRE Transactions on Information Theory*, IT-8 (Oct. 1962), 372–8.

[Rai87] ———, "First and second passage times of Rayleigh processes," *IEEE Transactions on Information Theory*, IT-33 (May 1987), 419–25.

[Rai88] ———, "Origin of Rice's Formula," *IEEE Transactions on Information Theory*, IT-34 (Nov. 1988), 1383–7.

[Rao45] Rao, C. R., "Information and the accuracy attainable in the estimation of statistical parameters," *Bulletin of the Calcutta Mathematical Society*, 37 (1945), 81–91.

[Ree62] Reed, I. S., "On a moment theorem for complex Gaussian processes," *IRE Transactions on Information Theory*, IT-8 (Apr. 1962), 194–5.

[Ree63] ———, and I. Selin, "A sequential test for the presence of a signal in one of k possible positions," *IEEE Transactions on Information Theory*, IT-9 (Oct. 1963), 286–8.

[Res57] Resnikoff, G. J., and G. J. Lieberman, *Tables of the Non-Central t-Distribution*. Stanford, CA: Stanford University Press, 1957.

[Ric44] Rice, S. O., "Mathematical analysis of random noise," *Bell System Technical Journal*, 23 (July 1944), 282–332; 24 (Jan. 1945), 46–156. Reprinted in N. Wax, *Selected Papers on Noise and Stochastic Processes* (New York: Dover Publications, Inc., 1954), 133–294.

[Ric58] ———, "Distribution of the duration of fades in radio transmission: Gaussian noise model," *Bell System Technical Journal*, 37 (May 1958), 581–635.

[Ric68] ———, "Uniform asymptotic expansions for saddle point integrals—Application to a probability distribution occurring in noise theory," *Bell System Technical Journal*, 47 (Nov. 1968), 1971–2013.

[Ric73] ———, "Efficient evaluation of integrals of analytic functions by the trapezoidal rule," *Bell System Technical Journal*, 52 (May–June 1973), 707–22.

[Ric77] Rickard, J. T., and G. M. Dillard, "Adaptive detection algorithms for multiple-target situations," *IEEE Transactions on Aerospace & Electronic Systems*, AES-13 (July 1977), 338–43.

[Ric80] Rice, S. O., "Distribution of quadratic forms in normal random variables—Evaluation by numerical integration," *SIAM Journal of Scientific and Statistical Computing*, 1 (Dec. 1980), 438–48.

[Ric82] ———, "Envelopes of narrow-band signals," *Proceedings of the IEEE*, 70 (July 1982), 692–9.

[Rie53] Riemann, B., "Sullo svolgimento del quoziente di due serie ipergeometriche in frazione continua infinita," (Oct. 1863). Reconstructed from notes by H. A. Schwarz. In *Collected Works of Bernhard Riemann*, ed. H. Weber (New York: Dover Publications, Inc., 1953), 424–30.

[Rih64] Rihaczek, A. W., "Radar resolution properties of pulse trains," *Proceedings of the IEEE*, 52 (Feb. 1964), 153–64.

[Rih65] ———, "Radar signal design for target resolution," *Proceedings of the IEEE*, 53 (Feb. 1965), 116–28.

[Rih69] ———, *Principles of High Resolution Radar*. New York: McGraw-Hill Book Co., 1969.

[Rit86] Ritcey, J. A., "Performance analysis of the censored mean-level detector," *IEEE Transactions on Aerospace & Electronic Systems*, AES-22 (July 1986), 443–54.

[Rob69] Robertson, G. H., "Computation of the noncentral chi-square distribution," *Bell System Technical Journal*, 48 (Jan. 1969), 201–7.

[Rob76] ———, "Computation of the noncentral F distribution (CFAR) detection," *IEEE Transactions on Aerospace & Electronic Systems*, AES-12 (Sept. 1976), 568–71.

[Roo63] Root, W. L., "Singular Gaussian measures in detection theory," *Proceedings of the Symposium on Time Series Analysis*, ed. M. Rosenblatt (New York: John Wiley & Sons, Inc., 1963), 292–315.

[Roo87] ———, "Remarks, mostly historical, on signal detection and signal parameter estimation," *Proceedings of the IEEE*, 75 (Nov. 1987), 1446–57.

[Rud61] Rudnick, P., "Likelihood detection of small signals in stationary noise," *Journal of Applied Physics*, 32 (Feb. 1961), 140–3.

[Rud62] ———, "A signal-to-noise property of binary decisions," *Nature*, 193 (Feb. 10, 1962), 604–5.

[Sal78] Saleh, B., *Photoelectron Statistics, with Applications to Spectroscopy and Optical Communications*. New York: Springer-Verlag, 1978.

[Sam48] Samuelson, P. A., "Exact distribution of continuous variables in sequential analysis," *Econometrica*, 16 (1948), 191–8.

[Sar91] Sarwate, D., "Computation of binary integration detection probabilities," *IEEE Transactions on Aerospace & Electronic Systems*, AES-27 (Nov. 1991), 894–7.

[Sch56] Schwartz, M., "A coincidence procedure for signal detection," *IRE Transactions on Information Theory*, IT-2 (Dec. 1956), 135–9.

[Sch65] Schweppe, F. C., "Evaluation of likelihood functions for Gaussian signals," *IEEE Transactions on Information Theory*, IT-11 (Jan. 1965), 61–70.

[Sch66] Schwartz, M., W. R. Bennett, and S. Stein, *Communication Systems and Techniques*. New York: McGraw-Hill Book Co., 1966.

[Sch69] Schwartz, C., "Numerical integration of analytic functions," *Journal of Computational Physics*, 4 (1969), 19–29.

[Sch71] Scharf, L. L., and D. W. Lytle, "Signal detection in Gaussian noise of unknown level: An invariance application," *IEEE Transactions on Information Theory*, IT-17 (July 1971), 404–11.

[Sch75] Schwartz, M., and L. Shaw, *Signal Processing: Discrete Spectral Analysis, Detection, and Estimation*. New York: McGraw-Hill Book Co., 1975.

[Sch91a] Scharf, L., *Statistical Signal Processing: Detection, Estimation, and Time-series Analysis*. Reading, MA: Addison-Wesley Publishing Co., 1991.

[Sch91b] Schleher, D. C., *MTI and Pulsed Doppler Radar*. Boston: Artech House, 1991.

[Sen85] Senior, J. M., *Optical Fiber Communications: Principles and Practice*. Englewood Cliffs, NJ: Prentice Hall, 1985.

[Shi57] Shinbrot, M., "A generalization of a method for the solution of the integral equation arising in optimization of time-varying linear systems with nonstationary inputs," *IRE Transactions on Information Theory*, IT-3 (Dec. 1957), 220–4.

[Shn89] Shnidman, D. A., "The calculation of the probability of detection and the generalized Marcum Q-function," *IEEE Transactions on Information Theory*, IT-35 (Mar. 1989), 389–400.

[Sie51] Siegert, A. J. F., "On the first-passage time probability problem," *Physical Review*, 81 (Feb. 15, 1951), 617–23.

[Sie56] Siebert, W. M., "A radar detection philosophy," *IRE Transactions on Information Theory*, IT-2 (Sept. 1956), 204–21.

[Sie57] Siegert, A. J. F., "A systematic approach to a class of problems in the theory of noise and other random phenomena—Part II, Examples," *IRE Transactions on Information Theory*, IT-3 (Mar. 1957), 38–44.

[Sil49] Silver, S., *Microwave Antenna Theory and Design*, M.I.T. Radiation Lab. Series, no. 12. New York: McGraw-Hill Book Co., 1949.

[Sko62] Skolnik, M. I., *Introduction to Radar Systems*. New York: McGraw-Hill Book Co., 1962.

[Sle58a] Slepian, D., "Fluctuations of random noise power," *Bell System Technical Journal*, 37 (Jan. 1958), 163–84.

[Sle58b] ———, "Some comments on the detection of Gaussian signals in Gaussian noise," *IRE Transactions on Information Theory*, IT-4 (June 1958), 65–8.

[Sle61] ———, and H. O. Pollak, "Prolate spheroidal wave functions, Fourier analysis, and uncertainty (I)," *Bell System Technical Journal*, 40 (Jan. 1961), 43–63.

[Sle62] ———, "The one-sided barrier problem for Gaussian noise," *Bell System Technical Journal*, 41 (Mar. 1962), 463–501.

[Sle65a] ———, "Some asymptotic expansions for prolate spheroidal wave functions," *Journal of Mathematics & Physics*, 44 (June 1965), 99–140.

[Sle65b] ———, and E. Sonnenblick, "Eigenvalues associated with the prolate spheroidal wave functions of zero order," *Bell System Technical Journal*, 44 (Oct. 1965), 1745–59.

[Sle69] ———, and T. Kadota, "Four integral equations of detection theory," *SIAM Journal on Applied Mathematics*, 17 (Nov. 1969), 1102–17.

[Sny75] Snyder, Donald L., *Random Point Processes*. New York: John Wiley & Sons, Inc., 1975.

[Sri86] Srinivasan, R., "Distributed radar detection theory," *Proceedings of the IEE*, pt. F, 133 (Feb. 1986), 55–60.

[Ste66] Stein, S., "Decision-oriented diversity for digital transmission," Ch. XI in [Sch66], 490–584.

[Stu64] Stutt, C. A., "Some results on real-part/imaginary-part and magnitude/phase relations in ambiguity functions," *IEEE Transactions on Information Theory*, IT-10 (Oct. 1964), 321–7.

[Swe59] Swerling, P., "Parameter estimation for waveforms in additive Gaussian noise," *Journal of the Society for Industrial and Applied Mathematics (SIAM)*, 7 (June 1959), 152–66.

[Swe60] ———, *Probability of Detection for Fluctuating Targets*, Rand Corp, Report RM-1217. Reprinted in *IRE Transactions on Information Theory*, IT-6 (Apr. 1960), 269–308.

[Swe64] ———, "Parameter estimation accuracy formulas." *IEEE Transactions on Information Theory*, IT-10 (Oct. 1964), 302–14.

[Tan38] Tang, P. C., "The power function of the analysis of variance tests with tables and illustrations of their use," *Statistical Research Memoirs*, 2 (1938), 126–49 + 8 tables.

[Tei81] Teich, M. C., "Role of the doubly stochastic Neyman Type-A and Thomas counting distributions in photon detection," *Applied Optics*, 20 (July 15, 1981), 2457–67.

[Tik61] Tikhonov, V. I., "The Markov nature of the envelope of quasi-harmonic oscillations," *Radio Engineering & Electronic Physics*, 6, 7 (1961), 961–71.

[Tit66] Titlebaum, E. L., "A generalization of a two-dimensional Fourier transform property for ambiguity functions," *IEEE Transactions on Information Theory*, IT-12 (Jan. 1966), 80–1.

[Tol34] Tolman, R. C., *Relativity, Thermodynamics, and Cosmology*. New York: Oxford University Press, 1934; reprinted by Oxford University Press, 1946.

[Tur60a] Turin, G. L., "An introduction to matched filters." *IRE Transactions on Information Theory*, IT-6 (June 1960), 311–29.

[Tur60b] ———, "The characteristic function of Hermitian quadratic forms in complex normal variables," *Biometrika*, 47, 1/2 (1960), 199–201.

[Tur61] Turyn, R., and Storer, J., "On binary sequences," *Proceedings of the American Mathematical Society*, 12 (1961), 394–9.

[Urk53] Urkowitz, H., "Filters for detection of small radar signals in clutter," *Journal of Applied Physics*, 24 (Aug. 1953), 1024–31.

[Van68] Van Trees, Harry L., *Detection, Estimation, and Modulation Theory*, vol. 1. New York: John Wiley & Sons, Inc., 1968.

[Vit64] Viterbi, A. J., "Phase coherent communication over the continuous Gaussian channel," in *Digital Communications with Space Applications*, ed. S. W. Golomb (Englewood Cliffs, NJ: Prentice Hall, 1964) 106–34 and Appendix Four, 196–204.

[Vit66] ———, *Principles of Coherent Communication*. New York: McGraw-Hill Book Co., 1966.

[Wal39] Wald, A., "Contributions to the theory of statistical estimation and testing of hypotheses," *Annals of Mathematical Statistics*, 10 (1939), 299–326.

[Wal47] ———, *Sequential Analysis*. New York: John Wiley & Sons, Inc., 1947. Reprinted by Dover Publications, New York; 1973.

[Wal50] ———, *Statistical Decision Functions*. New York: John Wiley & Sons, Inc. 1950. Second ed., Chelsea Publishing Co., Bronx, NY; 1971.

[Wal57] ———, *Selected Papers in Statistics and Probability*. Stanford, CA: Stanford University Press, 1957.

[Web68] Weber, C. L., *Elements of Detection and Signal Design*. New York: McGraw-Hill Book Co., 1968; and New York: Springer-Verlag, 1987.

[Wha71] Whalen, A. D., *Detection of Signals in Noise*. New York: Academic Press, Inc., 1971.

[Wie60] Wiener, N., *Extrapolation, Interpolation, and Smoothing of Stationary Time Series with Engineering Applications*. New York: John Wiley & Sons, Inc., 1960.

[Wil45] Wilcoxon, F., "Individual comparisons by ranking methods," *Biometrics*, 1 (June 1945), 80–3.

[Wil60] Wilcox, C. H., The Synthesis Problem for Radar Ambiguity Functions, University of Wisconsin, Mathematics Research Center, Madison, WI, Technical Summary Report no. 157 (April 1960). Reprinted in [Bla91], 229–60.

[Wil73] Wilcoxon, F., S. K. Katti, and R. A. Wilcox, "Critical values and probability levels for the Wilcoxon rank sum test and the Wilcoxon signed rank test," in *Selected Tables in Mathematical Statistics*, eds. H. L. Harter and D. B. Owen, American Mathematical Society, 1 (1973), 171–259.

[Win72] Winkler, R. L., *An Introduction to Bayesian Inference and Decision*. New York: Holt, Rinehart & Winston, 1972.

[Wol59] Wolfe, P., "The simplex method for quadratic programming," *Econometrica*, 27 (July 1959), 382–98.

[Wol62] Wolff, S. S., J. B. Thomas, and T. R. Williams, "The polarity-coincidence correlator: a nonparametric detection device," *IRE Transactions on Information Theory*, IT-8 (Jan. 1962), 5–9.

[Woo50] Woodward, P. M., and I. L. Davies, "A theory of radar information," *Philosophical Magazine*, 41 (Oct. 1950), 1001–17.

[Woo53] ———, *Probability and Information Theory, with Applications to Radar*. New York: McGraw-Hill Book Co.; Oxford: Pergamon Press Ltd., 1953. Second edition, Oxford: Pergamon Press, 1963.

[Wor68] Worley, R., "Optimum thresholds for binary integration," *IEEE Transactions on Information Theory*, IT-14 (Mar. 1968), 349–53.

[Yag63] Yaglom, A. M., "On the equivalence and perpendicularity of two Gaussian probability measures in function space," *Proceedings of the Symposium on Time Series Analysis*, ed. M. Rosenblatt (New York: John Wiley & Sons, Inc., 1963), 327–46.

[Yen64] Yengst, W. C., *Procedures of Modern Network Synthesis*. New York: Macmillan, Inc., 1964.

[You54] Youla, D. C., "The use of the method of maximum-likelihood in estimating continuous-modulated intelligence which has been corrupted by noise," *IRE Transactions on Information Theory*, PGIT-3 (Mar. 1954), 90–105.

[You57] ———, "The solution of a homogeneous Wiener–Hopf integral equation occurring in the expansion of second-order stationary random functions," *IRE Transactions on Information Theory*, IT-3 (Sept. 1957), 187–93.

[Yue78] Yue, O. C., R. Lugannani, and S. O. Rice, "Series approximations for the amplitude distribution and density of shot processes," *IEEE Transactions on Communications*, COM-26 (Jan. 1978), 45–54.

[Zad50] Zadeh, L. A., and J. R. Ragazzini, "An extension of Wiener's theory of prediction," *Journal of Applied Physics*, 21 (July 1950), 645–55.

[Zad52] ———, and J. R. Ragazzini, "Optimum filters for the detection of signals in noise," *Proceedings of the IRE*, 40 (Oct. 1952), 1223–31.

[Zet62] Zetterberg, L.-H., "Signal detection under noise interference in a game situation," *IRE Transactions on Information Theory*, IT-8 (Sept. 1962), S47–52.

[Zie59] Zierler, N., "Linear recurring sequences," *Journal of the Society for Industrial and Applied Mathematics*, 7 (Mar. 1959), 31–48.

[Ziv69] Ziv, J., and M. Zakai, "Some lower bounds on signal parameter estimation," *IEEE Transactions on Information Theory*, IT-15 (May 1969), 386–91.

Index

A

Airy pattern, 160
Ambiguity, parameter estimation, 259
Ambiguity function:
  chirp-modulated signal, 382
  clear area, 388
  complex, 253–5, 285
  discrete, 389
  Gaussian pulse, 382
  general properties, 379
  generalized, 243, 299
  Hermite signals, 384
  moments, 385
  pulse train, 386
  reduced, 251
  restrictions, 380
  self-transform property, 380
  single pulse, 381
Ambiguity matrix, 363
Ambiguity surface, 380
Amplifier, 74
Amplitude:
  standard, 126
  unknown, detection, 125–7
Amplitude modulation, 89
Analytic signal, 91
Antenna, effective area, 73
Antipodal signals, 85
Arrival time:
  estimation, 233
    mean-square error, 235
    narrowband signal, 246–9, 257
  unknown:
    maximum-likelihood detector, 272
    threshold detector, 266–71
Asymptotic relative efficiency, 165–9
  definition, 167
  linear and quadratic detectors, 171
  polarity coincidence counter, 332, 334
  rank test, 328, 545
  sign test, 321
  Wilcoxon two-sample statistic, 330
Autocovariance function, 29
  bandlimited noise, 98
  complex, 96, 97–100

Autocovariance function *(cont'd)*
  rectangular spectral density, 424
  stationary noise, 30
  white noise, 35
Autocovariance matrix, complex, 100
Autoregressive moving-average
      process, 459, 462
Avalanche photodiode, 491
Average power, stochastic process, 466
Average sample number, sequential
      test, 344

B

Balanced channel:
  binary, 116
  M-ary, 133
  Rayleigh fading, 134
Bandwidth:
  effective, 270
  root-mean-square (rms), 235, 280
Barker sequence, 390
Bayes cost, 122
  in estimation, 228
Bayes criterion, 46
  in estimation, 227
Bayes strategy, 7
  binary decisions, 8–9
  sequential detection, 337
  signal resolution, 358
Bayes's rule, 3–5, 7
Beam pattern, 160
Beamforming, 137, 156–64
  point source of noise, 161–4
Bessel function, modified, 108
  asymptotic form, 112
  trigonometric integral, 108, 527
Beta distribution, 307
Bias, 222
Bilateral exponential distribution,
      321, 324
Binary channel, 114
  incoherent:
    balanced, 116
    unilateral, 117

Binary communication system,
      multiple signals, 137
Binary decisions, 8
Binary integration, 176
  detection probability, 179
Blind velocities, 378
Boltzmann distribution, 70
Boltzmann's constant, 69, 481
Bose–Einstein distribution, 478

C

Campbell's theorem, 36, 499
Carrier, 89
Carrier frequency, unknown, 282–6,
      295
  least favorable distribution, 300
Cauchy distribution, 24
Cauchy–Schwarz inequality, 50
  traces, 412
Causal quasi-estimate, 455, 480
Censored mean-level detector, 310
Characteristic function, 189
  circular Gaussian, 103
Characteristic values, 37
Chebyshev inequality, 245
Chernoff bound, 211–3
Chirp modulation, 382
Chi-squared distribution, 18, 428
Circular Gaussian density function,
      102
Circular Gaussian processes, 100–3
  characteristic function, 103
  joint moments, 103
Clutter, 356, 395
  detection of pulse trains in, 377
  detection of single pulses in,
      370–7
  overspread, 376
  spectral density, 370
  underspread, 376
Cofunction, 40
Coherence, complete, 138–40
Coherent light, detection in
      background light, 482–4

Coherent signal plus stochastic signal, detection, 451–6
Col, 198
Colored noise, 37, 42
Comb filter, 260
Comb function, 260, 377, 387
Completeness relation, 32
Complex envelope, 90
Composite hypothesis (*see* Hypothesis, composite)
Concentration ellipsoid, 242
Conditional expected value, 229
  stochastic signal, 445
Conditional probability, 3
Conditional risk, estimator, 228
Constant-false-alarm rate (CFAR) receiver, 306, 316
  detection probability, 308
    graphs, 310–1
Convex function, 197, 289
Correlation coefficient, 29
  sample, 331
  signals, 117
Correlation receiver, 45
Cost, average, composite hypotheses, 121
  distributed detection, 180
  minimization, 6–8
Cost function, 227
  quadratic, 228
Cost-likelihood ratio, 121
Cost matrix, 7
Counting functional, 276
Covariance, time and frequency estimates, 257
Covariance matrix, 29
  state vector, 459
Cramér–Rao inequality, 239–42
Cross-ambiguity function, 392
Crossing rate, 276
  Gaussian process, 277
  rectified Gaussian process, 280
Cumulants, 190
Curvature, 204
Cutoff frequency, 372

D

Decision level, 9
  approximate computation:
    Bayes, 217
    Neyman–Pearson, 215
  random:
    detection probability, 538–40
    false-alarm probability, 540
Decision surface, 9
  Neyman–Pearson criterion, 10
Decision theory:
  binary, 8
  references, 8
Deflection, 168
Delta function, 32, 35
  derivative, 62, 512
  periodic, 366
Detection:
  maximum-likelihood (*see* Maximum-likelihood detection)
  nonparametric, 315
  probability of, 9
Detection probability:
  average, 124
  known signal, 44
    graph, 47
  randomized strategy, 21
Detection statistic, stochastic signal, 402
Detector, photoelectric, 471
Differentiation, functions of functions, 489
Diffraction pattern, 388
Dirac delta function, 32, 35 (*see also* Delta function)
Discrete random variables, decisions based on, 18–23
Discrete-time processing:
  known signal, 58–9
  stochastic signal, 456
Discriminator, 92
Distributed detection, 137, 174–85
Distribution-free receiver, 316
Diversity communications, 137

Doppler shift, 250, 544
    effect on detection in clutter, 376,
        378
Dummy hypothesis, 4
Duration, mean-square, 254
Durbin algorithm, 460, 464
Dynode, 486

E

Eckart filter, 406
Edgeworth series, 194
Effective signal-to-noise ratio (*see*
        Signal-to-noise ratio, effective)
Efficacy, 168
    threshold detector, 297
Efficient estimator, 240
    asymptotic, 241
Eigenfunctions, 37
    orthonormality, 38
Eigenvalues:
    homogeneous integral equations,
        37, 516
    sum of, 81
Energy detector, 268
Energy of signal, 47, 69
    narrowband signal, 92
Epoch, 220
Equal-eigenvalue approximation,
    429
Equipollent receivers, 167
Equipotentials, 202
Error:
    first and second kinds, 9
    probability of, 9
Error covariance matrix, 227, 231, 239
    signal parameters, 257
Error function, 191
Error-function integral, 6
    asymptotic form, 85
Estimanda, 221
Estimation, signal parameters, 242
Estimator, 221
    arrival time, narrowband signal,
        246–9

Bayes, 227–8
efficient, 240
    asymptotic, 241
jointly Gaussian parameters, 226
linear, 227, 230
maximum-a-posteriori-probability,
    223
maximum-likelihood, 223–4, 232
    signal amplitude, 232
    variance, 236
mean of Gaussian distribution,
    224–6
minimum-mean-square-error, 228
signal arrival time, 233
stochastic signal, causal, 429
unbiased, 222
Estimator-correlator interpretation,
    410
Estimator-correlator representation,
    causal, 438
Expected value, conditional, 229
Exponential distribution, 303
    bilateral, 321, 324
    discrete, 476

F

F distribution, 541
Factorial moments, 208
Fading (*see also* Swerling cases)
    balanced channel, 134
    distributed detection, 176, 184
    maximum-likelihood receiver, 150
    multiple signals, 142
        detection probability, 147–51
    Rayleigh, 126, 152
    Swerling cases, 152
Fading signals, detection probability,
    537–8
False alarm, probability of, 9
    known signal in Gaussian noise, 44
    randomized strategy, 21
    resolution:
        multiple signals, 365
        two signals, 360

stochastic signals, 417
    threshold detector, 418
False-alarm number, 154
False-alarm rate, 275 (*see also* Crossing rate)
False dismissal, 9
Filter, narrowband, 93
Filter bank:
    outputs, 285
    signals in clutter, 378
First-passage-time probability density function, 274
Fisher formula, estimation error, 237
    generalized, 238
Fisher information matrix, 238
    signal parameters, 243, 256
Fixed-sample-size test, 336
Fluctuating point target, 396
Flux lines, 202
Fourier series, 31
Frank code, 393
Fredholm determinant, 399, 416, 474, 480
    finite, 462, 465
    Lorentz spectral density, 433
    rational spectral density, 436
    stochastic signal, 431–3
      in white noise, 416, 448
    Toeplitz approximation, 422
    rational spectral density, 423
Fredholm integral equation, homogeneous, 37 (*see also* Homogeneous integral equations)
Frequency deviation, mean-square, 249, 254
Frequency modulation, 89
Fusion center, 174

G

Gain:
    antenna, 73
    avalanche photodiode, 492
    photomultiplier, 488

Gain pattern, 160
Gamma distribution, 146
Gamma function, 18
Gaussian approximation, 211
Gaussian noise:
    bivariate density function, 29
    multivariate density function, 28
    univariate density function, 5, 29
Gaussian pulse, ambiguity function, 382
Geometric distribution, 20, 476
Gram–Charlier series, 191
Gram–Schmidt orthogonalization procedure, 31, 33–4, 75
Ground mapper, 357

H

Hermite polynomial, 191
Hermite signals, 383
Hermitian kernel, 37
Hermitian matrix, 398
Heterodyne receiver, 140
    optical, 497
      output signal-to-noise ratio, 501
Hilbert space, 37
Hilbert transform, 92
Hohlraum, 69
Homodyne detector, 90
    optical, 503,
      output signal-to-noise ratio, 503
Homogeneity, statistical, 165, 317, 337
Homogeneous integral equations, 37
    solution, 65–8, 516
Hypergeometric distribution, 478
Hyperspherical shell, volume, 17
Hypothesis:
    alternative, 8
    composite, 22, 107, 120
    nonparametric, 315
    null, 8
    parametric, 315
    simple, 120

I

Impulse noise, 316
Impulse response:
  complex, 94
  matched filter, 49
  matrix, 443
  narrowband filter, 93
Incoherent detection, 115
Incoherent light:
  as narrowband Gaussian process, 471
  detection, absence of background, 481
  distribution of photoelectron emission times, 506–9
Incoherent signals, multiple, 140–1
Independent-increment process, 339
Inhomogeneous integral equations, solution, 60–5, 514
Innovation process, 440
Interference, unknown spectral density, 311–4
Interpulse interval, 350
Invariable parameters, 165
  unknown, 172
Irrelevance proof, 57–8
Iterates, 398
Itô integral, 439

J

Jamming (*see* Interference)

K

Kalman–Bucy equations, 445–9
  derivation, 553
Kalman equations, 458–60, 464
Kalman gain, 447
  discrete-time processing, 459
Karhunen–Loève expansion, 37–9, 42
Kernel:

Hermitian, 37
Hilbert space, 41
of integral equation, 37
positive definite, 39
symmetrical, 37
triangular, 65
Kronecker delta, 33

L

Laguerre polynomials, 476, 524
  associated, 478, 526
Laplace distribution, 24 (*see also* Bilateral exponential distribution)
Laser, ideal, 471
Least favorable distribution, 122, 125, 287, 290
  criterion, 292, 294
  weak-signal approximation, 297
Least squares, 232
Legendre polynomials, 31
Length of function, 33
Leucogenic noise, 61
  nonstationary, 64, 442
Light:
  polarized, 471
  unpolarized, 475
Likelihood functional, 46
  average, 124, 129
  coherent narrowband signals, 139
  coherent signals, 138
  incoherent signals, 141
  narrowband input, 104–6
  random phase, 108
  signal of unknown arrival time, 267
    and carrier frequency, 283
  stochastic signal in white noise, 438
Likelihood ratio, 4
  average, 289
  composite hypotheses, 122
  detection in Gaussian noise, 42–4
  discrete variables, 19
  emission times, 504

power series expansion, 127
sequential detection, 338
stochastic signals, 401
Linear detector, 171
Linear filtering, stationary noise, 30
Lines of force, 202
Lorentz contraction, 543
Lorentz–Fitzgerald equations, 542
Lorentz spectral density, 54, 63, 66,
    405, 424, 448, 477
  eigenvalues, 66, 548
  Fredholm determinant, 433, 547

M

M-ary channel, incoherent, 118
Mann–Whitney test, 330
Marcum's Q function (*see* Q function)
Markov process, 274, 344
Matched filter, 32
  colored noise, 53
  examples, 51, 54
  impulse response, 49
  known signal, 49–57
  maximizes signal-to-noise ratio, 50
  robust, 314
  signal resolution, 358
  transfer function, 52
Matrix impulse response, 443
Maximal-ratio combining, 139
Maximum-a-posteriori-probability
    estimator, 223, 431
Maximum-likelihood detection,
    128–32, 359
  signals in clutter, 371
Maximum-likelihood detector:
  detection probability, 282
  signal of unknown arrival time, 272
    and carrier frequency, 284
Maximum-likelihood estimator, 223–4
  signal parameters, 232
  stochastic signal, 410
  variance, 236

Maximum-likelihood receiver, 130,
    147, 150
  performance (graph), 132
Mean-square error, 222
  arrival-time estimation, 235
  estimation of stochastic signal, 433
Mercer's formula, 39, 399
Minimax strategy, 23
Minimum detectable signal, 46, 73
Minimum-mean-square-error
    estimator, 228
  signal sample, 458
Mixer, 90
Moment-generating function, 25, 189
  CFAR receiver, 309
  coherent plus stochastic signal, 555
  output current of photodetector,
    495, 496
  quadratic form, 535
  quadratic functional, 413
  sequential statistic, 341
  sum of quadratically rectified
    outputs, 144
    fading signals, 152
Moments, signal envelope, 253

N

Narrowband noise:
  amplitudes, joint distribution, 133
  complex representation, 95–7
  nonstationary, 98
  phase difference, distribution, 133
  quadrature components, 96
  stationary, 99
Negative binomial distribution, 478
Neutral point, 203
Newton's method, 199, 206
Neyman–Pearson criterion, 10–2, 46
  decision levels, approximate, 215
  distributed detection, 183
  extended, 123–5, 287
  randomized strategy, 21
  sequential detection, 339

Neyman type A distribution, 490
Noise:
  colored, 37
  Gaussian (*see* Gaussian noise)
  leucogenic, 61, 64, 442
  narrowband (*see* Narrowband noise)
  stationary, 30
  thermal, 27
  white (*see* White noise)
Noise figure, 74
Nonary communication system, 86
Noncentral Rayleigh distribution (*see*
    Rayleigh–Rice distribution)
Noncentrality parameter, 541
Nonparametric receiver, 316
  asymptotically, 320
Norm, 40
Normal distribution (*see* Gaussian
    noise)
Nyquist's law, 71

O

Observation interval, 2
  infinite, 49
    stochastic signals, 404
On-off system, 85
Operating characteristic, 12–4
  discrete data, 21
  known signal (graph), 48
  slope, 13, 26
Orthogonal signals, 75, 79
Orthogonality:
  principle of, 230–2, 433, 446
  random variables, 230
Orthonormal functions, 31, 40
Osculatory parabola, 205

P

Parameter space, 120
Parameters:
  invariable, 165
    sequential detection, 338
  unknown, 172
  signal, 120
    estimation, 246
Paraxial approximation, 159
Periodic codes, 390
Permanent, 508
Phase, random, 107
Phase difference, distribution, 133
Phase modulation, 89
Phased array, 159
Phasor, 91
Photocount distributions, 472
Photodetector, output current, 494
Photoelectrons, 471
  emission times, distribution, 504
Photomultiplier, 486
  single-stage, 490
Photon counting, 472
Photon energy, 471
Planck law, 480
Planck's constant, 70, 471
Poisson distribution, 25, 209, 479
  probability-generating function,
    209, 473
  saddlepoint integration, 209
  secondary electrons, 489
Poisson limit, 485
  photoelectron emission times, 505
Poisson point process, 35, 97, 472,
    494, 504
Polarity coincidence correlator, 332
Polyà distribution, 489
Polyphase codes, 389
Positive-definite kernel, 39
Posterior probability, 3, 19
Power of test, 9
Principle of orthogonality (*see*
    Orthogonality, principle of)
Prior probability, 3
Prior probability density function, 120
Probability:
  conditional, definition, 3
  of correct decision, 3
    multiple signals, 78
    orthogonal signals, 80
    simplex signals, 80

of detection (*see* Detection
    probability)
of error, 3
  multiple signals, 78
of false-alarm (*see* False-alarm
    probability)
posterior, 3
prior, 3
Probability density function, 2
  nominal, 317
Probability generating function, 207
  binomial distribution, 325
  output of avalanche diode, 492
  output of photomultiplier, 488
  photocount distribution, 473
  Poisson distribution, 209, 473
  Polyà distribution, 489
  rank statistic, 327
  rank-sum statistic, 330
Probability mass function, 19
Probability of detection (*see*
    Detection probability)
Proper values, 37
Pulse-amplitude modulated
    communication, 5, 75, 85
  error probability, 6
Pulse-compression radar, 383
Pulse-position modulation, optical,
    509
Pulse-repetition rate, 155
Pulse train, 259
  ambiguity function, 386
  coded, 389
  detection in clutter, 377

Q

Q function, 113, 523
  asymptotic forms, 525
  Mth order, 146, 526
    computation by recurrence,
      528–30
    Edgeworth series, 294
    saddlepoint approximation, 214,
      216

    saddlepoint integration, 200, 205
    trigonometric integrals, 527
    trigonometric integral, 534
Quadratic detector, comparison with
    linear, 171–2
Quadratic form, moment-generating
    function, 535
Quadratic programming, 299
Quadratic rectifier, filtered output, 414
Quadratic threshold detector, 143
  performance, 144–7
Quadrature components, 90
  narrowband noise, 96
Quantization, 176
Quantum efficiency, 471
Quasiharmonic signal, 91
Quaternary communication system, 75

R

Radar:
  pulse-compression, 383
  range measurement, 249
  receiver, 272
Radiometer, 268, 406
Radon–Nikodym derivative, 48
Rake radiometer, 407
Random avalanche gain, 491
  moments, 493
Randomization, 21
  rank test, 327
  sign test, 321
Range bins, 156, 305
Rank test, 326
  asymptotic relative efficiency, 545
Rational spectral density, 61, 422, 511
  discrete data, 459
  incoherent light, 479
Rayleigh distribution, 111
Rayleigh fading (*see* Fading)
Rayleigh–Rice distribution, 112, 523
  moments, 524, 526
Rayleigh–Ritz method, 67
Receiver:
  circuit model, 71–4

Receiver *(cont'd)*
  Neyman–Pearson, 318
  nonparametric, 316
  unbiased, 317
  with reference input, 330
Receivers, equipollent, 167
Rectangular spectral density:
  autocovariance function, 424
  incoherent light, 478
Rectifier, 92
  linear, 92
  quadratic, 92
Regularity domain, Laplace
      transforms, 189
Relative efficiency, sign test, 322
Relative frequency, 3
Reliability, 167
Repetition period, 259
Reproducing-kernel Hilbert space,
      40–1, 57
Resolution, 356
  many signals, 363–8
  signals in time, 365–8
  two signals, 357–62
Resolvability of signals, 356
Resonant circuit, 94
Riccati equation, 447
Rice–Nakagami distribution, 112 (*see
      also* Rayleigh–Rice distribution)
Risk, 228
  conditional, 7, 228
RKHS (*see* Reproducing-kernel
      Hilbert space)
Robust detection, 317

S

Saddlepoint, 198
  locating, 199, 209
Saddlepoint approximation, 213–5
Saddlepoint integration, 199
  curved path, 205
  detectability of stochastic signal,
      420
  Lorentz spectral density, 549

Sample mean, 16, 123, 225
Samples, 166
Sampling, 269
  by orthonormal functions, 31–3, 397
Scalar product, 33, 77, 397
  in RKHS, 40, 82
Scatter-multipath communication, 97,
      395
Schwarz inequality, 258 (*see also*
      Cauchy–Schwarz inequality)
  expectations, 170, 239
Schweppe likelihood-ratio receiver,
      449–50
Secant method, 199
Seismology, 396
Self-transform property, ambiguity
      function, 380
Semi-invariants, 190
Sensors, distributed detection:
  identical, 175–80
  nonidentical, 180–5
Sequential detection, 337
  signals of random phase, 347–50
  targets of unknown distance, 350–4
Sequential probability ratio test, 339
  average sample number, 344
  detection probability, 342
  variance of number of tests, 346
Shot noise, 493–7
  heterodyne detector, 500
Sign test, 320
  asymptotic relative efficiency, 321
Signal space, 74–8
Signal-to-noise ratio, 44, 45
  effective, 82, 168, 296
    heterodyne detector, 501
    homodyne detector, 503
    likelihood-ratio detector, 170
    threshold detector, 170
  loss, narrowband signals (graph),
      114
  maximum attainable, 49, 54, 70
  narrowband signal, 106
  signal in white noise, 47
Signal vectors, 75
  displacement, 79

Signals:
    analytic, 91
    antipodal, 85
    bandlimited, 373
    minimum detectable, 46, 73
    multiple coherent, 138–40
    multiple incoherent, 140–1
    narrowband, 91
    orthogonal, 75, 79
    quasiharmonic, 91
    random phase, 107–9
        detection probability, 111
        sequential detection, 347
    simplex, 80
    stochastic, 394
    unknown amplitude, detection,
        125–7
    unknown sign, detection, 293
Simplex, 80
Singular detection, 48
    stochastic signals, 411
Size of test, 9
Sodium, wavelength of chief line in
        spectrum, 481
Spectral density, 30
    bilateral, 35
    least favorable, 312
    Lorentz (*see* Lorentz spectral
        density)
    matrix, 444
    narrowband, 96
    rational, 61, 422, 511
        discrete data, 459
        incoherent light, 479
    unilateral, 34
Specular component, 451
Spread-spectrum communication
        system, 128
State equations, 441
State-space model, generation of
        stochastic signal, 442
State vector, 441
    covariance matrix, 443
Steepest descent, 201
    tracing path of, 206
Stochastic signal, 394

detectability:
        Lorentz spectral density, 424
        rectangular spectral density,
            424
        in white noise, 415–8
    discrete data, 456
        detectability, 462–5
    optimum detector, 402
    plus coherent signal, detection,
        451
    threshold detector, 405, 418–20
Strategy, 2
    minimax, 23
Student's t statistic, 319
Sufficient statistic, 14
    detection:
        known signal, 44–9
        signal of random phase, 109
    estimation, 225, 241
Swerling cases, detector performance,
        153–6
Symmetrical kernel, 37

T

t-test, 319
Temperature, effective, of noise, 69
Temporal coherence function, 471
Temporal modes, 475
Ternary communication system, 85
Thermal noise, 27, 69
Thévenin equivalent of receiver, 71
Threshold approximation, validity,
        420
Threshold detector:
    multiple inputs, 143
        performance, 144–7
    narrowband signals, 298
    noise of unknown strength, 305
    sequential test, 349
    signal of unknown arrival time, 268
    signal with unknown parameters,
        296
    stochastic signals, 405
        in white noise, 418–20

Threshold effect, 246
Threshold statistic, 127
Time-bandwidth product, 385
Toeplitz approximation, 422, 477
Toeplitz matrix, 365
Trace, 399
Transducer array, 137, 156
　　beam pattern, 160
Transfer function, 30
　　matched filter, 52
　　narrowband filter, 93
Transition probability density
　　function, 274
Transmission line, 69
Trapezoidal rule, 200
Two-input system, 331

U

Uncertainty ellipse, 258, 382
Uniform asymptotic expansion, 217
Uniformly most powerful test, 123,
　　126
Union bound, 365
Unit step function, 212

V

Vector-space representation, 397
Velocity of target, measurement, 250

W

Weak-signal approximation, 127
　　multiple signals, 142–4
Weak-signal detector (*see* Threshold
　　detector)
White noise, 34–7
　　autocovariance function, 35
　　linear filtering, 36
　　narrowband, 103–4
　　spectral density, 34
Whitening filter, 54
Wilcoxon signed rank statistic, 326
Wilcoxon two-sample test, 330

Z

Zero-crossing problem, 275
Ziv–Zakai bound, 244–6